Innovation and Competition Policy

Innovation and Competition Policy

Alan Gutterman

KLUWER LAW INTERNATIONAL
LONDON – THE HAGUE – BOSTON

Published by Kluwer Law International Ltd
Sterling House
66 Wilton Road
London SW1V 1DE
United Kingdom

Sold and distributed in the USA and Canada
by Kluwer Law International
675 Massachusetts Avenue
Cambridge MA 02139
USA

Kluwer Law International Ltd incorporates the
publishing programmes of
Graham & Trotman Ltd,
Kluwer Law & Taxation Publishers
and Martinus Nijhoff Publishers

In all other countries sold and distributed
by Kluwer Law International
P.O. Box 85889
2508 CN The Hague
The Netherlands

First published in 1997

ISBN 90 411 0945 5

British Library Cataloguing in Publication Data and Library of Congress Cataloging-in-Publication Data is available

Typeset by On Screen, West Hanney, Oxfordshire
Printed and bound in Great Britain by Hartnolls Ltd, Bodmin, Cornwall

Contents

1 Introduction

I. Challenges and Problems of Innovation

Society thrives on the continuous development of goods, services and ideas which can be applied to enhance the welfare and livelihood of its members. The creation of these new goods and services is often a long and complex process which calls for a variety of skills and resources in a number of areas, such as research, engineering, manufacturing, marketing, and service, as well as significant amounts of financial and human capital. This creative process, which brings novel and useful ideas from the mind to the marketplace, has been referred to as innovation (i.e. the introduction of "something new" and "novel"), and is the subject of this work.

There are several different ways to encourage innovation. For example, if the state has the resources, it may decide to allocate funds to invention and innovation, perhaps in the form of direct grants or prizes. Until recently, this has been a common strategy in the United States and the European Community ("EC"), where governments would appropriate funds to support research and development programs in areas considered to be important to the country's national and economic security interests. However, as noted elsewhere in this work, state-directed provision of innovation is not always effective. Moreover, the projects financed by the state in these countries were only a small portion of the new goods and services created within the economy.

In fact, it is far more typical for innovation in most developed countries to be left to the private sector; however, in order for entrepreneurs and firms to respond to the challenges of innovation, they must have a reasonable expectation that they will be rewarded for their efforts and investment by returns that are commensurate with the level of risk involved in a particular project. In order for these expectations to be realized, it is necessary to create an environment in which the entrepreneur or firm can both appropriate a material portion of the benefits from their innovative activities and, if necessary, join with other entrepreneurs and firms to share the often substantial risks that are associated with developing and commercializing any new idea or innovation. For this reason, this work focuses on the recognition and use of one special state-created property right – the patent – which can be used by an entrepreneur or firm to appropriate the benefits of their work by excluding use of new inventions by others, and on collaborative arrangements (e.g. joint ventures) focused primarily on the risky business of research and development.

II. Intellectual Property Rights

A. Patents

Private innovators, while appreciative of the intangible rewards of contributing to social progress, need to have the economic incentive to finance the innovation process. One method for establishing that incentive is the recognition of a state-created property right – the patent – which can be used by the owner thereof to exclude others from using novel inventions covered by the patent.

The patent owner has a number of strategic options. The primary purpose of the patent right is to provide the owner with sufficient protection from copying by, and competition from, others to allow the owner to commercially exploit the invention and appropriate a sufficient level of revenues from consumers to recoup the initial investment in innovation and incentivize the owner to engage in further research and development activities. In other cases, the patent owner may decide to supplement the revenues obtained from the owner's own commercial exploitation of the patent by licensing the right to practice the invention to others in exchange for royalties or some other consideration.[1] Still another strategy is for the patent owner to simply use the patent defensively to prevent others from making, using, or selling the patented product or process, even though the subject matter of the patent may not be exploited by the patent owner.[2] Finally, the patent owner may contribute the patent rights to a collaborative arrangement with one or more other firms (e.g. joint venture or contractual relationships, such as a license or manufacturing agreement) formed for the purpose of commercially exploiting the patented invention.

There are obviously a number of important legal, economic, and social issues associated with the creation, enforcement, and use of patent rights. In this work, we are concerned with three of those issues. First, we wish to examine some of the legal and economic justifications for intellectual property rights as a whole, and for patent rights in particular. Second, assuming the existence of adequate justification for exclusive private rights in new inventions and ideas, we seek to analyze the effect of such rights on competition in the marketplace and how the state might properly regulate the actual use of such rights under its antitrust or competition laws. Third, we review how antitrust and competition regulators deal with collaborative arrangements (e.g. joint ventures) dedicated to research and development activities, which are the cornerstone of creating new inventions eligible for intellectual property protection. In order to gain the benefits from comparing and contrasting how these issues have been approached in different countries with divergent political systems and concerns, we will focus both on the United States and the EC.

B. *Legal and Economic Nature of Intellectual Property Rights*

General framework of intellectual property rights For many years, "capital," in the form of machinery, land, buildings and other tangible assets, and "labor," in the form

[1] Normally, a patent owner in the United States is not required to license the patent rights to others as a condition to retaining all of the legal rights embodied in the patent. In 1988, the Congress passed amendments to the United States patent law statute (Title 35 of the United States Code, referred to herein as the "Patent Act") which added two new categories to Section 271(d) of the Patent Act (35 U.S.C. § 271(d)), which lists certain acts of patent owners and licensors that will not be deemed to be patent misuse. The relevant provision of Section 271(d) now reads as follows: "No patent owner otherwise entitled to relief for infringement or contributory infringement of a patent shall be denied relief or deemed guilty of misuse or illegal extension of the patent right by reason of his having done one or more of the following ... (4) refused to license or use any rights to the patent." However, many national patent laws outside of the United States provide for the issuance of compulsory licenses to third parties in the event that a patent is not adequately exploited. See, e.g. the United Kingdom Patents Act 1977 §§ 48–59.

[2] As discussed in note 1, however, defensive use of a patent without actually exploiting the patented invention may not be allowed in countries with compulsory licensing schemes.

of human resources measured on both a quantitative and qualitative basis, were perceived as the most important factors of production in the economic analysis of any business enterprise. However, particularly in light of the literal "globalization" of business activity in the last quarter century, it is clear that a new set of assets, denominated as "knowledge" and "innovation," have now moved to the forefront of business strategy for both large and small enterprises around the world.[3] This new factor of production is often referred to as "technology," which has competitive significance. The use of this term connotes a set of tangible assets which can be used by their owner and, like any other type of asset, may be susceptible to obsolescence, misappropriation and duplication. However, the legal framework that has been constructed to contain, define, and allocate ownership and control of these technological assets provides an owner with far more than a set of property rights. Instead, these legal rights, which form the substance of what is referred to as intellectual property rights, provide a tool which assist the owner to forge and maintain an overall competitive advantage based, in part, on its ability to use, and prevent others from using, the valuable innovations and ideas which meet the requirements of law around the world. Intellectual property can take a number of different forms, each of which entitles the owner thereof to certain legal rights in relation to third parties with regard to the specified subject matter.[4] Specifically, property rights may be available for the following:

(i) Any new or useful process, machine, manufacture, or composition of matter, as well as new or useful improvements thereof, may be eligible for protection under the patent laws.

(ii) Any formula, pattern, device or compilation of information which is used in one's business, and which gives the person in possession an opportunity to obtain an advantage over competitors who do not know or use it can be protected as a trade secret.

(iii) Original works of authorship fixed in any tangible medium of expression, such as books and records, can be protected under the copyright laws.

(iv) Words, names, symbols or devices or any combination thereof adopted and used by a manufacturer or merchant to identify and distinguish his goods from the goods manufactured or sold by others can be protected under the trademark laws.[5]

The scope of the property rights granted under the laws of patents, trade secrets, copyrights, and trademarks is not uniform. The holder of patent has the right, for a fixed pe-

[3] See, on this subject, Smith and Parr, *Valuation of Intellectual Property and Intangible Assets*, John Wiley & Sons, New York (1989), where the authors assert that "Intellectual property is fast becoming the most important asset possessed by corporations. Various forms of intellectual property are the foundation of market dominance and continuing profitability for many companies ... Companies are licensing, selling, joint venturing, and trading intellectual property around the world. Our global economy is based, in many ways, upon this commerce in intellectual property."

[4] See Chapter 2 for a detailed discussion of patent laws in the United States and the EC.

[5] The objects and creations which might be considered to be subjects of intellectual property protection are changing over time. For example, intellectual property rights have now been extended to cover new plant varieties in a number of countries and special protection has been developed for the powerful circuits in a silicon chip. Moreover, the law relating to the protection of computer programs is rapidly evolving. Also, one can expect a number of future developments with regard to the protection of the products of biotechnological research.

riod of time, to prevent others from making, using or selling the item, usually referred to as an "invention," without the consent of the holder of the patent, even if the other party independently discovers the invention. A trade secret owner cannot prevent third parties from independently developing the protected information; however, the owner does have a cause of action against any persons who may misappropriate the information. The owner of a copyright or a trademark has the right to prevent others from copying or using the copyrighted material or the protected word, name, symbol or device without the consent of the owner.

Patent protection is not available without filing an application with the appropriate governmental authority which contains a full description of the invention and various other items of information designed to demonstrate that the invention is sufficiently novel and useful to qualify for a patent. Trade secret protection does not require any filing; however, the owner must take steps to maintain the confidentiality of information for which trade secret protection will be requested. Copyright and trademark protection begins with the creation, in the case of copyrights, or first use, in the case of trademarks, of the work or "mark." However, registration of copyrights and trademarks with the appropriate governmental authority is required in order for the owner to take certain actions against infringers.

Patent, copyright and trademark protection extend only to the borders of the country in which the property right has been granted. Accordingly, the holder of a United States patent will generally have no right to prevent persons outside of the United States from making, using or selling the invention in a foreign country unless such person also obtains a patent in that foreign country. Trade secret laws are largely a function of state law in the United States and many countries fail to provide any trade secret protection. Moreover, the patent, copyright and trademark laws around the world differ substantially as to such things as the subject matter which might be eligible for protection and the scope of the rights granted to the owner of the property right.

Legal justifications for intellectual property rights Intellectual property rights have caused a great deal of controversy among legal scholars, economists and social philosophers. In particular, the debate has focused upon the legal rights granted by the state to the holder of a patent to effectively exclude all others from making, using, or selling the invention covered by the patent for a fixed period of time. Probably the most vexing issue surrounding the patent right lies in the nature and purpose of the right itself. Speculation has centered on whether a patent is intended to grant a "property-like" right to those who contribute to the common fund of knowledge, or is it simply an economic policy tool designed to provide an incentive to inventors and innovators to reveal their ideas to the public in order to promote and foster the creation and dissemination of new knowledge. Some of the characteristics of a patent, such as the universal right to exclude and the power to transfer, are similar to those associated with traditional property rights. On the other hand, the monopolistic elements in the genealogy of patent rights, as well as the relative difficulties associated with appropriating exclusive control over the use of ideas and knowledge, have served as the basis for severe criticism of the notion of intellectual "property."

While patents, copyrights, trademarks, and trade secrets are often referred to as property rights, there are actually a number of problems in reconciling those legal rights with the customary characteristics of the objects of traditional property rights. The law of property recognizes two different forms of legal relations with regard to property. The first form covers the relationship between a person and a thing and is referred to as a real relation, or a relation *in rem*.[6] When a person states that he or she owns a particular good, such as a car, a house, or a piece of land, the owner describes a relation that can be understood without reference to any other ascertainable person. On the other hand, a person's right to receive a certain sum of money from another person, while perhaps just as valuable as real rights in a thing, is actually a specific relationship between one person and another person (i.e. debtor and creditor) and is usually characterized as a personal relationship or relation *in personam*.[7]

Real rights imply a universal obligation on others to refrain from any acts which would interfere with the control of the good by the owner.[8] Real rights are based upon the satisfaction of two key conditions: physical control of the object of the rights and the compatibility of this physical control with the given property rules within the legal system.[9] In contrast, the factual condition of physical control generally does not exist in a relationship that is based upon personal rights. An obligee does not exert physical control on the obligor, nor does the obligee own the obligor or even the object of the obligation, such as the money that is to be paid or the services which are to be rendered.[10] Instead, the only thing that the obligee actually possesses is a claim against the obligor, which is an expectation that the courts and the state will be ready to enforce.[11] Moreover, the holder of a personal right can only enforce the claim against specific persons, not against all others, as is the case with a real right.

A patent is similar to a real right in that it creates a universal obligation on others to refrain from making, using or selling the patented invention during the term of the patent. However, critics of the concept of intellectual property are quick to point out that knowledge, which is the creation that is embodied in the patented invention, lacks most of the essential attributes of traditional real property, such as appropriability or natural scarcity.[12] Moreover, it has been argued that patent rights cannot be properly included among the various personal rights, since the patent holder has a right of exclusion which can be en-

[6] Real property rights can be traced back to the sixteenth-century writings of philosophers and jurists in France and Spain. They defined property as "the right to keep a good, to use it, to benefit from its yields, to exclude anybody else from its use, to alienate it, and even to destroy it." See Bouckaert, "What is Property?," *Harvard Journal of Law and Public Policy* 13 (1990), 775, 788 (referring to the writings of Vazquez, *Controversarium Illustrium*, Lib. I, Cap. XVIII (1559); Hotman, *Commentarius de Verbis Juris* (1569); Doneau, *Commentarius de Jure Civili,* Lib. 9, Cap. 8–12 (1595)).

[7] There are several different sources of personal rights: contract, tort, factual ("quasi-contract"), and direct imposition by the state. See Nicholas, *French Law of Contract* (1982), 28–55 (discussing the sources of obligations in French law). Contract and tort are generally considered to be the normal sources of personal rights. The ability of a person to create personal rights and obligations by contract is considered to be an expression of personal liberty. Quasi-contractual duties and obligations are normally created for reasons of natural justice or equity, rather than as a direct consequence of strict judicial doctrine. Bouckaert, supra note 6, at 804.

[8] Ibid. at 794.

[9] Ibid. at 794–6. The clearest example of real rights is the relationship recognized by the law between corporeal goods and their owners. The owner of land or a vase is recognized as having a proprietary interest therein. If the land should be occupied by a trespasser, or the vase is stolen by a thief, the landowner or the owner of the vase is entitled to full and exclusive possession and enjoyment of the property. This entitlement is not just based upon a personal obligation owed by the trespasser or the thief to the owner of the subject matter to vacate the land or give back the vase, as the case may be, but rather upon the property rules which have been developed by society and incorporated into its legal system based upon physical control of the specified property.

[10] Ibid. at 794–7.

[11] Ibid.

[12] For example, Renouard, a nineteenth-century legal authority, argued that traditional property theories were not adequate to protect literary and artistic production. He pointed out that artistic property had neither the same base, the same object, the same rules, the same consequences, nor the same limits as ordinary

forced even against those who are not in privity of contract with the patent holder.[13] While there are those who believe it is possible to find analogies between patent rights and traditional property law, as a general rule legal scholars have found it difficult to reconcile protection of intellectual creations with classical notions of property rights.[14]

Commentators have argued that if the obligations imposed on third parties as a result of intellectual property laws cannot be qualified as real or personal rights, then they can only be viewed as obligations which are directly imposed by the state and, as such, are exogenous to the inner logic of private law.[15] As one scholar has written, intellectual property rights "belong to the realm of macro-economic measures by which government pretends to stimulate particular activities to foster the general welfare."[16] For these writers, the question becomes whether or not the overall social welfare is enhanced by such a policy, particularly since the public is being asked to surrender its "legitimate right" to freely use information received outside of any contractual framework in order to subsidize the creation of new intelligence.[17]

Others have also objected to characterizing ideas and information as property, although they do not share the same aversion to some form of protection for ideas as those who would abolish intellectual property protection altogether. They concede that ideas in general and confidential information in particular have characteristics which cannot be accommodated within the general concept of real rights, since information which is "taken" actually becomes shared information due to the fact that the person from whom the information was taken still has the ability to use it and enjoy it. In fact, they note that most jurisdictions have actually rejected the notion of confidential information as property, and thus have refused to convict persons who have misappropriated trade secrets of classical property related offenses such as theft or criminal damage.[18] These authors similarly

cont.
property and, therefore, could not qualify as a category of classical property. Instead, Renouard believed that any recognition of an artist's right to exercise control over his creation was actually a legal privilege which fell outside the scope of civil law doctrine. However, he also considered the privilege to be justified as a legitimate reward for a valuable social service: the creation of a work of beauty (ibid. at 792). It has also been pointed out that the communication of knowledge does not deprive the owner of possession of the knowledge, as is the case when tangible items of property are transferred, but rather merely ends the owner's monopoly of natural ownership. As such, the use of knowledge is "non- rivalrous" and knowledge is not subject to "scarcity" in the same way as tangible goods. Ibid. at 802. However, some types of information may be rival in that the first person to possess the information may gain an advantage over his competitors. See Mackaay, "Economic Incentives in Markets for Information and Innovation," *Harvard Journal of Law and Public Policy* 13 (1990), 867, 892. The foregoing is the basis for trade secret law protection.

[13] For a general discussion of how contract, tort and quasi-contract law relates to patents, see Bouckaert, supra note 6, at 804–6.

[14] For example, Picard, a Belgian legal theorist, claimed it was futile to attempt to fit artistic and industrial property into the classical subdivision of rights. Instead, he argued that a new separate category, which he referred to as "intellectual rights," should be created to encompass intellectual creations such as artistic concepts, the plot of a novel, or a technological formula. While Picard was unable to reconcile intellectual property rights with classical property rights, he did believe that the creator of an intellectual product was entitled to claim against the world an obligation to refrain from copying or imitating the creation without the consent of the creator. Ibid. at 793 (citing Ricard, "Le Droit Pur," in *Cours d'Encyclopedie du Droit* (1899), 119–29). See also Roguin's "la théorie de monopole," which was based on the argument that protection of the artist or the inventor under the classical civil law framework could only be justified if the protection was referred to as a "universal obligation" based upon the act of the legislature in setting up a negative obligation for all others to abstain from copying or imitating the work. Id. at 792. (See Roguin, *La Règle de Droit. Systeme des Rapports de Droit Privé* (1989).)

[15] Ibid. at 805.

[16] Ibid.

[17] Ibid. at 806.

[18] Coleman, "Protection of Innovation Under the Common Law and In Equity," in Phillips (ed.), *Patents in Perspective* (1985), 18, 30.

come to the conclusion that the resolution of the issues surrounding the protection of ideas and information call for a balancing of a complex web of public and private interests, for which the traditional property-based analysis is ill-equipped.[19]

Given the apparent shortcomings of dealing with patents and other forms of intellectual property rights in the same manner as land and tangible goods, it is not surprising that a good deal of the debate in this area focuses upon other justifications for the creation and maintenance of the patent system. One alternative set of arguments for patent rights, as well as for the other forms of intellectual property rights, have been referred to as the "labor" or "reward" theory. In essence, advocates of this theory argue that people own their bodies and the products of the labors which are performed thereby, and are entitled to receive the fruits of their labors.[20] If one accepts this notion, then intellectual property rights serve as a means of providing a reward to inventors and others who might be involved in creative activities.[21] The labor theory has been rejected, however, by those who would take issue with granting to the inventor a right to receive the total value which the market places on the new invention, given that the actual value of the activities of the inventor may be much less[22] and that the value of the invention may be largely attributable to work which may have previously been performed by others.[23]

Another popular theory is the "natural law," or "moral rights," argument, which holds that individuals have a natural property right in their own ideas and that society has a moral obligation to protect these rights from being usurped or stolen by others.[24] It is not surprising that natural rights in ideas are a matter of great controversy. For example, the concept of moral rights was specifically rejected by drafters of the United States Constitution, as Thomas Jefferson wrote: "[i]nventions then cannot, in nature, be a subject of property. Society may give an exclusive right to the profits arising from them, as an encouragement to men to pursue ideas which may produce utilities, but this may or may not be done according to the will and convenience of society, without claim or complaint from anybody."[25] Moreover, even if one accepts the notion, it is difficult to jus-

[19] Ibid. at 49.

[20] Hettinger, "Justifying Intellectual Property," *Philosophy and Public Affairs* 18 (1989), 31, 37.

[21] Nance, "Forward: Owning Ideas," *Harvard Journal of Law and Public Policy* 13 (1990), 757, 764; Cornish, *Intellectual Property: Patents, Copyright, Trade Marks and Allied Rights* (1989), 79; Bainbridge, *Intellectual Property* (1992), 245–6. The "labor theory" of property is generally associated with John Locke. See, e.g. Locke, *Two Treatises of Government* (P. Laslett, ed., 1970).

[22] Hettinger, supra note 20, at 37. There also may be proportionality problems in matching the effort expended to the value of the rights which might be given to the laborer. For example, some inventions of extraordinary market value result from flashes of genius, while others with little market value, but perhaps with great social value, will require a significant amount of effort. Ibid. at 43.

[23] Ibid. at 37 and 40. Moreover, it has been pointed out that it is far from clear that what the laborer deserves to receive is property rights in the objects created. Alternatives include fees, awards, acknowledgment, gratitude, praise, security, power, status, and public financial support (ibid.). On balance, commentators note that rewarding inventive activity is little more than an incidental consequence of the modern system of patents. See, e.g. Cornish, supra note 21, at 79. See also Rawls, *A Theory of Justice* (1971); and Rachels, "What People Deserve," in J. Arthur and W. Shaw (eds.), *Justice and Economic Distribution* (1978), 150–63.

[24] See, e.g. Machlup and Penrose, "The Patent Controversy in the Nineteenth Century," *Journal of Economic History* 10 (1950), 1, 10–20. Machlup and Penrose have attempted to describe the natural rights position as follows: "A man has a natural property right in his own ideas. Their appropriation by others must be condemned as stealing ... Property is in essence exclusive. Hence enforcement of exclusivity in the use of a patented invention is the only appropriate way for society to recognize this right ... Justice requires that a man receive ... reward for his services in proportion as these services are useful to society ... The most appropriate way to secure to inventors rewards commensurate with their services is by means of exclusive patent rights in their inventions."

[25] See Hamilton, *Investigation of Concentration of Economic Power: Patents and Free Enterprise* 21 (TNEC Monograph No. 31) (1941).

tify the manner in which the enforceability of patent rights has been extended beyond those who may have been involved in usurping or stealing new inventions.

One last important set of theories in this area are the so-called "consequentialist" theories, which attempt to justify property rights on the basis of the good consequences flowing from their recognition.[26] With regard to intellectual property rights, consequentialists argue that recognition of exclusive legal rights in invention and the other products of creative activity facilitates the "good consequences" of the advancement of knowledge, industrial progress and economic efficiency. As one commentator explained, "[i]ndustrial progress is desirable to society. Inventions and their exploitation are necessary to secure industrial progress. Neither ... will be obtained to any adequate extent unless inventors and capitalists have hopes that successful ventures will yield profits which make it worth their while to make their efforts and risk their money. The ... most effective way for society to hold out these incentives is to grant exclusive patent rights in inventions."[27]

There are two major branches of consequentialist thought: utilitarianism and teleology.[28] Both branches focus upon the incentives which the law can establish; however, utilitarian and teleological theories differ in how the "good" which is to be promoted should be identified. Utilitarianism, with which the modern theories of "law and economics" are often identified,[29] tends to measure the good in terms of the satisfaction of human preferences, with measurement being allowed to occur through the aggregation of preferences that are registered by the operation of the market.[30] The utilitarian theory makes no judgment about whether the preferences of the market are, in some objective sense, appropriate or correct.[31]

In contrast, teleological theories are based upon some judgment regarding which is good or necessary for individuals, as well as for communities, notwithstanding any consumer preferences which may run contrary to these judgments.[32] In fact, this general school of thought is referred to as "teleological" because it involves various judgments regarding the proper ends of human life.[33] With respect to patents and other forms of intellectual property rights, teleological arguments are based on the intrinsic values of knowledge and aesthetic experience, which are considered to be values which deserve the support of the coercive powers of the government despite, and indeed because of, the insufficiency of consumer demand, even in a well-functioning market.[34] However, merely securing exclusive rights to authors and inventors does not guarantee that the market will produce the objectively "correct" amount of knowledge and aesthetic experience which the advocates of the teleological theory would think is desirable.[35] As such, teleologists may be more likely to propose non-market mechanisms, such as subsidies and prizes offered by the government, to encourage creative efforts.

[26] Nance, supra note 21, at 763.

[27] See, e.g. Machlup and Penrose, supra note 24, at 10.

[28] Nance, supra note 21, at 764.

[29] See, e.g. Posner, "Utilitarianism, Economics, and Legal Theory," *Journal of Legal Studies* 8 (1979), 103; Landes and Posner, "An Economic Analysis of Copyright Law," *Journal of Legal Studies* 17 (1989), 325.

[30] Nance, supra note 21, at 764.

[31] Ibid.

[32] Ibid. at 765.

[33] Ibid. at 765, n35. Teleological arguments are especially associated with classical "natural law" thought of Aristotle or Thomas Aquinas. See Finnis, *Natural Law and Natural Rights* (1980).

[34] Nance, supra note 21, at 765.

[35] Ibid. at 765–6.

Consequentialist arguments take patents out of the realm of property law and place them squarely in the economic arena. The desirability of patents was far from a settled question among the leading economists of the eighteenth and nineteenth centuries.[36] For example, Bentham [37] and Mill[38] were advocates of patents, and Adam Smith apparently was not opposed to their use.[39] Clark believed that innovation would nearly cease without patents.[40] On the other hand, Taussig and Pigou argued that patents made little difference[41] and Lowe was active in a movement to abolish the use of patents.[42] However, since the middle of this century, most commentators have expressed their support for the patent system on basic utilitarian grounds,[43] although there are those who continue to express concerns regarding externalities, and scholarly writings have included a variety of proposals to achieve a more efficient or optimal result, such as by providing a tax or subsidy, modifying the duration of patents, or changing the scope of patents. Economic justifications for the patent system are discussed in greater detail in Chapter 3.

C. Patent Rights and Regulation of Competition

Exploitation of patents by licensing Patent-based exclusivity can be used not only to exclude others from using the patented invention, but also to induce others to engage in activities which may be necessary for the fullest exploitation of the invention. For example, if the patent owner is unable to internally manufacture the number of units required to satisfy demand in the marketplace, the owner can enter into a contract with a third party to have the excess units manufactured. These relationships are forged through

[36] See, e.g. Machlup and Penrose, supra note 24, at 7–9; Dutton, *The Patent System and Inventive Activity During the Industrial Revolution 1750-1852* (1984), Chapter 1. Other sources of interest relating to the history of patents include Plant, "The Economic Theory Concerning Patents for Inventions," *Economica* 30 (1934); Scherer, *Industrial Market Structure and Economic Performance* (1970) 379; Machlup, *An Economic Review of the Patent System* (1958).

[37] See Bainbridge, supra note 21, at 246 (noting that Bentham argued that the exclusivity provided by the patent laws must be reserved for inventors due to the fact that invention involved a great deal of time, money and effort).

[38] Ibid. Mill appeared to be unconcerned about the possible problems of "allocative inefficiencies" which some believe are caused by patents. One scholar said of Mill that he "adopted the rhetoric of the free market economy by suggesting that the reward depended on the invention proving to have economic value and that, in any event, only the users of the commodity created were paying for the increased price caused by the patent monopoly." See Smit, *The Social Creation of a Legal Reality: A Study of the Emergence and Acceptance of the British Patent System as a Legal Instrument for the Control of New Technology*, PhD thesis, University of Edinburgh (1980), as quoted in Bainbridge, supra note 21, at 246.

[39] Meiners and Staaf, "Patents, Copyrights, and Trademarks: Property or Monopoly?," *Harvard Journal of Law and Public Policy* 13 (1990), 911.

[40] Plant, *Selected Economic Essays and Addresses* (1974) 39–40.

[41] Ibid.

[42] Plant, "The Economic Theory Concerning Patents for Inventions," *Economica* 47 (1934).

[43] O'Brien, "Patents: An Economist's View," in Phillips, supra note 18, at 32 and 41. See also, Mansfield, *The Economics of Industrial Change* (1968) 209 (" there are few leading economists, if any, who favour abolition of the patent system. Even those who publish their agnosticism with respect to the system's effects admit that it would be irresponsible, on the basis of our present knowledge, to recommend abolishing it.") There are some exceptions. See, e.g. Palmer, "Intellectual Property Rights: A Non-Posnerian Law and Economics Approach," *Hamline Law Review* 12 (1989), 261 (arguing that the primary justification of property is the priority of self-ownership and liberty and that recognition of intellectual property rights is contrary to that priority on the grounds that such rights interfere with other persons' use of their intellect, bodies, and otherwise lawful property). For further discussion of Palmer's theories, see Mackaay, supra note 12, at 868–9.

the use of licensing agreements, which are essentially contractual exceptions to the patent owner's ability to otherwise prevent the other party from using or manufacturing the licensed invention.

Competitive effects of patent licensing arrangements While it has not always been so, patent licensing agreements are now generally considered to be desirable because of the role that they play in the development of new products and processes, and in disseminating innovation. However, common elements of the licensing agreement create a number of antitrust and competition law issues. For example, in exchange for the affirmative rights to use the licensed subject matter, the licensee not only agrees to make certain payments to the licensor, but usually also will consent to be bound by various restrictions and limitations with respect to the use of the licensed patent(s) and other aspects of its business activities. In turn, the licensor may also be made subject to various restrictions during the term of the licensing arrangement, such as prohibitions on granting other licenses in the territories in which the original licensee intends to use the licensed patent.

The concern of antitrust and competition law regulators is that these ancillary restrictions, as well as the contact between the parties which necessarily follows from a close collaborative relationship, will impair competition. For example, exclusive patent licensing agreements can raise significant competition law problems, since such arrangements presumably foreclose the use of the licensed subject matter by others, including the licensor, in the licensed territory,[44] and heightens the risk of potentially anticompetitive collusion between the parties. In the absence of a patent, an arrangement in which two or more companies agree not to compete in a given geographic territory or product market would, under traditional principles of antitrust and competition law analysis, be struck down as a *per se* violation of applicable law.[45] On the other hand, when a patent

[44] A "pure" exclusive license includes the licensor's agreement that it will not itself license any undertaking within the particular territory apart from the chosen exclusive licensee, nor will it exploit that particular right itself in that territory. On the other hand, a sole license refers to an agreement under which the licensor reserves for itself the right to compete in the licensed territory, but is restricted from licensing any third parties situated there. Non-exclusive licenses, which allow the licensor to grant comparable licenses to a number of other persons or firms, generally do not raise significant problems, unless they contain other terms or conditions that unduly restrict the business activities of the licensees. For an extensive discussion of each of the various types of licenses, and the terms generally included therein, see Chapter 6.

[45] See, e.g. *Engine Specialties, Inc.* v. *Bombardier Limited*, 605 F.2d 1 (1st Cir. 1979) (exclusive distributorship between manufacturer and potential entrant held *per se* unlawful as market division under the United States Sherman Act). Determinations as to *per se* illegality are generally reached without reference to any surrounding circumstances which may mitigate the effect of the provision or justify its use. The *per se* rule was described by the Supreme Court in *Arizona* v. *Maricopa County Medical Society*, 457 U.S. 332, 343–4 (1982):

> "The elaborate inquiry into the reasonableness of a challenged business practice entails significant costs. Litigation of the effect or purpose of a practice often is extensive and complex ... Judges often lack the expert understanding of industrial market structures and behavior to determine with any confidence a practice's effect on competition ... And the result of the process in any given case may provide little certainty or guidance about the legality of a practice in another context ...
>
> The costs of judging business practices under the rule of reason, however, have been reduced by the recognition of per se rules. Once experience with a particular kind of restraint enables the Court to predict with confidence that the rule of reason will condemn it, it has applied a conclusive presumption that the restraint is unreasonable. As in every rule of general application, the match between the presumed and the actual is imperfect. For the sake of business certainty and litigation efficiency, we have tolerated the invalidation of some agreements that a fullblown inquiry might have proved to be reasonable."

is involved, a finding of *per se* liability would be inconsistent with the exclusivity that lies at the very essence of the patent law.

The relationship between intellectual property and antitrust laws Intellectual property law and antitrust law are each intended to achieve certain public policy objectives, many of which are, in certain circumstances, somewhat inconsistent. The uneasy relationship between intellectual property rights, which are exclusive rights recognized to insulate the holder thereof from market competition as a means for promoting innovation,[46] and the antitrust laws, which seek to promote consumer and economic welfare by fostering and preserving competition, is illustrated by the following passage:

> "The patent laws reward the inventor with the power to exclude others from exploiting his invention for a period of seventeen years. 35 U.S.C. § 154 (1976). In return, the public benefits from the disclosure of inventions, the entrance into the market of valuable products whose invention might have been delayed but for the incentives provided by the patent laws, and the increased competition the patented product creates in the marketplace. The antitrust laws, on the other hand, were enacted to protect competition in the market. The antitrust laws are based upon the fundamental premise that the public benefits most from a competitive marketplace ...The conflict between the antitrust and patent laws arises in the methods they embrace that were designed to achieve reciprocal goals. While the antitrust laws proscribe unreasonable restraints of competition, the patent laws reward the inventor with a temporary monopoly that insulates him from competitive exploitation of his patented art. When the patented product, as is often the case, represents merely one of many products that effectively compete in a given product market, few antitrust problems arise. When, however, the patented product is so successful that it evolves into its own economic market, as was the case here, or succeeds in engulfing a large section of a preexisting product market, the patent and antitrust laws necessarily clash. In such cases the primary purpose of the antitrust laws – to preserve competition – can be frustrated albeit temporarily, by a holder's exercise of the patent's inherent exclusionary power during its term."[47]

While it certainly true that the specific mechanisms which are employed by the intellectual property and antitrust laws may well lead to conflict, it has been argued that the two bodies of law can be reconciled by emphasizing their common goal of promoting overall consumer welfare. For example, one court has observed:

> "The patent system, which is rooted in the United States Constitution [Art. I, § 8, cl. 8], serves a very positive function in our system of competition, i.e., "the encouragement of investment based on risk ... [By so doing, it] encourages innovation and its fruits: new jobs and new industries, new consumer goods and trade benefits ... In that sense, therefore, and because the underlying goal of the antitrust laws is to promote competition ... the patent and antitrust laws are complementary."[48]

As a practical matter, the difficulties surrounding the interface of intellectual property law and antitrust law flow not so much from their ultimate objectives, as from the

[46] See e.g. *Kewanee Oil Co.* v. *Bicron Corp.*, 416 U.S. 470, 480-81 (1974) ("The patent laws promote ... progress by offering a right of exclusion for a limited period as an incentive to risk often enormous costs in terms of time, research and development."); *Sears, Roebuck & Co.* v. *Stiffel Co.*, 376 U.S. 225, 229–30 (1964) ("Patents ... encourage invention by rewarding the inventor with the right, limited to a [fixed term], to exclude others from use of his invention").

[47] *SCM Corp.* v. *Xerox Corp.*, 645 F2d 1195, 1203 (2d Cir. 1981), *cert. denied,* 454 U.S. 1016 (1982).

[48] *Loctite Corp.* v. *Ultraseal Corp.*, 781 F.2d 861, 876–7, 228 U.S.P.Q. 90 (Fed. Cir. 1985).

fact that the courts are often called upon to determine when, and why, one body of law should dominate over the other in determining the legality of a specific transaction. For example, in some situations, the principles of antitrust law will control so as to limit the range of permissible alternatives that can be used to commercialize new technology or new ideas even when the area of competition in question falls wholly within the scope of the exclusive rights recognized under the intellectual property laws.[49] On the other hand, intellectual property law sometimes legitimizes business transactions which in the absence of such rights would be *per se* violations of the antitrust laws.[50] As such, what is, on its face, an apparent conflict, may be better described as a process of reconciliation. Even if regulators concede that patent rights do, in specified circumstances, provide the patent owner with a legal right to enter into contracts that might be illegal in other transactional contexts, they have remained concerned about the possibility that the patent owner might use the strategic value of the patent-based exclusivity in a manner that goes beyond the "proper scope" of the patent. These issues typically come up when the licensor attempts to impose restrictions on the business activities of the licensee in areas related to the subject matter of the licensed patent, or appears to condition the grant of the license on the licensee's agreement to do one or more things that might have economic value to the licensor beyond any royalties or other consideration being paid for the license (e.g. purchasing or licensing products from the licensor which are not covered by the licensed patent(s)).

Apart from pure licensing arrangements, actual or potential competitors may enter into various other forms of collaborations which involve the development and/or use of intellectual property rights. For example, a research and development arrangement or a joint venture might be used to develop and commercialize a new technology or product. As noted above, the sharing of resources and risks may be advantageous to the participants, and may well lead to the creation of a products and processes that would not have otherwise been available in the absence of the arrangement. However, as with licensing agreements, antitrust regulators will be concerned about the possibility that cooperation may turn into collusion that inhibits competitiveness in the marketplace. In addition, since the participants may seek to further protect the investment that they might make in the collaboration through the use of restrictions on the research and business activities of the others (e.g. covenants not to engage in independent research relating to competitive products), an analysis of the effect of these so-called "collateral restraints" will also be part of the antitrust regulator's review.

Regulation of contracts involving patents and other intellectual property rights

Countries follow a variety of regulatory strategies in regulating contracts involving the monopoly rights provided to the owners of patents and other intellectual property rights. The most common form of regulation in developed countries are antitrust or competition laws, which typically deal with a variety of business strategies and collaborations involving two or more firms which might have a specified impact on competition in the relevant market. These laws not only cover licensing arrangements, but also other types of business collaborations (e.g. mergers and joint ventures) that might result in some com-

[49] See, e.g. *International Wood Processors* v. *Power Dry, Inc.*, 792 F.2d 416 (4th Cir. 1986) (attempt by licensor to conspire with banks, investors and other licensee to eliminate a licensee in order to improve the overall commercial attractiveness of the licensed product found to be a violation of the antitrust laws, even though the only competition restrained fell wholly within the scope of the licensed patent).

[50] See, e.g. *United States* v. *General Electric Co.*, 272 U.S. 476 (1926) (licensor of a patent was permitted to control the prices charged by his licensee for products covered by the patent).

bination of the business assets and resources of actual or potential competitors in a manner which reduces competition.[51] Competition may also be regulated or controlled by foreign investment laws, which focus upon the role that foreign firms and capital will have in the domestic economy of the regulating country, and by laws which are designed to vet specific forms of business collaborations (e.g. regulation of technology transfer agreements).[52]

As to regulatory content, countries may adopt rules which impose outright restrictions on the use of certain licensing practices (e.g. a rule that no patent owner may fix the prices to be charged by a licensee) or joint venture restrictions (e.g. a restriction on the ability of a venturer to prohibit the other venturer from engaging in competitive activities). Rules of this type declare certain practices to be *per se* illegal, without any detailed analysis of the commercial or economic circumstances surrounding the use of the provision. Alternatively, the applicable regulatory body may apply a "rule of reason" and assess a provision in a licensing agreement in relation to things such as the intent of the provision, the market power of the parties, and the effect of the provision on the relevant market. In most situations, substantive law is a combination of *per se* rules and rule of reason analysis.

Procedural regulation of licensing arrangements generally takes one of three distinct forms, based on the timing and purpose of the review. Some countries, such as Canada, Germany, the United Kingdom, and the United States do not require pre-effective review of licensing provisions and limit regulation to an after the fact analysis, generally in the context of potential or actual litigation proceedings, of anticompetitive effects. In contrast, the EC and Japan have adopted administrative policies which anticipate the review of license agreements in advance of effectiveness, and have promulgated detailed lists of acceptable and unacceptable provisions to serve as guidelines for parties and regulators. A third method, generally used in the developing countries, is to condition the legality of technology transfer agreements on review and approval by the government.

In many cases, antitrust and competition law issues are initially raised by the parties to the licensing agreement in the course of litigation regarding the arrangement, rather than by regulators. Litigation with respect to licensing arises largely as a function of the uncertainties and risks associated with using and valuing the subject matter of the transaction. Unlike tangible assets, the value of intellectual property rights, and thus any license with respect thereto, can change dramatically over time. Licenses to new technology are often entered into at the outset of the development process of a new product or a new idea, at a point when the ultimate shape of the marketplace is obscure or nonexistent. This means that if and when commercial success becomes a reality, the burdens and benefits of thc license are likely to be out of proportion with the real economic value represented by the license, and one or both of the parties may look to disengage from the licensing arrangement. For example, if the royalty is either significantly above or below its real economic value, one the parties has a strong incentive to void the license through litigation.

Antitrust issues are also often a factor in any litigation respecting patent licensing practices because many licensees are direct competitors of their licensors positioned to argue

[51] See Chapter 4 for a general summary of the regulatory framework relating to antitrust and competition law issues in the United States, the EC, and Japan.

[52] While technology transfer agreements, including technology licenses, are often regulated under separate laws in developing countries, such agreements are typically regulated as part of the general regime of antitrust and competition laws in the developed countries.

that particular provisions of the license amount to *per se* violations of applicable laws. Antitrust and competition law attacks on licensing practices involving patents can lead to extremely harsh results for the patent owner, since the patent can be challenged and invalidated in an antitrust proceeding. Moreover, challenges to patent validity in antitrust and competition law proceedings have been used in the past by regulators to compel patent holders to grant what are in effect "compulsory" patent licenses to third parties in order to ease actual or prospective monopolies.

National practices regarding the regulation of joint ventures are somewhat uniform, at least in those cases where the proposed collaboration involves participants of a specified size in relation to the effected market(s). For example, in the United States, joint ventures may be subject to review prior to the consummation of the transaction, although recent statutory changes have led to a more liberal attitude toward joint ventures which will be limited to research and development and production of the results of the joint research. The EC also reviews joint ventures and other collaborative arrangements in the same manner as it deals with licensing agreements. Finally, countries that extensively regulate direct foreign investment will generally be involved in reviewing any proposed joint venture between a local party and a foreign investor.

D. Organization of Review of Patents and Patent Licensing

The review of the relationship of patents and patent licensing to innovation in this work proceeds as follows. In the next chapter, we present an overview of the legal aspects of patent rights, including the requirements for obtaining patent protection in the United States and the EC. In Chapter 3, we review the extensive theoretical and empirical work that has been carried out with respect to economic aspects of patent rights, including their role as an incentive for innovation. After we review the general framework of antitrust and competition law in the United States and the EC, and the relationship between innovation and collaboration, we describe the basic components of a licensing agreement. Finally, in Chapters 9 through 11 will examine in detail the regulation of patent licensing in the United States and the EC.

III. Joint Ventures

While innovation can be pursued individually, by a single person or business entity, it has been argued that the most effective and efficient way to organize for innovation is through a collaborative effort with one or more other participants.[53] In this context, innovative collaboration brings together the complementary assets and other resources of the participants in a long-term or intermediate-term relationship dedicated to pursuing the promotion of invention, development or diffusion of new products or processes relating to production.[54] The use of a collaborative business relationship has been referred to as the pure private enterprise solution for the preservation of market selection and responsiveness, particularly in those cases where the industry is otherwise fragmented.[55] In addi-

[53] Jorde and Teece, *Innovation and Cooperation: Implications for Competition and Antitrust, Collaboration Among Competitors* (hereinafter *Collaboration*) (1990) 887, 898.

[54] See Brodley, "Antitrust Law and Innovation Cooperation," in *Collaboration*, supra note 53, 911–12.

[55] See, e.g. Jorde and Teece, supra note 53, at 898.

tion, the use of private agreements among firms as a means of coordination has been favorably contrasted to grand schemes of industrial planning and policy.[56]

Due to international competitive conditions which threaten the viability of national industries in the face of competition from abroad, collaborative business arrangements, particularly joint ventures, have become an increasingly common occurrence. For example, in recent years, the United States has seen the use of joint ventures expand beyond the traditional areas of metal industries, energy, and chemicals, into new high technology areas, such as computers, electronics, communications, pharmaceuticals, and financial services.[57] Significant joint ventures have been formed by large, leading firms in different industries as a means for taking advantage of complementary assets and to allow those firms to enter new areas. Xerox and duPont formed a new joint venture to develop and produce improved copying equipment,[58] and Westinghouse and Siemens AG got together to work on the manufacture of factory automation equipment. Joint ventures have even been formed by firms which were direct competitors, including the joint venture between General Motors and Toyota for the production of a limited class of automobiles, the collaboration between Merck and Johnson and Johnson for the development and marketing of new over-the-counter drugs, and the agreement between General Motors and Chrysler to produce manual transmissions for automobiles.[58]

Apart from the social efficiencies and public benefits that may flow from collaborative business relationships, the parties themselves generally realize substantial private benefits from such alliances. For example, firms often see collaborative relationships as a means for reducing the business risks associated with the necessarily large investments required for innovation in these vigorous industries.[60] Joint ventures also provide a mechanism for combining the complementary tangible and intangible assets necessary for the complete development and commercialization of new products. In addition, joint ventures provide a structure for managing some of the potentially adverse externalities which threaten the return on investment made with respect to innovation. Finally, collaborative relationships can, in many instances, achieve beneficial scale economies for the participants and provide the parties with transactional benefits that would not be possible with other contractual structures.

In this work, we begin our review of joint ventures in Chapters 7 and 8, where we examine the basic terms of free-standing joint ventures (e.g. partnerships or corporations in the United States) and contractual arrangements which generally require extensive integration of the skills and resources of the parties. Then, in Chapters 12 through 14, we examine in detail the regulation of joint ventures in the United States and the EC. In particular, we focus on those joint ventures dedicated to some of the earlier stages of the innovative process, particularly research and development.

[56] Ibid. Interfirm agreements are viewed as a means of preserving some of the benefits of the less hierarchical structures associated with smaller enterprises, and Jorde and Teece argue that the ability of a firm to rapidly respond to market needs, which is essential to the concept of innovation, is reduced whenever an organizational structure is excessively hierarchical.

[57] Ibid. Shapiro and Willig, "On The Antitrust Treatment of Production Joint Ventures," in *Collaboration*, supra note 1, at 929, 937.

[58] Ibid. at 933–4.

[59] Ibid. at 934.

[60] Ibid.

2 Patent law in the United States and the European Community

License agreements, and other forms of technology-based transactions, involve trading in various forms of "property rights" which have been recognized under the law as a means for rewarding innovative and creative activities. Intellectual property can take a number of different forms, each of which entitles the owner thereof to certain rights with regard to the use of the property. For purposes of this work, the essential form of intellectual property right is the "patent," which is intended to provide certain legal rights with respect to new or useful processes, machines, manufactures, compositions of matter, as well as new or useful improvements thereof, which have some minimum level of commercial value;[1] however, it should not be forgotten that most technology-based transactions involve the transfer and use of several forms of intellectual property.[2]

While there are sometimes significant variations among the patent laws of various nations, it is generally the case that the holder of a patent has the right, for a fixed period of time, to prevent others from making, using, or selling the item, usually referred to as an "invention," without the consent of the holder of the patent, even if the other party independently discovers the invention.[3] The patent right is an exclusionary right, and is usually not dependent upon the patent owner's affirmative use of the invention. In order to obtain these rights, however, the inventor or his or her assignee must file an application with the appropriate governmental authority which contains a description of the invention and various other items of information designed to demonstrate that the invention meets the legal requirements for patentability (e.g. novelty, utility, etc.). The contents of this application will eventually be published, and will be available to enhance the public knowledge during the term of the patent-based exclusivity, even though the public would not be able lawfully practice the invention until the patent expires.

[1] Intellectual property protection is reserved for items which have commercial value. See, e.g. Cornish, *Intellectual Property: Patents, Copyright, Trade Marks and Allied Rights* (1989), at 5 ("[I]ntellectual property protects applications of ideas and information that are of commercial value").

[2] For example, a party seeking the right to manufacture and distribute a product developed by another party will generally need to enter into a licensing agreement which covers (i) any *patents* owned by the developing party which cover any aspect of the products products which are the subject of the arrangement; (ii) any *trade secrets* possessed by the developing party which might be useful in manufacturing the product or in developing an appropriate market strategy for the sale and distribution of the product; (iii) any *copyrights* owned by the developing party covering tangible works of authorship, including any computer software or advertising copy, relating to the product; and (iv) any *trademarks* owned by the developing party which might be used in connection with marketing the product.

[3] In contrast, while the owner of a trade secret has a cause of action against others who misappropriate the secret, a trade secret owner cannot prevent others from practicing the trade secret if they have independently developed it without breaching any duty to the trade secret owner.

Although the exclusionary rights granted to the owner of a patent can have significant strategic value, there are a number of consequences which the inventor must take into account in deciding to file for a patent, particularly if the invention can be protected under trade secret law.[4] First, the information in the patent application will be available to permit the public to learn about the subject matter, thereby minimizing any advantage that the patentee may have had with regard to the general level of knowledge. Second, competitors can use the information in the patent application in their own research programs, sometimes to allocate resources away from further work on products which would merely infringe upon the patented invention. Finally, it is often possible to use the information to develop non-infringing products which improve upon, or even replace, the original item.

Patent protection extends only to the borders of the country in which the patent right has been granted. Accordingly, patent owners in the United States generally have no right to prevent persons outside the United States from making, using, or selling the invention in a foreign country unless they have also obtained a patent in the foreign country.[5] While there have been ongoing efforts to harmonize the patent laws of the world, significant variations can still be found in national patent laws on such things as the subject matter which is eligible for protection and the scope of the rights granted to the owner of the patent.

I. United States Patent Laws

A. Overview

In the United States, the federal patent law[6] protects any "new and useful process, machine, manufacture, or composition of matter, or any new and useful improvement thereof",[7] which could not "have been obvious at the time the invention was made".[8] An owner of a patent, sometimes referred to as a "patentee," has the right, for a determinate period of time, "to exclude others from making, using, or selling the invention throughout the United States."[9] These exclusionary or "monopoly" rights extend even to those who create the same invention completely independently (i.e. without copying from or even knowing about the existence of the patentee's invention). However, the grant of a patent does not necessarily provide the owner with the full legal right to practice

[4] Trade secrets can include items which might otherwise be eligible for patent protection, like formulas, patterns, devices, or compilations of various pieces of information which afford a competitive advantage to the owner by virtue of the fact that the item is confidential and not known to, or used by, competitors of the owner.

[5] The United States patent will, however, give the patent owner a right to prevent the sale in the United States of products covered by the patent which may have been manufactured outside of the United States, even if the owner does not possess equivalent patent rights in the country of manufacture.

[6] The patent statutes were enacted by Congress under US Const. Art. I, § 8, which grants to Congress the power "[t]o promote the Progress of Science and useful Arts, by securing for limited Times to Authors and Inventors the exclusive Right to their respective Writings and Discoveries." The principal statutory prerequisites to patentability in the United States are codified in Title 35 of the United States Code, which is referred to herein as the "Patent Act."

[7] 35 U.S.C. § 101.

[8] 35 U.S.C. § 103.

[9] 35 U.S.C. § 154.

the invention, since practice may also require the use of other technology patented or owned by someone else.

A fundamental condition to the limited monopoly rights provided to an inventor is the need for the inventor to publicly disclose the subject matter of the patent with particularity in the patent application. The patent application, which must be filed with the Patent and Trademark Office ("PTO") in the United States, must contain a full description of the invention and the specific inventive "claims" that one is seeking to patent. The level of detail that must be disclosed in a patent application must be sufficient to allow one skilled in the art to make and use the invention. The information set out in the application will eventually become available to the public during the term of the patent, thereby contributing to the advance of knowledge, even though practice of the invention described remains subject to the rights of the patentee.

Inventions fulfilling the general statutory requirements referred to above are called utility patents.[10] Protection will also be available for the ornamental characteristics of designs. The design of an object consists of the visual characteristics or aspects displayed by the object. The protection provided by these design patents is for the appearance of the object rather than any functional characteristics.[11] A utility or a design patent may be based on the same subject matter.[12] In addition, a plant patent may be available for certain new varieties of plants that have been asexually reproduced.

In the United States, patent protection has historically extended for a term of seventeen years (fourteen years in the case of design patents); however, recent amendments to the patent laws provide for a term of twenty years, measured from the earliest effective date of a completed application in the United States, for all patents issuing on applications filed on or after June 8, 1995.[13] At the end of this period, the exclusionary right of the owner expires and the right to use the invention passes into the public domain. As a rule, the protection period cannot be extended; however, patents on human drug products, medical devices, food additives, and color additives, which may be subject to regulatory review before marketing, may be extended for up to five years beyond the initial patent term, if certain conditions are met, in order to compensate the inventor for that period of review.[14]

B. Conditions to Patentability

Introduction An inventor must be overcome several hurdles in order to receive a patent for an invention. First, the invention must be within the subject matter stated in

[10] *Manual of Patent Examining Procedure*, US Department of Commerce, pp. 200–1 (1987).

[11] Ibid. at 1500–1.

[12] Ibid. at 1500–3.

[13] The change in the length of the patent term was included as part of the implementing legislation for the GATT (General Agreement on Tariff and Trade) Uruguay Round Agreement (Public L. No. 103-465), which was signed into law on December 8, 1994 (the "1994 Intellectual Property Legislation"). Title V of the 1994 Intellectual Property Legislation implements the Uruguay Round agreements on trade-related aspects of intellectual property rights (TRIPs) (se text below), although other sections thereof also focus on intellectual property matters. The changes with respect to the patent term actually implement a bilateral agreement between the United States and Japan reached in 1994, rather than the provisions of the TRIPs agreement.

[14] 35 U.S.C. § 156. Under the 1994 Intellectual Property Legislation, extensions of the patent term for up to five years are also available to compensate for delays in the issuance of the patent caused by an interference, a secrecy order, or a successful appeal to the Board of Patent Appeals and Interferences or the Court of Appeals for the Federal Circuit.

the statute. With regard to utility patents, this means the invention must be a process, a machine, an article of manufacture or a composition of matter. Second, the invention must be useful, new and not obvious, and these requirements are generally formalized and discussed as the conditions of utility,[15] novelty[16] and nonobviousness.[17]

Subject matter requirement In order to qualify for utility patent protection, an invention must be one of the following: a process, a machine, a manufacture or a composition of matter. In practice, each of these categories have been defined as follows:

(i) *Processes* A process has been defined as a mode of treatment of certain materials to produce a given result. It is an act, or a series of acts, performed upon the subject matter to be transformed and reduced to a different state or thing.[18] For example, patent protection might be available for new or improved processes of manufacturing an existing product.
(ii) *Machines* A machine is a device which have relatively moveable parts and which is capable of performing some new and useful function or operation.
(iii) *Articles of manufacture* Articles of manufacture have been defined as tangible objects, other than machines or compositions of matter, that are man-made and not found in substantially the same form in nature.[19]
(iv) *Compositions of matter* The term composition of matter includes the physical or chemical combination of two or more ingredients in order to produce a mixture or compound. It is possible to obtain protection for liquids, solids or gases. New chemicals, plastics or pharmaceutical mixtures are covered under this category.

The visual characteristics of an object, such as its configuration, shape or surface ornamentation, may be eligible for design patent protection provided that the design is also new and original.[20] Novelty in a design patent is present when an average observer perceives the new design to be different than, rather than just a modification of, an existing design. However, a design patent will not be granted for mere changes which might be made in the finish, color or workmanship of the object.

Finally, a plant patent may be available for distinct and new varieties of plants that have been asexually reproduced by the inventor, including cultivated sports, mutants, hybrids and newly found seedlings, but not "tuber propagated" plants or plants found in an uncultivated state.[21] An application for a plant patent must fully describe plant, the methods of asexual reproduction and the characteristics which distinguish the plant from other known varieties.

Utility requirement The subject matter of a utility patent must be useful.[22] This requirement is generally easily meant by establishing that the invention can be operated

[15] 35 U.S.C. § 101.
[16] 35 U.S.C. § 102.
[17] 35 U.S.C. § 103.
[18] *Cochrane* v. *Deener*, 94 U.S. 780, 788 (1876).
[19] *Diamond* v. *Chakrabarty*, 447 U.S. 303 (1981).
[20] 35 U.S.C. § 171.
[21] 35 U.S.C. § 161.
[22] 35 U.S.C. § 101.

in a manner that accomplishes at least one of its intended purposes. However, patent protection is not available when the intended purpose is immoral, frivolous or mischievous or when the intended purpose is not in some way beneficial to the progress of mankind.

Novelty requirement The novelty requirements under the Patent Act are met by applying Section 102 of the Patent Act.[23] Section 102 sets out various types of disclosures by, or activities of, the inventor or others, occurring before the making of the invention by an applicant or before the filing of a patent application for the invention, which may result in the denial of patent protection for an invention.[24] Specifically, an inventor may not receive patent protection for an invention in the following instances:

(i) *Known or used prior to invention* An invention cannot be patented if it was known or used by others in the United States, or patented or described in a printed publication in the United States or any foreign country, before the invention thereof by the applicant.[25]
(ii) *Patented, published or used prior to application* An invention cannot be patented if was patented or described in a printed publication in the United States or a foreign country or in public use or on sale in the United States, more than one year prior to the date that the application for patent is filed in the United States.[26]
(iii) *Delay in filing patent application after prior foreign patent* An inventor who has obtained a patent or an inventor's certificate with respect to the invention in a foreign country prior to the date of filing a patent application in the United States cannot obtain a United States patent on the invention if the foreign application was filed more than 12 months prior to the filing of the patent application in the United States.[27]
(iv) *Competing patent application filed prior to invention* An invention cannot be patented if was described in a patent granted to another inventor on the basis of an application which has been filed prior to the date that the first inventor actually invented the subject matter of the second application.[28]
(v) *Prior invention by another* The so-called "first-to-invent" requirement of the Patent Act denies patent protection if it is shown that the subject matter of the application was invented by another prior to the date that the subject matter was invented by the applicant.[29]

These events are sometimes collectively referred to as "statutory bars." In some cases, the inventor actually has some degree of control over the occurrence of a particular event,

[23] *Nickola* v. *Peterson*, 580 F.2d 898, 905 (6th Cir. 1978), *cert. denied*, 440 U.S. 961 (1979).
[24] 35 U.S.C. § 102(a)–(b).
[25] 35 U.S.C. § 102(a).
[26] 35 U.S.C. § 102(b).
[27] 35 U.S.C. § 102(d).
[28] 35 U.S.C. § 102(e).
[29] 35 U.S.C. § 102(g). The "first-to-invent" requirement is unique to the United States. In almost all other countries, a patent is granted to the person who is the "first-to-file" a patent application covering the invention, even if such person was not the first inventor of the subject matter.

such as the decision to publish information regarding the invention in a domestic or foreign journal or the choice of filing a patent application outside of the United States without making a concurrent filing in the United States. Nonetheless, even in those cases where a disqualifying event is caused by the inventor, the statutory bars will operate to deny patent protection in the United States.

The date of an invention is a key consideration in the determination of whether several of the statutory bars might apply to a given application. The invention date is either the date that the invention is reduced to practice by the inventor or the date that the invention is conceived if it can be shown that the inventor exercised reasonable diligence following the date of conception to reduce the invention to practice. Reduction to practice may be actual, such as producing a tangible embodiment of the invention, or constructive, which occurs if the invention is fully described in a patent application filed with the PTO.

If two inventors file applications which cover the same subject matter, priority must be determined in an "interference proceeding" conducted by the PTO. In such a proceeding, which is also used when an inventor files an application with claims that are similar to those in an issued United States patent, the party with the latest effective application filing date, referred to as the "junior" party, must bear the burden of proof in showing that the second application should be given priority over the filing made by the inventor, referred to as the "senior" party, with the earlier effective filing date.

In an interference proceeding, the trier of fact will examine each of the aforementioned factors in order to determine the respective dates of invention for both the junior and the senior party. Inventors can increase their changes of prevailing in an interference proceeding by maintaining written records as to the progress of the invention. This record, which can take the form of a laboratory notebook, should include a description of the invention, the results of any experiments and any further matters which might be relevant to the inventor's efforts to try and reduce the invention to practice. Entries should be dated, signed by the inventor and countersigned by a non-inventor who have observed the invention and is qualified to understand the invention.

Nonobviousness requirement In addition to the concepts of utility and novelty, inventions will only receive patent protection if they are not "obvious ... to a person having ordinary skill in the art to which said subject matter pertains."[30] The inquiry that will be made in satisfying the nonobviousness requirement will take into account some of the same factors that determine if the invention is sufficiently novel. Specifically, an examination must be made of various types of disclosures regarding the subject matter that may have been made prior to the filing of the patent application in order to determine the "state-of-the-art" at that time. These disclosures are sometimes referred to as the "prior art."

The United States Supreme Court[31] has stated that the question of nonobviousness requires a determination of the scope and content of the prior art, an analysis of the differences that may exist between the prior art and the claims at issue in the instant patent or patent application, resolution of the level of ordinary skill in the relevant area and, finally, an examination of specified secondary indicia of obviousness. In the same case,

[30] 35 U.S.C. § 103.
[31] *Graham* v. *John Deere Co.*, 383 U.S. 1, 86 S.Ct. 684, 15 L.Ed.2d 545 (1966).

the Supreme Court stated that commercial success of the subject invention, long-felt but unsolved needs, failure of others, and evidence of copying of the claimed invention are all objective indicia of nonobviousness.[32]

Inventorship requirement In the United States, patent applications can only be filed by, and the initial ownership of an issued patent may only vest in, the person or persons who have actually invented the subject matter claimed in the patent application.[33] An "inventor" must be an individual, therefore legal entities, such as a corporation or partnership cannot apply for a patent in the United States. In those cases where the invention resulted from the work of two or more persons, the application must be jointly made by all of the co-inventors.

Many inventions which qualify for patent protection are conceived by employees during the course of their employment by a corporation or other business entity. In the United States, an employer usually has the legal right to have the patent on an employee's invention assigned to it if the employee was hired to invent, provided that the parties have previously entered into a valid, written assignment-of-inventions agreement satisfying the requirements of applicable state statutes. In the event that an assignment-of-inventions agreement has not been executed by the employee, the employer has only a "shop right," or royalty-free license, to use the invention, but does not own the patent or have the right to grant licenses with respect thereto.

Written assignment-of-inventions agreements are now quite commonplace in the United States. In general, such agreements require an employee who conceives an invention which is related to the company's business to assign or agree to assign to the company all of the employee's rights in any patents or patent applications relating to the invention. The employee will agree to execute any and all documents and perform all acts necessary or appropriate to perfect the company's rights in such patents. In order for the company to determine whether there are any inventions which might be covered by the agreement, provision will be made for disclosure by the employee to the company of any and all inventions conceived while using the company's resources.

C. Patent Office Procedures

Introduction The Patent Act sets out the procedures for applying for a patent and patent applications are reviewed by examiners at work within the PTO. Patent applications must satisfy a number of formal requirements, including the need to adequately describe all the distinctive elements of the purported invention. After a patent has been issued, it may be reexamined for various reasons in order to ascertain the validity of the allowed claims.

Contents of the patent applications In the United States, patents are issued by the PTO after review and approval of a complete patent application containing: (i) a speci-

[32] *Graham* v. *John Deere Co.*, supra; *Specially Composites* v. *Cabot Corp.*, 845 F.2d 981, 991 (Fed. Cir. 1988).

[33] 35 U.S.C. § 102(f).

fication including one or more claims; (ii) an oath or declaration;[34] (iii) drawings if necessary; and (iv) the prescribed filing fee.[35] The specification must include a full written description of the invention, and of the manner and process of making and using the invention. The description must contain a detailed description of the preferred embodiment or embodiments of the invention in sufficient detail to enable any person of ordinary skill in that art to make and use the invention.[36] In addition, the specification must set forth the best mode that is contemplated by the inventor of carrying out the invention.[37]

The enablement requirement for the specification does not mean that the inventor must disclose data that would enable mass production of the invention or that the specification must set forth solutions to problems that may be encountered if an attempt is made to apply the invention to a particular use. Also, a specification will not be found to be deficient simply because some experimentation will be required to adapt the invention to a particular use, provided that the amount of experimentation that may be needed is not unreasonable, nor unduly extensive in either quality or quantity.[38]

The best mode requirement for the specification does not mean that the inventor must disclose the absolute best mode of practicing the invention, which may not be known by anyone at the time the application is filed, but rather than the inventor must simply describe the best mode of the invention that is known to him or her. This requirement really goes to the integrity of the disclosure process and is designed to ensure that inventors do not disclose an inferior version of the invention in a patent application simply to obtain the monopoly rights and then use a better mode which may have been known at the time of the filing to obtain a competitive advantage which extends beyond the term of the patent. An issued patent may be invalidated if it can be shown that the applicant knew of and concealed a better mode than that disclosed in the application.

The specification will include a series of numbered paragraphs which set out the various claims regarding the subject matter which the inventor believes distinguishes the invention from prior art. The claims, if allowed by the patent examiner, will serve as the basis for any future determination regarding whether or not another party may have infringed upon the patent covering the invention. The description and claims will usually be supplemented by drawings, such as graphed data and simplified mechanical representations, which are intended to illustrate all of the important features of the invention otherwise discussed in the specification.

Examination of the patent application Once a patent application has been properly filed at the PTO, it is assigned to an examining group within the PTO having particular knowledge regarding the subject matter covered by the application. An examiner within that group reviews the prior art relevant to the application in order to determine whether

[34] In general, the oath or declaration must (i) identify the specification to which it is directed; (ii) list each inventor and the residence and country of citizenship; (iii) state whether the inventor is a sole or joint inventor of the claimed invention; (iv) state that the inventor or inventors believe themselves to be the original and first inventors of the claimed subject matter; (v) state that each inventor has reviewed and understands the content of the specification and any specific amendments thereto; (vi) acknowledge the duty of candor to disclose information material to the examination of the application; and (vii) identify any related applications. See 37 C.F.R. § 1.63.

[35] 35 U.S.C. § 111; 37 C.F.R. § 1.51.

[36] 35 U.S.C. § 112.

[37] 35 U.S.C. § 112.

[38] See *Atlas Powder Co.* v. *E.I. duPont de Nemours & Co.*, 750 F.2d 1569 (Fed. Cir. 1984).

or not the claims are sufficiently distinct so as to satisfy all of the conditions for issuing a patent. The examiner will also make a determination regarding the sufficiency of the specification in light of the formal statutory requirements as to disclosure.

In the event that the examiner makes a determination that the invention is not sufficiently distinct from the prior art to qualify for patent protection, the applicant will be advised that the application has been "rejected." The notification will state the reasons for the rejection. If examiner finds fault with the application itself, the examiner will "object" to the application and advise the applicant of the perceived deficiencies which may have been noted in the application. In either case, applicants will have specified period of time to respond to any comments made by the examiner, either by making such arguments as might be necessary to convince the examiner that the initial comments were incorrect or, if necessary, amending the application itself in an attempt to alleviate the examiner's concerns. If the examiner is satisfied, a patent will be issued covering the claims approved in the examination process.

If the examiner twice rejects any of the claims which are filed by the applicant, or the examiner designates a rejection as "final," the applicant can appeal the examiner's decision to the Board of Patent Appeals and Interferences.[39] If an appeal is made, both the appellant-applicant and the examiner will be given the opportunity to file written arguments which support their positions and, if requested, both parties may make oral arguments before a panel of senior examiners.[40] Once the written and oral arguments have been completed, a decision will be rendered by the panel. If the appellant is still dissatisfied with the opinion of the panel, it may request review by the Federal Circuit or the United States District Court for the District of Columbia.[41]

Reexamination procedures After a patent has been issued, any person, including the patentee, may request reexamination of the patent by the PTO in order to verify its validity.[42] In addition, reexamination can be ordered by the PTO itself.[43] Reexamination can be a valuable procedure when one wants to test the validity of a patent prior to bringing an action for infringement, or before taking some action (e.g. commencing use of a similar process or machine) which might precipitate an infringement action by the patentee. When reexamination takes place, reference may only be made to patents or printed publications which are relevant to determining the patentability of at least one of the claims in the patent.[44] The results of any reexamination will be set out by the PTO in a certificate which describes the new content of the patent. The patentee will have the right to appeal any adverse decision to the Federal Circuit or seek *de novo* review of the matter by the United States District Court for the District of Columbia.

Provisional patent applications The 1994 Intellectual Property Legislation created a new procedure for the filing of provisional patent applications with more limited formal requirements and lower filing fees. Such applications:

(i) are generally considered to be equivalent to foreign priority applications;

[39] 35 U.S.C. § 134; 37 C.F.R. § 1.191.
[40] 37 C.F.R. §§ 1.192–1.194.
[41] 35 U.S.C. §§ 141, 145.
[42] 35 U.S.C. § 302; 37 C.F.R. § 1.510.
[43] 35 U.S.C. § 304; 37 C.F.R. § 1.520.
[44] 35 U.S.C. § 301.

(ii) do not require claims, or an oath or declaration;
(iii) become abandoned if a complete application claiming its priority is not filed within one year;
(iv) have a prior art effect date to defeat the patent of another in a patent issuing on a complete application claiming priority from the provisional application;
(v) may not be involved in an interference, although a subsequent complete application may;
(vi) establish a priority date for subsequent foreign filings under the Paris Convention, which must be filed within one year of the provisional application filing date unless the provisional application is abandoned; and
(vii) may be filed by converting a complete application to a provisional application.

D. Scope of the Patent Right

Introduction A patent right is a form of property. The owner can use a patent for a variety of strategic purposes in the course of its business activities. Like any other property right, a patent can be assigned in whole or in part. The patent owner may protect the exclusive rights embodied therein by legal actions against persons, referred to as infringers, who attempt to make, use or sell the patented invention without the consent of the owner.

Strategic uses of the patent right The owner of a patent right has a variety of strategic options with regard to the use and control of the patent rights, including the following:

(i) In those cases where the holder does not have the financing and production skills to use the invention itself, the patent can be used to induce investors to contribute the capital required for the commercial exploitation of the invention, since the investors can derive some comfort that commercialization of the invention can proceed, at least for some limited period of time, without legal competition in the country where a patent has been issued.
(ii) Obviously, in those cases where the holder does have sufficient functional capabilities (e.g. production, financing and distribution) to commercialize the invention, an aggressive program of patent enforcement can be used to discourage competitors and increase and maintain market share.
(iii) If the holder is unwilling or unable to use, make or sell the underlying invention itself, it can license the right to use the invention to others, a form of "exception" to the monopoly rights inherent in the patent grant, in return for a stream of royalty income based in some way on the licensee's use of the patent rights.
(iv) Patent licenses can be used as a contribution by the holder to a new business enterprise, such as a domestic or foreign joint venture. The ability of the owner to contribute patent rights in such situation and, in some cases, to receive royalties from the use of the patent, will depend upon local laws relating to permissible forms of investment and transfers of technology.

Assignment of patent rights Patents, applications for patents, partial interests in patents and any of the exclusive rights granted to the owner of a patent may all be assigned. An assignment must be in writing and while no specific form of writing is required it is generally good practice to identify the patent or patent application in the instrument by number and date and to identify the name(s) of the inventor(s) and the title of the invention.[45] While it is not necessary to record assignments, grants or conveyances in the PTO, it is advisable to do so promptly, since the law provides that any such assignment will be void as against any subsequent purchaser or mortgagee for a valuable consideration, without notice, unless recorded within three months from their date or prior to the date of the subsequent purchase or mortgage.[46]

Right to prevent infringement A patent owner has the right to commence legal proceedings against persons who infringe upon the owner's exclusive right to make, use or sell the invention claimed in the patent or prevent others from so doing. The patent owner may also have a cause of action against anyone who induces another to infringe the rights of the patent owner, or contributes to that infringement. As noted below, patentees may maximize the amount of damages which might be collectable in an infringement action by appropriately marking the patented invention.

E. Infringement of Patent Rights

Introduction A person who violates the legal monopoly granted to a patentee with respect to using, making and selling the invention which is the subject matter of the issued patent will be liable for patent infringement, as will any person who induces another to commit acts of infringement, or any person who contributes to the acts of infringement. Infringement is generally classified as either direct or indirect. A direct act of infringement is the actual use, production or sale of the subject invention and, as a legal matter, can take one of two different forms: literal infringement or infringement under the "doctrine of equivalents." Indirect acts of infringement generally refer to acts by persons other than a direct infringer to induce, or contribute to, the direct infringement.

Acts of infringer and burden of proof A person need not have notice or knowledge of the patent in order to be liable for direct infringement. The simple act of making, using or selling an invention[47] constitutes infringement regardless of the infringer's intent and knowledge of the patent. The "doctrine of equivalents" has been developed by the courts to reach those cases where literal infringement cannot be proved, but it can be shown that the product produced by the infringer performs substantially the same function, in substantially the same way, to achieve substantially the same result as the claimed invention. In any case, the burden of proof with respect to the issue of infringement, literal or otherwise, is a preponderance of the evidence and always rests with the patentee. Proof of knowledge or intent is required for a finding of liability for indirect infringement.

Defenses A defendant may defend a patent infringement action by denying that the acts in question constituted infringement and by attempting to demonstrate noninfringement. Alternatively, the defendant may raise several affirmative defenses based either

[45] 35 U.S.C. § 261.
[46] 35 U.S.C. § 261.
[47] Under the 1994 Intellectual Property Legislation, offers to sell and importation into the United States are stated to be infringing activities, like "make, use, and sell" under previous law.

upon the underlying invalidity of the patent in suit or specified conduct by the patentee which, under the law, mitigates against enforcement of the patent monopoly. For example, a patentee may be prevented from enforcing the patent rights if it is found that the patent engaged in inequitable conduct while applying for the patent, such as by failing to disclose all known prior art to the examiner. Similarly, a patentee who uses the patent in a manner intended to extend the permitted monopoly beyond the scope of the original claims may lose the ability to enforce the patent under the doctrine of "patent misuse."[48]

Injunctive relief A patentee may obtain preliminary and permanent injunctive relief with respect to an act of infringement. A preliminary injunction will be granted when the patentee is able to show a likelihood of success on the merits with respect to the validity of the patent and infringement, provided that he is also able to show that he will be irreparably harmed absent a preliminary injunction, that a balance of hardships between the parties militates toward the grant of a preliminary injunction and that the grant of a preliminary injunction is not a disservice to the best interests of the public.[49] As a general rule, injunctive relief is now routinely granted once a clear showing has been made that infringing acts have occurred. Permanent injunctive relief for the remaining term of a patent is generally available to a patent owner prevailing in an infringement action.[50]

Damages If a patentee prevails in a suit for infringement, he will be entitled to damages in an amount that is deemed adequate to compensate him for the infringement but in no event less than a reasonable royalty for the use of the invention by the infringer, together with interest and costs as fixed by the court.[51] In addition, an infringer who is shown, by "clear and convincing" evidence, to have intentionally, or "willfully," infringed an issued patent may be liable for treble damages to the patentee.

The measure of damages in an infringement action will be determined with reference to the lost profits of the patentee. Lost profits in an infringement action may take three different forms: (i) lost profits from potential sales which were lost to actual sales made by the infringer; (ii) lost profits from price reductions which the patentee was forced to make in response to competition from the infringer; and (iii) lost profits from any projected lost sales, including the inability of the patentee to raise its prices due to competition from the infringer and a reduction in the rate of the patentee's sales growth after the initial infringement.[52]

Monetary damages for infringement cannot be recovered by patentees unless an appropriate notice or marking has been placed on the invention or, if no notice or marking has been used, the patentee has actually notified the infringer that the patent is being infringed, in which case the damages will be limited to those which are attributable to acts following the notification. Accordingly, once a company has a patent on an invention, it

[48] The inability of the patentee to enforce the patent will generally not extend beyond the period during which the particular use continues to occur. Upon purge of the misuse, the patentee may thereafter regain the right to enforce his patent. *U.S. Gypsum Co.* v. *National Gypsum Co.*, 352 U.S. 457, 465 (1957).

[49] *T.J. Smith & NeRhew Ltd.* v. *Consolidated Medical Equip.*, 821 F.2d 646, 647 (Fed. Cir. 1987).

[50] 35 U.S.C. § 283. *Roche Prods. Inc.* v. *Bolar Pharmaceutical Co.*, 733 F.2d 858 (Fed. Cir.), *cert. denied*, 469 U.S. 856 (1984).

[51] 35 U.S.C. § 284.

[52] *Lam Inc.* v. *Johns-Manville Corp.*, 718 F.2d 1056, 1065–8 (Fed. Cir. 1983). Also, damages may be awarded if the patent owner can show that the price of the patented article would have increased had infringement not occurred. Ibid. at 1065. An award of projected lost profits is proper and not speculative in a two-supplier market if the patent owner can show its sales growth rate decreased during the infringement period and that it had a higher growth rate before and after the infringement period. Ibid. at 1067.

should place a "patent notice" on the invention, which consists simply of the word "patent" or the abbreviation "pat.," together with the number of the patent. Also, in the case of inventions for which an application has been filed but is still pending, it is prudent to mark the invention with the words "Patent Pending" or "Patent Applied For." Although "patent pending" notices have no real legal effect, they may dissuade competitors from further efforts to develop a similar product.

II. International Patent Law Regimes

A. Introduction

The protection provided by the United States patent laws extends only to its borders. Each country has its own patent laws and requires its own national patent to be granted in order for the inventor or his assignee to secure patent coverage therein, and thus the grant of a patent in the United States does not prevent a party in another country from making, using or selling the invention in that country. If the United States patent owner is interested in exploiting the invention in a foreign market, it must apply for and receive the patent rights covering the invention in the foreign market. As it is likely that overseas operations will reach two or even more countries, the prospect of multiple patent applications must be considered with almost every invention with broad market appeal.

B. Similarities and Differences in National Patent Laws

Introduction The International Convention for the Protection of Industrial Property, originally enacted in Paris in 1883 (the "Paris Convention"),[53] provides for equal treatment toward nationals among member states with respect to patent rights and establishes procedures to ease the process of filing multiple patent applications in different countries.[54] However, the equal treatment provisions do not guarantee harmonization of all of the substantive provisions of the various national laws. For one thing, there are a number of different forms of patent rights recognized under foreign patent laws, including utility models, petty patents, inventors' certificates, patents of importation, certificates of addition and patents of improvement. Moreover, substantial variations exist with respect to certain of the key substantive provisions of national patent laws.[55]

[53] The Paris Convention has been supplemented in some respects by the Patent Cooperation Treaty. The latest revisions of the Paris Convention appear at 21 U.S.T. 1629, 24 U.S.T. 2140, 828 U.N.T.S. 305.

[54] The Paris Convention is administered by the World Intellectual Property Organization (WIPO), a branch of the United Nations based in Geneva. This Convention established the general principle of non-discrimination by requiring signatory States to treat non-nationals who applied for a patent within their territory in the same way as they would their own nationals. Moreover, under the Convention, if an individual lodged an application within one signatory State, he could, within one year, lodge applications in other signatory States, and the later applications would be backdated to the date of the original application.

[55] The discussions within the Uruguay Round, referred to below, represent one effort at harmonizing the patent laws of the various countries. The World Intellectual Property Organization has also been attempting for many years to structure a treaty that would achieve harmonization, although substantial differences between developed and developing countries have hampered talks in that area.

Eligible subject matter Each country is free to establish its own definition of the types of products for which a patent may be granted. While the United States has argued that patent protection must be available for all products and processes which satisfy the criteria of novelty, utility, and nonobviousness, a number of countries have explicitly precluded patentability for products in specified areas, such as pharmaceuticals. Also, even if patent rights are available, the government may sometimes reserve the right to license the patented invention to others in the name of the "public interest," thereby undercutting the utility of the monopoly usually associated with patent rights.[56]

Scope of patentable claims Even though an invention is fully described in the application, the scope of protection is determined by the claims which are allowed in the examination process. Thus, even in cases where the scope of available subject matter protection is fairly broad, various administrative practices of the patent authority may force claims to be applied so narrowly that others can easily avoid claim coverage and obtain their own patents based on very slight variations of the original invention that lack true innovation.

Term of protection The term of patent protection around the world is not uniform. The United States and other developed countries seek a minimum term for patent protection of at least twenty years from the date that the patent application is filed. However, in many cases, the patent term in developing countries is far less than the norm in the developed countries, sometimes extending no longer than five years.

Persons entitled to patent protection While the United States issues patents to the "first inventor," assuming that the application has been filed in a timely fashion, almost all other countries give priority to the inventor who is the "first to file" an application, even if the another conceived the invention at an earlier date.

Working requirements Patent rights in some countries may be subject to early termination because the invention has not been used, or "worked," in the country during a specified period.[57] Alternatively, failure to work an invention may result in the grant of a license to third parties at low royalty rates without the consent of the patent holder.[58] In some developing countries, products covered by a patent may still be imported until the patent is actually worked in the country.

Enforcement The ability of a patent holder to enforce the patent rights effectively varies substantially around the world. Among the factors that must be considered regarding the potential patent litigation in any foreign jurisdiction are the time limits for bringing an infringement action, how far back in time the patent owner can go in collecting past damages,[59] whether or not injunctive relief is available, whether a bond must be posted

[56] In other cases, the government may, in the "public interest," issue a new patent which infringes upon some of the claims of an existing patent.

[57] Importation of the patented goods may not be sufficient to satisfy the working requirements in some countries. In such cases, the patent holder must attempt to locate a local partner to manufacture the goods, which may not be practical in light of the local resources.

[58] Compulsory licensing and "working" requirements are not confined to developing countries. Such provisions exist in Japan, Canada and portions of Western Europe.

[59] Under some laws, no infringement action may be brought against persons who, in good faith, may have begun to produce or sell a patented article prior to the date that the patent application is published.

by the plaintiff, the use of discovery proceedings, the amount of time that it takes to prosecute an infringement action and the attitude of local courts toward patent holders, particular when the patent is held by a foreigner.

C. Patents and GATT Negotiations

Patents, and other forms of intellectual property rights, were considered as part of the recently completed Uruguay round of negotiations for amendment of the General Agreement on Tariffs and Trade ("GATT"). In those discussions, consideration was given to so-called trade-related intellectual property issues ("TRIPs"), and an agreement reached in April 1994 included the following provisions:

1. Patents will be available and enjoyable without discrimination as to the place of invention, the field of technology, and whether the product is imported or produced locally.
2. The only types of inventions that countries can exclude from patentability are those whose exploitations would prejudice public order or morality and those involving diagnostic, therapeutic, or surgical methods for the treatment of humans or animals, new strains or races of plants or animals, or essentially biological processes for the production of plants and animals.[60]
3. Compulsory licenses or other "official licenses" are to be permitted only after consideration of the individual situation in which such a license is requested and, except in cases of national emergency, are subject to a number of conditions, including the following:
   (a) The requester of the license has used its best efforts to obtain a voluntary license on reasonable commercial terms.
   (b) The license will be terminated if the circumstances leading to its grant have ceased and are unlikely to recur.
   (c) The holder of the compulsory license pays adequate compensation for its right to use the invention.
   (d) Determination of the amount of adequate compensation is subject to independent review.
   (e) Where such a license is granted in order to enable the use of the subsequent patented invention, the license shall be granted only if the later invention is "an important technical advance of considerable economic significance" relative to the document patent and the owner of the document patent is entitled to a cross-license under the secured patent.
4. The minimum duration of a patent is to be twenty years from its filing date.
5. A requirement that in patent infringement trials involving patents granted for processes, there will be a presumption that a product that could have been made by the process was made by the patented process if either (a) the product itself is new or (b) there is a substantial likelihood that the product was made by the

[60] Countries taking advantage of this provision to preclude the grant of patent for new plants must, however, provide some alternative means of protection for such plants.

patented process and reasonable attempts made by the patentee have been unable to find out exactly what process was used.[61]

III. EC Patent Laws

A. Background

Patent rights have a long history within the territories that now comprise the EC. Evidence indicates that patents were being granted in Venice in the fifteenth century. In England, patents arose out of the practice of the Crown of granting to its favored subjects royalty monopolies or exclusive privileges or franchises with respect to trade in commodities, as well as over the use of new discoveries. This practice generated a great amount of controversy and opposition in England, leading to the passage of the Statute of Monopolies in the seventeenth century, which voided "all monopolies, and all commissions, grants, licenses, charters, and letters-patent."[62] However, a limited exception to the Statute of Monopolies was made for "letters patents and grants of privilege for the term of fourteen years or under ... of the sole working or making of any manner of new manufacture within this realm, to the true and first inventor or inventors of such manufactures ...".[63]

The existence of a number of separate national regimes of intellectual property in the EC constitutes an obvious barrier to the creation of a truly integrated market. Law and policy in the EC relating to intellectual property, since the coming into force of the EC Treaty, has reflected an attempt to reconcile two opposing principles, on the one hand, the free movement of goods and services, and, on the other hand, the national protection of intellectual property rights.[64] Early EC initiatives attempted to minimize this conflict through the development of intellectual property rights with a uniform effect throughout the EC (e.g. an EC patent or trademark). However, actual harmonization of EC intellectual property rights has only been pursued in earnest by the Commission since publication of the White Paper in 1985.[65]

In the White Paper the Commission noted that "Differences in intellectual property laws have a direct and negative impact on intra Community trade and on the ability of enterprises to treat the common market as a single environment for their economic activities."[66] The Commission has since taken a series of initiatives aimed at harmonizing, as far as possible, national laws with respect to patents, trade marks, and copyrights. At

[61] As described in note 13 supra, the United States has recently completed substantial amendments to its intellectual property laws (i.e. the 1994 Intellectual Property Legislation) in order to comply with the TRIPs provisions.

[62] See 1 Lipscomb III, *Walker on Patents* § 1.15 (ed. 1985) (quoting English Statute of Monopolies, 21 Jac. 1 ch. 3 (1623)).

[63] 21 Jac. 1, ch. 3 (1623), reprinted in Schwartz, Flynn and First, *Free Enterprise and Economic Organization: Antitrust* 1–2 (6th ed. 1983). For a description of the evolution of patent law in the United Kingdom, see, e.g. Cornish, supra note 1, at 6–13; Bainbridge, *Intellectual Property* (1992), 244–5; and Dutton, *The Patent System and Inventive Activity During the Industrial Revolution 1750–1852* (1984).

[64] While Art. 30 of the Treaty provides that all "quantitative restrictions" and "measures having equivalent effect" are prohibited between member states, Art. 36 of the Treaty goes on to provide that restrictions on imports or exports may be justified on grounds relating *inter alia* to "the protection of industrial and commercial property". For detailed discussion on the application of Art. 36, see Chapter 4.

[65] *Completing the Internal Market*, White Paper from the Commission to the Council, COM(85)310 Final (June 14, 1985) (the "White Paper").

[66] Ibid. at ¶ 145.

the same time, renewed efforts have been made by the EC at a political level to facilitate the introduction of an EC patent and trade mark. Also, the Green Paper on industrial design, published in June 1991, recommended the introduction of an EC design together with a minimum harmonization of design laws of the member states. In addition, since the publication of the Green Paper on copyright in 1988, the Commission has proposed a number of directives relating to the harmonization of copyright and neighboring rights.

B. Harmonization of EC Patent Laws

Introduction Harmonization of patent laws in the EC has, in large part, been achieved outside the scope of the Treaty through a variety of international agreements, including the Paris Convention, the Strasbourg Convention,[67] and the European Patent Convention (the "EPC"), which came into force on October 7, 1977.[68] Although the Strasbourg Convention set out much of the substantive law necessary for the emergence of a harmonized patent system, significant differences continued to exist between the patent laws of the various member states into the mid-1970s, and persons wishing to acquire patent protection were obliged to make individual patent applications with the competent patent authority of each member state, a costly and time consuming exercise. However, the EPC created a mechanism for rationalizing and simplifying the procedure for granting patents by enabling a person to file a single patent application which would be subject to a uniform examination procedure, and to obtain a patent having the same effect as a national patent in any, or all, of the Contracting States. Also, in 1975, a Community Patent Convention (the "CPC") was adopted to establish a single Community Patent, having a unitary character throughout the Community, and having a common system of law applied to it in such matters as revocation and infringement.[69]

European Patent Convention The EPC established the European Patent Organization, which consists of a European Patent Office ("EPO") and a managing body called the Administrative Council. The task of granting European patents is carried out by the EPO, under the supervision of the Administrative Council.[70] The European Patent Organization is financed through the imposition of fees and through payments and special contributions by the contracting states. The purpose of creating the European Patent Organization as a centralized system was to avoid the duplication of effort and improve the overall quality of the patent inspection.

[67] Council of Europe Convention on the Unification of Certain Points of Substantive Law on Patents for Inventions, E.T.S. No. 47, opened for signature on November 27, 1963. The Convention entered into force on August 1, 1980.

[68] Convention for the Grant of European Patents.

[69] Convention for the European Patent for the Common Market, OJ Eur. Comm. (No. L17) 1 (1976). A number of amendments have been made to the Convention over the years. For example, the CPC was supplemented by three Protocols which are part of the Community Patent Agreement (89/65JEEC) signed on December 15, 1989. These protocols covered: the settlement of litigation concerning the infringement and validity of EC patents; privileges and immunities of the Common Appeal Court; and the statute of the Common Appeal Court. Despite the existence of the EPC, the Commission has gone on record as regarding the establishment of a unitary EC patent as fundamental to the completion of the internal market. See, e.g. the White Paper at ¶ 148 (Commission emphasized that the EC patent would offer important advantages to industry and favored its immediate entry into force, if only among the member states in a position to ratify it).

[70] EPC, Art. 4.

Much of the substantive law in the EPC is based on the Strasbourg Convention of ten years earlier, particularly as regards the subject matter of a patent application. Under the EPC, a European patent will be granted for any inventions which are susceptible of industrial application, which are new, and which involve "an inventive step."[71] Excluded from the relevant definition are scientific theories, aesthetic creations and methods for medical and veterinary treatment.[72] An invention is considered to be new if it does not form part of the state of the art[73] and is considered as involving an inventive step if, after having regard to the state of the art, it is not obvious to a person skilled in the art.[74]

An application for a European patent may be filed in the EPO, or branch office, or at the competent patent authority of a member state, which will forward it to the EPO's main office.[75] Once the application has been filed, it is then subject to an examination and search procedure, after which the EPO will decide whether to refuse or grant the European patent.[76] Applications filed under this procedure are published eighteen months from the priority date. Within nine months of the publication of the fact of the grant of the European patent, any person can give notice of opposition to the patent so granted.[77] Once granted, the European patent takes effect as the equivalent of a national patent in all, or if requested, a designated number, of the member states which adhere to the Convention. Any infringement of the European patent is thereafter dealt with by the national courts.[78] The term of the European patent is twenty years from the date of filing of the application.[79] The EPC, which allows an applicant to acquire a "bundle of rights" with obvious savings in costs and administrative time, has been quite successful, although the lack of a common appeal structure from decisions which are made by national courts regarding patent litigation can, in theory, lead to variations in the treatment of a European patent among member states.

Community Patent Convention The CPC establishes the procedures for the EPO to grant a "Community Patent" which, rather than taking effect as a bundle of national rights, would have a unitary character in the EC and a common system of law provided for it.

[71] The criteria for the grant of a European patent are comparable to that in many of the member states. For example, in the United Kingdom, Patents Act (1977) § 1(1) provides that "A patent may be granted only for an invention in respect of which the following conditions are satisfied, that is to say – (a) the invention is new; (b) it involves an inventive step; (c) it is capable of industrial application ...". With regard to the relevant law in the United Kingdom, see Cornish, supra note 1; Bainbridge, supra note 64; and White, Jeffs, Jaboc, Cornish and Vitoria, *Encyclopedia of United Kingdom and European Patent Law* (1977).

[72] EPC, Art. 52. The Commission has also published a draft directive on biotechnological innovation that seeks to harmonize patent protection in this rapidly developing field, which includes genetic engineering. It may be possible to patent "products of nature" but only if the human intervention required is beyond the level of merely selecting biological material and allowing it to perform an inherent biological function under natural conditions. A draft regulation on Community Plant Variety Rights has also been prepared, although completion thereof will require resolution of the issues raised by the farmer's privilege in relation to the seeds they produce.

[73] EPC, Art. 54.

[74] EPC, Art. 56.

[75] EPC, Art. 75(1).

[76] EPC, Arts. 90–8.

[77] EPC, Art. 99.

[78] EPC, Art. 64.

[79] EPC, Art. 63(1). Procedures have been implemented for extending the effective period of patent protection for certain pharmaceutical to take into account the delays in commercialization generally caused by regulatory reviews of the safety and efficacy of the pharamaceuticals. The extension is effected through the issuance of "supplementary protection certificates." The measures are similar to regulations previously adopted in the United States and Japan.

A Community Patent would have equal effect throughout the Community and could only be granted transferred, revoked or allowed to lapse with respect to the EC as a whole.[80] Moreover, it would only be subject to the provisions of the CPC and, in appropriate cases, those conditions of the EPC which are binding on every European patent.[81] Therefore, the criteria included in the EPC with respect to the conditions for the grant of a European patent are equally applicable to a patent to be granted under the CPC. The CPC system is not exclusive, and when and if the CPC comes into force, an applicant would have the choice of: (i) securing patents individually in the member states; (ii) availing of the EPC and acquiring a bundle of rights in different member states; or (iii) acquiring a Community patent which is valid throughout the EC.

No separate EC patent office would be established in order to administer the CPC, rather a patent administration division would be set up within the EPO for the purpose of making decisions with respect to entries in the register of Community Patents.[82] The preparation of applications for a Community Patent will be somewhat complicated by the requirement that the specifications must be translated into all official languages of the EC, and lodged in the EPO which will, on request, forward a copy of the translations to the competent authority of each member state.[83] A Community Patent will be considered to be "property" and, as such, may be licensed or transferred in the same way as national patents.[84]

While it has been many years since the CPC was first proposed and adopted, the Community Patent has yet to come into existence. One of the major problems with a Community Patent is the expense associated with complying with the requirement that the patent application must be translated into all the official languages of the EC,[85] since failure to file all translations will render the Community Patent invalid.[86] In fact, industry representatives claim that the application preparation burdens are so substantial, that it is unlikely that the CPC will ever come into effect, despite the opinions of the Commission and various regulators.

EC design laws The law of designs is concerned with the protection given to the visual form or features of industrial or consumer products.[87] The degree of variation among national systems for the protection of industrial designs in the EC is far greater than is the case with patents and other forms of intellectual property rights. While the White Paper failed to address the issue of the legal protection of industrial designs, the Commission produced a Green Paper on this subject in July 1991, which emphasized that the diver-

[80] 1975 CPC, Arts. 1–2.
[81] 1975 CPC, Art. 2(3).
[82] 1975 CPC, Art. 8.
[83] 1975 CPC, Art. 33.
[84] 1975 CPC, Arts. 39–48.
[85] 1975 CPC, Art. 30.
[86] 1975 CPC, Art. 30(6).
[87] Lahore, "Harmonization of Design Laws in the European Communities: The Copyright Dilemma," *Common Market Law Review* 20 (1983), 233. A "design" has been defined as the two-dimensional or three- dimensional features of the appearance of a product which are capable of being perceived by the human senses as regards form and/or color and which are not dictated solely by the technical function of the product. Thus the internal parts of a product, which are not visible to the eye, will usually not be protectable, save in exceptional circumstances. See Art. 3 of the preliminary draft proposal for a Regulation on the Community design (the "Draft Design Regulation"), which is contained in Annex 1 to the Green Paper on the Legal Protection of industrial design, Brussels, June 1991, at 3 (Unpublished, Doc 111/F/5131/91) (the "Design Green Paper").

sity of protection available in the member states was incompatible with the creation of internal market conditions.[88] In an effort to remedy the situation, the Green Paper envisaged a two-pronged approach: first, the creation of a "Community Design" system, comprising both registered and unregistered protection, which would be implemented through a Regulation,[89] and second, the limited approximation of the design laws of the member states, which would be implemented through a Directive.[90]

The Community Design would involve a two-tier regime of protection. First, a registered Community Design granted by a Community Design Office would provide up to twenty-five years protection. Second, an informal, unregistered Community Design would provide protection for a period of three years from the publication or public use of a design within the EC. A design will be protected as a Community Design to the extent that it has a "distinctive character," which means that: (i) it is unknown to other specialists in the sector concerned operating in the EC; and (ii) through the overall impression it displays in the eyes of the relevant public, it distinguishes itself from any other design known to such circles.[91] Once registered, a Community Design provides protection with a blocking or monopoly effect which will extend to any design which in the eyes of the relevant public displays an overall impression of "substantial similarity" to that of the Community Design.[92]

The Draft Design Directive contains many of the same substantive features included in the Draft Design Regulation, including the definition of a design, the test of distinctive character, and the term of protection. In most member states protection under specific design protection law can be cumulated with possible protection under copyright law. The Draft Design Directive proposed that designs registered in accordance with the Directive must also be eligible for copyright protection. The extent and conditions under which such protection is conferred, including the level of originality required, would fall to be determined by each member state.[93]

[88] Ibid.

[89] The Draft Design Regulation, supra note 87.

[90] See the preliminary draft proposal for a Directive on the approximation of the legislation of the member states on the legal protection of design (the "Draft Design Directive"), which is contained in Annex 1 to the Design Green Paper, supra note 87.

[91] Draft Design Regulation, Art. 4.

[92] Draft Design Regulation, Art. 18.

[93] Draft Design Directive, Art. 14(1).

3 Economic Aspects of Patent Rights

I. Introduction

As we saw in Chapter 1, it is now generally accepted that recognition of patents and other forms of intellectual property rights is based primarily on economic, rather than traditional property rights, factors. Accordingly, it becomes important to closely examine each of the main economic justifications for creation and maintainance of these exclusive legal rights. In this chapter, we describe four of the main economic theories of patent rights, and review and analyze some of empirical data that has been collected in connection with each of the theories. In addition, taking patent rights as a given, we explore some of the economic consideration relating to the legal boundaries of the patent right itself.

II. Economic Justifications of the Patent System

A. Introduction

There are several important consequentialist theories with respect to intellectual property rights. These theories have been generally referred to as the incentive-to-invent theory, the incentive-to-disclose theory, the incentive-to-innovate theory, and the prospect theory. Each of theory raises its own unique issues and have been the subject of scholarly debate and empirical examination.[1] Moreover, consequential justifications for patents and other forms of intellectual property rights have been the focus of general criticism, including skepticism regarding the need for the state to create special incentives for in-

[1] While discussed herein as separate theories, in most cases the debate regarding the economic justification of patent rights simultaneously includes stimulation of inventive activities, the need to facilitate disclosure of valuable information, and the development of incentives for creating new products and markets. See, e.g. Cornish, *Intellectual Property: Patents, Copyright, Trade Marks and Allied Rights* (1989), at 79 ("[Patents] are intended to encourage the making of inventions and the subsequent innovative work that will put those inventions to practical use; and they are expected to procure information about the invention for the rest of the industry and the public generally, which otherwise might be withheld, at least for a period that could be critical."). See also the Report of the Banks Committee on the British Patent System 1970 (Cmnd. 4407, 1970, 1): "The basic theory of the patent system is simple and reasonable. It is desirable in the public interest that industrial techniques should be improved. In order to encourage improvement, and to encourage the disclosure of improvements in preference to their use in secret, any person devising an improvement in a manufactured article, or in machinery or methods for making it, may upon disclosure of his improvement at the Patent Office demand to be given a monopoly in the use of it for a period of ... years. After that period it passes into the public domain; and the temporary monopoly is not objectionable, for if it had not been for the inventor who devised and disclosed the improvement nobody would have been able to use it at that or any other time, since nobody would have known about it. Furthermore, the giving of the monopoly encourages the putting into practice of the invention, for the only way the inventor can make a profit from it (or even recover the fees for his patent) is by putting it into practice; either by using it himself, and deriving an advantage over his competitors by its use, or by allowing others to use it in return for royalties."

ventive activities,[2] concern about extending the role of law beyond the distribution of rights into the highly sensitive and political area of promoting some social policy which an interest group may perceive as being necessary or desirable,[3] and hostility toward any attempt by the state to override people's revealed preferences in the production of goods and services to impact the rate of innovative activity, at least when compared to "more basic issues" in production related to health, safety and education.[4] In addition, some commentators, while recognizing the good consequences which flow from inventive activity, continue to resist patents as unreasonable restraints on personal liberties.[5]

B. Incentive-to-Invent Theory

Statement of the theory The incentive-to-invent theory is based on the premise that the patent system, by providing the inventor with the legal right to exclude others from making, using, and selling his invention, provides the requisite economic incentive to encourage inventors to incur the costs necessary for the development of new inventions. The incentive-to-invent theory is based on the belief that without a patent system, competitors who have not contributed to the costs of invention will "free ride" upon the efforts of the inventor, imitate the invention, and drive down the price of the invention to the marginal cost of production, thereby precluding the original inventor from recovering any portion of his original development costs. Over time, inventors would begin to reduce the expectations regarding their anticipated rate of return on investments in the research required to generate new inventions. If the anticipated return falls far enough, the incentive for investment in research may disappear, and the resources (e.g.financial capital and labor) may well be redirected to other areas. As a result, society will underinvest in new inventions, thereby preventing the discovery of important knowledge or, at a minimum, substantially delaying its creation and use.[6]

[2] It is known that universities and similar institutions will continue to invent without the need for any special economic inducement and that many business enterprises will engage in research and innovation as a matter of competitive necessity, whether or not patent protection is available. A number of writers have argued that new inventions will arise independent of any state incentives or property right awards whenever the state of basic knowledge or other social conditions are favorable. See, e.g. Usher, *A History of Mechanical Inventions* (1929) 1–31; Gilfillan, *The Sociology of Invention* (1935) 71–8; Ogburn, *Social Change* (1923) 86; Machlup, *An Economic Review of the Patent System*, Subcommittee on Patents, Trademarks, and Copyrights of the Senate Committee on the Judiciary, Study No. 15, 85th Cong. 2d Sess (1958), 23 n120 and 24 n127; Kahn, "Fundamental Deficiencies of the American Patent Law," *American Economic Review* 30 (1940), 475, 479–81.

[3] See Nance, "Forward: Owning Ideas," *Harvard Journal of Law and Public Policy* 13 (1990), 757, 765–71.

[4] Ibid.

[5] Ibid. See also Palmer, "Intellectual Property Rights: A Non-Posnerian Law and Economics Approach," *Hamline Law Review* 12 (1989), 261, 281 (referring to Legget, *Democratick Editorials: Essays in Jacksonian Political Economy* (1984) 399) (arguing that the right of a free exercise of our thinking faculties is given by nature to all mankind and these faculties cannot be the object of exclusive rights).

[6] See, generally, Bowman, *Patent and Antitrust Law* (1973), 2–3; Scherer, *Industrial Market Structure and Economic Performance* (1970), 379–99; McGee, "Patent Exploitation: Some Economic and Legal Problems," *Journal of Law and Economics* 9 (1966), 135; Usher, "The Welfare Economics of Invention," *Economica* 31, 279 (1964); Arrow, "Economic Welfare and the Allocation of Resources for Invention," in National Bureau of Economic Research, *Rate and Direction of Inventive Activity* (1962) 609; Nelson, "The Economics of Invention: A Survey of the Literature," *Journal of Business* 32, 101 (1959); supra, note 2; Polanyi, "Patent Reform," *Review of Economic Studies* 11 (1944), 61; Plant, "The Economic Theory Concerning Patents for Inventions," *Economica* 1 (1934), 30.

While recognizing that important inventions sometimes are created without any economic incentive,[7] the proponents of the incentive-to-invent theory argue that inventive activity will usually only occur if the inventor is provided with some assurance that he will be able to appropriate a sufficient amount of the social value of the invention in order to justify the research and development expenditures necessary for the invention to be completed.[8] They also point out that subsidies for invention are particularly important given the high levels of risk which are involved in research.[9] As a result, they conclude that the protection from competition which are afforded by the patent system, which allow the inventor to charge prices which correspond to the value of the invention to the users, can be justified as a means for achieving the requisite match of the private benefits of invention and the value of the invention to society.[10] It has also been suggested that patents, to the extent that they serve to protect the inventor's investment of otherwise scarce resources, are no different than any other set of legal rules which serve to protect the products of risk capital.[11]

Advocates of the incentive-to-invent theory usually point to the language of the Patent Clause of the United States Constitution as support for their view of the underlying purpose of the patent system. The framers of the Constitution, mindful of the experience of their predecessors in England, still provided the Congress with the power "To promote the Progress of Science and useful Arts, by securing for limited Times to Authors and Inventors the exclusive Right to their respective Writings and Discoveries."[12] Since that time, the Supreme Court, while recognizing that the patent laws did stimulate the efforts of genius by providing a reasonable reward to inventors, has consistently reiterated that patents are primarily designed as a means for promoting scientific progress.[13]

Analysis of the theory In analyzing the incentive-to-invent theory, consideration must be given to the effect that such an incentive might have on research activities which might otherwise occur naturally in the marketplace, as well as the real need to provide such incentives in the face of other non-patent competitive factors which might induce firms

[7] See, with regard to inventors and inventive activities, Dutton, *The Patent System and Inventive Activity During the Industrial Revolution 1750–1852* (1984), 103–21; Meinhardt, *Inventions, Patents and Monopoly* (1950), 3–11.

[8] See, e.g. Machlup and Penrose, "The Patent Controversy in the Nineteenth Century," *Journal of Economic History* 10 (1950), 1, 20–5 ("Industrial progress is desirable to society. Inventions and their exploitation are necessary to secure industrial progress. Neither ... will be obtained to any adequate extent unless inventors and capitalists have hopes that successful ventures will yield profits which make it worth their while to make their efforts and risk their money. The ... most effective way for society to hold out these incentives is to grant exclusive patent rights in inventions.") See also Machlup, *The Production and Distribution of Knowledge in the United States* (1962), 162–70; Jewkes, *The Sources of Invention* (1958); Scherer, supra note 6, at 384; Baxter, "Legal Restrictions on Exploitation of the Patent Monopoly: An Economic Analysis," *Yale Law Journal* 76 (1966), 267, 268–9; Machlup, supra note 6, at 57–8; Arrow, "Economic Welfare and the Allocation of Resources for Invention," in *The Rate and Direction of Inventive Activity* (1962).

[9] Ibid. at 614–15; Nelson, "The Simple Economics of Basic Scientific Research," *Journal of Political Economy* 67 (1959), 297.

[10] See Baxter, supra note 8, at 270.

[11] See, e.g. Schwartzman, *Innovation in the Pharmaceutical Industry* (1976), 164. If this is so, it is possible to argue that the patent system is appropriate without any particular analysis of its efficacy.

[12] U.S. Const., Art. I, § 8, cl. 8.

[13] See *Pennock* v. *Dialogue*, 27 U.S. (2 Pet.) 1 (1829), as quoted in *Motion Picture Patents Co.* v. *Universal Film Mfg. Co.*, 243 U.S. 502, 510–11 (1917). In *Motion Picture Patents*, the Supreme Court also said that "this court has consistently held that the primary purpose of our patents laws is not the creation of private fortunes for the owners of patents but is 'to promote the progress of science and useful arts". Ibid. at 510–11. But, see below for a discussion of how the language used in the Constitution has been cited by some as restricting the use of patents as an incentive for activities beyond "invention," such as "innovation."

to engage in inventive activity even if patent protection was not available. In addition, it may well be important to consider whether or not the patent system actually provides an effective incentive to produce the intended increase in inventive activity.

Potential adverse effects on research activities Any incentive to engage in a specified activity must be weighed against the effect that it might have upon the level of other activities which involve the same resources. The existence of the patent incentive has been criticized on a number of grounds for possible distortions on the overall efficiency of development activities. For example, many commentators claim that recognition of patents leads to over-investment in simply "inventing around" an existing patent, thereby wasting the scarce resources available for use in research and development. Others worry about the effect that patents might have on the incentive to develop improvements to existing inventions, as well as the possibility that patents might cause rivals to unwisely shift the focus of their own research and development programs.

Over-investment in substitute inventions One alleged problem with the patent system is the risk that the exclusive rights might lead to over-investment in substitute inventions, at the expense of investment in creating new knowledge that would build upon, rather than duplicate, the work of prior inventors. For example, rivals may have incentives to produce substitute inventions rather than pay for the right to use the patented invention if the estimated costs of producing substitutes are lower than the royalty rate which might be charged by the holder of the original patent. Some economists have argued that the expense of producing substitute inventions is a social cost of the patent system, since these expenses could have been avoided if rivals were simply allowed to use the patented item without paying tribute to the original inventor.[14]

In response, various commentators have cited the benefits which might be derived from the products or processes which result from inventing around patents. For example, a number of judicial opinions have looked with favor on the incentive to invent around existing patents as a mechanism for promoting further technical progress.[15] Moreover, a strong argument has been made that it is reasonable to assume that competitors which elect to incur the costs of inventing around the original patent would not do so unless they believed that the research and development work would allow them to create superior substitutes for the prior invention.[16] If, in fact, the substitutes are of greater value, then inventing around may not necessarily be socially wasteful.

Effect on incentives to develop improvements A number of commentators have expressed the concern that a patent, once granted, reduces the incentives to third parties to engage in any further research toward improving upon the invention covered by the patent, since the improvements presumably could not be exploited without the approval of the owner of the original patent.[17] Any reduction in incentives to develop

[14] See Scherer, supra note 6, at 386–7 (noting that while inventing around has probably had a positive effect upon the rate of technological progress in some fields, it may well be that the resources used in such efforts might be better spent on activities with "higher social incremental payoffs").

[15] See, e.g. *Yarway Corp.* v. *Eur-Control USA*, 775 F2d 268, 277 (Fed Cir 1985); *State Indus.* v. *A.O. Smith Corp.*, 751 F.2d 1226, 1236 (Fed Cir 1985); *Kimberly-Clark Corp.* v. *Johnson and Johnson*, 745 F.2d 1437 (Fed. Cir. 1984).

[16] Bowman, supra note 6, at 21–2.

[17] Scherer, supra note 6, at 392; Baxter, supra note 10, at 270; Machlup, supra note 6, at 64; Kahn, supra note 2, at 482; Plant, supra note 6, at 46.

improvements to patented inventions must be seriously considered, since it is often the case that the improvements prove to be just as useful, if not more so, than the original invention. However, an improvement may be afforded patent protection in its own right and the owner of the patent on the improvement may find himself in possession of an asset that is of significant value to the original patent owner and her licensees. As such, the owner of the improvement may be able to negotiate with the original patent owner for a cross-license which allows each party to exploit both the old and the improved technology on a more competitive basis.[18]

Shift in focus of research and development activities Another set of arguments regarding the adverse effect of patent rights concerns the possibility that firms will shift the focus of their research and development activities away from those areas in which the return on investment appears to be limited by the existence of a large number of existing patents,[19] or away from fields in which patent protection is not available.[20] If rivals have already established a strong patent position in a given area, firms may be reluctant to risk becoming involved in costly litigation with respect to alleged infringement. Also, since basic research is unlikely to produce knowledge that would qualify as inventions, it might be expected that research resources would be shifted from basic research to applied research,[21] creating the possibility that the flow of valuable new knowledge would be retarded.

Availability of non-patent barriers to competition A large body of literature suggests that patents may not be necessary to induce firms to invest in new inventions, since they will be adequately protected by other barriers to competition. For example, in many cases, potential competitors will be dissuaded due to the difficulty of acquiring the technical knowledge necessary to manufacture a complex product and the expenses which are associated with developing a service organization to support the product.[22] In addition, other valuable non-patent barriers to market entry include production facilities, managerial experience, and marketing and distribution channels, each of which may have a sufficient deterrent effect upon competition so as to allow the inventor to achieve a sufficient return on investment in invention without the need for patent protection.[23]

A number of empirical studies have been conducted on the importance of patent protection in allowing firms to secure the benefits of research and development activities. Mansfield found that patents were not essential to protecting inventions in many industries.[24] Levin and his colleagues conducted a survey of research and development personnel at United States corporations and concluded that in most industries, advantages associated with a head start, including the establishment of new production and distrib-

[18] Eisenberg, "Patents and the Progress of Science: Exclusive Rights and Experimental Use," *University of Chicago Law Review* 56 (1989), 1017 at 1027 n43.

[19] Machlup, supra note 6, at 45–50 and sources cited therein.

[20] McGee, supra note 6, at 137.

[21] Turner, "The Patent System and Competitive Policy," *New York University Law Review* 44 (1969), 450, 454–5.

[22] Taylor and Silberston, *The Economic Impact of the Patent System: A Study of the British Experience* (1973), 26.

[23] Scherer, supra note 6, at 387; Arrow, supra note 8, at 619–22.

[24] Mansfield, "Patents and Innovation: An Empirical Study," *Management Science* 32 (1984), 173, 176.

ution facilities, and moving rapidly down a learning curve, were significantly more effective than patents in enabling a firm to reap benefits from innovation.[25] In a separate study, Levin also found that patents were ineffective in relation to other types of entry barriers in protecting the competitive advantages of new technologies in most industries.[26]

Other researchers have attempted to evaluate the role of the patent system by estimating the level of inventive activity which might have occurred without a patent system or under a system that included patents, yet significantly reduced the apparent efficacy of the patent as a barrier to practice of new inventions. In one case, a comparative analysis was made of the industrialization rate of several countries with and without patent systems during the late nineteenth and early twentieth centuries. The researcher found little evidence that the lack of a patent system hampered industrial development.[27] However, it was pointed out that the countries which did not have patent protection may have been free riding on inventions which were stimulated by patent protection abroad, including new inventions developed by citizens of those countries and protected under foreign patent laws.[28]

Another large study along this same vein was conducted by researchers from the Department of Applied Economics at the University of Cambridge, who examined the economic impact of the patent system in the United Kingdom for the survey year 1968.[29] The researchers analyzed approximately forty firms in the United Kingdom, all of which were in industries in which there was rapid technological change and considerable amounts of patenting and research activity. The firms were asked to make various comparisons between the traditional patent system and an alternative system in which patents could be obtained in the usual way, but that patent would be granted under a regime which called for almost automatic compulsory licensing of patents to any party offering reasonable commercial terms to the patent owner.

One of the first questions asked by the researchers is what amount of research which was actually conducted in 1968 under the traditional patent system would not have been conducted under the alternative compulsory licensing system. Overall, expenditures on research and development would have decreased by 8 percent; however, responses varied widely by industry, with the effect being almost negligible for electrical engineering, while expenditures in the pharmaceutical industry would have fallen by 64 percent.[30] A second question examined the amount of production which would be lost if the alternative system were used. While a few industries, such as pharmaceuticals, would have suffered a significant drop in sales, the impact in most industries was quite modest.[31]

[24] Ibid.

[25] Levin, Klevorick, Nelson and Winter, "Appropriating the Returns from Industrial Research and Development," *Brookings Papers on Economic Activity* 783 (1987).

[26] Levin, "A New Look at the Patent System," *AEA Papers and Proceedings* 76 (1986), 2, 200. Levin used the results of his survey to make several suggestions as to appropriate patent policy. First, he noted that in a majority of nonchemical industries, there would appear to be little to gain from lengthening the patent life, since the effect on research and development would be negligible. Ibid. at 201. Moreover, even in those areas where patents are effective, discounting the stream of anticipated returns implied that lengthening the life of the patent beyond seventeen years would not have much impact upon incentives unless there is a substantial lag between the grant of a patent and the years during which the patentee stands to obtain the maximum commercial benefit, such as is the case in the pharmaceutical area. Ibid.

[27] See Schiff, *Industrialization Without National Patents* (1971) 34–41, 96–106.

[28] Ibid. at 23–4, 102–4.

[29] Taylor and Silberson, supra note 22.

[30] Ibid. at 199 (Table 9.2).

[31] Ibid. at 202. Pharmaceutical sales would have dropped by 68 percent.

[32] See Scherer, *The Economic Effects of Compulsory Licensing* (1977), 9–10.

In general, the Cambridge Study appears to indicate that in most industries the importance of patents with respect to research and development may well be overstated by critics of the current patent system. It is worth noting that a similar study was also conducted of United States firms and it was found that respondents also did not consider patents to be of particular importance in making decisions regarding research and development projects.[32] However, one must be careful of making broad generalizations, particularly since, as described below, there appear to be several important industries where patents seem to play a significant role in innovation and in the decision-making relative to research and development activities.

Another method of assessing the effect of patents on the competitive activities of rivals is to examine how difficult it might be for other firms to imitate new patented products. Scherer found that for seven major process innovations in the United States, the imitation period, which was defined as the period between first use and use by 60 percent of the producers, varied from one year to twenty years, with a majority of these innovations being insulated from widespread imitation in the marketplace for at least four years.[33] If the delays in imitation are related to some of the technical difficulties associated with understanding and using the new process, rather than the need to "invent around" patents taken out on the process, than it might be argued that patents are not necessary to protect the investment made by the innovator.[34] It would appear that any inferences in this area will depend upon the degree to which the process is protected by non-patent means, such as by trade secret law.

Industry-based variances in importance of patents Even if it is conceded that patents do provide some significant incentive for inventive activities, it appears that the actual importance of patent protection will vary depending on the industry sector. For example, the Cambridge Study found that patents are relatively unimportant in industries where insulation from competition is provided by other non-patent barriers, such as economies of scale or the underlying complexity of the technology used in the industry.[35] In a later study, Von Hippel found that patent protection was "relatively ineffective" in most industries in capturing the competitive advantages of new and improved production processes and products in relation to other factors.[36] Specifically, he concluded that patents appeared to be unnecessary to bring about technological change in industries where the firms may be able to maintain new ideas as trade secrets, entry barriers are high, the imitation lags are long, and market concentration is high.[37]

[33] Scherer, supra note 6. Figures estimated from Mansfield, *Industrial Research and Technological Progress* (1968). See also the results of a similar study conducted by Mansfield and others as described in Mansfield, Schwartz and Wagner, "Imitation Costs and Patents: An Empirical Study," *The Economic Journal* 91 (1981), 907, 913 (60 percent of all of the patented successful innovations included in the sample were imitated within four years of their introduction). See also Bainbridge, *Intellectual Property* (1992), 248 (noting the importance of lead time as an alternative to patent protection).

[34] However, see Horstmann et al., "Patents as Information Transfer Mechanisms: To Patent or (Maybe) Not to Patent," *Journal of Political Economy* 93 (1985), 837, 848 (noting that expected profits from imitation decline and the frequency of patenting increases from an expansion of patent coverage, thereby providing evidence of the importance of patent protection in relation to imitative competition).

[35] Taylor and Silberson, supra note 22.

[36] Von Hippel, *The Sources of Innovation* (1988), 47–8. These conclusions were based on a fairly low rate of return from licensing, as well as responses from firms in many industries indicating that levels of research and development expenditure would be only marginally effected by the absence of patent protection. Ibid. at 48.

[37] Ibid. When trade secrets are an important element of protection for the innovation, patents become relatively less important in the commercialization process. In such cases, a system of compulsory patent

There are, however, some industries where patent protection appears to a significant factor in innovation. For example, the Cambridge Study also found that in some industries, like finished and specialty chemicals, including pharmaceuticals, patents had a substantial effect on competition, prices and profits, since in those industries, lead time and manufacturing know-how are rarely important factors, relatively strong patents can be acquired, and research expenses are fairly large,[38] and similar conclusions have been reached by in other studies.[39] Von Hippel, for one, describing the importance of patents in the broader chemicals area, which includes pharmaceuticals, noted that the patents are unusually "strong," in that it is possible to patent not only the actual molecule found to have useful medical properties, but also its analogs.[40] This allows the inventor to claim a large number of specific molecules without having to synthesize all of them. In addition, since it is often impossible to fully understand the mechanisms by which pharmaceuticals can achieve their medical effects, it is difficult for competitors to use the information disclosed in a drug patent to actually invent around the patent, although this may change as biochemists are able to enhance their understanding of the underlying biology.[41]

Schwartzman has also studied the pharmaceutical industry extensively, and has argued that pharmaceutical firms are generally averse to competing on the basis of price (i.e. price-cutting), and that the changes in market share which do occur in the drug industry are based upon the introduction of innovative new products, a phenemenon which further enhances the importances of patents.[42] In addition, he has described that pharmaceutical firms have almost always indicated that patent protection was quite important to them in their decisions regarding research and development. This is due, in large part, to the fact that the duplication of compounds is fairly simple, since each label contains the formula which can be readily replicated by any number of manufacturers. As a result, when patent protection expires, the original producer presumably will lose a significant amount of its prior market share to generic manufacturers that compete on the basis of price, even though the original producer has established a well-known brand name and gained the confidence of physicians.[43]

Researchers studying the costs of imitation, as well as the time required to complete the process, have also noted variations among industries.[44] For example, in the pharmaceuticals area, the ratio of imitation cost to innovation cost for a significant number of

cont..
licensing (e.g. forced licensing of the patents to third parties under certain conditions, such as failure to exploit the patents) would not be problematical to innovating firms, since they would still retain the required know-how, leaving the compulsory licensee with merely the right to try and produce the product without fear of an infringement claim.

[38] Taylor and Silberson, supra note 22.

[39] See, e.g. Levin, supra note 26, at 200; and Mansfield, Schwartz and Wagner, supra note 33, at 915 (one-half of the innovations in the sample would not have been introduced without patent protection and the bulk of these innovations were in the pharmaceutical industry in which almost all innovations would not have been introduced if no patent protection was available). However, it has been pointed out that even in the pharmaceutical industry the "strength" of a patent arises from a combination of factors; in particular, the fact that many countries now require that new pharmaceutical products undergo lengthy clinical testing procedures, thereby increasing the "lead-time" of a novel new drug product. See Cornish, supra note 1, at 82.

[40] Von Hippel, supra note 36, 47–8.

[41] Ibid.

[42] Schwartzman, supra note 11, at 165.

[43] Ibid. (noting that over 30 percent of the total number of prescriptions of leading multiple-source drugs written in 1973 were for generic brands).

[44] See, e.g. Mansfield, Schwartz and Wagner, supra note 33, at 907–18.

new products deviated substantially from the average ratio among all industries studied, and a similar phenomenon was noted with respect to the ratio of the imitation time to the innovation time.[45] Overall, the study concluded that patents did increase the costs of imitation, with pharmaceutical patents increasing imitation costs by around 30 percent, while the increases in other industries ranged from 7 to 10 percent.[46]

Importance of patents to certain types of inventors Just as the importance of patent protection varies between industries, one must remember that patents may prove to be more valuable incentives to certain groups of inventors. It should not be forgotten that inventive activity is undertaken by a broad range of actors, from large corporations with diverse research organizations to the "one-man band" operating in the back of his garage and that, as such, it may well continue to be necessary to provide a diverse range of possible incentives.[47] Jewkes has noted that small organizations centered around individuals of outstanding quality have historically generated a significant number of the important technical advances[48] and that patents may be very important for this group, since their size makes it unlikely that they will be able to construct many of the non-patent barriers which are typically available to larger firms (e.g. economies of scale, broad distribution networks).[49]

The ex ante *incentive effect of patent protection* Although various studies have found that other factors, such as head start and the availability of specialized functional skills (e.g. distribution and marketing), may have been more important, with the benefit of hindsight, in commercializing new technologies, this is not to say that patent protection is unimportant. It is useful to bear in mind the *ex ante* incentive effect that patents may well have on research and development decisions. At the time that a commitment is made to fund the development of a new product, there is no way of knowing how rapid imitation will be or just what effect other non-patent barriers will have upon competition. As such, many inventors will look to patents to provide them with a form of insurance that they will have sufficient time to combine the resources necessary to exploit their invention. Also, if the inventor believes that the invention will be highly profitable, he will expect others to quickly enter the market by the production of substitutes;[50] therefore, inventors who believe their invention will be profitable will seek patent protection so as to undermine the efforts of other to encroach upon his expected profits.

[45] The researchers found that, on average, the ratio of imitation cost to the innovation costs was about 0.65, and the ratio of the imitation time to the innovation time was 0.70. Ibid. at 908–10. However, there was substantial variation in many cases and results tended to differ among industries. For example, in the pharmaceutical industry, 54 percent of the new products fell within the 0.60–0.80 ratio with respect to the costs and 52 percent fell within the 0.30–0.70 ratio with respect to the time period. Ibid.

[46] Ibid. at 913.

[47] Cornish, supra note 1, at 80.

[48] Ibid. (citing Jewkes, Sawers and Stillerman, *The Sources of Invention* (1969), Chapter 9). However, Cornish does point out that it does not necessarily follow that *all* the individual inventors are necessarily spurred by the patent system, noting that academic scientists will continue to engage in research more out of a desire for knowledge and recognition in the academic community. Ibid. n93.

[49] For example, the availability of patent protection has provided smaller firms with an asset which allows them to raise the venture capital necessary to expand their business activities to the point where they can begin to develop some of the non-patent barriers to entry for their own use. However, it has been noted that larger organizations may need patents in order to protect their ability to receive a return on the massive amounts of investment that such firms typically make in order to develop new technology-based products. See Bainbridge, supra note 33, at 246–7.

[50] See Scherer, supra note 6 (the rate of imitation is linked to the profitability of the innovation). See also Beck, "Patents, Property Rights and Social Welfare: Search for a Restricted Optimum," *Southern*

Alternative competitive incentives for inventive activity It has been argued that there are generally a variety of alternative competitive incentives for inventive activity that will cause firms to engage in research and development without regard to the availability or strength of patent protection for the resultant products. For example, investment in research and development may be justified simply by the opportunity to be first in the market with a new invention[51] or by competitive requirements which mandate that the firm attempt to keep pace with the technological progress of its rivals.[52] In addition, firms may well be able to derive independent profits from using their advance knowledge of new technologies to invest and speculate in assets which they may anticipate will be increase in value upon the general release of the new technology.[53]

Overall efficacy of the incentive effect While it is generally thought that the patent system does provide some amount of incentive in certain areas for inventors to engage in activities that they might otherwise forgo if patent protection was not available, it is important to note that a number of experts have opined that recognition of intellectual property rights has not demonstrably achieved a more optimal rate of social investment in innovation and creativity than what would have occurred without intellectual property rights.[54] In fact, skepticism regarding the efficacy of exclusive rights in ideas and other intellectual creations has quite a long history. For example, in the early years of the new American nation, Thomas Jefferson wrote:

> "It is a fact, as fair as I am informed, that England was, until we copied her, the only country on earth which ever, by a general law, gave a legal right to the exclusive use of an idea. In some other countries it is sometimes done, in a great case, and by a special and personal act, but, generally speaking, other nations have thought that these monopolies produce more embarrassment than advantage to society; and it may be observed that the nations which refuse monopolies of invention are as fruitful as England in new and use devices."[55]

It can already be seen that the patent system has not proven to be an effective means for stimulating all forms of research and development activity. For example, in the United States and in other major developed countries, significant amounts of research,

cont.
Economic Journal October (1976); and Kemp, "The Follow-On Development Process and the Market for Diuretics," in Helms (ed.), *Drug Development and Marketing* (1975) (describing evidence to the effect that patented innovations that are profitable encourage the production of substitutes and their patenting in glass containers, oil refining, shoe machinery and pharmaceuticals).

[51] Scherer, supra note 6, at 384–7; Machlup, supra note 2, at 23 n121, 24 n128, 38–9 and sources cited therein, 59–60; Turner, supra note 21, at 452; Cornish, supra note 1, at 82–3 (expressing doubts about the role of patents in exploiting major inventions and noting that with major inventions the advantage of being first in the field may well be an adequate incentive).

[52] Scherer, supra note 6, at 386; Machlup, supra note 6, at 38–9.

[53] See, e.g. Hirshleifer, "The Private and Social Value of Information and the Reward to Inventive Activity," *American Economic Review* 61 (1971), 561 (citing Whitney as an example and arguing that Whitney could have used his information about a new and superior method of processing cotton to make investments in cotton and cotton-bearing land and in businesses engaged in the selling, warehousing, and transporting of cotton, as well as in other speculative activities concerning cotton, substitutes for cotton (wool), and goods which complimented cotton (textiles and machinery)).

[54] See, e.g. Machlup, supra note 6; Jewkes et al., supra note 48, at 25–54.

[55] Jefferson, Letter to Isaac McPherson, Monticello, August 13, 1813, in *The Writings of Thomas Jefferson* XIII, 326–8 (A. Lipscomb ed. 1904).

particularly basis research, are already supported by government financing, either through direct subsidies, support of government research institutes, or financing of university research. This is partly because the areas in which government financed research is undertaken are characterized by a high level of uncertainty and the very real possibility that the results of the research effort will be some sort of fundamental principle, rather than a specific product which qualifies as a candidate for a patent. Therefore, any practical discussion of the utility of patents may effectively be limited to areas in which the time and risk discounts may be made acceptable to firms and individuals, with or without the existence of some form of patent protection.

Conclusions The verdict on the incentive-to-invent theory is somewhat mixed. There appears to be a good deal of empirical evidence to the effect that patent protection is relatively unimportant, at least in relation to other non-patent barriers to competition, in many industries in providing any substantial additional incentive with respect to creating potential new products and processes. As to whether the patent system adversely impacts the subsequent research activities of other firms, the answer would appear to depend on the scope of the original patent, an issue that is taken up separately below. On the other hand, patents clearly remain important in selected industries, and are likely to be useful incentives for smaller firms and new market entrants.

C. Incentive-to-Disclose Theory

Statement of the theory The incentive-to-disclose theory emphasizes the utility of the disclosure requirements which are a condition for obtaining a grant of a patent right.[56] Advocates of the theory argue that without patent protection inventors would conceal the knowledge embodied in their inventions so as to prevent competitors from appropriating them and exploiting them without compensation to the inventor.[57] As a result, the public is not only denied access to the new information and the benefits associated with wider distribution, alternative uses and price competition in the marketplace,[58] but also incurs the social costs associated with duplicative research effort.[59] Therefore, the patent system, by conditioning the patent grant on specified public disclosures by the inventor, serves as a means of discouraging secrecy and wasteful duplicative research and

[56] In order to obtain a patent, an inventor must publicly disclose the subject matter of the patent with particularity in the patent application. The patent application must contain a full description of the invention and the specific inventive "claims" that one is seeking to patent. In the United States, the level of detail that must be disclosed in a patent application must be sufficient to allow one skilled in the art to make and use the invention. 35 U.S.C. § 112. The information set out in the application will eventually become available to the public during the term of the patent, thereby contributing to the advance of knowledge, even though practice of the invention described remains subject to the exclusive rights of the patentee. In the United Kingdom, see Patents Act (1977) § 14 relating to patent applications and requiring that the application include a specification which discloses the invention in a manner which is clear enough and complete enough for the invention to be performed by a person skilled in the art.

[57] Bowman, supra note 6, at 12–13; Scherer, supra note 6, at 440.

[58] Ko, "An Economic Analysis of Biotechnology Patent Protection," *Yale Law Journal* 102 (1992), 777, 795.

[59] Adelman, "Property Rights Theory and Patent-Antitrust: The Role of Compulsory Licensing," *New York University Law Review* 52 (1977), 977, 982.

also increases the public's access to valuable information, thereby insuring that industrial progress will continue at a sustained rate over time.[60]

Analysis of the theory The incentive-to-disclose theory has often been cited in judicial opinions.[61] However, outside of the courtroom, there has been a good deal of skepticism regarding the importance of the disclosure incentive as a motive for granting patent rights.[62] One group argues that the inventions that are disclosed under the patent system would have been disclosed in any event, since if they could be practiced in secret then the inventor would have relied on the protection provided under trade secret law and contract. Also, others have focused upon the fact that the disclosures in patent filings are generally inadequate to allow others to begin to understand and practice the invention. On the other hand, a number of studies have demonstrated that the information generated through patent filings is substantial, albeit sometimes quite difficult to access effectively.

Ability to protect new inventions by alternative means Critics of the incentive-to-disclose theory take issue with the presumption that the patent system actually promotes disclosure of inventions that would otherwise be maintained in secrecy.[63] They contend that patent protection will only be sought in those cases where it is impractical to protect the invention through the use of alternative means, such as trade secrets.[64] For example, if the new invention takes the form of an article or chemical compound which can be easily duplicated as soon as it is placed into the general stream of commerce, it is likely that patent protection will be sought; however, in such situations, the world does not gain access to knowledge that would otherwise have been maintained in secrecy. Alternatively, if an inventor chooses to patent a new manufacturing process which could have been practiced as a trade secret, then presumably the patent system has had some role in inducing the disclosure of information which might otherwise have been kept secret by the inventor.

The foregoing suggests that the incentive-to-disclose theory only makes sense when certain conditions are satisfied. First, a firm is likely to apply for a patent only when it has some concerns about the viability of trade secret protection, as might be the case when exploitation of the invention requires dissemination of the technology to a large number

[60] See Machlup and Penrose, supra note 8, at 10. ("To secure [industrial progress] at a sustained rate it is necessary that new inventions become generally known as parts of the technology of society. In the absence of protection against immediate imitation of novel technological ideas, an inventor will keep his invention secret ... Hence it is in the interest of society to induce the inventor to disclose his secret for the use of future generations. This can best be done by granting exclusive patent rights to the inventor in return for public disclosure of his invention.") The disclosure aspects of the patent application process have been characterized as the consideration provided by the customer in what is, in effect, a contractual relationship with the state. See Bainbridge, supra note 33, at 242.

[61] See, e.g. *Universal Oil Prods. Co.* v. *Globe Oil & Ref. Co.*, 322 US 471, 484 (1944); *Grant* v. *Raymond*, 31 U.S. 218, 247 (1832); *Sinclair & Carroll Co.* v. *Interchemical Corp.*, 325 U.S. 327, 331 (1945); *Cross* v. *Iizuka*, 753 F2d 1040, 1046 (Fed. Cir. 1985); *Flick-Reedy Corp.* v. *Hydro-Line Mfg. Co.*, 351 F.2d 546, 550–1 (7th Cir 1965).

[62] Gilfillan, "The Root of Patents, or Squaring Patents by Their Roots," *Journal of the Patent Officers Society* 31 (1949), 611, 612.

[63] See Scherer, supra note 6, at 381; Machlup, supra note 6, at 32, 33, 53 and 76.

[64] See, e.g. Easterbrook, "Intellectual Property is Still Property," *Harvard Journal of Law and Public Policy* 13 (1990), 108, 109–10; Cornish, supra note 1, at 84 ("Only if the invention is one that need never be revealed to the rest of the industry in the course of exploiting it does the patent provide a clear long-term gain in terms of publicity.").

of firms.[65] Second, patents may be considered to be valuable if it is believed that the technology can easily be lawfully discovered, through reverse engineering techniques, by competitors.[66] Third, patent protection may well be the preferred form of protection if the inventor believes that the fixed term of the patent exceeds the period of time which she reasonably believes the invention could have been maintained as a trade secret.[67] Finally, the decision to seek patent protection for inventions which could otherwise be secretly practiced might well depend upon the inventor's assessment of how difficult it may be to prove misappropriation by others.[68]

Adequacy and utility of disclosures in patent applications Another practical criticism of the incentive-to-disclose theory is that the disclosures which are made by the inventors in their patent applications often are insufficient to provide the public with enough information to accomplish the goal of effective diffusion of knowledge. A number of commentators have noted that inventors generally withhold important parts of the invention in order to continue to protect them as trade secrets or to otherwise reduce the possibility of imitation.[69] In fact, this practice has become so widespread, that it has been estimated that about half of all the inventions covered by a patent cannot be used without supplementary information,[70] in spite of the statutory requirement that the specification provide an enabling disclosure.

Even in those cases where inventors have made adequate and full disclosures in their patent applications, the utility of the disclosures will vary substantially depending upon the subject matter of the patent. For example, mechanical inventions typically produce predictable results; accordingly, rivals are able to learn a good deal from observing the specifications in an application which covers a mechanical item. On the other hand, a patent on a chemical compound teaches very little about how variations in the structure of the chemical or reactive conditions would alter its effects.[71] This explains, to some extent, why courts generally grant a narrower scope to chemical patents than they do with mechanical patents.[72] Courts reason that the breadth of claims a patent's disclosure supports "varies inversely with the degree of unpredictability of the factors involved."[73]

Another variable which is becoming increasingly important as inventors seek patent protection in a number of different countries is the disparate policies that may be followed in various national patent systems regarding the specific information which must be included in the patent applications. Failure to require that the patent application contain

[65] See Bowman, supra note 6, at 13. See also Bainbridge, supra note 33, at 243 ("where the invention relates to a process, the owner might still be well advised to seek patent protection because of the uncertainty of the law of confidence").

[66] Plant, supra note 6, at 44.

[67] See Bowman, supra note 6, at 13.

[68] Eisenberg, "Proprietary Rights and the Norms of Science in Biotechnology Research," *Yale Law Journal* 97 (1987), 177, 215.

[69] See, e.g. Machlup, supra note 6, at 32–3; Nordhaus, *Invention, Growth, and Welfare: A Theoretical Treatment of Technological Change* (1969), 89. See also *Brenner* v. *Manson*, 383 U.S. 519, 533–4 (1966).

[70] Sanders, Rossman and Harris, "Attitudes of Assignees Toward Patented Inventions," *Patent, Trademark and Copyright Journal of Research and Education* 2, 463, 467–8. See also Turner, supra note 21, at 454.

[71] *In re Cook*, 439 F.2d 730, 734 (C.C.P.A. 1971).

[72] *In re Fisher*, 427 F.2d 833, 839 (C.C.P.A. 1970).

[73] *Hormone Research Found.* v. *Genentech, Inc.*, 708 F Supp 1096, 1107 (ND Cal 1988) (summary judgment of noninfringement granted), *aff'd in part, vacated in part*, 904 F.2d 1558 (Fed. Cir. 1990) *cert. dismissed*, 59 U.S.L.W. 3687 (U.S. April 3, 1991) (No. 90-1052) (quoting *In re Fisher*, supra note 72).

enough information to meet the objective of disseminating new knowledge will reduce the patent system to an index of sources from which further information may be obtained only by negotiating a license with the patent holder.[74] However, it is clear that each country adopts it own policies regarding the content of the specifications in the application. For example, in the United States, specifications have been characterized as being "noteworthy for [their] dogged attention to pedestrian detail," while in Germany the specifications are "hazy about practical steps but ... more likely to reveal basic concepts."[75]

One of the difficulties with obtaining information from the patent system is that there are a number of problems relating to accessing the information. For example, it has been pointed out that relatively few libraries maintain patent information, that patents are poorly covered by indexing and abstracting journals, that some countries' patent specifications are only available in microform, and that the patents of many countries appear in semi-random order.[76] Moreover, some countries fail to print patents, and in other situations the material which is available from the patent office is barely legible.[77]

The utility of patent disclosures also depends upon the ability of the relevant inventive community to absorb all of the new knowledge which might be contained in the patent application. While this issue is probably not as important in the developed countries, it is relevant to debates regarding the proper role of the patent system in developing countries where, in most cases, the local scientific and business communities may lack the resources and skills to absorb and apply the technical information which may be set out in patent applications. These deficiencies are often the reason that developing countries impose obligations upon any foreign inventors seeking to license or transfer technology into the country to provide technical assistance to the local licensee or transferee in order to raise the level of technical competence in the country.

Disclosures and diffusion of knowledge Another important benefit of disclosures made through the patent system is the positive effect that the information might have upon the general diffusion of knowledge. Presumably, firms and individuals can use information in patent applications not only as a means for learning to practice new technologies, but also as a tool for understanding the current areas of research interest in a given area. With this knowledge, firms should be able to avoid the waste caused by duplicative research and make better decisions regarding the creation and maintenance of their own portfolio of technical assets. For example, information regarding patents that have been obtained by other inventors can be used to structure future research projects, perhaps focusing upon improvements or enhancements to prior patented products or developing alternative solutions to problems addressed by other inventors.[78] Also, the information can be used to make appropriate decisions regarding the possible need to obtain licenses from other patent holders.

Unfortunately, evidence on the actual utility of patent disclosures as a source of competitive information has been mixed and somewhat difficult to interpret. For example, one study that was conducted in the early 1960s estimated that the wastage from duplicative

[74] Cornish, supra note 1, at 84.

[75] Ibid. at 84. See also Oppenheim, "Information Aspects of Patents," in Phillips (ed.), *Patents in Perspective* (1985), at 54 (noting that accessing the international pool of patent information requires familiarity with a vast range of different languages).

[76] Ibid.

[77] Ibid. at 55.

[78] See, e.g. Higgins in collaboration with Silberston, *Invention and Economic Growth in Britain and the United States 1853–1963* (1969).

research efforts in generating information that already existed in published specifications in the United Kingdom amounts to just under 1 percent of the total annual research expenditures in that country.[79] Moreover, inquiries regarding the content of patent specifications have been considered to be extremely low, leading to tentative conclusions that others do not find them to be of much utility. However, the absolute number of requests for review of specifications may be an unreliable indication of the knowledge which flows therefrom, since the results of a single inquiry may well be transmitted to many others, particularly when the reviewer works for a large company.[80]

The British Library funded two studies in the early 1980s which looked into the uses which were made of the patent system by industry[81] and by academics.[82] The studies point out that among those persons who did a good deal of reading in their technical areas, patents were not considered to be anything special.[83] For those persons who did very little reading, patents were considered to be less important than other sources of information, such as journal articles.[84] As a general matter, use of patent information as a means for solving problems was rare, and only occurred in those cases where the technology was completely outside of the experience of the design or research team.[85] Participants in the studies expressed their belief that patent literature was drafted in terms of deliberate vagueness, of "calculated misdirection."[86] In fairness, it should also be pointed out that surveys revealed that there was a fair amount of ignorance regarding the potential uses of patent information.[87]

Disclosure and the market for information As noted in the previous discussion of some of the modern economic theories of property rights, patents provide a means for facilitating market transactions in intangible items, such as ideas and knowledge, which are not easily susceptible to the types of physical control generally associated with corporeal goods. In this regard, it has been argued that patents may have a positive effect upon the disclosure not only of the information which is set out in the patent application, but also of the technical know-how which is often required to actually use the technology covered by the patent. This may be due to the fact that since patents define proprietary interests and are enforceable at law, inventors may be more willing to disclose ancillary know-how to a licensee when the transaction also includes patents, since the inventor has a firmer base for litigation when a patent is involved then if the inventor was simply transferring the know-how.[88]

[79] See Martyn, *Literature Searching by Research Scientists* (1964).

[80] For a general review of patent data sources in the United Kingdom, see Chapter 3 in Bosworth, *Intellectual Property Rights* (1986).

[81] See Stephenson and Riley, *Use of Patent Information in Industry*, Library and Information Research Report 4 (1982); Stephenson, "The Use of Patent Information in Industry," *World Patent Information* 4 (1982), 164–71.

[82] Jalloq, *Use of Patent Literature by Academics in Chemistry and Physics, Electrical and Mechanical Engineering*, British Library R&D Report 5770 (1982).

[83] Eisenschitz, "The Value of Patent Information," in Phillips (ed.), supra note 75, at 47.

[84] Ibid.

[85] Ibid.

[86] Ibid.

[87] Ibid.

[88] Cornish, supra note 1, at 85. The registration procedure provided by the patent system may also serve as a means for alerting potential licensees to new product ideas which ultimately result in a license covering not only the publicized information, but also the associated trade secrets. Ibid. With regard to the importance of "know-how" as a source of information, see Kingston, "Innovation Patents and Warrants," in Phillips, supra note 75, at 80 ("[S]tudies of licence agreements show that it is *know-how*, not what is disclosed in the patent specification, that is important.").

Conclusions On balance, the incentive-to-disclose theory appears not to provide a compelling justification for implementation of a patent system, although disclosure is certainly a valuable byproduct of the registration that is required in order to secure the valuable exclusivity afforded under the patent rights.[89] It may well be that disclosures could be improved by broadening the permitted research uses of patented inventions, a topic which is discussed below. It is also important to remember that patent disclosures do provide a source of valuable information to the marketplace regarding the research activities of rival firms and potential new problems that might profitably be pursued in future periods.

Debate in the disclosure area continues as to methods that might be used in order to improve the quality of information that is generated by the patent system.[90] For example, it has often been argued that patent examiners should not allow claims which are not the subject of very specific descriptions in the application, including an insistence that inventors identify and describe the best example of the manner in which the subject invention actually works.[91] Another suggestion is to insure that all known prior art is listed in reasonable detail in the patent specification.[92] Finally, some commentators point out the disclosure benefits of the patent system would be enhanced if the system is expanded in order to provide legal protection for certain activities, such as research and development programs and the refinements necessary to ready an invention for commercialization, which do not presently qualify for patent protection.[93]

D. Incentive-to-Innovate Theory

Statement of the theory While the incentive-to-invent and incentive-to-disclose theories have proven to be appealing to the courts in supporting the use of the patent system, a number of commentators advance the argument that the patent monopoly is of greatest importance in encouraging firms to invest in "innovation," which includes all of the steps required to actually commercially exploit a new product or process during the patent term.[94] As opposed to the two theories previously discussed, the inventive-to-innovate the-

[89] See, e.g. Aubrey, "A Justification of the Patent System," in Phillips, supra note 75, at 1 (The information from the patent system "forms a valuable source of information for those who understand it and know how to use it, but to many it deserves the Scotsman's comment on the dictionary: 'fine meaty reading, but a wee bit disconnected'."). See also O'Brien, "Patents: An Economist's View," in Phillips, supra note 75, at 33 (the dissemination of technical knowledge through the public disclosures in the patent application is certainly a "beneficial externality" of the patent system). Others have a much higher opinion of the information available through the patent system. See, e.g. Oppenheim, "Information Aspects of Patents," in Phillips, supra note 75, at 55–6 (noting that the patent literature is the same order of size as the journal literature, that the patent literature frequently represents the only source, and that patents provide more information more quickly than alternative sources).

[90] See, e.g. ibid. at 56–7; Eisenschitz, supra note 85, at 48–51.

[91] See, e.g. ibid. at 49 and 50.

[92] Ibid. at 50.

[93] Ibid. at 51–3. Also, note that the innovation warrant scheme advocated by Kingston, and described elsewhere, would presumably improve the level of disclosure to the extent that it facilitates the publication of information currently protected under the law of trade secrets. Ibid. at 52–3.

[94] See, e.g. Scherer, supra note 6, at 441. The courts have not completely ignored innovation in the context determining the role of the patent system. See, e.g.the concurring opinion of Judge Frank in *Picard* v. *United Aircraft Corp.*, 128 F.2d 632 (2d Cir. 1942), in which he states: "The controversy between the defenders and assailants of our patent system may be about a false issue – the stimulus to invention. The real issue may be the stimulus to investment." Ibid. at 643. Realization of the importance of innovation is evidence by the title of the 1983 Green Paper examining the national intellectual property rights system in the United Kingdom: *Intellectual Property Rights and Innovation.* See Cmnd. 9117, HMSO 1983. See also Staines, "Some Thoughts on the 1983 Green Paper," in Phillips, supra note 75, at 100–8.

ory, like the prospect theory described below, focuses upon post-inventive activities, such as further research and development which might be required in order to determine commercial feasibility, construction of new plants and equipment, refinements which may be necessary in order to satisfy the tastes and demands of consumers, and the promotional activities that might be need to induce prospective purchasers to buy the product.[95]

The incentive-to-innovate theory is similar to the incentive-to-invent and incentive-to-disclose theories in that it attempts to justify patent rights as a means for inducing economic actors to engage in certain socially beneficial activities. However, while the first two theories assume that the purpose of the patent system has been achieved at the time the patent is granted, and that the succeeding period of exclusivity is merely the price that society must pay to fulfill its promise to prospective inventors,[96] the incentive-to-innovate theory holds that patents have an ongoing role during the patent term to induce firms to make investment in technological innovation and incur additional expenses which often far exceed the initial costs of inventing.[97]

Innovation as the primary objective of the patent system has created some amount of legal controversy. In the United States, some have noted that the use of patents for the promotion of innovation, as opposed to the act of "invention," reaches outside of the original constitutional objective of promoting "the Progress of Science and useful Arts."[98] However, one of the stated purposes of the modern British patent system is to encourage innovation and the improvement of industrial techniques.[99] Moreover, support for consideration of the notion of innovation and commercialization as objectives of the patent system can be found in the emphasis that early English patent statutes placed upon providing a reward for those who brought "any new trade into the realm."[100]

The incentive-to-innovate theory is partially grounded in the works of Joseph Schumpeter relating to economic development.[101] Schumpeter was among the first to make a

[95] Ko, supra note 58, at 799.

[96] See Machlup, supra note 6, at 55, where he explains: "If one accepts the theory that patent protection has the social function of serving as an incentive for inventive activity, one accepts, by implication, that the beneficial effects of this incentive system must flow, not from existing patents, but from the hope for future profits from future patents; this hope may induce people to undertake certain investments and useful activities – to wit, financing and arranging industrial research–which they might not undertake otherwise .... [E]xisting patents impose a burden on society, a burden which it has decided to carry in order to hold to people the chance of obtaining future profits from future patents on future inventions."

[97] See Mansfield et al., *The Production and Application of New Industrial Technology* (1977); Scherer, supra note 6, at 381 (noting that the costs of post-invention development work constitute more than three-quarters of all the industrial expenditures on research and development).

[98] Machlup, supra note 6, at 56.

[99] See the White Paper on Patent Law Reform 1975 (Cmnd. 6000, 1975 §§ 1-4) and the Report of the Banks Committee on the British Patent System 1970 (Cmnd. 4407, 1970, § 1).

[100] See *Darcy* v. *Allein* (Court of Queens Bench) 1 WPC 1 (1602), where the following was written: "When any man by his own charge and industry, or by his own with and invention doth bring any new trade into the realm or any engine tending to the furtherance of a trade that has never been used before; and that for the good of the realm; in such cases the King may grant him a monopoly patent for some reasonable time, until the subjects may learn the same, in consideration of the good that he doth bring by his invention to the Commonwealth."

[101] See Schumpeter, *Capitalism, Socialism and Democracy* (1950) 81–110; Schumpeter, *Business Cycles* 1 (1939), 84–192; Schumpeter, *The Theory of Economic Development* 61–94 (Transaction reprint, Redvers Opie trans. 1983). See also, generally, Scherer, *Innovation and Growth: Schumpeterian Perspectives* (1984); Kamien and Schwartz, *Market Structure and Innovation* (1982); Ruttan, "Usher and Schumpeter on Invention, Innovation, and Technological Change," *Quarterly Journal of Economics* 73 (1959), 596; Solo, "Innovation in the Capitalist Process: A Critique of the Schumpeterian Theory," *Quarterly Journal of Economics* 65 (1951), 417.

distinction between invention, which he said itself produces "no economically relevant effect at all,"[102] and innovation, which is the basis for revolutionary changes in the economic system through a "process of creative destruction."[103] Schumpeter argued that progress arises by the creation of new firms to carry out new innovations, thereby driving out old firms that are no longer able to compete with their obsolete goods and services. In this model, competition is based primarily on new commodities and technologies, rather than on price competition between firms which simply offer similar goods and services.[104]

While Schumpeter's work does not focus exclusively on the patent system, he does assert that in his dynamic model of the capitalist system, monopoly conditions, such as those which are said to exist upon the grant of a patent, may promote the desired innovation and growth more effectively than competition.[105] Schumpeter holds that monopoly conditions are desirable in order to provide firms with the time and space required for further development[106] and that protection from competition provides the requisite hedge against losses which may be necessary in order to induce investment in innovation.[107] Also, the prospect of monopoly returns promotes the transfer of financial resources to innovative firms and away from alternative uses which only offer investors the promise of ordinary returns.[108]

The incentive-to-innovate theory predicts that the monopoly conditions created by the patent system will actually increase, rather than restrict, the use of new technologies by facilitating the activities of innovators. In turn, any attempt to reduce the strength of the patent right would have the effect of undermining the incentives to expend the capital and effort which is needed in order to commercialize inventions once they have been discovered. Patent rights, whether measured in terms of their scope or otherwise, should be broad enough to facilitate innovation in the Schumpeterian sense, which means not only reducing a new technological invention to practice, but also carrying out any new combination of productive resources that amounts to "the setting up of a new production function,"[109] or development of new consumer goods, new markets, and new forms of industrial organization.[110]

Analysis of the theory The incentive-to-innovate theory has drawn many of the criticisms which were earlier discussed in the context of the incentive-to-invent theory. Moreover, many economists have taken issue with the assumption that the process of innovation proceeds more rapidly when the level of rivalry and competition is reduced through patent protection. Also, it is far from clear that patents provide sufficient protection for further innovation, since patents currently only provide legal protection for the initial in-

[102] Schumpeter, *Capitalism, Socialism and Democracy*, supra note 101, at 84. See also ibid. at 88–9 ("As long as they are not carried into practice, inventions are economically irrelevant. And to carry any improvement into effect is a task entirely different from inventing it.")

[103] Ibid. at 83.

[104] Ibid. at 84.

[105] Ibid. at 87–106. Schumpeter also observed that, in his opinion, economic advances were more frequently traced to big business (i.e. monopolists) than to firms engaged in atomistically competitive industries. Ibid. at 82.

[106] Ibid. at 89.

[107] Ibid. at 88 (arguing that the monopoly profits received by the firm are in part a charge to consumers for the imputed cost of self-insurance by the firm against losses which might be incurred in future innovative activities).

[108] Schumpeter, supra note 101, at 67–74.

[109] Schumpeter, *Business Cycles*, supra note 101, at 87.

[110] Ibid. at 84.

vention. Finally, the very idea of attempting to distinguish between invention and innovation has come under attack.

Role of monopolies and competition in inducing innovation Schumpeter's key assumption – that monopolistic conditions created by patents more readily induce innovation that competitive conditions – has been challenged by economists.[111] For example, it has been argued that rivalry normally accelerates the pace of technological research, development, and innovation, as long as the number of firms competing is not excessive.[112] While rivalry of this type will not maximize joint profits due to the compressed development time and duplication, it will confer a net benefit or consumer surplus upon society.[113] The empirical studies which have been conducted on the proposition have been inconclusive;[114] however, a number of commentators have expressed concerns about the potential adverse effects that patents might have upon competition and the ability of new firms to enter the marketplace.[115]

It has been suggested that support for the Schumpeterian preference for monopoly can be found in some industries. One writer has pointed to the growing biotechnology area in which many small firms with broad patents covering important new products and processes have been able to prosper in the shadows of the larger pharmaceutical companies with superior innovative resources.[116] However, while it is true that a number of small and specialized firms have attempted to enter the market in this industry over the last few years, it remains to be seen whether they will survive for any significant length of time before being absorbed by one of the larger companies which possess the complementary manufacturing and distribution capabilities.

Efficacy of patents to exclude competition Commentators have often expressed skepticism about the efficacy of patents for excluding competition with respect to innovation. For example, Von Hippel has offered the following analysis of the apparent lack of value of patents to innovators in many fields. First, he claims that patents will only be sought if the cost of application and maintenance remains relatively small in relation to other costs associated with the product or process.[117] Second, the owner of a patent may still be unable to exploit the invention as a result of another blocking patent, a situation which often occurs in rapidly developing technologies.[118] Third, actual enforce-

[111] See, e.g. Arrow, supra note 8 (arguing that incentive to innovate should be greater for competitive firms than for monopolists); Villard, "Competition, Oligopoly and Research," *Journal of Political Economy* 66, 483 (1958) (arguing that "competitive oligopoly," characterized by a small number of big firms, promotes research better than either pure competition or monopoly).

[112] Scherer, "Research and Development Resource Allocation Under Rivalry," *Quarterly Journal of Economics* 82 (August 1967), 359, 392.

[113] Ibid.

[114] See Scherer, supra note 6, at 169–255 and 363–4 and sources cited therein; Kamien and Schwartz, supra note 101, at 49–104 and sources cited therein.

[115] See, e.g. Machlup, *Production and Distribution of Knowledge in the United States* (1962) 170 (expressing concerns that the patent position of big firms may make it almost impossible for new firms to enter the industry); and Hettinger, "Justifying Intellectual Property," *Philosophy and Public Affairs,* 18 (1989), 50 (referring to Noble, America by Design (1982), Chapter 6) ("patents are frequently bought up in order to suppress competition")).

[116] See Kenney, "Schumpeterian Innovation and Entrepreneurs in Capitalism: A Case Study of the US Biotechnology Industry," *Resources Policy* 15 (1988), 21 ("The role of small biotechnology firms in reducing these innovations to practice and in their ability to continue to grow demonstrates that the independent entrepreneur ... has been very active in the biotechnology industry.")

[117] Ibid. at 51–2 (citing the semiconductor industry and the electronics industry).

[118] Ibid.

ment of a patent against alleged infringers can be quite time-consuming and expensive.[119] Finally, in many cases, it still may be more cost effective for a competitor to attempt to invent around the patent, rather than to take out a license, thereby reducing the potential royalty income to the patent owner.[120]

However, other economists have argued that the various non-patent barriers to imitation are not necessarily sufficient, in and of themselves, to provide the type of protection required in order to incentive innovation.[121] For example, while it is true that an innovator may be able to achieve some sort of competitive advantage by being first in the marketplace and developing the requisite economies of scale, in general an innovator seeking to assess the possibility of imitation is confronted with an extra decision variable with a wide and unknown variance.[122] This dilemma is particularly acute when the innovator is a small firm, since it is even more unlikely that the firm can rely upon the some of the non-patent advantages generally associated with market power. As such, patents may have value as a means for reducing the variable degree of risk that is typically associated with innovation.[123]

The role of patent protection in determining the competitive characteristics within an industry appears to vary from field to field. For example, as noted previously, there are certain industries, such as pharmaceuticals, in which firms appear to be so averse to price competition that rivalry has taken the form of product or innovative competition.[124] In these industries, the primary engine for continuing research and development is the need for companies to defend their positions in the market by continuing to develop further products. It is true that those industries in which innovative competition occurs do tend to make greater use of patents. However, it has been argued that the use of patents may simply reflect the type of industry competition, as well as relatively higher costs of research and development, rather than defining the cause of the competition. In those cases where technological change is rapid, patents will generally become obsolete in the face of further advances. While a patent may allow the innovator to more quickly recover his investment, and may serve to reduce some of the risks involved in development to the extent that they may slow imitation, patents generally do not amount to a crucial factor of technological change.[125]

Sources of innovative activity One commentator who has made a study of the sources of innovation has noted that the original inventor is not necessarily the one who innovates.[126] Instead, innovation will be undertaken by the firm which anticipates the greatest return on investment from the discovery. Innovation may come from a variety of sources, such as users or suppliers, with the point of origin varying depending upon the technological field. The foregoing has been cited as evidence that granting a monopoly

[119] Ibid. See also Cooper, "Patents and Innovation," in Teeling-Smith (ed.), *Innovation and the Balance of Payments: The Experience in the Pharmaceutical Industry* (1967).

[120] Von Hippel, supra note 36, at 51–2.

[121] See, e.g. O'Brien, supra note 89, at 37.

[122] Ibid.

[123] Ibid.

[124] See, e.g. Cornish, supra note 1, at 81; and Schwartzman, supra note 11, at 103.

[125] However, there are examples where the existence of a major patent position did affect the technology in an industry. For example, in the radio and petroleum cracking industries, the existence of a basic patent has stimulated companies to make improvements on the basic patent rather than merely imitate. See Scherer, supra note 6. In other industries, the broad patent position of a single firm (e.g. duPont's position in nylon) has probably done a good deal to frustrate technological change.

[126] Von Hippel, supra note 36, at 11–42.

to the inventor does not lead to the desired innovation in all cases;[127] however, the real answer probably lies in assessing the transaction costs that might impact the sale or license of the patent by the inventor to other firms who might be in a position to more fully commercialize the subject matter of the patent.

Distinguishing innovation from invention Some commentators have challenged the incentive-to-innovate theory on the fundamental grounds that it is often not possible to distinguish between invention and innovation. For example, it has been argued that the distinction may only be critical with those handful of inventions which truly constitute a major technological breakthrough and therefore may require a substantial amount of costly and time-consuming development work. In the typical case of a routine improvement on a technology that has already been largely developed, the additional difficulties of making the improvement marketable may be relatively small. As such, the patent system may be better suited to supporting small advances and improvements, rather than in inducing the rapid development of larger commercial products.

It has also been pointed out that it is very rare that a new discovery is made and then developed into a marketable product such that one clearly identify two discrete and identifiable processes: conception and commercialization. In most cases, what is referred to as innovation is instead a continuing process of technology building on technology, all the time leading to further patents as innovators improve the original product to the point where it is commercially viable.[128] Therefore, patents are really part of the process of the development of knowledge in a particular area of technology, a view which is consistent with the fact that most patented inventions are small improvements which have been made on existing technology.[129]

Others have championed the distinction between invention and innovation as part of a broader argument to the effect that patents are most necessary as a means for protecting innovation, rather than just the creation of a new idea which has yet to be reduced to practice as a commercial product. For example, Kingston argues that patent protection for an invention does not provide the patent holder with the ability to realize the required investment return unless their is a strong and direct link between the invention and the ultimate product which consumers are willing to purchase.[130] He notes that while in the chemicals area, there generally is such a link, such is not the case with respect to most mechanical and electrical products.[131] Therefore, Kingston concludes that an effort must be made to protect innovation, which he defines as "a new idea, appropriately realized," if the patent system is to serve as any meaningful form of economic incentive.[132]

[127] Ko, supra note 58, at 800.

[128] See Langrish, Gibbons, Evans and Jevons, *Wealth from Knowledge* (1972).

[129] However, others have specifically pointed out the importance of patent to the process of innovation, as opposed to the act of initial invention. See, e.g. O'Brien, supra note 89, at 37 ("the very process of innovation, involving a substantial investment under conditions of considerable uncertainty, would suggest that any device which reduces the uncertainty will increase the willingness of firms to innovate.").

[130] Kingston, supra note 88, at 68–9. As evidence of the fact that the current form of patent protection, which emphasizes "inventions," may not be sufficient to provide innovators with the required return, Kingston cites the empirical evidence developed by Von Hippel in Von Hippel, *Increasing Innovators' Returns from Innovation,* Working Paper No. 1192-81 under National Science Foundation Grant No. PRA 80-119244 (noting that social returns from investment in innovation far exceed private returns and arguing that a significant amount of the innovator's investment "leaks away" to others, including competitors of the innovator).

[131] Kingston, supra notc 88, at 69.

[132] Ibid. at 68–9.

Potential adverse consequences of broad patent scope The incentive-to-innovate theory, like the prospect theory discussed below, appears to have a preference for the creation of patent rights which are relatively broad in scope, since these are the types of patents which presumably will afford the holder of the patent with the requisite exclusivity to achieve the innovation anticipated by the theory. A number of commentators have voiced their discomfort with the use of broad patents, particularly as to the potential adverse effects that broad patents might have upon further development of the technology. For example, broad pioneer patents may cause important future improvements to be ignored or withheld from the market and protected via secrecy or some other manner, resulting in a diminution in the rapidity of technological changes. Similarly, the breadth of the initial patent may serve to dissuade rivals from undertaking valuable "follow on" projects or from incurring the costs necessary to "invent around" the patent.

Conclusions There would seem to be little doubt that patents can be an important part of the process of innovation, given that they serve to reduce a portion of the risks that firms must consider in deciding to make the substantial investments in innovation. However, it is not clear that innovation cannot occur in less than monopolistic markets, and in many industries the rate and direction of technological change is determined by a variety of non-patent factors. The issue is whether these other factors are sufficient to produce the desired level of innovation. If not, then the object of the patent right would need to be modified to include developmental steps beyond conception of the invention. It is this type of change which forms the foundation for several proposed alternatives to the patent system discussed in the last chapter of this work.

E. The Prospect Theory

Statement of the theory Like the incentive-to-innovate theory, Edmund Kitch's "prospect theory" of patent protection focuses upon the events which occur following the initial creation of an invention.[133] Kitch has argued that the patent system's award of exclusive rights in new inventions shortly after their discovery allows the inventor to invest in development without fear that others will appropriate the results of her work, thereby encouraging the inventor to seek to coordinate further innovative activities with other firms.[134] Under the prospect theory, the patent owner's rights will prevent others from investing in improvements to the patented technology without negotiating with the patent owner, thereby promoting the efficient allocation of innovative resources by permitting the patent owner to force other researchers to share information and thereby

[133] Kitch, "The Nature and Function of the Patent System," *Journal of Law and Economics* 20 (1977), 265, 267–1. The term "prospect" is derived from analogies between the function of the monopoly rights granted by patents and the award of exclusive mineral claims in government-owned lands in the American West. See Forst, "Legal Incidents of Non-Use of Patented Inventions Reconsidered," *George Washington Law Review* 14 (1946), 273, 279 n24 ("An interesting analogy may be drawn between the law relating to patents for inventions and the mining law, an analogy which emphasizes the fact that patents are only one of the many situations where an exclusive grant is provided to encourage effort and capital investment.") Kitch views the patent system as a means for awarding exclusive, publicly recorded ownership in new technological "prospects" shortly after discovery. Kitch, supra at 268.

[134] Ibid. at 276. He also argued that the patent system defines a framework of legal relations among firms to facilitate disclosure, licensing, etc. Ibid. at 277–8.

avoid the social waste of duplicative research which is likely to occur if different investigators attempt to develop the improvements independently.[135]

In many respects, the prospect theory is consistent with the broader theories of property rights expounded by Demsetz and Posner in that it concerns itself with the problems which might arise from inefficient exploitation of resources. Kitch's arguments are, in fact, an attempt to reconcile general property rights theory to the creation and use of knowledge and intangible resources. He notes that while information does not present the same difficulties of exhaustion and scarcity that might apply to the use of any tangible resource, the resources available for the use and exploitation of the information are limited, and therefore property rights in the inventions which embody that information are needed in order to improve the efficiency with which these resources are managed by placing the patent owner in a position to monitor and control post-invention research and development activities.[136]

Kitch makes a number of arguments to the effect that the patent system's own incentives for seeking patent protection at an early stage in the development of a new invention is an indication of the role that patent rights were expected to play in the ongoing development of the invention.[137] He emphasizes that inventions are commonly patented long before they reach the stage where it becomes commercially feasible to exploit them, since any delay in filing a patent application in order to continue to develop the invention may jeopardize the inventor's ability to obtain the patent rights in the future.[138] Therefore, Kitch asserts that it is likely that further research will usually be required in order for an invention to be fully developed during the patent term[139] and that the patent system is designed to promote efficiency in the development effort.

Kitch also bases his arguments on the belief that the patent rights cover not only the original, somewhat primitive, version of the invention, but also any subsequent improvements or refinements in the invention.[140] This is because improvements which fall within the scope of the claims in original patent, as well as any newly discovered uses of the invention, cannot be used by others without a license from the original inventor which covers the underlying patent.[141] Therefore, the patent owner is placed in a position where he can benefit from the further research of others, while other researchers will only have an incentive to conduct further development work if they are able to reach an accommodation with the patent owner. Accordingly, the patent system is an effi-

[135] Ibid. at 276–9.

[136] Ibid. at 275–6.

[137] Ibid. at 267–71. A patent may be awarded simply upon a showing that the invention works. The inventor need not demonstrate that the invention works well or that the invention provides the best means of accomplishing a specified purpose. Therefore, the utilitarian criteria underlying the patent grant are somewhat crude.

[138] For example, in the United States a delay in filing a patent application may bar a future claim if in the interim the inventor makes a public use of the invention (35 U.S.C. § 102(b)); if the invention is described in the literature or used by others (35 U.S.C. § 102(b)); if a competitor files an earlier patent application on the same invention and is able to demonstrate that he was the first-to-invent (35 U.S.C. § 102(g)); or if there is further technical progress in the field which destroys the "nonobviousness" of the original invention (35 U.S.C. § 103).

[139] Kitch, supra note 133, at 269–71.

[140] Ibid. at 268–9.

[141] Even though the subsequent inventor might be able to acquire a patent on the improvement or new use of the prior invention, he would still need license on the underlying invention in order to practice the improvement or new use. In turn, the original inventor would need a license from the subsequent inventor to practice the improvement or new use. Therefore, practice of the results of the further research would depend upon cooperation between the two patent holders. See, e.g. *Marconi Wireless Tel. Co.* v. *DeForest Radio Tel. & Tel. Co.*, 236 F 942 (S D NY 1916), *aff'd*, 243 F 560 (2d Cir. 1917).

cient mechanism for controlling subsequent research, whether conducted by the patent owner himself or by his licensees.

Finally, Kitch asserts that the exclusive rights which the patent system grants in technology prospects are necessary in order to provide an incentive to the original inventor to continue the work required in order to develop the invention without fear that competitors will appropriate the results of the research work.[142] In addition, he argues that patent rights facilitate the transfer of technology to other firms who can make specific contributions to the development effort, since the patent owner can make disclosures without sacrificing his exclusive rights. Therefore, the research work can be conducted more efficiently and without any unnecessary duplication.[143]

The notion of a "pioneer" invention has long been a part of the patent law's equitable "doctrine of equivalents," which finds infringement when a device is essentially the same as the one that has been patented.[144] A "pioneer" invention in this context has been defined by the courts as "a patent covering a function never before performed, a wholly novel device, or one of such novelty and importance as to make a distinct step in the progress of the art,"[145] and such inventions have been held to be "entitled to a broad range of equivalents."[146] Kitch's own definition of a prospect appears to be much broader than any which has been used by the courts. Kitch would apply to prospect theory in any situation where a new invention creates room for further development, whether or not the prospect represents a significant step to advance the art.[147] In some ways, the Kitch definition compliments those decisions which would award pioneer status to an invention which anticipates a new branch of industry.[148]

Analysis of the theory By focusing upon research activities following the initial discovery of an invention, the prospect theory contributes to a greater appreciation of the possible social effects which follow from the grant of patent rights. As with the incentive-to-innovate theory, the prospect theory embraces the role that patents can play in creating incentives and efficiencies *after* the invention has been created and the requisite disclosures made into the broader public domain. However, the courts have shown little interest in the theory,[149] economists have criticized the theory in general, as well as several of its basic underlying assumptions, and the theory is also subject to a number of the criticisms which have been made regarding the incentive-to-innovate theory, including the realism of distinguishing invention and innovation and the real need for substantial amounts of post-patent development work in order to prepare patented inventions for commercial use.[150]

[142] Kitch, supra note 133, at 276–9.
[143] Ibid.
[144] See, e.g. *Graver Tank & Mfg. Co.* v. *Linde Air Prods. Co.*, 339 U.S. 605, 608 (1950).
[145] *Westinghouse* v. *Boyden Power Brake Co.*, 170 U.S. 537, 561–2 (1898).
[146] Chisum, *Patents* (1984), at § 18.04[2].
[147] Kitch, supra note 133, at 268, 269, 278 and 279.
[148] See, e.g. *Ludlum Steel Co.* v. *Terry*, 37 F.2d 153, 160 (N.D.N.Y. 1928).
[149] Eisenberg, supra note 18, at 1043.
[150] See, e.g. Sanders, "Speedy Entry of Patented Inventions into Commercial Use," *Patent, Trademark and Copyright Journal of Research and Education* 6 (1962), 87 (studying assigned patents issued in 1938, 1948, and 1952, and finding that of the estimated 10 percent of patented inventions ever put to commercial use, about 90 percent were first put to use either before the patent application was filed or during the period that the application was pending and prior to the issuance of the patent).

Continuing inefficiencies from pre-patent competition While it would appear that the prospect theory attempts to justify the patent system as a means for improving the efficiency associated with research and development activities, it has also been argued that granting a broad patent for use in coordinating post-patent work does not eliminate, and may actually increase, the amount of rivalry which occurs in order to be the one who receives the prospect patent. According to this analysis, a system of granting patents broad enough to serve as an exclusive right to commercialize the underlying technology would simply shift all of the "wasteful" competition back one stage, since there is no pre-patent right to get a patent.[151]

Ability of prospect patent holders to coordinate research It has been pointed out that the control over research activities hypothesized in the prospect theory is not present in the real world of competition.[152] One group of researchers looking at the prospect theory in the context of the optimal patent scope has reported that they have not found a single case where the holder of a broad patent, which would seem to be a very important feature of the type of control advocated by Kitch, was able to use it effectively through tailored licensing to coordinate research and development activities of others.[153] Moreover, even if a patent holder attempted to control development work through some system of selective licensing, the high transaction costs which are typically associated with those types of licenses might severely diminish the overall benefits of coordination.[154]

Critics have also questioned Kitch's assumptions about the original inventor's ability to exercise control over future technological developments.[155] The ability of the patent owner to exercise effective "blocking" control over the exploitation of subsequent patents on improvements to the original invention may well be somewhat problematic and will depend, in large part, on the scope of the original patent itself. In many cases, the scope of the patent claims in the original patent may well have been drawn quite narrowly in order for a distinction to be drawn between the invention and prior art.[156] Moreover, in

[151] See McFetridge and Smith, "Patents, Prospects, and Economic Surplus: A Comment," *Journal of Law and Economics* 23 (1980), 202. Kitch's response to McFetridge and Smith appears in Kitch, "Patents, Prospects, and Economic Surplus: A Reply," *Journal of Law and Economics* 23 (1980), 205, 206. An alternative to the present patent system offered by Kingston, which is described below, would handle the "race-to-invent" problem in a number of different ways. For example, if another party demonstrates that it has already has made a significant investment in any product for which protection is being sought, it might be provided with a limited period of additional time to complete the investment. If this is done, the original applicant is spared the expense of any further investment which might only lead to an infringing product and it may even be possible for the two parties to cooperative in the development of the product. Alternatively, protection might be awarded to the original applicant, provided that the other party withdraws its objection. This should lead to negotiation between the two parties which results in the rights being awarded to the party which values them the highest, thereby achieving the goal of an efficient exclusive rights system. See Kingston, *The Political Economy of Innovation* (1984), 221–2.

[152] See, e.g. Scherer, supra note 6, at 447 n30 (stating that Kitch's views seem "little influenced by any concern for reality").

[153] Merges and Nelson, "On the Complex Economics of Patent Scope," *Columbia Law Review* 90 (1990), 839, 872.

[154] See Caves et al., "The Imperfect Market for Technology Licenses," *Oxford Bulletin of Economics and Statistics* 45 (1983), 249, 260–62.

[155] See, e.g. Beck, "The Prospect Theory of the Patent System and Unproductive Competition," *Research in Law and Economics* 5 (1983), 193 (noting that contrary to Kitch's assumptions, patents do not protect all or even many future developments of a technology).

[156] See, e.g. Merges, "Commercial Success and Patent Standards: Economic Perspectives on Innovation," *California Law Review* 76 (1988), 803, 840–1 and sources cited therein.

some cases, subsequent researchers might find it worth their while to attempt to develop improvements to the patented invention without the consent of the original patent holder if the improvement itself could be patented since the improver could sell licenses to the improvement to the original patent holder and his licensees, as well as strengthen his bargaining position with regard to obtaining a license to use the underlying patent.[157]

Effect of underlying technology The ability of the prospect patent holder to coordinate the research work associated with further development of the technology disclosed in the pioneer patent may also be impaired by the nature of the technology itself. Presumably, in order for the pioneer to exercise effective control over technological progress, the scope of the pioneer patent must be drawn broadly enough to encompass all the foreseeable products of coordinated research and development. However, in many cases, particularly in the pharmaceutical area, the underlying science and technology is extremely unpredictable, making it close to impossible to anticipate how, where and when the next breakthroughs might occur. In such situations, any attempt to craft a broad pioneer patent might simply have the adverse effect of discouraging other researchers from entering the field, a result which is even more problematical given that the disclosures set out in the pioneer patent are generally, in the case of unpredictable technological areas, of little value to the technical community.

Strategic objectives of pioneer firms One other fundamental issue regarding the prospect theory relates to the willingness and the ability of the prospect firm to actually serve the coordinating function envisioned by Kitch. For example, it has been argued that elimination of rivalry may lead to complacency on the part of the pioneer firm since it "diminishes the threatened costs of inaction."[158] It has also been observed that after the initial inventor has fully developed one part of the "prospect," it may have little interest in pursuing other potential uses which may be of particular interest to other firms.[159] This problem has been described as follows:

> "The firm presently realizing monopoly profits may be less motivated to seek additional profits than one earning only normal profits. It may, in other words, be less hungry for additional profits than the firm without a monopoly position. Several reasons for this are possible. First, it may begin to regard additional leisure as superior to additional profits ... Second, it may become more concerned with protecting its current monopoly position than in acquiring a new one."[160]

While it may be that the pioneer firm eventually loses the strategic motivation to develop particular areas which have been "staked out" in the pioneer patent, another difficulty may well be that the pioneer firm lacks the resources (e.g. financing, engineering skills, marketing channels) to adequately pursue each of the potential opportunities which logically follow from the subject matter of the pioneer patent. For example, it may well be that other firms have experience and expertise which can be used to develop a specific application of the technology far more rapidly than would be the case if the pioneer firm attempted to take on the project by itself. The foregoing argues for some form of arrange-

[157] Eisenberg, supra note 18, at 1044.
[158] Merges and Nelson, supra note 153, at 872.
[159] Ibid. at 873.
[160] Kamien and Schwartz, *Market Structure and Innovation* (1982), 29 and 30.

ment which divests the pioneer firm of exclusive control as to identified uses and applications in the event that it does not exploit the opportunity in a timely fashion.[161]

Conclusions The potential utility of the prospect theory lies in its emphasis upon the steps which are required to commercialize many inventions after their initial development. However, most of the basic assumptions underlying the theory which attempt to predict enhanced efficiencies from cooperative efforts following the grant of the patent are subject to a fair degree of criticism and it would appear unlikely that any one firm would be able, or willing, to exercise the degree of control over various possible uses and applications of the patented technology that Kitch envisions. In fact, it would appear that the grant of a broad patent simply for the purposes advocated by Kitch would create serious risks that the technology would be underutilized and that the research efforts of rivals would be diverted to other, perhaps less promising, areas of activity.

III. Economic Aspects of the Legal Boundaries of the Patent Right

A. Introduction

Most economic analysis regarding the legal boundaries of the patent grant has focused upon the balance between the incentives which the patent provides to the inventor or innovator on the one hand and the social costs resulting from the underuse of the new invention as a result of the monopoly granted under the patent.[162] For example, in examining the effects of increasing the duration of the patent rights, researchers have tended to focus upon whether the increased incentives to inventors and innovators exceeds the losses associated with a longer monopoly.[163] Similar models have been used in analyzing patent licensing practices,[164] as well as in determining the optimal breadth of patent claims.[165]

Other economists have criticized the so-called "trade-off" analysis between incentives and "deadweight" loss as ignoring at least one other important issue: the effect that the boundaries of the patent right might have upon the creation of "substitutes,"[166]

[161] While this approach might take the form of compulsory licensing, it might well be argued that it is socially wasteful to remit any royalty to the pioneer firm in those situations where the patent scope was broadly drawn in order to facilitate coordination unless the pioneer firm actually is able to serve in that function. See, e.g. Merges & Nelson, supra note 153, at 907. Patent "working" requirements, which are common in many developing countries, are designed to insure that the patent owner actually uses, or "works," the patent and any failure to work the patent within a specified period of time would result in cancellation of the patent or the award of non-royalty bearing compulsory licenses.

[162] Ibid. at 868.

[163] See, e.g. Nordhaus, *Invention, Growth, and Economic Welfare* (1969).

[164] See, e.g. Kaplow, "The Patent-Antitrust Intersection: A Reappraisal," *Harvard Law Review* 97 (1984), 1813, 1855–67.

[165] See Gilbert and Shapiro, *Optimal Patent Length and Breadth* University of California, Berkeley, Department of Economics, Working Paper No. 89-102, January 1989.

[166] Gilbert and Shapiro, as well as Klemperer, do recognize that broadening the scope of an initial patent right may increase social costs by precluding a correspondingly wider range of potential substitutes. Ibid. at 4–5 (effect of substitutes on price patentee can charge); Klemperer, *How Broad Should the Scope of Patent Protection Be?*, cited in n2 of Merges and Nelson, supra note 153 (modeling cost of precluded substitutes by picturing consumers traveling along a product distribution line). However, Merges and Nelson are critical of Gilbert, Shapiro and Klemperer for treating substitutes "as if they were already in existence or could be made so trivially." Ibid. at 869–70. But cf. Gilbert and Shapiro, supra note 165 (acknowledging that a policy of adjusting the rewards to inventors solely be varying the scope of the patent might "retard subsequent innovation by establishing monopoly rights to an entire line of research").

including subsequent inventions that not only substitute for the initial invention, but also improve upon it in some way.[167] In this alternative analytical approach, emphasis is placed on the belief that there are usually a number of different firms and individuals involved in inventive activities in any given technological area and that these activities may continue well beyond the date that a patent is awarded on an initial invention. Accordingly, it is argued that policy decisions regarding the legal boundaries of the patent right must specifically take into account the effect upon this multi-actor portfolio of inventive efforts[168] as well as the efficiency of that effort over the long run.[169]

The debates regarding the proper legal boundaries of the patent right have been concentrated in several areas. First, one of the most interesting areas of analysis is the "scope" of the patent rights, as measured by the breadth of future enhancements and improvements which might arguably be covered by the claims in the original patent. Evaluations of patent scope are analogous to the breadth of real property rights and necessarily draw upon some of the basic underlying theorems of law and economics. Duration of the patent right is another area of interest, as is the effect of adopting a compulsory licensing policy which may serve to diminish the economic strength of patent rights which may have previously been granted. Finally, the uncertainty regarding the impact that patent rights might have on the ongoing search for new knowledge can be illustrated in the struggle to develop an exemption from a claim of infringement for legitimate experimental use of patented inventions.

B. Patent Scope

The scope of a patent is determined by the nature of the claims contained in the original patent application which actually are allowed at the time that the patent is granted. The breadth of the permitted claims presumably has an effect upon the degree of exclusive control which the patent holder will be able to exercise over subsequent inventions. The broader the original patent, the more likely it is that new inventions in closely related areas

167 Merges and Nelson are critical of other researchers for ignoring "subsequent inventions which not only substitute for the initial invention, but also improve on it in some way." Merges and Nelson, supra note 153, at 870. On the same topic, they comment as follows: "The article by McFetridge and Rafiquzzaman ["The Scope and Duration of the Patent Right and the Nature of Research Rivalry," *Research in Law and Econmics* 8, 91 (1986)] raises the same general problem. These authors argue that the greater the degree of post-patent competition, or potential competition, the greater the price discipline on the innovator. The degree of competition is dependent on the scope of protection given to the innovator. In this model competition eats into the percentage of cost savings brought about by the innovator's invention ... But note that it considers only the effect of patents on the development of substitutes for that patented technology and not the effect of the patent on improvements and extensions of the patented technology. It is in this sense still more static than dynamic in its approach and, hence, different from the view taken here." Ibid. at 869 n129.

168 For example, the breadth of the initial patent right may well have a significant effect upon the ability or desire of different parties to stay in the inventing competition over time. Ibid. at 869. See, generally, Kamien and Schwartz, *Market Structure and Innovation* (1982), 105–12.

169 See Merges and Nelson, supra note 153, at 869. See, e.g. Lippman and McCardle, "Dropout Behavior in R&D Races with Learning," *Rand Journal of Economics* 18 (1987), 287. Kaplow also refers to changes in "market structure" as a potential social cost of permitting a particular restrictive licensing practice: "The possibility of adverse effects from long-run changes in market structure occurring over the patent life adds another element to aggregate social cost and reinforces [my] conclusion about the relation of private to social benefits [i.e. that private benefits will exceed social benefits due to the presence of social costs not borne by the patentee]." Kaplow, supra note 164, at 1828 n35.

will infringe upon the original patent by relying upon all or a portion of the technology described in one or more of the claims.

Issues of patent scope arise in a number of contexts. First, the agency responsible for the issuance of patents addresses the question of patent scope whenever it reviews a patent application to determine the claims that will be allowed. While the decision of the patent administrator is governed by a number of legal rules and principles, there are many situations where the examiner still has a great deal of discretion regarding which claims should be admitted and which claims should be narrowed or rejected. After the patent is issued, the courts will focus on scope issues whenever a question of infringement arises. In any infringement action, the patent holder will argue that the product produced by the defendant falls within the boundaries of her patent or that any differences between the two products are insignificant. The defendant, in turn, will argue that the original patent was invalid[170] or that his product is materially different from the patent holder's invention.

Patent scope is particularly important in the context of the incentive-to-innovate and prospect theories. In each case, it is argued that the innovator or pioneer, as the case may be, should be granted patent rights which are sufficiently strong so as to permit her to control the advance of innovation in the specified market. For example, the pioneer's ability to coordinate research work in the technical fields covered by the pioneer patent will depend, in large part, on her ability to assert patent rights which are broad enough to dissuade others from independently developing any new inventions which improve or enhance the original invention out of fear that the new inventions will infringe upon the claims set out in the pioneer patent.

Various economists have attempted to address the debate regarding patent scope by relying upon Coase's theories of property allocation, which suggest that parties will bargain to a Pareto-superior solution regarding the allocation of property rights no matter how entitlements are initially assigned. If this is true, the "size" of the right, however that is defined, will affect only the individual parties' personal gain, not the overall efficiency of the allocation in benefiting society.[171] In the context of patents, the Coasean theory implies that the scope of patents has no impact upon technological and scientific progress as long as the market in question is perfectly competitive.

While the Coasean theorem is widely referred to in law and economics literature, *ex hypothesi* this will only occur if transaction costs are low relative to the parties' expected gains from trade and if there are few opportunities for extracting rents through strategic behaviour.[172] As far as the negotiation of property rights between technological pioneers and improvers is concerned, in practice it is extremely difficult for the par-

[170] The issue of the validity of a patent turns upon whether the underlying invention satisfies the essential conditions that are imposed for the grant of the patent right. While in many cases, issues in this area are resolved by reference to various objective standards having to do with the state of the "prior art," it is important to remember the historical context in which the patent system operates. This appears in a number of cases. For example, the Supreme Court has noted that "The [Patent] clause is both a grant of power and a limitation. This qualified authority, unlike the power often exercised in sixteenth and seventeenth centuries by the English Crown, is limited to the promotion of advances in the 'useful arts." *Graham* v. *John Deere Co.*, 383 U.S. 1, 5 (1966). See also Irons and Sears, "The Constitutional Standard of Invention: The Touchstone for Patent Reform," *Utah Law Review* (1973) 653, 654.

[171] See Demsetz, "Toward a Theory of Property Rights," *American Economic Review* 57 (1967), 347, 359 (discussing bargaining process by using example of patents). See also Coase, "The Problem of Social Cost," *Journal of Law and Economics* 3 (1960), 1.

[172] See Cooter and Ulen, *Law and Economics* (1988) 105 n15; Kelman, "Consumption Theory, Production Theory, and Ideology in the Coase Theorem," *Southern California Law Review* 52 (1979), 669.

ties to agree on the value of the pioneering invention and the expected value of improvements, and to find an acceptable division of the profits from the surplus created whenever a new improvement is combined with a pioneering invention.[173] These high informational costs, coupled with the opportunities for hold-up, may well militate against an optimal solution being achieved by private contracting alone. Bargaining in this area may be further complicated by the steep transactional costs of technology licensing.[174]

This suggests that patent scope can be an extremely important issue is designing a patent system in order to achieve a desired level of economic efficiency and technological output. As noted above, it is highly unlikely that inventors would be able to appropriate the value of their inventions without patents or some other form of universal protection, given the transaction costs that would be associated with relying on individual contracts based on the common law duties of confidentiality and secrecy. In fact, the Coasean theorem would actually predict the emergence of patent-like rights in that situation. However, the importance of patent scope is limited to the objects of the patent right; therefore, it is important to remember that the efficacy of the Coasean theorem with respect to patents may be further limited by the degree to which the value and utility of the subject product is a function of characteristics which can only be protected outside the existing patent system.

A number of commentators have attempted to study the economic significance of patent scope. For example, Merges and Nelson claim that issues of patent scope are of particular importance since they are subject to a great degree of discretion.[175] They have analyzed patent scope in a manner that goes beyond simply looking at a trade-off between the incentives to inventors on the one hand and the "deadweight loss" which the patent monopoly purportedly causes to society.[176] This is done by focusing upon the effect that the scope of the initial patent grant has on the post-invention environment for further research and development and subsequent improvements;[177] however, contrary to Kitch, they treat as an open question whether or not granting the initial inventor a patent with a broad scope induces most effective development and future invention.[178]

Merges and Nelson attempt to show that patent scope can be used as a means for "tailoring" the patent grant to achieve the proper balance between incentives to the initial inventor and the social gains which arise from retaining opportunities for others to "stay

[173] Merges and Nelson, supra note 153, at 877 n160. For some background on the strategic aspects of licensing transactions, see F. Bidault, *Technology Pricing: From Principles to Strategy* (B. Page and P. Sherwood trans. 1989) 83–137; see also Meurer, "The Settlement of Patent Litigation," *Rand Journal of Economics* 20 (1989), 77, 77 ("patent validity disputes are not always resolved with licensing agreements because of incentive problems created by private information about validity").

[174] See, e.g. Contractor, *International Technology Licensing: Compensation, Costs, and Negotiation* (1981) 104–5 (calculating the transaction costs in licensing deals studied); Teece, *The Multinational Corporation and the Resource Cost of International Technology Transfer* (1976), 44 (indicating that transfer costs constitute over 19 percent of total project costs in international projects studied).

[175] See Merges and Nelson, supra note 153, at 839. On this topic, see also McFetridge and Rafiquzzaman, supra note 167; Gilbert and Shapiro, supra note 165; and Ko, supra note 58.

[176] See Merges and Nelson, supra note 153, at 842.

[177] Ibid. at 843. Explaining their approach in terms of "the two-dimensional analysis of incentives and deadweight loss," Merges and Nelson state that it is their view that "the concept of incentives ... should embrace post-invention conditions favorable to the inventor, such as extension of an initial patent to cover subsequently-developed versions of the invention ... [and] ... the notion of a patent's social costs should include its potential to reduce competition in the market for improvements to the patented technology." Ibid. at 842–3.

[178] Ibid. at 843. Kitch's "prospect theory" is based on the presumption that granting broad scope to an initial inventor induces more effective development and future invention. Ibid.

in the invention game."[179] In particular, they conclude that while an effort should be made to avoid any extensive reduction in the incentives for the pioneering firm, decisions regarding patent scope should at the margin favor the creation of a competitive environment for improvements, rather than an environment which is dominated by the pioneer firm, since in many industries the efficiency gains which might be achieved from the coordination advocated by Kitch are likely to be outweighed by the loss of competition with respect to improvements to the basic invention.[180] They also emphasize that patent scope decisions must take into account the nature of technical advance in the specific industry.[181]

While the efficacy of the incentive provided by a patent is related to the scope of the patent rights, it is also important to consider the degree to which the patent holder is able to rely upon the legal system to uphold the validity of the patent against legal challenges by competitors. In turn, actual and potential rivals will base their responses to the patent position achieved by the initial holder on the risks associated with engaging in research and development activities which arguably infringe upon the claims allowed by the examiner at the time that the patent was issued. As such, serious consideration must be given to the procedures which are used to determine the initial scope of the patent, as well as the costs associated with subsequent infringement actions.

The value of a property right is a function of the costs associated with establishing and maintaining its boundaries. For example, the owner of a large parcel of land must be diligent in protecting its borders through the use of physical barriers as well as the tools provided by law. Recognizable and clear boundaries are also important to preserve the exchange value of the property, since potential transferees will want assurances regarding the legal scope of the transferred property. In summary, the incentive embodied in the value of the patent right will depend upon such things as the examination procedures used by the registry and the manner in which any disputes regarding the validity of the rights granted to a particular applicant are to be resolved.[182]

Patent scope is a key consideration for all of the economic justifications offered for the maintenance of the patent system. Patent scope determines the value of the incentive provided to the inventor with respect to inventive and innovate activities, and also determines the impact that the exclusive right will have upon the subsequent actions of competitors in the marketplace. Also, the value of the patent right depends upon the

[179] Ibid. at 916.

[180] Ibid. at 843–4. Narrow patent scope is consistent with the arguments of others who emphasize the social utility associated with the ability to imitate and utilize the creative products of others. See, e.g. Rahl, "The Right to 'Appropriate' Trade Values," *Ohio State Law Journal* 23 (1962), 56, 70–2. ("Imitation is inherent in any system of competition and it is imperative for an economy in which there is rapid technological advance. It follows that preservation of a high degree of freedom to copy, imitate, mimic, and 'appropriate' the published or marketed trade values of others is not merely defensible, but is of vital importance ... It is not freedom of competition which requires apologies. It is interference with freedom which must always be explained.")

[181] See Merges and Nelson, supra note 153, at 916. Merges and Nelson have also written on the relationship between patent scope decisions and appropriate antitrust policies. See, e.g. Merges and Nelson, "Market Structure and Technical Advance: The Role of Patent Scope Decisions," in Jorde and Teece (eds.), *Antitrust, Innovation and Competitiveness* (1992), 185–232.

[182] The high and uncertain costs of patent litigation are frequently cited and would appear to diminish the efficacy of any incentive provided by the patent right. See, e.g. White, Jeffs, Jacob, Cornish and Vitoria, *Encyclopedia of United Kingdom and European Patent Law* (1977). A number of proposals have been made to reduce these costs. See, e.g. Livesley, "Patent Law Reform: A Patent Agent's Personal View," in Phillips, supra note 75, at 111–12 (advocating the use of arbitration as a means for "fairly" resolving patent disputes). See also O'Brien, supra note 89, at 36 (noting that the costs associated with patent litigation can "crush" smaller firms).

attitudes of the law, as expressed through patent administrators and the courts, as to enforcement of the rights originally granted to the patentee. As such, it may be beneficial to explore how it may be possible to enhance the manner in which patent scope is expressed and defined, perhaps with the effect of reducing uncertainties among various competitive actors as to those activities which can be pursued without encroaching upon the rights granted to others.

C. Patent Duration

A number of economists have analyzed the question of the optimal length of the patent grant.[183] In many cases, the concern has been that the duration of the patent term serves as a means for measuring the period of allocative inefficiency that troubles the critics of the patent system. The current patent term is largely a function of historical factors which generally no longer have any relevance. As such, an effort has been made to identify each of those factors which are truly important in making any decision with respect to the duration of the patent right. Among the most cited factors are: the importance of the invention, the speed at which the invention yields its advantages, and the complexity and length of the development period.

Establishing the optimal patent term may well prove to be particularly difficult. For example, one commentator has argued that the patent term must at least be long enough to persuade new investors that imitation will be deferred long enough to allow for recovery of an acceptable rate of return.[184] Since empirical work has demonstrated that the rate of imitation is hardly constant and varies substantially between industries, one immediately runs into problems with attempting to establish patent terms which are not only consistent with actual imitation rates, but also with the *ex ante* expectations of investors.

Wholly apart from the issue of imitation, another economic factor of some significance is the period of time which might be required in order for the inventor to recover her investment in developing the new product or process. Arguably, in those cases where a major invention yields its various advantages quickly and achieves rapid acceptance in the marketplace, the period of patent protection should be shortened, since the inventor has received an adequate reward for her efforts and any further extension of the exclusive rights might simply exacerbate the potential allocative problems of the original patent. On the other hand, inventions which have greater difficulty in achieving rapid acceptance would have the benefit of a longer period of patent protection in order to allow sufficient time to recoup the development costs.[185]

[183] See, e.g. Nordhaus, supra note 163; Scherer, "Nordhaus's Theory of Optimal Patent Life: A Geometric Reinterpretation," *American Economic Review* 62 (1972), 422; Tandon, "Optimal Patents with Compulsory Licensing," *Journal of Political Economy* 90 (1982), 470.

[184] Scherer, supra note 183, at 426.

[185] It has been calculated that for a process innovation which reduces costs by approximately 10 percent, the optimum life of a patent ranges between three and seven years for a demand elasticity of 0.7 to 4.0. See Nordhaus, supra note 163. Alternatively, if the invention has a less dramatic effect on the level of costs and savings (e.g.1 percent), and the elasticity of demand is less or equal to one, than the optimal patent life would range from fifteen to twenty years. Ibid. However, it is important to remember that the present value of a return on investment in the later years of an extended patent life may be too variable and uncertain to have any real effect upon the incentive provided by the patent right. Cornish, supra note 1, at 84.

An important policy question with respect to attempting to adjust the period of patent protection to take into account the economic characteristics of inventions is whether or not it would be possible to do the job efficiently. Presumably, such a system would require a sophisticated administrative structure utilizing personnel who have considerable technical knowledge of all of the industries involved, as well as the ability to forecast economic conditions during the term of the patent and afterwards. Also, the administrators might need to be given some power to adjust their early decisions in the event that one or more of the assumptions prove to be in error. The foregoing would not only be quite expensive, but might also result in a good deal of uncertainty in the minds of prospective inventors.[186]

It may well be that the actual life of a patent may only be important to a limited number of inventions, since various studies have found that the majority of patents actually become obsolete well before the patent term expires as new developments make them economically useless.[187] However, it would be dangerous to conclude that the term of protection should be shortened based upon the number of patents which survive their full term, since such an analysis does not take into account the quality and economic impact of the patents. A single important patent which runs for its entire term may yield far more economic and technical benefits than hundreds of minor patents which last only a few years, thus patents still may be a stimulus to truly innovative forms of invention. Also, the rate of invention may be influenced by the patent term, since a long period of patent protection may induce competitors to work rapidly to invent around a patent, rather than simply waiting until the patent expires.

It is worth noting that there have been initiatives to vary the duration of the patent grant in appropriate circumstances. For example, both the United States and the European Community have recently amended their patent laws to extend the term of pharmaceutical patents to take into account the delays in commercialization caused by the need to comply with product safety testing regulations. At the other end of the spectrum, a number of countries have introduced the concept of "petty patents" or "utility patents" which are available to provide limited protection for minor technical advances without the expense and delay of a preliminary examination.[188] In 1983, a proposal was made in the United Kingdom for recognition of a "registered invention," which would provide ten years' protection on completion of certain formalities without the need for prior examination.[189] While it is intended that "abbreviated" rights of this type will allow certain groups, such as individual inventors and small firms, to attempt to commercialize their inventions on a cost-effective basis, others have noted that any new rights must be carefully examined for their effect upon the utility of the current system of patent protection.[190]

[186] Ibid. at 83 ("A single period for all patents has an arbitrary appearance; but the British experience of granting extensions in deserving cases demonstrated the practical hurdles in the way of introducing individual variations.").

[187] For example, in countries where the holder has to renew his patent, typically only 50 percent last more than six years and only 10 percent survive to their final year. See, Federico, *Renewal Fees and Other Patent Fees in Foreign Countries*, Study of Sub-Committee on Patents, Trade Marks and Copyrights (1958). A 1970 report on the British patent system disclosed that the proportion in commercial production of all patents in force was about 30 percent. *The British Patent System*, Report of the Committee to Examine the Patent System, Cmnd. 4407, HMSO, 1970. Another study conducted by the Department of Applied Economics at Cambridge University found that only one-fifth to one-quarter of British patents are renewed for their complete term. Taylor and Silberston, supra note 22.

[188] See the general discussion in Cornish, supra note 1, at 83, including nn. 4 and 5.

[189] Staines, supra note 94, at 106.

[190] Ibid. at 107 (noting the danger that a system of unexamined rights could provide a deterrent to investment on the part of large companies).

D. Compulsory Licensing Requirements

Compulsory licensing might be considered as a means for reducing some of the adverse costs of the patent system. For example, requiring compulsory licensing at reasonable rates may reduce the underutilization costs of the patent system; however, uses of lesser value than the royalty rate would still not be covered. Compulsory licensing might also reduce to some extent the contribution of the patent system to monopoly power, the wasteful duplication of research efforts, the problem of blocking patent strategies, and the concerns about research in areas that may well already be covered by patents.[191] On the other hand, balanced against the possible utilities of compulsory licensing is the very real risk that it would substantially reduce the potential rewards to the inventors to the point where the patent no longer operates as a viable incentive to innovation.[192]

Studies have indicated that future patenting activities of firms subjected to the possibility of compulsory licensing declines significantly in relation to the control group.[193] However, such a result does not necessarily mean that the overall expenditures of the firms on research will decline, since firms may be able to rely on alternative methods (e.g.trade secret protection) for protecting their investments in the creation of the new knowledge.[194] As to the market effects of compulsory licensing, the studies indicate that compulsory licensing decrees had no statistically significant deconcentrating consequences in the relevant broadly defined product markets.[195]

Another important factor to consider is the patent licensing, compulsory or otherwise, is generally only effective when the associated know-how is transmitted. Accordingly, it may well be that a compulsory licensing scheme would only achieve the desired social benefits if the patent owner was also required to transfer and disclose whatever trade secret information might be necessary in order to practice and commercialize the product covered by the patent. Such a requirement may well substantially endanger the overall efficacy of the patent system. As at least one commentator has pointed out while commenting on the proposal in the 1983 Green Paper that know-how be included in any compulsory license required under the patent laws, the net result of any such requirement might be a withdrawal by many companies from the patent system and increased reliance on secrecy in research and development as a means of protection.[196]

[191] Some might advocate that failure to affirmatively exploit a patent right might be ground for outright termination of the patent. Compulsory "working" requirements would appear to be consistent the underlying objectives of both the incentive-to-innovate and pioneer theories, since both attempt to justify the patent system as providing the patent owner with the means for achieving a return on her investment by commercializing the patented product. Use of patents to merely "block" the development efforts of others, while certainly providing some degree of "competitive value" to the patent owner, would not appear to provide the benefits to society that are associated with further activities designed to distribute new products and, perhaps, create new markets.

[192] Turner, supra note 21, at 456.

[193] Scherer, *Innovation and Growth: Schumpeterian Perspectives* (1984), 207–8.

[194] Ibid., Table 10.2. If innovators substitute secrecy for patents, society loses the disclosure benefits associated with patenting.

[195] Ibid.

[196] See Staines, supra note 94, at 107. See also O'Brien, supra note 89, at 39.

E. Experimental Use of Patented Inventions

The tension between private "property-like" rights in new inventions and the public interest in permitting the scientific community to enjoy free access to these discoveries can be seen the debate regarding the propriety of exempting experimental use of patent inventions in research from infringement liability. The benefits of creating and enforcing an experimental use exemption have been concisely articulated by Professor Eisenberg, who has written:

> "free access to prior discoveries performs a critical function in the system of social control in science by permitting scrutiny of research claims by scientific peers. It thereby allows the scientific community to challenge the validity of prior discoveries and to establish the value of the research contributions of its members. Free access also facilitates scientific progress by permitting subsequent investigators to use prior discoveries in their research without having to go to the trouble and expense of obtaining permission for their predecessors, thereby lowering the costs of research and permitting researchers to embark upon the pursuit of divergent and even inconsistent approaches to further research. Finally, free access facilitates independence in research, which might be compromised by the need to obtain approvals from other scientists who might disagree with their approaches or to disclose their theories."[197]

Eisenberg's attempts to balance the interests of the initial inventor and the desirability of using patented inventions in order to continue technical advance led to three main recommendations. First, infringement liability would not be imposed in those cases where the patented invention is used to check the adequacy of the specification and the validity of the patent holder's claims.[198] Second, in those cases where there is actually a market of research users for the patented invention, research use of the invention would not be exempt from infringement liability if the user is an ordinary customer of the patented invention.[199] Finally, a patent holder would not be entitled to enjoin the use of an invention covered by a patent in subsequent research in the specific field; however, it may be appropriate in some cases to award a reasonable royalty after the fact to be sure that the patent holder receives an adequate return on the initial investment in the development of the patented invention.[200]

[197] Eisenberg, supra note 18, at 1065–6.
[198] Ibid. at 1078.
[199] Ibid.
[200] Ibid.

4 General Framework of Antitrust and Competition Law and Regulation in the United States, the EC and Japan

I. General Framework of United States Antitrust Laws

A. Introduction

The federal antitrust laws of the United States are set out in several different statutes. The Sherman Act[1] prohibits "contracts," "combinations" and "conspiracies" in "restraint of trade," where United States commerce is affected. The Clayton Act[2] prohibits "acquisitions" of stock or assets where the effect thereof "may be" to lessen competition substantially, or to create a monopoly, in any line of commerce.[3] The Federal Trade Commission Act ("FTC Act") declares unlawful any "unfair methods of competition in or affecting commerce, and unfair or deceptive acts or practices in or affecting commerce."[4] The Hart–Scott–Rodino Antitrust Improvements Act, which is actually included in the Clayton Act as Section 7A,[5] forbids certain acquisitions of voting securities or assets unless a prior notification has been filed with the government and the specified waiting period has expired.

The federal antitrust laws are generally enforced by the Antitrust Division of the Department of Justice ("DOJ") and the Federal Trade Commission ("FTC"), which was actually established under the terms of the FTC Act, although individuals may also have private rights of action for damages suffered as a result of the antitrust violations of another. Injunctive relief is available against actions which violate the antitrust laws and courts may declare any relevant contract, including any license agreement and the underlying patent rights, to be unenforceable. Damages of up to three times the actual damages sustained may be awarded and criminal penalties may also be assessed.

B. General Operation of Federal Antitrust Laws

The Sherman Act Section 1 of the Sherman Act prohibits "contracts," "combinations" and "conspiracies" in "restraint of trade," where United States commerce is affected.[6] Section 2 of the Sherman Act makes any attempt to monopolize any part of interstate or foreign trade a criminal offense.[7] While the Sherman Act, as well as the Clayton Act, does not explicitly mention patents or any form of intellectual property rights, it is settled that

[1] 15 U.S.C. §§ 1–2.

[2] 15 U.S.C. §§ 12–18.

[3] Section 2 of the Clayton Act (15 U.S.C. § 13) actually includes another major piece of federal antitrust legislation, the Robinson–Patman Price Discrimination Act, which was added in 1936 to specified discriminatory pricing practices which injured competition among purchasers of the products.

[4] 15 U.S.C.§ 45.

[5] 15 U.S.C. § 18a.

[6] 15 U.S.C. § 1.

[7] Ibid.

a license agreement may be a contract which runs afoul of the prohibitions set out in Section 1 of the Sherman Act.

Certain types of contracts and agreements are deemed illegal *per se* "because of their pernicious effect on competition and lack of any redeeming virtue ..."[8] As a general rule, the *per se* doctrine applies only to agreements that are among competitors at the same distribution level, referred to as horizontal agreements, and is used to prohibit price fixing, allocation of markets, boycotts and similar agreements among competitors. Most other types of agreements, including those which contain vertical restraints involving persons at different levels in the chain of distribution (i.e. manufacturers and distributors), are judged under the "rule of reason," which attempts to balance the business justifications and any procompetitive effects of the arrangement against any perceived anticompetitive effects.[9]

The Clayton Act Section 7 of the Clayton Act prohibits certain acquisitions of stock or assets, including intellectual property rights (e.g. exclusive licensing agreements), where the effect of such acquisition "may be" to substantially lessen competition, or to create a monopoly, in any line of commerce in any section of the country.[10] In determining if a transaction violates Section 7, the relevant "product" and "geographic" markets must be defined and the effect of the transaction on competition in those markets must be determined.[11] The parameters of the relevant product market are determined by the degree of interchangeability of use, or cross-elasticity of demand, between the particular product at issue and the substitutes for it. Accordingly, it is unlikely that a product will be deemed to be in a product market of its own if other products could be readily substituted by end users. Competitive effect is measured by reference to the market shares of the parties to the specific transaction and the concentration of the market.

C. Regulation of Joint Ventures

General standards of review Section 7 of the Clayton Act applies to the formation of joint ventures, assuming that the joint venturers acquire stock in the joint venture and/or the joint venture acquires assets from one or both of the joint venturers. However, joint ventures raise special considerations because they have a dual propensity for both procompetitive and for anticompetitive effects. For example, joint venturers may be able to realize transactional efficiencies, share the costs and risks of the endeavor, thereby pro-

[8] *Northern Pacific Railway Co.* v. *United States*, 356 U.S. 1 (1958) (contracts and agreements falling within the *per se* doctrine are conclusively presumed to be "illegal without elaborate inquiry as to the precise harm they have caused").

[9] The rule of reason requires the fact finder to weigh "all of the circumstances of a case in deciding whether a restrictive practice should be prohibited as imposing an unreasonable restraint on competition." *Continental* v. *GTE Sylvania*, 433 U.S. 36, 49 (1977). The factors to be considered in evaluating the restraint include: "[Tlhe facts peculiar to the business to which the restraint is applied; its condition before and after the restraint was imposed; the nature of the restraint and its effect, actual or probable. The history of the restraint, the evil believed to exist, the reason for adopting the particular remedy, the purpose or end sought to be attained, are all relevant facts." *Chicago Board of Trade* v. *United States*, 246 U.S. 231, 238 (1918).

[10] 15 U.S.C. § 18.

[11] The parameters of the relevant product market are determined in general by the degree of interchangeability of use, or cross-elasticity of demand, between the particular product at issue and the substitutes for it. Accordingly, it is unlikely that a product will be deemed to be in a product market of its own if other products could be readily substituted by end users.

moting competition by providing a means for new competition by companies that may not have entered the market on their own. On the other hand, if the joint venturers would have entered the market separately, without the formation of the joint venture, then the affiliation might serve to eliminate the competition that otherwise would have existed between them. Moreover, there are always concerns that the cooperation between the parties in the joint venture may lead to collusion or collateral restraints which are unrelated to the main purpose of the joint venture.

Analyzing the competitive effects of a new joint venture at the time of creation requires a balancing of the procompetitive and anticompetitive consequences of the transaction. If the venturers have previously been either competitors or potential competitors in the same line of commerce, it may be necessary to evaluate a wide variety of factors. This requires an economic analysis of the companies involved, including their size and relationship, as well as consideration of the length of time, scope of activities and other elements of the venture. Other factors to be taken into account include the number and power of the competitors in the relevant market; the reasons and the necessities for the creation of the joint venture; the joint venture's line of commerce and the relationship thereof to that of its parents; the potential power of the joint venture in the relevant market; and an appraisal of what the competition in the relevant market would have been if one of the joint venturers had entered it alone instead of through the joint venture.[12]

The Hart–Scott–Rodino Antitrust Improvements Act Section 7A of the Clayton Act, which was added by the Hart–Scott– Rodino Antitrust Improvements Act of 1976 (the "HSR Act"), provides that certain acquisitions of voting securities or assets may not take place unless prior notification has been filed with DOJ and the FTC, and the specified waiting period has expired. The waiting period ordinarily is thirty days, but may be either extended or shortened under certain circumstances by the DOJ or the FTC. The DOJ and the FTC have the right to request additional information or documents regarding the proposed transaction. If such a request is made, the waiting period is extended until twenty days following delivery of all of the supplemental information or documents to the appropriate agency.

The notification and review requirements of the HSR Act are applicable to acquisition and corporate joint ventures which meet certain thresholds. For example, a party involved in forming a corporate joint venture must make the required filing under the HSR Act if:

- (i) the party is engaged in commerce, or the joint venture itself is engaged in commerce;
- (ii) the party has total assets or annual net sales of $100 million or more;
- (iii) at least one of the other parties to the joint venture has total assets or annual net sales of $10 million or more;
- (iv) the party will hold 15 percent or more of the joint venture's stock or assets (or the party's interest in the joint venture is valued at more than $15 million); and
- (v) the joint venture will have total assets of $10 million or more.

[12] For a comprehensive list of some of the factors which are to be considered, see *United States* v. *Penn-Olin Chemical Co.*, 378 U.S. 158 (1964).

If a proposed venture is organized as a partnership or other non-corporate joint venture, filings under the HSR Act are not required, even if one or more corporations are participants in the venture.

National Cooperative Research and Production Act Courts have long wrestled with balancing the factors which must be considered in analyzing the net competitive effects of a joint venture. In addition, attempts have been made to enact various statutes that relate specifically to joint ventures. For example, the National Cooperative Research and Production Act of 1993 ("NCRPA") addresses certain cooperative arrangements that are focused on research and development and/or early-stage commercialization. The NCRPA is intended to address concerns about the risks which parties to various research and development and production joint ventures might otherwise face under the antitrust laws. Specifically, the NCRPA clarifies that joint ventures that were engaged in certain activities which fell short of marketing and distribution (e.g. basic research work, testing and experimental production, and production of a product, process or service) would be analyzed under the rule of reason standard, and that a successful plaintiff in any antitrust action which involved a covered venture could only recover actual damages, and not treble damages, provided that the venturers make certain disclosures to the DOJ and the FTC.

D. Role of Regulatory Agencies in Antitrust Policy

While a good deal of the law and policy in this area is essentially a function of case law interpretations of relevant statutes, the attitude of the government, as reflected by the DOJ and the FTC can provide reliable clues as to the direction in which regulation will proceed. This is true for several reasons. First, many companies conform their conduct to the views of the Antitrust Division merely to avoid challenges and the attendant cost of litigation. Second, courts tend to rely on government views for guidance, since the inherent ambiguity of the antitrust laws, particularly at the interface between the exclusive rights created under the patent laws and the preference for competition under the antitrust laws, often makes it difficult for courts on their own to discern the law from established precedents.

Unlike the EC and Japan, where antitrust regulators have promulgated specific standards for determining if a particular transaction will run afoul of the antitrust laws, the DOJ has adopted a policy of periodically publishing various guidelines in order to state the DOJ's antitrust enforcement policy regarding such transactions. These guidelines are not intended to be an exemption from the antitrust laws, rather they are designed to assist those who need to predict whether the DOJ will challenge a practice as anticompetitive.[13] In general, the publication of any new guidelines is preceded by significant periods of public comment, as well as consultation with business firms, economists and members of the antitrust bar, and new guidelines typically are drafted in a manner which attempts to conform with relevant case law in the area.

[13] See, e.g. Section 1.0 of the 1995 Antitrust Guidelines for the Licensing of Intellectual Property, which are described in greater detail below. Therein, the DOJ makes its clear that any guidelines which might be issued by the DOJ cannot remove judgment and discretion from the enforcement of the antitrust laws, and that each case must be evaluated in light of its own facts. The DOJ also made it clear that its guidelines will be applied reasonably and flexibly.

One area in which the DOJ and the FTC have been active in attempting to provide some form of administrative guidance is with respect to mergers and other transactions which might be covered by Section 7 of the Clayton Act. The DOJ published a set of Merger Guidelines in 1982. In general, the FTC endorsed the Guidelines, although it published its own Statement Concerning Horizontal Merger Guidelines in 1982. In 1984, the DOJ released revised Guidelines which contained a mathematical formula, commonly referred to as the Herfindahl–Hirschman Index or "HHI," that combines market share data and concentration ratios, resulting in a numerical index that is used by DOJ and the FTC in deciding whether to challenge a transaction. The DOJ and the FTC jointly issued revised Horizontal Merger Guidelines (the "1992 Merger Guidelines") on April 2, 1992.

Another influential set of administrative guidelines were the DOJ's 1988 Antitrust Enforcement Guidelines for International Operations" (the "1988 International Guidelines").[14] While the 1988 International Guidelines were intended to provide guidance on all aspects of the DOJ's internal enforcement practices and procedures relating to international operations, they also set out a good deal of the information related to the license and transfer of intellectual property rights, as well as analyses of criminal violations of the Sherman Act, monopolies, mergers, joint ventures, and vertical non-price distribution arrangements. The portions of the 1988 International Guidelines dealing with intellectual property rights have recently been replaced by the 1995 Antitrust Guidelines for the Licensing of Intellectual Property, which were finally adopted on April 6, 1995 (the "IP Guidelines"),[15] and the remainder of the 1988 International Guidelines will be superseded by new joint guidelines regarding the policies of the United States government with respect to enforcement of federal antitrust laws in the international arena which were first published for comment in October 1994 ("1994 International Guidelines").

E. Extraterritorial Application of United States Antitrust Laws

Courts in the United States have long taken the general position that the Sherman Act applies to all conduct that has any actual or probable effect on United States commerce, including conduct in foreign countries that is intended to and does effect United States imports or exports.[16] Subsequent cases have attempted to provide some further definition to the general assertion of jurisdiction. In one important case, the Ninth Circuit Court of Appeals held that United States courts should consider issues of "international comity and fairness" in deciding whether or not to extend the United States antitrust laws into the international arena, as well as "whether the interests, of, and links to, the United States – including the magnitude of the effect on American ... commerce – are sufficiently strong, vis-a-vis those of other nations, to justify an assertion of extraterritorial authority."[17]

[14] See 4 Trade Reg. Rep. (CCH) § 13,109.

[15] The IP Guidelines were first published for comment in August 1994, and were ultimately adopted, with some modifications from the initial draft, by the DOJ and the FTC on April 6, 1995. The new guidelines explicitly supersede section 3.6 in Part I, "Intellectual Property Licensing Arrangements," and cases 6, 10, 11, and 12 in Part II of the 1988 International Guidelines.

[16] *United States* v. *Aluminum Co.* of America, 148 F.2d 416, 443–5 (2d Cir. 1945).

[17] This "jurisdictional rule of reason" test, which was set out in *Timberlane Lumber Co.* v. *Bank of America*, 549 F.2d 597 (9th Cir. 1976), requires the court to take into account a number of different factors, including: "the nationality of the parties; the degree to which application of American antitrust law will cause conflicts with foreign law or policy and therefore adversely affect the United States foreign relations; the relative importance of conduct that occurred here rather than abroad; the existence of an intent to affect United States commerce, the substantiality of those effects, and their foreseeability; the pendency of related

The courts and the DOJ have each recognized that the antitrust laws of the United States should only be applied to foreign conduct that has a "substantial and foreseeable effect" on United States commerce. This condition would appear to be satisfied whenever an agreement eliminates a foreign party, such as a foreign partner in a joint venture, as a competitor or potential competitor in the United States, or otherwise restricts the degree to which either party will compete for business in the United States, since in such cases United States consumers will presumably be deprived of the benefits associated with such competition. In addition, an effect on United States commerce will be found if the agreement restricts competition in a foreign market and the restriction will reduce the export opportunities of other United States companies in that foreign market.

Even in those cases where the United States antitrust laws apply to extraterritorial activities, the activity may be exempted by operation of either the Webb–Pomerene Export Trade Act[18] or the Export Trading Company Act of 1982.[19] The Webb–Pomerene Act permits the creation of export trading associations and the Export Trading Company Act allows the creation of export trading companies. Also, certain joint research and development and/or production activities may be protected under the National Cooperative Research and Production Act. Both the Webb–Pomerene Act and the Export Trading Company Act require advance registration or certification by the relevant federal agencies.

As noted above, the 1994 International Guidelines are intended to set out the policies of the United States government with respect to extraterritorial enforcement of federal antitrust laws. The 1994 International Guidelines address the subject matter jurisdiction of the DOJ and the FTC over conduct and entities outside of the United States and the considerations and concerns that will govern decisions to exercise jurisdiction; mutual assistance on international antitrust enforcement matters; and the effects of the involvement of foreign governments on the antitrust liabilities of private parties. The 1994 International Guidelines are largely based upon the general proposition that "[a]nticompetitive conduct that affects US domestic or foreign commerce may violate the US antitrust laws regardless of where such conduct occurs or the nationality of the parties involved," and point to the recent Supreme Court statement that "the Sherman Act applies to foreign conduct that was meant to produce and did in fact produce some substantial effect in the United States."[20] Foreign conduct which may fall within the United States antitrust laws includes not only imports into the United States market, but also mergers of foreign firms accounting for substantial United States sales, conduct of foreign firms with the requisite effect on exports of United States firms, and anticompetitive conduct by foreign firms selling to the United States government.

cont.
foreign litigation; whether the plaintiff has a remedy abroad; whether court-ordered relief would place a party under obligations to perform acts illegal in another country or would otherwise subject him to conflicting legal requirements; whether court-ordered relief could be made effective; whether an order for relief would be acceptable here if made by a foreign court under similar circumstances; and whether an inter-governmental agreement with the other nation has addressed the subject matter."

[18] 15 U.S.C. §§ 61–6.

[19] 15 U.S.C. § 4001 et.

[20] *Hartford Fire Ins. Co.* v. *California*, 113 S.Ct. 2891, 2909 (1993).

II. General Framework of EC Competition Laws

A. Introduction

The competition policies of the EC are central to the overriding objectives of the EC itself, since they are designed to eliminate practices which interfere with the integration of the separate economies of the EC member states into a single European market. The antitrust laws of the EC are set out in the EC Treaty (the "Treaty"), as well as in the national laws of each of the members of the EC. Article 85 of the Treaty provides that, unless an exemption applies, any agreement which prevents, restricts, or distorts competition within the EC is prohibited and automatically void and is subject to investigation and penalties by the EC Commission, and Article 86 of the Treaty prohibits abuses of a dominant position by a single firm.[21]

B. Article 85 of the EC Treaty

Introduction Article 85 of the Treaty is divided into three main sections. Article 85(1) lists examples of agreements which would be objectionable, such as agreements which fix prices, control or share markets, or contain tie-in provisions. Article 85(2) provides that agreements which are prohibited under Article 85(1) are automatically void and may not be enforced. However, Article 85(3) opens the possibility of an exemption for agreements which might otherwise be prohibited by Article 85(1) in those cases where the agreement satisfies certain specified conditions.

Article 85(1) Article 85(1) prohibits as incompatible with the common market "all agreements between undertakings, decisions by associations of undertakings and concerted practices which may affect trade between member states [*to an appreciable extent*] and which have as their object or effect the prevention, restriction or distortion of competition within the common market [*to an appreciable extent*]." Article 85(1) applies to both horizontal agreements, including cartels and joint ventures affecting principally interbrand competition, and to vertical agreements, like distribution and licensing agreements affecting interbrand competition. The Treaty itself provides as illustrative examples of anti-competitive types of agreements ones which:

> "(a) directly or indirectly fix purchase or selling prices or any other trading conditions [i.e. price-fixing and resale price maintenance];
> (b) limit or control production, markets, technical development, or investment [i.e. non-competition covenants and exclusive agreements];
> (c) share markets or sources of supply [i.e. cartels, joint purchasing agreements, quota arrangements];
> (d) apply dissimilar conditions to equivalent transactions with other trading parties, thereby placing them at a competitive disadvantage [i.e. cartels applying discriminatory terms to non-members];
> (e) make the conclusion of contracts subject to the acceptance of supplementary obligations which, by their nature or according to commercial usage, have no connection with the subject of such contracts [i.e. tying arrangements]."

[21] References herein to the "Commission" and to Art(s). 85 or 86 refer to the EC Commission and Art(s). 85 or 86 of the EC Treaty unless the context indicates otherwise.

Agreements falling outside of Article 85(1) The Commission has declared that certain agreements will be considered to fall outside of Article 85(1), regardless of the fact that such agreements may be of the type referred to in Article 85(1). For example, in its Notice of December 3, 1986 relating to so-called "Agreements of Minor Importance" the Commission stated that it would not challenge agreements which provide for distribution of goods, provision of services, and/or certain cooperative efforts if:

(i) the goods or services which are the subject of the agreement, together with the participating companies' other goods or services which are considered by users to be "equivalent" in view of their characteristics, price, and intended use, do not represent more than 5 percent of the total market for such goods or services in the area of the Common Market affected by the agreement; and

(ii) the aggregate annual turnover of the participating undertakings in all goods and services in the last financial year does not exceed 200 million ECUs, (a) the goods or services of the participating companies do not represent more than 5 percent of the total market value of such goods or services in the area of the affected market, and (b) the aggregate annual sales of the participating firms do not exceed 200 million ECU.

The Commission has also issued similar notices within which it has set forth the terms and conditions under which it considers that the following other types of agreements would be exempt from the scope of Article 85(1): (i) exclusive sales agency agreements; (ii) agreements for cooperation between small- and medium-sized enterprises; and (iii) subcontracting agreements.

Exemptions under Article 85(3) As noted above, Article 85(1) will not apply to agreements which are exempted pursuant to Article 85(3). In order for any agreement to be eligible for exemption, it must be able to satisfy two positive conditions and one negative condition, as follows:

(i) it must contribute to the improvement of production or distribution of goods or to the promotion of technical progress;
(ii) it must allow consumers a fair share of the resulting benefit;
(iii) it must not impose on its parties restrictions which are not indispensable to the attainment of such beneficial objectives; and
(iv) it must not give the parties an opportunity substantially to eliminate interbrand competition in the relevant product market.

There are two kinds of exemptions. First, an exemption may be had by way of individual decision granted by the Commission upon the parties' application in the form of a notification. Whenever an agreement is "notified," the parties submit the agreement and related information to the appropriate regulators prior to the effectiveness of the agreement along with a request an individual exemption or, in the alternative, a "negative clearance," which is basically a ruling that the agreement does not meet the conditions for prohibition.[22]

[22] See EC Regulation 17 of 1962.

The second kind of exemption is the "block exemption," which is automatically granted, without any notification or other formal action, to categories of agreements which are set out in a series of regulations issued by the Commission. Among those block exemptions which may apply to collaborative arrangements are those covering exclusive distribution agreements, exclusive purchasing agreements, specialization agreements, research and development agreements, patent licensing agreements, know-how licensing agreements, and franchise agreements.

In most cases, the block exemptions define by general characteristics and by the type of restriction those practices which need not be notified and those practices which either should be notified and justified or avoided completely since they will, in most cases, be successfully challenged if they are brought to the attention of the Commission by notification or otherwise. Most of the block exemptions can be understood as creating a "white list," a "grey list," and a "black list," each of which have the following general effect:

(i) Practices on the "white list" are not considered to have any restrictive competitive effect and therefore fall outside of Article 85(1) altogether and need not be notified or justified.
(ii) Practices on the "grey list" admittedly restrict competition and, as such, would violate Article 85(1); however, there is generally a high likelihood that such practices will be viewed favorably in the event that the agreement is notified.
(iii) Practices on the "black list" clearly violate Article 85(1) and cannot be automatically exempted under the regulation, nor is it likely that such practices will be permitted even if a specific exemption is requested by notification.

If the agreement includes a provision which is on the black list, it will make the entire agreement potentially unenforceable and void, including the restrictions which might otherwise be on the applicable white list. In the event that a national court is called upon to enforce the agreement in the future, it will have to apply its own rules of severance, with the effect being that there may be no terms or provisions left in the agreement to enforce. For example, a licensor in this situation may find that not only is it unable to enforce any of the restrictions in the agreement, such as prohibiting the licensee from selling products outside of the specific territory referred to in the agreement, but it also is unable to collect any royalties from the licensee.

Penalties for violations of Article 85(1) Article 85 is of crucial commercial importance as it affects the enforceability of ordinary contracts and agreements, even where one or more of the contracting parties are located outside of the EC. As a general matter, there are three consequences to any violation or infringement of Article 85(1):

(i) Unlawful provisions may be stricken from the agreement by the courts, which may have a significant impact upon the business and commercial relationship between the parties.
(ii) Parties to unlawful agreements may be subject to fines imposed by the Commission which, in theory, can run up to 10 percent of the total turnover of each party to the contract and all other associated companies in the same group, regardless of whether they operate in the same market.

(iii) Third parties may claim damages from the parties to an unlawful agreement for the losses caused to them as a consequence of that unlawful agreement, and may also apply for injunctive relief.

C. Regulation of Collaborative Business Arrangements

Commission's approach to collaborative arrangements The Commission has expended a good deal of effort in the development of law and regulatory policy with respect to the myriad of forms of collaborative business arrangements, which are quite often referred to generically as "joint ventures." The Commission has repeatedly taken note of the fact that there are situations where collaboration can have a number of advantages which, taken together, may override any anticompetitive effects from the collaboration, especially when the collaboration occurs one or more steps (e.g. research and development) away from the market for the final products.[23] On the other hand, the case law in this area indicates a real concern about the potential anticompetitive effects of collaborative arrangements, such as the diminution, if not outright elimination, of actual or potential competition between the joint venturers; the risk that the collaboration in the joint venture area will ultimately "spillover" into areas outside of the joint venture; and the danger that the joint venture may foreclose possibilities for other parties to cooperate with the joint venturers.

In analyzing any particular cooperative arrangement under Article 85, the Commission attempts to strike a balance between, on the one hand, the benefits of the arrangement and, on the other hand, the anticompetitive effects of the arrangements, with special emphasis on the potential restrictions which the joint venture might place upon trading within the EC. In addition to attempting to assess the net competitive effect of the specific arrangement, it will be incumbent on the Commission to take steps designed to insure that any restrictions on the activities of the parties that are included in the joint venture agreement are not excessive, but rather are limited to those that are indispensable to the creation of the joint venture and its attainment of the positive benefits of cooperation.[24]

A threshold determination with regard to any cooperative arrangement is whether the collaboration falls under the general rubric of Article 85, or is a "concentration" or "concentrative joint venture" of the type which should be analyzed under the EC's Merger Regulations, which are described below. A joint venture which is considered as being "cooperative" usually is confined to less than all the functional steps in the development and commercialization of a new product or technology (e.g. research and development, production, and/or distribution and marketing), or is limited to products or technologies which involve substantially less than all the assets of each of the participants. In turn, any joint venture which amounts to a concentration involves what is essentially a combination of all or a part of the old businesses of the participants.

Characterization of a joint venture as "collaborative" or "concentrative" has historically been quite important due to the substantial differences between the manner in which collaborative joint ventures have been handled and the treatment of mergers and concentrations under the Merger Regulation. As a general matter, collaborative joint ven-

[23] Bellamy and Child, *Common Market Law of Competition* 4th edn., London (1993) § 5-002. See also, e.g. Fourteenth Report on Competition Policy (1984), 37–8; and Fifteenth Report on Competition Policy (1985), point 282.

[24] Ibid. at § 5-004.

tures have typically been regarded as falling under Article 85(1), unless the competitive effects of the joint venture were so minor that it fell under the *de minimis* exceptions to Article 85(1). On the other hand, under the Merger Regulation, a concentrative joint venture will be approved unless the concentration creates or strengthens a dominant position as a result of which effective competition will be significantly impeded in the whole or in any substantial part of the EC.[25]

The broad application of Article 85(1) to collaborative arrangements meant that the participants were likely subject to potentially lengthy review,[26] as well as the uncertainties associated with an implicit finding of infringement. It also meant that the best that could be hoped for was the grant of a time-limited and revocable exemption under Article 85(3), rather than the final exemption that was provided to concentrative joint ventures under the terms of the Merger Regulation. These problems with an Article 85(3) exemption were particularly acute when a substantial capital investment was required in order to commence joint venture operations. The disparity of treatment is also evident with "structural" joint ventures, which are those situations where the joint venture effectively creates a new and permanent or quasi-permanent force in the market, but is not "concentrative" enough to fall under the Merger Regulation.[27]

In recent years, the Commission has appeared to revise its approach to analyzing and processing cooperative joint ventures due, in part, to the apparent anomalies created by the manner in which mergers are regulated. One area of change has been with respect to the Commission's economic analysis of joint ventures under Article 85. It has been noted that the Commission now seems to be making a more realistic and fact-based economic assessment regarding potential competition and of the extent to which any of the restrictions on competition which may be identified in the arrangement will have an "appreciable" effect on competition.[28] Additional changes have been made in order to expedite the review of "structural" joint ventures,[29] expand the scope of the specialization and research and development block exemptions, which are described below, to cover arrangements

[25] See Art. 6(1)(b) of the Merger Regulation. A holding by the Commission that a notified concentration does not raise serious doubts as to compatibility with the common market is, to some extent, analogous with the grant of a negative clearance under Art. 85(1) (i.e. the arrangement does not even fall under the scope of Art. 85(1)), as opposed to an approval of the agreement under Art. 85(3) by exemption or otherwise, which carries the implied belief that the agreement does infringe Art. 85(1).

[26] Interestingly, the lengthy review for an individual exemption under Art. 85(3) should be contrasted to the limited period of time, no more than five months, that the Commission has to make a decision under the Merger Regulation. See below.

[27] A number of joint ventures extend beyond the temporary and limited coordination of commercial behavior between independent market participants, but do not amount of a concentration of all of the resources of the participants. Note the Commission's statement regarding the assessment of cooperative joint ventures pursuant to Article 85, OJ 1993 C43/2: "The creation of a JV which combines wholly or in part the existing activities of the parents in a particular economic area or takes over new activities for the parents brings, over and above [the co-ordination inherent in a concentrative JV] a change in the structure of the participating enterprises."

[28] See, e.g. Bellamy and Child, supra note 23, at 224 n14. Some commentators have characterized these changes as an attempt to apply some form of "rule of reason" in this area. See, e.g. Venit, *Oedipus Rex: Recent Developments in the Structural Approach to Joint Ventures under EEC Competition Law* (1991) World Competition Law 5. But, see also Whish and Sufrin, "Article 85 and the Rule of Reason," *1987 Yearbook of European Law*, Oxford (1988).

[29] The Commission's new "fast track" system for notifications of "structural" joint ventures anticipates that the parties will be informed within two months of the Commission's receipt of a "complete notification" whether the joint venture raises "serious doubts as to its compatibility with the EEC competition rules." The Commission's response may take the form of either a comfort letter or a formal negative clearance or

which extend to sales activities, and clarify that intellectual property licenses granted within the scope of a joint venture will fall within the block exemptions for patent and know-how licenses described below.[30]

While the changes in the Commission's approach generally have been welcomed, concerns remain about the potential for over-regulation through the application of Article 85(1). One alternative that has been suggested is that the Commission might approach the question of whether or not an arrangement infringes Article 85(1) by using a "rule of reason" standard similar to that which is applied in the United States, with the possible consequence being that any given joint venture might fall outside of Article 85(1) altogether after the Commission balances the various procompetitive and anticompetitive effects. While the use of a "rule of reason" standard has its advantages, in general the Commission has opted to recognize a broad application of Article 85(1) and then temper its judgment by permitting parties to pursue "desirable" collaboration through block and individual exemptions and comfort letters granted under Article 85(3).[31]

Group exemptions for collaborative arrangements While collaborative arrangements other than intellectual property licenses are often generally referred to as "joint ventures," the Commission has, in fact, chosen to analyze such arrangements by reference primarily to functional activities in which the participants are involved. This is evidenced by the Commission's election to implement a series of block exemptions dealing with collaborations focusing on research and development, production, and distribution, regardless of the actual legal form of the collaboration arrangement (e.g. joint venture, research contract, distribution agreement). In each case, the block exemption is based on the Commission's experience in reviewing agreements for purposes of determining whether Article 85(1) was applicable and, if so, whether an Article 85(3) exemption was warranted in the particular circumstances.

Research and development block exemption The research and development block exemption, Regulation 418/85, applies to three general types of research and development arrangements:

(i) joint research and development of the products or processes without any subsequent joint exploitation of the results;

cont.
exemption. However, the Commission is not bound to follow this procedure, it is intended only to provide internal guidance as to how such arrangements will be handled. See [1993] 4 CMLR 238. Even before the Commission announced the new procedure, it appeared that the Commission was making an effort to review "structural" joint ventures within in a timetable similar to that which applied to mergers. See, e.g. Bellamy and Child, supra note 23, at 224–5 n15. The Commission has attempted to summarize its administrative practices regarding the assessment of joint ventures in a recent Notice on the subject. See Commission Notice concerning the assessment of cooperative joint ventures pursuant to Art. 85, OJ 1993 C43/2.

30 Reg. 151/93, OJ 1993 L21/8.

31 It is worth noting that the ECJ has, in some instances, applied what seems to be a "rule of reason" approach to the issue of whether or not a "restriction" on competition in an agreement necessarily involves an infringement of Art. 85(1), particularly if the restriction is merely ancillary to achievement of an aim which is not in itself anticompetitive or is not an "appreciable" restriction on competition. See, e.g. Case 258/78 *Nungesser* v. *Commission* [1982] ECR 2015 [1983] 1 CMLR 278; Case 161/84 *Pronuptia* [1986] 1 CMLR 414. However, the ECJ has yet to apply this type of analysis to a specific joint venture or to the provision of a joint venture agreement.

(ii) joint research and development of the products or processes with joint exploitation of the results;
(iii) joint exploitation of the results of research and development carried out as the result of earlier research and development agreements between the same undertakings.

The Regulation carefully defines "research and development" itself to cover "the acquisition of technical knowledge and the carrying out of theoretical analysis. systematic study of experimentation, including experimental production. technical testing of products or processes. the establishment of the necessary facilities and the obtaining of intellectual property rights for the results.

The exemption will not be available where the parties are competing manufacturers and their combined market shares of the products "capable of being improved or replaced by the contract" exceed 20 percent in the EC or a substantial part thereof; moreover, if the distribution of the products concerned is carried on by one of the contracting parties, a joint venture, or a commonly designated third party, the above market share limit amount is reduced to 10 percent.

Regulation 418/85 permits the use of provisions which restrict: research by the parties in related areas, agreements with third parties to conduct competing research, purchases from non-parties, the manufacture of contract products, and/or sales in territories reserved for the parties and sales in reserved territories for a period of five years. The block exemption also permits the use of minimum quantity and quality provisions, as well as confidentiality agreements and restrictions on the grant of exclusive distribution rights to specified parties. On the other hand, the parties may not include restrictions relating to: independent research and development activities with third parties in connected fields; challenges to the validity of intellectual property rights; or prices, quantities or customers.

Specialization block exemption A block exemption is provided under Regulation No. 417/85 for so-called "specialization agreements," which are designed to bring about specialization at the production level between potential or actual competitors. One situation in which the exemption might be applicable is when each of the parties agree to specialize in the production of certain products and to refrain from producing other products which are to be left for manufacture by the other parties. In most cases, the agreement regarding production is accompanied by additional agreements from each party to obtain exclusively from the other its requirements of those products which it has itself agreed not to manufacture and to supply to the other party the products in the manufacture of which it has specialized. Another type of specialization agreement is when one party ceases to manufacture certain products and simply agrees to purchase all of its requirements exclusively from the other party.[32]

Specialization agreements between actual or potential competitors necessarily involve a restriction on the competitive freedom of the parties and, subject only to the *de*

[32] Some of the principal cases in this area illustrate the scope of specialization arrangements: *Clima-Chappee*, OJ 1969 L195/1 [1970] CMLR D7 (specialization and exclusive cross-distribution of heating equipment); *Sopelem/Langen*, OJ 1972 L13/47 [1972] CMLR D77 (specialization and *de facto* exclusive cross-supply arrangement); *Prym/Beka*, OJ 1973 L296/24 [1973] CMLR D250 (transfer of production assets with obligations not to re-enter production and to source exclusively from acquiror); *Bayer/Gist- Brocades*, OJ 1976 L30/13 [1976] 1 CMLR D98 (specialization and cross-supply underpinned by joint investments); and *Rheinzink* v. *Commission* [1984] ECR 1679 [1985] 1 CMLR 688 (market sharing with cross-supply to cover production shortfalls).

minimis rules, will generally constitute an infringement of Article 85(1). This does not mean, however, that the Commission will strike down such agreements, particularly when it can be shown that the agreement rationalizes production, creates greater efficiencies, and lowers costs. In fact, the Commission is often inclined to exempt such arrangements under Article 85(3), absent unduly restrictive provisions in the agreement or a finding that the parties are actually using the agreement as a means for dividing the EC market territorially.[33]

Agreements covered by this block exemption may allow for exclusive distribution rights,[34] may restrict the parties from entering into specialization agreements with third parties relating to similar products,[35] and may allow for exclusive procurement arrangements involving the other contracting party or a joint venture formed for the purpose of manufacturing the product.[36] The block exemption applies to companies which do not account for more than 20 percent of the market for such products in the EC or in a substantial part thereof, and which have total annual sales not in excess of one billion ECU; however, where the distribution of the products concerned is carried on by one of the contracting parties, a joint venture, or a commonly designated third party, the market share limit is reduced to 10 percent.[37]

Distribution and agency agreements There are basically five types of distribution arrangements:

(i) exclusive distribution arrangements, under which the distributor is granted an exclusive territory in which it can concentrate its sales activities without competition from the supplier or any other distributor appointed by the supplier;
(ii) sole distribution arrangements, under which the distributor still is the only distributor in a given territory, but the supplier reserves the right to supply some or all of the customers in the territory;
(iii) exclusive purchasing arrangements, under which the distributor is obligated to obtain all supplies of the specified products from the supplier;
(iv) selective distribution arrangements, under which a manufacturer of specialized consumer goods is allowed to impose certain quality standards on its distributors;
(v) franchising, under which franchisees are granted franchise to operate specified businesses in accordance with strict standards established by the franchisor as a condition of the arrangement.

33 See, e.g. *SNPE/LEL*, OJ 1978 L191/41 [1978] 2 CMLR 758.

34 Regulation 417/85, Art. 2(1)(c). However, such rights may only be granted if intermediaries and users can also obtain the products from other suppliers and the parties do not render it difficult for intermediaries or users to obtain the products.

35 Ibid. at Art. 2(1)(a). Similar products include not only identical products, but also products which are considered by users to be equivalent in view of their characteristics, price and intended use.

36 Ibid. at Art. 2(1)(b). The provision is not exempt, however, where the products are obtainable on more favorable terms elsewhere and the other party, the joint undertaking or the undertaking charged with manufacture is not prepared to offer the same terms.

37 Art. 4 of the block exemption provides for an opposition procedure whereby undertakings exceeding the size limits provided by the exemption may notify their agreements to the Commission and thereby benefit from the exemption if not opposed by the Commission within a period of six months after notification. In practice, however, the use of this procedure remains exceptional.

The Commission has implemented block exemptions and other guidelines relating to each of the aforementioned arrangements. In addition, a number of the member states have their own laws and regulations with respect to the content of distribution and agency agreements.[38]

Exclusive distribution agreements The regulation concerning exclusive distribution agreements (Regulation No. 1983/83) applies to "agreements to which two undertakings are parties whereby one party agrees with the other to supply certain goods for resale in the whole or a defined area of the Common Market only to that other." The only other obligation that may be imposed on the supplier is to refrain from supplying the contract goods to users in the contract territory. The block exemption allows the following obligations to be imposed upon the distributor:

(i) the obligation of the principal to not sell the product to any other end user in the territory;
(ii) the obligation of the distributor to not manufacture or distribute competing prod ucts;
(iii) the obligation of the distributor to purchase the products for resale only from the other party to the contract;
(iv) the obligation of the distributor to not prospect (actively market, establish a subsidiary, or maintain a warehouse for the products outside of the contract territory);
(v) the obligation of the distributor to buy complete sets of the products or to buy minimum quantities of them;
(vi) the obligation of the distributor to sell the subject products under the trademark or in the trade dress provided by the principal; and
(vi) the obligation of the distributor to undertake any promotional sales measures, and in particular: (a) to advertise; (b) to insure the maintenance of a sales network or an inventory; (c) to insure service to the clientele; and (d) to employ personnel which have specialized or technical training.

However, distributors must remain free to set their own prices and sales conditions, and may not be constrained as to their choice of customers. Additionally, the warranties of the manufacturer must be honored throughout the EC.

A reciprocal exclusive distribution agreement between competitors will not benefit from the block exemption, while a non-reciprocal distribution agreement will, provided at least one of the parties has a total annual turnover below 100 million ECU. Other agreements excluded from the scope of the block exemption include exclusive distribution agreements in markets where users are unable to obtain the contract goods from alternative sources, even outside the contract territory, and agreements to which the parties render it difficult for intermediaries or users to obtain contract goods from other dealers inside the EC, including through the exercise of intellectual property rights.

[38] For example, under the French national competition law, a party: (i) cannot impose on its French distributor the obligation to engage in engage in predatory pricing; (ii) cannot impose a minimum resale price, but it may advise its French distributor regarding any recommended resale prices; (iii) cannot refuse to sell its own products; and (iv) should treat all of its intermediaries fairly and avoid discriminatory practices (i.e. discounts, rebates, bonuses, or other conditions of sale should be accorded equally to all intermediaries).

Exclusive purchasing agreements The block exemption for exclusive purchasing agreements (Regulation No. 1984/83) applies where the reseller agrees with the supplier "to purchase certain goods specified in the agreement for resale only from the supplier or from a connected business or from another source which the supplier has entrusted with the sale of its goods". The exemption applies only to agreements whose duration does not exceed five years. In addition, no more than two firms must be involved.

The principal difference with exclusive distribution agreements is that the reseller is not provided with any exclusive distribution area. Similar to the exclusive distribution agreement block exemption, the exclusive purchasing agreement block exemption is not applicable to reciprocal exclusive purchasing agreements between manufacturers of identical or equivalent goods or to non-reciprocal exclusive purchasing agreements between manufacturers of such goods, unless one of them has a total annual turnover of no more 100 million ECU. This Regulation also contains specific provisions applicable to petroleum and beer supply agreements.

Selective distribution agreements Selective distribution agreements apply mainly to consumer goods which are sold through high class department stores and other specialty outlets. It therefore affects perfumes, hifi equipment and other such products which require some sort of specialist expertise and handling. Manufacturers who wish to maintain high standards for their products may wish to lay down minimum quality criteria for their retail outlets, such as ensuring that the sales and service personnel are properly trained, that the outlets are properly laid out and maintained to specified standards. EC law makes the somewhat arbitrary distinction between such qualitative restrictions (which are deemed to fall outside Article 85(1)) and any quantitative restrictions (e.g. a restriction on handling of competing products, or a restriction or quota on the number of resellers in a particular district). As a general rule, selective distribution is only acceptable if one can show that the goods have certain special qualities that require a particular standard of premises, specially trained staff or special facilities for post-sale servicing of the products.

Franchising agreements The block exemption for franchise agreements (Regulation No. 4087/88) is available where a franchisor agrees to restrain from exploiting its own franchise, marketing the goods or services which are the subject matter of the franchise, supplying the goods to third parties, or granting the right to exploit all or part of the franchise to third parties. As to franchisees, they may be made subject to restrictions with respect to: (i) exploiting the franchise beyond the contract premises, (ii) seeking customers outside the contract territory for the goods or services which are the subject matter of the franchise, and (iii) manufacturing, selling, or using goods that compete with the franchisor's goods in the course of providing services.

The Regulation does not apply where competitors enter into franchising agreements, in the case of provisions that prevent franchisees from continuing to use licensed know-how after the termination of the agreement although it has became public, price fixing provisions, clauses which prohibit the franchisees from challenging the validity of the industrial or intellectual rights of the franchisor, or provisions which bar franchisees from supplying goods or services within the EC to end purchasers based solely on the latters' places of residence. As with several other block exemptions, a "fast-track" opposition procedure is available to parties seeking exemption in cases where certain provisions are not expressly barred or prohibited by the regulation.

D. Merger Controls

Originally, the Commission believed that the EC Treaty provisions on competition should not apply to mergers. However, due to the growing merger and acquisition activity within the EC, the EC Council felt that it was important for the Commission to be given the explicit power authority to oppose or modify large scale mergers which were likely to effect the EC market. This grant of authority was accomplished through the adoption of the Merger Control Regulation (Regulation 4064/89 with respect to the control of concentrations between undertakings) which applies to mergers and acquisitions with a "Community dimension." Those transactions which are subject to the Merger Control Regulation are mergers and acquisitions where the sales of the parties (both worldwide and within the Community) are above the following thresholds:

(i) ECU 5,000 million group turnover of all parties worldwide;
(ii) plus ECU 250 million group turnover of at least two parties in the EC;
(iii) no more than two thirds EC turnover in the same member state.

Mergers and acquisitions, which are referred to in the Regulation as "concentrations," which fall within the scope of the Merger Control Regulation must be notified to the Commission within one week of the signing of the agreement with respect thereto. A concentration will be prohibited if the Commission finds it to be incompatible with the common market, which will be deemed to be the case in the event that the concentration strengthens a party's dominant position in the EC such that effective competition is likely to be impeded. The Commission has one month to notify the parties as to whether it will review the transaction.[39] If the Commission, upon initial enquiry, has concerns about the proposed transaction it must launch a detailed investigation (a so-called "Phase 2 investigation"). The Commission then has four months to complete its investigation and reach a final decision.[40]

E. Monopolization and Abuse of Dominant Position

Article 86 of the Treaty prohibits any abuse of a dominant position by a single firm in the EC. In determining whether or not a particular transaction falls within the scope of Article 86, the Commission must first decide if there exists a "dominant position" in the EC or in a substantial part thereof. Article 86 does not define what exactly is meant by a "dominant position"; however, the case law has stated that it is present where an entity can conduct business while disregarding its competitors, suppliers, or customers. Although the size of an entity is important in making this determination, the test is one of market control in a substantial part of the EC. Accordingly, the Commission will taken into account not only the percentage of the relevant market that an enterprise may control, but also the relative strength of its competitors and their respective access to know-how, raw materials, and capital, as well as any barriers to entry.

While it is generally assumed that Article 86 will apply almost exclusively to firms of larger size, it is important to remember that a smaller firm could fall within its scope

[39] Art. 10.1 of the Merger Regulation.
[40] Ibid.

where it introduces a product onto the market for which no significant alternatives exist. Also, a dominant position may be found to exist in limited markets, such as a market for replacement parts. Finally, firms with a product line that essentially forms its own market, such as the products manufactured by IBM as part of its computer equipment business, may be found to exercise a dominant position over its own products and may be required to provide competitors with technical information that would allow them to manufacture compatible and competitive products.

Article 86 operates independently of Article 85, and arrangements which may escape the need for notification with the Commission for purposes of Article 85 by virtue of complying with one of the group exemptions may nonetheless be covered by Article 86 of the Treaty in situations where the agreement constitutes an abuse of a dominant position. The ECJ considered this question with respect to patent licensing agreements in *Tetra Pak* v. *Commission.*[41] In that case, Tetra Pak acquired a United States company and, in so doing, acquired an exclusive license for a technique for manufacturing and filing sterile cartons. While the actual terms of the license brought it within the group exemption for patent licensing agreements, the Commission went on to issue a decision condemning Tetra Pak for its acquisition of the license on the basis that it constituted an abuse of a dominant position which infringed Article 86. The ECJ upheld the decision, stating that the grant of an exemption under Article 85(3), whether individually or by virtue of the patent licensing block exemption, did not provide protection against the possible effects of Article 86.

Assuming a firm is found to hold a dominant position in a substantial part of the EC, Article 86 prohibits that firm from abusing that position. Specific instances of such an abuse might include (i) imposing unfair buying or selling prices or other trading commissions; (ii) limitations on markets, production, or technical development which result in prejudice to consumers; (iii) imposing tying arrangements; or (iv) discriminating in commercial transactions that would lead to a competitive disadvantage over trading partners. In addition, a recent case raised the issue of whether or not a dominant firm's decision not to grant licenses might be challenged as a violation of Article 86.[42]

F. National Competition Laws in the EC

In addition to EC competition law, parties to distribution agreements in the EC must also comply with the competition law regimes of the various EC member states. For example, under the French national competition law:

(i) a party cannot impose on its French distributor the obligation to engage in en gage in predatory pricing;
(ii) a party cannot impose a minimum resale price, but it may advise its French distributor regarding any recommended resale prices;
(iii) a party cannot refuse to sell its own products; and
(iv) a party should treat all of its intermediaries fairly and avoid discriminatory practices (i.e. discounts, rebates, bonuses, or other conditions of sale should be accorded equally to all intermediaries).

[41] Case T-51/89 *Tetra Pak* v. *Commission* [1990] II ECR 309 [1991] 4 CMLR 334.
[42] 89/205/EEC (IV/31.851– *Magill TV Guide/ITP, BBC, and RTE*) OJ 1989 L78/43.

III. General Framework of Japanese Antitrust Laws

A. The Antimonopoly Law and the Fair Trade Commission

In Japan, antitrust issues are considered under the terms of that country's Antimonopoly Law which, among other things, will prohibit any agreement or contract between foreign and Japanese entrepreneurs which contain any terms which amount to either an unreasonable restraint of trade or an unfair trade practice.[43] The Antimonopoly Law is administered by the Fair Trade Commission ("FTC"), which is part of the Prime Minister's office but retains full independence in performing its duties. Among other things, the FTC (i) receives reports or notifications; (ii) designates unfair practices; (iii) investigates suspected violations of the Antimonopoly Law; and (iv) eliminates any unfair practices chiefly through private consultation with the parties to an agreement. International agreements will also be subject to review under the country's foreign investment laws.

B. International Licensing and Distribution Agreements

Notification and FTC examination The Antimonopoly Law includes a system of notification and examination for international licensing and distribution agreements which has been established in order to protect and promote fair and free competition. The Antimonopoly Law requires that all such international agreements must be filed with the FTC after execution and the FTC is given broad authority to review the validity of the agreement in terms of the Antimonopoly Law. If the FTC determines that a provision in an agreement constitutes a "private monopoly," "unreasonable restraint," or "unfair trade practice" under the Antimonopoly Law, the FTC will either request that the contracting parties voluntarily delete such a provision or make an official recommendation to urge the Japanese party to do so. If the parties refuses to accept the recommendation, the FTC may commence a formal proceeding to enforce the recommendation and the decision resulting from this proceeding can be appealed to the Tokyo High Court.

FTC Guidelines on international licensing agreements As in the EC, the FTC has promulgated various guidelines regarding the content of international know-how license agreements which are intended to improve the transparency of its process of review and assist private parties in drafting agreements which do not run afoul of the Antimonopoly Law. In 1989, the FTC announced a new set of patent and know-how licensing guidelines which apply to both domestic and international license agreements, although the compulsory registration requirements of the Antimonopoly Act apply only to international agreements.[44] As a general rule, the system avoids litigation and monetary penalties, instead the FTC simply denies registration to the agreement until the parties remove any unacceptable provisions from the license agreement.

The structure, as well as the content, of the know-how guidelines appears to have been borrowed from the EC, although it is believed that, in some cases, the Japanese rules

[43] See Law Concerning Prohibition of Private Monopoly and Maintenance of Fair Trade, Law No. 54 of 1947, as amended.

[44] The know-how guidelines cover only technological know-how related to industrial use and not non-secret know-how.

may be more permissive than in the EC. As in the EC, the Japanese guidelines establish three categories to evaluate the terms of any license agreement:

(i) license terms likely not to be considered unfair trade practices since they have only a negligible effect on competition ("white list");
(ii) license terms that might be unfair trade practices ("grey list"); and
(iii) license terms that are "highly likely" to be considered unfair trade practices ("black list").

The following licensing agreement restrictions are among those that have been placed on the white list in the know-how guidelines:

(i) granting a license for a limited period insofar as licensed know-how remains secret;
(ii) restricting exploitation of licensed know-how to a specific field of technology;
(iii) requiring minimum production, minimum sales volume of goods manufactured exploiting licensed know-how, or minimum use of licensed know-how;
(iv) making it obligatory for the licensee not to handle substitutable goods or similar goods that are in competition with licensed goods or not to employ substitutable technology or similar technology for a short period after the expiration or termination of the licensing agreement, provided that such obligation is confined to the extent necessary for preventing unauthorized exploitation of the licensed know-how;
(v) making it obligatory for the licensee to inform the licensor of knowledge or experience newly obtained by the licensee regarding licensed know-how, or to grant the licensor a non-exclusive license with respect to an invention improved or applied by the licensee, provided that the licensor has similar obligations to the licensee and that the obligations of both parties with respect to informing the other or granting a non-exclusive license are roughly balanced in substance;
(vi) making it obligatory for the licensee to maintain specified quality standards, provided that the obligation is confined to the extent necessary to guarantee the effectiveness of licensed know-how or maintain the goodwill of a licensed trademark;
(vii) making it obligatory for the licensee to procure raw materials or components from the licensor or a person designated by the licensor in those cases where the use of specified quality standards are not sufficient in order to guarantee the effectiveness of the licensed know-how, maintain the goodwill of a trademark, or such obligation is vital for protection of the secrecy of licensed know-how, provided that such obligations are confined to the extent necessary to guarantee the effectiveness of licensed know-how, maintain the goodwill of a trademark, or for protecting the secrecy of licensed know-how;
(viii) restricting the ability of the licensee to export goods into any area in which (a) the licensor has restricted its patent rights on the licensed goods, (b) the licensor has continuously marketed the licensed goods, or (c) the licensor has assigned exclusive sales rights to a third party;
(ix) restricting the licensee's export price or export volume of licensed goods or making it obligatory for the licensee to export through the licensor or a person designated by the licensor if the restrictions or obligations pertain to areas described in (viii) above and are confined to the extent necessary;
(x) making it obligatory for the licensee to use production or sales volume or price of the finished product as a basis for royalty, provided that the finished

product cannot be produced without using the licensed know-how, or making it obligatory for the licensee to use the consumption of raw materials or components that are necessary for producing licensed goods as a basis for royalty;

(xi) making it obligatory for the licensee to accept licensing of more than two items of know-how as a package, provided that such an obligation is necessary for guaranteeing the effectiveness of licensed know-how;

(xii) providing that royalties continue to accrue after licensed know-how has become publicly known due to reasons for which the licensee is not responsible, insofar as it constitutes installment payment or extended payment of royalties, or insofar as royalties for use are charged for a short period after licensed know-how has become publicly known during the term of the licensing agreements;

(xiii) providing that the licensor can terminate the licensing agreement if the licensee challenges whether licensed know-how has become publicly known;

(xiv) making it obligatory for the licensee not to disclose licensed know-how to a third party so long as licensed know-how remains secret;

(xv) making it obligatory for the licensee to use his best efforts to exploit licensed know-how.

Unlike the situation in the EC, grey list items, those which may be deemed to be unfair trade practices, will be the subject of further review by the FTC in light of the position of the parties to the agreement in the relevant market, the conditions of the relevant market, the duration of restrictions imposed, and the restrictions taken as a whole.

Restrictions on handling competing goods An attempt to restrict the licensee's ability to handle competing goods or to use competing technologies during the term of the license agreement will be struck down if competition is reduced due to the fact that competing companies will be denied access to important customers or the licensee is deprived of the freedom to select its preferred goods or technologies.

Restrictions on sale of licensed goods Licensees cannot be required to sell the licensed goods only through the licensor or a person designated by the licensor, nor can licensees be prevented from selling to persons designated by the licensor. A provision of this type will be attacked in those cases where depriving the licensee of the freedom to select its sales outlets will result in a reduction of competition.

Non-reciprocal grant-back requirements While reciprocal grant-back provisions are included on the white list, a licensee cannot be forced to inform or license the licensor with respect to improvements of the licensed know-how unless the licensor is under a comparable obligation to the licensee during the term of the license agreement.

Required use of licensor's trademarks A licensee generally cannot be forced to use a trademark designated by the licensor for licensed goods if the result is that the business activities of the licensee are unjustly restricted by its inability to freely select a trademark for the goods. Concerns will also arise in those cases where the licensee is required to use the licensor's trademarks even after the licensed know-how has become public.

Restrictions on quality Restrictions on the quality of licensed goods, raw materials, components and similar items which are not necessary to guarantee the effectiveness of the licensed know-how or maintain the goodwill of a trademark are not permitted if such restrictions would reduce competition in the market for the licensed goods or the market for the raw materials, components or similar items.

Mandatory procurement requirements A licensor may not obligate a licensee to procure raw materials or components from the licensor or a person designated by the licensor other than for the reasons specified on the white list, since such a restriction might either deprive the licensee of the freedom to select its sources for the items, thereby impairing efficient competition, or reduce competition in the markets for the raw materials or components.

Export restrictions Geographic export restrictions beyond those permitted under the white list (i.e. areas in which the licensor has restricted patent rights, has engaged in continuous marketing, or assigned exclusive sales rights to another) will generally not be allowed if the effect of the restrictions is to reduce competition in the export market. Similarly, restrictions on the licensee's export price or export volume of licensed goods, or any obligation for the licensee to export through the licensor or a person designated by the licensor, will be disallowed to the extent that they go beyond those permitted under the white list.

Royalties unrelated to licensed goods Other than the formulae included in the white list for computing royalties on the basis of finished products or consumption of raw materials, a licensee cannot be required to pay royalties based on products or services other than the licensed goods if the result would be to establish transaction terms which are unduly disadvantageous to the licensee.

Tying arrangements Tying arrangements, which impose an obligation on the licensee to accept licenses for more than two items of know-how as a package, are forbidden unless confined to the extent necessary to guarantee the effectiveness of the licensed know-how, since the licensee is deprived of the freedom to select its technology and competition in the technology market is likely to be reduced. Moreover, tying arrangements will be disallowed in those cases where the licensee is forced to pay extra royalties or the duration of royalty payments is extended.

Trade secret no-contest clauses A licensee cannot be prevented from contesting whether or not the licensed know-how is in the public domain. Concerns regarding such a provision are that the licensee would have to continue paying royalties with respect to technology that otherwise would be freely available, thereby resulting in transaction terms which are unduly disadvantageous to the licensee.

Termination provisions The license agreement cannot include unjustly disadvantageous termination conditions, such as allowing for unilateral or immediate termination without appropriate notice by the licensor, except in those cases where the license agreement become unenforceable (e.g. due to the insolvency of the licensee).

Restrictions on the black list are generally considered as unfair business practices and must be removed from the agreement unless a specific justification is presented. Among the clauses on the black list are those which: (i) restrict the sale or resale prices of licensed goods; (ii) prohibit the licensee from handling competing goods or employing any competing technology after the expiration or termination of the license agreement, unless the prohibition is for a short time to prevent unauthorized use of the licensed know-how; (iii) restrict the use of licensed technology or require the licensee to pay a royalty for such use after licensed know-how becomes public[45]; (iv) restrict research and development activities by the licensee; or (v) obligate the licensee to assign to the licensor the rights to any improvements on the licensed technology or to grant the licensor an exclusive license with respect to an improvement.

FTC Guidelines on international distribution agreements The Antimonopoly Law also applies to the terms of any international distribution agreement relating to the distribution of imported goods by a Japanese firm. The Antimonopoly Law generally prohibits the acquisition of a dominant market position through the terms of any such agreement. In its "General Designation" of unfair practices in the context of distribution agreements, the FTC has stated its intent to challenge unduly exclusive dealing, resale price maintenance, and customer and territorial restrictions which might be employed as a tool of resale price maintenance.[46]

In addition, the FTC has taken the position that agreements for the distribution of well-known brands of imported goods will be challenged if a Japanese party would, by reason of its entering into an agreement pursuant to which it would become a distributor of imported goods, acquire a market share of 25 percent or more.[47] The FTC has also stated that the following practices may be deemed to be "unfair," and therefore illegal, in the context of any such agreement: (i) restricting the resale prices of the goods covered by an agreement; (ii) restricting the person to whom the goods covered by an agreement are resold; (iii) imposing the obligation to purchase parts, etc., for the goods covered by an agreement from the foreign party or a person designated by such foreign party; (iv) unduly hindering parallel importation of the goods covered by the agreement; and (v) imposing an unduly disadvantageous condition for the termination of the agreement.[48]

C. Other Antitrust Regulatory Policies

Japan's Antimonopoly Law, which was originally modeled after the antitrust laws of the United States, states its purpose as the prevention of "excessive concentration of economic power" and the promotion of free and fair competition. However, the most important characteristic of the Japanese law of competition has been the set of exemptions provided under the Antimonopoly Law, which includes depression and rationalization cartels, industrial property rights, designated cooperatives, regulated industries, "natural monopolies," and legitimate acts under legislation adopted with respect to the development of specific industries. Moreover, from time to time, additional exemptions have been made available that complement overall industrial policies designed to focus upon specific industries.[49]

[45] Royalties may be required for a short time after the know-how has come into the public domain provided that the royalty is based on the use of the know-how after public disclosure and during the term of the license agreement.

[46] See FTC Notice No. 15, General Designation Items 11–13.

[47] See the Antimonopoly Guideline for Sole Import Distributorship Etc. Agreements, which was published in 1972.

[48] Ibid.

[49] For example, in the early 1970s, exemptions were, in place which covered expert trade, land transpoltation, shipping conferences, and insurance, as well as particular industries such as coal, machinery, fertilizers, nonferrous metals, and textiles. According to the annual reports of the FTC, cartels have often been legally sanctioned in dozens of industries.

The effect of these exemptions has often been to promote, rather than deter, industry concentration at levels far in excess of those that might be tolerated in nations such as the United States. Yet, the FTC has pursued prevention of acts that would constitute "unfair business practices." As a result, a system of vigorous competition has developed among firms in Japan. It is, however, quite clear that potential new entrants have a great deal of difficulty in penetrating industries and markets in which the existing competitors have already developed the "firm-level" practices necessary for effective competition. Moreover, as the economy has developed, antitrust relief to various industries has been gradually abandoned or successfully challenged,[50] and an effort has been made to concentrate the influence of regulation upon those firms and industries that have become uncompetitive.

The Antimonopoly Act does not contain a specific legislative exemption for joint innovation arrangements. However, the FTC has adopted a basic policy which has led to the exemption of most joint innovation arrangements from the scope of the Antimonopoly Act.[51] In fact, the Japanese FTC has stated that the evaluation of the anticompetitive effect of joint research and development at the product market stage will depend significantly "on the competition and market shares among the participants and the market structure of the industry to which the participants belong ... In cases where the market shares of the participants are small ... the effects will be small."[52] Commentators have noted that they believe that "small" in this context is likely to include cooperative contractual arrangements where the parties have a combined market share of less than 25 percent, meaning that such arrangements are not likely to be closely examined by FTC officials.[53]

The FTC explicitly states that if anticompetitive effects are being alleged with respect to a proposed cooperative arrangement, then the procompetitive benefits associated with any innovation arising from the arrangement must also be balanced in assessing the transaction.[54] Moreover, when this balancing process takes place, it must occur not only with a particular market, but also across markets in light of the "possibility of the emergence of competition at the intersection of industrial sectors as a result of joint R&D between firms in different sectors."[55] Among the procompetitive justifications which are recognized by the FTC are: (i) the difficulty of single-firm innovation; (ii) the faster innovation created by cooperation and specialization between joint participants; (iii) the pursuit of innovation in new fields by utilizing shared technology and know-how; and (iv) enhancement of the technological level of each participant through the interchange of technology.[56]

[50] For example. in the early 1980s, MITI's "administrative guidance" with respect to the oil industry was challenged and reversed in the courts, suggesting that the FTC may be gaining greater latitude in seeking to limit MITI's discretion with respect to the structuring of domestic markets.

[51] Jorde and Teece, "Innovation and Cooperation: Implications for Competition and Antitrust," in *Collaboration Among Competitors* (1990) (hereinafter *Collaboration*), 887, 901.

[52] Ibid. at 901 n14 (citing Federal Trade Commission (Japan), *Research and Development Activities in Private Enterprises and Problems They Pose in the Competition Policy* (1984), 37–9.

[53] Ibid. The "safe harbor" for arrangements involving parties with less than 25 percent combined market share is similar to the safe harbor applicable to horizontal mergers contained in Japan's Merger Guidelines. Ibid.

[54] Ibid.

[55] Ibid. (citing Federal Trade Commission (Japan), *Research and Development Activities in Private Enterprises and Problems They Pose in the Competition Policy* (1984), 37–9).

[56] Ibid.

Teece and Jorde believe that the basic Japanese attitude regarding joint research and development activities is that they are procompetitive and therefore should not be restricted under that nation's competition laws.[57] They have pointed out that the literal Japanese translation of research and development – *kenkyu kaihatsu* – implicitly includes commercialization, and that here is no semantic distinction between the concepts of research and development and commercialization.[58] They have also praised what appears to be a Japanese system of cooperation and consultation among its governmental agencies with respect to state-sponsored cooperative research and development projects.[59] For example, they point out that whenever Japan's powerful Ministry of Trade and Industry ("MITI") wishes to promote a cooperative research and development project, the Japanese FTC will be consulted in advance and once the Japanese FTC has cleared a project or activity, it is unlikely that there will be any subsequent antitrust actions by the government.[60]

Other commentators take issue with Teece and Jorde's interpretation of competition policies in Japan. Brodley has written that the apparent close collaboration between research and commercialization in Japan has occurred primarily within a single firm, not through the use of jointly owned production or marketing facilities,[61] and that Japan actually rarely permits joint production ventures[62] or marketing collaborations.[63] Shapiro and Willig note that while Japanese policy has been to promote cooperative research and development efforts, which are structured as "study groups," collaboration generally terminates when the research and development has been completed and the parties are left to their own independent production efforts.[64] They also point out that joint production ventures involving domestic firms are not commonly used in Japan, although there has been some increase in production collaboration between Japanese firms and foreign firms.[65]

There are also other differences between legal and business practices in Japan and the United States which must be taken into account in making any comparison. First, the Japanese attitude toward mergers and acquisitions, particularly in industries that are in the process of reorganization and consolidation, appears to be much more liberal

[57] Ibid. Jorde and Teece claim that Japanese antitrust policies are based on an appreciation that if the policies are to err as between protecting innovation and competition, innovation is the preferred goal and objective. Ibid. at 905.

[58] Ibid. at 901.

[59] Ibid.

[60] Ibid. (citing examples of activities authorized by the Japanese Act for Facilitation of Research in Key Technology, or the Research Association for Mining and Manufacturing Technology Act). Jorde and Teece also point out that treble damages are not available to private parties seeking to enforce Japanese antitrust laws and that, as general rule, private suits for single damages under Japan's antitrust laws are very rare and usually unsuccessful. Ibid.

[61] See Brodley, "Antitrust Law and Innovation Cooperation," in *Collaboration*, supra note 51 (1987), 887, 913 (citing Freeman, *Technology Policy and Economic Performance: Lessons from Japan* London: Pinter Publishers; Ouichi, "The Logic of Research and Development," *California Management Review* 30 (1988), 9–33; Porter, *Antitrust Policy and Economic Competitiveness*, Hearings before Committee on the Judiciary, 100th Cong., 1st Sess., 1987: Washington: USGPO (1988) 154–71).

[62] Ibid. (citing Barfield, Congressional Testimony, in *Antitrust and Trade Regulation Reports*, Bureau of National Affairs (1989), 57, 370).

[63] Ibid. at 920.

[64] Shapiro and Willig, "On The Antitrust Treatment of Production Joint Ventures," in *Collaboration*, supra note 51, 929, 939.

[65] Ibid. at 939.

than in the United States.[66] Second, antitrust enforcement in certain industries has been almost non-existent, leading to bid-rigging and other practices which never would be tolerated by United States regulators.[67] Finally, while formal production joint ventures are rarely seen in Japan, it may well be that the Japanese practice of having a number of firms organized within a single "business group" facilitates a sharing of technology and complementary assets with little free-riding and transactions costs in much the same manner as a joint venture might do for unaffiliated firms.[68] In effect, as these firms are not direct competitors, since competitors are generally not in the same business group, the benefits or risk and asset sharing which are normally achieved through joint ventures that involve horizontal competitors are pursued by an organization based on vertical and conglomerate relationships.[69]

[66] Japan's powerful Ministry of International Trade and Industry has often recommended that certain Japanese industries should be reorganized and consolidated, generally when it becomes apparent that there is too much capacity for the industry to remain competitive in the face of competition from developing economies, such as Korea. In such cases, the consolidation of capacity takes the form of mergers, rather than joint ventures, and it has been pointed out that the horizontal combinations would like not be permitted under United States antitrust laws. Ibid. at 939–40.

[67] See ibid. at 940 (citing an article in the *New York Times* (November 29, 1989, page D1), which reported: "[there is] a long-established system in Japan known as 'Dango,' under which construction companies consult in secret before submitting bids and decide in advance which ones will submit the winning bids ... it had been years since the [Japanese Fair Trade Commission] was willing to challenge the politically powerful industry on charges that it violated Japan's Anti-Monopoly Act.").

[68] Ibid. (referring to Japanese business groups organized around a major bank which include cross-shareholding relationships among firms in many industries, a multiproduct trading company, and liberal lending policies and overall monitoring from the bank).

[69] Ibid.

5 Innovation and Collaboration

I. The Innovation Process

A. Introduction

There are a variety of ways to define the ultimate goals and objectives of the business enterprise. While it is true that the firm is generally dedicated to maximizing the investment return on the cash and other assets contributed to the enterprise, it is perhaps more useful to develop a definition of the output from the firm's business activities that is more closely aligned to the actual uses of the firm's resources. A term that has become quite popular recently is "innovation," which has been defined as the search for, and the discovery, development, improvement, adoption and commercialization of, new processes, new products, and new organizational structures and procedures.[1] The search for innovation is extremely complex and costly, and involves a good deal of uncertainty, risk taking, probing and reprobing, experimenting, and testing.[2]

B. Stages of the Innovation Process

Introduction While it is easy to oversimplify the innovation process, it is generally true that the creation, development and commercial exploitation of new products or services usually requires a good deal of cooperation among researchers, scientists, engineers, and marketing and sales personnel, as well as the ongoing involvement of senior management. From an engineering perspective, the key stages of innovation include product conceptualization, technical feasibility, product development, commercial validation and pre-production preparations, full-scale product production, and the ongoing support of the product. The first four stages are often referred to generally as "research and development." The fifth stage is commonly known as "manufacturing," and the final stage is both "service and support" and, to the extent that it involves improvements and enhancements to the original product or service, a form of ongoing research and development. Non-engineering innovation stages include distribution and marketing strategies; however, decisions in these areas inevitably influence, and are influenced by, the engineering-based activities.

Engineering stages of new product development The engineering stages of new product development can be briefly described as follows.

[1] Jorde and Teece, "Innovation and Cooperation: Implications for Competition and Antitrust," in *Collaboration Among Competitors* (1990) (hereinafter *Collaboration*), 887, 889.

[2] Ibid. See also Dosi, "Sources, Procedures, and Microeconomic Effects of Innovation," *Journal of Economic Literature* 26 (1988), 1120–71.

Product conceptualization During this stage, an effort is made to verify that a concept for a potential new product is viable, to define the key performance specifications and objectives of the new product, and to identify the major potential barriers to development, manufacturing and marketing of the new product.

Technical feasibility Technical feasibility work is designed to demonstrate that it is technically and economically feasible to continue the development of the product concept into a commercial product. This conclusion requires completion of various experiments and engineering tests. At the end of this stage, there should be sufficient information to prepare plans for the next stage, product development, including a cost and time schedule.

Product development In the development stage, engineers and designers perform the necessary work on materials, processes and design to test and prove that the product is commercially producible.

Commercial validation and preparation for production Once the product development work is completed, the next stage is to complete development of the manufacturing techniques and to demonstrate the validity of the new product in the marketplace. It is typical to prepare a preproduction prototype or process at this stage which is tested by limited placements with potential users (i.e. "betasites").

Full-scale production During the stage of full-scale production, the manufacturing or process facility is built (or capacity is otherwise acquired) and full-scale production runs are made. At this point, no further major design changes are made to the product, and the focus is on optimizing the process of manufacturing to meet customer demands.

Product support Product support involves not only warranty services, other repairs, and spare parts, but also the design and development of minor improvements and enhancements and new applications. In some cases, ideas for new product spin-offs or major design changes may be identified, at which point the focus would return once again to the beginning stages of the innovation process.[3]

Non-engineering stages of new product development The process of new product development is not limited to engineering tasks. For example, no product development effort should be launched without careful consideration of how all the required activities are to be financed. There are obviously a vast array of financing strategies which can be used to fund new product development, although the alternatives available at any one point in time will depend on a number of factors, including the size of the firm, cash flow from other activities, the type of product, the creditworthiness of the firm and/or its backers, and general economic conditions.

In addition, the success or failure of the innovation process will ultimately be tied to the quality and execution of the company's marketing and sales strategies for the particular product or service, including the plans developed with respect to distribution, pricing and packaging. In recent years, firms have begun to bring marketing personnel

[3] Adapted from *Engineering Stages of New Product Development*, National Society of Professional Engineers (1990).

into the product development process at a much earlier stage, and many new products actually are conceived as responses to perceived customer needs. It is this phenomenon that is responsible, to some extent, for a good deal of the revision to the traditional model of innovation.

C. Models of the Innovation Process

It is certainly difficult to make generalizations regarding the stages of the innovation process. Traditional models of the so-called "commercialization" process took a serial approach to the process, with the new products or services beginning with the identification of a customer need which was turned into some form of specification by marketing personnel and then turned over to the company's development and design engineers for further work. The resultant prototype was then adapted for production in large-scale manufacturing processes, and the final product eventually was incorporated into the company's portfolio and distributed through its sales and distribution channels. In most cases, there was little or no feedback or overlap between and among the various stages. The serial model lends itself to looking at innovation as a vertical process.[4]

The economics of innovation has recently become a very active area of economic inquiry, although the results are far from settled.[5] There has been increasing recognition of the need for constant interaction between various specialists and other participants, including potential customers, during the product development process. The traditional serial model of innovation has been criticized for its inability to capture how innovation actually proceeds in a number of technology-driven industries.[6] For example, it has been argued that the traditional model applies only, if at all, to scale-intensive industries, and does fit for an industry, such as semiconductors, computers, and automobiles, in which technological change flows from a number of small, but cumulatively important, incremental innovations.[7]

A new model of innovation, the "simultaneous" model, has been advanced as the proper basis for gaining an understanding of innovation.[8] The simultaneous model recognizes that innovation often depends upon the existence and smooth operation of tight linkages and feedback mechanisms, including linkages within firms, and often between firms and other organizations. Under the simultaneous model, innovation is not a serial process. Even though the model conceptually concedes the existence of a general flow of activity similar to that in the traditional model (i.e. development to production to marketing), its main contribution is its ability to recognize that innovation requires the involvement of a number of economic actors who do not have a simple upstream–downstream relationship to one another. For example, in the simultaneous model of innovation, research personnel must be able to work closely with manufacturing personnel

[4] Jorde and Teece, supra note 1, at 889. Characterizing innovation as a vertical process enables economists and legal scholars to fit innovation into the traditional antitrust analysis of vertical restraints. See also Grossman and Shapiro, "Research Joint Ventures: An Antitrust Analysis," *Journal of Law and Economics* 2 (1986) 315, 319; Tirole, *The Theory of Industrial Organization*, Cambridge: MIT Press (1988), 389.

[5] See, e.g. Shapiro, "Patents and R&D Rivalry," *American Economic Review* 75 (1985), 25–30; Levin, "A New Look at the Patent System," *American Economic Review* 76 (1986), 199–202.

[6] Jorde and Teece, supra note 1, at 889. See also Teece, "Inter-organizational Requirements of the Innovation Process," *Managerial and Decision Economics* 10, 35–42.

[7] Jorde and Teece, supra note 1, at 889.

[8] Ibid. at 890–1.

and marketing staff, as well as with outside suppliers of raw materials and new complimentary technologies, in order to rapidly design products that meet customer needs quickly and efficiently.[9]

The simultaneous model of innovation has a variety of important consequences.[10] First, it implies that innovation does not always begin with research and development, as it does in the serial model of innovation. Second, pioneer companies may be able to achieve market leadership through rapid product design, rather than supremacy in basic research and development, since they are often able to "design in" science and technology that is either already in the public domain or available from other firms. Third, an innovative firm need not have the internal capacity to independently perform all of the steps included in the serial model, rather they can develop and assemble leading edge products by contracting for the needed inputs and complimentary assets in a variety of lateral and horizontal linkages with outside firms and organizations, including universities.

II. Innovation and the Business Planning Process

A. Introduction

The stages of innovation should be closely linked to the company's overall business planning process, and this normally is evidenced by the company's approach to each of the key functional areas involved in identifying, developing and commercializing new products and services. While organizational schemes may vary somewhat from firm to firm, it is fairly common to differentiate among research and development, manufacturing and production, distribution, marketing, and service and support activities in assessing the strengths and weaknesses of the firm. Once the firm has evaluated and described its resources in each of these areas, the key step in the business planning process is to identify the competitive advantages and distinctive competencies of the firm. In so doing, management can then determine those areas in which collaborative relationships with third parties might be most beneficial to the firm and its business.

B. Analysis of Key Functional Areas

Research and development activities Research and development activities cover each of the activities enumerated under the first four stages of the new product development process. In most cases, a company will be involved in several different research and development projects simultaneously, each of which will its own distinct set of risks and milestones. A company's research and development activities should emphasize those new products which the company anticipates will be needed in order to respond to changing market needs, and the company should develop a strategy for meeting the competitive challenges created by new technologies or scientific approaches which may become practical over the next few years. While a good amount of the company's research and development may be conducted by its own personnel, it is not uncommon for a company to enter into contracts with third parties (e.g. universities) in order to gain access to their special expertise.

[9] Ibid. at 890.
[10] Ibid. at 891.

Manufacturing and production activities Manufacturing and production involves a wide variety of different activities dedicated to the commercial production of the company's products in quantities, and at rates, which are consistent with market demand for the products. While clearly research and development can be quite expensive, manufacturing will also represent a significant cost to the company, and every effort will be made to reduce manufacturing costs as way for the company to obtain a pricing advantage over competitors. Some of the activities during the manufacturing/production stage include not only the construction of manufacturing facilities and the related equipment, but also the design and implementation of quality control procedures, the development of a distribution system (i.e. shipment and warehousing processes), the creation of customer assistance schemes, and contracting for procurement of parts, raw materials and other production items.

The company's business plan should address how it intends to manufacture its products and get them into the hands of customers or into the appropriate distribution channels. In particular, the following issues should be covered:

(i) Will the company manufacture and produce its products internally or will it subcontract with others for all or a portion of the manufacturing process? If the company will be subcontracting, are there any plans to develop an internal manufacturing capacity at some time in the future?
(ii) Does the company, or its subcontractors, have any particular production advantage in relation to the company's competitors? If so, how long can that advantage be exploited and how is it protected?
(iii) What is current production capacity? Is it sufficient to meet the requirement of the company in the future? If not, what plans does the company have for expanding such capacity and what costs and risks are associated with such plans? What are the critical parts or components in the production process? How are these parts/components acquired? Are any of these parts/components "single-sourced," or does the company have other suppliers? What lead times are required in order to acquire such parts/components, and how does this impact the ability of the company to meet increases in demand?
(v) What are the standard costs for production at various volume levels?

Distribution activities The key distribution issue confronting the firm is how it intends to make its products and services available to eventual end users. There are a number of strategies which might be used, often in combination with one another. For example, the company could develop its own distribution force; however, the cost of such a network may be prohibitive, especially if the company is relatively new. In other cases, the company might look to enter into one or more sale representative or distributor agreements with third parties under which the third party would assume the responsibility for selling the company's products in designated demographic and geographic markets. While the company would have to pay the representative/distributor a commission, or grant them a substantial discount on sales of the products for the purpose of resale, the use of third parties is seen as a quick and cost-effective mechanism for entering a new market. On the other hand, if a distributor fails to vigorously market the company's products in the manner originally contemplated, any competitive advantage embodied in the company's technology may be lost as competitors are allowed to catch-up and enter the market with similar products.

Distribution strategies may change as time goes by. For example, as the company acquires the resources necessary for it to develop its own direct distribution capabilities, it may be able to reduce its dependence on outside distributors. In other cases, outside distributors will be brought into to assist the company enter new markets which are beyond its existing skills and resources. This strategy is particularly popular whenever a company embarks on large-scale international sales activities. Another possibility is the expansion of an existing relationship with a distributor to include development of new products that compliment those which the distributor might otherwise have in its product line.

Marketing activities Marketing activities include a wide variety of issues. Certainly, the company's marketing plans should cover traditional issues such as promotional strategies, pricing strategies, and market identification and selection. However, modern marketing techniques recognize the factors that form the basis for the new serial model of innovation, and properly applied should extend to distribution, and to field service and product support, each of which are important elements of the value provided by the company to its customers. In any event, the company's marketing plan should consider the following.

Promotional strategies The company must determine the proper mix of its promotional activities, including advertising, trade shows, and direct sales efforts.

Pricing strategies In some cases, the company may be able to take advantage of demand pricing, particularly if the product or service is particularly innovative or perceived by customers as providing a significant advantage over existing offerings. In other cases, cost-based pricing may be more appropriate in order to achieve market share objectives.

Market selection In many cases, the company will elect to concentrate on one or more specific demographic and/or geographic markets. For example, the company might choose to focus on sales to larger firms in metropolitan areas, as opposed to expending resources on promotional activities in remote areas in which customers are not likely to place orders of any significant size. The company may also decide to attempt penetration in foreign markets, perhaps using a distribution strategy (e.g. local distributors) which differs from that used in domestic markets.

The company's marketing strategies will likely change substantially over time. For example, while it may well be possible to pursue demand-based pricing fairly early in the life of a particular product, at some point it may be necessary for the company to revert to cost-based pricing, particularly when the market becomes more competitive or demand begins to decline for other reasons. Also, entry into specific markets may be deferred until the company has achieved its initial objectives, particularly if further product development work is required in order to adapt the basic product to the requirement of another market (e.g. foreign markets with specific technical needs).

Service and support operations Support and service, which focus on activities that occur after the production and sale of the product or service, are part of both the company's engineering and marketing strategies. From an engineering perspective, support and service means the ability of company to provide technical instructions, warranty services and other repairs, spare parts and, in some cases, improvements and enhancements. As a marketing tool, support and service is a means for maintaining customer satisfaction, building loyalty, and developing an ongoing relationship which allows the company to further understand the changing needs of its customer base. In fact, post-sale contacts with customers serve as the basis for identifying new product spin-offs or design changes which would eventually become part of the company's product portfolio. Also, ongoing field service and support can be an important source of additional revenues to the company.

As with manufacturing and distribution, the key choices for the company with respect to support and service is whether or not the company should conduct such activities directly using its own personnel or subcontract with third parties. The business plan should describe the company's plans in this area. The company should be mindful of the risks associated with using third party support and service providers. Obviously, any failure by such parties to perform their duties will likely lead to substantial customer dissatisfaction and harm the company's goodwill. Also, the company may be deprived of the opportunity to maintain regular contact with the end users of its products. Finally, in those situations where the company must transfer a good deal of its proprietary technical information to the third party in order for the third party to carry out its duties, there should be some legitimate concern that the third party will use the technical information to reduce the company's competitive advantage.

C. Competitive Analysis

Once the company has described its assets and resources in each of the key functional areas relevant to innovation in its line of business, it must then identify its own competitive advantages and distinctive competencies, and then analyze the relationship between its strengths and the resources and the tangible and intangible assets and capabilities which will be required in order to achieve success in the marketplace. For example, while a company may well have extraordinary engineering personnel, it may discover that its viability is dependent upon gaining access to specific distribution channels in order for its products to reach the customers with the most to gain from the engineering-based advantages. If the company identifies a gap between its own resources and the skills required for profitable innovation, it must then decide whether to attempt to developed the needed skills independently or to turn to a relationship with a third party in order to bolster its competitive profile.

One method that can be used in order to identify these advantages and competencies is to ask a series of questions regarding the manner in which the company completes the tasks necessary for it to develop, produce, and distribute its products and services. This exercise, which is sometimes referred to as "value-chain analysis," is designed to establish the basis upon which the company will compete in the future and, to some extent, to identify those areas in which the company's skills fail to match the expertise of competitors. In any event, the following questions should be considered:

(i) Is the company able to compete on the basis of its strengths in procurement, processing or manufacturing?
(ii) Is the company able to compete on the basis of its practices with respect to inventories, packaging or production flexibility?
(iii) Is the company able to compete on the basis of its strengths with respect to sales, distribution, customer service, delivery, promotion, maintenance or field engineering?

(iv) Is the company able to compete on the basis of the characteristics and capabilities of its products and services?
(v) Is the company able to compete on the basis of its operational flexibility to accept its products to meet the needs of different customer "niches"?
(vi) Is the company able to compete on the basis of its human resources strategies?
(vii) Is the company able to compete on the basis of its strengths in the area of research and development?
(viii) Is the company able to compete on the basis of a unique technological advantage, which is either embodied in its product line or is utilized in the course of its manufacturing?
(ix) Is the company able to compete on the basis of its financial assets or strategic business relationships?
(x) Is the company able to compete on the basis of its managerial skills, internal controls or innovative business goals and philosophies?

Once the key components of the relevant value chain for the company's products and services are identified, a similar series of questions should be asked for each of its major competitors. Thereafter, the company can choose the basis upon which it wishes to compete by reference to the important competitive factors in the industry and specific customer "needs." For example, while a given product may be technically "inferior," it may flourish due to the exceptional distribution capabilities of the manufacturer. Some companies may be able to become the lowest cost entry in the marketplace by virtue of its ability to use low-cost production facilities in an offshore location. In other cases, a company's unique technology can be used to develop a variety of new tools or applications which can be used to by customers to solve their own engineering or development problems.

III. Innovation and Business Collaborations

A. Private Benefits from Business Collaborations

Reduction of partners' business risks Innovation, particularly in high technology industries, is often a costly and time-consuming process. As suggested by the simultaneous model of innovation described above, innovation may require extensive changes, modifications and experimentation at each stage of the development and production process. Often, it is assumed that potential competitors may be unable or unwilling to assume the entire range of business and financial risks which may be associated with innovation. However, two or more firms might be willing to enter the marketplace in some form of joint venture or other collaboration with other similarly situated firms; provided that the participants believe that the return on investment is attractive taking into account the dilution which necessarily follows from shared control of the project. From an antitrust and competition law perspective, the reduced business risks for the participants may translate into a procompetitive effect in the marketplace if a new competitive force enters into what may have been a concentrated and less than fully competitive market.

Access to complementary assets Technological superiority is not sufficient to ensure that a firm will be able to appropriate all or most of the benefits from its investment in research, particularly if the technology can be easily imitated by the firm's competitors. Appropriation of the benefits may only be possible if the firm has access to the assets and capabilities necessary to package the new technology so that it is valuable to the end user.[11] The assets are often referred to as "complementary assets," and include complementary technologies, manufacturing, marketing, distribution, sales, and service. However, those assets that have the highest strategic value are not the generic assets of a firm, such as general purpose facilities and equipment and nonspecific skills, which are typically disembodied and codified and, as a result, easily transferable, but rather are specific assets, including highly differentiated system and firm-specific assets and skills.[12] Specific assets are, in essence, a firm's particular assemblage of physical assets and prior learning, all of which are tightly embedded in the firm's organizational structure and accordingly are difficult for the firm's competitors to replicate.[13]

Ideally, the firm itself will be world-class in those complementary assets required for successful exploitation of its technologies. However, in many cases, the firm will attempt to build various contractual relationships with partners able to perform the key activities which are complementary to the easily imitatable activities of the initial firm.[14] The value chain analysis that occurs in the course of developing the strategic plan of the firm is not only useful in identifying the unique competitive assets and skills of the firm, it can also be used to highlight those areas in which the firm might profitably pursue some sort of business collaboration with an outside party. For example, weaknesses in manufacturing or distribution can cripple the firm's ability to exploit its competitive advantage with regard to technology and research, and may dictate the need to bring in one or more outside partners to perform all or a portion of those functions for the firm. Correspondingly, if the firm is having difficulties in the area of new product development, a research collaboration with a third party may be able to provide it with access to new technologies, as well as to the talents of engineers and scientists currently working with other firms.

The firm's competitive strengths may also be exploited through business collaborations, and may also serve as a basis for expanding the firm's activities into new functional and product-related areas. For example, one firm with strengths in the development of new technologies and products might only be able to realize a full required return on investment if it is able to build a link with another firm capable of adapting and utilizing the technology in a superior manufacturing process, particularly since it is generally more difficult to replicate the manufacturing process than it is to reverse engineer the actual product.[15] Also, firms with substantial manufacturing and/or distribution expertise may be able to expand their own line of internally-developed products by licensing the rights to new technology and products from other firms, and may even be able to use the knowledge gained through such licenses to develop its own new products, provided that the license does not prohibit such activities.

As in other areas, the benefits from the integration of complementary assets may vary depending on the functional area of collaboration. For example, it is likely that research

[11] See, e.g. Jorde and Teece, supra note 1, at 896.
[12] Ibid.
[13] Ibid.
[14] Ibid.
[15] Ibid. (pointing out that once competitors reverse engineer a product, success in the marketplace, which is determined by the amount of the profits from the product which can be appropriated, essentially depends upon manufacturing costs and quality, and that many firms in the United States and Great Britain have failed to capture any substantial value from their inventions due to weaknesses in the manufacturing area).

and development activities provide greater opportunities for the realization of creative synergies between the parties than might be the case with manufacturing, since it is likely that the research backgrounds and skills of the parties will be far more diverse than their skills in the production area.[16] However, diversity among the participants may lead to complications, at least in those cases where the business cultures of the parties are sufficiently different so as to create difficulties with respect to communicating and interpreting the results from the research work.

Managing externalities In economics, an "externality" exists when benefits are created by an investment which cannot be fully captured by the firm which originally made the investment. A common example of an externality is the inability of an innovator to capture all of the value of the social benefits which may be created from the risks and investments made by the innovator, even though it may be possible to obtain some form of intellectual property rights with respect to components of the innovation. The concern in the situation where the products of the investment, which include know-how and other items with the characteristics of a "public good" when the investment activity is innovation, cannot be fully appropriated by the innovator is that the innovator, frustrated by seeing the benefits from the investment redistributed to competitors and users, will lose the incentive to make further investments in innovation.[17]

Although there are economists and others who would still disagree, it is generally accepted that there is some need to maintain adequate incentives to invest in innovative activity and that, absent government subsidies, this is best done through the development of methods for curtailing free-riding. One obvious legal and regulatory strategy in this area has been recognition of intellectual property rights, particularly patent rights.[18] However, as we have seen, patents have often been shown to be of limited utility in providing innovations with a material degree of protection from imitation.[19] As a result, innovators holding patent rights continue to run the risk that prospective users will nonetheless continue to develop inventions which prove to be comparable to, or substitutes for, the original innovation, often at a cost which is substantially less than the investment made by the first party to venture into the field.

Faced with the prospect that imitators would substantially reduce their investment returns on innovation, firms have turned to other strategies. Among the most common of these is the use of new organizational forms, such as joint ventures among the major prospective users of the innovation, which force the users to contribute to the costs associated with the innovation in proportion to the likely benefits each of the users expect to receive from the creation of the innovation,[20] thereby creating a real incentive for the participants to commit to protecting the results from the joint venture activities rather than attempting to independently replicate the results for their own account.[21]

[16] Shapiro and Willig, supra note 15, at 937. See also Grossman and Shapiro, supra note 4.

[17] See, e.g. Jorde and Teece, supra note 1, at 895 (noting that innovation has well-known free rider and public good characteristics from know-how leakage and other spillovers).

[18] Ibid.

[19] Ibid. (citing studies appearing in Levin et al., "Appropriating the Returns from Industrial Research and Development," *Brookings Papers on Economic Activity* 3 (1987), 783–820, and Mansfield et al., *Technology Transfer, Productivity, and Economic Policy*, New York: W.W. Norton (1982)).

[20] See, e.g. Kitch, supra note 23, at 871.

[21] Shapiro and Willig, supra note 15, at 937. It has been pointed that if a technological leader could secure adequate legal protection, and if technology and knowledge can actually be transferred at zero cost, as is assumed in many of the theoretical models, then the transactional structures which would exist in the marketplace for new technology would generally be quite simple. For example, the developer of the

There is no general consensus, however, that these strategies will be effective in reducing concerns about the appropriability of the results of innovation. For example, some commentators point out that a research joint venture may not be sufficient to overcome many of the problems, unless there are many potential competitors involved as part of the joint venture.[22]

Even if an innovator is not able to use a joint venture or similar organizational form to effectively reduce the number of potential imitators, such arrangements can produce a variety of other benefits that could increase the share of economic benefit that the innovator receives from the innovation. For example, by entering into one or more collaborative relationships in order to build, acquire, or rent, on an exclusive basis, the complimentary assets which are required to package and commercialize the new technology, the innovator may be able to efficiently increase the value for the end user and capitalizes on the very significant advantage of being a "first-mover."[23] The type and structure of these relationships will depend on the firm-specific assets of the innovator, the process of innovation that is specific to the product under consideration, and the research appropriability characteristics in the specific industry.[24]

Economies of scale Even if the parties could have entered the market on their own, it may be that the cooperative efforts conducted within the joint venture may permit operation at a more efficient scale, or at an earlier date. For example, a small firm may not be able to support research at an efficient scale; however, a joint venture with one or more larger firms may allow the project to continue, particularly since it is generally thought that the horizontal linkages implicit in a joint venture are useful in overcoming scale barriers in research, and can also be useful in defining technical standards for the products developed during the course of the research.[25] Also, even if a firm is able to finance or manage a new project, it may not be efficient from a risk management perspective for it to take on the project-specific risk of the particular venture in relation to its other investments. However, by joining forces in a joint venture with other firms, the firm may be comfortable with a more limited investment and risk exposure.

Economies of scale, as with any other form of "efficiency" from a collaborative business relationship, can be quite difficult to measure. In general, it is assumed that the benefits of economies of scale in the research area might be far greater than when the joint venture relates to production or marketing, since the marginal cost of an additional firm using a given research finding is far less than the average cost incurred by the discoverer,[26] meaning that it would be quite wasteful for other firms to conduct parallel re-

cont.
technology could either sell the technology to others for cash or license it to downstream firms who can then engage in the activities which might be required to extract value from the technology. See Jorde and Teece, supra note 1, at 892. See also Teece, "Profiting from Technological Innovation," *Research Policy* December (1986), 285–305. However, the inadequacies of patent protection has arguably led to the use of more complex collaborative arrangements.

[22]See, e.g. supra note 1, at 895.

[23]Ibid. at 895–6. See also Levin et al., supra note 34 (noting that first-mover advantages, such as lead time and learning curve advantages, secrecy, and investment in sales or service support were more effective in appropriating the rents from innovative activities than intellectual property protection in many cases).

[24]See, e.g. Brodley, "Antitrust Law and Innovation Cooperation," in *Collaboration*, supra note 1, 911, 913. With regard to variations in appropriability between industries, see Mowery (ed), *International Collaborative Ventures in US Manufacturing*, Cambridge: Ballinger (1988).

[25]Jorde and Teece, supra note 1, at 894.

[26]Shapiro and Willig, supra note 15, at 937. See also Grossman and Shapiro, supra note 4.

search projects.[27] However, several commentators have urged that caution be used in assuming beneficial economies from large-scale research efforts, since the advantages of scale has not been positively established.[28]

Economies of scale have also been studied in relation to identifying the optimal antitrust policy with regard to mergers and similar consolidations. While empirical research indicates that the incidence of plants operating at suboptimal scale (i.e. too small to realize all the known economies of scale) was greater in industries which were less concentrated;[29] it is not evident that antitrust policies that permit greater concentration (i.e. higher volume of collaborative business relationships) would necessarily lead to anything more than quite modest reductions in costs attributable to scale efficiencies.[30] Moreover, as to non-plant economies, including organizational expenses,[31] it appears inefficiencies are more likely, or of greater magnitude, when competitive constraints are the weakest (i.e. highly concentrated industry structures).[32]

Transactional benefits Even in those cases where the competitive effects of a proposed joint venture are merely neutral, or the formation of the joint venture is only marginally related to competitive considerations, the structure may offer significant transactional efficiencies over other forms of arrangements.[33] The notion of transactional efficiencies is extremely important in any economic analysis of structural forms and has its grounding in the work of economists such as Coase and Williamson.[34] From an antitrust and competition law perspective, transactional efficiencies come into play as regulators examine whether there are any other forms of contractual arrangement with equal efficiencies, yet which present less of a threat to competition.[35] As a practical matter, however, it is quite difficult to measure the relative efficiencies of two or more alternative transactional forms, although economists may refer to the presence or absence of particular institutional forms in the industry as empirical evidence of the efficiency of a given structure.[36]

[27] Shapiro and Willig, supra note 15, at 937.

[28] See, e.g. Brodley, supra note 39, at 913. See also Dasgupta, "The Theory of Technological Competition," in Stiglitz and Mathewson (eds), *New Development in the Analysis of Market Structure* (1986), 399–446 (finding no clear evidence regarding the advantages of large size in research activities).

[29] Scherer, "Antitrust, Efficiency, and Progress," in *Collaboration*, supra note 1, 33, 37 (citing Weiss, "Optimal Plant Size and the Extent of Sub-Optimal Capacity," in *Essays on Industrial Organization in Honor of Joe S. Bain* (1976) 123; and Scherer, "The Determinants of Industrial Plant Sizes in Six Nations," *Review of Economics and Statistics* 23 (1973), 125).

[30] Scherer, *Antitrust, Efficiency, and Progress*, supra note 44 (citing Scherer, *Industrial Market Structure and Economic Performance* (2nd edn. 1980), 469–70.

[31] Organizational expenses include capital raising, central office functions, field sales representation, advertising, and other activities typically spanning the operation of multiple plants.

[32] Ibid.; Siegfried and Wheeler, "Cost Efficiency and Monopoly: A Survey," *Quarterly Review of Economics and Business* 21 (1981), 11). See also Pickford, "A New Test for Manufacturing Industry Efficiency," *International Journal of Industrial Organization* 3 (1985), 153 (study found that New Zealand manufacturers protected from foreign competition by tariffs and other trade barriers had unit costs exceeding those of their most efficient overseas peers by 74 percent on average).

[33] See, e.g. Pitofsky, supra note 25, at 841–2.

[34] See, e.g. Coase, "The Nature of the Firm," *Economica* 4 (1937), 386 and Williamson, *Markets and Hierarchies, Analysis and Antitrust Implications: A Study in the Economics of Internal Organization* (1975).

[35] See, e.g. Kitch, supra note 23, at 872.

[36] Ibid. An economist would not feel constrained to looking simply at arrangements in the United States. For example, while an efficient arrangement may not be used in the United States due to concerns about antitrust liability, evidence that the arrangement had been successfully utilized in similar economies (e.g. the European Community or Japan) would be quite important to the economist's assessment of efficiencies. Ibid.

While there may be a number of transactional benefits from using a joint venture or other form of collaborative arrangement, the most common of these include the resolution of valuation problems, which is important when one or more of the parties is to contribute tangible and intangible assets that have a value which is speculative and based on future events; the reduction of managerial uncertainties, since the collaborative arrangement is better suited to management participation by more than one of the parties than a single, unilateral contract; the minimization of contractual uncertainties; and enhancement of the efficiency of technology transfer.

Resolution of valuation problems Complex collaborative relationships, such as a joint venture, can be particularly useful in situations where there would otherwise be difficulties in measuring the value of the contribution to be made by one or more of the parties to the project. For example, as noted above, collaborative business relationships might be attractive in those situations where the parties own or control complementary assets. While one form of transaction which might be used is for one party to sell the relevant asset to the other party in a one-time contractual sale arrangement, such a transaction may be problematic for a variety of reasons. First, the parties may be unable to agree on the purchase price, since many desirable assets, like new technology, are extremely difficult to value until they are incorporated into their proposed usage. Second, it may not be possible for the purchaser to fully utilize the asset without the continuing cooperation of the seller. Third, the specialized assets may be impossible to separate from the other activities of the selling firm.

If any or all of the foregoing factors exist with respect to the use of a sales contract, a joint venture, with its continuing requirement of cooperation, with investment returns based on results, and its ability to adapt as the project develops, may be the preferred transactional structure.[37] Alternatively, the parties may structure the transaction in a manner which, while less integrative than a free-standing joint venture entity, still ties the amount and timing of the payments for the subject asset to actual performance over the term of the relationship. For example, a licensing arrangement, coupled with a commitment on the part of the transferror to provide technical assistance and further research work over the term of the agreement, is often an effective transactional structure.

Managerial uncertainties In those cases where the future development of the market, or the products and services to be supplied in the collaboration, is somewhat uncertain, the partners may prefer the greater degree of managerial control that can be provided through the use of the joint venture form, as opposed to an arrangement pursuant to which one of the parties cedes to the other full decision-making power regarding key aspects of the innovation process. For example, while a firm with technological competence may simply license the technology to other firms with particular competence in manufacturing and distribution, a joint venture, in which the technology-provider works hand in hand with the other parties, allows the provider an opportunity to closely observe the manner in which the technology is being exploited, as well as the ability to learn new skills and capabilities in areas in which the provider has less experience.

Minimizing contractual uncertainties Just as different transactional structures may be used to resolve anticipated valuation problems, the form of transaction can be a useful tool in establishing procedures for the parties to discuss and resolve unexpected

[37] Ibid. at 872–3.

problems which may arise over the term of the relationship. In general, it is often argued that the joint venture form may be simpler to administer than a contractual arrangement focusing on the same objectives, since a long-term contract will generally require that the parties to the contract attempt to anticipate all sorts of contingencies and deal with them in the contract. For example, it may be difficult to negotiate the purchase of research and development under a long-term contract, since it may be impossible for the parties to clearly predict or measure the costs of the development work or to specify the output or results which are to be purchased. In such cases, the partial integration offered by the joint venture form may be a more efficient transaction form than attempt to use a research and technology purchase contract.[38]

Enhanced efficiencies for technology transfer Transactional efficiencies are particularly important in situations where one of the main purposes or objectives of the arrangement is efficient technology transfer. A number of commentators have noted that technology transfer among the various activities constituting innovation is far from costless. For example, it has been argued that the market for knowledge is, in fact, far from perfect,[39] that simple unilateral contracts are likely to be inefficient,[40] and that the effective transfer of technology will actually require the use of complex bilateral and multilateral contracts, internal organization, or some other sort of hybrid structure.[41] Accordingly, firms are often forced to seek protection for their investments in innovation through various organizational forms for undertaking their innovative activities.[42] This is particularly true in cases where the development and commercialization of new technology follow the simultaneous model of innovation, since the knowledge which is to be transferred generally does not lend itself to be bundled and shipped in one lot, and the close relationship between "research" activities and "commercialization" activities requires ongoing interaction between parties from the upstream development work to the downstream interfaces with the actual end users.[43]

As a general matter, there are many situations in which technology transfer cannot be effectively achieved in the context of contractual relationships that are arm's-length and non-exclusive.[44] In such cases, complex cooperative relationships, such as joint ventures, may provide the most effective means for exploiting the technology. In fact, the simultaneous model of innovation, with its emphasis on rapid market interface between the technology and the end user, has been cited as evidence for regulators to adopt a more liberal attitude toward the scope of the joint venture's activities on the basis that restrictions on the ability of firms in a joint research arrangement to continue to cooperate in

[38] Ibid. at 872.

[39] Arrow, "Economic Welfare and the Allocation of Resources for Invention," in National Bureau of Economic Research (ed.), *The Rate and Direction of Inventive Activity*, Princeton: Princeton University Press (1962) 609–25.

[40] See, e.g. Teece, "Economies of Scope and the Scope of the Enterprise," *Journal of Economic Behavior and Organization* 1 (1980), 223–47; and Teece, "Towards an Economic Theory of the Multiproduct Firm," *Journal of Economic Behavior and Organization* 3 (1982), 39–63.

[41] Jorde and Teece, supra note 1, at 892–3; Williamson, *The Economic Institutions of Capitalism: Firms, Markets, Relational Contracting* New York: Free Press (1985); Teece, "Profiting from technological innovation," *Research Policy*, December (1986), 285–305.

[42] Jorde and Teece, supra note 1, at 895; and Teece, "Profiting from technological innovation."

[43] Jorde and Teece, supra note 1, at 893.

[44] Ibid. See also, on the characteristics of knowledge, Winter, "Knowledge and Competence as Strategic Assets," in Teece (ed.), *The Competitive Challenge*, Cambridge: Ballinger (1987).

the commercialization of the technology might impose an unnecessary technology transfer burden,[45] as well as a loss of effectiveness and timeliness, and much higher costs.[46]

IV. Basic Forms of Business Collaborations

A. Introduction

As we have seen, innovation is a long and complex process which extends from the moment that the new product is conceived to the time that the product is in full-scale production and is being distributed in quantities sufficient to satisfy market demand. Each stage of the innovation process requires various specialized skills, assets and other resources which may not be owned or controlled by a single firm. In fact, it is the rare firm that is ever able to internally provide all the functions necessary for successful innovation, particularly when the business has just been organized. As such it is common for firms to seek out one or more business collaboration arrangements in order to assist them in executing their business plans and in maximizing the value of their own firm-specific assets.

While economists and other academics tend to speak of business collaborations as "joint ventures" regardless of their form, it is more common in the business world to refer to such arrangement by reference to the principal functions covered in the arrangement. The most popular forms of collaboration are licensing agreements, which are designed to facilitate the transfer of the legal right to use intellectual property from one party to another; research and development arrangements, supply arrangements; manufacturing and distribution arrangements; and actual joint ventures, such as newly formed corporations or partnerships owned and operated by two or more entities for a specified business purpose. In some cases, the collaboration actually takes the form of an outright acquisition of a firm or all of its assets and legal rights.

These basic forms of business collaboration can, of course, be combined into more complex arrangements covering a broad range of functional activities. For example, a joint venture may be established for the development, manufacture and distribution of one or more related products. License agreements often extend beyond mere technology transfer to include further research and development or the manufacture and/or distribution of products for the benefit of the licensor. Moreover, in some cases, the functional arrangements covered in a business collaboration will be coupled with financial support, in the form of an investment or long-term loan, from one party to another.

B. Licensing Arrangements

The most basic form of strategic business relationship is a licensing arrangement, which may cover patents, trade secrets, copyrights and/or trademarks. A license conveys to the licensee the right to use the licensed intellectual property right for the purposes described in the agreement without fear of being sued for infringement or misappropriation by the owner of the right. In effect, a license is a contractual exception to the otherwise exclu-

[45] Jorde and Teece, supra note 1, at 894.
[46] Ibid.

sive rights of the owner. In some cases, the license will extend to subsequent improvements or enhancements to the original licensed subject matter.

While a license is appropriate when the licensee simply wishes to exploit its own technology without being "blocked" by the rights of the licensor, it is quite common to see a license used as a fundamental tool in one of the other forms of strategic business relationship. For example, in order for a party to manufacture and distribute the products developed by another party, it must generally be granted a license to make and sell the products. In such cases, the arrangement between the parties includes the basic technology license and an agreement covering the terms and conditions upon which the products will ultimately be sold to consumers, and the licensor will typically receive a front-end payment plus royalties based on the revenues received by the licensee from the use of the licensed technology.

C. Research and Development Relationships

Research and development arrangements contemplate a coordinated effort by the parties to make fundamental changes in, or improvements to, specified core technologies. The research may be conducted under an agreement which provides for one party to fund the research work of another party in return for the rights to use the resultant technology in various applications. If the parties have complimentary technical skills and assets, the development program may include cross-licensing agreements and sharing of scientific and engineering personnel. In recent years, many domestic firms have turned to foreign firms as new strategic partners to conduct basic research and development work on promising technologies. While a research and development agreement can be structured as a one-time "fee-for-service" type of arrangement, such a collaboration is often the first step toward a broader arrangement covering the sale and distribution of products arising from the research work.[47]

D. Supply Arrangements

Almost every company enters into one or more important supply arrangements with respect to some or all of the components for its products and services, including the raw materials and other goods and services necessary for the manufacture of the company's products and services. Supplier relationships can raise a variety of concerns. For example, while the company might welcome the opportunity to enter into a long-term supply contract with one or more parties in order to "lock-in" pricing and component availability parameters during periods of rapid growth and product development, care must be taken to guard against the possibility that the company may become too dependent on the supplier and that the existence of such a long-term supply contract might foreclose the company from pursuing other supply and/or manufacturing strategies, including "in-house" production, that may be more cost-effective. Excessive dependence on outside suppliers also raises the possibility of adverse consequences in the event the company loses access to the products and services of a given supplier, either due to the termination of a specific contract or discontinuation of business by the supplier.

[47] Collaboration with respect to research and development can, and often does, extend to joint sales and distribution activities; however, the legality of such downstream collaboration depends upon the characteristics of the market and the participants, and the relationship, if any, of the sales and distribution activities to upstream collaboration.

E. Manufacturing and Distribution Arrangements

Introduction In order to exploit technical advantage gained through its research and development activities, a firm must be able to effectively manufacture and distribute the products which embody the technology. Although it is certainly possible for one firm to possess the technical, financial and business resources to permit internal production of the products and direct sales of the products to end users without engaging intermediaries, it is more likely that one or more partners will be sought to assist in manufacturing and distribution and that the firm will be called upon to transfer a portion of its technical information to these third parties.

Manufacturing and product purchase arrangements One common and important type of technology transaction involves the manufacture and purchase of technology-based goods. For example, a firm may be able to realize greater efficiencies in production by utilizing low-cost manufacturers overseas to make their products. The firm may achieve these cost savings by granting a license to the manufacturer to make the products and agreeing to repurchase the products at a fixed price, perhaps at some multiple of the manufacturer's actual cost. The license must include some agreement as to technical assistance, since the manufacturer must be trained in a manner which guarantees that its products are similar in quality to those that would otherwise be produced by the licensor.

Sales representative arrangements In many cases, a firm may lack the sales and distribution network and resources to fully realize the commercial potential for one or more of its products. One basic form of contractual relationship in this area is the sale representative arrangement, which is often used as an entry strategy in foreign markets. A manufacturer will contract with another party to represent the manufacturer in locating potential customers for the products of the manufacturer, and the sales representative will be entitled to a commission based upon the volume of sales generated from its activities on behalf of the manufacturer. Sales representative arrangements are one of the loosest forms of strategic business relationships.

Distribution arrangements A more complex transactional structure is a distribution arrangement, which involves a sales of goods by the manufacturer to the other party (the "distributor") for eventual resale by the distributor for its own account. A distribution arrangement often includes a license covering all the patents, trade secrets and trademarks required in order for the distributor to market and service the products. This type of arrangement will allow the manufacture to take advantage of the sales network of the distributor. In turn, the distributor is able to gain access to new products without incurring the costs of internal development. Distribution arrangements are quite common, and the terms thereof have become somewhat standardized. However, more complex types of distribution arrangements involve upstream collaboration by the parties on the development and/or manufacture of new products which ultimately become the subject of the distribution efforts.

Other manufacturing and distribution arrangements Another well-known form of manufacturing and distribution arrangement is the so-called "original equipment manufacturer" or "OEM" agreement, under which a firm manufactures products embodying the technology according to the specifications given by another firm and sells them to the second firm for incorporation by the purchaser into certain of its products which are to be sold to end users. Alternatively, a firm may simply grant another firm a list to manufacture and sell certain products that have been developed by the licensor, sometimes with a right to purchase the finished products from the licensee on special terms and conditions. Manufacturing license arrangements will often be used in situations where the licensee actually provided funding for the development of the licensed technology, and the licensor lacks the internal manufacturing capabilities to produce the number of units necessary for successful commercialization.

F. Joint Venture Arrangements

Rather than relying upon one or more of the contractual arrangements referred to above, the parties may choose to create a formal, commonly-owned joint venture company to conduct some or all of the activities relating to the technology. For example, a joint venture may be formed to conduct additional research and development projects, as well as to manufacture and distribute the various derivative products. A joint venture arrangement serves a number of purposes and, obviously, tends to bring the interests of the parties into much closer alignment than would be the case if the technology transfer simply consisted of a license agreement between two essentially unrelated parties.

The parties to a joint venture must resolve most of the issues that would otherwise arise in any of the basic contractual relationships. For example, if one party is perform research and development work for the joint venture, a research agreement will probably be used which sets out the objectives of the work and the compensation which the researching party will receive for the performance of the work. In addition, the parties must agree on a number of other issues which are particular to the use of the joint venture form, such as the control mechanisms for the joint venture, the form of contribution which each party must make in the form of patent licenses, technical information, technical and management assistance, training, cash, capital, labor and other similar items, and the procedures for terminating the venture.

G. Negotiated Acquisitions

Business collaborations may, in appropriate situations, actually take the form of a negotiated acquisition of one firm, or its assets and legal rights, by another firm. For example, a research and development or distribution arrangement may include an option in favor of one of the parties, typically the larger partner commissioning the research or distributing the product of a smaller partner, to ultimately acquire control of the other on specified economic terms. Even when a party is not given an option to purchase in advance, an acquisition may well be the appropriate extension of a relationship which originally began through the use of one of the functional types of arrangements described above.

V. Corporate Partnering Arrangements

A. Introduction

One way to illustrate the concept of business collaboration is to examine "corporate partnering." These relationships usually involve one or more ongoing arrangements between a larger company with relatively mature functional capabilities in areas like product development, manufacturing or distribution (the "senior partner") and a small, entrepreneurial technology firm with strong skills in research and new product development, but lacking the capital and other functional resources to bring its products to market (the "junior partner"). However, corporate partnering also occurs between companies of comparable size if the parties possess complimentary assets, although these relationships generally raise greater concerns under antitrust and competition laws due to the potential effect that the collaboration might have on competition in the marketplace.

B. Basic Forms of Corporate Partnering Arrangements

There are a wide variety of potential corporate partnering relationships, some of the most basic of which can be described as follows:

(i) Senior partner makes an equity investment in the junior partner on terms similar to those typically associated with a venture capital-type investment. The senior partner would have the right to designate one or more representatives on the junior partner's board of directors and junior partner would make various covenants regarding the operation of its business activities; however, the parties do not enter into any other specific functional agreements (e.g. research and development, distribution, etc.).

(ii) Senior partner engages junior partner to conduct a contractual research and development project. The junior partner is paid a research fee and, perhaps, a royalty based on specified use of the technology developed during the research project, which will be owned by the senior partner.

(iii) Senior partner agrees to act as a distributor or an original equipment manufacturer with respect to one or more products developed by the junior partner and also makes an equity investment or technology development funding contribution to junior partner.

(iv) Senior partner provides funding to junior partner for development of new products using base technology owned by junior partner. Senior partner agrees to purchase new products from junior partner at a multiple of the junior partner's manufacturing cost and junior partner agrees to pay royalties to the senior partner on any product sales the junior partner makes to others.

(v) Senior partner makes an equity investment in junior partner on the terms described above and also agrees to provide additional funding for the junior partner to conduct research and development work on one or more specified technologies. The senior partner would have a right to participate in managing the research project and to license the resultant technology for use in various areas and applications.

(vi) Senior partner provides funding for the junior partner to conduct research and development work on a new product. The results of the research project would be owned by the junior partner; however, senior partner would be entitled to purchase rights to the product for a front-end payment and future royalties based on product sales.

(vii) Senior partner makes an equity investment in junior partner on the terms described above, agrees to fund research and development work of the junior partner and/or purchase products from the junior partner and receives an option to purchase the junior partner in the future at a price based on a specified formula.

Every specific transaction within a more general structural format involves a number of additional decisions. For example, a distribution-based corporate partnering relationship requires the parties to consider the duration of the arrangement, the scope of the products (e.g. existing products and future enhancements) and markets (e.g. domestic and/or international) to be covered by the arrangement, and the degree of "exclusivity" associated with the arrangement. Any research and development relationship similarly raises issues of technical scope, as well as usage of the various applications of the technology developed during the relationship. The decisions made with respect to these issues will determine, in large part, the strategic importance of the relationship to each party.

C. Characteristics of the Parties

Introduction Each of the parties to a corporate partnering arrangement bring a distinct set of strengths, needs and concerns with them to the bargaining table. The senior partner looks to a junior partner as a means for filling in a gap in the senior partner's own resources, such as the ability to rapidly develop a desired new technology or portfolio of products. The senior partner may also wish to gain access to certain management and technical personnel of the junior partner. In turn, the junior partner is looking for additional capital and functional support in areas that it has been unable to fully develop, such as marketing and distribution. However, regardless of the complimentary skills that the parties bring to the relationship, tension will always exist between the senior partner's desire to manage and control the activities of the junior partner and the junior partner's desire to retain its independence and autonomy.

Senior partner characteristics As noted above, the senior partner is generally larger than the junior partner and is able to provide substantial amounts of new capital and financing, either directly or through its ability to influence commercial lenders to make loans on favorable terms to the junior partner, as well a strong distribution network and, hopefully, and installed base of customers and end users of some of the products which fall within the junior partner's areas of interest and expertise. In addition, the senior partner usually has established support services, including packaging design, advertising, and customer support (e.g. engineering, service and maintenance).

In looking for potential corporate partners, the senior partner is seeking the ability to accelerate its production of new products and services, either to supplement its own existing products or to expand into new or related product areas. The senior partner is hoping that a new relationship with a junior partner will provide the senior partner with access to future new products and technology and, in those cases where the senior partner is actually attempt to enter a new market, with market expertise from the junior partner in the specific niche in which the junior partner's technology will be most attractive. Also, the senior partner may be seeking to build a relationship with the managers and employees in a more aggressive, entrepreneurial environment, which might not be possible within the culture of the senior partner itself.

The concerns of the senior partner with respect to any corporate partnering relationship are similar to those which might exist in an internal development project. For example, if the senior partner is looking for the junior partner to provide it with a stream of new products, the senior partner will want to make sure that it is able to achieve timely delivery of that amount of products which are sufficient to meet the needs of the senior partner and that the products delivered satisfy the senior partner's expectations with regard to quality and performance. The senior partner will also be concerned about the costs of the relationship, which might be in the form of an investment and/or funding for a particular project. Finally, as with any other type of investment, the senior partner will want to be sure that it is able to obtain a satisfactory return on the contributions, financial and otherwise, which it makes to the resource base of the junior partner.

Junior partner characteristics The junior partner's strengths will typically lies in its technological base, which serves as the foundation for each of its existing and future products. The junior partner also brings an aggressive, entrepreneurial corporate culture to the specific relationship with the senior partner, which presumably supports a tolerance for risk-taking and an ability to bring new products to the marketplace more quickly than would have been possible if the senior partner had attempted the development effort on its own. Finally, properly-managed and focused junior partner candidates should have an understanding of where and how its products and technical advances will fit into the general market and satisfy the needs of potential customers.

A junior partner obviously looks to corporate partnering relationship as a means for obtaining additional capital which might be required in order to fund research and development work and to expand its manufacturing and distribution capabilities. However, beyond capital, which presumably could be obtained on some terms from non-corporate investors, such as venture capital investors and financial institutions, a corporate partnering arrangement should also provide the junior partner with an avenue for enhanced sales of its products, the establishment of market credibility, and immediate access to necessary functional skills, such as engineering, marketing and distribution. Also, a solid relationship with a reputable senior partner should ultimately enhance the value of the junior partner, which will be helpful if the junior partner undertakes a public offering in the future.

Investments made by senior partners in the context of a corporate partnering relationship often are made on terms which are similar to those that are used in venture capital financings, and raise many of the same issues of management control and valuation. However, while venture capital investors are often unable to demonstrate the requisite patience in the development and progress of the junior partner's business, and generally will seek valuations which fit with their own required returns on investment, a corporate partner may willing to accept a lower return on investment due to other non-financial benefits which may result from the relationship. This results in a lower level of dilution for the existing shareholders, since the corporate investor is focusing on marginal profits and returns in relation to alternative uses of resources, rather than upon the return on investment from the entire operations of the junior partner.

The concerns of a junior partner regarding a potential corporate partnering relationship are significant. First, the disparity in size and scope of activities between the parties will cause the junior partner to have concerns regarding the level of the senior partner's commitment to the relationship, not only financially, but also in terms of the attention that will be received from management and other personnel within the senior partner's organization. Second, the junior partner may have some reservations regarding its ability to meet any product delivery obligations, particularly if the pricing, payment, and delivery schedules are negotiated before the product has been developed and tested. Third, the junior partner will want to be sure that it is able to protect its technology in the event that something happens to undermine, or even terminate, the relationship before it reaches the anticipated objectives. Finally, the ability of the junior partner to retain its independence may be the greatest problem for the junior partner, which will want to be sure that it is able to make its own decision regarding future development and marketing strategies, as well as financing and liquidity strategies (e.g. public offering or acquisition by third party other than senior partner).

D. Analysis of Specific Corporate Partnering Arrangements

Research and development arrangements Research and development arrangements are among the most common forms of corporate partnering relationships. A decision by a senior partner to enter into a "development-driven" relationship will depend upon its assessment of its own internal technical capabilities. For example, it may be apparent to management of the senior partner that the development of the technology rights necessary to compete in a given market requires the skills and abilities of technical employees and scientists who would have little interest in becoming directly involved in the senior partner's own corporate culture. Accordingly, the senior partner's access to that type of talent must come through the strategic business relationship with a junior partner.

The concern for the senior partner regarding its own internal capabilities may also lie in its own inability to effectively initiate and manage the creative process necessary to complete the development of the new technology. The senior partner may be unable to recruit and retain research managers who are capable of replicating an entrepreneurial and informal environment of the type that might exist within the junior partner and which would facilitate the development of the technology. Also, the focused efforts of the junior partner may permit each of the parties to complete the development effort on a far more rapid basis than would be possible if the project were competing with other projects within the senior partner's own research staff.

A "development-driven" relationship certainly provides the junior partner with a stream of revenues which may not otherwise be available at the early stages of its own internal development. However, absent any right to use the developed technology or to receive royalties from the senior partner's use of the technology in the future, the junior partner will have effectively "capped" its own internal growth and development in that technological area. Whether or not this is acceptable really depends upon the place of the technology within the junior partner's overall development plans including any other planned products, the availability of capital, and the amount of human resources that must be devoted to completion of the development program.

Product development and purchase arrangements An advanced form of "development-driven" relationship involves applied research and development work by the junior partner for the purpose of creating a new product which can be used by the senior partner either for internal consumption or for inclusion in the senior partner's product line. A relationship of this type is not unlike the traditional "original equipment manufacturer" arrangement. Usually, the agreement will provide for detailed specifications from the senior partner with respect to products to be developed. Thereafter, the junior partner will sell the products to the senior partner at a multiple of the manufacturing cost and, in the event that the junior partner sells the products to others, it may also be liable to the senior partner for a royalty payment. The arrangement takes advantage of the junior partner's technological expertise and ability to rapidly product innovations and improvements which may be of use in the senior partner's products.

A product development arrangement may take a number of different forms. For example, the agreement may call for the senior partner to provide funds for research and development of the new product by the junior partner and, upon completion of the development program, the senior partner would have the right to purchase the product from the junior partner by making a lump sum payment at the time of purchase and agreeing to make additional royalty payments as the product is sold. Due to the costs and risks involved, the senior partner will usually insist upon some period of exclusivity with respect to the new product, however the junior partner's concerns regarding such a relationship are alleviated by its ability to use the amounts paid for development of other products and, presumably, its expectation regarding a satisfactory royalty return due to the senior partner's financial commitment to the product. Also, the junior partner is able to maintain its small organization while reaping large rewards from the marketing efforts of the senior partner.

In addition to the funded new product development effort, the parties may consider various other arrangements, including: the distribution of the junior partner's existing product by the senior partner; distribution of the junior partner's existing product and future enhancements by the senior partner; development of new products by the junior partner for offshore manufacturing and distribution by the senior partner; joint development of new products, manufacture by the junior partner and ultimate distribution by the senior partner; joint development and distribution of a single product by the senior partner; and joint development and distribution of a series of new products by the senior partner.

As always, a relationship of this type is risky for the junior partner, particularly if a concern exists that the senior partner will modify its own product line or take similar steps that will eliminate the need for the junior partner's products. The junior partner should retain the right to use the underlying technology for different applications in order to alleviate any substantial dependence upon the relationship with the senior partner. Also, while the junior partner can take comfort from the fact that the senior partner's internal use of the products eliminates any immediate competition, care must be taken that the underlying technology is adequately protected.

The senior partner's use of the junior partner's products in its own products creates a good deal of dependency between the parties. As such, this contractual relationship requires a substantially greater degree of communication between the parties than might be the case in other situations. The junior partner should ensure that sufficient support exists within the senior partner for using the junior partner's products and regular meetings should be scheduled to ensure that the delivered products conform to the expectations of the senior partner. In some cases, the relationship should be supplemented by an equity investment by the senior partner in the junior partner.

Investment relationships An investment relationship involves an equity or debt investment by the senior partner directly in the junior partner. While in most cases the investment will be accompanied by one or more functional arrangements, such as an agreement by the senior partner to distribute certain of the junior partner's products or a commitment by the junior partner to conduct various research and development activities on behalf of the senior partner, it is often the case that the senior partner will limit its initial involvement to taking a minority position in the junior partner, generally with the intent of entering into further agreements as the relationship matures and evolves.

Investment relationships, with or without an accompanying functional agreement, will often be attractive to both parties for a variety of reasons. The senior partner may be seeking a window on new technologies, a good return on its investment and the benefits of an "entrepreneurial" environment without the dislocations that might be caused by attempting to modify the partner's own corporate culture and internal development efforts in order to replicate the atmosphere necessary to development the innovative technology within the timeframe necessary for market acceptance. Also, a direct, non-controlling, investment should provide an opportunity for the senior partner to evaluate the opportunities for a long-term relationship or an acquisition.

As for the junior partner, an investment by a reputable senior corporate partner can prove to be a most desirable development. For example, the junior partner receives capital to fund its development efforts; credibility for its development efforts in the marketplace, if the senior partner permits appropriate publicity of the relationship; and credibility with venture capitalists and other potential institutional and public investors. More importantly, the junior partner should ensure that the investment brings a thoughtful sharing of the senior partner's expertise and experience with respect to business, financial, manufacturing and technical expertise, through board participation or otherwise.

Equity investments can be tremendously risky for the senior partner and, as such, may not be suitable for the particular corporate culture adopted by the senior partner. The investment is usually made at a time when the junior partner is still in the development stage, without any track record of success at product or technical development. The senior partner has only indirect control over the incentives and motivations of the key employees and managers of the junior partner and the lack of any specific development program may result in the junior partner undertaking one or more projects which may prove to be of little commercial value for the senior partner. Moreover, the risk that the junior partner's development efforts may be focused in areas that are outside of the expertise of the senior partner tends to dilute the attractiveness of the senior partner's expertise to the junior partner.

For the junior partner, a senior partner can prove to be a very attractive alternative financing vehicle, provided that the junior partner is able to retain control over the key business decisions which might effect the company in the future. The ability of the junior partner to seek alternative strategic business relationships may be adversely impacted by the presence of the senior partner and the senior partner may wield a significant amount of influence over corporate activities, either through the position of its representatives on the company's board of directors, or under the terms of any special voting rights or shareholders' agreement. Also, unlike venture capitalists, the senior partner may be reluctant to provide future funding in order to undertake business objectives that were not contemplated at the time of the original investment.

6 Licensing Arrangements

I. Overview

A. Introduction

As discussed in the previous chapter, there are a wide range of business collaborations, and the best choice will depend on the particular facts and circumstances, including the goals and objectives of the parties and the importance of the collaborative relationship to their overall strategic plan. However, with the possible exception of sales agent and representative agreements, licensing agreements may well be the most prevalent form of business collaboration, either as "stand-alone" agreements amounting to a temporary lease of the licensed legal rights or as part of a more complex functional arrangement.

A licensing arrangement is created when one party, the "licensor," which owns, or otherwise controls the right to specify the uses of, a valuable legal right, grants to the other party, the "licensee," the right or license to utilize the legal rights for the purposes specified in the contract between the parties.[1] In consideration for the grant of the license, the licensee agrees to compensate the licensor, perhaps by the payment of a flat fee or by payment of an amount determined by reference to amounts received by the licensee from the use of the licensed rights. However, in some cases, the compensation may be "in-kind," such as when the licensee agrees to deliver finished goods to the licensor for resale.

A license is not an outright assignment of the legal rights from the licensor to the licensee, since the licensor will retain actual ownership of the licensed subject matter. When the term of the license expires, the licensed rights will revert back to the licensor and the licensee will generally not be able to use any of the elements of the licensed subject matter which may still be eligible for legal protection. As such, a license actually amounts to a form of rental arrangement with respect to the subject matter and the amount of compensation requested by the licensor is often computed with the intent of arriving at a fair rate of return in relation to amounts expended by the licensor in developing the licensed technology. For its part, a licensee compares the rental costs of the license to the burden of independently developing legal and viable alternatives to the licensed subject matter.[2]

[1] As used herein, the term "valuable legal right" includes not only intellectual property rights, such as the authority to lawfully practice a patented invention without fear of being sued for infringement, but also the right to use technology, knowledge, trade secrets and technical data, all of which are considered to be a form of property held and controlled by the owner thereof.

[2] A potential licensee may attempt to develop substitutes for a patented product which do not infringe upon the claims set out in the patent which might be the subject of a license or, as to trade secrets, may undertake to independently develop the process, knowledge or other information through inspection of the licensor's products or reverse engineering.

B. Classification of Licensing Agreements

Introduction Licensing agreements can be characterized by the subject matter of the license, such as a patent or trademark license, or by the permitted functional uses of the licensed subject matter, such as a manufacturing or distribution license. While referring to the subject matter of the license is useful in insuring that the license covers all the legal rights necessary to accomplish the objectives of the agreement, the permitted functional uses of the licensed subject matter is determinative of the strategic role that the license is expected to plan in the business plans of the parties thereto.

Classification by licensed subject matter When classifying licensing agreements by their licensed subject matter, the parties may refer to a patent license, a trademark license or a hybrid license, which includes two or more of the basic forms of intellectual property rights (e.g., patents and trade secrets). It is quite common for the parties to use hybrid licenses, since the grant of a mere patent license is generally not sufficient to allow the licensee to make and use the patented invention. In such cases, the license will also include all of the trade secrets of the licensor which might be necessary in order for the licensee to exploit the patented invention in the manner set out in the license agreement.

Patent licenses The holder of a patent has the right to exclude others from the making, using or selling of an invention during a specified time in the country in which the patent has been issued. A patent license is an exception to the monopoly rights of the licensor and allows the licensee to utilize the subject matter of the patent without liability to the licensor for infringement. A mere patent license, without access to further trade secret information owned by the licensor, will only be useful to a licensee with technical resources that permit it to independently develop products which might fall within the claims of the licensed patents.[3]

Trademark license A trademark license allows the licensee to use the licensor's trademarks in the course of the licensee's business, generally in connection with the manufacture and distribution of the licensor's products. Since, in effect, a trademark license amounts to a lease of the licensor's "goodwill," the licensor will impose a number of special obligations on the licensee calculated to closely monitor and control the "quality" of the goods or services marketed by the licensee under the scope of the licensed trademark. A trademark license is usually accompanied by the transfer of other information which might be useful in the marketing process, such as advertising materials, market surveys and customer data, all of which might be eligible for protection under copyright or trade secret law.

Copyright license Copyright protection extends to written works, and even software, and affords to the holder of the copyright the exclusive right to publish the copyrighted work or to determine who may publish the work. As such, a copyright license permits the use by the licensee of copyrighted materials developed by the licensor, in-

[3] Patent licenses may be sought by licensees who believe that the patent may "block" their efforts to exploit their own new patented invention or products, since it is possible to obtain a patent on inventions which improve or enhance an existing patented invention or which build upon elements of technology which are covered by a broad basic patent right.

cluding written materials prepared in order to assist users in understanding the operation of the licensor's products.

Trade secret license While patent, trademark, and copyright licenses generally involve items which can be easily identified by reference to publicly-available information, a trade secret license involves business information and technical data which, by its nature, have been maintained in confidence by the licensor and, as such, must be effectively "transferred" from the licensor to the licensee. A trade secret license creates a fiduciary relationship between the parties, in which the licensee agrees to insure that information is not disclosed in a manner which undermines its commercial value. In many cases, trade secret licenses are referred to as technology or "know-how" licenses.

Hybrid licenses A license arrangement may cover more than one of the basic types of legal rights described above. For example, when a license is entered into in the course of a relationship which contemplates that the licensee will manufacture and distribute one or more of the licensor's products, it is generally necessary for the license to cover all of the patents and trade secrets which may be required in order for the licensee to manufacture and service the products, as well as the trademarks and copyrights relating to the marketing and operation of the products. Even in the cases where a licensee only intends to use a patented invention for its internal purposes, the license will include any trade secrets required to use the invention and permission to reproduce and distribute internally any copyrighted materials relating to the operation of the invention.

Classification by permitted functional uses Since most license agreements in a strategic business relationship are "hybrids," it may be useful to characterize the licensing arrangement by reference to the specific functional uses of the transferred technology. In so doing, the parties can focus upon the activities to be performed by the licensee in the context of the broader relationship. Even in those cases where trade secrets form the largest part of the licensed technology package, categorization tends to parallel the exclusive rights granted to the owner of a patent or a copyright under applicable statutes.

Use licenses A use license simply provides the licensee with the right to use the licensed subject matter. Use licenses do not necessarily include the right to copy, modify, sell or distribute the licensed subject matter or the right to use the licensed subject matter to manufacture or produce goods or products for sale or distribution by the licensee. A common example of a use license is the end-user license which is granted to users of consumer software. Another form of use license arises when the licensee is given a right to use a patented process or trade secret information strictly for internal purposes and not for marketing of products and services.

Manufacturing licenses A manufacturing license is a common form of licensing arrangement and includes the right to manufacture the technology covered by the license, manufacture products which include the licensed technology, or manufacture products through the use of the licensed technology. A manufacturing right should be explicitly granted in the license agreement and may have different meanings depending upon the nature of the technology that is covered by the license. For example, the holder of a patent is granted the exclusive statutory right to "make" the products which are referred to in the patent, a manufacturing license with respect to a patent provides the licensee with the right to make the products. If a copyright is licensed, the licensee will have the right to "copy" the specific work. As discussed below, the right to "manufacture" will generally be limited to a specific type of product.

Modification license A modification license allows a licensee to correct, modify, enhance or improve the licensed technology for use in one or more specific applications. As a general rule, when a licensee is given the right to modify the original technology, the licensor will have the right to use any of the modifications, including any statutory rights which may arise as a result of the licensee's activities, under a "grant-back" provision in the original license agreement.

Distribution license Although a licensee may be given the right to copy and manufacture the licensed subject matter, or make products using or including the licensed subject matter, it does not normally have the right to distribute and market the licensed subject matter or its byproducts to third parties in the absence of a specific sale and distribution right. In some cases, the manufactured products can only be sold to the licensor. However, in many cases, the licensee will have some right to distribute one or more of the products manufactured with the licensed subject matter to third parties in a specified geographic or demographic market.

Sublicenses The rights of the licensee to use the licensed subject matter may, in some cases, include a right to sublicense to third parties all or a portion of the various legal rights originally granted to the licensee. Sublicensing may often be appropriate when the licensing arrangement is intended to facilitate the distribution of products into a given market and the licensee wishes to engage the services of third party sales agents or distributors. However, broad sublicensing rights should raise concerns on the part of the licensor regarding its ability to protect the licensed subject matter, particularly trade secret information, and to retain some degree of control over its distribution channels and practices.

Special commercial situation licenses Licenses may be granted in a number of special commercial situations. One common example is the grant of a license under an "OEM," or "original equipment manufacturer," or "VAR," or "value added reseller," agreement. In these cases, the licensee is given a right to purchase or license certain or the licensor's products, incorporate them into products which are to be developed by the licensee, perhaps with the benefit of technology licensed from the licensor, and market the licensed products as part of its own product line. The manner in which the licensed products are to be incorporated into the licensee's products may vary. For example, the licensed products may not be identified in the new products, as is the cases with a so-called "private label" arrangement. On the other hand, the licensee may be allowed to resell or relicense the licensed products in an identifiable form with the "value added" items developed by the licensee.

C. Advantages and Disadvantages of Licensing Agreements

Introduction Every licensing arrangement has a number of advantages and disadvantages to both the licensor and the licensee. This may be largely attributed to the fact that the rental of the intellectual property rights covered by the license agreement creates a form of reliance between the parties and, to some extent, restricts their independence with regard to commercial activities in a particular area. As a result, it is important for both parties to attempt to assess the objectives and skills of the other party prior to the time that the license agreement is executed.

Potential advantages of a licensing arrangement The parties to a licensing arrangement may have a number of different objectives. Certainly, for a licensor, one of the goals of a licensing arrangement is to realize revenues from the use of the technology by the licensee which can be offset against any of the expenses associated with the development or acquisition of the subject technology. From the licensee's perspective, a licensing agreement often represents a cost-effective method for accessing new technologies which fit into the licensee's overall strategic business plans. For both parties, the license should facilitate the creation of a relationship in which each party is permitted to contribute its specific strengths while, at the same time, taking steps to minimize any weaknesses or functional limitations in its own competitive capabilities.

Access to vertical capabilities The "life-cycle" for any product requires successive focus upon such disparate capabilities as research and development, the refinement of products for production, acquisition and processing of raw or semi-finished materials, manufacturing, distribution, sales and marketing and service. In those cases where a firm may lack the financial or managerial resources to independently fulfill each of the tasks needed to bring a product to market, a licensing arrangement with one or more parties can provide the firm with the required access to other vertical capabilities to allow the firm to fully exploit it current strengths.

For example, a firm may possess the technical skills for the development of new products, yet lack the financing and facilities required for large-scale production and manufacturing. However, if the firm is able to license the technology to a firm the resources capable of producing the new products on the requisite scale, it is then able to exploit its technical skills without taking on the costs and risks of developing its own production capability. The firm might consider using the proceeds from the sale of the various new products to finance the acquisition of its own manufacturing function. If this is the case, the licensor will seek to structure the original license so that the manufacturing rights revert back to the licensor when it is ready to take over production itself.

Technology acquisition and exchange A firm may look at a licensing arrangement as a means for acquiring new or existing technology which may be needed in order for the firm to lawfully continue its own development and product sales efforts. Given the rapid technical changes that sweep across almost every industry, even the largest firm may no longer be able to keep pace with the new frontiers of research in all areas. As a result, firms will maintain their technological portfolio through licensing arrangements with other commercial entities, universities and research organizations, including entities located outside of the United States, without subjecting themselves to the costs and risks associated with new development efforts.

Licensing also presents an attractive vehicle for exchanging technology and marketing information between the licensor and the licensee. For example, a licensor may grant a license to a licensee which is actively involved in research and development activities which may result in enhancement and improvements to the licensed technology. By using "grant-back" clauses, which provide for a license from the licensee to the licensor of any improvements made to the basic technology, it is possible for the original licensor to minimize the costs associated with further development of the technology. Similarly, the grant of a license for the sale and distribution of the licensor's products may also permit the licensor to gain valuable information about new markets.

Market penetration Market penetration is one of the primary attractions of a licensing arrangement for both parties. For a firm which lacks the distribution network necessary to sell its new products in a given market, a licensing agreement with an experienced distributor will facilitate rapid market entry and allow the licensor to generate revenue from sales in a new market without the need to commit any significant new capital to the development of a direct sales force in the market. From the distributor's perspective, the licensing agreement may allow it to use its sales channels to distribute a new line of products which might not have otherwise been available through its internal development efforts or allow it to exploit the reputation and goodwill associated with the licensor's products.

Even in those cases where the licensor may have some sales capability, there will be situations where it is more efficient to engage licensees with specific expertise in a given geographic or product market. However, the success of this type of strategy depends, in large part, upon the skills of the licensee. As such, the licensor may want to retain the ability to alter its sales strategies in the market, including the option to make a direct investment, either in the form of a wholly-owned operation or as a more formal joint venture with the licensee or a new partner.

Neutralizing blocking patents A firm may develop technology which, if properly exploited, might conflict with the existing patent rights of other firms. In those cases, the firm may consider entering into a cross-licensing agreement with the firms holding the blocking patents in order to insure that the technology can be utilized to its fullest extent without concern about an infringement action. An arrangement of this type may be quite expensive and the firm should investigate the possibility of blocking patents before any significant amount of resources are expended on research and development activities. Moreover, cross-licensing arrangements raise a number of antitrust issues, depending upon the size of the market and the relative positions of the firms.

Satisfying local regulatory requirements In order to enter a new foreign market, licensing, rather than direct investment, may be necessary in order to satisfy local customs and to comply with any governmental decrees regulating the participation of foreign firms in a specific market. For example, a number of countries have enacted laws to encourage development and expansion of local industry and reduce the degree of dependence on the use of imported technology. In such cases, licensing will be the only way to penetrate the local market, since any direct investment by foreign firms in those sectors which have been targeted for "self-sufficiency" will be limited or restricted. Moreover, even in those cases where a firm is permitted to make a direct investment, it may be prudent to enter into various license agreements which provide access to existing distribution channels of local firms, many of whom may receive, directly or indirectly, some form of government assistance.

Protection of intellectual property rights Licensing may be required in order to preserve the protection of inventions and other items under the intellectual property laws of a number of foreign countries. For example, many countries will require that the holder of a patent granted under local law must actually work or exploit the patented invention in the local market or forfeit the exclusive rights generally associated with the grant of a patent. In such cases, the holder of the patent will license a local firm to work the patent in order to avoid any action by the local authorities to grant a compulsory license of the patent to local firms, thereby severely undercutting the value of the patent rights to the original inventor.

Cost reduction Licensing is now a common strategy for firms seeking to control or reduce their manufacturing costs with respect to any new products which utilize technology developed by the firm. In recent years, firms have licensed foreign firms to manufacture products at a fraction of the cost that would be incurred by the licensor. A strategy of this type allows the firm to maintain its market share, lower its overall costs and improve its margins, while avoiding the need to constantly reconfigure and rehabilitate its own production capabilities.

Enhancing cash flow and harvesting While licensing agreements are usually entered into for at least one of the broader strategic reasons referred to above, it often makes sense to grant a license simply to generate additional cash flow from the technology in the form of royalties and fees. These types of licenses are particularly appropriate when the firm is seeking to "harvest" its previous investment in a given area and has no further desire to continue with development work, perhaps due to the costs associated with keeping up with new technologies or a broader change in the strategic objectives of the firm.

Potential disadvantages of a licensing arrangement Just as it is essential to understand the specific goals and objectives which are being pursued within the licensing strategy, notice needs to be taken of the various disadvantages to the licensor and the licensee which may arise out of the relationship. Obviously, licensing, like any other form of strategic business relationship, creates a material degree of interdependence between two firms with distinct cultures and business objectives. However, certain other factors must also be considered, including the very real possibility of creating a new competitor.

Dependence A licensing arrangement is generally chosen when it fulfills a specific strategic business need of one or both of the parties. For example, a licensor may enter an arrangement with a licensee to distribute the licensor's products in a specified market when the licensor believes that it would be too costly or risky for it to attempt to engage in any direct sales efforts in the market. As such, the licensor depends upon the licensee to assist the licensor in recovering its initial investment and achieve market acceptance for its products. If, for whatever reason, the licensee is unable to fulfill its obligations under the license, the licensor may lose the market opportunities to another firm or suffer a reduction in revenues such that the entire investment in the development of the licensed products cannot be recovered. Therefore, it is essential that the licensor exercise caution in the selection of licensees, since mistakes can result in material harm to its business.

The risks and uncertainties of dependence on the performance of the other party to the licensing arrangement can be particularly acute when the parties differ substantially in size and maturity. For example, a start-up firm with a single new product may well be forced to rely on the manufacturing and distribution capabilities of a larger, established firm in the industry. Unfortunately, what is clearly the highest priority for the start-up firm may not be as important for its larger partner. In many cases, the larger firm decide to take a different strategic direction, often using one or more products which compete with or replace the products that were developed by the start-up firm. Accordingly, the start-up firm will attempt to build in mechanisms for retaining the interest of the larger firm, including prohibitions on competitive activities and milestones which must be achieved in order for the large firm to retain its position as the exclusive distributor.

Expropriation of licensed technology When the licensee is an actual or potential competitor of the licensor, a real concern arises as to whether or not the licensed technology will be utilized in a manner which minimizes its value to the licensor. In some cases, it may be extremely difficult for the licensor to prevent a licensee from exploiting the licensed technology following the expiration of any statutory rights or the termination of the license agreement. In any event, negotiation of any licensing agreement requires that the parties carefully focus upon any territorial limitations or other contractual restraints which may be necessary in order to alleviate the concerns of the licensor regarding the use of the technology. In the context of a trade secret license, this means specific provisions covering the protection of the information by the licensee during the term of the license and a requirement that the information be returned to the licensor upon termination of the license.

Risks to value of goodwill A party to a licensing arrangement may have a concern about protecting the value of the goodwill associated with its products or activities. For example, if a licensor licenses trademarks to the licensee for use in connection with the distribution of the licensor's products, it will want assurances that the licensee will maintain the quality of the licensed products in accordance with the expectations of the licensor's existing customers. Similarly, when a firm licenses a new product from another firm for inclusion in its product line, it will want to insure that the new products perform at the same level as the licensee's other products. In either case, a failure of one of the parties to meet the reasonable expectations of the other party has the potential for diminishing the value of the goodwill associated with that party's business.

Licensee's costs of exploiting licensed rights For the licensee, the potential costs associated with fulfilling its obligations under the licensing agreement are not insignificant. Although the licensed technology may permit the licensee to pursue its plans to expand its existing product line or increase market penetration in the face of competitive pressures, the use and exploitation of licensed technology may require substantial outlays of capital. Moreover, royalty payments based upon the volume of production or gross sales must also be taken into account in assessing the costs of marketing the new products.

D. Conditions for Successful Licensing Arrangements

Selecting licensing partners Licensing partners can be used to assist the licensor in conducting further research and development work, production and manufacturing of products which embody the licensed technology and the distribution of products which may have previously been manufactured by the licensor or another licensee or manufactured by the distributor-licensee. Among the various characteristics which might be considered in evaluating a potential licensee are the following:

(i) *technical and functional experience*. The potential licensee should be able to demonstrate experience in specific technical area and in the particular function or functions which are at the essence of the particular relationship. For example, if the relationship is being established for the distribution of products, the licensee/distributor should be able to provide proven skills in marketing similar or complimentary products in the target market(s). If possible, the experience of the licensee should compliment, rather than merely duplicate, that of the licensor.

(ii) *formal strategic plan*. The potential licensee should be able to provide a formal strategic plan setting out in detail the steps that will be taken to fully exploit the licensed technology. The plan should include firm estimates of the licensee's proposed expenditures and the anticipated revenues and income, which estimates might be incorporated into the license agreement in the form of covenants.

(iii) *management skills and support*. The potential licensee should have one or more senior managers with the skill and experience to ensure that the licensee is able to execute its strategic plans. Such persons should be able to serve as in-house "champions" for the business relationship in the face of competing opportunities. In addition, the licensee should have strong management personnel in the functional area(s) which are to be the subject of the licensing arrangement.

(iv) *financial resources*. The potential licensee should be able to provide sufficient capital to fund the formal strategic plan, as well as to finance additional work which might be required on the licensed technology.

(v) *compatibility*. The potential licensee should be compatible with the firm in a number of areas, such as the level of commitment to the licensing arrangement, the size and structure of the organization and the underlying national and corporate cultures. Under the best of circumstances, the parties will have had some sort of prior relationship with one another.

(vi) *governmental and regulatory acumen*. There are cases, such as when the licensing arrangement pertains to a product or technology which is subject to government testing and approval, when the skill and experience of a prospective licensee in dealing with the government can be extremely important. For example, in a number of foreign countries, the government exercise actual or *de facto* authority over local distribution.

(vii) *reputation*. If possible, an effort should be made to find a licensee with a solid reputation in the market, as well as the requisite functional skills.

Barriers to successful collaborations While business collaborations can obviously be enormously valuable to a firm, it is by no means certain that a particular relationship will be successful and that the anticipated goals and objectives will be realized. Over the years, a good deal of "lore," as well as empirical data and information, has been collected regarding the factors that might be most useful in attempting to explain the performance of business collaborations, particular joint ventures. While it is difficult to draw any generalizations from this information in identifying some of the barriers to a successful licensing arrangement, the following can nonetheless serve as a useful guide for managers in negotiating and implementing such agreements:

(i) In many cases, the technological objectives of the relationship were never realized due to the loss of interest on the part of one of the participants in utilizing the relationship as the vehicle for the development and exploitation of the specified technology.
(ii) The pre-formation planning for the relationship was often inadequate and, perhaps, too hasty. For example, a distribution relationship might fail if the product was inappropriate for the target market or the parties failed to adequately assess the relevant competitive factors in the marketplace. Unsuccessful relationships the planning effort and the key operational activities of the relationship.
(iii) The participants failed to communicate effectively prior to the formation of the relationship, perhaps due to language, cultural or geographic constraints. In addition, relationships suffer from the unwillingness of managers to effectively contribute and share generic and industry-specific knowledge with representatives of the other participant, leading to mistrust between personnel from both participants.
(iv) The participants failed to reach a consensus on the fundamental objectives of the relationship. In many cases, this phenomenon is evidenced by undue concern regarding methods or procedures, thereby limiting the flexibility of the relationship to utilize alternative methods to pursue the overriding objective. Similarly, the participants may impair the chances of success for the relationship by limiting its activities to a narrow product or customer group.
(v) Even when the participants did appear to reach an agreement regarding the fundamental objectives of the relationship, confusion often occurred when the participants failed to recognize the importance of acquiring knowledge and process experience in the course of the relationship which could be used in a diverse set of product and market areas. In some cases, it can be misleading to measure performance of the relationship as simply of function of revenues or market share.
(vi) Incorrect or inflexible objectives often led to poor strategic decisions regarding the operation of the relationship. For example, the participants may be driven to ill-advised cost-cutting efforts in order to increase demand and the rate of growth for the products being manufactured and sold by the venture. However, as a result, competitive production facilities of one or both of the participants may be eliminated or conflicts may arise due to business objectives of the participants arising outside of the scope of the relationship.
(vii) While the relationship is based, in part, on the desire of the participants to effectively utilize their respective comparative advantages, the prospects of the relationship often floundered due to the unwillingness of one or both of the participants to share control or provide the needed support to the relationship. For example, one of the participants may be unwilling to commit senior managers and technical as a subsidiary, rather than an independent business enterprise.
(viii) In some cases, adverse local business conditions doomed the relationship. For example, local regulation of the relationship may be extremely burdensome. Also, economic conditions may change rapidly, based on local political and environmental factors beyond the control of the participants.

Any number of additional factors have been cited for the failure of any business collaboration, such as general incompatibility between the participants due to disparities in size, goals and culture, burdensome communication costs due to the unwillingness of one of the participants to delegate control over the day-to-day operations of the business, impasses on major issues without adequate dispute resolution procedures, conflicts with other internal functional and financial objectives of one of the participants, inability of the participants to agree on the proper allocation of responsibilities and the rights upon any termination of the relationship and, finally, the lack of skill and experience in one or more areas required for successful pursuit of the objectives of the relationship.

II. Definition and Use of the Licensed Technology

A. Introduction

Obviously, the clause granting the licensee the various legal rights to exploit the technology embodied in the licensed subject matter is one of the most important terms of the license. As such, it is important for the parties to clearly understand all of the legal rights possessed by the licensor which may be relevant to the proposed transaction, as well as the intended use of each of these rights by the licensee. This information allows the "granting clause" of the license to be drafted in a manner which precisely enumerates the rights which are to be given to the licensee and the permitted uses of the technology.

The key issues which need to be considered in formulating a granting clause are the definition of the actual legal rights that are to be the subject of the license, the permitted uses of such rights, whether or not the licensee is to have the exclusive right to practice and use the legal rights granted in the license for the purposes specified and whether the licensee itself will have the right to grant further licenses (i.e. sublicenses) of the licensed rights to third parties. The answers to each of the aforementioned issues will determine the scope and utility of the license grant, as well as the amount of consideration which a licensee might well be asked to pay for the use of the license.

B. Definition of the Licensed Technology

The agreement should define the current and contemplated subject matter of the license, including the specific information to be transferred from the licensor to the licensee and any other forms of intellectual property, such as issued patents, pending patent applications, and trademarks, which might be necessary in order for the licensee to use the trade secrets in the manner set out in the agreement. Also, since a licensing arrangement is an ongoing relationship, the parties must address the effect, if any, of changes in the nature of the licensed rights after the term of the license has begun (e.g. filing of patent applications covering products and technology previously protected as trade secrets), as well as the consequences of any improvements or enhancements to the underlying technology.

C. Scope of Use of the Licensed Subject Matter

In general Once the legal rights which are to be included as part of the licensed subject matter have been properly defined, attention turns to determining the limitations, if any, which are to be imposed on the licensee's ability to use the licensed rights. Restrictions on a licensee can take a number of different forms; however, the two most common types of restrictions typically relate to the permitted functional uses of the licensed subject matter and the demographic or geographic markets in which the licensee is allowed to exploit the licensed subject matter.

Functional uses of the licensed subject matter A license transfers to the licensee a set, or "package," of statutory and nonstatutory legal rights with respect to a specific technical process or product. For example, in its broadest form, a patent license allows the licensee to do all of the things that would otherwise be precluded by the patent monopoly: make, use or sell the invention covered by the patent. The issues with respect to the permitted functional uses of the licensed subject matter relate to defining which of the legal rights originally granted to the holder of the patent, trademark, copyright or trade secret will be passed on to the licensee. A specific description of each of the permitted functional uses of the licensed subject matter is extremely important, since it is unlikely that a permitted use will be implied other than from the actions of the parties.

The scope of the permitted functional uses of the licensed subject matter depends upon the intended purposes of the business relationship. For example, if the licensee is going to manufacture and sell the products covered under a licensed patent, the license should permit the licensee to "make, use and sell" the patented invention. Moreover, if the licensee is to sell the patented products under the licensor's brand name, the license should cover the requisite trademarks. On the other hand, if a licensee merely wishes to use the patented invention as part of its own production process, or the licensee's activities are confined to manufacturing products which are to be sold by others, the right to "sell" the invention will not be included as part of the licensing package.

Field of use restrictions: products and applications In many cases, the licensed rights may be used to manufacture a number of different "products," each of which may be used for one or more distinguishable applications. For example, in the computer and electronics area, the technology may be used to manufacture one or more products which might be suitable for sale into both the "consumer" market and into the "business" market. Similarly, in the pharmaceutical area, prescription and "over-the-counter" drugs form two distinct markets. Since a single licensee may not have the distribution skills for accessing more than one market, the licensor may wish to select two licensees whose activities will be limited to those applications which are best suited to their sales network, thereby increasing the probable royalty stream associated with the technology.

Field of use restrictions: geographic and territorial The licensor will often attempt to impose geographic and territorial restrictions upon the licensee's ability to make and/or distribute the licensed products outside of the region or countries specified in the license. A restriction of this type is designed to allow the licensor to maintain control over the manner in which the licensed products are distributed and to permit the licensor to select the best mode for distribution in each geographic market. However, licensees will usually resist these limitations, since they generally see the license as a means for expanding their own distribution capabilities outside of their traditional markets.

D. Exclusive and Non-Exclusive License Rights

In general In addition to the functional, product and geographic restrictions that may be imposed upon the ability of the licensee to utilize the licensed subject matter, the value and utility of the licensed rights will also depend upon its ability to exploit the technology with little or no lawful competition from others, including the licensor itself. There are three general types of licensing categories which describe the license grant in terms of competitiveness: "exclusive" licenses, which provide the licensee with the right to exploit the rights without competition from the licensor or any other party, "semi-exclusive" licenses, which limit competition to the licensor, and "non-exclusive" licenses, which are used when one or more other licensees are to be granted the right to exploit the licensed subject matter in a manner which may compete with the uses of the licensee.

Strategic considerations in exclusive licensing Obviously, the licensee perceives an exclusive license as being the most advantageous form of technology transfer, even though it can be anticipated that the "cost" of such a license will exceed the amount required for a non-exclusive license. An exclusive license provides the licensee with access to all of the results of the licensor's research and development work in the specific area, without the need to incur the substantial "up-front" costs associated therewith. Moreover, an exclusive grant allows the licensee to achieve its strategic objectives of being "first to market" with respect to the specified products.

For its part, a licensor will be understandably reticent to divest itself of control over the exploitation of the technology rights by granting an exclusive license, particularly when it has invested a significant amount of funds and time in developing the technology. However, in many cases, the grant of an exclusive license to a specific licensee may be advantageous to the licensor given the perceived skills and resources of the licensee in the areas of manufacturing and distribution. Accordingly, it is more than likely that any grant of an exclusive license will include a number of conditions calculated to insure that the licensee will perform in a manner that allows the licensor to realize the optimal benefits from the transferred technology.

Contractual restrictions on exclusivity One method that is often used to control the risks that a licensor might feel come with the grant of an exclusive license is to restrict the scope of the exclusive rights in some manner. For example, a licensor may restrict exclusivity by limiting it to that period of time which may be necessary for the licensee to achieve the desired advantage of being the first in the "market" with the product or technology covered by the license. This strategy may be appropriate when it is anticipated that the licensee will expend substantial amounts on promoting and distributing the product or technology. Once this initial investment has been recovered, or the licensee has had sufficient time to realize the benefits of its distribution strategies, the licensor will hopefully be able to enter the market directly or to obtain additional royalties through the grant of other licenses.

Performance conditions for exclusivity In almost all cases, the ongoing exclusivity of a license grant will be conditioned upon continued satisfactory performance of the licensee in relation to specified measures included as part of the license agreement. While the conditions will vary depending upon the nature of the licensed rights and the objectives of each of the parties, the following types of tests might be considered:

(i) The licensee may be obligated to reduce the licensed rights to practice, or complete development of an improvement to the technology, by a certain date.

(ii) When the license pertains to a distribution relationship, the licensee may agree to establish a certain number of dealerships or agents by a certain date.
(iii) The licensee may be required to expend certain minimum amounts for further development or improvement of the licensed technology or for the promotion of the various products and services included in the license.
(iv) The licensee is often required to meet specific revenue objectives or minimum royalties or, alternatively, capture a given minimum share of the target market for the licensed products.

E. Sublicensing and Assignment

A person who rents a property right from its owner will sometimes want to rent, or even sell, its rights as a renter to a third party. In the licensing context, the grant of a license by a licensee to a third party is referred to as a "sublicense" and a sale or full conveyance of the licensee's rights under the original license is usually referred to as an assignment. Sublicenses raise all of the same issues as the original license. For example, the licensee may sublicense all, or only a portion, of the legal rights which the licensee received from the licensor. In contrast, the assignment of the licensee's rights will generally terminate its interest in the licensed subject matter.

Obviously, the licensor will have some concern about any transaction, whether a sublicense or an assignment, which will give a third party the right to use and exploit the licensed technology. As a general rule, absent an express or implied agreement to the contrary, a license is not assignable and the licensee has no right to grant sublicenses to third parties. However, are often a number of situations where sublicensing may be essential to the success of the overall business relationship. For example, when the original arrangement contemplates that the licensee will manufacture and/or distribute the licensed technology or products which embody such technology, it may be necessary for the licensee to hire various subcontractors, consultants, distributors, dealers or agents in order to properly exploit the licensed rights. In each case, it may be necessary to sublicense a portion of the licensed right to these persons or entities.[4]

If the licensee is to be given the right to sublicense, the licensor will want to insure that it has the ability to exercise adequate control over the number of licensees, the uses which can be made of the licensed rights by the sublicensees, and the ability of the sublicensees to adequately protect the technical information embodied in the licensed rights. In addition, the licensor will want to understand how the grant of a sublicense will impact the overall economic relationship between the parties, particularly with respect to royalties which might be paid to the licensee by a sublicensee. In many cases, the licensor can effectively manage sublicensing activities by either insisting on the right to review and approve the terms of every proposed sublicense or by providing that sublicenses may only be consummated in accordance with terms set out in a form sublicense agreement which might be appended to the original license agreement.

[4] It may also be necessary to sublicense a portion of the technology to end users i4 connection with the final use of the licensed products. In general, this type of sublicense does not create material issues between the licensor and the licensee and the right to grant these types of sublicenses will be routinely granted to licensees engaged in distribution of the licensed products.

III. Warranties and Guarantees

A. Introduction

Under United States law, the grant of a license does not include any implied warranties regarding the subject matter, such as a guarantee that the technology has commercial utility or that patents or copyrights included in the subject matter do not infringe upon the rights of third parties. However, in other countries, a license actually imposes a number of affirmative rights and obligations on each of the parties, including the rights of the licensee to use any inventions which may be covered by the license and the duty of the licensor to facilitate the lawful use of the licensed subject matter by the licensee. For example, in Germany a patent license carries an implied warranty that production using the patented invention is technically possible and that all of the representations set forth in the license agreement with respect to the functionality of the invention are true and correct.

In most cases, the parties to a license agreement will not rely upon any implied warranties regarding the licensed subject matter. Instead, the license agreement will generally contain a number of representations and warranties from the licensor with regard to the legal status and utility of any of the inventions and technical information being transferred to the licensee. A common representation requested by licensees is an affirmative statement that the licensed subject matter does not infringe upon the rights of any third party and has been lawfully developed or acquired by the licensor. However, the licensee will usually attempt to secure broader representations covering the performance characteristics of the licensed subject matter. In turn, the licensor will seek to limit or disclaim any express or implied warranties regarding the licensed subject matter.

B. Infringement or Misappropriation of Third Party Rights

In general Almost every license agreement will contain some form of representation or warranty by the licensor that the licensee is free to use the licensed subject matter in the manner contemplated under the agreement without fear of a claim from a third party that the licensed rights infringe upon their statutory legal rights or that any trade secrets included in the license were misappropriated from the third party.[5] With regard to trade secrets, this basic representation may be supplemented by additional representations from each person involved in developing the target company's trade secrets that he or she has not violated any non-disclosure or non-competition agreement with a former employer or any other party as a result of such person's activities for the target company.

Defending claims by third parties Representations regarding infringement and misappropriation are invariably accompanied by the licensor's agreement to defend the licensee against, and hold the licensee harmless from, any claims regarding infringement or misappropriation made by third parties, including the amount of any reasonable attorneys' fees incurred by the licensee as a result of any such claim. In most cases, the licensor will already have an independent interest in vigorously defending any infringe-

[5] Under the Uniform Commercial Code, every merchant implicitly warrants that the goods shall be free from any claim for infringement of patent or trademark. See Uniform Commercial Code § 2-312(3).

ment or misappropriation action brought by a third party. However, the specific provisions in the license agreement are intended to make it clear that the licensor will have the right to manage the defense and settlement of any such claim, even if the licensee is named as a defendant.

The principle of allocating control of the litigation to the licensor may be modified by the parties in certain circumstances. For example, it may be that an exclusive licensee has a substantial economic interest in the outcome of the litigation, apart from the prospect of remitting damages to the third party claimant. This may be due to the fact that the licensee considers the license to be essential to its new product strategies. In such cases, the licensee can be given the option to defend any such litigation. In addition, the costs associated with defending any litigation might be shared between the parties, as could the right to share in any awards. However the costs and duties are to be allocated, both parties should agree to cooperate with one another to the extent necessary to properly defend any such action.

Right to obtain license or modify technology One alternative to incurring the expense of defending an infringement or misappropriation action is to obtain a license from the third party claimant or to modify the licensed subject matter in a way that avoid the claims of the third party. Although this option is often provided in the license agreement, usually at the request of the licensor, it may be impractical for a number of reasons. First, the costs of obtaining a license may actually be quite high, particularly if the third party is aware of the potential exposure that the licensor might have to the licensee. Second, while modification of the licensed subject matter to avoid the third party's claims may be possible when patent infringement is at issue, it may simply be impossible to redesign a product to remove all remnants of a misappropriated trade secret. Finally, any change may so alter the functionality of the licensed subject matter that the expectations of the licensee are no longer met.

Remedies for breach of representation Acknowledging that a certain amount of risk is always associated with any representation relating to infringement or misappropriation, the parties must carefully craft the remedies which might be available to the licensee in the event that the representation turns out to be untrue. As noted above, a licensor may agree to indemnify the licensee for any damages suffered in the course of an infringement action. However, the licensor will want to limit the amount of its liability in some fashion, often through a stipulation that the liability of the licensor cannot exceed some percentage, or multiple, of the royalties which have actually been paid by the licensee.

C. Performance Warranties and Guarantees

Performance warranties and guarantees relating to licensed technology present some of the most difficult issues in the entire licensing area. Warranties can cover a number of business issues relating to the commercial utility of the technology, including warranties that the licensed process can be used to manufacture specified products and that the process will be able to operate under specified conditions. In addition, the license agreement might include warranties that the performance characteristics of the licensed subject matter will conform to standards represented by the licensor. For example, the licensor may be required to provide a representation that licensed patents and related trade secrets under a manufacturing license will reduce the specified undesirable byproducts of the manufacturing process by a certain amount or percentage.

As a general rule, performance warranties are a matter of negotiation between the parties, although in some countries local law may require some representation to the effect that licensed technology is suitable to produce the result contemplated by the parties, such as the manufacture of the specified product. One of the key difficulties in drafting warranties of this type is the fact that, to some extent, the performance of the technology often depends upon the skills of the licensee itself. As such, great care must be taken in this area to clearly understand the scope and duration of the warranties; the parties or persons along the distribution chain who will be entitled to the benefits of the warranties, including any customers of the licensee; and the remedies available to the beneficiaries with respect to any breach of warranty.

If, for whatever reason, the licensed technology fails to perform in the specified manner, some provision must be made to provide the licensee with an appropriate remedy. For its part, the licensor may wish to limit the remedies to such actions as may be commercially reasonable in order to insure satisfaction on the part of the beneficiary. On the other hand, the licensee may bargain for very specific remedies, such as repair, replacement, refunds or, in the case of certain defects in the technology, a commitment to undertake further research and development. In addition, unless the license agreement states that the remedies contained therein are exclusive, it may be possible for the licensee, as well as any other beneficiaries, to obtain monetary damages.

It is common for the agreement to include language which limits the prospective liabilities of the licensor and set outs the specific procedures which are to be followed in the event that the products fail to comply with the express warranties provided in the agreement. Typically, the licensor will explicitly disclaim any liabilities for any general, special or consequential damages which may arise of out of the sale, manufacture or use of the products. In addition, the remedies for any warranty will be limited to those set out in the agreement and may range from repairing, replacing or correcting any defects to refunding the purchase price and removing the defective products. The choice of remedies is generally given to the licensor.

IV. Rights and Obligations of the Parties

A. Introduction

The license agreement will usually contain a number of very specific rights and obligations of the parties which must be meant over the term of the relationship. A few of these obligations, such as the duties which relate to the protection of the licensed subject matter, are shared by both parties. The obligations of the licensor, aside from those emanating from the representations and warranties, generally run to continuing technical assistance to the licensee and honoring any contractual rights of the licensee with respect to exclusivity. As to the licensee, it may have a number of responsibilities, the most important of which will relate to the duty to actively exploit the licensed rights and refrain from engaging in competing activities. Of course, the licensee will also have the separate contractual obligation to pay the consideration set out in the license agreement.

B. Protection of the Licensed Subject Matter

In almost every case, the license agreement will contain various agreements and procedures which obligate both parties to protect the licensed subject matter from infringement or, in the case of trade secrets and technical information, misappropriation by third parties. Among the matters which might be covered in a license agreement are the manner in which the parties will deal with any misappropriation of the licensed information by a third party and the methods which will be used to maintain the confidentiality of such information. If statutory intellectual property rights are also included in the license, procedures will be included with respect to establishing and maintaining each of the various rights, such as preparing, filing and prosecuting any patent applications which might be required and insuring that all fees are paid on a timely basis, and provision will be made as to managing infringement actions involving the licensed rights.

C. Technical Assistance and Commercial Support

Transfer of technical information The license agreement must clearly describe the manner in which documents and other information evidencing the technology rights are to be transferred to the licensee. In the case of trade secrets, the agreement might contain a list, which need not be exclusive, of the documents embodying the trade secrets which will be delivered to the licensee. The licensor may wish to include a recital to the effect that the delivery of the listed documents completely fulfills all of its obligations with respect to transmittal of the trade secrets.

Technical assistance and training Technical assistance and training are an important part of ensuring that the technology will be effectively transferred to, and utilized by, the licensee. Information contained in a patent will generally not be sufficient to allow the licensee to begin production and manufacturing and trade secrets, by their nature, will require "hands-on" training in addition to the information included as part of any documents which may be provided to the licensee. Accordingly, unless the licensee has the requisite technical skills and experience in the use of the licensed subject matter of the license, the parties should anticipate that the licensing relationship should include some ongoing technical assistance and training from the licensor to the licensee.

Technical assistance can be broken out into two stages. During the first stage, which is sometimes referred to as "initial" technical assistance, the licensor is working with the licensee to provide a sufficient amount of training to enable the licensee to begin using the licensed subject matter on the scale contemplated by the parties. The initial technical assistance usually extends for a fixed period of time as set out in the license agreement. Although the parties may lay out the obligations of the licensor in some detail, such as the number of consultants that the licensor will make available for training, the licensor will generally want to avoid an overly broad set of duties, preferring instead to have the flexibility to provide a program of technical assistance which it determines, at its own discretion, is reasonably sufficient to effect the initial transfer of technology.[6]

The initial technical assistance is usually either provided "free" or at an amount which reflects only the actual costs that are incurred by the licensor (e.g. reasonable travel and living expenses of trainers) with respect to the training. While the licensor might wish to reduce its expenditures by training the licensee's personnel at the licensor's facilities, it is often not possible to effectively transfer the technology without working at the

[6] However, some licensees will insist that the licensor at least provide representations regarding the quality of the required training and the availability of certain personnel during the training program.

licensee's facilities. In any case, the license agreement should clearly specify the situs of the training and the manner in which any travel expenses will be borne. The agreement should also make it clear that employees of either party that might be sent to the other party will remain employees of their original employer and will abide by the rules, regulations and customs which govern the activities of the party at which they may be located.[7]

Technical assistance after the initial stage of training is generally provided on an "as needed" basis. The licensor may agree to make its personnel available for ongoing support, training and general "troubleshooting" during the term of the agreement. Any additional technical assistance will be billed at a per hour or per diem rate established in the license agreement, with the rates being subject to cost of living adjustments over the term of the agreement. The licensee will also bear any travel expenses for its own employees as well as for any employees of licensor who may be asked to travel to the licensee's facilities. The licensee will be required to provide the licensor with advance notice of any request for additional technical assistance and the licensor's obligations will be subject to availability of appropriate personnel. In the event the licensor does not wish to provide additional assistance of any type, these exceptions should be noted in the agreement.

Other technical and commercial support In addition to training in the use of the transferred technology, the parties may also provide for assistance in the operation of equipment, the "start-up" of plant and factory, the maintenance and service of products and special techniques with respect to marketing and sales of products manufactured through the use of the licensed technology. In those cases where both parties are involved in the manufacture of products covered by the license agreement, the licensor may provide the licensee with information about the best sources for necessary components and raw materials. The parties may either develop procedures for the joint purchase of supplies, equipment and raw materials, or the licensor may purchase supplies on behalf of the licensee or provide assistance to the licensee in dealing with suppliers to insure that a sufficient amount of supplies needed for exploitation of the license will be available at reasonable prices and on a timely basis.

D. Licensee's Improvements and Grant-Back Provisions

While the primary element of the technology package in any licensing relationship is the legal rights owned and developed by the licensor, in many cases the licensee will possess both the technical skills, and the economic motivation, to develop various improvements to the originally licensed technology which conducting its activities under the license. If it is anticipated that the licensee will develop improvements to the original technology, the license agreement should clearly describe the respective rights of both parties with respect to the ownership and use thereof, as well as any contemplated changes that the creation of such improvements might have upon the basic economic structure of the relationship.

The first issue which should be addressed is the ownership of the improvements. The answer to this question generally lies with the original intent of the parties. For ex-

[7] If work permits or other immigration authorizations are required in order for training to be completed, responsibility for such matters should be set forth in the agreement.

ample, the licensor may be willing to accept a lower royalty rate for the license of the original technology if it is contemplated that the licensee will be developing new improvements which will be owned by the licensor at the time that the development work is completed. However, if the improvements are of the type that could only be developed by the licensee in light of its particular skills and resources, or are of a type that are unexpected or unrelated to the original licensed subject matter, it may be appropriate for the licensee to be vested with full title to such improvements, perhaps with an appropriate adjustment to the amount of royalties for the original technology or, as described below, an appropriate agreement regarding use of the new technology by the licensor.

One possible solution which might be used in cases where the parties are working together on developing various improvements to the original technology is to provide that each party will own any improvements which they develop independently and that the parties will jointly own any improvements that arise from their cooperative efforts, with cross-licenses to one another. In order to implement this alternative, the licensee would submit each improvement to the licensor for review. The licensor would agree to maintain any such submission in confidence and to advise the licensee if it agrees with the licensee's claim of independent development. In the event of any disagreement, the matter might be submitted to some form of alternative dispute resolution mechanism.

Once ownership issues have been resolved, consideration must be given to the rights of each party to use the new technology in similar or new applications or fields of activity. In the most extreme case, if the improvements are owned by the licensor, the licensee will be required to assign all of its rights in the improvements to the licensor and would have no right to use the improvements under the original license agreement, although in some cases the licensee might be provided a right of first offer on any license of the improvements which the licensor may intend to give to third parties. Alternatively, the licensee may be given the right to use the improvements only for purposes limited to the objectives of the original licensing relationship.

If the improvements are owned by the licensee, it may simply retain all of the rights in the new discoveries, without any further obligation to the licensor. This may occur when the licensee has made a large lump-sum payment to the licensor at the time that the license agreement was originally signed. In effect, the licensor is taking the position that it purchased the original technology in order to serve as a starting point for further research and development work. However, in most cases where the improvements are owned by the licensee, it will have an obligation to grant a license to the licensor with respect to the use of the improvements. These types of provisions, commonly called "grant-back" clauses, can have a significant impact on the economic incentive to the licensee to undertake research which might lead to improvements in the original technology.

E. Licensee's Duty to Exploit Licensed Subject Matter

The ability and willingness of the licensee to exploit the licensed technology is a key factor in the success or failure of the business relationship, particularly when the licensee is to be given, for some period of time, the exclusive right to use the technology for a specific application or product or in a designated market. Since an exclusive license represents what amounts to a wholesale delegation of a commercial opportunity to the licensee, the licensor develops a keen dependence upon the capabilities of the licensee to successfully utilize the licensed technology and products. In the event that the licensee is unable, or unwilling, to adhere to its obligations under the arrangement, the licensor may suffer substantial economic losses and, in the worst case, the licensor will be unable to market and sell the new products in a timely fashion once the rights have reverted back to it.

The legal duties and obligations of the licensee to exploit the licensed rights will depend upon the type of license. In the United States, there is no implied duty or obligation under a non-exclusive license which requires that a licensee actively exploit the license. However, at least in the United States, the grant of an exclusive license may create an implied obligation on the part of the licensee to use its "best efforts" to exploit the licensed subject matter in "good faith". In most cases, the parties will incorporate any expectations as to the licensee's performance into the license agreement itself, rather than relying upon any implied standards under the law. This is particularly important in light of the fact that it is often difficult to define such concepts as "best efforts" and "good faith."

F. Restrictions on Licensee Engaging in Competitive Activities

Given the desire of the licensor to maximize the flow of revenue from the licensing relationship, an attempt may be made to restrict the ability of the licensee to engage in any activities relating to the development or commercialization of technologies or products which might serve as a form of substitute for the licensed subject matter. In effect, the licensor hopes to be the licensee's "sole supplier" of the relevant technology, at least with respect to the commercial activities contemplated at the time the license is executed.

For its part, the licensee will be reluctant to agree to any substantially limitations on its activities, especially in those cases where the licensing relationship is seen as an opportunity to reduce the time and expenditures required to absorb new forms of technology and know-how which will serve as a basic for new product development efforts by the licensee in the future. Nonetheless, it is probably not unreasonable for the licensee to be asked to agree to certain restrictions. For example, the licensee will usually be prohibited from using the licensed technology to manufacture and sell new products which may compete with the licensed products. Also, the licensee will be restricted from acting as a manufacturer or distributor of competing products engineered by third parties. Finally, in appropriate circumstances, a specific prohibition on the use of "reverse engineering" techniques by the licensee should be included.

G. Quality Control and Products Liability Claims

Although the parties are free to specify the scope of the various expressed and implied warranties regarding the performance of the licensed technology, statutory and common law principles with respect to strict or products liability expose manufacturers to the risk of claims for damages arising out of injuries suffered due to defects in any products manufactured with the use of the licensed technology. The specter of potential products liability claims makes it important for the parties to consider including one or more provisions designed to insure quality control in the use of the licensed technology and to allocate the burden of products liability claims. In addition, since the courts have maintained that control over the nature and quality of goods is essential to a valid trademark license, a trademark licensor may have a duty to monitor the actions of the licensee in order to protect its own trademark rights.

H. Breach of the Terms of the Licensing Agreement

The parties will usually attempt to specify certain of the consequences associated with a breach of a representation or term set out in the license agreement. Obviously, one remedy would be to permit the non-breaching party to terminate the agreement. In many cases, however, the non-breaching party may elect to continue the relationship and look to the breaching party for a specified remedy or for monetary payments based on the damages suffered by the non-breaching party. As a general matter, the agreement will contain various limitations on the amount of the liabilities which might arise out of any breach of the agreement or the remedies that might be available to the non-breaching party.

V. Compensation for Use of Licensed Rights

A. Introduction

In most cases, the license will call for payment of some form of compensation by the licensee to the licensor for the use of the licensed subject matter. The form of compensation, be it cash or "in-kind," as well as the amount and timing of the payments will be determined by negotiation between the parties and each side will seek to satisfy their own financial and accounting objectives with respect to cash flow management, cost recovery and minimization of tax obligations. While industry standards are helpful in arriving at an appropriate "value" for the license, each transaction will be unique and the compensation package will ultimately depend upon the needs of, and alternatives available to, the parties.

Cash payments are the most common type of compensation paid under a licensing arrangement. Cash payment can be classified in one of two ways: "royalties," which are measured by reference to the actual use of the licensed technology by the licensor; and "lump sum" or other periodic forms of payments, which may be made in addition to, or in lieu of, royalties and which are typically unrelated to the licensee's use of the licensed technology. The two types of cash payments are not mutually exclusive and, in many cases, the license agreement calls for some lump sum payment at the time of execution, which covers the licensor's costs of negotiating the agreement and transferring the technology; additional minimum periodic royalty payments to insure that the licensee is serious about using the technology; and a "performance-based" royalty based on the licensee's actual use of the technology over the term of the license.

B. Royalties

In general Royalties are periodic payments which bear a specified relationship to one or more objective standards of consumption or performance with respect to the licensee's use of the technology. The royalty formula can be a percentage of the revenues derived by the licensee from the sale of products using the licensed subject matter or can be established as a fixed amount per unit or product. The royalty percentage or amount, which is generally referred to as the "royalty rate," can vary over the license term, depending upon the requirements of the parties.

Royalty measurement standard Private and public economists have developed various formulae for establishing royalty rates. Among the common measures are the gross production of the licensee, the gross or net sales price of the licensed products, the net or gross profits from the licensed products, the number of units manufactured or sold by the licensee, or the cost of the licensed products. The proper measurement standard will depend upon the particular circumstances surrounding the transaction, as well as the custom in the industry in which the parties are active.

Establishing the royalty rate as a fixed or, in many cases, varying percentage of the gross or net sales price is probably the most common measurement standard when "products" are to be produced by the licensee using the licensed subject matter. Obviously, the choice between "gross" and "net" sale price will have a significant effect upon the relative interests of the parties. The licensor will prefer to base royalties on gross revenues, while the licensee will prefer to use a net revenue standard which takes into account one or more deductions from the amounts actually received from the sale of the licensed technology or products.

If the "net" sales price is utilized, the parties must carefully consider the various items which are to be deducted from the invoice price for the licensed products. Although the items may vary with the system of accounting utilized by the licensee, among the items to be considered are duties, tariffs and sales, value added and turnover taxes; discounts and returns actually allowed; transportation and insurance expenses; charges for installation; amounts paid by the licensee for products purchased from the licensor or any other party recommended by, or affiliated with, the licensor; fees associated with the maintenance of any statutory rights included in the license; and revenues received by the licensee with respect to service, maintenance and spare parts.

An alternative to using net sales or similar price-related milestones for computing royalties is to base payments on the number of units manufactured or sold by the licensee. The use of a "volume" measure shifts the risks and benefits of marketing efficiencies to the licensee, since the amount of the royalty on a given unit is fixed while the actual sales price for the unit may vary upward or downward. While a licensor might consider this type of measurement scheme to be attractive, it may actually be counter-productive if the per unit royalty burden on the licensee becomes so high that it abandons exploitation of the licensed rights.

When the license covers a process, rather than products, production volume is generally used as the royalty base. In such cases, the royalty rate is expressed as a percentage of unit of measure which is applied to the product made by using the licensed process, such as weight, volume, square footage, or linear feet. However, it is possible to relate the royalties in this type of situation to the net sales price of the products made using the licensed process. For example, if the licensee use a licensed process to produce paint and then sells the paint to consumers, the royalty would be based on the net sales price of the final paint products.

Royalty rate The royalty rate can be a certain amount per unit or product or a percentage of revenues, profits or sales. Whatever standard of measurement is used, the parties may consider varying the royalty rate to be paid by the licensee in relation to the level of use or performance. For example, at the beginning of the license period, the licensee may argue for a lower royalty rate in order to provide an incentive for using the new technology and incurring the "start-up" costs associated with the distribution and promotion of the products. On the other hand, the licensor may prefer a higher initial royalty rate in order to recover the costs of developing the licensed technology. Once the initial goals have been met, the rate can be increased, recognizing that the licensee has achieved the requisite economies of scale, or decreased, assuming that the licensor has recovered its costs and the licensee then requires an incentive to expand its sales and production efforts.

Minimum and maximum royalty provisions In order to provide the licensor with assurances that the licensing arrangement will allow it to recover a portion of the costs associated with the development of the licensed technology, the licensee may be required to remit a specified minimum royalty amount on a periodic basis during the license term, whether or not the royalties are actually earned under the standard specified in the agreement. Failure of the licensee to make the minimum royalty payments may lead to the termination of the agreement or an action for damages. Also, even if the licensee continues to make all the minimum payments, if any underlying performance milestones are not being achieved, the licensor may seek the right to terminate the license or, in the event that the license is exclusive, convert the license to a non-exclusive arrangement.

Minimum royalty payments may be particularly appropriate in those cases, such as in the pharmaceutical area, where regulatory requirements may create long delays in the commencement of sales of royalty-producing products. The licensor may require the licensee to make either minimum periodic payments during the development and testing of the new product, or to make benchmark payment whenever the licensee has passed certain objective milestones in the process of bringing the product to market. While the licensee may agree to such demands, it may argue for the right to credit amounts paid in the development stage against royalties which might become due once the product is sold in the market.

On the other hand, a licensee may seek to limit the total amount of its obligations to make royalty payments to the licensor. For example, the parties may agree that the royalty obligation will terminate when a specified amount of royalties have been paid by the licensee, presumably because the licensee has fully performed on its obligations to exploit the licensed technology. Thereafter, the licensee would have a perpetual, "paid-up" license to use the technology in the manner specified in the license agreement. The royalty "cap" will correspond to the level of sales anticipated by the parties at the time that the license agreement is executed. A maximum royalty provision will override any royalty period included in the license agreement.

C. "Lump Sum" or Non-Royalty Payments

Although royalty payments are the most typical form of compensatory arrangement under a licensing agreement, in certain cases the parties may dispense with any requirement of royalty payments, preferring instead to limit the amount of consideration to a single fixed amount, payable at one time upon execution of the agreement or in two or more installments over the period of the license. Lump sum payments, whether paid in advance or in several installments over the period of the license, can be advantageous in a number of circumstances, such as when the technology is mature or subject to obsolescence or the licensor seeks to recover the costs of developing the licensed technology or provide an incentive for the licensee to exploit the technology.

Lump sum or non-royalty payments are not limited to those instances where no royalty obligation is imposed under the license. While it is certainly possible to require both lump sum payments and royalties, in most cases the amount of any lump sum payments will be fully or partially credited against any future royalties which may be owing as a result of the actual use of the licensed technology. If lump sum payments are treated as a form of advance against future royalties, they may or may not be refundable if the amount of any future royalties does not exceed the initial payment amounts.

VI. Term and Termination of the License Agreement

A. Introduction

Although the parties are typically most concerned about establishing the guidelines necessary to insure the success of the relationship, great care must be taken with respect to determining the circumstances, if any, which might lead to termination of the licensing agreement. First of all, the parties should determine the natural duration of the licensing arrangement, absent an event which might lead to an early termination. The agreement should then specify the events which might lead to early termination, the consequences associated with any termination of the agreement, the effect of termination upon the then-existing rights of the parties, any post-termination covenants or restrictions with respect to the licensed rights and, if appropriate, any additional remedies and liabilities relating to termination.

B. Term of the Licensing Agreement

Each license agreement will generally have a specified duration or term, although it is possible to provide that the term of the license will be perpetual, either by its terms or by virtue of additional payments made by the licensee. Obviously, a licensor will wish to limit the duration of the license, particularly when the specified rights are exclusive, in order to minimize the risk that the royalty rate was too low or that the licensee is unable to adequately exploit the licensed rights. On the other hand, any licensee will want to insure that the duration of the license is sufficient to allow it to recover the amount of expenditures which must be made in order to adequately exploit the technology.

When the license pertains to patent rights, the term should generally extend until the last of the patent rights expires. When the license includes the rights to use trade secrets, the licensor will want payments to continue for so long as the products continue to manufactured by the licensee, while the licensee will argue that the obligation to make payments should terminate after some period of time. As a general rule, trade secret licenses extend for five to ten years; however, the term of the license ultimately depends on the relative bargaining strengths of the parties and period of time that secrecy can be maintained. A trade secret license does not, at least as a matter of law, terminate merely due to the dissemination of the information in a manner which reduces or eliminates it secrecy, at least in those cases where the trade secret information was confidential at the time that the license agreement commenced.

The term of the license agreement is generally strictly regulated in developing countries. For example, the length of a patent license may be limited to a period which is substantially less than the duration of the patent. The restrictions on trade secret or "know-how" licenses may be even more onerous, in some cases prevention the term of any royalty obligation from extending beyond the period of time required in order for the licensee to be able to independently utilize the technology. Also, trade secret licenses in developing countries may automatically terminate upon release of the information into the public domain, a problem that may be exacerbated when local laws and practices with respect to the protection of any trade secrets or similar non-statutory rights are more lenient than those which might exist in the United States.

C. Early Termination Provisions

Both the licensor and the licensee may be given a right to terminate the license agreement prior to the expiration of the term specified in the agreement. Early termination provisions should be carefully tailored to suit the circumstances which may surround the specific arrangement. As a general rule, termination provisions which are usually inserted into commercial agreements, such as the right to terminate the agreement in the event of the bankruptcy or insolvency of one of the parties or a material breach of the terms of the agreement, should also be included in the license agreement. In addition, both of the parties may have the right to terminate or modify the arrangement upon the occurrence of certain events.

From the perspective of the licensor, the license should be terminated in the event that the licensee defaults in making any required royalty payments, if the licensee fails to take such steps as are necessary to adequately exploit the licensed technology in the product or market area to which the license applies or if the licensee fails to abide by the limitations imposed upon the use of the technology rights. Given the substantial investment of time and money in the technology rights on the part of the licensee, the license will generally require that the licensor deliver a notice to the licensee of the event constituting a default and a licensee will have a specified period of time following delivery of the notice in which to cure any such default.

A licensee may want to terminate the license agreement in order to relieve itself from any burden to remit minimum royalties or to make further expenditures necessary to meet its obligations to adequately exploit the technology rights, particularly if the licensee determines that the technology has become obsolete. The licensee may also want to be able to terminate the agreement upon any default by the licensor or when the licensor has breached one of the representations and warranties relating to the technology and the conduct of the relationship. Any attempt by the licensee to terminate the license in order to avoid future obligations may result in the imposition of some form of damage payment, since the licensor will then need to replace the licensee with a third party in the product or market area to which the agreement relates.

When an exclusive license is involved, the parties may consider various alternatives to an outright termination of the license agreement. For example, rather than terminate the rights of the licensee upon the failure of the licensee to meet certain minimum royalty milestones, the licensor may simply convert the license to non-exclusive and appoint other licensees in the same field of activity. Similarly, an exclusive licensee wishing to reduce its obligations under the license agreement may, in lieu of terminating the entire relationship, allow the licensor to convert the license to non-exclusive. In each case, the compensation that is to be paid under the licensing agreement may be adjusted.

D. Effect of Termination and Post-Termination Covenants

The license agreement should clearly describe all of the consequences associated with any termination of the agreement, including the nature of any covenants which are to survive the termination of the agreement. In addition to providing for a full and final accounting of amounts owed under the agreement, it will be important to specify the procedures to be taken for protecting and returning any technical information used in the relationship and the manner in which responsibilities for any statutory rights are to be allocated between the parties. Also, the parties must consider how ongoing relationships with customers and sublicensees will be handled, particularly since the licensor has an interest in ensuring that such parties are not unduly disadvantaged by the termination of the arrangement with the licensee. Finally, the parties may want to provide for damages in those cases where the termination arises due to a default by one of the parties.

7 Joint Ventures

I. Introduction

A. Definition and Classification

One of the most common forms of organizational strategic business relationship is the "joint venture." The joint venture utilizes a separate business entity (e.g. a corporation or a partnership in the United States), to conduct the specified business activities.[1] The use of a separate entity allows the parties to limit the liabilities associated with the relationship and, in fact, the structure may be required in order for the parties to avail themselves of incentives and concessions offered under any local foreign investment programs when the joint venture is used as a means for selling and distributing products and services into a foreign market.

Joint ventures can be classified by the various functions which may be performed by the enterprise. For example, a joint venture may be formed to conduct research and development work on new products or technical application, to manufacture or produce various products, to market and distribute products and services in a specified geographic area or to perform some combination of the aforementioned functions. The function of the joint venture will be linked to the overall objectives of the parties and will dictate to a large extent the substantive terms of the joint venture arrangement.

B. Motivations for Joint Venture Relationships

A joint venture carries with it a number of advantages and disadvantages. On the one hand, a joint venture can provide a party with access to resources and skills which are unavailable to it at any reasonable cost. However, on the other hand, use of a joint venture can be quite risky given the reliance that must be placed upon the ability and willingness of the other party to perform its obligations during the term of the arrangement. In considering whether or not to enter into a joint venture, notice should be taken of some of the following motivations which might drive a party toward such a relationship:

[1] Joint ventures which involve the creation and use of a separate legal entity are referred to generally as "equity joint ventures," since each party will contribute capital to the joint venture, participate in the management of its operations and share the profits from, and risks associated with, the conduct of the joint venture activities. In contrast, it has now become quite common to refer to "contractual joint ventures," which do not involve the creation of a separate legal entity that is jointly owned and managed by the parties, but instead consists of one or more contractual agreements which obligate the parties to provide various operations or services on a long-term basis which retaining ownership of the tangible and intangible assets which are needed to achieve the objectives of the relationship.

(i) The party may not have sufficient financial resources to take on a particular project by itself and may seek a partner to assist in sharing the financial burden and other risks of the project. Projects which require a substantial amount of investment in development work and product testing may be good candidate for the joint venture structure. In some cases, a partner may even be willing to provide most, if not all, of the funding in exchange for gaining access to the intangible assets of the party.
(ii) The party may want to gain access to the technical resources and skills of another party. A joint venture which brings together managers and scientists from each party will facilitate the rapid exchange of information regarding existing and new technologies.
(iii) The party may seek a joint venture with a local party in a new foreign market as a means of accelerating the pace of market penetration. The local party should be able to provide the requisite knowledge of local tastes and customs and providing for equity involvement by the local partner will ensure that it will provide the best services to the joint venture. A joint venture may also be required in order to satisfy host country rules which require local participation.
(iv) The party may use a joint venture to decrease the costs associated with the manufacture of its products while retaining some control over the quality of the process and the technology used in the course of completing the manufacturing activities.
(v) A party may choose toa joint venture structure, rather than a network of contractual relationships, in order to ensure that it is in a position to directly manage the specific functional process, be it research work, manufacturing or distribution. Also, a joint venture may be appropriate when the party believes that it will need to provide personnel to facilitate rapid transfer of trade secrets and other information for use in the collaborative venture.

C. Basic Structural Components of Equity Joint Ventures

The formation and use of an equity joint venture, referred to hereafter simply as a "joint venture," must take into account all of the same issues which are generally encountered with any new business enterprise. For example, each of the joint venture partners will contribute various resources and skills to the new enterprise, including products, financing, personnel, facilities, raw materials and marketing expertise. These contributions may take the form of direct investment in the joint venture or may be provided under the terms of one or more ancillary agreements between the joint venture entity and the partners. The partners must also agree on a number of issues regarding the management and operation of the enterprise.

The basic structural components of the joint venture are as follows:

(i) *formation*. The parties, A and B, agree to form a separate business entity (e.g. a corporation or a partnership in the United States). The procedures to be followed regarding the formation of the entity are typically specified by relevant laws in the jurisdiction in which the entity is to be organized.
(ii) *governing documents*. The law relating to the specific form of business entity used for the joint venture will always specify one or more legal governing documents which need to be prepared in order for the legal existence of the entity

to be recognized. For example, when a joint venture is organized in the corporate form in the United States, articles of incorporation and bylaws must be drafted and approved by the parties. These basic charter documents describe the capital structure of the entity as well as any specific rights granted to the shareholders with respect to voting, distributions and liquidation of the entity. Articles are filed with the appropriate regulatory body at the time the entity is formed, while the bylaws are approved by the board of directors of the entity and ratified by the parties in their capacity as shareholders.

(iii) *capital contributions*. A and B agree to make various contributions of cash, tangible assets and other intangible rights to the joint venture. A and B are the sole owners of the joint venture and receive shares, if the corporate form is used , or partnership interests, in the partnership form is used, which evidence their interests in the venture.

(iv) *joint venture management agreement*. While the legal governing documents will often contain various provisions with respect to management of the venture, it is also typical for the parties to enter into some form of management agreement relating to the joint venture's activities. For example, in the United States, A and B might enter into a shareholders, or partnership agreement which will govern their respective rights and obligations with respect to the operation of the joint venture. This agreement is the key operational document for the joint venture and sets forth the understanding of the parties with respect to a number of matters: the initial capitalization of the venture, additional financing for the venture, management and control of the venture, the business and operational activities of the venture, transfers of ownership interests, allocations and distributions of income and other assets, and the term and termination of the venture. In addition, the agreement will include various representations and warranties by each of the parties, agreements with respect to confidentiality of information exchanged in the course of the venture, procedures for the resolution of disputes and a description of any internal controls required to monitor the progress of the venture.

(v) *ancillary agreements*. A and B may also both enter into separate contractual agreements with the joint venture. These agreements will cover any services to be provided to the entity by the parties, as well as any agreements with respect to the purchase of products developed or manufactured by the joint venture or the use of the assets of the joint venture by the parties in activities that may, or may not, be related to the specific purpose of the joint venture. Among the various agreements are an administrative services agreement, a supply agreement, an equipment purchase agreement, one or more licenses with respect to technology or manufacturing rights, one or more distribution agreements and, in some cases, an agreement with respect to the lease, acquisition or construction of appropriate facilities.

(vi) *termination provisions*. Although the joint venture is established as a separate legal entity, the enterprise will only survive for that period which is required in order for each of the parties to achieve the specific goals and objectives of the relationship. Upon termination of the joint venture, the entity will either be liquidated, with the parties each receiving their agreed allocation of the remaining assets and resources of the venture, or one of the parties will buy the interest of the other party and continue to operate the business of the venture outside of the original joint venture relationship. Most of the termination provisions will be described in great detail in the joint venturer agreement and, in the case of a corporate joint venture, the charter documents.

D. Regulation of Joint Venture Relationships

The formation and operation of a joint venture generally raises many of the same legal issues as confront any other new business entity. For example, the parties must always be mindful of the requirements imposed by relevant law relating to formation and legal operation of the business entity (e.g. partnership or corporations laws in the United States). In addition, any of the contributions to, or distributions from, the joint venture by the parties will be subject to specific rules under domestic and international tax laws. Joint ventures often involve an exchange of sensitive business information between the parties and raise a number of concerns with respect to the protection of intellectual property and related rights. Also, as describe in great detail elsewhere in this work, joint ventures raise unique issues in the antitrust and competition law areas. Finally, in the case of international joint ventures, there may be a need to obtain approval from local regulators of an investment by a foreign party and/or for the transfer of technology to the joint venture under applicable foreign investment laws.

E. General Issues in Structuring the Joint Venture

Structuring a new joint venture requires consideration of a wide variety of legal and business issues. Among those which are the most important are the following:

(i) defining the objectives of the joint venture;
(ii) identifying the contributions to the joint venture by the parties;
(iii) constructing the capital structure of the joint venture;
(iv) defining the rights of the parties with respect to management and control of the joint venture;
(v) "day-to-day" functional operation of the joint venture;
(vi) procedures for transferring joint venture interests or withdrawing from the joint venture; and
(vii) duration and termination of the joint venture.

II. Defining the Objectives of the Joint Venture

At the outset of every proposed joint venture, it is necessary for the parties arrive at some understanding of the basic objectives of the proposed enterprise. This includes an identification of the nature and scope of the proposed activities of the joint venture and a description of the expectations of each of the parties as to the ultimate financial and technical performance of the joint venture. The process of defining the basic objectives of the joint venture is usually facilitated by the preparation of a summary of the key business terms of the transaction, as well as a formal business plan.

In defining the objectives of the joint venture, it is necessary for the parties to take into account the scope of the joint venture business, the size of the joint venture, the way in which the parties will allocate management responsibility for the day-to-day operations of the joint venture, the decision-making process for major decisions regarding the venture, the terms of any ancillary agreements between the joint venture and either of the parties and the term of the joint venture and any events that might lead to an early termination of the business relationship.

Once the parties have reached general agreement on the purpose and objectives of the relationship, it is then necessary to give consideration to the preparation of the initial business plan for the venture. In some cases, the business plan will be completed prior to the formation of the venture. However, it is not uncommon for discussions to continue in earnest during the first few months following the initial capital contributions to the venture. The business plan should set forth in reasonable detail the goals and objectives of the venture with respect to research and development, production, distribution, licensing and such other matters as may be determined by the parties. Agreement should be reached on the procedures for updating the plan, usually on a semiannual or annual basis.

III. Contributions to the Joint Venture

A. Introduction

The first step in establishing any joint venture is determining the contributions which each party will be making to the enterprise. In this area, consideration should be given not only to the amount of cash which each of the parties will be expected to contribute to fund the various activities of the venture, but also the content and valuation of any "in-kind" contributions, the relative rights which are to be associated with the stock instruments to be issued to the parties in connection with their capital contributions, the obligations of the parties to contribute additional capital to the venture or to guarantee any borrowings by the venture from commercial lending institutions and, finally, the nature of any other "contributions" to the venture by the parties which are made other than in the form of a capital contribution.

B. Tangible and Intangible Assets

In general The planning and negotiation process prior to the actual formation of the joint venture should include an analysis of the financial, technical and functional requirements of the venture throughout the proposed term of its existence. Although this type of analysis should initially be conducted without regard to the specific capabilities of the parties in each of these areas, it will ultimately be necessary for the parties to focus upon each of the specific contributions will are to be made by the parties.

Cash and cash equivalents Cash contributions from the parties provide most of the initial working capital for the venture, although it may also be possible for the venture to secure working capital through loans from public agencies and commercial banks at the time that the venture is formed. The total amount of cash to be contributed really depends upon the relevant planning horizon, but should be sufficient to carry the venture through one or more milestones which are perceived to be relevant to the overall success of the venture.

Services The laws of each jurisdiction will vary as to whether or not services rendered to the venture by one of the parties will be valid consideration for the issuance of shares. Usually, if it is anticipated that one or both of the parties will provide managerial or technical services to the venture, the nature of those services, and the related amount of compensation, will be specified in a separate agreement between the party and the venture. Compensation received by the parties under those arrangements will bear no relation to their capital contribution to the venture.

Tangible property Tangible property, in the form of equipment or facilities, is often part of the initial capital contribution to the venture. Relevant issues include the appropriate value to be placed on the contributed property and the need for assurances from the contributor that the property is in good working order and suitable for use in the activities of the venture. As always, one or both of the parties may lease tangible property to the venture, rather than making an outright capital contribution.

Intangible property Intangible property can take many forms, such as statutory intellectual property rights, trade secrets and certain technical information deemed essential to the activities of the venture, as well as unregistered trademarks, tradenames and the goodwill that is associated therewith. Often the very purpose of the venture is to conduct activities relative to development and commercialization of intangible property rights or to facilitate the distribution of "known" and reputable products by a local party in a new market deemed to be unattractive for direct selling activities by the foreign producer. As such, the key issues relate to properly defining the scope of the intangible property which is to be transferred to the venture and the value which is to be assigned to the contribution.

C. Other Forms of Contributions to the Joint Venture

In addition to cash, services, tangible property and intangible property, each of which might be contributed to the venture in the form of capital, consideration must be given to various other contributions which are to be made by each of the parties to the venture. Among the items that might be included are the following:

(i) *research and development expertise*. In those cases where the venture is to be engaged in significant research and development activities relative to the production and commercialization of specified products, one of the parties may be engaged to conduct a specified program of research and development. The venture may enter into a research and development agreement with the party providing for various payments to be made to the party, while allowing the venture to retain sole ownership of the results of the development program.

(ii) *administrative and management services*. Although it is often the intent of the parties to build the joint venture into a self-sustaining business enterprise, it is not uncommon for one or both of the parties to contract with the venture to provide the venture with administrative and management services for a portion of the contemplated term of the venture.

(iii) *supply arrangements*. A joint venture is an appropriate business form when one of the parties is in a unique position with respect to its ability to supply one or more essential components to the venture. As such, the venture will often enter into a supply agreement with the party designed to ensure the ready availability of needed components for the products which are to be manufactured and distributed by the venture.

(iv) *lease or licensing arrangements*. As noted above, both tangible and intangible property may either be transferred to the venture in the form of a capital contribution or may be loaned or rented, in the form of a lease or license, to the venture in return for a specified rate of return to be set forth in the contract.

(v) *manufacturing arrangements*. A joint venture usually aspires to develop its own manufacturing capabilities, often to take advantage of perceived benefits associated with the local labor supply. However, one of the parties may agree to provide its own manufacturing capabilities to the venture for some portion of its term and, in some cases, the local party may undertake all of the manufacturing responsibilities, a role which itself should raise concerns regarding capacity, the relation to other unrelated activities of the party and proper allocation of overhead costs.

(vi) *distribution and marketing agreements*. Developing an independent distribution and marketing capability is generally perceived to be far too expensive for the joint venture form. Usually, one of the parties is believed to be able to provide adequate capabilities through its existing distribution channels such that the successful marketing within the territory of the products developed by the venture is assured. Therefore, the terms of any distribution agreements between the venture and the parties are key elements of the negotiation process and the performance of the parties under these agreements often dictates the long-term success or failure of the relationship.

(vii) *regulatory expertise*. The relationship that might exist between one of the parties and the relevant regulatory authorities can be an important factor in the selection of the joint venture party. Local parties may be able to secure preferred financing arrangements from public lending agencies and, in those cases where regulatory approvals are required for commercialization of specified products, the party may have significant experience in dealing with the bureaucratic processes.

IV. Capital Structure of the Joint Venture

A. Introduction

Whatever form that the initial capital contribution might take, a corporate joint venture actually commences operations upon the sale and issuance of shares to the parties. The parties become the owners, as shareholders, of the corporation with such rights as may be created under the laws of the jurisdiction in which the corporation is organized, as well as any contractual agreements between the shareholders. The shares of capital stock to be issued to the parties carry a number of rights with respect to management and control of the enterprise as well as the manner in which the holder can expect to receive a return with respect to its invested capital.

B. Alternative Forms of Capital Structure

In general The parties may choose among various alternative forms of capital structure in organizing the economic and voting rights with respect to the corporate joint venture. The simplest form of capital structure use a single class of common stock, thereby assuring that each party has identical rights with respect to voting, interim distributions and liquidation, assuming that the parties are issued an equal number of shares of common stock. However, it is common to use a multi-class capital structure that takes advantage of the flexibility provided by the corporate form to tailor the shareholder rights to the objectives of each of the parties.

Single class of common stock The simplest capital structure calls for a single class of stock, typically referred to as "common stock," which is issued to the parties in proportion to the value of their original capital contributions to the venture. The shares would have indistinguishable rights with respect to dividends, liquidation and voting, although it is not unlikely that the parties may enter into supplemental agreements governing the management of the venture which effectively modify the voting rights imposed by statute.

Multi-class capital structure A more complex capital structure utilizes two classes of stock and illustrates some of the issues to be borne in mind when some degree of uncertainty exists regarding the timing and amount of future income from business activities. Separate classes can be denominated in any number of different ways, such as "Class A" and "Class B" common stock or as common and "preferred" stock, and a different class is issued to each of the parties. In most cases, some mechanism will be created which permits or "forces" conversion of shares of one of the classes into shares of the other class, thereby placing each of the parties on an equal footing following the completion of this so-called "conversion event." Once the conversion event has occurred, the parties will share future distributions of cash and assets from the enterprise in direct proportion to the original value of their contributions to the capital of the venture.

C. Accounting for Additional Capital Contributions

It is often necessary to provide additional capital to the venture following the development stage. As part of the periodic budgeting process, the parties must consider the need for additional financing and reach agreement regarding the timing and amount of the new contributions. If additional capital is necessary, each of the parties will usually be obligated to make the contribution in proportion to their then-existing interest in the income and assets of the venture. Should one of the parties be unable or unwilling to make another capital contribution, the interests of the parties will be modified to reflect the actual amounts of capital contributed. However, in some cases, the failure to make an additional contribution may have more dire consequences, perhaps providing the non-defaulting party with the right to terminate the venture or purchase the interest of the other party.

D. External Financing of the Joint Venture

In addition to funding from the parties, it may be necessary for the venture to borrow money from commercial lending institutions or public financing agencies. External financing of this type obviously creates a number of issues relative to the operation of the business: a burden on current cash flow, a security interest on the assets and properties of the venture and, in some cases, various restrictions on the business activities of the venture and the prerogatives of the parties as shareholders and managers. In addition, external financing may be contingent upon the availability of guarantees from each of the parties, a factor which should be taken into account at the outset of the venture.

Guarantees create the greatest difficulties when one of the parties is unable to provide a guarantee in proportion to its equity interest in the venture. In those situations, the other party may guarantee all or a disproportionate portion of the external financing and, in return, will be entitled to receive a fee from the non-guarantor calculated as a percentage of the outstanding borrowings. Alternatively, the guarantor's fee may be paid out of income generated by the venture and the other party may agree to indemnify the guarantor for any losses suffered as a result of its assumption of the obligations under the guarantee.

The parties should seriously consider the availability of external financing before entering into the joint venture. For example, it is important to understand whether the joint venture will be able to obtain acceptable banking and lending facilities from domestic, foreign or governmental sources. Among the factors to be considered would be the conditions and terms of any credit facilities, as well as the interest rate and associated fees and the nature of any guarantees or security interests that might be required. In addition, external financing may impair the ability of the parties to remove profits from the joint venture in the form of dividends.

V. Management and Control of the Joint Venture

A. Introduction

Devising an appropriate structure for the management and control of the venture is one of the most important matters to be negotiated between the parties. Assuming that the choice is made to utilize the corporate form of business entity, the parties must consider various issues relating to the election of the board of directors, the selection of the officers and key managers, the respective voting rights of each of the parties on matters deemed to be material to the existence and operation of the enterprise and the specific duties of the board of directors, the officers and any committees created by the parties to manage one or more of the business functions of the enterprise. Also, while hopefully the parties will not reach a situation where they are unable to agree on a certain matter, the parties should agree upon a procedure for the orderly resolution of disputes.

B. Methods for Allocating Control of the Joint Venture

In general The relative interests of the parties in the profits of the corporation, which is usually determined by their ownership of those classes of shares which are entitled to share in any distributions made by the corporation, need not dictate the degree of actual control exercised by the parties. While a majority interest in the profits of the enterprise might well mandate the assumption of control, there are situations where the nature of the party's contribution to the enterprise or local laws will require that the party having a minority interest in the profits of the corporation be able to control certain of its activities.

Factors for consideration in allocating control Several different factors need to be considered in the course of allocating control of the joint venture, including the functional and operational objectives of the venture, the level within the corporate structure at which decisions should be made and responsibilities placed, the degree to which the success of the venture is depending upon the use of "shared control," or unanimous voting requirements and the key commitments of each of the parties and the major business and technical milestones that must be achieved in order for the venture to prosper.

Functional and operational objectives The parties should focus upon the key functional and operational objectives of the venture and attempt to allocate responsibilities in a manner that takes into account which party possesses the requisite expertise necessary for achieving those objectives. In most cases, control should be allocated in relation to their respective substantive contributions to the venture, although the non-expert party must be given the opportunity to review major actions proposed by the controlling party. Therefore, a party contributing technology to the enterprise should have the primary right to control its use in development efforts and a party with responsibility for sale and distribution of the joint venture's products should be given primary authority over various marketing matters.

Level of decision-making Serious consideration must be given to the appropriate scope of authority to be exercised by the shareholders, the directors and the officers and managers of the venture. If the success of the venture depends upon its ability to move rapidly to exploit opportunities in the local marketplace, it probably makes sense to delegate a good deal of discretion of the officers and managers of the venture, with the role of the directors and shareholders being confined to periodic performance reviews. Alternatively, in those cases where the development of the venture involves completion of a series of tasks or the purpose of the venture is the exchange of functional knowledge and expertise, an effort should be made to establish shared decision-making procedures.

Need for shared control The parties must consider how the use of shared control requirements (i.e. provisions that require that specified actions must be approved by both parties) will impact the day-to-day actions of the venture and its ability to achieve the goals and objectives specified in its business plan. While shared control might appear to be an attractive means for easing any concerns that might exist at the commencement of the relationship, the practical effect of such an arrangement might be to divert attention from the objectives of the venture to burdensome procedural measures which are of little consequence to effective management of the business. As a general rule, it is probably better for the parties to limit the areas in which the shared control requirement will apply to those which really do have a material impact on their ability to protect their interest in the joint venture.

Key commitments and performance milestones Control is best understood in the context of identifying the crucial financial and technical commitments of the venture as well as the manner in which the performance of the venture is to be monitored and assessed. In the course of developing the overall strategic business plan for the venture, care must be taken to identify material and largely irreversible commitments of cash and other resources, as well as the key technical and financial objectives of the venture. Each item provides an opportunity to assess the management and control of the venture and should require close consultation and agreement between the parties. In some cases, it may be appropriate to abandon or modify the original business plan or transfer control of the venture from one party to another.

Allocating control in "50–50" joint ventures The parties are free to establish and maintain any appropriate distinctions between their interest in the assets and profits of the venture on the one hand and their right to exercise control over the operations of the venture on the other hand. For example, parties that each contribute a relatively equal amount of cash and assets to the joint venture and agree that profits and distributions will be shared equally (i.e. "50–50") may nonetheless allocate control over certain decisions in a manner that departs from shared control. For example, one of the following procedures may be used by the parties:

(i) Each party might be issued a separate class of stock in the corporation with distinct rights to vote on the election of directors as well as on certain other matters relating to the corporation. For example, one party might be issued a class of stock having the right to elect a majority of the members of the board of directors while the other party might be issued a class of stock providing the right to approve certain major actions relating to the affairs of the corporation.

(ii) The parties might enter into a contractual agreement providing for the right of each of the parties to designate specific officers and managers of the corporation. For example, one party might be given the right to designate the chief executive officer of the corporation and the other party might be given the right to designate the chief financial or technical officer of the corporation.

(iii) Even in those cases where the parties share nominal control of the board of directors, agreement might be reached as to the delegation of decisions in specified areas to a subcommittee of the board of directors controlled by the party having the expertise in that area. For example, decisions regarding research and development activities might be left to the chief technical officer of the corporation, appointed by the party with expertise in that area, and a subcommittee of the board of directors controlled by representatives of the appropriate party.

Allocating control in "non-50–50" joint ventures While we have assumed that the parties have an equal interest in the profits of the corporation, it is often the case that one of the parties will have a greater interest in the profits or that the corporation will be owned by more than two parties, at least one of which has a profit interest which is less than that of one of the other parties. Even in those cases, it is possible to utilize one of the mechanisms described above to provide a party with a minority profit interest with effective control of the corporation. In addition, local corporate law may provide certain voting rights to minority shareholders which effectively grant them the right to veto actions taken by the majority shareholder, either directly or by virtue of the right of the minority shareholders to demand an appraisal of the value of their shares and the repurchase of their interest at the specified value by the corporation.

C. Board of Directors

Introduction The procedures for selection of the board of directors may be set forth in the charter documents of the corporation or in the agreements between the parties which describe their rights with respect to the voting of their shares. Among the issues which should be considered are the actual size and composition of the board, the procedures for allocating control of the board in those cases when the parties do not choose equal representation, and the circumstances under which control of the board may shift to one of the parties.

Size and composition of the board of directors The charter documents will specify the number of authorized directors, the requirements regarding the number of directors necessary to constitute a quorum and, if appropriate, any special requirements with respect to the composition of the board of directors, including the rights of the respective parties to designate a specified number or percentage of the members of the body. Although the parties may impose supermajority voting requirements with respect to certain matters, actions by the board of directors generally require the vote or consent of a majority of its members. Accordingly, the ability of one of the parties to designate a majority of the members of the board of directors provides it with the ability to effectively control the day-to-day operations of the venture.

Control of the board of directors As described above, any number of mechanisms exist for allocating effective control of the venture between the parties. With respect to control of the board of directors, the parties may choose among the following structural forms:

(i) One of the parties will be given the right to designate such number of directors as might be necessary in order to control the board of directors. For example, the party would have the right to designate three members of a five person board of directors.
(ii) As above, one of the parties will be given the right to control the board of directors, subject to the requirement that certain actions cannot be taken without the consent of all of the directors. Many commentators consider this to be the most appropriate structure for the joint venture form, since it vests in one of the parties the ability to initiate actions and manage the venture while preserving appropriate protections for the non-controlling party.
(iii) Although one of the parties is given the right to control the board of directors pursuant to one of the aforementioned procedures, provision may be made for a shift in control upon the occurrence of one of the events specified by the parties.
(iv) The parties may opt for shared control of the board of directors, although provision may be made for a shift of control to one of the parties at some point during the development of the venture.
(v) The parties may wish to provide for one or more mutually selected independent directors and implement a voting procedure which vests final decision-making authority in the independent directors in those situations where the parties are unable to reach a consensus. However, it may be difficult to locate persons willing to arbitrate disputes between the parties and it may be necessary to resort to the use of arbitration procedures.

The changing needs of the venture mandates that the parties give serious consideration to providing for a mechanism for shifting effective control of the board of directors to one party after the passage of a specified period of time or upon the occurrence of one of several events to be agreed upon by the parties. While such "vote switch" procedures will allow one of the parties to elect a majority of the board of directors, they need not alter the respective interests of the parties in the profits of the venture. Depending upon the circumstances, a change in control of the board of directors may be accompanied by corresponding changes in the scope of authority provided to the body in the charter documents of the venture.

D. Matters Subject to Unanimous Shareholder Approval

The parties need to strike an appropriate balance between permitting the officers and managers of the joint venture to make appropriate decisions regarding the operation of the enterprise and reserving the right, as the owners of the venture, to review and approve certain matters. The matters which are subject to the "shared control" of the owners, thereby requiring approval of both joint venture partners, should be limited to those items that are material to the performance of the venture, since making numerous actions subject to a unanimous vote will diminish, or even eliminate, the ability of the venture to quickly respond to appropriate business opportunities and changes in competitive and other environmental conditions.

While there are seemingly endless possibilities for the listing of items which might require the unanimous consent of the shareholders, or a supermajority vote of the board of directors, the following matters are typically considered:

(i) capital expenditures or indebtedness in excess of a specified dollar amount at any one time, other than as provided in the annual operating budget;
(ii) sale of asset the value of which exceeds a specified dollar amount at any one time or the aggregate value of which exceeds a specified dollar amount within any six month period;
(iii) creation of material liens, mortgages, encumbrances or other charges of any kind on any asset of the venture or guarantees of payment by or of performance of the obligations of any third party;
(iv) approval of the annual operating budget and annual capital expenditure budget of the joint venture, as well as any amendments thereto;
(v) any commitment of the joint venture which either creates a liability in excess of a specified dollar amount or which obligates the joint venture to perform under any arrangement, contract or agreement which is not within the ordinary course of its business;
(vi) any amendment to the articles of incorporation or bylaws of the joint venture corporation;
(vii) any issuance of additional shares of the joint venture corporation, declaration of dividends or reinvestment of earnings at a time when the corporation is otherwise permitted by law to pay dividends;
(viii) any agreement or material transaction between the joint venture and either (a) a shareholder or any affiliate of a shareholder or (b) a director or executive officer of the joint venture;
(ix) adoption of or change in any major policy of the joint venture relating to operations, manufacture, sale of products, or changes in product lines;
(x) a dissolution, liquidation, merger, consolidation, other business combination or sale of substantially all the assets of the joint venture.

E. Dispute Resolution Procedures

Whether the joint venture is a "50–50" company or one party owns more than 50% of the venture but has agreed to supermajority voting provisions, the possibility of "deadlock" must be taken into account in structuring the enterprise. Although it may be somewhat awkward to spend inordinate amounts of time before the joint venture is even formed in debating the consequences of any failure of the parties to agree, some procedures for resolving a dispute between the parties without having to resort to the costs and aggravations of litigation are usually desirable.

When an event occurs which is considered to be a "deadlock" under the terms of the agreements between the parties, a number of different dispute resolution mechanisms can be used. Among the more common methods are mediation and arbitration, bilateral discussions, swing-vote directors, put–sell options and, in some extreme cases, termination and dissolution of the venture.

Mediation and arbitration Some parties will rely upon mediation and arbitration involving independent third parties as a means for resolving disputes. This procedure may work fairly well when the dispute involves questions of fact or law or can be resolved by reference to the precise terms of the joint venture agreement itself. However, when the dispute arises over issues of policy or business judgment, mediator or arbitrators can only assist the parties in communicating, but are not competent to render a decision as to the proper business strategy.

Bilateral discussions The parties may agree that any dispute among representatives of the parties at the venture level will be passed up successive levels of the internal "ladder" of each organization, eventually reaching to the office of the chief executive officer of each party. Hopefully, the embarrassment of bringing more and more people from the parent organization into the affairs of the joint venture will encourage the officers and directors of the venture to reach some accommodation.

Swing-vote directors One easy method of resolving a deadlock is by providing for the presence of a tie breaking vote at meetings of the directors and shareholders. The tie breaker can be an officer or director elected alternately each year by one side under a bylaw requirement that the officer be elected from among the directors representing a different class each year. The swing-vote director may also be a third party in whom the partners have confidence.

Put–sell options Both parties may be given a right to "put" their shares to the other party at a fixed price and the other party must either agree to purchase the shares at the price fixed by the party making the "put" or sell its own shares to the first party at the same price. Options of these types are more fully discussed below in the context of "termination," however, they are usually only viable when both parties have the financial resources to fund a purchase or when the parties are able to reach some agreement as to extending credit on the transaction.

Termination and dissolution In extreme cases, the parties may agree to terminate the joint venture, dissolve the company and liquidate and distribute the assets in the manner provided in the articles of incorporation.

VI. Operating the Joint Venture

A. Introduction

Although a good deal of the emphasis during the period of negotiating the terms of the joint venture is placed on issues relating to capitalization and management, it is important for the parties to remember that the venture company will become an independent operating entity with a life of its own. As such, the joint venture documents should cover the key points relating to operation of the venture, including the responsibilities for the functional activities of the company (e.g. research and development, manufacturing and/or distribution); operational matters, such as legal compliance, insurance, staffing and other similar issues; accounting and financial reporting matters; and the allocation and distribution of joint venture income.

B. Functional Activities of the Joint Venture

The key functional activities of the joint venture, which might include research and development, manufacturing, sales and distribution and service, may either be conducted by personnel employed directly by the joint venture or by one or both of the parties pursuant to ancillary contractual agreements between the party and the joint venture. For example, one party may agree to contribute a portion of its existing technology to the venture as consideration for its shares and then enter into a development agreement with the joint venture entity to perform the additional work with the technology required in order to create new products which can be marketed by the joint venture.

When the joint venture structure uses ancillary agreements with the parties to fulfill the functional needs of the venture, the negotiation of each agreement generally raises the same sorts of issues as would arise in any other context. However, if the functions are to be performed by joint venture personnel, many of whom may be loaned or assigned to the venture by the parties, it becomes important for the parties to lay out their expectations in the joint venture documents. For example, if joint venture employees are to be involved in development work, it is usually a good idea for the parties to agree upon a development plan and to include covenants in the joint venture agreement regarding those matters which typically arise in a technology agreement, such as ownership of inventions and duties to perfect and protect all of the legal rights of the joint venture in the technology.

C. Operational Activities of the Joint Venture

The operational activities of the joint venture include a number of matters relating to the conduct of the internal affairs of the enterprise, as well as compliance with all applicable laws and regulations that may be relevant to the continued existence of the entity. As with any other business enterprise, matters such as staffing, acquisition and leasing of real and personal properties, insurance, business licensing and working conditions will need to be considered. One of the parties may be given the primary responsibility for managing a specific operational area, or the joint venture agreement may simply set out a number of covenants and agreements which obligate both parties to cause the joint venture to take certain actions.

D. Accounting Matters and Financial Reporting

Accounting and financial reporting is an essential part of monitoring the progress and management of the joint venture. The parties must reach agreement on the establishment of accounting and internal controls, including the selection of an independent public accountant to audit the books and records of the venture. In addition, the shareholders' or joint venture agreement should describe all of the financial information that is to be prepared and distributed to the parties during the term of the venture, since these reports will provide the basis for budgetary and capital allocation decisions, permit discovery of deviation from goals and objectives specified in period business plans and, finally, allow the parties to evaluate the performance of various product and marketing strategies or the progress of technology development efforts.

E. Allocations and Distributions of Joint Venture Income

Compensation to the parties in a joint venture can take a number of tangible and intangible forms. While certainly a good number of joint ventures are pursued in order to assure access to new technologies, improved production methods, new distribution channels and innovative management skills, it is the promise of exceptional cash returns on invested capital that often serves as the overriding consideration in analyzing any specific business opportunity.

The activities of the joint venture will presumably generate income to the enterprise. Moreover, any sale of assets which might be owned by the joint venture will also provide cash which would be available for distribution to the shareholders or for reinvestment in the activities of the business. Since external financing opportunities for the joint venture may be limited, it is important for the parties to carefully consider the policies that are to be followed with respect to distributions of operating income and the internal application of funds. As such, it is important to maintain an ongoing capital budgeting process which ensures that the enterprise will not distribute funds which are better utilized for expansion and related purposes.

While it is generally the case that the shareholders will share in all distributions made by the enterprise in proportion to the value of their respective capital contributions, agreement may be reached with respect to the timing of distributions and any priorities that might exist between the shareholders in the event that the amounts available for distribution are less than originally contemplated. For example, two classes of securities may be issued containing different provisions with respect to "dividends" paid by the enterprise. One class of stock might be entitled to a preferred return out of operating income for a given period prior to any distribution to the other shareholder. Similarly, distributions upon liquidation of the enterprise may also reflect a preference for one of the shareholders.

Since shareholders of a corporation generally have limited liability with respect to the debts and obligations incurred by the corporation, the law will carefully guard the rights of third party creditors to prevent distributions of cash and other assets to shareholders which impair the solvency of the corporate form and prejudice the rights of creditors. Dividends and redemptions are not permitted unless the corporation meets certain financial tests immediately following the distribution. Accordingly, it is important to consider the requirements of the jurisdiction in which the corporation is organized and, in some cases, directors of the corporation will be held liable for improper distributions of corporate assets.

VII. Restrictions on Transfers of Ownership and Withdrawals

A. Introduction

In most cases, the choice of the joint venture format for the specific business relationship is driven, at least in part, by the belief that the parties have the compatibility and financial stability which are expected to be the major elements contributing to the prospects of future success for the venture. Accordingly, the parties will wish to ensure that each of them remains committed to the venture by establishing appropriate procedures relating to the transfer of shares evidencing equity ownership in the enterprise. Similarly, the parties will need to consider the possibility that one of them will want to withdraw from the joint venture prior to the end of the specified term.

B. Restrictions on Transfers of Ownership

As a general rule, the parties will generally agree to a strict prohibition on any sale or transfer of the shares for a specified period of time, usually corresponding to the initial term of the venture. Precluded transfers will include not only an outright sale of the shares, but also any pledge or other encumbrance of the shares, although in some cases a party will be permitted to transfer the shares to a successor corporation or a wholly-owned subsidiary of the party. Waiver of the restriction on transfer would require the consent of the other party.

After the restriction period has ended, the parties will usually agree to a "right of first offer" or a "right of first refusal." The right of first offer obligates the party wishing to find a purchaser for its shares to first offer them to the other shareholder on the terms at which the potential seller would like to complete the transaction. If the other shareholder is unwilling to purchase the shares on those terms, the seller would them have a specified period of time to find a purchaser on terms no less favorable to the seller than those originally offered to the other shareholder. The right of first refusal is slightly different, since the selling shareholder must first have an offer to purchase the shares from a third party at the time the transaction is proposed to the other shareholder.

C. Pre-Termination Withdrawals of a Joint Venture Partner

There may be a variety of situations in which one of the joint venturers will be allowed, or required, to withdraw from the venture prior to the date upon which the parties originally contemplated that the relationship would terminate. For example, a fundamental change with respect to one of the parties, such as a change in ownership of the party, may cause a sufficient amount of uneasiness for the other, or "non-changing" party, to trigger a right in the hands of the non-changing party to purchase the shares of the other party.

Another situation where withdrawal may be appropriate is when a party which is not actively involved in the management of the venture is dissatisfied with the level of performance which the venture has achieved in the hands of the other party and seeks to have its interest purchased by the party that is in control of the venture.

Withdrawal of one of the parties and continuation of the business by the other party is obviously an alternative to simply dissolving and liquidating the entity. In some cases, continuing the business without both of the original parties is simply not a viable alternative, particularly when the withdrawing party takes with it certain resources, technical or otherwise, which are of fundamental importance to the success of the venture. However, since dissolution and liquidation often diminishes the value of the interests of both parties, some effort will generally be made to establish equitable procedures for liquidating the interests of the parties without terminating the business itself.

A number of procedures might be used for facilitating the withdrawal of one of the parties from the joint venture, either by the sale of shares of the withdrawing party to the remaining joint venturer or by permitting the withdrawing party to secure some liquidity for its investment through transactions involving third parties. Among the provisions which might be considered by the parties are the following:

(i) granting the right to either party, at its option, to liquidate and dissolve the joint venture. If such a right is provided, the parties should also include a valuation mechanism and an option in favor of the other party to purchase the interest of party that wishes to withdraw, thereby allowing the business activities of the venture to continue.
(ii) permitting either party to sell its shares to a third party, perhaps in a public offering, subject once again to a right of first refusal in favor of the other party. An interesting variation on this alternative would permit either party to obtain an offer for the sale of the entire venture, with the other party having the right of first refusal to buy the interest of the withdrawing party on the same terms and conditions presented by the third party.
(iii) use a buy–sell option which permits one of the parties to set a price on its shares and then allows the other party to choose between buying the shares or selling its own shares at the stated price.
(iv) one of the parties may be given the right to sell its interest in the venture to the other party after a fixed period of time at a price to be fixed by an independent appraisal. The other party must either purchase the shares of dissolve the venture. This method may be utilized when a certain milestone has not been achieved by a date specified by the parties.
(v) in certain situations, one of the parties may be provided with the unilateral right to purchase the shares of the other party after a fixed period of time, with the price to be determined at the time the option is exercised.

VIII. Duration and Termination of the Joint Venture

A. Introduction

The joint venture documents must include provisions which specify the duration of the enterprise, the events which may lead to the termination and dissolution of the joint venture and the manner in which the assets and resources of the entity will be redistributed to the parties upon liquidation. The duration, or "term," of the joint venture will depend upon the specific goals and objectives of the parties and the amount of time which will be needed in order to achieve the required investment return. Termination and dissolution may occur either at the end of the fixed term or upon the occurrence of specified events prior to the termination date otherwise specified in the venture documents. The process of liquidation is generally governed by applicable law, although the parties do have some freedom to specify how the remaining assets of the joint venture will be allocated between them after the claims of creditors have been satisfied.

B. Duration of the Joint Venture

A number of issues need to be considered with respect to establishing the duration of the joint venture. In concept, a successful joint venture may enjoy perpetual existence and the parties will seek ways in which the basic relationship can be expanded to include new functional, technical, product and market opportunities. However, at the outset the venturers may well be satisfied with establishing an initial term for the enterprise which fits within the realistic business horizon for each of the particular activities to be undertaken and provides the parties with a reasonable opportunity to achieve a proper return on their investment of time, capital and other resources.

C. Election to Terminate the Joint Venture

In spite of the best intentions of the parties, it is often necessary to consider the circumstances upon which one or both of the parties may elect to terminate the enterprise. As noted above in the context of describing the events which might trigger the right of one of the parties to "buy-out" the interest of the other party, it is likely that the parties will contemplate the termination of the venture upon the occurrence of a fundamental change relating to one of the parties, a material breach by one of the parties of its obligations to the relationship or the failure of the venture to attain specified technical or financial milestones.

D. Liquidation and Dissolution of the Joint Venture

Termination of the joint venture leads to liquidation and dissolution of the business entity, a process which is largely dictated by applicable law and any agreements between the parties regarding the distribution of specified assets of the joint venture. As noted above, the liquidation provisions can be tailored to suit the specific activities of the joint venture. For example, when one or more of the assets of the joint venture are best suited to use by one, but not both, of the parties and sale of the asset and distribution of the cash proceeds makes little sense, then an effort will be made to ensure that such assets are distributed to the party to whom it has the most value.

When the assets of the joint venture consist not only of cash, but also includes such things as equipment, property and intangible property rights, it is important to establish an appropriate valuation procedure. Tangible assets may easily be valued by reference to book or replacement values. However, intangible property rights often require the use of independent appraisers, although it may be difficult to value those rights in isolation and their value to one of the parties may be greater than the amount that might be received if the rights were sold to a third party and the proceeds distributed to the parties.

The business of the joint venture will largely cease during the period required to liquidate and dissolve, although the management of the enterprise, or any other persons appointed to supervise the liquidation of the venture, will continue to collect any receivables and pay any liabilities. Appropriate provision will be made for the contingent claims of creditors and a final accounting will be made which records each of the venture's assets and permits the preparation of all final income tax returns. In cases where a dispute arises regarding the final accounting, it may be necessary for the parties to engage third party experts and mediators to assist in valuation and similar matters.

Even though the formal joint venture relationship is terminated and the entity is liquidated and dissolved, certain agreements between the parties will survive the completion of the formal business relationship. Most importantly, the parties will remain obligated under the terms of any agreement with respect to the confidentiality of proprietary information that was exchanged or otherwise received in the course of the joint venture. In most cases, each of the parties will be required to return all documentation, including copies thereof, received from the other party which may constitute confidential information. Also, it is important for the parties to reach agreement as to how customers of the joint venture are to be handled in the future.

In addition, consideration must be given to satisfying the claims of creditors and other third parties which may arise following the dissolution of the venture. Although reference must be made to applicable law regarding the liability of shareholders for the return of liquidating distributions, it is appropriate to consider maintaining insurance to cover any liabilities associated with defective products, labor claims or, if appropriate, environmental matters. In the event that claims are successfully asserted, the parties are presumably free to allocate responsibilities between them in accordance with various indemnification and contribution agreements.

8 Collaborative Business Relationships other than Joint Ventures

I. Introduction

While contractual joint ventures are the primary focus of this part of the work, it is important to remember that collaborative business relationships may take a number of different forms. As mentioned in the previous chapter, contractual joint ventures may create one or more complex relationships between the parties even though no separate legal form business entity is used to carry out the specified activities. Among the more common areas in which a business collaboration may be created is research and development, manufacturing, and sales and distribution.

II. Research and Development Arrangements

A. Introduction

Research and development arrangements can take a variety of different forms. In some cases, the research agreement is simply a "fee-for-service" arrangement which calls for the sponsor to pay a fee to the researching party to conduct specified work over a fixed period of time. In other situations, the research agreement is one of several agreements in a much more complex set of economic relationships between the parties. Examples of when a research agreement may be used are as follows:

(i) Research agreements between private sponsors and universities provide funding for basic research in areas of mutual interest. Universities receive fee income for the research work and, in some cases, royalties from the commercialization of products derived from the research work. The private sponsors gain access to "cutting-edge" basic research in new areas and the skills of researchers in the academic community.
(ii) A larger firm engaged in the distribution of technology-based products may enter into a research agreement with a smaller firm to provide funding for the smaller firm to develop new products which can be incorporated into the large firm's product line. The smaller firm will receive a research fee and revenues from the sale of the products to the larger firm.
(iii) Joint venture arrangements may include a program of research activities which are to be jointly engaged in by the joint venturers for the benefit of the joint venture and each of the participants. For example, both parties may conduct re-

search to develop products which will be marketed by the joint venture in a specified geographic territory. In addition, the joint venture might license the technology to each of the joint venturers for use in areas which fall outside of the joint venture's mission.

(iv) A large firm may make an investment in a smaller firm to permit the smaller firm to continue its existing research and development activities. In addition, the smaller firm may enter into a specific research agreement with the larger firm to develop products of specific interest to the larger firm, with the larger firm providing any required additional financing in the form of loans or advances against projected royalties.

In negotiating any research and development arrangement, the parties must give consideration to the following issues: the scope and content of the research program, the budget for conducting the research program, the manner in which the research program will be staffed by the researching party, the contributions of technology, services and materials which may be made by the parties in order to properly complete the research program, the allocation of ownership and usage rights with respect to the technology developed during the course of the research program and any other arrangements with respect to further work relating to the commercialization of the technologies and products created during the research program.

B. The Research Program

Introduction The central element of any research arrangement is a full description of the proposed research program. This should include not only a statement of the scope and content of the program, but should also deal with specific items such as budgeting and payment, staffing considerations, recordkeeping procedures, mechanisms for transferring technical information between the parties and, in many cases, the formal provisions relating to management of the research program.

Scope and content of the research program The research agreement generally begins with a statement of the proposed scope and content of the research program. This will usually consist of, at a minimum, an identification of the field of activity in which the research is to be conducted (e.g. research relating to the treatment of cancer using a specified technology); a listing of the goals and objectives of the research program, such as development of products suitable for clinical testing, and some general specification of the major tasks that will be undertaken in the course of the research program. Budgetary and payment matters are also included, since the amount of funds available will bound the scope of the research program.

The research budget The research budgeting process, including the profit element which may be built into the compensation scheme, can be one of the most complex components of structuring the research agreement. It is important to consider the allocation of the direct and indirect costs of the project, the amount of any profit which may be earned by the researching party, the effect of any overruns in spending on the research project and accounting and payment procedures.

The research budget, which defines the amounts which will be paid to the researching party under the agreement, may be divided into several different categories. The first item would be those direct costs incurred by the researching party in connection with the research activities. The second item would be a charge for a proportion of the general overhead of the researching party which should be allocated its activities in connection with the research project. These indirect costs are usually stated as a percentage of the direct costs. The third item would the costs of acquiring specialized capital equipment and facilities. Finally, the budget may include a "profit" figure, which is often expressed as some percentage of one or more of the aforementioned items.

Preparing an accurate estimate of the anticipated expenses associated with the proposed research work can be one of the most difficult tasks facing the parties, since it is quite difficult for them to predict the cost of completing all of the required research work, particularly when the project may extend over a long period of time. The research agreement may either contain a preliminary research budget for the initial stage of the project or may leave it to the parties to agree upon a budget within a specified period of time following execution of the agreement. While some research agreements will actually include a budget for the entire term of the project, it generally makes better sense to review the budget on an annual basis in multi-year projects.

The research budget can either be quite general, leaving the details to the discretion of the researching party, or set out in great detail with a number of different line items. Whatever the method, the parties must consider how the costs associated with the following types of items should be allocated: salaries and wages, contract labor, fringe benefits, facilities and equipment related expenses, supplies, development and prototype materials, freight and transportation, training and education, travel expenses, data processing costs, license fees, insurance, professional services, depreciation and amortization of capital acquisitions, sales and use taxes, and periodic lease payments under capital or financing leases. In addition, the research budget may include an allocation for indirect costs, such as overhead, incurred by the researching party in performing under the contract.

Staffing of the research project The research agreement should describe the understanding of the parties as to how the research project will be staffed. In a number of cases, the research agreement will contain an explicit statement regarding the minimum amount of time, often expressed as "person-years," that personnel of the researching party will devote to the research project, as well as the percentage of such time that is to come from internal resources. For example, the research agreement might state that the researching party agrees to devote at least twenty person-years each year to conducting the research work, at least 75 percent of which will be from internal resources, rather than from outside consultants. The research agreement might also refer to a list of the specific persons who are to work on the project, including their respective time commitments.

Two additional issues that often arise with regard to the staffing of research projects are the availability of certain key researchers and the need for the researching party to recruit other scientists with experience and skills in areas that may not then be available among the researching party's existing personnel. If the sponsor is relying on a particular researcher to manage, or do work on, the research project, the terms of participation should be set out in the research agreement. Also, if the researching party is required to recruit new personnel, the costs associated therewith, as well as the consequences of failing to locate and hire any of the necessary persons, should be dealt with in the agreement.

The research agreement should also address whether the researching party will have the right to subcontract any or all of the research work which it is obligated to perform.

In some cases, the right to subcontract may be conditioned upon the prior approval of the sponsor. If the researching party contracts with persons, such as outside consultants or other business organizations, for services relating to the research project, the research agreement should provide for procedures to ensure that subcontractors will agree to forfeit any rights that they might have in the technology which they might develop during the course of their work. Also, the sponsor may seek indemnification from the researching party as to any claims made by subcontractors arising out of their work on the research project.

Management and reporting procedures Given the complexity of most research and development arrangements, particularly if an effort is being made to discover new commercial applications through refinements of untested basic technology, it is usually important to develop various procedures relating to management of the overall project and reporting the progress of the research effort to the sponsor. The actual content of the management procedures will depend upon the persons involved in the project, as well as the overall importance of the project to the parties.

A relatively simply management procedure is to require that each party designate a person to serve as the project manager with respect to the specific research program. Each project manager would have authority on behalf of his or her respective party to make day-to-day management decisions with respect to the research project. The project managers would meet frequently during the research project to discuss the progress being made, as well as the problems being encountered, by the researching party. As a general rule, the research agreement should provide for meetings no less frequently than monthly or quarterly.

Another common method for managing a research project is a committee comprised of representatives named by each of the parties who are reasonably acceptable to the other party. The committee will usually meet at least once each year during the research program and at least once during the year following the completion or termination of the program. However, it is more than likely that the committee will meet more often, particularly during the early stages of the research program. At its meetings, members of the committee will formulate the overall objectives of the research program, monitor the costs and expenses incurred by the researching party in conducting the program and, as appropriate, consider the progress of the researching party through the various stages of the research program.

Technology exchange and technical assistance Although coordination at the administrative level is always important to the research effort, technology exchange between the parties, including technical assistance in understanding the new technology conceived during the research program is absolutely key to the success and value of the work. In addition to the high-level technical meetings referred to above, the parties should establish a regular program for ensuring that the results of the research work are properly documented and smoothly communicated between the two organizations.

The research agreement generally contains a broad covenant regarding the intent of each party to exchange all the technical information developed during the course of the research program. In addition, the parties will agree to provide all the technical assistance and training to the other party which may be required in order to ensure that the exchange of technical information has been successful. In particular, the parties will agree to exchange all of the information necessary to obtain governmental and regulatory approvals for new products which use or incorporate the technology developed during the research program. If information which is owned by third parties is required in order to effectively make the exchange of technology, the parties will undertake to obtain the permission of such third parties, generally licensors, to have the information disclosed.

The research agreement should establish a schedule that anticipates the disclosure to the sponsor of all information that relates to new inventions, technical information and other types of materials which the researching party may conceive or discover in the course of the research project. Such disclosures, which are to be made in writing, should be made no less frequently than on a quarterly basis, although any material developments should probably be conveyed to the sponsor as soon as practicable. The obligation to disclose should be made subject to any third party rights which might restrict the researching party's ability to disclose any such information to the sponsor; however, the researching party should disclose such limitations to the sponsor.

C. Contributions to the Research Program

Introduction The conduct of the research program often depends upon the availability of certain resources, technical and otherwise, which are under the ownership or control of one of the parties prior to the commencement of the research work. The research agreement will normally include a description of the various contributions which each of the parties will make to the research program with respect to existing technologies, raw materials and other items.

Contributions of existing technologies to the project The research work often builds upon technology that has been previously developed by one or both of the parties. If this is the case, provision should be made for contributing such technologies, either by assignment or license, to the research program. This is typically done by granting a non-exclusive license relating to the technology to the other party, with the proviso that the technology can only be used by the licensee to perform its obligations under the research agreement. The licensor should also agree to provide technical assistance to the licensee with respect to the use of the licensed technology and, if appropriate, access to any improvements or enhancements of the basic technology which may be developed by the licensor during the research program. As noted below, at the end of the research program, the licensed technology will usually revert back to the owner, along with any improvements which might have been developed during the course of the research program.

Participation of sponsor personnel in research program There may be situations where the sponsor wants to supplement the resources of the researching party by makings its own personnel and support services available for use in the research program. In such instances, the research agreement might provide the sponsor with the right, but not the obligation, to supply its personnel and related services to the research program if it believes that such actions will facilitate the timely and successful completion of the research work. If the sponsor does choose to supply personnel, the researching party may be given a right to review the qualifications of the persons who are to work on the project and such persons will be required to execute non-disclosure agreements. The expenses of providing such personnel and services will usually be borne by the sponsor, unless the assistance is made necessary by a failure in performance by the researching party.

D. Representations and Covenants

Introduction As with any commercial agreement, the research agreement will include various representations and warranties from both parties as to the performance of their obligations under the agreement. Also, the research agreement will set out a variety of covenants from the parties with regard to their activities in relation to the research program.

Representations and warranties Each of the parties will make representations and warranties regarding a variety of matters. For example, both parties will be called upon to make representations and warranties regarding its right and authority to enter into and perform its obligations under the research agreement and the absence of any conflict between its performance under the agreement and any provision of any statute or contractual agreement relating to the party's activities. Also, if a party is contributing technology for use in the research program, it will usually make a representation relating to its good title to such technology. Another common representation relates to securing any governmental approvals required for a party to complete all of its various obligations under the research agreement.

Researching party's obligation to conduct research work While there will no doubt be a number of economic inducements for the researching party to perform all of its obligations under the research agreement, the sponsor will nonetheless insist upon a specific covenant from the researching party regarding the conduct of the research program. As a general rule, the researching party will represent and warrant that it will use its "best efforts" or "reasonable efforts" to perform all of the research work necessary to complete the research project on a timely basis. This general representation may be supplemented by objective milestones which can be used as a basis for permitting the sponsor to terminate the research agreement.

Restrictions on competitive activities One or both of the parties may have concerns regarding the possibility that the other party may enter into activities which conflict with their obligations to fulfill the objectives of the research relationship. For example, as noted above, the sponsor will have concerns that the researching party not engage in any other research project which may have an adverse effect upon the staffing expectations of the sponsor or which may result in the transfer of the sponsor's technology to third parties. In turn, when the researching party is relying upon the sponsor's interest in further commercialization of the products developed during the research project as a source of further revenue (e.g. the sponsor has the right to an exclusive royalty-bearing license), it will want assurances that the sponsor will not contract with others for research which might result in a competitive product.

Restrictions on competitive activities begin with agreement from both parties that the research program will be the exclusive mechanism for research work designed to achieve the stated goals and objectives. In addition, if the researching party is relying upon royalties from the sponsor's commercial exploitation of the research work and the sponsor begins selling competitive products, the researching party may be protected by imposing an obligation on the sponsor to pay a royalty based on sales of the competitive products; however, the sponsor's obligation will usually only apply to competitive products which are introduced within a stated period of time and only if the sales of the original product have actually declined.

Confidentiality Any technical information exchanged under the research agreement will be received by the other party under an obligation to maintain it in confidence and not to use the information other than for purposes which are consistent with the objectives of the research relationship. The amount of detail which is used in the research agreement to describe this obligation of confidentiality will vary depending on the circumstances. In some cases, the party will simply agree to use the same degree of care that it uses to protect its own confidential information of like character. But, in other situations, extensive procedures for protecting any trade secret information and patentable inventions will be included in the research agreement or in a separate confidentiality agreement. For example, the party may be required to have each of the persons involved in the research project execute a non-disclosure agreement relating to information developed during the project.

E. Ownership and Protection of the Technology

Introduction The research agreement should document the understanding of the parties regarding the ownership of any inventions, technical information or other items which may be created or discovered in the course of the research program. The agreement should also set out the procedures for protecting the technology, including the steps that will be taken to perfect any statutory rights and for ensuring that trade secrets will be maintained in confidence. It is important that both parties cooperate in this area, particularly since it is common for the non-owner to be given various rights to use the technology following the completion of the research work.

Ownership of project-related technology As a general rule, the research work will produce a broad variety of tangible and intangible products which may be eligible for protection under the intellectual property laws. For example, the research program may result in the creation of inventions that would be eligible for patent protection, know-how and technical information which might be covered under the trade secret laws, and new or improved biological materials which can be used in various applications within and outside the field of activity. In any case, the research agreement should clearly describe the manner in which ownership of the inventions, technical information and any other rights arising out of the research project will be vested.[1]

Perfection of intellectual property rights Whatever decision is made regarding ownership of any of the technology arising out of the research project, the agreement must provide for cooperation between the parties relating to perfecting and maintaining the rights of the owner in the technology. This generally means that the non-owner will be asked to provide all of the necessary assistance in preparing and filing applications for statutory protection (e.g. patents, copyrights and trademarks), as well as implementing procedures designed to protect the secrecy of technical information and other trade se-

[1] Failure to provide for ownership of the research products will expose the parties to the uncertainties of litigation. Courts are likely to focus upon the "cooperative efforts" of each of the parties to the research agreement and apportion the rights between the parties on an equitable basis, often leading to joint ownership of inventions and information contrary to the intent of one of the parties.

crets which were developed during the research program.[2] Also, procedures should established for protecting the intellectual property rights from infringement by others.

F. Uses of Project-Related Technology

Introduction Issues relating to ownership of project-related technology should be distinguished from the permitted uses of such technology by the parties. In almost all cases, the party that does not have ownership rights in the technology will be granted the right, in the form of a license, to use the technology for specific purposes described in the research agreement. The license may simply allow the licensee to use the technology for internal work, without the right to develop and sell any commercial products which embody the technology. On the other hand, the owner may be willing to grant the licensee the exclusive right to use the technology in the sale and distribution of new products.

Uses of sponsor-owned technology by researching party As noted above, in most cases the sponsor will be granted ownership rights in the technology developed using the funds that the sponsor provided to the researching party. Obviously, it is possible for the arrangement to be structured as a simple "fee-for-service" contract, in which case the researching party would have no residual rights in the new technology and its compensation would be limited to the research fees paid by the sponsor. However, it is more likely that the research fees will be supplemented by some additional rights with respect to the new technology and/or the sponsor's attempts to commercialize any new products.

In almost all cases, the researching party will be given the right to use the project-related technology for internal purposes which are not related to developing commercial products. This type of right allows the researching party to use the technology in its laboratories for experimental purposes and serves as a means for the researching party to advance its technological capabilities. These rights, which take the form of a non-exclusive license, will generally be granted on a royalty-free basis.

In some limited cases, the sponsor may be uninterested or unable to use the technology outside of the field of activity set out in the research agreement. In such a situation, the sponsor might be willing to grant the researching party a license to use the technology in areas which are not of interest to the sponsor. For example, a sponsor from the United States who contracts with a Japanese company to develop technology which the sponsor intends to use in enlarging its domestic product line might be quite happy to license the technology to the Japanese company for exploitation in Japan, provided that the Japanese company agrees to pay royalties and to limit its activities to Japan.

Uses of project-related technology owned by researching party University research arrangements often provide that the university will retain ownership of all the intellectual property rights developed during the research program. However, since the sponsor would not have been willing to fund the research unless it perceived some commercial

[2] In most cases, the party which is to own the technology will have responsibility for perfecting patent rights; however, there may be situations in which the other party is given responsibility in a particular area. For example, when the researching party is in Japan and the sponsor is in the United States, the sponsor may retain ownership of the technology but allow the researching party to apply for patents in Japan, perhaps as part of an arrangement by which the researching party receives an exclusive license in Japan.

opportunity, the research agreement will generally contemplate an ongoing relationship between the parties in which the sponsor contributes additional funds and resources in order to commercialize the products of the research program. The resultant relationship can take a number of different forms, with the actual structure being a function of the types of products and technologies developed by the researching party and the skills and objectives of the sponsor.

One of the most common methods for providing a sponsor with access to the technology developed during the research project is to provide the sponsor with a right of first refusal to obtain an exclusive license to the technology. The procedure begins with a requirement that the researching party provide the sponsor with all of the technical information regarding the technology which may be necessary for the sponsor to determine whether or not to exercise the right and obtain the license. The sponsor would then have a fixed period of time in which to make a decision. If the sponsor decides not to exercise the right, the researching party would be free to pursue other licensing arrangements and, in some cases, the sponsor would be precluded from marketing or selling competitive products.

Another structure that is commonly used when the researching party is primarily engaged in product development activities and the sponsor is primarily engaged in distribution activities is to provide that the sponsor would have the option to elect to act as the sole distributor of certain new products developed as a result of the research program. This type of arrangement is common in the software area, where certain companies have the skills to write new programs and other companies serve as distribution and marketing organizations. The distribution agreement would contain whatever terms and conditions are agreed upon by the parties and would also include such licenses of the technology as may be necessary for the sponsor-distributor to market the products.

Rather than providing for a license to the new technology between the parties for the purpose of commercialization, another structure which might be used is the creation of a joint venture between the parties for the purpose of manufacturing and marketing the products developed during the research program. In this case, the party with ownership rights to the technology would license or assign the technology to the joint venture and one or both of the parties would agree to provide additional financing, as well as the required manufacturing and distribution capabilities, to exploit the technology. Both parties might be given the right to terminate the joint venture by purchasing the interest of the other party for a price based on a predetermined formula.

G. Termination of the Research Arrangement

Introduction In general, the research program will continue until the development work has been successfully completed, the research budget has been expended or the specified term of the research project expires. If the research program has been properly planned and conceived, the work will be successfully completed during the original term of the research agreement and without requiring the expenditure of amounts in excess of the original budget. However, as with any commercial agreement, consideration must be given to the possibility that the research arrangement will, for a variety of reasons, be terminated before the research work is completed.

Events causing termination of the research arrangement Perhaps the most common reason for terminating the research program is the realization that the research party will not be able to successfully complete the project or that the products which are to produced from the development work will fail to have sufficient commercial value to warrant further investment by the sponsor. In such cases, the sponsor will have a right to terminate the research program. However, the sponsor's right of terminate may not be used arbitrarily. In most cases, the sponsor will be obligated to use its reasonable business judgment in determining that the research project is no longer viable and will almost always be required to consult with the researching party before formally terminating the research agreement. Often the parties agreement that a third-party will be engaged to mediate any controversy regarding the continued viability of the project.

If, for any reason, the sponsor is unable to fund all of its obligations to finance the research program, one or both parties will have the right to terminate the development work. The manner in which accrued expenses and ownership rights to any technology are to be allocated between the parties in such circumstances will depends upon the understanding of the parties at the time that the research agreement was first executed. For example, if the failure to provide additional financing amounts to a default by the sponsor with respect its original agreement to fulfill the requirements of the original budget, the researching party may have some cause of action against the sponsor, as well as the right to retain title to any new technology developed during the course of the abbreviated research program.

Depending upon the circumstances, the research agreement generally includes various other events which might lead to the termination of the research program. Most agreements provide for termination upon the expiration of the term originally specified in the agreement or the expenditure of all of the funds provided for the research program by the sponsor, although these events really are tantamount to concluding that the research program has failed. Other events of termination include the occurrence of a bankruptcy or similar proceeding with respect to one of the parties, which will usually create a right in the non-effected party to terminate the agreement. Similarly, a material breach of the agreement by one of the parties will create a right of termination in favor of the non-breaching party. Finally, the agreement will permit the parties to terminate the agreement by mutual consent.

Rights and obligations of the parties upon termination The research agreement should describe the various rights and obligations of the parties upon termination of the agreement. In most cases, the following matters will be covered:

(i) As a general rule, termination of the research agreement should not effect the right of either party to receive payment of accrued amounts owed to it, either in the form of a research fee or a royalty, at the time the agreement is terminated. However, in many cases, parties will have a right to offset any accrued obligations against amounts which might be owed to such party by the other.

(ii) Ownership rights and permitted uses of the technology contributed to, or developed during, the research program should be allocated in the manner otherwise provided for in the research agreement. All patents and technical information arising from the conduct of the research work should be delivered to the owner. The agreement should specify the delivery date and cover how the expenses associated with such deliveries are to be borne.

(iii) Covenants of the parties with respect to protecting the confidentiality of information developed over the course of the research work should survive termination for the period provided in the agreement. Typically, the parties will be obligated to maintain the confidentiality of such information for periods extending anywhere from three to five years following termination, subject to any prior lawful disclosure by either party.

(iv) If the arrangement involves a license to sell products which may have been developed over the course of the research program, the licensee will generally be given the right to sell any remaining products held in stock at the time of termination and will have an obligation to continue to pay royalties to the licensor. The licensee must continue to sell the products at the price at which products were offered prior to termination. If service arrangements regarding the products were in place, the agreement should address how customers will be serviced following termination.

III. Manufacturing and Distribution Relationships

A. Introduction

Once the research and development work has been completed, attention turns to those functional activities, manufacturing and distribution, which are necessary for the commercial exploitation of the underlying technology. Although it is certainly possible for one firm to possess the financial and technical resources to permit internal production of the products and direct sales of the products to end users without the use of intermediaries, it is likely that one or more partners will be sought to assist in the manufacturing and/or distribution areas.

As in the research and development arena, manufacturing and distribution relationships can take a number of different forms. As to manufacturing, the relationship can take the form of a product purchase arrangement pursuant to which the manufacturer produces products which conform to the specifications provided by the buyer/purchaser. In many cases, the buyer/purchaser actually licenses technology to the manufacturer to allow it to produce the specified products. In the distribution area, the most common type of relationship involves the situation where a particular firm arranges for the manufacture of the products and the distributor purchases the products from the manufacturer for subsequent resale to various end users. Other variations include the original equipment manufacturer, or "OEM," form of relationship, which involves the purchase and distribution by the OEM of products that are manufactured by the seller to meet the precise specifications of the OEM, and the manufacturing and distribution licensing arrangement, which involves a license by the licensor of all the legal rights and technical information which might be required in order for the licensee to manufacture and distribute the products.

B. Basic Manufacturing Relationship

The basic manufacturing relationship is used in those situations where the firm that developed or otherwise acquired the particular product chooses to have it manufactured by another firm and then distributes the manufactured products, either alone or in combination with one or more distributors. In addition to any technology license from the innovating firm to the manufacturer which may be required for the manufacturer to actually produce the products, the parties will enter into some form of product purchase agreement. Such agreements will typically contain terms relating to the form and acceptance of orders, change orders, cancellation of orders, transfer pricing (i.e. the price at which the products will be "sold" or transferred to the innovating firm), terms of payment, shipping terms and risk of loss, the consequences of any nonconforming or defective shipments, warranties, and defaults. The quantity of products to be produced and transferred will also be a key negotiating item.

C. Basic Distributorship Relationship

General structure of the relationship The basic distributorship relationship involves a purchase of goods by the distributor from the principal and the resale of the purchased goods by the distributor for its own account to its customers. The distributor's sole income from the transaction is derived from the markup of the price of the goods which are sold to the customers. There are a number of different subcategories of the basic distribution relationship. For example, one common distinction is between a "stocking" and a "non-stocking" type of distributor, with the stocking distributor maintaining a stock of the goods in its inventory for sale to its various customers and a non-stocking distributor foregoing the costs of maintaining an inventory and simply purchasing goods from the principal at the time an order is received from a customer.

The principal's price to the distributor will typically reflect a discount from the suggested retail price. The amount of the discount will vary depending upon the size of the orders and the scope of the activities which are to be performed by the distributor. Since the distributor will often need to maintain sales offices, keep inventory on hand and train its salespersons to demonstrate the products, it will want to ensure that there will be sufficient "margin" for it to make an adequate profit, after taking into account its own costs and investments relating to marketing.

The contract in a distribution relationship will take on many of the aspects of a direct sales agreement. For example, it will generally cover the description of the goods and services, the manner in which the subject goods are to be priced and delivered and allied requirements with respect to shipping and insurance; the form and timing of payment and related credit arrangements and any warranties and limitations which might be imposed on the respective liabilities of the parties to the agreement. However, the usual terms covering the sale of goods will be supplemented by additional provisions on how the distributor will pursue the manufacturer's principal objective of penetrating a market in which direct sales efforts are either inappropriate or economically infeasible.

Obligations of the distributor Among the obligations of the distributor which are often included in the distribution agreement are the following:

(i) The distributor will agree to promote the products, identifying them by their correct names as products of the manufacturer. The distributor will also agree to maintain a qualified sales force and distribution organization, although it may be necessary for the manufacturer to supply appropriate training with respect to the uses and functions of the products.

(ii) The distributor will maintain a staff of trained technicians and a stock of spare parts and technical literature adequate to provide technical support and service to its customers. Once again, it may be necessary for the manufacturer to provide training in the appropriate service techniques, particularly when concerns exist regarding the overall quality and reputation of the product offerings.

(iii) The distributor will establish and maintain a place or places of business in areas throughout the territory adequate to provide good customer support and marketing coverage.

(iv) The distributor will agree not to sell or distribute products which are competitive with the products that are produced by the manufacturer.

(v) The distributor will be obligated to maintain an inventory of products sufficient to satisfy customer orders on a timely basis; however, the obligations of the distributor in this regard must be made subject to any other agreement covering the allocation of limited products supplies among other distributors.

(vi) The distributor should provide the manufacturer with periodic reports regarding product sales, competitive goods and prices and additional information relating to the markets in which the distributor's activities are being conducted.

(vii) The distributor will be obligated to conduct its business in a manner that will reflect favorably on the manufacturer and will agree not to engage in any deceptive, misleading, illegal or unethical business practices. It is important to understand the importance of local regulations and customs regarding the sale and distribution of goods, particularly if they differ materially from those that might exist in the home market of the manufacturer.

(viii) Warranty obligations with respect to the products should be clarified, as should the allocation of the costs and responsibilities associated with any warranty repairs. The distributor should agree to promptly inform the manufacturer of any complaints regarding the products.

Distributor review One of the fundamental goals and objectives of any form of distribution arrangement is to generate revenues from sales of products and, in turn, to build and maintain market share. While in some cases the obligation of a distributor under the agreement will be limited to the use of its "best efforts," it is generally most desirable for the manufacturer to impose various standards under which the performance of the distributor will be measured. However, the ability of the manufacturer to effectively monitor the performance of the distributor will depend upon a number of factors, including the relative size and experience of the parties and the consequences attached to any failure by the distributor to meet the expectations of the manufacturer.

Performance criteria can take a number of different forms, such as sales quotas, minimum inventory purchases, levels of staffing, shelf or floor space requirements, number and geographic coverage of dealer appointments, warehouse locations and the amount of expenditures on promotional activities. Whatever the form of criteria, it is important to provide ongoing incentives for the distributor and to take into account anticipated changes in market conditions and in the various strategic objectives of each of the parties. One method for ensuring that a distributor will diligently attempt to meet any performance measures is to condition any ongoing right of exclusivity on the achievement of the objectives. If the distributor does not attain the goals set forth in the agreement, the manufacturer may have the right to engage other distributors and, in some cases, to terminate the entire relationship.

In addition to any objective performance measures, the agreement may include a number of other procedures calculated to build and reinforce communication between the parties regarding the conduct of the relationship. For example, it generally makes sense to provide for periodic financial and marketing reports, on-site inspections of distributor facilities, training sessions and other meetings or telephonic conferences between representatives of each of the parties. Ongoing contact of this sort will give the manufacturer a good sense of the distributor's commitment to the relationship and the feasibility of expanding the scope of the arrangement into other areas.

Term and termination Although each of the parties may certainly contemplate that the distribution arrangement will continue for a long period of time, it is necessary to give careful consideration to the term of the arrangement, any conditions with respect to renewal of the arrangement and, most importantly, the circumstances under which it will be possible for one or both of the parties to terminate the arrangement prior the end of any specified term. Also, the agreement should describe the effect of any termination of the agreement, as well as any procedures that may be established to regulate the resolution of disputes among the parties with respect to their respective rights following any termination of the agreement.

Termination of the distribution agreement may expose the terminating party to potential claims with respect to damages that might be suffered by the other party due to termination of the relationship. While both parties will attempt to limit their exposure with respect to such claims by including disclaimers of any liability for damages or restitution arising out any decision to terminate the agreement, such provisions may not be effective under the laws of various nations. Also termination generally does not effect the rights of the parties to receive payments that may be owed as a result of activities which occurred prior to the date of termination.

A number of countries now have laws which provide various protections for local agents and distributors upon termination of an international distribution relationship. For example, it may be possible for the terminated distributor to obtain compensation for the value of the goodwill which may have been developed with respect to the manufacturer's products, for the loss of expected profits during the remaining term of the agreement, for the value of capital investments which the distributor may have made in the course of marketing the products and, in some cases, for the cost of employee severance benefits paid to those employees which the distributor need to lay off as a result of the termination of the distribution agreement. The manufacturer can avoid liability if the termination is "justified," however, most laws contain a very narrow definition of "just cause," limiting it to criminal acts and gross incompetence.

The distribution agreement should provide for settling outstanding accounts and payment obligations between the parties. For example, subject to any agreement which the parties might have regarding the repurchase of inventory which might be held by the distributor at the time of termination, the distributor will remain obligated to remit to the manufacturer any amounts which might be due for products purchased prior to termination. In turn, the distributor may have the right to "set off" against any amounts due to the manufacturer any expenses for which it might be entitled to reimbursement by the manufacturer, such as for copies of promotional materials.

Since termination will generally occur while each of the parties are still involved in various activities relating to the arrangement, the more difficult issues usually arise out of the need to ensure that existing customers are properly serviced, that outstanding orders are filled, that existing inventories are accounted for and that warranty obligations are satisfied. In some cases, "termination" of the agreement actually commences a period of "winding-down," in which both parties seek to properly allocate ongoing obligations to customers within the territory.

D. OEM Relationships

Introduction An OEM relationship actually combines elements of the manufacturing and distribution functions. The principal will agree to sell goods to the distributor which, in turn, will add "value" to the goods, such as enhancements or additional applications, and then sell the "value-added" products to its customers. An OEM agreement usually involves detailed negotiations regarding the specific requirements of the OEM with regard to the products, as well as the terms upon which the products will be sold to the OEM. Properly structured, an OEM relationship can provide smaller firms with a significant customer for its products at a time when it may be struggling to develop its own distribution network and stable of customer accounts.

The legal structure of an OEM relationship combines elements of three distinct strategic business arrangements: a manufacturing and supply agreement wherein one party, the "manufacturer," actually makes products which are to be supplied to the other party, the "OEM", a product purchase agreement, which includes the terms that will govern the sale of the products by the manufacturer to the OEM, and a distribution agreement under which the OEM distributes the products purchased from the manufacturer as part of the line of products otherwise sold by the OEM. The parties to any form of OEM arrangement must deal with most of the issues which exist in separate product purchase and distribution relationships. In addition, the parties must deal with the unique considerations that arise when the products are to be customized to the special needs of the OEM.

Product specifications One of the key threshold issues in an OEM relationship is reaching agreement on the specifications of the products which are to be produced and sold by the manufacturer to the OEM. The process begins with the OEM's submission of design requirements which are used by the manufacturer to develop a prototype version of the anticipated product. While the manufacturer will often agree to assume the costs of developing the prototype, it will require reimbursement from the OEM for any additional expenses relating to subsequent adaptations or changes which might be made in the original specifications. In some cases, the OEM may even be required to pay for the cost of the original prototype if it fails to purchase a sufficient volume of products.

As with any form of distribution arrangement, the OEM generally bargains for the right to sell all of the current and future products of the manufacturer. Obviously, there will be a good deal of reluctance on the part of the manufacture to grant such broad rights to the OEM. One compromise is to include certain specified enhancements as part of the original agreement, but provide that any other products will only be covered by a "right of first negotiation" which obligates the manufacturer to negotiate in good faith with the OEM regarding a new agreement on additional products.

Product manufacture and acceptance procedures Once the product design has been agreed upon, the parties must then agree upon the procedures which are to be followed as to manufacture of the products and acceptance of the finished items by the OEM. The agreement will set out the requirements with respect to production, inspection and testing of products, as well as the rights of the OEM to audit the quality control procedures utilized by the manufacturer. The manufacturer may seek limitations on the OEM's ability to delay acceptance of products by requiring that delivered products which have not been rejected within a specified period will be deemed to have been accepted for purposes of the agreement.

The OEM may demand a license to manufacture the products itself in the event a substantial amount of the manufacturer's products fail to conform to the specifications established by the OEM. The manufacturer will be reluctant to grant such a license in situations other than bankruptcy or insolvency; however, it may be necessary to grant the OEM a non-exclusive license which comes into effect upon the occurrence of certain events. For example, a non-exclusive royalty bearing license may be granted if the manufacturer is unable to adhere to the quality control standards relating to the products. If the manufacturer is not able to meet its commitments due to a cessation of its business activities, the manufacturing license may be royalty free. In most cases, the license only extends to the manufacture of the number of products which were originally contemplated by each of the parties and terminates at the time that the agreement would have otherwise expired.

Pricing and ordering procedures The issues relating to pricing and ordering in any OEM relationship are similar to those in any other type of product purchase arrangement. The OEM will seek retroactive discounts based on the volume of products purchased and both parties will seek adjustments in the purchase price to take into account any increases or decreases in the cost of components. Firm orders will be placed in accordance with an agreed-upon monthly or quarterly schedule and forecasts will be used in order to manage the long-term production schedule and allow the manufacturer to take advantage of volume purchases of components. Cancellations and changes in delivery dates are handled in the same manner as in any distribution agreement.

Product warranties Warranty issues in the OEM area are somewhat complicated because the products are actually incorporated into other goods which are manufactured or assembled by the OEM. For example, manufacturers will be unwilling to assume any responsibility for the components or parts which may be supplied by the OEM. Also, the obligations of the manufacturer with respect to infringement claims by third parties will not extend to components supplied by the distributor, compliance any specifications or designs which may have been supplied by the OEM, the OEM's modifications to the products supplied by the manufacturer, unless approved by the manufacturer, or any combination of the manufacturer's products with those of the OEM.

When the manufacturer's products involve software, it is generally customary for the manufacturer to provide the OEM with updates, which are mandatory engineering changes made to correct errors that lead to product failure, at no extra charge. On the other hand, if the manufacturer develops software enhancements which are not required to prevent product failure, but do result in increased functionality or improved performance, the OEM will usually have the right to purchase the enhancements at a price that is to be negotiated when they are available. In some cases, the pricing of the enhancements may be made subject to a "most favored nation" clause.

Termination Termination issues in the OEM area are unique in that the manufacturer's products usually are an essential feature of other products or services provided by the OEM to its end users. The OEM will want to be sure that it has some access to the products, service and technology provided by the manufacturer prior to the date of termination, assuming that the termination is not due to a material default by the OEM. The parties often agree that the manufacturer will continue to provide some support for a minimum period of time following termination, including field services, spare parts and software updates. If, due to bankruptcy or some other event which results in the cessation of the manufacturer's business activities, the manufacturer is unable to provide all of the required support, the OEM be given a license to manufacture the required products and spare parts.

E. Manufacture and Distribution Licensing

Introduction Companies which have developed products based on new technologies, but which lack the manufacturing capacity and distribution channels to adequately exploit the innovations, may look to larger companies as partners to assist in the production and distribution of the products. One form of relationship takes the form of a non-exclusive license that is granted by the party which has developed the products and the accompanying technology, referred to herein as the "developer," to the larger company, referred to herein as the "senior party," that allows the senior party to manufacture the products and then sell the products to end users. The developer would be entitled to a royalty on sales made by the senior party and would also have a right to purchase additional products from the senior party. The developer would remain free to manufacture the products itself and the senior party will usually be obligated to purchase some products from the developer over the term of the agreement.

Royalties received by the developer from sales which might be made by the senior party can be used for the development of new products, as well as to allow the developer to begin to "ramp up" its own manufacturing capabilities. Also, use of the senior party as a manufacturer allows the developer to have a sufficient volume of products available to permit it to begin to construct its own distribution network. Moreover, the senior party becomes a guaranteed purchaser for some number of products over the term of the agreement. Finally, the solid relationship with a larger company enhances the credibility and reputation of the developer and allows it to gain access to new financing.

Manufacturing activities In order for the senior party to commence and continue manufacturing the licensed products, the agreement must provide for a license of the developer's patents and other intellectual property rights to the senior party for the purposes described in the agreement, as well as for the actual transfer of technical information, such as trade secrets, which is need to build and manufacture the products. In addition, the developer will agree to provide some training to senior party's engineers regarding the use of the technical information and the parties will attempt to work together to minimize the costs of raw materials and other inputs into the manufacturing process.

Scope of manufacturing and distribution licenses The developer will grant the senior party a non-exclusive license to use the developer's patents and technical information to manufacture, use, lease and sell the licensed products in a specified territory. In turn, the senior party will generally grant the developer to use senior party's patents as they relate to improvements or enhancements which the senior party might make to the licensed products during the term of the agreement. As noted above, the senior party would be required to pay a royalty to the developer based on a specified percentage of the net sales price of products sold by the senior party. The senior party's license to the developer is usually royalty-free.

Senior party's obligation to purchase products In addition to the royalties received from the sale of products by the senior party, the developer will be able to receive income from the sale of products to the senior party. The parties should specify the number of products which must be purchased, the price to be charged for the products, the manner of payment and any requirements as to forecasted purchases or minimum orders. The price which the senior party will be charged for the products will generally be established by reference to the prices the developer is charging to similar customers for any like amount of products at the time of delivery. For example, the agreement may state that the purchase price for the products shall be the developer's OEM list price for comparable quantities at the time of delivery. In most cases, the senior party will receive the benefits of a "most favored customer" clause that obligates the developer to charge the senior party the best price offered to any customer, irrespective of discount.

Developer's right to purchase products The developer will also have the right purchase products from the senior party. For example, the agreement might provide that the developer shall have the right to purchase from the senior party, under the developer's own private label, products manufactured by the senior party pursuant to the license. The developer would be entitled to the best price, including all discounts, terms and conditions on which the senior party offers such products to any other customer, irrespective of quantity.

9 Collaboration and Competition Regulation

I. Introduction

The actual regulatory practices in the United States and in the EC with regard to patent licensing arrangements are discussed elsewhere in this work. This chapter focuses on the core ideas and principles which regulators have applied in their attempts to review specific transactions and develop general guidelines which can be relied on by innovative firms in the marketplace. We begin by reviewing some of the theoretical and empirical work of various economists regarding the impact of collaborative business relationships on the level of competition. This is important, since much of this work forms the basis for the decisions which courts and regulators have made in this area. We then turn to a description of the basic principles driving regulatory and legal policies in each of the United States and the EC with regard to patent licensing arrangements, including a summary of some of the relevant policies which have been applied to joint ventures and other business collaborations in general.

II. Collaboration and Competition

A. Introduction

While, as a general matter, the participants in any particular collaborative business relationship can be expected to derive substantial private benefits from the arrangement, public policy considerations demand that such arrangements must also be examined for their "social" effects. While there may well be a number of factors that should be taken into account in this type of analysis, the most common area of consideration tends to be the effect that the relationship might have on the competition in the relevant marketplace, with the presumption being that any arrangement with substantial potential for restricting the level of competition will lead to a level of social harm that would require regulators to step in and modify, if not prohibit, the particular arrangement.

In this section, we examine some of the work that has been done by economists with respect to identifying and quantifying the social benefits and costs of joint ventures and other forms of collaborative business arrangements. We begin by reviewing the pro-competitive effects of collaborative relationships, which include several of the factors considered to be private benefits to the participants, as well as enhanced efficiencies from the integration of complementary assets and resources and the reduction of duplicative research activities. We then analyze the potential anticompetitive effects of joint ventures, many of which have occupied antitrust and competition regulators in both the United States and in the EC. Thereafter, we look at the difficult process of balancing the benefits and costs of a joint venture, including various proposals to assist regulators and judges in developing a more consistent approach to rule of reason analysis.

B. Procompetitive Effects of Collaborative Relationships

Introduction The study and analysis of the procompetitive effects of collaborative business relationships is, somewhat surprisingly, a relatively new phenomenon. This might be explained by the fact that, until recently, courts and regulators tended to view joint ventures in the same manner as mergers. As a result, they placed a good deal of weight on high concentration, collateral restraints and information sharing, all of which were considered to be per se bad and unlawful, and failed to give appropriate consideration to the potential the joint venture might have for producing economic benefits.[1]

It is now generally assumed that collaborative business relationships can have a variety of procompetitive effects, particularly with regard to improving efficiencies, enhancing the rate of innovation, and reducing the waste of scarce resources in duplicative research activities. In addition, certain of the purported private benefits of business collaborations may also translate into benefits for society as a whole.[2] For example, the ability of parties to obtain economies of scale with respect to logistics, productions, and procurement[3] may reduce the costs of production and, ultimately, the price of goods to consumers. Similarly, consumers may benefit from the ability of firms to use collaborative relationships to share the business and financial risks associated with investments in products which will serve uncertain demand or which involve uncertain technologies. Also, collaborative relationships can serve a purpose similar to that of patents in reducing the effect of externalities on innovators, thereby retaining or enhancing incentives to innovate.

Enhanced operational efficiencies One of the most commonly cited benefits from the use of a collaborative business relationship, such as a joint venture, is the level of efficiency that the participants may be able to achieve in relation to conducting the activities independently. Efficiencies may flow from a variety of sources; however, it is generally believed that the benefits are derived from integrating the complementary assets and resources of the participants, as well as from economies of scale. While operational efficiencies obviously lead to private benefits for the participants, they may also lead to public benefits to the extent that the efficiencies accelerate the pace and dissemination of technological advance.

One area of debate in this regard has been devising a procedure for actually measuring the level of efficiency associated with any particular structural relationship. Some commentators argue that courts have rarely made any real inquiry into the efficiencies of an arrangement once some threshold finding has been made that the arrangement was

[1] See, e.g. Weston and Ornstein, "Efficiency Considerations in Joint Ventures," in *Collaboration Among Competitors* (1990) (hereinafter *Collaboration*) 875, 884.

[2] See, e.g. Kitch, "The Antitrust Economics of Joint Ventures," in *Collaboration*, supra note 1, 865. Kitch argues that the legal documentation used for a joint venture can, and should, always be structured in a manner which accentuates the manner in which the efficiency features that result in private benefits to the parties may also lead to public benefits. Kitch also notes that joint venture agreements should include features that limit the possibilities that the arrangement will simply become a device to facilitate collusion. Id. at 873.

[3] Shapiro and Willig, "On The Antitrust Treatment of Production Joint Ventures," in *Collaboration*, supra note 1, 929, 930.

a joint venture, rather than an unlawful cartel. These critics maintain that, given the wide range of structures that might be characterized as joint ventures, the balancing process which is part of the rule of reason analysis of joint ventures must explicitly measure the efficiencies and other sorts of procompetitive effects of a joint venture.[4]

There obviously are substantial difficulties in attempting to measure the efficiencies associated with a joint venture. For example, there are a wide array of potential efficiencies which might be taken into account, and there really is no agreement as to the relative weights that ought to be afforded to each type of efficiency in the rule of reason analysis. Moreover, there are few, if any, reliable methods which can be used to quantify the efficiencies that might be present in a given transaction.[5] Nonetheless, it would appear that some attempt must be made to assess the level of efficiencies in a joint venture, even if only in terms of "high", "low", or "in-between", in order for their to be a full and effective analysis of the pro- and anticompetitive effects of joint ventures.[6]

One method that might be used for measuring the levels of efficiency associated with a joint venture is to assume that such levels can be predicted on the basis of extent of the integration between the venturers. Accordingly, high levels of integration which might occur with a project involving joint construction of new productive assets will typically indicate a substantially more efficient operation than what might occur in the absence of any joint arrangement.[7] This approach would make no attempt to draw finer distinctions between the various different kinds of efficiencies which might occur in a joint venture, such as the economies of scale, synergies from the joint use of complimentary resources, and the capital formation advantages which might flow from the risk sharing in the joint venture context.[8]

Reduction of duplication of research activities Debate has focused upon whether collaborative research activities are beneficial from the perspective of reducing the amount of duplication which might occur if firms simply were to proceed independently[9] or whether horizontal cooperation may be harmful in the research area from a resultant loss of diversity. On the one hand, commentators have noted the existence of what has been referred to as "shameful and needless duplication of effort" in the research area,[10] and have pointed out that waste often occurs from the fact that independent research activities often proceed down identical or near-identical technological paths.[11] These commentators suggest that the coordination of research plans will minimize the welfare losses arising from the duplication in this area,[12] a result which may become all the more important in light of the relatively scarce resources which might be available for research and development.

[4] See, e.g. Pitofsky, "A Framework for Antitrust Analysis of Joint Ventures," in *Collaboration*, supra note 1, 839, 850–1.

[5] See, in this regard, Fisher and Lande, "Efficiency Considerations and Merger Enforcement," *California Law Review* 71 (1983), 1580, 1604.

[6] See, e.g. Pitofsky, supra note 4, 862.

[7] Ibid. at 851.

[8] Ibid.

[9] See also Kitch, supra note 2.

[10] See David, "R&D Consortia," *High Technology*, October (1985), 42 (quoting William Norris, then-CEO of Control Data Corporation).

[11] Jorde and Teece, in *Collaboration*, supra note 1, 887, 894.

[12] Ibid.

On the other hand, some economists have argued that the gains from competition may more than offset the losses arising from duplicative research activities,[13] and raise legitimate concerns that a system of innovation which converges on a single view of the technological possibilities, as may be the case when cooperative multi-firm structures are used, may needlessly close off productive avenues of inquiry.[14] Other economists have noted that there is no real agreement as to what constitutes the optimal level of research rivalry.[15]

In response, advocates of policies designed to facilitate cooperative research point out that a private enterprise economy without any coordination and communication among horizontal participants does not, in and of itself, guarantee that the market for innovation will achieve the desired level of diversity at the lowest cost.[16] Also, they suggest, without elaboration, that cooperation need not be the enemy of diversity, and that coordination among firms can occur in ways which minimize the amount of duplication without the industry converging to a single technological approach.[17]

C. *Anticompetitive Effects of Business Collaborations*

Introduction While it is now often argued that there actually is no necessary conflict between promoting cooperation and competition, provided that the cooperation itself serves to improve efficiency or to advance innovation,[18] there is a large body of case law and commentary[19] regarding the potential anticompetitive effects of joint ventures and other types of collaborative relationships. Among the problems most commonly discussed are the following:

(i) reduction of potential competition by non-venturers, perhaps by removing potential entry (sometimes referred to as the "foreclosure" effect);
(ii) reduction of competition between the venturers, which may have the effect of restricting the independent future growth of the venturers;
(iii) reduction in the overall pace of innovation or technical change in change in the marketplace, again attributable to a loss of competitive edge from the collaboration;
(iv) the spill-over effects with respect to competition among the venturers, often referred to as "collusion";

13 See, e.g. Nalebuff and Stiglitz, "Information, Competition and Markets," *American Economic Review* 72 (1983), 278–84.

14 Jorde and Teece, supra note 11, 894.

15 See, e.g. Brodley, "Antitrust Law and Innovation Cooperation," in *Collaboration*, supra note 1, 911, 913. See also Baldwin and Scott, *Market Structure and Technological Change*, New York: Harwood (1987).

16 Jorde and Teece, supra note 11, 894.

17 Ibid. at 894–5 (noting also that Bell Laboratories achieved a very considerable level of internal diversity during the pre-divestiture period).

18 Ibid. at 888 n 2 (citing Schumpeter, *Capitalism, Socialism and Democracy* (1942), 85, who pointed out, when compared to competition among firms with similar products and technologies, the competition that counts "comes from the new commodity, the new technology, the new source of supply ... This kind of competition is ... so much more important that it becomes a matter of comparative indifference whether competition in the ordinary sense functions more of less promptly.").

19 See, e.g. Pitofsky, supra note 4, 843; and Weston and Ornstein, supra note 1, 875.

(v) the use of collateral restrictive agreements which unduly restrict the activities of the joint venture and/or the participants in a manner beyond that which may be necessary to realize the efficiencies from the arrangement.

Reduction of potential competition The most commonly cited area of concern with respect to collaborative business relationships, especially joint ventures, is the possibility that the arrangement will reduce the level of competition in the marketplace, and that the firms participating in the joint venture will occupy a position in the market such that they are able, when acting in concert, to charge more than a competitive price. A full analysis of this question requires an evaluation of the ease with which firms which are already in those markets can expand their production as well as the ease with which firms which are not presently in the market would be able to enter it.[20] Since it is often difficult to objectively analyze and quantify these variables, regulators have tended to rely upon concentration measures in downstream markets (e.g. the distribution and sales) as a proxy for the relevant factors.

There are a variety of problems with the use of downstream concentration measures. For example, it has been argued that a collaborative relationship at the research and development stage is better evaluated by reference to markets for future research and innovation, rather than the present product market.[21] While the logic behind the use of such so-called "innovation markets" is quite compelling, many commentators have dismissed the idea on the basis that it is impossibly speculative to attempt to define markets for invention, research, and innovation.[22] Nonetheless, recent regulatory efforts to develop guidelines for analyzing technology-based collaborative relationships have included formal recognition of innovation markets, although product markets will still be an important factor in situations where it is difficult to properly define an innovation market.

Another challenge to the use of concentration measures follows from a critical reassessment of the model of competition which should be used by regulators in antitrust analysis. For many years, antitrust regulators relied upon the validity and utility of the "static structural model" of competition, which predicts that large companies will become progressively larger and more powerful, and that concentrated industries will become even more concentrated over time, due to the existence of monopoly profits[23] to justify antagonism to any concentrative structure, including joint ventures. However, new theoretical developments and empirical evidence have been used to challenge the structural-based theories with respect to concentration. For example, investigators have attempted to show that even the most concentrated industries are subject to the rigors of competition, and that even though industries with high levels of concentration showed above-average profitability, over time those industries reverted to a more normal or average level of profitability.[24]

[21] See Ordover and Willig, "Antitrust for High-Technology Industries: Assessing Research Joint Ventures," *Journal of Law and Economics* 28 (1985), 311.

[22] See, e.g. Kitch, supra note 2, 869 ns 8 and 9 (criticizing the approach taken in Ordover and Willigs, supra note 21, as "abstract analysis").

[23] See Weston and Ornstein, supra note 1, 882.

[24] See Brozen, "The Antitrust Task Force Deconcentration Recommendation," *Journal of Law and Economics* 13 (1970), 279; Brozen, "Bain's Concentration and Rates of Return Revisited," *Journal of Law and Economics* 14 (1971), 351; Brozen, "The Persistence of 'High Rates of Return' in High-Stable Concentration Industries," *Journal of Law and Economics* 14 (1971), 501.

Other investigators have stressed the importance of other variables, such as capital intensity, in the analysis of the influence of concentration on profitability.[25] In addition, Landes and Posner have rejected the structural models and have chosen to focus upon the importance of demand and competitive supply elasticities (i.e. substitutes) in assessing the degree of market power in their antitrust analysis.[26] Also, Weston and Ornstein have developed an alternative model of competition, sometimes referred to as the "dynamic competition model".[27] Under this model, which bears some similarities to the underlying assumptions for the simultaneous innovation model described above, competition involves more than simplified pricing and output decisions, which are the basis of the classical model.[28] Instead, the profitability of firms depends on a range of factors that have become the foundation for the new strategic planning literature: technological leadership, product quality, product variety, promotional efforts, and relative cost efficiencies.[29]

Some commentators have suggested that the risk that collaboration will inhibit competition in a given industry is a function of the rate of technological change in that industry. Specifically, they suggest that collaboration between firms in industries in which there is a rapid rate of technological change is not likely to have an adverse effect on competition due to the "diversity of sources of innovation" and global competition.[30] If this is true, than antitrust policy in high technology-based industries should be less restrictive toward collaboration, a conclusion which also follows from the fact that new innovation projects in those areas tend to be more risky and costly than in industries in which technical standards are more settled. Others have noted, however, that there is far from any clear economic or legal consensus for these conclusions.[31]

Reduced competition between venturers While the possibility of reduced competition between the participants in a collaborative relationship is essentially part of the large issue of reduced competition discussed above, it is often discussed separately for analytical purposes. Once again, this area raises measurement problems. One economist has argued that the possibility that a joint venture will result in reduced competition can be analyzed by examining the relationship between the activities of the joint venture and the rest of the participant's businesses.[32] For example, if the joint venture is small in relation to the other businesses of each of the participants, there should be little incentive to refrain from competing in the other businesses since there is more for them to

[25] See Comanor and Wilson, "Advertising Market Structure and Performance," *Review of Economics and Statistics* 49 (1967), 423.

[26] Landes and Posner, supra note 20. See also Liebeler, "Market Power and Competitive Superiority in Concentrated Industries," *University of California Law Review* 25 (1978), 1231; Demsetz, "Industry Structure, Market Rivalry, and Public Policy," *Journal of Law and Economics* 16 (1973), 1; Fisher, "Diagnosing Monopoly," *Quarterly Review of Economics and Business* 19 (1979), 7; Baumol, "Contestable Markets: An Uprising in Theory of Industry Structure," *American Economic Review* 72 (1982), 1; *Industrial Concentration: The New Learning* (H. Goldschmid et al. eds. 1974); J.S. McGee, *In Defense of Industrial Concentration* (1971); H. Demsetz, *The Market Concentration Doctrine* (1973); and J.F. Weston, *Concentration and Efficiency; The Other Side of the Monopoly Issue*, Special Issues in the Public Interest No. 4 (Hudson Institute, May 1978).

[27] See Weston and Ornstein, supra note 1, 883.

[28] Ibid.

[29] Ibid.

[30] See, e.g. Jorde and Teece, "Innovation, Cooperation and Antitrust: Balancing Competition and Cooperation," *High Technology Law Journal* 4 (1989), 1–113.

[31] See Brodley, supra note 15, 912.

[32] See, e.g. Kitch, supra note 2, 870.

lose in the other businesses than in the joint venture; however, if the joint venture is large, and the other businesses are small, the reserve implication arises.[33]

Reduced pace of innovation Another potential problem with research and development joint ventures stems from the possibility that by replacing the competitive independent decision-making of the parties by the joint decision-making associated with collaborate activities in the joint venture, the pace of innovation may be slowed due to the fact that the parties will no longer be engaged in a patent race to discover the solution to identified problem. The problem becomes even more acute if the collaboration does not at least significantly reduce the costs of research and development, since the overall rate of return on research investment will decrease. Also, the anticompetitive effects will be even greater to the extent that the parties would have otherwise been the principal competitors with respect to the specific research and development activities.[34]

Spill-over effects and collusion Concerns regarding "spill-over" effects relating to a collaborative relationship go to the possibility that the joint activities within the context of the arrangement will "spill-over" into other business areas, thereby leading to the potential for collusion and unlawful cooperation between the participants. Critics of explicitly liberalizing antitrust treatment of collaborative arrangements concede that such collaboration will not generally harm competition if alternative suppliers for the innovation exist, if the downstream markets are effectively competitive, and if the collaboration is limited to a bounded fraction of innovation suppliers.[35] However, such critics argue that innovative collaboration presents a danger of injuring competition through collusion when the collaboration includes substantially the entire market, when entry barriers exist, and when the joint enterprise attempts to deny necessary inputs to rival firms.[36]

Many economists would agree that the possibility of anticompetitive collusion is somewhat related to the activities of the joint venture.[37] Probably most suspect would be joint ventures relating to the current marketing and pricing of the products procedures by the venturers, while joint ventures which are related to research and development of future products, the promotion of the product type, or some phase of the production process would be less likely to result in collusion.[38]

Collateral restrictive agreements Collaborative business relationships generally involve a network of contractual duties and obligations, many of which are focused on activities of the participants which might otherwise have an impact on the market in which the products and services which are the subject of the arrangement would otherwise be sold or distributed. As a general matter, these obligations, which are referred to as "col-

[33] Ibid.

[34] Shapiro and Willig, supra note 3, 937. See also Katz, "An Analysis of Cooperative Research and Development," *Journal of Economics* 17 (1986), 527–43; Grossman and Shapiro, "Dynamic R&D Competition," *Economic Journal* 97 (1987), 372– 87; and Ordover and Baumol, "Antitrust Policy and High-Technology Industries," *Oxford Review of Economic Policy: Technical Progress* Winter (1988). See also Shapiro, "Patent Licensing and R&D Policy," *American Economic Review, Papers and Proceedings* 75 (1985), 25–30 (discussion patent licensing, which is sometimes considered as an alternative to joint research and development).

[35] See, e.g. Brodley, supra note 15, 912.

[36] Ibid.

[37] See, e.g. Kitch, supra note 2, 870.

[38] Ibid. at 870–1.

lateral restrictive agreements," are designed to insure that the participants devote the expected level of assets and resources to the joint venture, rather than diverting those resources to other activities make it less likely that the objectives of the joint venture will be achieved. While certain restrictive agreements entered into in connection with a joint venture will be anticompetitive, it is important to look beyond the mere legal form of the arrangement to the reasons and effects thereof, since it may be that the agreement actually is procompetitive, at least in the specific context.[39]

If the joint venture is otherwise found to be an efficient transactional solution to market imperfections, an economist may be willing to tolerate certain restrictive agreements between the parties which would normally be illegal under the antitrust law if such collateral agreements are necessary in order to create the incentives for the formation of the joint venture. A common example of a collateral restraint which might satisfy this sort of criteria is the practice of assigning exclusive territories to the participants in a joint venture, since the grant of such rights might well be the best method for allocating the benefits of the joint venture, which may have formed to take advantage of the scale efficiencies associated with a multi-party venture, among the participants.[40]

On the other hand, a non-competition agreement among the participants which extends outside of the field of activities of the joint venture may be viewed as an unnecessary restraint on the business prerogative of the participants, particularly in those cases where the effect of such an agreement would be to insulate one of the participants from the competitive efforts of other firms involved in the joint venture. Similar concerns are often raised with respect to obligations imposed on participants with regard to the procurement of raw materials and other items.

While, in theory, courts should, as part of the overall rule of reason analytical process, determine whether the collateral restrictions are reasonably related to achieving the efficiencies associated with the particular joint venture, in practice, this has proven to be difficult.[41] Commentators continue to champion the importance of such an analysis, and also argue that an effort should be made to examine whether the same efficiencies could be achieved through a less restrictive alternative.[42] It is important to remember, however, that a joint venture should not be found illegal simply if the same or similar efficiencies can be achieved in a less restrictive manner, instead it is only necessary that those efficiencies be discounted in some manner in the total rule of reason analysis if they could have been achieved with a substantially lesser level or even the complete absence of restraint.[43]

[39] See Weston and Ornstein, supra note 1, 883.

[40] See, e.g. Kitch, supra note 2, 873.

[41] There are some exceptions to the general practice. For example, in *National Bancard Corp. (NaBANCO)* v. *VISA USA, Inc.*, 779 F.2d 592 (11th Cir. 1986), the court examined and upheld a discount arrangement between various classes of banks within the VISA credit card system. The court specifically analyzed a price arrangement between different categories of banks and concluded that the arrangement should be sustained because the "restraint at issue potentially could create an efficiency enhancing integration to which the restraint is ancillary." Ibid. at 599. The court also stated that the arrangement was not just reasonably related but essential to achieving substantial efficiencies in the operation of the credit card system. Ibid. at 601–2. See also the opinion of Judge Posner in *General Leaseways* v. *National Truck Leasing Ass'n*, 744 F.2d 588 (7th Cir. 1984) (rejecting after careful analysis defendant's claim of efficiencies).

[42] See, e.g. Pitofsky, supra note 4, 861.

[43] Ibid. at 858.

D. Balancing Procompetitive and Anticompetitive Effects

Introduction Once a determination has been made regarding the various procompetitive and anticompetitive effects of a particular joint venture or other collaborative relationship, rule of reason analysis inherently demands some form of balancing process to determine whether the positive social benefits from the venture outweigh any adverse impact on competition. Obviously, this can be a very difficult process, and it appears that courts have been illl prepared and unable to articulate and apply objective and consistent standards in this area.[44] Among the problems which are commonly discussed are the unwillingness of the courts to give adequate consideration to efficiencies from a joint venture and the need to take into account various limitations on the possible anticompetitive effects of the arrangement. These problems have, in turn, led to some controversial proposals for statutory codification of the factors which should be considered in applying rule of reason analysis.

Consideration of joint venture efficiencies As noted above, serious consideration of the efficiencies from the use of a joint venture or other form of collaborative business relationship is a relatively new factor in the analysis of such arrangements. While a good deal of theoretical work has been done by economists which may assist in carrying out the balancing process, it still remains difficult to compute the cost savings from a joint venture.[45] As a result, it is widely believed that the possible efficiency contributions from the use of collaborative relationships are not adequately considered in the joint venture analysis.[46]

One method which has been suggested for the balancing process is the use of a presumption regarding the presence of efficiencies based on the level of integration associated with the joint venture, with substantial efficiencies presumed from high levels of integration and insubstantial efficiencies presumed from low levels of integration.[47] Once the appropriate presumption was established, the challenger of the transaction would have the burden of showing that substantial efficiencies were not present regardless of high levels of integration, and the defender of the transaction would, in turn, have the task of demonstrating that low levels of integration actually understate the efficiencies from the relationship.[48]

Limitations on anticompetitive effects The balancing process should also consider certain limitations on the possible anticompetitive effects of a particular collaborative relationship. For example, while an arrangement may involve integration, as well as collateral restrictions, which may substantially impair the ability of the participants to independently compete in a market, antitrust regulators may nonetheless approve the ven-

[44] See, e.g. Posner, "The Next Step in Antitrust Treatment of Restricted Distribution: Per Se Legality," *University of Chicago Law Review* 48 (1980), 6 (criticizing results of use by courts of a standard "so poorly articulated and particularized, applied by tribunals so poorly equipped to understand and apply it").

[45] An early example of some of the theoretical work in this area is Williamson, "Economics as an Antitrust Defense: The Welfare Tradeoffs," *American Economic Review* 58 (1968), 18.

[46] See, e.g. Weston and Ornstein, supra note 1, 875. Another commentator has noted that courts sometimes take steps to simplify the analysis by merely treating the efficiencies from a collaborative relationship as being neutral in deciding whether the arrangement should be permitted. Pitofsky, supra note 4, 863. See also *FTC* v. *Procter & Gamble, Co.*, 386 U.S. 568 (1967).

[47] See, e.g. Pitofsky, supra note 4, 863.

[48] Ibid.

ture in situations where the arrangement is between two small non-competing companies involved in an unconcentrated market. In such situations, the anticompetitive effect of the relationship are said to be "insubstantial," since there is likely to be little or no effect on price.[49]

It has also been pointed out that most joint ventures generally have inherent limitations which tend to provide further safeguards against anticompetitive effects.[50] These include: (i) the complexity and rigidity of long-term contracts, (ii) moral hazard if contracts are loosely constructed to meet changing conditions, and (iii) disagreements arising from the mismatching of expectations with regard to the relative inputs and performance of the joint venture.[51] In addition, concerns about a possible inhibition on competition between the venturers as a result of the joint venture tend to become less importance when one begins to realize some of the operational difficulties that occur with respect to internalizing the learning of strategically valuable information for each party while limiting such benefits to the joint venture itself.[52]

E. Empirical Studies on Competitive Effects of Joint Ventures

Theoretical arguments regarding the balance between procompetitive and anticompetitive effects of joint ventures have been supplemented by attempts to empirically study the actual competitive impact of joint ventures. For example, one major empirical study of joint ventures concluded that there was weak, if any, potential for augmentation of market power in joint ventures even when the parents are horizontally related.[53] Other studies have uncovered a variety of positive effects from joint ventures. In one case, it was found that joint ventures produce increases in the market values of the participating firms, which reflects the positive expectations of investors regarding the efficiency gains from joint venture activity.[54] Joint ventures have also been found to stimulate innovation by increasing technological rivalry.[55]

III. Principles for Review of Licensing Arrangements under United States Antitrust Law

A. Introduction

As with joint ventures and other forms of collaborative business relationships, antitrust regulators in the United States attempt to identify, and balance, the various procompetitive and anticompetitive effects of licensing arrangements. However, the subject matter of a patent licensing agreement raises a number of unique issues relating to the monopoly-like characteristics of a patent. For example, in determining the market impact

[49] Ibid. at 842.
[50] See Weston and Ornstein, supra note 1, 883.
[51] Ibid.
[52] Ibid.
[53] See Berg, Duncan, Friedman, *Joint Venture Strategies and Corporate Innovation* (1982).
[54] McConnell, *Joint Venture: An Alternative to Merger*, Purdue University (1983).
[55] Berg and Friedman, "Joint Ventures, Competition and Technological Complementarities: Evidence from Chemicals," *Southern Economic Journal* (1977), 1330.

of the licensing agreement, the nature and market power of the patents involved must be taken into consideration. Also, the appropriate standard of review and analysis with respect to licensing arrangements must be selected. In addition, patent licensing has produced two doctrinal issues that have dominated how courts and regulators have acted in this area: the "first sale," or "exhaustion," doctrine; and the doctrine of "patent misuse." Finally, a brief overview of United States antitrust law review of joint ventures and other collaborative business arrangements is presented.

B. Procompetitive Benefits and Anticompetitive Effects of Licensing Arrangements

Procompetitive benefits of licensing arrangements The procompetitive benefits that can be derived from intellectual property licensing were explicitly recognized in the 1995 Antitrust Guidelines for the Licensing of Intellectual Property (the "IP Guidelines").[56] The IP Guidelines noted that licensing activities can facilitate the combination of each of the complementary components of production (e.g. manufacturing and distribution facilities, workforces, and intellectual property rights owned by others)[57] necessary for the licensor to realize the commercial value of its own intellectual property rights. While it sometimes possible for intellectual property rights owners to develop or purchase each of these complementary inputs on their own, it is often more efficient to contract with others for these inputs through licensing arrangements or joint ventures which involve the license or transfer of intellectual property rights. These enhanced efficiencies not only benefit consumers through the reduction of costs and the introduction of new products, they also increase the incentive for innovation, and the expected level of research and development, by increasing the expected returns from development and use of intellectual property rights.

The IP Guidelines also point out that many of the common limitations negotiated as part of a licensing arrangement, such as field-of-use and territorial restrictions, may actually serve procompetitive ends by allowing licensors to manage and exploit their intellectual property rights as efficiently and effectively as possible. For example, when licensees are granted various forms of exclusive rights with regard to the exploitation of the licensed intellectual property rights, an attempt is being made to give the licensee an incentive to make an investment in commercializing the products embodying the licensed intellectual property by protecting the licensee against free-riding on the licensee's investments by other licensees or by the licensor. The IP Guidelines also acknowledge that such restrictions may serve to promote the licensor's incentive to grant licenses, and thereby disseminate the products of its research activities, by protecting the licensor from competition in the licensor's own technology in a market niche it prefers to keep to itself.[58]

[56] See IP Guidelines § 2.3. The IP Guidelines were first published for comment in August 1994, and were ultimately adopted, with some modifications from the initial draft, by the DOJ and the FTC on April 6, 1995.

[57] Sometimes the use of an item of intellectual property requires access to another, as is the case when a product covered by a patent right cannot be practiced without infringing another patent. In such situations, licensing of the right to use the "blocking" patent can promote the coordinated development of those technologies that are in a blocking relationship. IP Guidelines § 2.3.

[58] Intellectual property rights owners are under no obligation to grant licenses to others, even in situations where they have market power. See, e.g. *SCM Corp.* v. *Xerox Corp.*, 645 F.2d 1195 (2d Cir. 1981), *cert. denied*, 455 U.S. 1016 (1982).

Anticompetitive effects of licensing arrangements Although regulators have come to recognize the welfare-enhancing proclivities of licensing arrangements, they still have significant concerns regarding situations in which licensing arrangements impede competition which likely would have taken place in the absence of the license.[59] Specifically, the IP Guidelines make it clear that regulators will closely scrutinize agreements between actual or likely potential competitors[60] which may reduce or eliminate competition in the antitrust market(s) in which they compete or are likely to compete,[61] and that action will be taken against arrangements which facilitate market allocation or price-fixing for products or services that might be supplied by the licensees.[62]

The IP Guidelines include several examples of how licensing arrangements might, under the circumstances, be anticompetitive. For example, a license agreement which transfers little or no useful intellectual property, but which imposes restraints upon entities which otherwise might be able to compete with the licensor by using alternative technologies (e.g. restrictions on the use of goods or technologies other than the licensed subject matter and/or contractual provisions that penalize licensees for dealing with suppliers of substitute technologies), might have significant adverse effects in downstream goods markets or in other markets. Similarly, an arrangement that effectively merges the research and development activities of two of only a few entities that could plausibly engage in research and development in the relevant field might harm competition for development of new intellectual property. Also, restrictions in the license relating to one market may actual impair competition in another market, as might be the case if the restrictions foreclose access to, or raise the price of, an important input.[63]

C. Nature and Market Power of Intellectual Property Rights

Comparable treatment of intellectual property While it is clear that intellectual property has important characteristics that tend to distinguish it from more traditional forms of tangible and intangible property, the IP Guidelines make it clear that regulators will apply the same general antitrust principles to conduct involving intellectual property that it applies to conduct involving any other form of property.[64] This comparable treatment is based on the acceptance of the principal that the exclusionary rights embodied in patents, copyrights and trade secrets are similar to the rights enjoyed by owners of other forms of private property. As such, antitrust laws will have the same interest in whether or certain acquisitions of intellectual property, and certain types of agreements respecting such intellectual property, will have anticompetitive effects.

[59] IP Guidelines § 3.1.

[60] In note 14 of the IP Guidelines, it is stated that a firm will be treated as a likely potential competitor if its entry is likely under the standards of section 3.3 of the 1992 Merger Guidelines, or if there is evidence of likely actual entry by that firm. The IP Guidelines note that competitive concerns are more likely to arise in those situations where the number of actual and likely potential competitors is not large.

[61] Language in the 1988 International Guidelines focused on the scope of the intellectual property rights, stating that "[T]he owner of intellectual property is entitled to enjoy whatever market power the property itself may confer." Feeling that this condition did not provide firms with clear signposts to distinguish market power conferred by the intellectual property right from market power that is not, the IP Guidelines state the principle that "antitrust concerns may arise when a licensing arrangement harms competition among entities that would have been actual or likely potential competitors in a relevant market in the absence of the license." IP Guidelines § 3.1.

[62] See Example 1 in the IP Guidelines.

[63] IP Guidelines § 3.1.

[64] Ibid. § 2.1.

Market power aspects of intellectual property rights The IP Guidelines make it clear that the exclusionary rights embodied in patents, copyrights and trade secrets will not, in and of themselves, create a presumption that the owner of such rights has "market power."[65] The IP Guidelines note that there will often be sufficient actual or potential close substitutes for a product, process, or work covered by the intellectual property rights to prevent the exercise of market power.[66] Even in those cases where the owner of the intellectual property right does acquire some form of market power, and thus is able to obtain a significant level of profits above and beyond that which would have been anticipated in the presence of competition, the antitrust laws would not be violated to the extent that such market power is merely a product of superior skills or products, business acumen or industry.[67] On the other hand, market power achieved through the exploitation of intellectual property rights is not exempt from antitrust review when such market power is illegally acquired or maintained, or when the owner engages in unreasonable conduct in connection with such property.[68]

D. Rule of Reason Analysis of Licensing Arrangements

For a number of years, courts and regulators in the United States took a decidedly hostile approach to the restrictions which typically are included in a licensing arrangement, often concluding that such restraints were per se unlawful without any meaningful inquiry into the surrounding circumstances or the commercial motivations of the parties. However, as recognition of the importance of technology and innovation has grown, along with the appreciation of the importance of intellectual property rights in encouraging research and development, attitudes have changed, and the IP Guidelines make it clear that restraints in vertical licensing relationships should now be evaluated under the rule of reason whenever the arrangement creates significant integrative efficiencies.[69]

The IP Guidelines include examples of when the rule of reason analysis will be appropriate. For example, licensing arrangements which facilitate the combination of the licensor's intellectual property rights with complementary factors of production owned

[65] In IP Guidelines § 2.2, "market power" is defined as the ability profitably to maintain prices above, or output below, competitive levels for a significant period of time. The DOJ also notes that "market power" is often exercised in other economic dimensions, such as quality, service and innovation. The definition of "market power" in the IP Guidelines assumes that all competitive dimensions are held constant except the ones in which power is being exercised. For example, it would not necessarily be indicative of market power that a seller is able to charge higher prices for a higher-quality product. See IP Guidelines § 2.2, note 9.

[66] The case law remains unclear on this issue. For example, in *Jefferson Parish Hospital District No. 2* v. *Hyde*, 466 U.S. 2, 16 (1984), the view was expressed in dictum that if a product is protected by a patent, "it is fair to presume that the inability to buy the product elsewhere gives the seller market power". On the other hand, Justice O'Connor's concurring opinion in the same case noted that "a patent holder has no market power in any relevant sense if there are close substitutes for the patented product." See ibid. at 37 n7. See also *Abbott Laboratories* v. *Brennan*, 952 F.2d 1346, 1354–5 (Fed. Cir. 1991) (no presumption of market power from intellectual property right); and *Digidyne Corp.* v. *Data General Corp.*, 734 F.2d 1336, 1341–2 (9th Cir. 1984) (requisite economic power is presumed from copyright), *cert. denied*, 473 U.S. 908 (1985).

[67] See, e.g. *United States* v. *Grinnell Corp.*, 384 U.S. 563, 571 (1966) (market power (or even a monopoly) that is solely a "consequence of a superior product, business acumen, or historical accident" does not violate the antitrust laws). See also *United States* v. *Aluminum Co. of America*, 148 F.2d 416, 430 (2d Cir. 1945) (Sherman Act is not violated by the attainment of market power solely through "superior skill, foresight and industry").

[68] IP Guidelines § 2.2.

[69] Ibid. § 3.4.

by the licensee, align the incentives of the parties to promote the development and marketing of the licensed technology, or lead to a substantial reduction of transaction costs would normally be expected to have the requisite level of integrative efficiency.[70] In contrast to the sometimes strict and formalistic process of review in early licensing cases, regulators are now expected to take into account such divergent factors as the business of the parties to the license, the particular markets, and the purpose and effect of the restraint in question.[71]

While the IP Guidelines appear to signal that the future review of restrictions in licensing arrangements will take a more realistic approach to commercial practice, it does not mean that regulators will not challenge restraints in cases where there is no efficiency-producing integration of economic activity,[72] and the "nature and necessary effect" of the particular restraint "are so plainly anticompetitive" that it should be treated as unlawful *per se*.[73] For example, the IP Guidelines point out that *per se* illegality would be appropriate for naked price-fixing provisions, output restraints, and market division among horizontal competitors, as well as for certain group boycotts and resale price maintenance.[74]

E. Judicial Limitations on the Rights of Patent Owners

Introduction The monopoly rights created under the patent laws are recognized as a necessary exception to the general rules against monopolies as a means of providing an incentive for innovation and promoting the progress of science and the useful arts. However, the courts have developed several limitations on the rights of the patent owner. One limitation on the rights of the patent owner is expressed in the "first sale" or "exhaustion" doctrine, which provides that the rights of a patent owner under the patent laws with respect to controlling the use of the patented product expire upon the first sale of the product into the stream of commerce. Also, the courts have developed and applied the equitable doctrine of "patent misuse" to prevent a patent owner from using the patent to gain advantages beyond the scope of the patent monopoly, such as by imposing conditions which restrain competition in any unpatented goods or extend the patent beyond its term.

Resale restrictions and the "exhaustion" doctrine One of the most difficult areas in the regulation of licensing practices is the use of restrictions on the resale of patent products. Restrictions of this type arise in situations where a patent owner either manufactures the patented products himself, or has them manufactured by others under his direction, sells those products to a licensee for consumption, incorporation into other goods, or direct resale, and conditions are imposed on the manner in which the licensee can attempt to resell the licensed products. While these resale restrictions can take a number of different forms, the most common relate to the resale price, the field of use, and the territory of resale, and the courts have analyzed such restrictions under both the patent laws and the antitrust laws.

[70] Ibid.

[71] Ibid.

[72] An ordinarily suspect restraint incorporated in a licensing agreement will not escape *per se* treatment, however, if the putative integration itself is a sham or if there is an insufficient relationship between the restraint and an efficiency-producing integration. Ibid.

[73] Ibid. See, e.g. *National Society of Professional Engineers* v. *United States*, 435 U.S. 679, 692 (1978).

[74] Ibid.

The "exhaustion of monopoly" doctrine for patent law purposes was first announced by the Supreme Court in the seminal case of *Adams* v. *Burke,*[75] where the Court declared that "[When a patentee] sells a [patented article] whose sole value is in its use, he receives the consideration for its use and he parts with the right to restrict that use."[76] A number of other cases have since upheld the exhaustion doctrine, concluding that attempts to restrict the manufacture, use or sale of a patented item after it has first been "sold" are *per se* illegal as a matter of patent law, and must therefore be tested under the antitrust laws.

The impact of the exhaustion doctrine in the antitrust analysis of resale restrictions has varied over the years. In general, there is little question that sellers are not allowed to impose direct restrictions on the price at which the licensed products can be resold in downstream markets, regardless of whether the products are patented.[77] With regard to nonprice restraints on resale, the case law developments have been somewhat complex. While there were some isolated exceptions,[78] almost all nonprice resale restrictions were found to be *per se* illegal under the antitrust laws up to the mid-1970s.[79] However, in 1977 the Supreme Court made a striking reversal in this trend when, in *Continental TV, Inc.* v. *GTE Sylvania*, Inc.,[80] it adopted rule of reason analysis for such restraints.

Patent misuse doctrine The "misuse doctrine" originally arose strictly as a matter of patent law.[81] For patent law purposes, "misuse" is an equitable defense to an infringement action and does not constitute a separate cause of action which can be affirmatively brought by a licensee. Under the doctrine, the patent owner will be precluded from any attempt to enforce patent rights against all persons who may be infringing the patent for so long as the patent owner engages in activities which are deemed to be patent misuse. Patent misuse does not cause the patent to be invalid, it simply makes it unenforceable during the period of misuse. Enforceability may be reinstated at any time, even during a pending proceeding, if the misuse is "purged," which generally requires a showing that the offensive restriction or condition has been canceled and its effects have dissipated. Mere nonenforcement of an offensive restriction or condition will not, without more, be sufficient to purge the misuse.

[75] 84 U.S. 453 (1873).

[76] Ibid.

[77] See, e.g. *United States* v. *Univis Lens Co.*, 316 U.S. 241 (1942). The Supreme Court, relying on *Adams* v. *Burke*, held that price restrictions on the resale of a patented lens violated Section 1 of the Sherman Act, stating: "The patentee may surrender his monopoly in whole by the sale of his patent or in part by the sale of an article embodying the invention. His monopoly remains so long as he retains the ownership of the patented article. But sale of it exhausts the monopoly in that article and the patentee may not thereafter, by virtue of his patent, control the use or disposition of the article." 316 U.S. at 250.

[78] See, e.g. *General Talking Pictures Corp.* v. *Western Electric*, 304 U.S. 175, *aff'd on reh'g* 305 U.S. 124, 127 (1938) (a field of use restriction imposed by a patent owner on a manufacturing licensee, rather than a purchasing licensee, was not unlawful). The Supreme Court's prior decision in *Adams* v. *Burke* apparently was not relevant in this instance, since the Court treated the initial sale of the patented goods in that case as outside of the scope of the license given by the licensor.

[79] The *per se* illegality of resale restrictions reached its height in the Supreme Court's decision in *United States* v. *Arnold, Schwinn & Co.*, 388 U.S. 365 (1967).

[80] 433 U.S. 36 (1977).

[81] The Supreme Court has never applied the misuse doctrine to copyrights and trade secrets, but many of the lower courts have assumed that an analogous doctrine exists, at least in the case of copyrights. See, e.g. *Tempo Music, Inc.* v. *Myers*, 407 F.2d 503, 507 (4th Cir. 1969) (a holding of infringement was reversed on the grounds that copyright owners had failed to notify licensees of certain rights accorded those licensees by a prior judgment against the copyright owners and was thereby guilty of unclean hands).

Over the years, courts began to apply the misuse doctrine in the antitrust area to invalidate licensing provisions which, in the view of the court, amounted to an attempt by the patent owner to use the bargaining power of his patent to gain a commercial advantage in some line of commerce not covered by the patent. The doctrine was, for some period of time, widely used to strike down attempts by licensors to require the purchase of unpatented items as a condition for the license of the patent (i.e. tying arrangements),[82] impose price and territorial restrictions on the sale of patented products[83] and territorial restrictions on the sale of unpatented products made with a patented process,[84] require royalty payments after the expiration of a patent,[85] and require a licensee to refrain from dealing in competing products not covered by the patent.[86] Resale restrictions have also been found to be patent misuse.[87]

The misuse doctrine has been subject to criticism stemming from the fact that the doctrine is generally applied without any showing of an actual violation of the antitrust laws,[88] and without any attempt to analyze the economic consequences that would generally be considered to be important in other types of antitrust cases. In fact, the sole inquiry with respect to the defense has been whether or not the patent owner has attempted to "extend his monopoly" beyond the scope of the claims recognized in the subject patent.[89] Dissatisfaction with the continued rigid application of the doctrine has led to a variety of legislative proposals that would attempt to address, in broad terms, concerns regarding the misuse doctrine and the application of antitrust laws to technology-based transactions. While these initiatives have been largely unsuccessful,[90] the Patent Act has been amended to provide that certain tying arrangements will not be considered patent misuse for purposes of establishing a defense to an infringement action unless there is an explicit determination that the patent owner does, in fact, have "market power" in the "relevant market."[91]

Practical Guidelines for Assessing Licensing Arrangements The current analytical framework for assessing licensing arrangements in the United States has been nicely summarized as follows:[92]

[82] See, e.g. *Morton Salt Co.* v. *G.S. Suppiger Co.*, 314 U.S. 488 (1942).

[83] *Ansul* v. *Uniroyal*, 448 F.2d 872, 880 (2d Cir. 1971).

[84] *Robintech* v. *Chemidus Wavin*, 628 F.2d 142 (D.C. Cir. 1980).

[85] *Rocform Corp.* v. *Acitelli-Standard Concrete Wall*, 367 F.2d 678 (6th Cir. 1966).

[86] *National Lockwasher Co.* v. *George K. Garrett Co.*, 137 F.2d 255 (3d Cir. 1943).

[87] See, e.g. *Baldwin Lima Hamilton* v. *Tattnal Measuring Sys. Co.*, 169 F.Supp. 1, 28-30 (E.D. Pa. 1958), *aff'd per curiam* 268 F.2d 395 (3d Cir. 1958); *Hensley Equipment Co.* v. *Esco Corp.*, 383 F.2d 252, 262–4, *mod'fd* 386 F.2d 442 (5th Cir. 1967).

[88] See, e.g. *Morton Salt Co.* v. *G.S. Suppiger Co.*, where the Supreme Court stated: "It is unnecessary to decide whether respondent has violated the Clayton Act ... maintenance of the present suit to restrain ... manufacture or sale of the alleged infringing machines is contrary to public policy ..." 314 U.S. at 494.

[89] See, e.g. *Morton Salt Co.* v. *G.S. Suppiger Co.*, where the Supreme Court stated: "[I]t thus appears that respondent is making use of its patent monopoly to restrain competition in the marketing of unpatented articles, salt tablets, for use in the patented machines, and in aiding in the creation of a limited monopoly not within that granted by the patent." 314 U.S. at 491.

[90] For example, the National Productivity and Innovation Act of 1984 contained provisions which would have changed the law with respect to the use of per se rules to analyze antitrust and misuse issues. See H.R. 3878 and S. 1841 in the 98th Congress. While those provisions were not adopted, a portion of the bill dealing with joint ventures was enacted as the National Cooperative Research and Production Act, which is described below.

[91] See "Judicial and Administrative Analysis and Regulation of Patent Licensing Arrangements and Provisions – Tying Arrangements," in Chapter 6.

[92] Preovolos, "Intellectual Property Licensing: The Practitioner's Perspective," *Competition: The Journal of the Antitrust and Trade Regulation Law Section of the State Bar of California* 4 (1994), 8.

(i) What is the relevant market? This issue has been complicated somewhat by the IP Guidelines' recognition of so-called "technology" and "innovation" markets.
(ii) What is the nature of the relationship between the licensor and the licensee? If the licensor and the licensee are actual or potential competitors, then it is more likely that the licensing arrangement will be viewed as horizontal in nature, which means the restrictions in the arrangement will be more closely scrutinized than if the parties were in a vertical arrangement.
(iii) Is the particular restriction likely to have anticompetitive effects? A restriction generally will only have an anticompetitive effect if it precludes competition that actually would have taken place in the absence of the restriction.
(iv) What are business reasons and efficiency justifications for the use of the particular restraint? The business reason for using a particular restrict is important not only to determining if it is anticompetitive, but also to whether the procompetitive effects of the restrictions outweigh any anticompetitive effects.
(v) Does the arrangement restrict competition outside of the lawful scope of the licensed intellectual property right? This is the traditional "extension of monopoly" test that might be applied when the licensing arrangement has an effect on goods and services which fall outside of the licensed subject matter.
(vi) Who is likely to be harmed by the arrangement? This inquiry goes to the real competitive significance of any restriction, and the impacted parties might include the licensor, the licensee, consumers, and/or competitors.

F. Regulation of Joint Ventures and other Collaborative Business Arrangements under United States Antitrust Laws

Introduction The United States antitrust laws apply to the formation of joint ventures, assuming that the joint venturers acquire stock in the joint venture and/or the joint venture acquires assets from one or both of the joint venturers.[93] Joint ventures raise special considerations because they have a dual propensity for both procompetitive and for anticompetitive effects. For example, joint venturers may be able to realize transactional efficiencies, share the costs and risks of the endeavor, thereby promoting competition by providing a means for new competition by companies that may not have entered the market on their own. On the other hand, if the joint venturers would have entered the market separately, without the formation of the joint venture, then the affiliation might serve to eliminate the competition that otherwise would have existed between them. Moreover, there are always concerns that the cooperation between the parties in the joint venture may lead to collusion or collateral restraints which are unrelated to the main purpose of the joint venture.

Based on recent cases and statutory developments, the type of joint venture that is least likely to pose antitrust problems is a research and development joint venture, since its activities are typically at least a step or two removed from the level where the most sig-

[93] *See United States* v. *Penn-Olin Chemical Company*, 378 U.S. 158 (1964) (§ 7 of the Clayton Act applies to the formation of joint ventures, assuming that the joint venturers acquire stock in the joint venture and/or the joint venture acquires assets from one or both of the joint venturers).

nificant amounts of competition might be expected to occur. Production joint ventures should also be the subjected to relaxed treatment, particularly in light of the recently adopted National Cooperative Research and Production Act of 1993.[94] On the other hand, marketing joint ventures are the most likely to raise antitrust problems, whether or not the joint venture is involved in research and development and production activities. Moreover, the analysis may turn on the scope of activities of the proposed joint venture, with "one-shot" joint ventures with a limited focus (e.g. a joint venture to build a particular, multibillion dollar hydroelectric plant) much less likely to be a cause for concern as opposed to a joint venture in which the venturers integrate all their activities in a particular field.

IV. Principles for Review for Licensing Arrangements under EC Competition Law

A. Introduction

In evaluating patent licenses or similar agreements, the threshold issue under EC competition law is whether or not the provisions thereof, when taken as a whole, contravene Article 85(1).[95] In general, the test for the application of Article 85(1) in this area is the same as it is with other agreements, namely whether the license or any of its provisions may adversely affect trade between member states and prevent, restrict or distort competition within the common market to an appreciable extent. If it is concluded that Article 85(1) does apply to the license, reference is made to the block exemptions promulgated by the Commission for technology licenses. If relief is not available under the block exemptions, application can still be made to the Commission for an individual exemption.

The evaluation of patent licensing agreements under EC competition law requires consideration of a number of different legal and public policy issues, some of which are unique to the EC, but most of which are similar to those which are considered by courts and regulators in the United States. Regulators in both the United States and the EC are concerned with practices that might impair competition in the marketplace; however, EC competition laws also have a special concern about agreements that impair the free movement of goods between countries within the EC. As such, it is not surprising that the EC has paid a good deal of attention to regulating patent licensing practices, such as exclusive licenses, which use national patent rights as a means for closing territorial markets to competition from goods and technologies produced elsewhere in the common market.

Regulators have also focused on various aspects of the procompetitive and anticompetitive aspects of patent licensing. For example, while the "limited license" theory is no longer widely followed, it effectively justified any restriction imposed by the licensor on the grounds that the licensee would otherwise have no right to use the subject matter absent the license. In addition, EC competition policy is increasingly recognizing

[94] 15 U.S.C. §§ 4301–5.

[95] In Case 193/83 *Windsurfing International* v. *Commission* [1986] ECR 611 [1986] 3 CMLR 489, the ECJ held that the effect of an agreement on trade between member states should be judged as a whole rather than from individual clauses.

the need to encourage the use and dissemination of new technologies through licensing arrangements. On the other hand, the EC has consistently grappled with the concept of "the specific subject matter" of the patent as a means for determining the boundaries of the restrictions and limitations which can be lawfully imposed by a licensor. Such concerns are similar to those that underlie the exhaustion and patent misuse doctrines in the United States. The EC also has a "restriction of competitiveness theory," which explicitly focuses on the effect that provisions in the license might have on the licensee's ability to compete. Recently, cases have raised the possibility that a dominant firm's refusal to license may amount to an abuse of a dominant position that would infringe Article 86 of the Treaty. Finally, it is useful to consider the statements that the Commission has made regarding the advantages and disadvantages of collaboration in the context of a joint venture, since many of the same factors might apply to a licensing arrangement.

B. National Patent Rights and the Free Movement of Goods

In addition to the general problems associated with balancing the exclusive rights of a patent owner and the desire for free market competition,[96] EC policies in this area must take into account the need to promote the establishment of a single market among all of the member states. Since intellectual property rights currently remain as national, as opposed to EC-wide, rights,[97] the owner of an intellectual property right in one member state has the right, under certain conditions, to attempt to prevent the importation of products lawfully marketed in another member state by bring a suit for the infringement of his right under national law. However, infringement actions of this type will have the effect of obstructing the free movement of goods across the borders of the member states, thereby undermining the single market objectives of the EC.[98]

In attempting to resolve the apparent conflict between national intellectual property rights and the free movement of goods within the EC, the European Court of Justice ("ECJ") has relied primarily on an analysis of Articles 30 and 36 of the Treaty. Article 30 prohibits all quantitative restrictions and measures having equivalent effect on imports and exports between member states, a limitation which is broad enough to encompass the rights of an intellectual property owner to bring an infringement action against unauthorized importers.[99] In turn, Article 36 establishes sets out an exception to the restrictions set out in Article 30 for measures that can somehow be "justified on grounds of... the protection of industrial and commercial property."

[96] While competitive aspects of the existence and use of intellectual property rights are typically addressed as a matter of competition law, certain intellectual property law statutes also grapple with similar issues. Scc, c.g. the United Kingdom Patents Act 1977: § 44 (tying provisions), § 45 (determination of certain contracts on expiry of the patents), and §§ 48 et seq. (compulsory licenses).

[97] A Community Patent Convention aimed at creating a Europe-wide patent was signed in December 1975, but did not enter into force for want of ratification. An agreement for the second Community Patent Convention replacing the 1975 agreement was signed in Luxembourg in December 1989 (OJ 1989 L401/1), but has not yet been ratified.

[98] See also Art. 222 of the Treaty, which provides that the Treaty "shall in no way prejudice the rules in Member States governing the system of property ownership."

[99] An injunction, which generally would be awarded in the event of a successful infringement action, is, in itself, "a measure" within Art. 30 of the EC Treaty. See Case 6/81 *Industrie Diensten Groep* v. *Beele* [1982] ECR 707 [1982] 3 CMLR 102, § 7.

Article 36, being a derogation from the otherwise fundamental principle of free movement of goods, is not only limited by the very words of the Article, but also has been narrowly construed by the ECJ. For example, Article 36, by its own terms, will not save restrictions that "constitute means of arbitrary discrimination or a disguised restriction on trade between member states."[100] In addition, the ECJ has held that it is the "existence" of intellectual property rights, including patents, which is protected by Article 36, and not the actual "exercise" of such rights, which is otherwise subject to the general limitations arising from the rules of the Treaty. Also, as to the "exercise" of intellectual property rights, the ECJ has held that such exercise can only be justified for purposes of Article 30 if the action is taken for the purpose of protecting what the ECJ has come to refer to as the "specific subject matter" of the particular form of intellectual property right, a concept described in greater detail below.[101]

The case of *Centrafarm* v. *Sterling Drug*[102] was where the well-known "exhaustion of monopoly" doctrine was first developed to deal with the purchase of patented products in one member sand their resale in another member state even when the original seller is the owner of patents covering the products in the second member state. In that case, the defendant had purchased patented products from a licensed subsidiary of the plaintiff in the United Kingdom and imported the products into the Netherlands, even though the plaintiff held a corresponding Dutch patent on the products. The plaintiff's efforts to use his patent rights in the Netherlands to bar resale of the products by the defendant were unsuccessful, with the ECJ stating:

> "An obstacle to the free movement of goods may arise out of the existence, within a national legislation concerning industrial and commercial property, of provisions laying down that a patentee's right is not exhausted when the product protected by the patent is marketed in another Member State, with the result that the patentee can prevent importation of the product into his own Member State when it has been marketed in another State.
>
> Whereas an obstacle to the free movement of goods of this kind may be justified on the ground of protection of industrial property where such protection is invoked against a product coming from a Member State where it is not patentable and has been manufactured by third parties without the consent of the patentee and in cases where there exist patents, the original proprietors of which are legally and economically independent, a derogation from the principle of the free movement of goods is not, however, justified where product has been put on to the market in a legal manner, by the patentee himself or with his consent, in the Member State from which it has been imported, in particular in the case of a proprietor of parallel patents.
>
> In fact, if a patentee could prevent the import of protected products marketed by him or with his consent in another Member State, he would be able to partition off national markets and thereby restrict trade between Member States, in a situation where no such restriction was necessary to guarantee the essence of the exclusive rights flowing from the parallel patents".[103]

[100] E.g. Case 35/87 *Thetford Corporation* v. *Fiamma* [1988] ECR 3585 [1988] 3 CMLR 549 and Case 434/85 *Allen & Hanburys* v. *Generics* (U.K.) Limited [1988] ECR 1245; Cases C-235/89 and C-30/90 *Commission* v. *United Kingdom and Italy* [1992] 1 ECR 777 and 829 [1992] 2 CMLR 709; Case C-191/90 *Generics* v. *Smith Kline & French* [1992] 1 ECR 5335 [1993] 1 CMLR 89.

[101] Case 15/74 *Centrafarm* v. *Sterling Drug* [1974] ECR 1147 [1974] 2 CMLR 480.

[102] Ibid.

[103] Case 15/74 *Centrafarm* v. *Sterling Drug* [1974] ECR 1147 [1974] 2 CMLR 480.

The ECJ's decision has come to stand for the principle that Articles 30 to 36 of the Treaty provide a complete defense to an infringement action if it is shown that the products in question have been previously marketed in another member state by the proprietor of the right or with his consent. Put another way, previous marketing activities "exhaust" the ability of the owner of the right to prevent the subsequent free circulation of the products concerned throughout the EC, and no action can be brought thereafter under the patent or any parallel patent that would impede movement of the goods. The doctrine has since been expanded to apply even in those situations where the particular product was imported from a member state where patent protection was not available, provided that it had been put on the market in that member state by the patentee or with its consent.[104]

In light of the ECJ's decisions in *Centrafarm* and other cases regarding the relationship between patent rights and the free movement of goods within the EC, it was apparent that where a license was limited to permitting sales only in one member state, it was not the patent rights of the licensor which served as the basis for restricting exports by the licensee, but rather the contractual obligations included in the license itself.[105] Based on the language in the cases, license provisions of this type would not be permissible under the competition laws, which meant that it would be difficult, if not impossible, to protect a licensee in one member state from imports by licensees in other territories. However, there are still some limitations which a licensor can impose on the conduct of its licensees which can impact the economic incentives for such licensees to export the licensed products into other territories, such as requiring that licensees must manufacture only in a particular factory or in part of a member state.[106]

Predicting the effect of allowing licensees in other territories to export the licensed products into the territory of a particular licensee can be extremely difficult, given that it will depend upon a variety of market conditions and logistical factors. For example, other licensees may be unwilling to compete with a licensee in the subject territory in situations where the products are cheap or bulky, making it uneconomical for the other licensees to incur the transportation costs. On the other hand, transportation costs may be an insignificant barrier in cases where the subject matter is particularly costly or valuable to the consumer. In cases where licensees from other territories have incentives to export the licensed products into the territory, consumers therein may well be able benefit from lower prices. However, if adequate protection is not available in a particular territory, perhaps due to weaknesses in national patent laws, then no local firm will seek a license for the specified products and consumers may therefore lose access to the products, especially if the costs of transport and distribution make it unattractive for licensees in other territories to bring the products into the territory.

[104] See, e.g. Case 187/80 [1981] ECR 2063 [1981] 3 CMLR 463. On the other hand, the doctrine of exhaustion does not apply in the situation where the goods were originally marketed without the consent of the patentee in a member state where the was no patent protection. See, e.g. Case 24/67 *Parke, Davis* v. *Probel* [1968] ECR 55 [1968] CMLR 47.

[105] See, e.g. Korah, *Patent Licensing and EEC Competition Rules* Reg. 2349/84, ESC Publishing Limited: Oxford (1985), 2 (referring to Wright, "Patent and Knowhow Licensing," in Beckerleg (ed.), *EEC Competition Law* (1978), 1).

[106] Ibid. at 2–3.

C. Limited License Theory

The "limited license" theory was originally adopted in the Commission's 1962 Notice on Patent Licensing Agreements, which was not officially withdrawn until 1984.[107] Under this theory, almost any restriction included as part of a license would be permitted on the ground that but for the license, the licensee would have no right to exploit or use the patented product or process at all. Therefore, the license, when granted, "opens the door" that would otherwise be closed, and restrictions on the use of the patented invention should be viewed simply as defining the general scope of the exception from the patent monopoly that the holder of the patent would otherwise be entitled to maintain. Taken to its fullest extent, the theory might be used to justify and permit restrictions which traditionally been struck down by competition regulators as *per se* violations of the competition laws, such as attempts to control the prices at which licensees might resell patent products.

The current position of the Commission, however, effectively amounts to at least a partial rejection of the limited license theory. For example, with respect to limitations on the quantity of the products a licensee might be permitted to manufacture under the terms of a license, the block exemption for patent licenses specifies that such limitations do infringe Article 85(1),[108] thereby taking the reverse position to that which was adopted in the 1962 Patent Notice. The Commission has also taken issue with licenses which impose territorial restrictions (e.g. a license that limits sales to only one member state), holding that a license that does not include a right to export the licensed products throughout the common market will generally be regarded as infringing Article 85(1).[109] In general, the attitude of the Commission with regard to the scope of patent licenses is that "the fact that an undertaking which has granted licenses to other undertakings is the proprietor of an industrial property right does not entitle it to control the market for the product under license."[110]

D. Encouraging Technology Transfer by Licensing

Another policy consideration in regulating the content of patent licenses is the fact that one of the parties, either the licensor or the licensee, may regard the restrictions placed on the freedom of the other party as a basic precondition for consummating the license in the first place. For example, the licensee may be unwilling to incur the heavy costs and risks associated with the commercialization of the licensed technology unless it is granted exclusive rights. Similarly, a licensor might be more willing to grant a license

107 See § 1 of the Patent Notice December 24, 1962, which was withdrawn in OJ 1984 C220/14.

108 See Art. 3(5) of Regulation 2349/84 (the "Patent Licensing Regulation"), which was finally approved on July 23, 1984 and took effect on January 1, 1985 (OJ 1985, L113/34). The Patent Licensing Regulation was amended by Regulation 151/93 (OJ 1993 L21/8), with effect from April 1, 1993. References herein to "Art." and "Recitals" refer to Articles and Recitals included in the Patent Licensing Regulation unless specified otherwise in the text.

109 See, e.g. *Raymond/Nagoya*, OJ 1972 L143/49 [1972] CMLR (license to sell in "Japan, Korea, Formosa, Continental China and the Philippines" held to contain a prohibition on export "which follows from the limitation of its sales rights to that territory"); *AOIP/Beyrard*, OJ 1976 L6/8 [1976] 1 CMLR D14 (license to export to territories where there were no other licenses or assignees construed as prohibition on exports to territories where there were other assignees or licensees).

110 Case 193/83 *Windsurfing International* v. *Commission* [1986] ECR 611 [1986] 3 CMLR 489, § 42.

if the licensee is prepared to accept various restrictions. In the either case, there is a conflict. On the one hand, the restrictions may appear to impair competition; however, allowing such restrictions may be the only way for society to gain the benefits of the collaboration, which presumably could include the dissemination of new technology, increased innovation, and better exploitation of the patented invention.[111]

The Commission's policy with regard to dealing with the potentially pro-competitive effects of licensing restrictions was recently summarized in the following terms:

> "The protection from competition which certain cartels, certain monopoly positions or State-granted exclusive rights may provide reduces the incentive for the firms concerned to adapt to technological change, let alone to lead the way in introducing innovation. Competition policy measures in such areas should, by improving the efficiency of structures, free the resources required for technological progress. Similarly the breaking down of barriers and the resulting broadening of markets should create sufficient demand to warrant R & D investment and enable technological innovation to be financed.
>
> The desire to ensure convergence between specific interest and the Community's general interest is at the heart of all competition policy. As far as technological research and development are concerned, this objective is reflected in an effort to balance the legitimate interests of firms in keeping a substantial part of the benefits of their investment (without which there would be no incentive to innovate) against the common interest in ensuring that the results of research are disseminated and that there are a number of different focal points of research."[112]

E. Specific Subject Matter of the Patent Right

The notion of the "specific subject matter" has become a guiding principle behind the Commission's analysis of various restrictions and limitations in licensing arrangements. In such cases, the Commission has generally attempted to distinguish between those provisions which flow from the inherent subject matter of the licensed right, and which therefore can be said to form part of the right itself and should be tolerated, and those provisions which are merely an attempt to extend the economic power of the licensor beyond those inherent rights. However, in spite of the efforts of the Commission, it remains far from clear where the line will be drawn between terms that relate to the "existence" of the patent right, and those which fall within the scope of the "exercise" of such rights.

In 1974, the Commission explained the distinctions underlying the notion of "specific subject matter" as follows:

> "any appraisal of particular patent licensing provisions requires prior differentiation between terms which are germane to the existence, and those which relate to the exercise, of patent rights, in order to establish upon which provisions the Commission may properly rule. While the differentiation remains to be more fully worked out by future decisions of the Court, it is clear that patent licensing agreements are not automati-

[111] The Commission has explicitly recognized the potential conflicts. See, e.g. *Tetra Pak* 1 (BTG License), OJ 1988 L272/27 [1990] 4 CMLR 47 (at §§ 56 and 57).

[112] Twenty-First Report on Competition Policy (1991), §§ 49–50.

> cally within Article 85(1) if the agreements simply confer rights to exploit patented inventions against payment of royalties, but that questions of applicability of Article 85(1) arise if a grant is accompanied by terms which go beyond the need to ensure the existence of an industrial property right."[113]

In that same year, the ECJ elaborated on the concept of "specific subject matter" in the *Centrafarm* case.[114] After declaring that Article 36 of the Treaty only permits derogations from the free movement of goods where such derogations are justified to safeguard the rights which constitute the specific subject matter of any intellectual property, the ECJ went on to define such specific subject matter as follows: "In relation to patents, the specific subject matter of the industrial property is the guarantee that the patentee, to reward the creative effort of the inventor, has the exclusive right to use an invention with a view to manufacturing industrial products and putting them into circulation for the first time, either directly or by the grant of licenses to third parties, as well as the right to oppose infringements."

The next year, the Commission refused to exempt a license agreement under which the licensor agreed to restrictions in its freedom to grant licenses to other parties without the consent of the existing licensees. The Commission argued that a restriction of this type was not in the essence of the licensor's rights as a patent holder, even though the existing licensees had invested substantial amounts in the improvement of the licensed invention. The Commission would, however, have allowed differences in the amount of royalties charged to the various licensees.[115]

The "specific subject matter" of patent rights was raised once again in *Windsurfing International Inc.* v. *Commission,*[116] where the Commission struck down a provision under which a licensor had the right to terminate the agreement if the licensee changed its production site. In that case, the licensor did not have patent protection in all territories of the EC, and the licensee could have lawfully manufactured the products in other territories without paying a royalty to the licensor. In the view of the Commission, the licensor could not rely on the "specific subject matter" of the patent in order to gain the protection normally afforded by the patent in territories where it did not also have patent protection, and the licensor was simply attempting to restrict the economic freedom of the licensee through the use of a contractual provision which had nothing to do with the patent.[117]

The *Windsurfing International* case also involved a requirement that the licensees of a patented rig for sailboards could only use the rig on certain types of board, none of which were covered by patents that had been licensed to the licensees. The Commission rejected the licensor's argument that such a requirement was necessary for quality control purposes, noting that such controls did not come within the "specific subject matter" of the patent unless they related to a product covered by the patent. Moreover, said the Commission, even if the boards were covered by licensed patents, such controls must be effected in light of quality and safety criteria that had been agreed upon in advance and which were based on criteria that were objectively verifiable.[118]

[113] Fourth Report on Competition Policy, (1974), point 20.
[114] Case 15/74 *Centrafarm* v. *Sterling Drug* [1974] ECR 1147 [1974] 2 CMLR 480.
[115] Decision 75/57 *Bronbemaling* v. *Heidemaatschappji*, OJ 1975 L249/27.
[116] Case 193/83 *Windsurfing International Inc.* v. *Commission* [1986] ECR 611.
[117] Ibid.
[118] Ibid.

F. Restriction of Competitiveness Theory

The restriction of competitiveness theory looks at whether particular contractual obligations, while not strictly restrictions on the freedom of the licensee, can nonetheless be considered as falling under Article 85(1) on the basis that such obligations are sufficiently onerous to as to adversely affect the "competitiveness" of the licensee. For example, on at least two separate occasions, the Commission has objected to various methods of computing royalties due under a patent license on the grounds that the methods could not be objectively justified, and adversely affect the licensee's "competitiveness" by placing it at a competitive disadvantage.[119] This approach has been partially upheld, at least in principle, by the ECJ,[120] although doubts remain about expanding Article 85(1) to include non-restrictive contractual provisions,[121] such as amount of the consideration to be paid by the licensee for the license.[122]

G. Regulation of Licenses Under Article 86 of the Treaty

Article 86 may, in light of recent developments in the *Magill*[123] case, apply to the licensing practices of dominant companies in the EC. In *Magill*, the Commission considered the issue of when a dominant company may reserve to itself the exclusive right to product and sell a product protected under intellectual property rights by refusing to grant licenses to third parties. *Magill* involved the refusal by two United Kingdom broadcasters and one Irish broadcaster to license programming information to Magill for inclusion in a new weekly television guide that Magill planned to publish in the United Kingdom and Ireland. The broadcasters had a factual monopoly over the information, which was also subject to copyright.

Each broadcaster published information concerning its own programs in its own weekly television guide. The broadcasters also allowed newspapers in the United Kingdom and Ireland to publish the comprehensive information for all television channels for one day or for the weekend, and allowed both newspapers and the other broadcasters to publish highlights, but not full and comprehensive information, on a weekly basis. Therefore, no one had the right to publish comprehensive program listings on a weekly basis, which meant that the United Kingdom and Ireland were the only two member states without a comprehensive weekly publication covering all of the channels.

The Commission held that the broadcasters had abused their dominant position in the market for program listings by preventing the introduction of a new product, namely

[119] *AOIP/Beyrard*, OJ 1976 L6/8 [1976] 1 CMLR D14 (royalties payable after expiry of patent); and *Windsurfing*, OJ 1983 L229/1 [1984] 1 CMLR 1 (royaltIes calculated on price of complete sailboards rather than the patented rigs only).

[120] Case 193/83 *Windsurfing International* v. *Commission* [1986] ECR 611 [1986] 3 CMLR 489. However, the ECJ basically rejected the Commission's findings on the facts of the case.

[121] See Case 320/87 *Ottung* v. *Klee* [1989] ECR 1177 [1990] 4 CMLR 915, where the ECJ found that an obligation to pay royalties which was unlimited in time and extended beyond the date upon which the patent expired was not *per se* a restriction of competition and could only infringe Art. 85(1) to the extent that the licensee's freedom of action was otherwise restricted under the terms of the license agreement.

[122] For example, English courts do not appear to be prepared to examine the adequacy of the consideration for a license. See, e.g. *Chemidus Wavin* v. *TERI* [1976] 2 CMLR 387 (Ch.D.) and [1978] 3 CMLR 514 (CA).

[123] 89/205/EEC (IV/31.851, *Magill TV Guide/ITP, BBC, and RTE*) OJ 1989 L78/43.

a comprehensive weekly television guide, for which there was significant consumer demand. In effect, the broader holding of the Commission was a dominant company would, depending on the specific circumstances, violate EC competition rules by reserving its exclusive rights through the refusal to license others if the refusal evidenced abusive behavior, and specifically if the refusal to license others was intended to reserve for the rightholder a derivative market. In the *Magill* situation, the Commission ordered the broadcasters to make their program listings available to each other and to third parties, subject to a reasonable royalty.

The broadcasters appealed the Commission's decision to the Court of First Instance ("CFI"), which upheld the decision.[124] Central to the CFI's findings was its conclusions concerning the application of the amorphous EC legal concepts of the existence and the exercise of IP rights, the specific subject matter of those rights, and the relationship between these concepts and the essential function of the rights. The CFI had to decide whether the broadcasters' right to refuse to license Magill fell within the existence/specific subject matter of their copyright. If so, the next question was whether this would *per se* preclude the Treaty's application or whether the Treaty could, in some circumstances, nevertheless apply to limit rights falling within the specific subject matter of the right.

The CFI held that "in principle, the protection of the specific subject matter of a copyright entitles the copyrightholder to *reserve the exclusive right to reproduce the protected work*... However, while it is plain that the exercise of the exclusive right to reproduce a protected work is not itself an abuse, that does not apply when, *in light of the details of each individual case*, it is apparent that right is exercised in such ways and circumstances as in fact to pursue an aim manifestly contrary to Article 86, and in particular if it involves evidence of abusive conduct."[125] The CFI held that the broadcasters' exercise of their copyright to preclude the emergence of a new product in competition with their products, and for which there was significant consumer demand, constituted abusive conduct because the broadcasters were using their dominant position in the market for television listings to gain a monopoly in the derivative market for weekly guides. The CFI also seemed to be influenced by the fact that the broadcasters had authorized free reproduction of daily/weekend listings and of highlights in the United Kingdom and Ireland and comprehensive listings in other member states.

The decisions of the Commission and the CFI were appealed to the ECJ, and in June 1994, Advocate General Gulmann ("AG") issued his opinion in which he advised the ECJ to overturn both prior decisions.[126] In the AG's view, the ECJ should properly hold that a dominant company generally would not violate the competition laws by reserving an exclusive right by refusing to license others provided (i) the products to be produced under the license would compete with the rightholder's products in the sense that they would fulfill roughly equivalent consumer needs and (ii) the rightholder has not licensed the intellectual property rights to any other licensee under comparable circumstances. On the other hand, the AG did believe that a dominant company would, depending on the circumstances, violate the competition laws by (i) refusing to license non-competing

[124] Case T-69/89 *RTE* v. *Commission* [1991] ECR II 485 and Case T-76/89 *ITP Limited* v. *Commission* [1991] ECR II 575 (hereinafter "CFI Opinion").

[125] CFI Opinion at point 71 (emphasis added).

[126] Opinion of Advocate General Gulmann, *Radio Telefis Eireann and Independent Television Publications Limited* v. *Commission of the European Communities*, Joined Cases C-241/91 P and C-242/91 P, Delivered on June, 1994 (hereinafter "AG Opinion").

products; (ii) engaging in discriminatory licensing (e.g. licensing some but not others under comparable circumstances); or (iii) charging unreasonable royalties.

The AG's Opinion is provocative in the sense that it asks the ECJ expressly to adopt the view that "it is possible pursuant to Article 86 to interfere with rights within the specific subject-matter"[127] where those rights are exercised in "special circumstances."[128] Thus, the specific subject matter concept would not constitute a bright line, mechanical standard, but rather would involve deciding in each case whether special circumstances exist justifying application of Article 86. The AG also asks the ECJ to confirm that it is within the power of the Commission to order compulsory licensing to redress competition law violations in appropriate cases.[129] The AG then carves out an exception for rights deemed to be "necessary" to preserve "the essential function" of the right. "[I]f an exercise of a given intellectual property right is necessary in order for that right to be able to fulfill its *essential function*, that exercise is *not affected* by the Treaty rules."[130] Thus, circumstances which might warrant application of Article 86 *per se* do not exist if the exercise of the right at issue is deemed to be "necessary" in order to fulfill the essential function of the right. However, as expressly acknowledged by the AG, the decision by the Commission and the courts as to what is "necessary" to fulfill the essential function of the right is "not an absolute concept but a relative one."[131]

To take an example, in the case of copyright, the AG agreed with the CFI that the essential function of the right is to reward the author for creative effort and to protect his/her moral rights in the work. However, in his view, the CFI was incorrect to hold that the broadcasters had abused their dominant position by exercising their exclusive rights to preclude a competing product from entering the market because requiring the rightholder to license competing products would too severely limit the reward for creative effort. Therefore, in this circumstance, "[e]ven if the market is limited to the prejudice of consumers, the right to refuse licenses ... must be regarded as necessary in order to guarantee the rightholder the reward for his creative effort."[132] On the other hand, the AG acknowledges that a dominant company could, depending on the circumstances, violate EC competition rules by refusing to license non-competing products;[133] engaging in discriminatory licensing practices;[134] and by charging unreasonable royalties;[135] notwithstanding that this too would limit the reward for creative effort.

The AG's Opinion in *Magill* is the first serious attempt to advise the ECJ essentially to do away with the specific subject matter concept as a *per se* standard, at least in competition law cases, in favor of an approach based on whether or not special circumstances exist warranting application of Article 86. By acknowledging the limitations of the existence/exercise distinction as a *per se* standard, the AG is expressing the view taken by most observers for some time, but which has yet to be acknowledged by the ECJ. Although opponents of any standard based on a case-by-case assessment have argued that such an approach would adversely affect legal certainty in the EC, certainty always comes at the expense of fairness in individual cases. These concerns about fairness are amplified when judicial institutions try to create before the fact absolute rules that will apply,

[127] Ibid. at point 58.
[128] Ibid. at point 62.
[129] Ibid. at point 84.
[130] Ibid. at point 80 (emphasis added).
[131] Ibid. at point 81.
[132] Ibid. at point 97.
[133] Ibid. at point 96.
[134] Ibid. at points 59–61, 113–17.
[135] Ibid. at points 59–61.

for example, to industries like the information technology industry that are technically complex and fast moving.

The AG is correct to point out, as he does in his Opinion, that it is in the nature of any Article 86 analysis that it be based on an after-the-fact factual inquiry by the Commission and the courts and that this should not raise more concerns in this context than it does in any Article 86 case. What he fails to see is that this approach should be applied equally to refusals to license others to produce competing products. In any event, history has shown that any approach adopted by the ECJ to the intellectual property/competition interface, even if coached in absolute terms, will in fact result in interest balancing, in which case certainty is actually increased by acknowledging this to be the case.

The AG's Opinion is problematical in one area, at least to the extent that he would carve out an absolute exception for refusals to grant licenses for competing products. Should this be the rule that is adopted by the ECJ, the rule that dominant companies always will have carte blanche to reserve markets to themselves by refusing to license others to produce competing products could have unforeseen and perhaps extremely unfortunate consequences, if it were to be applied in the categorical manner the AG seems at times to propose. It would be far better for the ECJ to hold (as the AG advises with regard to refusals to license for non-competing products, in discriminatory licensing and unreasonable royalties) that it is for the Commission and the courts to decide on a case-by-case basis, taking into account the fundamental importance of intellectual property rights, whether special circumstances justify a finding that a refusal by a dominant company to license a competing product violates the EC competition laws.

In April 1995, the ECJ held that, in some instances, a refusal to license intellectual property rights may actually be an abuse of a dominant position contrary to Article 86, if it can be shown that the refusal was "unjustified." The ECJ noted that, in this particular case, ownership of the copyright gave various television companies a *de facto* monopoly because they were the only source of listing information for which a significant amount of consumer demand existed. The ECJ explained that the companies were "excluding all competition from that market solely in order to secure [their] respective monopolies," and it was also pointed out that the effect of their actions was to block the emergence of a new product which they themselves were not providing. The effect of the ECJ's decision is difficult to measure at this time, although intellectual property owners may now be confronted with the burden of "justifying" their licensing practices, even though the ECJ has given no guidance as to what might constitute "justification" in a given circumstance.

H. Commission's Views Regarding Collaborative Arrangements

Introduction Collaborative arrangements, which are generally referred to herein as joint ventures, present difficult legal and economic issues for competition regulators in the EC. On the one hand, any agreement which restricts the freedom of the parties to conduct their own research and development activities or which otherwise make it more difficult for a party to gain some sort of competitive advantage over another party is likely to infringe Article 85(1), especially in cases where the parties are active in research-based or technologically dynamic industries.[136] On the other hand, however, the Commission

[136] The restrictions included in the agreements generally flow from the need of the parties to impose conditions which are designed to insure that each gets a fair share of the results, as well as the benefits of any exclusive rights which may be created during the joint work.

has often taken note of the fact that there are situations where collaboration can have a number of advantages which, taken together, may override any anticompetitive effects from the collaboration, especially when the collaboration occurs one or more steps (e.g. research and development) away from the market for the final products.[137]

For example, if joint research and development leads to quicker and more efficient results, or joint production leads to greater efficiency and reduced costs, then the collaboration will have achieved the objectives generally associated with insuring that markets remain competitive.

Advantages of collaboration in innovation The Commission has made a number of public statements regarding the potential advantages of collaboration in pursuing innovation. For example, in its Fourteenth Report on Competition Policy, the Commission declared that it cannot be denied in many cases that "the synergy arising out of co-operation is necessary because it enables the partners to share the financial risks involved and in particular to bring together a wider range of intellectual and mental resources and experience, thus promoting the transfer of technology. In the absence of such co-operation, the innovation may not take place at all, or otherwise not as successfully or as efficiently ... The present situation in the Community demands a more rapid and effective transformation of new ideas into marketable products and processes, which may be facilitated by joint efforts by several undertakings."[138]

The following year, in its Fifteenth Report on Competition Policy,[139] the Commission listed the following advantages with respect to research and development collaboration:

> "(i) Investment in R&D can be kept to a minimum. Economies of scale can be achieved.
> (ii) Research budgets can be made to go further and risks spread by sharing the costs and benefits of a project between several firms or spreading a given sum over a series of relatively independent projects.
> (iii) Cross-frontier R&D collaboration within the Community can help to open up national markets. The Community's poor performance in high technology is mainly due not to too low a level of expenditure on R&D, but to the low productivity of such expenditure because of the fragmentation of markets and supply. Governments' nationalistic support of national champions in their procurement policies checks or prevents the growth of new enterprises and limits the potential supply of high-technology products. International R&D collaboration can enlarge markets and supply for the products (high technology or otherwise) incorporating the results of the joint research to a Community or even world scale."

The Fifteenth Report also contained a summary of a number of the potential benefits that the Commission might look for as it analyzes a joint venture:

[137] Bellamy and Child, *Common Market Law of Competition* 4th edn, London (1993) § 5–002. The increasing competitive importance of the joint venture form in the EC in recent years can be traced by a review of the Commission's own annual reports on competition since the middle of the last decade. See, e.g. the section generally entitled "The Development of Concentration, Competition and Competitiveness," which usually appears as Part Four of the Commission's Annual Reports on Competition Policy.

[138] See Fourteenth Report on Competition Policy (1984), 37–8.

[139] See Fifteenth Report on Competition Policy (1985), point 282.

"joint ventures can contribute to a number of general economic objectives:
- (i) integration of the internal market, especially by means of cross-border co-operation;
- (ii) facilitation of risky investment;
- (iii) promotion of innovation and transfer of technology;
- (iv) development of new markets;
- (v) improvement of the competitiveness of Community industry;
- (vi) strengthening the competitive position of small and medium-sized firms;
- (vii) elimination of structural overcapacity."[140]

The potential benefits from collaboration have also been recognized in the EC's decision to implement several programs designed to increase its competitiveness in world markets for certain high technology products. One program, ESPRIT, which was adopted by the EC Council on March 9, 1984,[141] was designed to support pre-competitive research and development collaborations involving at least two undertakings not in the same member state in the areas of advanced microelectronics, software technology, information processing, integrated computer manufacture, and office systems. On March 25, 1985, the EC Council adopted BRITE,[142] which focused on research in the application of new technologies and development of such technologies in a variety of industrial techniques, including lasers, metal shaping and forming, and testing of raw materials. Finally, RACE, adopted on August 7, 1985,[143] was designed to enhance research relating to telecommunication technology.

Anticompetitive effects of collaboration in innovation While there may be significant benefits to the use of a joint venture or any other form of cooperative arrangement, there are obviously some potential anticompetitive effects which must be taken into account by regulators. In its Fifteenth Report on Competition Policy, the Commission cautioned that the economic effects of research and development collaborations are not always wholly beneficial. Specifically, the Commission noted that:

"(i) Powerful firms may enter into R&D agreements with potentially very innovative rivals in order to be able to control technological progress. In other cases, R&D collaboration may raise entry barriers to non-participating competitors.

(ii) R&D collaboration can also facilitate coordination of pricing and production and enable abnormal profits to be made from innovations. Such dangers are greatest where, as is frequently the case in the Community, there are non-tariff barriers between national markets. In such cases, cross-frontier collaboration at the R&D stage may give way to geographical division of the market on national lines for the product resulting from the R&D."[144]

The case law in this area indicates the development of a list of potential anticompetitive effects against which is used to assess each particular joint venture. For example, one of the most direct anticompetitive effects of a joint venture is the diminution, if not outright elimination, of actual or potential competition between the various parties to

[140] Ibid., point 26.
[141] OJ L67/54.
[142] OJ L83/8.
[143] OJ L210/24.
[144] Fifteenth Report on Competition Policy (1985), point 26.

the joint venture. In addition, concerns will arise that the collaboration in the joint venture area will ultimately "spillover" into areas outside of the joint venture, thereby adversely effecting competition in other markets. Also, joint ventures may foreclose possibilities for other parties to cooperate with the partners, a situation which is of particular concern when the industry is already concentrated. Finally, the Commission is concerned about what has been referred to as the "network effect," which essentially is a concern regarding the possibility that a party may seek to divide the single EC market by using two or more parallel joint ventures, all of which are under common control, but formed in order to act separately in particular geographic markets.[145]

Regulators in both the United States and the EC have often expressed their concerns regarding the "inherent anticompetitive effects" of joint ventures. Simply stated, this means that since joint venturers are unlikely, in most cases, to want to compete with the joint venture itself,[146] if the venturers are actual or potential competitors in a concentrated market the creation of the joint venture will reduce the number of competitors. This principle appears in the *Penn-Olin* case, which was decided by the United States Supreme Court in 1964[147] and which seems to have been followed by the Commission. The Supreme Court declared that if one of the parents of a joint venture could have entered the new market independently with the threat of the other or others doing so, then the market would be more competitive.[148]

One of the potential problems in applying the doctrine of inherent anticompetitive effects is the fact that it ignores the pro-competitive aspects of the joint venture, such as the fact that the combination of the complementary resources which are owned or controlled by the venturers may allow the joint venture to enter the market more rapidly and efficiently. Also, it may be difficult to assume that all the venturers would have entered the market independently, since the entry of the first firm may have increased supply in the market to the point where it may not make any sense for additional firms to take on the expenses that might be associated with market entry. Finally, the impact of the inherent anticompetitive effects of a joint venture will be depend upon the competitive characteristics of the market, with the greatest concerns regarding restrictions on competition generally arising where the parties compete with one another in oligopolistic markets.[149]

[145] See, e.g. Bellamy and Child, supra note 166, § 5–003. See also ibid. at § 5–005 ("The Commission's principal underlying aims in relation to joint ventures are to ensure that the affected Community markets remain sufficiently competitive and are not compartmentalised."). In its Fifteenth Report on Competition Policy (1985), at point 26, the Commission noted that while joint ventures may produce a number of desirable competitive effects, they may also be anti-competitive to the extent that lead to "market sharing, raising of barriers to entry, and the intensification of market power."

[146] The Commission has indicated that even if the there is no express prohibition placed on the venturers as to competition with the joint venture, their equity share in the joint venture may, in and of itself, create a disincentive for competition. See, e.g. *GEC/Weir Pump*, OJ 1977, L327/26; [1978] 1 CMLR D42.

[147] 378 US 158 (1964). See above.

[148] Ibid.

[149] V. Korah, *R&D and the EEC Competition Rules: Regulation 418/85*, ESC Publishing, (1986), 6.

10 Development of Judicial and Regulatory Policies Relating to Patent Licensing Arrangements

I. Early Developments

A. United States

In large part, the development of antitrust law policies regarding licensing arrangements in the United States reflects the tension that necessarily exists at the interface between the content and function of the intellectual property laws and the antitrust laws. Until the late 1930s, the courts generally took the position that the antitrust laws did not apply to a patent owner's exercise of his exclusive rights, and even allowed patent licensors to impose restrictions, such as limitations on the prices which licensees of patented products could charge when offering such products for resale, which would not be permitted if the transaction were viewed as a vertical contract under the antitrust laws. However, subsequent cases often evidenced a good deal of hostility toward the concept of exclusive rights in new inventions and technologies, and worked to create a variety of limitations on the scope and use of patents in the market.

One of the first significant Supreme Court decisions regarding the scope of patent rights was not an antitrust case; however, the decision would eventually influence the attitude of the courts in later cases. In *Adams* v. *Burke*,[1] the Supreme Court first announced the "exhaustion of monopoly" doctrine. In that case, the patent owner had appointed exclusive licensees in several geographic territories to resell patented coffin lids which they had originally purchased from the patent owner. The defendant lawfully purchased the coffin lid in a territory assigned to one licensee and resold the lid in a territory that had been assigned exclusively to another licensee. The patent owner sued for infringement, arguing that any license which may have accompanied the first sale was extinguished by removing the lid from the territory in which that sale was made. However, the Court ruled for the defendant, stating: "[When a patentee] sells a [patented article] whose sole value is in its use, he receives the consideration for its use and he parts with the right to restrict that use."[2]

During the early part of this century, the courts were apparently willing to grant a good deal of latitude to a patent owner's exercise of his patent rights. The tone was set in *United States* v. *United States Shoe Machinery Co.*, where the Supreme Court said: "Of course, there is restraint in a patent. Its strength is in the restraint, the right to exclude others

[1] 84 U.S. 453 (1873).
[2] Ibid.

from the use of the invention ... Its exertion within the field covered by the patent law is not an offense against the Anti-Trust Act."[3] This principle reached its zenith in the seminal case of *United States* v. *General Electric*,[4] which dealt with an arrangement in which General Electric granted a license to Westinghouse, its principal competitor, to manufacture light bulbs under patents owned by General Electric, provided that Westinghouse agreed not to sell the light bulbs other than at prices established by General Electric for its own sales of the light bulbs. In announcing what amounted to a *per se* rule of legality for such "resale price restrictions" when a patent was involved, the Supreme Court took note of the patentee's right to "secure pecuniary reward" for its patent "monopoly,"[5] including the right of the patent owner to impose restrictions reasonably designed to optimize the return on its patent, like the right to "acquire profit" through control over the "price at which the [patented] article is sold."[6]

By the end of the 1930s, however, courts, including the Supreme Court, had begun to take the position that the exercise of a patent owner's rights should be subject to, and generally limited by, antitrust rules. This was often done by applying a two-step process in analyzing any patent licensing arrangement. The first step consisted of the court's inquiry into whether the particular provision(s) in the license were beyond what the court viewed as being the proper scope or subject matter of the patent right. Courts tended to rely on the exhaustion of monopoly in rendering its decisions on these issues. The second step in the analysis, assuming that the provision was not fully justified by reference to patent law principles, was to apply the traditional antitrust rules to the licensing practice.

One of the results of this two-step analysis was the substantial reduction in the precedential value of the *General Electric* rule. In *United States* v. *Univis Lens Co.*,[7] the Supreme Court, relying on the exhaustion of monopoly doctrine, refused to apply the *General Electric* rule to price restrictions on the resale of a patent product, and held such restrictions to be *per se* unlawful violations of the Sherman Act. Specifically, the Court noted that "[s]ale of the [patented article] exhausts the monopoly in that article and the patentee may not thereafter, by virtue of his patent, control the use or the disposition of the article."[8] Having decided that the patent rules were not applicable to the purchaser's resale practices, the Court went on review the restrictions under normal antitrust rules[9] which, in general, dictate that resale price fixing arrangements are to be considered *per se* unlawful.

[3] 247 U.S. 32, 57 (1918).

[4] 72 U.S. 476, 490 (1926).

[5] Ibid. at 489–90.

[6] bid. at 490.

[7] 316 U.S. 241 (1942). See also, e.g. *Bauer & Cie* v. *O'Donnell*, 229 U.S. 1 (1913); *Ethyl Gasoline Corp.* v. *United States*, 309 U.S. 436, 44 U.S.P.Q. 614 (1940); *Edward Katzinger Co.* v. *Chicago Metallic Mfg. Co.*, 329 U.S. 394, 72 U.S.P.Q. 18 (1947); *MacGregor* v. *Westinghouse Electric & Mfg. Co.*, 329 U.S. 402, 72 U.S.P.Q. 21 (1947); *United States* v. *General Electric Co.*, 80 F. Supp. 989, 79 U.S.P.Q. 124 (S.D.N.Y. 1948).

[8] 316 U.S. at 250.

[9] See, e.g. *Adams* v. *Burke*, 84 U.S. 453, 456 (1873); *Hobbie* v. *Jennison*, 149 U.S. 355 (1893); *Keeler* v. *Standard Folding Bed Co.*, 157 U.S. 659 (1895); *Bauer & Cie* v. *O'Donnell*, 229 U.S. 1, 18 (1913); *United States* v. *Masonite Corp.*, 316 U.S. 265,53 U.S.P.Q. 396 (1942). See also *Atari Games Corp.* v. *Nintendo of America, Inc.*, 897 F.2d 1572, 14 U.S.P.Q. 2d 1034 (Fed. Cir. 1990) ("[A] patent owner may not take the property right granted by a patent and use it to extend his power in the marketplace improperly, i.e. beyond the limits of what Congress intended to give in the patent laws.").

Courts also began to transform the patent misuse doctrine, which originally arose strictly as a matter of patent law,[10] into a tool to overturn various patent licensing practices that the courts believed amounted to an attempt by a patent owner to use the bargaining power of his patent to gain a commercial advantage in some line of commerce not covered by the patent. For example, in *Interstate Circuit* v. *United States*, the Supreme Court refused to apply the *General Electric* rule to an agreement under which several distributors of copyrighted films, at the urging of two large exhibitors, had imposed minimum admission price restrictions on all exhibitors, basing the decision on the fact that the price restriction affected commerce outside the scope of the statutory grant (i.e. films other than pictures covered by the particular license).[11]

The lower courts were quick to follow the Supreme Court's lead in apply the patent misuse doctrine. For example, in one case, the owner of a patented process and patented machine was denied enforceability under the misuse doctrine for attempting to control the selling price of the unpatented article produced by the patented process and machine.[12] In another case involving the owner of a patent on parking meters, the court ruling that the owner could not establish prices for ancillary devices, such as coin counter, which were not covered by a patent.[13] Also, in *Cummer-Graham Co.* v. *Straight Side Basket Corp.*[14] restrictions on the price of unpatented baskets which were manufactured by patented machines were held to be unlawful.

This antitrust law version of the patent misuse doctrine proved to be particular popular with courts which drew analogies between certain patent licensing practices and unlawful tying arrangements. For example, *Morton Salt Co.* v. *G.S. Suppiger Co.*[15] involved what the court deemed to be an unlawful tying arrangement in which the licensor conditioned a license on patented salt injection machines on the licensee's agreement to purchase ordinary table salt. The Court noted: "The use of a patent to suppress competition in the sale of an unpatented article may deprive a patentee of the aid of a court of equity to restrain an alleged infringement."[16] Similarly, in *National Lockwasher Co.* v. *George K. Garrett Co.*,[17] the court concluded that the restrictions imposed by the patent owner on the ability of its manufacturing licensees to produce competing products amounted to misuse of the licensed patent,[18] thereby barring enforcement of the licensed patent against infringers. The Third Circuit drew an analogy between the restrictions at issue and patent tie-ins, finding that the patent owner in this situation had "[u]sed its patent

[10] The Supreme Court has never applied the misuse doctrine to copyrights and trade secrets, but many of the lower courts have assumed that an analogous doctrine exists, at least in the case of copyrights. See, e.g. *Tempo Music, Inc.* v. *Myers*, 407 F.2d 503, 507 (4th Cir. 1969) (a holding of infringement was reversed on the grounds that copyright owners had failed to notify licensees of certain rights accorded those licensees by a prior judgment against the copyright owners and was thereby guilty of unclean hands).

[11] 306 U.S. 208, 228 (1939) ("[T]he owner of a copyright [cannot] dictate ... the admission price which shall be paid for an entertainment which includes features other than the particular picture licensed.")

[12] *Barber-Colman* v. *National Tool Co.* 136 E2d 339 (6th Cir. 1943).

[13] *United States* v. *Vehicular Parking*, 54 F. Supp. 828 (D. Del. 1944).

[14] 142 F.2d 646 (5th Cir. 1944).

[15] 314 U.S. 488 (1942).

[16] Ibid. at 491.

[17] 137 F.2d 255, 58 U.S.P.Q. 460 (3d Cir. 1943).

[18] Specifically, the owner of a patent on a compression spring lock washer device had entered into a number of licenses with manufacturing firms which provided for the licensees to manufacture its patented form of washer device. The licenses also contained express restrictions on the ability of the licensees to produce competing washers, which were material since there were other types of lock washers that the licensees might otherwise have produced.

monopoly to suppress the manufacture of possible competing goods not covered by its patent."[19]

The tying analogy was also used in the Supreme Court's decisions in *Mercoid Corp.* v. *Mid-Continent Investment Co.*[20] and *Mercoid Corp.* v. *Minneapolis-Honeywell Regulator Co.*[21] Mid-Continent had developed and patented a furnace heating system which it had licensed to Honeywell. Neither Mid-Continent nor Honeywell had been active in the furnace system business, and it was therefore not possible for either of them to sell the exact product covered by the scope of the patent claims. However, Honeywell manufactured and sold a stoker switch which had been specifically designed for use in the patented system, and provided each purchaser of the switch with a license to build and use the system covered by the patent. Mercoid manufactured a switch identical to the switch manufactured by Honeywell whose only commercial purpose was to infringe Mid-Continent's patent. Mid-Continent had offered to give Mercoid a license to make the switch, similar to the license it had given to Honeywell; but Mercoid had declined to accept the license.[22] Mid-Continent and Honeywell subsequently brought an action against Mercoid for contributory infringement.[23]

The Supreme Court, using the tying analogy, found that the suit violated the antitrust laws as an attempt to enforce the patent on the furnace system on a line of commerce (i.e. stocker switches) not precisely covered by the claims of the patent. The Court stated:

> "The necessities or convenience of the patentee do not justify any use of the monopoly of the patent to create another monopoly. The fact that the patentee has the power to refuse a license does not enable him to enlarge the monopoly of the patent by the expedient of attaching conditions to its use ... The method by which the monopoly is sought to be extended is immaterial ... When the patentee ties something else to his invention, he acts only by virtue of his right as the owner of property to make contracts concerning it and not otherwise."[24]

The Court also took the opportunity to openly express its displeasure with the entire concept of patents:

> "The grant of a patent is the grant of a special privilege 'to promote the progress of science and useful arts.' Const., Art. I, § 8. It carries, of course, a right to be free from

[19] 137 F.2d 255, 256, 58 U.S.P.Q. 460 (3d Cir. 1943). See also, e.g. *Berlenbach* v. *Anderson and Thompson Ski Co.*, 329 F.2d 782, 141 U.S.P.Q. 84 (9th Cir.), cert. denied, 379 U.S. 830 (1964); *Compton* v. *Metal Products, Inc.*, 453 F.2d 38, 172 U.S.P.Q. 263 (4th Cir. 1971), *cert. denied*, 406 U.S. 968 (1972) (misuse defense found where license barred the patentee licensor from marketing competing items so long as it was receiving royalties under the license); *McCullough* v. *Kammerer Corp.*, 166 F.2d 759, 76 U.S.P.Q. 503 (9th Cir.), *cert. denied*, 335 U.S. 813 (1948) (misuse defense found where exclusive patent license prohibited the licensee user of a patented pipe cutting device from using other competing devices); and *F.C. Russell Co.* v. *Consumers Insulation Co.*, 226 F.2d 373, 107 U.S.P.Q. 131 (3d Cir. 1955) (misuse defense found where licenses to distribute the patentee's patented line of storm windows prohibited the licensees from carrying competing lines of storm windows); *Krampe* v. *Ideal Industries, Inc.*, 347 F.Supp. 1384, 175 U.S.P.Q. 688 (N.D. Ill. 1972) (misuse defense found where patent license prohibited licensee from marketing competing devices).

[20] 320 U.S. 661 (1944).

[21] 320 U.S. 680 (1944).

[22] This fact, which apparently was ignored by the Supreme Court in its ruling, is relevant in that it is generally accepted that tying cannot occur in situations where the party in question has indicated a willingness to sell the tied items independently of one another.

[23] The doctrine of contributory infringement was developed in order to allow patent owner's to enforce their rights against parties who, while not technically infringing the patent, sell an article which is specifically adapted for use by others to infringe the patent.

[24] 320 U.S. 661, 666 (1944).

> competition in the practice of the invention. But the limits of the patent are narrowly and strictly confined to the precise terms of the grant ... It is the public interest which is dominant in the patent system ... It is the protection of the public in a system of free enterprise which alike nullifies a patent where any part of it is invalid ... and denies to the patentee after issuance the power to use it in such a way as to acquire a monopoly which is not plainly within the terms of the grant ... The patent is a privilege. But it is a privilege which is conditioned by the public purpose."[25]

This theme was continued in *United States* v. *Line Material Inc.*,[26] where the Supreme Court held *per se* unlawful a price-restrictive cross-license on two "blocking" patents which represented a major advance in the handling of electrical power. The Court was concerned about the adverse competitive effects of price restrictions, as well as the potential that firms might use the cross-licensing arrangement to conspire to fix prices, thus reducing or eliminating competition. Justice Douglas explained the Court's reasoning that such an arrangement would be beyond the intended economic scope of the patent right as follows:

> "It is said ... that he, the patentee, has that monopoly anyway-that his exclusive right to make, use, and vend would give him the right to exclude others and manufacture the invention and market it at any price he chose. That is true. But what he gets by the price-fixing agreement with his competitors is much more than that. He then gets not a benefit inherent in the right of exclusion but a benefit which flows from suppression of competition by combination with his competitors. Then he gets the benefits of the production and marketing facilities of competitors without the risks of price competition. ... In short, he and his associates get the benefits of a conspiracy or combination in restraint of competition. That is more than an 'exclusive right' to an invention; it's an 'exclusive right' to form a combination with competitors to fix the prices of the products of invention. The patentee creates by that method a powerful inducement for the abandonment of competition, for the cessation of litigation concerning the validity of patents, for the acceptance of patents no matter how dubious, for the abandonment of research in the development of competing patents. Those who can get stabilized markets, assured margins, and freedom from price cutting will find a price-fixing license an attractive alternative to the more arduous methods of maintaining their competitive positions. Competition tends to become impaired not by reason of the public's preference for the patented article but because of the preference of competitors for price fixing and for the increased profits which that method of doing business promises."[27]

Up to the later part of the 1970s, courts and regulators remained focused on the actual or potential anticompetitive effects of licensing arrangements, and failed to give serious consideration to whether or not there were any positive economic benefits to be gained from the use of particular restrictions. An important illustration of this phenomenon was *United States* v. *Arnold Schwinn & Co.*,[28] where the Supreme Court ruled that all territorial restrictions imposed by a manufacturer on reseller distributors were *per se* violations of the Sherman Act, thus ignoring the utility of territorial exclusivity as an incentive to resellers. The Court, relying on the "ancient rule against restraints on alienation,"[29] applied the exhaustion doctrine in declaring: "Once the manufacturer has parted

[25] Ibid. at 665–6.
[26] 333 U.S. 287 (1948).
[27] Ibid. at 319.
[28] 388 U.S. 365 (1967).
[29] Ibid. at 380.

with title and risk ... his effort thereafter to restrict territory or persons to whom the product may be transferred ... is a *per se* violation of Section 1 of the Sherman Act."[30]

The exhaustion doctrine was also applied to strike down field of use resale restrictions. For example, in *Ansul Co.* v. *Uniroyal, Inc.*,[31] the court considered the situation where the patent owner sold a patented agrochemical product to dealers who placed under customer restrictions with respect to the resale of the patented products. Citing the various Supreme Court cases in which the exhaustion doctrine was articulated, the court noted "a valid patent does not authorize the holder to impose... customer restrictions upon purchasers of the patented product."[32] The court went on to declare the customer restrictions to be *per se* illegal under Section 1 of the Sherman Act.

The protections formally available to patent owners were also significantly reduced by the decision of Supreme Court in *Lear* v. *Adkins*,[33] where the Court abolished the doctrine of licensee estoppel on antitrust grounds. Accordingly, licensees could thereafter challenge the validity of the patent under which they were is licensed and, if successful, could avoid the various obligations and restrictions in the license and practice the invention without compensating the licensor. As explained by the Court:

> "Licensees may often be the only individuals with enough economic incentive to challenge the patentability of an inventor's discovery. If they are muzzled, the public may continually be required to pay tribute to would-be monopolists without need of justification. We think it plain that the technical requirements of contract doctrine must give way before the demands of the public interest in the typical situation involving the negotiation of a license after a patent has issued."[34]

The attitude in the courts was largely mirrored by the policies of the Antitrust Division of the Department of Justice. In general, the Antitrust Division was extremely antagonistic to the notion that the owner of intellectual property rights could impose, through licensing arrangements, limits and conditions on the use and exploitation of governmentally-enforced property rights in inventions and ideas, and generally took the position that a number of typical licensing agreement provisions were *per se* illegal. As one DOJ official explained in 1975, the following practices, generally referred to as the "Nine No-Nos," tended to almost invariably have an "adverse effect upon competition" and should, under the DOJ's enforcement policies at that time, be considered *per se* unlawful:

[30] Ibid. at 382.

[31] 306 F.Supp. 541, 163 U.S.P.Q. 517 (S.D.N.Y. 1969), *rev'd in part on other grounds*, 448 F.2d 872 (2d Cir. 1971), *cert. denied*, 404 U.S. 1018 (1972).

[32] Ibid. See also *In re Yarn Processing Patent Validity Litigation*, 541 F.2d 1127, 1135, 192 U.S.P.Q. 241 (5th Cir. 1976), *cert. denied*, 433 U.S. 910 (1977); *Munters Corp.* v. *Burgess Industries, Inc.*, 1978-2 Trade Cas. (CCH) §62,149 (S.D.N.Y. 1978); *Baldwin-Lima-Hamilton Corp.* v. *Tatnall Measuring Systems Co.*, 169 F.Supp. 1, 120 U.S.P.Q. 34 (E.D.Pa. 1958), *aff'd per curiam*, 268 F.2d 395 (3d Cir.), *cert. denied*, 361 U.S. 894 (1959); *United States* v. *American Linen Supply Co.*, 141 F.Supp. 105, 109 U.S.P.Q. 272 (N.D. Ill. 1956); *United States* v. *Consolidated Car-Heating Co.*, 1950 Trade Cas. (CCH) §62,656 (S.D.N.Y. 1950).

[33] 395 U.S. 653, 670-671 (1969). A full discussion of *Lear* is beyond the scope of this chapter. Some useful references are McCarthy, "Unmuzzling the Patent Licensee: Chaos in the Wake of Lear v. Adkins," *George Washington Law Review* 45, (1977) 429; Smith, "Foreclosure of Licensee Validity Challenges with Procedural Barriers: Federal Jurisdiction," *Idaho Law Review* 16, (1985) 177; Schlicher, "Judicial Regulation of Patent Licensing, Litigation and Settlement Policies Created in Lear v. Adkins." *AIPLA Papers* 3 (1985).

[34] Ibid.

(i) tying patented goods to unpatented goods;
(ii) grant-back clauses;
(iii) restrictions on a purchaser's resale of patented goods;
(iv) restrictions on a licensee's ability to sell goods outside the scope of the patent;
(v) agreements not to further license without consent;
(vi) mandatory package licensing;
(vii) conditioning royalties on products not related to the patent;
(viii) restrictions on the resale of products made under a process patent;
(ix) a requirement that the patentee adhere to resale prices.[35]

The "Nine No-No's" came to dominate the manner in which licensing arrangements will negotiated and structured at that time. Since a number of the cases brought by the government during the 1970s ended with abandonment of the challenged provisions as part of a consent decree, the DOJ's attitudes ascended largely without any actual tests in the courtroom.[36] However, with the change in the Administration at the beginning of the 1980s, the policies of the Antitrust Division began to change dramatically, as a new group of regulators, many with economic training, brought a renewed appreciation of the potential social benefits which may be realized from licensing.

B. European Community

Patent licensing agreements have always presented a particular administrative and regulatory challenge for drafters of EC competition laws, and the difficulties have continued through to the date of this writing. From the first time that the DG IV began to grapple with crafting exemptions covering patent licensing agreements, it realized that a number of the covenants, restrictions and limitations that the licensor was bound to impose on the licensee to protect the licensor's continuing interest in the licensed subject matter would almost always have the type of competitive effect which would bring the agreement within the scope of Article 85(1). This meant that some sort of procedure needed to be develop to assess whether or not the agreement would be eligible for exemption under Article 85(3), particularly since regulators were quickly confronted with the possibility that literally thousands of license agreements might need review in any given year.

In many ways, the fundamental issues confronting the Commission were little different than those which dominated the development of case law and administrative attitudes in the United States: balancing the need to allow licensors to exercise a sufficient amount of discretion in order to protect their legitimate interests with respect to the licensed subject matter, while avoiding a situation where the provisions in the license agreement extended beyond the rights normally associated with the underlying patent. However, there were a set of issues which were unique to the EC, and which have dominated

[35] Bruce Wilson, "Licensing Practices: Myth or Reality?", Remarks before the American Patent Law Association (January 21, 1975). See also, speeches by Richard W. McLaren, September 25, 1969, "Common Law Protection of Unpatented Ideas"; Richard H. Stern, May 22, 1971, "Place-of-Sale Restrictions in the International Licensing of Patents and Know-How – Are There Antitrust Hazards?".

[36] See e.g. *United States* v. *Wisconsin Alumni Research Foundation*, 1970 C.C.H. Trade Cases § 73,015 (1970) (grantback discontinued).

the debate between the Commission and industry for several decades. These issues centered on the licensor's desire to grant, and the need of many licensee to receive, exclusive territorial and customer rights in the use of the licensed subject matter. While the use of exclusive rights presumably provided a variety of private and public benefits, it clashed directly with the EC's mission of removing all territorial barriers to the free movement of goods and services within the emerging common trading zone.

The first public statement by the Commission with respect to the approach that it might take with respect to patent licensing provisions was set out in the "Christmas Message," published on December 24, 1962.[37] In the Notice, the Commission articulated its view that patent licensing agreements are desirable because of their role in the development of new products and techniques, and that it would also be desirable to promote the profitability of the new products and techniques and to disseminate innovation. The Notice, which included four sections, set out a list of those patent licensing provisions which the Commission regarded, at that point in time, as falling completely outside of Article 85(1), including the following:

(i) obligations imposed on the licensee having as their objects either limitation of the exploitation of the invention to manufacture, use, or sale, and limitations to certain fields of use;
(ii) limitations imposed also to the quantity of products to be manufactured, the period and territorial area for which they were permitted to be used, and limitations on the licensee's powers of disposal, e.g. on sub-licensing;
(iii) obligations as to marking of the product with a patent number and the observance of quality standards necessary for the technically perfect exploitation of the patent, also non-exclusive improvement undertakings on a reciprocal basis between licensor and licensee; and
(iv) undertakings to ensure that the exclusivity of the grant to the licensee was protected.

The Notice stated that patent licenses which contained only those clauses listed above would not need to be notified, basing that conclusion on the belief that such clauses "entail only the partial maintenance of the right of prohibition contained in the patentee's exclusive right in relation to the licensee who in other respects is authorised to exploit the invention." The Notice also declared that all other clauses would be analyzed and reviewed on their individual merits, and specified that a general "appraisal" was not feasible for agreements relating to the joint ownership of patents, reciprocal licensing, or multiple parallel licensing, or to clauses which purported to extend beyond the period of validity of the relevant patent.

The Notice's apparent acceptance of the legitimacy of exclusive licensing arrangements was consistent with the approach that was then taken under German competition law, and was also influenced by the limited case law experience in this area and the Commission's decision to devote its limited resources to the detailed analysis of notifications of exclusive distribution and purchasing agreements.[38] The Commission's position was welcomed by industry participants and, given that licensing agreements are a

[37] See Patent Notice December 24, 1962, which was withdrawn in OJ 1984 C220/14.
[38] Goyder, *EC Competition Law*, 2d ed, Clarendon Press: Oxford (1993), 290.

common form of strategic business relationship, it was not surprising that a significant amount of pressure was brought to bear on the Commission to move quickly to consider and articulate a group exemption for patent licensing arrangements.

The Council did adopt Regulation 19/65, which gave the Commission the power to promulgate regulations which would, pursuant to Article 85(3), provide an exemption from Article 85(1) for certain types of agreements and concerted practices involving only two parties which included restrictions relating to the acquisition or use of industrial property rights that would otherwise fall within the scope of Article 85(1).[39] Unfortunately, the process of developing such an exemption proved to be quite tedious, given the need to accumulate a sufficient amount of case law upon which a group exemption could be based, as well as the Commission's decision to devote its limited resources to the detailed analysis of notifications of exclusive distribution and purchasing agreements.[40] Moreover, the inherent exclusivity associated with the subject matter of a patent licensing agreement,[41] as well as the complexity of the various restrictions typically included in such agreements, made the development of an exemption more challenging than it may have been for other commercial agreements.[42]

Within several years following the publication of the "Christmas Message," the Commission's attitude regarding the competitive effects of exclusive licenses appeared to be changing substantially, and by the beginning of the 1970s, the Commission was taking the position that exclusive licenses may well infringe Article 85(1).[43] Specifically, the Commission stated:

> "A patent confers on its holder the exclusive right to manufacture the products which are the subject of the invention. The holder may cede, by licences, for a given territory, the use of the rights derived from its patent. However, if it undertakes to limit the exploitation of its exclusive right to a single undertaking in a territory and thus confers on that single undertaking the right to exploit the invention and to prevent other undertakings from using it, it thus loses the power to contract with other applicants for a licence. In certain cases. the exclusive character of a manufacturing licence may restrict competition and be covered by the prohibition set out in Article 85(1)."[44]

In 1972, the Commission elaborated further on the issues surrounding exclusive licenses in the *Davidson Rubber*[45] case. In that case, the licensor had developed a patented process for the efficient manufacture of padded elbow rests and seat cushions for automobiles, and had granted exclusive licenses on the patented process, and the related know-how, in several different countries within the EC. The licensees had agreed to exchange

[39] Regulation 19/65/EEC.

[40] Goyder, supra note 39, 290.

[41] Another complicating factor was the inherent tension between the exclusivity granted under national laws and the overall goals of the competition laws with respect to the free movement of goods and the reduction of the restrictions on competition.

[42] For example, the patent licensing block exemption which was ultimately adopted in Regulation 2349/84, see below, turned out to be much more detailed and lengthy than the block exemptions relating to exclusive distribution agreements (Regulation 1983/83) or exclusive purchasing agreements (Regulation 1984/83). Efforts to reconcile national patent rights with EC-wide competition law policies were also accompanied by initiatives to harmonize patent laws within the EC and to develop an "EC patent," which would be effective throughout the EC following a centralized application and review procedure.

[43] Korah, *Patent Licensing and EEC Competition Rules Reg. 2349/84*, ESC Publishing Limited: Oxford (1985), 1.

[44] [1972] CMLR D70.

[45] [1972] CMLR D52.

information with each other and with the licensor regarding the usage of the manufacturing process. The licensees, who were not prohibited from exporting their products, delivered the products directly to buyers in the automobile industry, many of whom had their own resources for manufacturing the particular items. The total aggregate output of the licensees represented approximately 20 percent and 40 percent of total production in Germany and France, respectively.[46]

The Commission, noting the substantial market shares of certain of the licensees, concluded that the exclusive licensing arrangement infringed Article 85(1), since it limited the number of potential suppliers and therefore produced a noticeable effect on the level and manner of competition in the industry.[47] The Commission then went on to exempt the arrangements under Article 85(3), explaining that the licensees would not have agreed to make the required investment in the new process if they could not have been sure of exclusivity within their particular territory. Also, the Commission pointed out that the adverse effects on the level of competition would be mitigated by the fact that the resources of the purchasers improved their bargaining position with the licensees, since they could choose to manufacture the items in question on their own, and that the licensees would be able to compete with each other through the export of the goods into other territories.[48]

The *Davidson Rubber* decision placed the Commission firmly on record that it would consider exclusive licenses which were limited to one member state as infringing Article 85(1), although exemption under Article 85(3) was likely to be available in situations where exclusivity was necessary into order to induce the licensee to take on the risks associated with tooling up for the manufacture of new products which had yet to be tried in the EC. The approach taken by the Commission, which amounted to the position that exclusive licenses were *per se* violative of Article 85(1), has been subject to criticism by commentators, some of whom take the position that exclusive arrangements with respect to untried products cannot be restrictive of competition, as no competition existed before the agreement was put into place. They argued that such arrangements fell outside of Article 85(1), and that there is no need for the Commission to even review those agreements, a requirement that they believed increased the uncertainty associated with licensing arrangements and deters investment in worthwhile projects.[49]

During the remainder of the 1970s, the Commission engaged itself in an increasingly critical and detailed assessment of the justifications offered for the creation of exclusive licenses, as well as in a thorough examination of the competitive effects of the various ancillary restraints generally imposed on the parties in the context of an exclusive licensing arrangement. In one case, the Commission held that restrictions on a licensor not to appoint other licensees in a particular territory fell outside of the essence

[46] Goyder, supra note 39, at 294.

[47] Korah, supra note 44, at 1 (Commission "decided that but for the exclusive rights Davidson might have granted a license for the same area to someone else and, if it had, the licensee might have exported to other member states.").

[48] Apparently, prohibitions on exports had originally been included as part of the licensing arrangements; however, they were removed at the insistence of the Commission.

[49] See, e.g. Korah, supra note 44, at 1–2. By applying an automatic prohibition under Art. 85(1) to restrictions on conduct likely to have significant effects on the market, as it did in *Davidson*, the Commission retained the power to vet all agreements itself according to uniform criteria, since it has exclusive power to grant exemptions under Art. 85(3). This is problematical, at least in the view of one commentator, given that the Commission is so short on staff that it cannot begin to grant enough individual exemptions. Ibid. at 3.

of a patent, and thus infringed Article 85(1).[50] In another situation, the Commission actually refused to grant an exemption for a license agreement which restricted the licensor's ability to grant further licenses without the consent of existing licensees, reasoning that the restriction impeded the promotion of technical and economic progress by limiting the number of potential competitors.[51] On the other hand, the Commission did grant a negative clearance to an exclusive patent license given by an EC manufacturer to a Japanese licensee which included an export ban on the licensee from selling the licensed goods within the EC, basing that decision upon its conclusion that the nature of the licensed product was such that it was unlikely that orders would emanate from the EC and that there would thus be no effect on the competitive structure in the EC.[52]

One of the strictest assessments of the potential effect of exclusive licensing arrangements was set out in the Commission's decision in *AOIP* v. *Beyrard*, which was decided in 1976.[53] In that case, the Commission took issue with a wide range of clauses that are almost routinely included in any exclusive license agreement. Specifically, the Commission held that the grant of exclusive territorial rights, as well as prohibitions on the ability of the exclusive licensee to export the licensed goods to any country in which the licensor has either licensed the patent or assigned its rights to third parties, infringed Article 85(1) and could not be exempted under Article 85(3). The Commission also took issue with a no-challenge clause, an obligation to pay royalties during the lifetime of the most recent original or improvement patent whether or not the patent was being exploited by the licensee, and an obligation on the licensee to refrain from any kind of competition with the licensor in the field covered by the agreement.[54]

The decision in *AOIP* raised the possibility that every exclusive license would be subject to review by the Commission, creating substantial uncertainty within the business community. As a result, pressure grew for the Commission to accelerate its effort to promulgate a group exemption dealing with patent licenses. Apparently, the Commission itself conducted extensive internal discussions from late 1974 through 1977,[55] and the earliest draft of a form of group exemption was made available for consultative purposes in late 1977. This draft appeared to indicate that a group exemption would not cover any agreement which contained any of the following provisions:

[50] *Kabel -und Metallwerke Neumeyer (KMN)/Luchaire* [1975] 2 CMLR D40. The Commission did grant an exemption for the exclusive licensing arrangement under Art. 85(3), but only after requiring the parties to delete grantback obligations on the licensee, an obligation on the licensee not to sell the licensed goods within the EC, and an obligation on the licensee not to contest the validity of the patented processes.

[51] *Zuid-Nederlandsche Bronbemaling en Grondboringen* v. *Heidemaatschappij Beheer* [1975] 2 CMLR D67.

[52] *Raymond/Nagoya Rubber Company* [1972] CMLR D45. The licensed goods were plastic attachment used in the construction of cars. The nature of such items was that they were manufactured only to the special order of car manufacturers, and therefore it was unlikely that orders would be received in Japan from the EC. The license itself was primarily for the incorporation of the items into Japanese vehicles, and the licensee was at the time the largest sub-contractor in the Japanese automobile industry.

[53] [1976] 1 CMLR D14.

[54] The field of the agreement was broadly defined to include any new processes or devices in the field which may be invented by the licensor during the term of the agreement, even if such new processes or devices were based on different principles but could be used for the same purposes. If, on the other hand, the licensee developed new processes or devices which could be used for the same purposes as the licensed items but which were based on different principles, the agreement purported to obligate the licensee to continue to pay royalties on such new processes or devices. These provisions were treated as restrictions which were inherently incapable of receiving an exemption.

[55] See, e.g. Sixth Report on Competition Policy at 17, where there is a reference to a December 1974 conference which involved the Commission and governmental expert regarding the main problems in the area of patent licensing.

(i) no-challenge provisions which would restrict the ability of the licensee to challenge the validity of the relevant patent;
(ii) non-competition provisions between the licensor and licensee in respect of matters not directly covered by the relevant patents and know-how;
(iii) extension of agreements beyond the life of the most recent patents covered by the agreement unless the agreement is terminable by the parties on reasonably short notice. The group exemption would also not be available where there was a requirement to pay royalties after the expiry of the last patent;
(iv) agreements requiring purchase of unpatented supplies for use with the patented process;
(v) restrictions on the manner in which the product made with the licensed patent is sold, such as requirements that the product must be packaged in small tubes or large bottles;
(vi) exclusivity restrictions with respect to sales or exports which exceeded the period necessary in order for the new product to penetrate its market.

Additional amendments were made to this early draft following the initial publication. One of the amendments was to introduce a condition that exclusivity would only be permitted in situations where the combined licensed territory for individual licensees (or licensees within the same group) was of a maximum population of one hundred million. Also, the Commission indicated that prohibitions on the export of licensed products outside the licensed territory might be allowed where the patented products are available for sale by independent third parties on the open market by way of parallel imports and the licensee either manufactured the products itself or subcontracted out the manufacturing to a third party. This type of condition effectively denied the exemption in cases where the licensee's role was limited to sales, rather than manufacturing, presumably on the basis that protection was only necessary for the licensee in those cases where it was being forced to take on the risk and expense associated with developing manufacturing capacity.

The amended draft was roundly criticized by industry. In particular, firms voiced concern regarding the population-based conditions on the use of territorial restrictions and claimed that they would make the proposed block exemption unattractive to many potential licensors. Firms considered the limits on the size of exclusive territories to territories of no more than one hundred million to be unduly arbitrary,[56] and the Commission responded by indicating that it would be willing to consider the use of a more generous qualification.

While the Commission was wresting the difficult issues involved in preparing the draft block exemption, it was also considering many of the same matters in its decision in the well-known *Maize Seed* case.[57] In that case, the licensor, which held plant breeders' rights in certain varieties of maize seed, had granted the licensee exclusive rights to produce and sell such varieties of seed in Germany. In addition, in order to provide protection to the licensee, the licensor had agreed in the license agreement to ensure that the seeds would not be exported into Germany other than through the licensee (i.e. a restriction covering parallel trading with respect to the seeds). The licensor had previously appointed licensees

[56] The Commission had attempted to impose a similar condition in its block exemption for exclusive distribution agreements, also meeting with heavy industry opposition.

[57] 3 CMLR 434 [1978].

in France, where the licensor was located, although the licensor itself did not itself produce seeds for commercial use.

Upon consideration, the Commission held that the following obligations included in the license agreement infringed the terms of Article 85(1): (i) the obligation on the licensor or on any of its licensees to refrain from having the seeds produced or sold by other licensees in Germany; and (ii) the obligation on the licensor or on any of its licensees to refrain from producing or selling the seeds in Germany themselves. The Commission believed that the grant of an exclusive license to a licensee restricted competition because the licensor could no longer supply products in the territory, nor could it license others to do so. The Commission also found that the obligation on the licensor or on any of its licensees to prevent third parties from exporting the seeds to Germany without the licensee's consent or authorization infringed Article 85(1).[58]

The licensee appealed the Commission's decision to the ECJ, which itself explained the Commission's reasoning as follows:

> "The Commission considered that, as in the case of a patent, exclusive propagation rights granted by the owner of breeders' rights to a licensee within the Common Market are, in principle, capable of being considered to have satisfied the tests for exemption under Article 85(3). There are even circumstances in which exclusive selling rights linked with prohibitions against exporting might also be exempted, for example when the exclusivity is needed to protect small or medium-sized undertakings in their attempt to penetrate a new market or promoting a new product provided that parallel imports are not restricted at the same time."[59]

In the particular case, however, the Commission concluded that the aforementioned tests were not satisfied, especially given that the prices paid for the seed in Germany were as much as 70 percent higher than those which were paid in France.

Extensive consultations, involving industry representatives and academic and practicing lawyers, followed the circulation of this initial draft, and the first official publication of a draft patent licensing block exemption finally occurred in 1979.[60] The draft was very similar in structure to Regulation 67/67, which dealt with exclusive distribution agreements. Article 1 covered exclusive rights to manufacture and distribute specified products, and indicated that such rights would be exempted from Article 85(1), provided that the annual turnover of the licensor or licensee did not exceed one hundred million units of account. The recitals to the draft exemption stated:

> "The territorial protection that arises from exclusive sales rights and related export bans can only be allowed if it is requisite for ensuring the expansion of technical progress. The Commission accepts that this protection is necessary for the majority of undertakings as a determining factor to facilitate decisions on investments relating to the development and marketing of new technologies. For undertakings with very high turnovers this protection would not on the other hand seem appropriate-having regard to their extensive financial resources ... the turnover limit set in the Regulation will ensure that most independent undertakings in the Community that grant or take licenses will be able to qualify for the exemption. But the exemption will not be available for a number of firms which have particularly large financial resources and which, moreover, hold the

[58] See also Cases 56 and 58/64 *Consten and Grundig* v. *Commission* [1966] ECR 299 [1966] CMLR 418.

[59] [1982] ECR 2023–4.

[60] OJ C58, 3.3.1979, p.12.

> bulk of the patents in force in the Common Market. Subject to this restriction, territorial protection may be allowed for the full duration of all patents extant at the time of the licensing agreement ... the Regulation assumes that the licensee himself undertakes investments for the manufacture of the licensed product. It does not, therefore, apply to mere sales licenses."

Article 2 of the draft exemption listed a number of ancillary restrictions that, if included as part of the license agreement, would not prevent the parties from claiming the exemption. Among the clauses that might be allowed were field of use restrictions, provided that the relevant products in each of the fields from which the licensee is excluded differed materially from the products for which the license was granted; an obligation to respect the specifications of the licensor relating to minimum standards of quality; and an obligations to provide non-exclusive rights to the licensor as to improvements which the licensee may have made to the subject matter of the license during the course of using the licensed patents.

Article 3 of the draft exemption was the so-called "black list," which included fourteen restrictions which, if present in the license, would take the agreement outside the scope of the block exemption. Among the restricted clauses were no-challenge and no-competition clauses, obligations to pay royalties after the expiration of the last relevant patent, maximum quantity restrictions, and restrictions relating to the price or condition upon which the licensee should sell including restrictions on classes of customer. In addition, the license agreement could not contain provisions that prevented the licensee using his own trade mark in parallel with that of the licensor.

While the actual publication of a draft exemption was a significant achievement, it met with substantial criticism from industry, particularly with respect to the conditions which the draft exemption imposed on exclusivity provisions. Moreover, progress toward a final block exemption was slowed once again by the ECJ's difficulty in scheduling hearings on the appeal in the *Maize Seed* case. The Commission clearly had an interest in having its own administrative procedures supported by the ECJ, and this could not be accomplished under after the ECJ rendered its decision in *Nungesser* v. *Commission*[61] in 1982. As a result of these factors, the EC entered the new decade with no ultimate resolution of the patent licensing issues, and this situation would remain in place until a revised draft of the exemption was once again published in late 1983.

II. Recent Developments: the Late 1970s to the Early 1990s

A. United States

The antagonism of courts and regulators to the use patent rights in licensing agreements began to shift in the late 1970s. One key milestone occurred when the Supreme Court revisited its prior decision in the *Schwinn* case in 1977 when it heard *Continental TV, Inc.* v. *GTE Sylvania*.[62] *Sylvania* involved a location restriction imposed by a manufacturer on one of its distributors with respect to the use of trademark rights. In a striking reversal from its earlier decision in *Schwinn*, the Supreme Court expressly overruled its hold-

[61] *Breeders' rights: Maize Seed*, OJ 1978 L286/23 [1978] 3 CMLR 434 on appeal Case 258/78 *Nungesser* v. *Commission* [1982] ECR 2015 [1983] 1 CMLR 278.

[62] 433 U.S. 36 (1977).

ing in the prior case and declared that such territorial restrictions, as well as other types of vertical restrictions, should not be considered *per se* unlawful, but rather should be analyzed under the rule of reason. As the Court said:

> "Certainly, there has been no showing in this case, either generally or with respect to Sylvania's agreements, that vertical restrictions have or are likely to have a 'pernicious effect on competition' or that they 'lack ... any redeeming virtue." Ibid. Accordingly, we conclude that the *per se* rule stated in *Schwinn* must he overruled. In so holding we do not foreclose the possibility that particular applications of vertical restrictions might justify *per se* prohibition under *Northern Pac. R. Co.* But we do make clear that departure from the rule-of-reason standard must he based upon demonstrable economic effect rather than –as in *Schwinn* – upon formalistic line drawing.
>
> In sum, we conclude that the appropriate decisioFn is to return to the rule of reason that governed vertical restrictions prior to *Schwinn*. When anticompetitive effects are shown to result from particular vertical restrictions they can he adequately policed under the rule of reason, the standard traditionally applied for the majority of anticompetitive practices challenged under § 1 of the Act."[63]

The DOJ's policies with respect to patent licenses, as well antitrust enforcement in general, also changed dramatically from the beginning of the 1980s.[64] In 1981, the Antitrust Division, under the direction of Assistant Attorney General William F. Baxter, specifically rejected the application of *per se* rules of antitrust illegality to license transactions,[65] and argued that protection of intellectual property rights by the courts was an important and essential element in the country's efforts to provide stimulation for productivity and innovation. The DOJ came to hold the belief that some licensing, even though on a restricted basis, is better than no licensing at all, and that intellectual property owners should be permitted the maximum possible degree of freedom in the commercialization of inventions and new ideas. Accordingly, the DOJ began to consistently take the position that licensing restrictions that induce the licensor to exploit patents were procompetitive, not anticompetitive, and explained that the "Nine No-No's" contained more error than accuracy.

The courts also responded to a growing recognition of the importance of innovation to competitiveness in dynamic global markets, and began to place some emphasis upon the role that intellectual property rights could play as an incentive in the development of new technology and the formation of capital for risky new ventures.[66] The significant and

[63] Ibid. at 58–9 (citations omitted).

[64] The change in attitude has been attributed to the Reagan Administration's policy of appointing DOJ officials who believed in the "Chicago School" of *laissez-faire* economics, rather than the interventionist approach of the "Harvard School." See, e.g. Charles F. Rule, "Antitrust Implications of International Licensing: After the Nine No-Nos," reprinted in Trade Regulation Reports 4 (CCH) § 13 (October 21, 1986) 131; Andewelt, "Antitrust Perspective on Intellectual Property Protection," *Patent Trademark and Copyright Journal (BNA)* 30, 319 (1985); McGrath, "Patent Licensing: A Fresh Look at Antitrust Principles in a Changing Economic Environment," *Trade Regulation Reports* (CCH) 4, § 13, (April 5, 1984), 130; Millstein, "The Role of Antitrust in an Age of Technology," *Cardoza Law Review* 9 (1988) 1175; Lipsky, Remarks to ABA Antitrust Section, November 5, 1981), reprinted in *Antitrust Law Journl* 50, (1981), 518.

[65] See, e.g. speech by Roger B. Andewelt, Chief of the Intellectual Property Section, December 2, 1981, before Houston Patent Law Association: speech by Abbott B. Lipsky, Deputy Assistant Attorney General, November 5, 1981, before A.B.A. Antitrust Section, National Institute.

[66] See, e.g. the 1985 report of the President's Commission on Industrial Competitiveness, entitled *Global Competition: The New Reality*, where it is stated: "If the developers of a new technology cannot be assured of gaining adequate financial benefits from its commercialization, they have few incentives to make the huge investments required." (Vol. 1, p. 21).

fundamental changes with regard to judicial attitudes can be seen in *Dawson Chemical Co.* v. *Rohm & Haas*, where the Supreme Court said:

> "The policy of free competition runs deep in our law. It underlies both the doctrine of patent misuse and the general principle that the boundary of a patent monopoly is to be limited by the literal scope of the patent claims. But the policy of stimulating invention that underlies the entire patent system runs no less deep. And the doctrine of contributory infringement, which has been called 'an expression both of laws and morals,' *Mercoid I*, 320 U.S., at 677 (Frankfurter, J., dissenting), can be of crucial importance in ensuring that the endeavors and investments of the inventor do not go unrewarded."[67]

The decision of the majority in *Dawson* represented a major retreat from the former policies of the Court in finding *per se* violations whenever a patent licensing practice could arguably be fit into the tying analog.[68] In addition, the Court indicated that it would respect the absolute right of a patent owner to choose not to license competitors, noting that to treat as misuse the refusal to license would essentially deprive the patent owner of the right to exploit the technology exclusively on its own.[69] Finally, the Court clearly acknowledged that patent owners must be permitted to attempt to recoup their investments through the use of commercially workable schemes it the public policy objective of encouraging the development of new technologies is to be achieved. Specifically, the Court wrote:

> "Development of new uses for existing chemicals is a major component of practical chemical research. It is extraordinarily expensive. It may take years of unsuccessful testing before a chemical having a desired property is identified, and it may take several years of further testing before a proper and safe method for using that chemical is developed.
>
> Under the construction of [principles of misuse] that petitioners advance, the rewards available to those willing to undergo the time, expense. and interim frustration of such practical research would provide at best a dubious incentive. Others could await the results of the testing and then jump on the profit bandwagon by demanding licenses to sell the chemical used in the newly developed process. Refusal to accede to such a demand, if accompanied by any attempt to profit from the invention through sale of the unpatented chemical, would risk forfeiture of any patent protection whatsoever on a finding of patent misuse. As a result, non-inventors would be almost assured of an opportunity to share in the spoils, even though they had contributed nothing to the discovery. The incentive to await the discoveries of others might well prove sweeter than the incentive to take the initiative oneself (448 U.S. at 221–2)."

Another interesting case illustrating the change in the attitudes of the courts was *United States* v. *Studiengesellschaft Kohle, m.b.H.*,[70] where the court considered the legality of a licensing strategy employed by the owner of a patent covering the only commercially feasible process for producing a specified unpatented product. In that case, the patent owner had granted an exclusive license to one licensee which covered production and

[67] 448 U.S. 175 (1980).

[68] Four justices would have continued the application of the doctrine. In his dissent, Justice Stevens stated: "[E]very license granted under this patent has been conditioned on the purchase of an unpatented product from the patentee. This is a classic case of patent misuse (448 U.S. at 240)."

[69] 448 U.S. at 215.

[70] 670 F.2d 1122, 212 U.S.P.Q. 889 (D.C. Cir. 1981).

marketing of the product, and had also granted a separate series of licenses, all of which were nonexclusive, to other firms which covered only the right to produce and use the patented process internally in the production of other unpatented products. In the view of the government, the fact that the restricted items were not patented meant that the prohibition as to sales permitted the owner to "extend the patent monopoly."

The DC Circuit, in overturning the lower court's finding that the arrangement amounted to a *per se* violation of the antitrust laws because of its effect upon competition in unpatented products,[71] pointed out that the actual "restraints" on the unpatented products were no greater than those which followed from the uniqueness of the patented process, which the patentee had the lawful right to license or withhold as it saw fit. The court rejected the arguments made by the government and, relying on the defense's showing that the purpose of the prohibition on sales was merely to harmonize the licenses with an earlier exclusive sales license, held that the restriction was to be tested by a rule of reason and was lawful under the Sherman Act. Not only did the court depart radically from the former strict application of the "extension of monopoly" doctrine, it chose to make a reasoned analysis of the commercial environment in which the transaction occurred. The court indicated that the restrictions on use were actually procompetitive and socially beneficial, since it encouraged the patent owner to give licenses to end users for their own consumption which he could not have given without the restrictions.[72]

The combined effect of the changes in DOJ policies and Supreme Court attitudes was a real shift toward the promotion of technology transfer through licensing. For the first time, courts began to abandon the use of those mechanical principals that had previously been used to strike down useful licensing provisions without any attempt to actually analyze their competitive impact or the realities of the commercial setting in which the parties to the license are operating. In addition, courts and regulators began to recognize that possession of a patent was not, in and of itself, tantamount to an ability to control a market, and that other factors influenced whether or not a particular party or group would be able to successfully commercialize a specific product, technology or innovation.

The importance of intellectual property issues was also demonstrated by the creation of the Court of Appeals for the Federal Circuit in 1982,[73] in order to bring about greater uniformity on issues of patent validity and patent infringement. The court was given exclusive jurisdiction over appeals from district courts in patent cases,[74] and once an appeal goes to the Federal Circuit, the court has jurisdiction over the entire case and not just the issues of validity and infringement.[75] This new court, partly due to the fact that several of the initial appointees were prominent members of the patent bar, brought a new energy to the enforceability of patent rights, and the Federal Circuit was expected to play an increasingly important role in reconciling patent and antitrust issues

[71] *United States* v. *Studiengesellschaft Kohle, m.b.H.*, 1978-2 Trade Cas. (CCH) § 62,291 (D.D.C. 1978), *rev'd*, 670 F.2d 1122 (D.C. Cir. 1981).

[72] However, the value of *Studiengesellschaft* as a precedent must be tempered by the decision of the same court in *Robintech* v. *Chemidus Wavin, Inc.*, 628 F.2d 142, 146–9 (D.C. Cir. 1980), just a year before. In *Robintech*, the court held that patent misuse had occurred when the licensor of an apparatus patent imposed territorial restrictions on the sale of unpatented pipe made by the apparatus, irrespective of any procompetitive or socially useful effects the restrictions might have had.

[73] P, L. 97–164.

[74] 28 U.S.C. § 1295(a)(1).

[75] *Atari, Inc.* v. *JS & A Group, Inc.*, 747 F2d 1422, 1437, 223 U.S.P.Q. 1074 (Fed.Cir. 1984).

which regularly arise during patent litigation. The timing and scope of the Federal Circuit's influences remains to be seen, however, given that it has been in existence for a relatively short period and the fact that its decisions often remain subject to policies developed and articulated by the Supreme Court[76] and the other circuit courts.[77]

By the end of the 1980s, the antagonism to patents and other intellectual property rights had been almost completely replaced by a more liberal and flexible approach to review and analysis of licensing arrangements. The changing attitudes of the DOJ were reflected in its pronouncements relating to the antitrust treatment of transactions involving intellectual property rights. The DOJ published, on November 10, 1988, its "Antitrust Enforcement Guidelines for International Operations" (the "1988 International Guidelines").[78] While the 1988 International Guidelines were intended to provide guidance on all aspects of the DOJ's internal enforcement practices and procedures relating to international operations, a good deal of the information related to licensing and transfer of intellectual property rights, including trade secrets and know-how.[79] The DOJ made it clear that the 1988 International Guidelines was not a restatement of the law as it was developed by the courts, rather the purpose of the Guidelines was to provide some general guidance as to the way in which the DOJ would analyze certain commonly occurring issues in relation to its own decisions with respect to enforcement.

According to the 1988 International Guidelines, a *per se* rule would be applied to unlawful "naked" restraints of trade contained in a license agreement that inherently restrict output or raise prices, and which are not a part of a contract or other economic integration that "may generate procompetitive activities." Thus, for example, price fixing and bid-rigging among competitors would be subject to criminal prosecution under the Sherman Act. On the other hand, license agreement restraints which are not *per se* unlawful will be examined under a rule of reason, pursuant to which the DOJ first determines whether the transaction would "likely create, enhance, or facilitate the exercise of market power in any relevant market." If the transaction would not be anticompetitive, the DOJ would not challenge it; however, if market power issues exist, the DOJ would analyze whether the "risk of anticompetitive harm is outweighed by procompetitive efficiencies that the parties claim will result from the transaction or conduct."

In the 1988 International Guidelines, the DOJ acknowledged that intellectual property rights do not, in and of themselves, raise a presumption of monopoly power. Specifically, the DOJ noted:

[76] See, e.g. *Senza-Gel Corp.* v. *Seiffhart*, 8O3 F.2d 661, 667–68 (Fed. Cir. 1986), where the Federal Circuit wrote explained the difficulties associated with reinterpreting old Supreme Court decisions with regard to patent misuse as follows: "Commentators and courts have questioned the rationale appearing in Supreme Court opinions dealing with misuse in view of recent economic theory and Supreme Court decisions in non-misuse contexts ... We are bound, however, to adhere to existing Supreme Court guidance in the area until otherwise directed by Congress or by the Supreme Court."

[77] For example, the Federal Circuit has apparently limited its ability to develop its own antitrust jurisprudence by electing to follow the appropriate regional circuit with respect to matters not required for a uniform construction of the Patent Act. See *Atari, Inc.* v. *JS & A Group*, 747 F.2d 142:, 1439 (Fed.Cir. 1984). Accordingly, the Federal Circuit follows the regional circuits with respect to antitrust analysis (see. e.g. *American Hoist & Derrick Co.* v. *Sowa & Sons*, 725 F.2d 1350, 1366 (Fed. Cir. 1984) (using Ninth Circuit precedents to decide a relevant market issue)).

[78] See Trade Regulation Reports 4 (CCH) § 13, 109.

[79] In additional to intellectual property licensing agreements, the 1988 International Guidelines also analyze: (i) criminal violations of Section 1 of the Sherman Act; (ii) monopolies; (iii) mergers; (iv) joint ventures; and (v) vertical non-price distribution arrangements.

> "At one time, the Department's enforcement policy proceeded on the assumption that intellectual property rights and certain practices in licensing of those rights conflicted with the goals of the antitrust laws because intellectual property rights were thought to create 'monopolies.' Today, however, the Department recognizes that intellectual property-even a patent-does not necessarily confer a monopoly or market power in any relevant market. A patent, for example, is merely the right to exclude others from making, using or reselling the product or process covered by the patent claims. The patented product or process remains subject to competition from products or processes outside the scope of the patent that are economic substitutes."

By 1990, the apparent tension between the goals of the patent laws and the antitrust laws, which had been manifest for much of the middle part of the century, had been almost completely mitigated. In fact, the Federal Circuit wrote that "the aims and objectives of patent and antitrust laws may seem, at first glance, wholly at odds. However, the two bodies of law are actually complementary, as both are aimed at encouraging innovation, industry and competition."[80] This view was to be echoed by the DOJ in its most recent pronouncement relating to patents and intellectual property, as described below, where it is confirmed that intellectual property laws and the antitrust laws, while apparently contradictory in their aims and objectives, actually share the common purpose of encouraging innovation, industry and competition and enhancing the level of consumer welfare.[81]

The changing attitudes toward licensing arrangements were also accompanied by corresponding modifications in antitrust policies with respect to other types of collaboration business arrangements, including joint ventures. In general, contractual arrangements for the advancement of innovation were subjected to "rule of reason" analysis in most cases, rather than to the strict merger analysis which theretofore had been applied to joint ventures. In particular, research joint ventures were treated with great permissiveness under antitrust law,[82] and the passage of the National Cooperative Research and Production Act, effectively removed most of the remaining restraints on the use of joint ventures in the areas of research and production.[83]

A few commentators have argued that there is still more which could be done in reforming United States antitrust laws to promote the formation and operation of interfirm agreements designed for the purpose of developing and commercializing new technologies.[84] Teece and Jorde, for example, have argued that United States antitrust law standards are ambiguous, claiming that while these contractual arrangements for the advancement of innovation appear to be subject to a "rule of reason" analysis in most

[80] *Atari Games Corp.* v. *Nintendo of America, Inc.*, 897 F.2d 1572, 1576 (Fed. Cir. 1990).

[81] See, e.g. the 1995 Antitrust Guidelines for the Licensing of Intellectual Property § 1 ("The intellectual property laws provide incentives for innovation and its commercialization by establishing enforceable property rights for the creators of new and useful products, more efficient processes, and original works of expression ... The antitrust laws promote innovation and consumer welfare by prohibiting certain actions by firms that deter those firms and others from competing with respect to either existing or new ways of serving consumers.").

[82] See Brodley, "Antitrust Law and Innovation Cooperation, in *Collaboration Among Competitors* (1990) (hereinafter *Collaboration*) 887, 914.

[83] See, e.g. Scott, "Diversification versus Cooperation in R&D Investment," *Managerial Decision Economics* 9 (1988), 173–86.

[84] See Jorde and Teece, "Innovation and Cooperation: Implications for Competition and Antitrust," in *Collaboration*, supra note 82, 887, 899.

cases, that the legal elements and standards involved in this analysis are quite muddled.[85] Teece and Jorde have also criticized the lack of a clear "safe harbor" for horizontal contractual arrangements among firms which, taken together, have a relatively small portion of the relevant market(s).[86] They hold that, given the importance of innovation to social progress and the fact that innovation is a powerful spur to competition, antitrust policy should err toward facilitating innovation.[87]

However, many commentators take issue with Teece and Jorde's conclusions regarding the effect of antitrust policies in the area of collaborative arrangements, citing the enactment of the National Cooperative Research and Production Act.[88] They also have pointed out that while a good deal has recently been learned about the economics of innovation, extreme caution should still be exercised in making any sweeping changes with respect to antitrust policies in this area.[89] In addition, it has been pointed out that overzealous enforcement of the antitrust law has been, at best, a very minor factor in the overall decline in the competitiveness and performance of United States industries, and that blame would be more aptly placed on actions or inactions of individual firms, including a general failure to engage in cooperative activities.[90] Shapiro and Willig also argued that there is a lack of evidence as to the assertion that innovation has been stifled, or that United States firms have been hindered in their efforts to compete in foreign markets, as a result of "overly strict" antitrust policy in the United States.[91]

B. European Community

The Commission's ultimate position on the content of the block exemption for patent licenses was greatly influenced by the ECJ's decision in *Nungesser* v. *Commission*,[92] where the ECJ heard various arguments appealing the Commission's earlier findings in

[85] Ibid.

[86] Ibid. The authors noted that the antitrust laws had previously clearly recognized a "safe harbor" for mergers and acquisitions between firms that would have less than a 20 percent market share after the transaction. Id.

[87] Ibid. at 905. Jorde and Teece claim that innovation yields social returns in excess of private returns. Ibid.

[88] It has been claimed that the Act was adopted not to redress any particular deficiencies in the actual enforcement of the laws in this area, but rather simply as a symbolic gesture intended to correct overdrawn perceptions of antitrust risk, particularly among smaller firms. See Brodley, supra note 82, at 914 n1 (citing Report of House Committee on the Judiciary, 98th Cong., 2d Sess., 1984, Rep. 98-656, 10–13: The National Productivity and Innovation Act and Related Legislation, Hearings before Senate Judiciary Committee, 98 Cong., 1st and 2d Sess. Washington: U.S.G.P.O., 1984, at 234, and also noting that the General Counsel of the National Science Foundation gave testimony which appears at 235, 237 of the Report to the effect that the most important effect of the Act would be "symbolic").

[89] Ibid. at 913.

[90] Dertouzos, Lester and Solow, *Made in America: Regaining the Productive Edge*, Cambridge: MIT Press (1986), 399–446 (citing a number of factors, including outdated strategies, short-term planning horizons, technological weaknesses, neglect of human resources, government and industry tension, and a general failure of persons and institutions to act in a cooperative fashion).

[91] Shapiro and Willig, "On The Antitrust Treatment of Production Joint Ventures," in *Collaboration*, supra note 82, 929, 946 (noting that at least some of the blame for "competitiveness" problems can be placed on such things as the low domestic savings rate, the high cost of capital relative to foreign markets, and the "inevitable" narrowing of any technical advantages which the United States may have previously enjoyed over its trading partners).

[92] *Breeders' rights: Maize Seed*, OJ 1978 L286/23 [1978] 3 CMLR 434 on appeal Case 258/78 *Nungesser* v. *Commission* [1982] ECR 2015 [1983] 1 CMLR 278.

the *Maize Seed* case. Before the ECJ, the licensee, and others, argued that exclusive licenses are often the sole means by which new goods and products developed in one member state will be introduced into the market in other member states, since the licensee rarely will be willing to undertake the costs and risks of attempting to launch the new product in a new market without some protection against the prospect of direct competition from the licensor or from other parties licensed for the same purpose in that market.[93] Arguments also raised the point that a policy of automatically holding that exclusive licenses fell under Article 85(1), thereby requiring an exemption under Article 85(3), conflicted with the pursuit of a sensible competition policy.

The ECJ's decision upheld the findings of the Commission with respect to the application of Article 85(1), as well as the Commission's decision to grant an exemption under Article 85(3), and also provided the basis for much of the Commission's future policies regarding exclusive licensing arrangements. First, the ECJ set out the important distinction between the licensor's obligations not to produce or sell the seeds in Germany, either itself or through licensees, and its obligations to prevent third parties (i.e. non-licensees) from exporting seeds into Germany. The ECJ said:

> "It should be observed that those two sets of considerations relate to two legal situations which are not necessarily identical. The first case concerns a so-called open exclusive license or assignment and the exclusivity of the license relates solely to the contractual relationship between the owner of the right and the licensee, whereby the owner merely undertakes not to grant other licenses in respect of the same territory and not to compete himself with the licensee on that territory. On the other hand, the second case involves an exclusive license or assignment with absolute territorial protection, under which the parties to the contract propose, as regards the products and the territory in question, to eliminate all competition from third parties, such as parallel importers or licensees for other territories."[94]

As to the relationship between the licensor and the licensee, the ECJ recognized that exclusivity for a limited period can provide the necessary incentive for the licensees to devote their resources to innovative efforts. Specifically, the ECJ observed in its decision that a licensee "which was not certain that it would not encounter competition from other licensees for the territory granted to it, or from the owner of the right himself, might be deterred from accepting the risk of cultivating and marketing that product; such a result would be damaging to the dissemination of a new technology and would prejudice competition in the Community between the new product and similar existing products."[95] The ECJ then concluded, after taking regard of the specific nature of the products involved in the case, that open licenses were not, in and of themselves, incompatible with Article 85(1).[96]

However, as to the restrictions which the license attempted to place upon the activities of third parties who were not bound under the particular contract, the ECJ ruled as follows:

> "As regard to the position of third parties, the Commission in essence criticises the parties to the contract for having extended the definition of exclusivity to importers who

[93] See [1982] ECR at pp. 2036 et seq.
[94] Case 258/78 *Nungesser* v. *Commission* [1982] ECR 2015 [1983] 1 CMLR 278 § 53.
[95] Ibid. § 57.
[96] Ibid.

> are not bound to the contract, in particular parallel importers. Parallel importers or exporters ... had found themselves subjected to pressure and legal proceedings by I.N.R.A. ... and the applicants, the purpose of which was to maintain the exclusive position of the applicants on the German market. The Court has consistently held . . . that absolute territorial protection granted to a licensee in order to enable parallel imports to be controlled and prevented results in the artificial maintenance of separate national markets, contrary to the Treaty."[97]

A new draft of the block exemption was finally published in late 1983, and an amended version was published in early 1984. The 1984 draft incorporated a concept which had been included as part of the block exemption for exclusive distribution agreements to the effect that licensees could be prevented from active sales outside the territory whilst keeping ability to respond if unsolicited sales orders happen to be received from outside its territory. Another draft included the right for the licensor to impose an absolute ban for a period of five years on the licensee against the making even of "passive sales" in respect of orders received from outside its territory. Finally, on July 23, 1984, Regulation 2349/84 (the "Patent Licensing Regulation") was approved, and the Regulation took effect on January 1, 1985.[98]

The Patent Licensing Regulation provided a block exemption under Article 85(3) for agreements that covered territories within the EC, and agreements outside the EC which had an effect within the EC,[99] which related to assignment,[100] acquisition, licensing and sub-licensing of patents.[101] In addition to patent licenses, the Regulation also was intended to apply to hybrid patent and know-how licenses, but "only in so far as the licensed products are necessary for achieving the objects of the licensed technology and as long as at least one of the licensed patents remains in force."[102]

Article 1(1) of the Patent Licensing Regulation set out various obligations which, although recognized as restrictive of competition, could be included in licensing agreements without losing the protection afforded by the block exemption.[103] The majority of the obligations listed in Article 1(1) were designed to ensure that licensees could be allowed to exploit the licensed subject matter in a particular territory without facing direct competition from the licensor or other licensees; however, the exemption was strictly limited to situations where the licensed products were protected by parallel patents in

[97] Ibid. § 60.

[98] OJ 1985, L113/34. The Patent Licensing Regulation was amended by Regulation 151/93 (OJ 1993 L21/8), with effect from April 1, 1993. References herein to "Art." and "Recitals" refer to articles and recitals included in the Patent Licensing Regulation unless specified otherwise in the text.

[99] See Recitals (3) to (5), and Art. 1(1)(1).

[100] Recital (6) ("The Regulation should also apply to agreements concerning the assignment and acquisition of the rights referred to in point 4 above [author's note: patents] where the risk associated with the exploitation remains with the assignor ...").

[101] The Patent Licensing Regulation applied to patent applications, utility models, applications for the registration of utility models, and agreements relating to the exploitation of an invention for which application was made within one year from the date of the agreement. Art. 10. The Regulation did not, however, cover agreements regarding plant breeders' rights. Art. 5(1)(4). In addition, the Regulation did not apply to agreements relating to sales alone; these were covered by the provisions of the block exemption for exclusive distribution rights (EEC Regulation 1983/83). Recital 7. While the Patent Licensing Regulation did not, pursuant to Art. 5(1)(2), apply to patent licensing agreements between competitors who held interests in a joint venture or between one of them and the joint venture if the agreements related to the activities of the joint venture, this exclusion was later qualified by Art. 5(2)(a), introduced by Regulation 151/93.

[102] Recital 9.

[103] The availability of the exemption was not dependent on the parties accepting all of the restrictions set out in Art. 1, but operates only if the arrangement also satisfies the requirements of Art. 3. Art. 1(1).

the territory, and also applied only for so long as the patents remain in force. In most cases, the obligations included as part of Article 1 were only eligible for the exemption if the licensee manufactured the licensed product itself or had it manufactured on his behalf,[104] and thus the exemption could not be used when the licensee merely acquired and sold the licensed products.[105]

Specifically, Article 1(1) allowed the licensor to afford its licensees substantial amounts of exclusivity be agreeing that a license would not be granted to others for the same invention within the licensed territory, nor that the licensor would itself exploit the invention within the licensed territory. In turn, the licensees could be placed under certain obligations to limit its exploitation of the licensed subject matter to the licensed territory by agreement that he would not exploit the invention in territories within the common market that were reserved for the licensor; that he would not manufacture or use the licensed product, or use the patented process or know-how, in territories that were allocated to other licensees; and that he would not engage in specified competitive activities in territories that were allocated to other licensees, other than certain passive competitive actions which could not begin for at least five years from the time the product was first put on the market.[106]

Other provisions in the Patent Licensing Regulation reflected the prior decisions of the Commission and the ECJ with respect to the boundaries of the specific subject matter of the licensed patent whereby the Commission and the ECJ attempted to determine which restrictions and obligations in a patent license merely reflected the licensor's rights to exploit his exclusive rights, and which restrictions or obligations actually amounted to an unlawful extension of such rights in a manner that adversely impacted competition. Among the permitted obligations were requirements as to mandatory procurement with the licensor, minimum royalty and quantity provisions, and non-assignment clauses.[107] Provisions that were perceived as being so anti-competitive that they would preclude any of the positive effects enumerated in Article 85(3), and thus needed to be avoided in order for the group exemption to apply, included non-competition clauses which went further than the restraints described above, provisions requiring the payment of royalties on non-patented products, and customer and price restrictions.[108]

According to the recitals included in the Patent Licensing Regulation, the restrictions listed in Article 3 "may fall under the prohibition of Article 85(1),"[109] thereby implying that an individual exemption might be available for such a restriction in unusual circumstances. As a practical matter, however, the Commission took a strict attitude toward the inclusion of black-listed clauses, often condemning them without any detailed explanation in individual decisions.[110] Commentators have been critical of the Com-

[104] Art. 1(2).

[105] Recital (7). Distribution agreements are separately regulated under the block exemption for such arrangements set out in Reg. 1983/83.

[106] Outright prohibitions could be placed on what is often referred to as "active" marketing, which presumably includes actively seeking customers (by advertising or promoting goods), establishing a branch (or a subsidiary), maintaining a distribution depot, or engaging in advertising specifically aimed at other territories. On the other hand, "passive" marketing, such as filling unsolicited orders received from a customers in another territory, could only be prohibited for up to five years from the date that the product was first marketed anywhere in the EC.

[107] Art. 2.

[108] Art. 3.

[109] Recital (19).

[110] See, e.g. *Velcro/Aplix*, OJ 1985, L233/22. For a criticism of the Commission's approach to the clauses listed in Art. 3, see Korah, supra note 44, at 111 ("Several blacklisted provisions in the patent

mission's approach to the items listed in Article 3, noting that antitrust regulators in the United States appeared to have substantially changed their views on such matters in recent years in recognition that the use of restrictions of the type listed can ultimately increase the level of competition by enhancing the incentives for innovation.[111]

Article 4 set out an accelerated procedure, known as the "opposition procedure," for obtaining an individual exemption for agreements which fall outside of the grey and white clauses, but did not contain any of the prohibited clauses on the black list.[112] Under this procedure, the parties could notify the clause to the Commission and gain the security that the clause would be covered by the exemption unless the Commission acted to open a full review within six months after the notification was filed. Of course, agreements which did not qualify for the exemption under the Patent Licensing Regulation could still be notified under Article 85(3), and the parties could also seek an individual exemption.

Even though the Patent Licensing Regulation was applicable in a specific case, Article 9 of the Regulation gave the Commission the authority to withdraw the benefits provided by the Regulation in any case where it found that an arrangement involved certain restrictions of competition that could not be justified under Article 85(3).[113] For example, the exemption could be withdrawn in situations where it appears the licensed product or process was not exposed to effective competition in the licensed territory from identical products or services, or from products or services considered by users to be equivalent as to their characteristics, price and intended use.[114] The exemption could also be withdrawn if the licensor did not have the right to terminate the exclusivity granted to the licensee within five years after the agreement was made, where the licensee had not made adequate use of the patent or had not used it at all.[115]

In addition, the exemption could be withdrawn if the licensee refused to meet demand from users or resellers in the territory of other licensees; if one or both parties refused to meet demand from users or resellers in their territories who would market the product on other territories; or if either or both parties made it difficult for users or resellers to obtain the product from other resellers, or where they exercise rights or take measures to prevent resellers or users from getting or selling products that have lawfully been put on the market by the patentee or with his consent.[116] Finally, the Commission could also withdraw the exemption in situations where an arbitration award renders an agreement incompatible with Article 85(3).[117]

cont.
licensing regulation are inserted to ensure that the patentee does not have an unjustified competitive advantage. Advantage, however, depends on the balance of the contract as a whole, not on particular clauses. This tendency is regretted.").

[111] See ibid. at 61. In particular, it has been argued that imposing limitations on the options available to potential licensors for the structuring of relationships with, and among, their licensees, may unduly deter innovators from granting licenses, thereby retarding the scope and speed of the dissemination of new technologies. Ibid.

[112] Art. 4(1). However, the amount of material required to support the request could be quite extensive.

[113] Art. 9.

[114] Art. 9(2). See also *Tetra Pak I* (BTG License), OJ 1988 L272/27 [1990] 4 CMLR 47 (Commission stated that, had the licensor not renounced exclusivity, it would have withdrawn the exemption pursuant to its Art. 9 authority on the grounds that there was no effective competition for the licensed products).

[115] Art. 9(3).

[116] Arts. 9(4) and (5).

[117] Art. 9(1).

The Patent Licensing Regulation could not be used when the licensee merely acquires and sells the licensed products, and does not actually manufacture the licensed products or have the product manufactured on its behalf.[118] Accordingly, the Patent Licensing Regulation did not cover the license of patents for the sole purpose of allowing the licensee to sell and distribute the products covered by such patents. However, the Commission developed several separate group exemptions to cover various types of distribution arrangements, including exclusive and sole distribution agreements,[119] exclusive purchasing agreements,[120] and franchising agreements.[121] The Commission also developed standards which it applied in assessing selective distribution arrangements. The Patent Licensing Regulation also did not cover license agreements between joint venture partners in the context of a joint venture, and license agreements between one or both of the joint venture partners and the joint venture were covered only if certain market share limitations are not exceeded (i.e. 20 percent in the case of a production license alone; 10 percent in the case of a production and distribution license).

The first decision of the ECJ regarding patent licensing following the effectiveness of the Patent Licensing Regulation was *Windsurfing International* v. *Commission*,[122] which involved a review of the Commission's findings regarding the terms of the patent licenses entered into by Windsurfing International and its German licensees. The licenses involved patents covering the rig (i.e. mast, sail, and spars) which was normally attached to a sailboard. There was a separate market for the rig; however, usually the rigs and the boards were sold as a completely integrated unit. The boards were not covered by the licensed patents; nonetheless, the license contained a number of provisions pertaining to the boards. The Commission had held that the following restrictions placed in the licensing agreement infringed Article 85(1):

- (i) the requirement that the licensor's approval must be obtained for any type of board on which the licensee intended to place the rig, even though the licensed product had been specified as a complete sailboard;
- (ii) the obligation on the licensee to sell only complete sailboards (i.e. the patented rig and an approved board);
- (iii) the licensee's obligation to pay royalties on the entire selling price of the complete sailboard, rather than only on the price of the patented right;
- (iv) the requirement on the licensee to fix a notice on the board stating that it was "licensed" by the licensor;
- (v) no-challenge clauses with respect to the licensor's trade mark; and

[118] Recital (7).

[119] Regulation 1983/83. The group exemption for exclusive and sole distribution arrangements permits the imposition of a number of obligations on the distributor, including non-competition covenants, agreements to purchase the products for resale only from the other party to the arrangement, restrictions on various sales activities outside of the contract territory, and obligations to undertake certain promotional activities (e.g. advertising, maintenance of a sales network, employ qualified and trained personnel). However, the group exemption does not apply to reciprocal exclusive agreements between competitors, nor to non-reciprocal agreements where there is not at least one part with a total annual turnover below 100 million ECU. Also, very importantly, an exclusive distribution arrangement only qualifies for the group exemption in situations where user are able to obtain the contract goods from alternative sources, including sources outside the contract territory.

[120] Regulation 1984/83.

[121] Regulation 4087/88.

[122] Case 193/83 [1986] 3 CMLR 489.

(vi) a restriction on the licensee against manufacturing the sailboard except at certain factories in Germany.

While Advocate-General Lenz strenuously argued that only the first three restrictions listed above infringed Article 85(1), the Court of Justice ultimately concurred with the findings which were made by the Commission on all matters other than item (iii). The ECJ concurred with the Commission's objections regarding the requirement that the licensor must approve the boards which were to be used with the rigs, noting that the licensor had failed to show that their control over the manufacture of non-patented components, such as boards, related to any objective criteria relating to the patented rig, particularly since no technical method of verifying the quality of the board was provided by the agreement. Specifically, the ECJ noted that the requirement did not come within "the specific subject matter of the patent."

As to item (iii), however, the ECJ found that the method of calculating royalties simply on the entire sailboard might have a small effect in restricting competition with regard to the board but no effect on the sale of rigs, particularly since the royalties levied on the sale of complete sailboards proved to have been no higher than that laid down for the sale of separate rigs. Finally, the ECJ did not accept the licensor's argument with regard to item (vi) that the restriction of manufacturing to one location was necessary in order to ensure quality control, particularly since numerous licensees throughout Germany were manufacturing components for the rig and the licensor could not have reasonably imposed conditions of quality control over all such manufacturers.

The ECJ also opined that it was not necessary for the Commission to examine each restrictive clause included as part of an agreement to see if the clause individually had an effect on trade between member states in order for Article 85(1) to become applicable to the agreement. Instead, it was sufficient if a finding could be made that the particular agreement itself, when taken as a whole, had the requisite effect on trade. In the case before it, the ECJ apparently was satisfied that the requisite effect existed due to the fact that the level of imports of complete boards into the EC amounted to about 20 percent of national markets and the level of imports of their individual components was about 10 percent of national markets.

The Patent Licensing Regulation was supplemented several years later by Regulation 556/89, which became effective on April 1, 1989. Regulation 556/89 was a block exemption which covered know-how licensing agreements which did not involve any other forms of intellectual property rights, as well as mixed licensing agreements (e.g. patent and know-how licenses) which are not covered by the terms of the Patent Licensing Regulation. Know-how was defined in Regulation 556/89 as non-patented technical information which is "secret," that is, not generally known or easily accessible, "substantial," which means the know-how is necessary, or of decisive importance, for the whole or a major part of the manufacturing process, or a product, or the development thereof, and identified in the license agreement or in other appropriate documentation.

Regulation 556/89 was similar in format and approach to the Patent Licensing Regulation, and reflected the Commission's realization that so long as know-how remains confidential and has some amount of intrinsic commercial value which licensees are prepared to pay for, restrictions can be placed on its use. Regulation 556/89 permitted the use of non-disclosure obligations and restrictions on the use of the licensed know-how which extend beyond the term of the agreement, provided that the know-how remained secret. On the other hand, licensees could be restricted from use know-how that comes into the public domain, nor can they be subject to price, quantity, or customer restrictions. Also, licensees could not be restricted from engaging in competing activities; however, it was permissible to provide for the loss of exclusivity in the event of competing activities.

III. United States: 1995 Antitrust Guidelines for the Licensing and Acquisition of Intellectual Property Rights

A. Introduction

The portions of the 1988 International Guidelines dealing with intellectual property rights have recently been replaced by the 1995 Antitrust Guidelines for the Licensing of Intellectual Property, which were finally adopted on April 6, 1995 (the "IP Guidelines").[123] In announcing the IP Guidelines, Richard A. Gilbert, the then-Deputy Assistant Attorney General in the Antitrust Division, noted that "[t]echnology licensing and related partnerships are essential in today's economy to remain globally competitive and to market the products that knowledge assets help to create."[124] He also pointed out that the IP Guidelines were intended to share the core principles that were expressed in the section on technology licensing included in the 1988 International Guidelines, namely:

> "the generally procompetitive nature of licensing arrangements, the absence of a presumption that intellectual property necessarily creates market power in the antitrust context, and the validity of applying the same general antitrust approach to the analysis of conduct involving intellectual property that [the DOJ and FTC] apply to conduct involving other forms of tangible or intangible property."[125]

The IP Guidelines' description of the procompetitive and anticompetitive effects of licensing arrangements, nature and market power of intellectual property rights, and the standard of analysis (i.e. rule of reason) for licensing relationships has already been discussion as part of the general guiding principles for regulatory and judicial review of licensing agreements. However, the IP Guidelines also provided valuable guidance regarding the identification and analysis of the relevant markets, including so-called "innovation markets",[126] for analyzing any particular licensing agreement; determining

[123] The IP Guidelines were first published for comment in August 1994, and were ultimately adopted, with some modifications from the initial draft, by the DOJ and the FTC on April 6, 1995. The new guidelines explicitly supersede section 3.6 in Part I, "Intellectual Property Licensing Arrangements," and cases 6, 10, 11, and 12 in Part II of the 1988 International Guidelines. The IP Guidelines cover licensing of patents, copyrights, trade secrets and know-how, but not trademarks. The IP Guidelines state that "[a]lthough the same general antitrust principles that apply to other forms of intellectual property apply to trademarks as well, these Guidelines deal with innovation-related issues that typically arise with respect to patents, copyrights, and trade secrets, rather than with product-differentiation issues that typically arise with respect to trademarks." Specific language was added to the final version of the IP Guidelines explaining that they applied to know-how also.

[124] The text of Gilbert's speech was released by the DOJ and the FTC on April 6, 1995 (the "Gilbert Speech").

[125] Ibid.

[126] The Gilbert Speech, ibid., stated: "An innovation market can be useful to identify competitive effects that cannot be adequately analyzed in markets for goods and services when the arrangement may affect the quantities, availabilities, or prices of products that do not presently exists, as in a research and development joint venture. Innovation markets also may be useful when an arrangement has competitive effects from research and development in geographic markets where product market competition is limited or nonexistent."

whether a particular arrangement should be analyzed as a horizontal or a vertical relationship; and the steps to be followed in conducting a rule of reason analysis of a licensing arrangement. In addition, the IP Guidelines establish, for the first time, an antitrust "safety zone" with respect to licensing transactions.[127]

B. Market Identification and Analysis

Introduction As with any transaction subject to antitrust scrutiny, market identification and analysis is a key issue. The IP Guidelines identify three distinct "markets" in which competition may be affected by a licensing agreement.[128] As to two of these markets, the "goods" and "technology" markets, the IP Guidelines state that market scope and market share measurement issues will be approached in a manner similar to that which the DOJ uses in analyzing horizontal relationships under its 1992 Merger Guidelines. However, in a move which created a good deal of controversy, the IP Guidelines also introduced a third market, referred to as an "innovation" market, as a tool for assessing if a particular licensing arrangement is likely to substantially reduce investment in research and development.

Goods markets Obviously, there may be numerous goods markets which are affected by a particular licensing arrangement. For example, a restraint set out in the licensing agreement may impact the market for intermediate and final goods produced with the licensed intellectual property. In addition, the licensing agreement may have effects in upstream markets, including markets for goods which are used as inputs, along with the intellectual property, to the production of other goods.[129]

Technology markets The concept of "technology" markets looks at intellectual property rights themselves as independent commodities, which can be licensed, transferred, or acquired in the same manner as any other form of asset. In a technology market, the licensed intellectual property will compete with other technologies which can be used as a means for achieving the same objective (i.e. substitutes), and by other goods that compete with the goods that are produced using the intellectual property and by other processes used to produce those other goods.[130] In many cases, there is little or no reason to attempt to analyze competitive effects in a technology market separately from those in a goods market. This is particularly true in situations where it is will be difficult to determine the actual prices for the technology and its substitute processes.[131]

The IP Guidelines contain a procedure for estimating market shares in "technology" markets, and indicate that forecasts of market acceptance will generally extend over a

[127] Specifically, neither the DOJ or FTC can be expected, absent some extraordinary circumstances, to challenge a restraint included in a licensing arrangement if "(1) the restraint is not facially anticompetitive, and (2) the licensor and its licensees collectively account for no more than 20 percent of each relevant market significantly affected by the restraint." IP Guidelines § 4.3.

[128] Ibid. § 3.2.

[129] Ibid. § 3.2.1.

[130] Ibid. § 3.2.2.

[131] As noted in IP Guidelines § 3.2.2, pricing estimates may be difficult when the technology is licensed royalty-free in exchange for the right to use other technology, or when it is licensed as part of a package license.

two-year period.[132] The DOJ intends to define technology markets by using the general approach described in the 1992 Merger Guidelines. The IP Guidelines include an acknowledgement that intellectual property may not be quantifiable in monetary terms, such as technology that is licensed royalty-free in exchange for the right to use other technology. If that is the case, the IP Guidelines provide for delineating the relevant technology market by identifying comparable substitutes to the licensed technology.

Innovation markets The concept of an "innovation" market includes those firms which possess the specialized assets or characteristics which are thought to be necessary in order to successfully conduct research and development which might result in new or improved products or processes. Participants in an innovation market include not only firms which are currently participating in the relevant goods or technology markets, but other firms which, while not active in such markets, nonetheless have the assets necessary to enter the market at some point in the future. In order to recognize an innovation market, the capacity to conduct the specific research and development activity necessary for innovation must be scarce, and one must be able to relate the capacity for innovation to specialized assets or characteristics of specific, and reasonably identifiable, firms.[133] While competition in the innovation market can be assessed independently, the IP Guidelines note that innovation markets can simply be used to identify competitive effects in relevant goods and technology markets.[134]

Defining an innovation market can be quite difficult, and generally serves no purpose if hypothetical coordination among the firms in the candidate market would not profitably retard or restrict innovation in the technology.[135] One factor that will be taken into account is the number of firms with the capabilities required to conduct the necessary research and development. If the number of firms with the required capability is small, it is likely that the innovation market concept will be evoked. On the other hand, if the number of firms with the required capability is very large (either because there are a large number of such firms in the particular industry, or because many firms in other industries have the required capability), then regulators will generally conclude that the innovation market is competitive, and that it is unlikely that any single firm or plausible aggregation of firms could acquire a large enough share of those specific assets which are necessary for innovation to have an adverse impact on competition.[136]

When a relevant innovation market is identified, the actual competitive significance of each firm will be assessed by referring to a number of factors, including the shares of the identifiable assets or characteristics upon which innovation depends, shares of research and development expenditures, shares of the related product, or equal shares

[132] When estimating the market share of a participant using new technology, the DOJ generally will forecast market acceptance over a two-year period using the best available information. For technologies not yet commercialized, the two-year period will begin with commercial introduction. When market shares or other indicia of market power are not readily available, and it appears that competing technologies are all equally efficient, the DOJ's analysis will treat each participant in the technology market as having an equal market share. IP Guidelines § 3.2.2. Two technologies will be regarded as being "comparably efficient" if they can be used to produce, at the same cost, goods perceived by consumers to be close substitutes. Ibid. n 22. See Example 2 of the IP Guidelines.

[133] IP Guidelines § 3.2.3. See also Complaint, *United States* v. *General Motors Corp.*, Civ. No. 93-530 (D. Del., filed November 16, 1993).

[134] See, e.g. Complaint, *United States* v. *Flow International Corp.*, Civ. No. 94-71320 (E.D. Mich., filed April 4, 1994).

[135] IP Guidelines § 3.2.3.

[136] Ibid. Example 3.

that are assigned to reflect the equal likelihood of innovating.[137] In evaluating competitive effects, consideration will also begin to other factors, such as the nature of the restraint, competitive harms resulting from the elimination of alternative research paths, the likelihood that other firms may eventually acquire the requisite capabilities, and any efficiency benefits from the integration of complementary research and development programs.[138] Where market data is unavailable to accurately identify a relevant innovation market, the IP Guidelines state that equal market shares may be assigned to entities with similar capabilities and incentives to pursue research and development as do the actual parties to the licensing arrangement.[139]

C. Horizontal and Vertical Relationships

Introduction According to the IP Guidelines, the antitrust analysis of a particular licensing arrangement will depend upon whether the relationship of the parties (i.e. the licensor and the licensee) is primarily horizontal or vertical in nature, or whether the relationship has substantial aspects of both types.[140] In most cases, licensing arrangements will be characterized as a form of vertical relationship, since a license is usually entered into as a means of integrating complementary skills and resources from different levels of the chain of production and distribution. On the other hand, when the parties to the licensing arrangement would be actual or likely competitors in a particular market without the agreement, regulators will closely scrutinize the transaction, since such agreements often create the potential for anticompetitive restraints.[141] However, according to the IP Guidelines, the existence of a horizontal relationship between a licensor and its licensees is not inherently suspect, and such a relationship need not give rise to an anticompetitive effect.[142]

Characteristics of horizontal arrangements The IP Guidelines point out that the DOJ will ordinarily analyze a licensing arrangement as horizontal with respect to a particular goods market in those situations where the parties would be actual or likely potential competitors in that market without the licensing arrangement.[143] With regard to technology markets, a horizontal relationship will be deemed to exist between the parties whenever the arrangement involves the acquisition of technologies that are economic substitutes for technologies which the licensee owns or controls.[144] As a general rule, the DOJ will continue to challenge certain types of horizontal restraints as *per se* unlawful in appropriate cases, including restraints in licensing arrangements that constitute price fixing, allocation of markets or customers, or agreements to reduce output. In other cases, horizontal restraints will be evaluated under the rule of reason.[145]

[137] Cf. 1992 Horizontal Merger Guidelines § 1.41 and n15.

[138] IP Guidelines § 3.2.3 and Example 3.

[139] Example 4 of the IP Guidelines illustrates how an innovation market can be used to evaluate the competitive effects of a research and development joint venture.

[140] IP Guidelines § 3.3.

[141] Ibid. § 5.1. The IP Guidelines note, however, that certain licensing arrangements among horizontal competitors, like joint ventures, often promote rather than hinder competition.

[142] Ibid. § 3.3.

[143] Ibid.

[144] Ibid.

[145] See, e.g. Example 9 of the IP Guidelines involving the joint assignment of patent rights on designs that are economic substitutes to a new joint venture which would have responsibility for licensing each of

Characteristics of vertical arrangements Vertical arrangements generally arise when a licensor and its licensees stand in a seller–buyer relationship, or when a licensor is primarily engaged in the business of research and development and the licensees are manufacturers interested in acquiring certain of the rights to the technology developed by the licensor.[146] Also, the licensor may be a component manufacturer owning intellectual property rights in a product that the licensee manufactures by combining the component with other inputs, or the licensor may manufacture the product, and the licensees may operate primarily in distribution and marketing.[147] While purely vertical relationships tend not to have anticompetitive effects, there is no presumption to that effect.[148] Also, although licensing arrangements typically have a vertical component, the licensor and its licensees may also have a horizontal relationship in the market containing the technology being licensed or in other markets in which they are actual or likely potential competitors.[149]

Identifying the form of arrangement A series of examples are provided in the IP Guidelines to illustrate horizontal, vertical and hybrid relationships between a licensor and its licensees. In one case, a manufacturer of farm equipment develops a new emission control technology for its tractor engines and licenses it to another manufacturer of farm equipment. Assuming that the new technology is far superior to the technology currently owned and used by the licensee, and that there is no likelihood that the licensee would have been able to develop the improved technology on its own, the IP Guidelines conclude that the relationship between the parties with respect to the supply and use of the technology is vertical, since the parties stand in a complementary relationship with respect to the new technology in that the licensee is a consumer and not a competitor with respect thereto.[150] On the other hand, in another example, a firm licenses new technology to a licensee which already owns an alternative form of technology that is an economic substitute for the new licensed technology. In this situation, the IP Guidelines conclude that the relationship between the parties in the relevant market including the licensed technology is vertical in part, since the parties are in a complementary relationship, and horizontal in part, since both of the parties own technologies that are economic substitutes for each other.[151]

cont.
the patents to others and establishing and collecting royalties. While regulators would be concerned that the arrangement amounts to horizontal price fixing, rule of reason analysis might be appropriate if there is some valid efficiency justification for the joint marketing arrangement. In the DOJ's words, "[t]he competitive issue with regard to a joint assignment of patent rights is whether the assignment has an adverse impact on competition in technology and goods markets that is not outweighed by procompetitive efficiencies such as benefits in the use or dissemination of the technology." Ibid.

146 IP Guidelines § 3.3.

147 Ibid.

148 Ibid. ("a purely vertical relationship [does not] assure that there are no anticompetitive effects").

149 Ibid.

150 Example 3 of IP Guidelines. The conclusion is based in large part on the fact that the technologies of the parties are not economic substitutes for each other, given the superiority of the licensor's new technology. Superiority is measured by the fact that the licensee's technology cannot discipline the prices that the licensor will be able to charge for the use of its new technology. The example does state that the parties may have a horizontal relationship in the goods market if they continue to sell farm equipment products that are economic substitutes for each other. See also Example 4 of the IP Guidelines.

151 See Example 4 of the IP Guidelines. Whether or not the firms license their technologies to others is not important for the conclusion that the firms have a horizontal relationship in this relevant market, nor is it

A final example in the IP Guidelines involves the case where a firm develops a new, patented pharmaceutical for the treatment of a particular disease and elects, rather than marketing the drug on its own, to grant exclusive manufacturing rights with respect to the drug to the only other company selling a drug for treatment of the subject disease.[152] Assuming that the licensee would manufacture and sell the licensor's patented drug, the relationship of the parties is in part vertical, because the licensee would be a customer of the licensor in the technology market. However, the relationship is also horizontal in part, because the licensor is a likely potential competitor of the licensee in the relevant goods market as well as in the relevant technology market.[153]

D. Procedure for Rule of Reason Analysis

Assuming that a restraint in a licensing arrangement is eligible for rule of reason analysis,[154] the IP Guidelines set out the preferred steps for conducting the analysis. First, an inquiry is to be made into whether the restraint has an anticompetitive effect.[155] If so, consideration will then be given to whether or not the restraint produces offsetting procompetitive effects, such as by facilitating the efficient development and exploitation of new technology.[156] If such offsetting benefits are established, an assessment will be made as to whether the restraint is reasonably necessary in order to achieve the efficiencies. If it is, and the efficiencies outweigh the anticompetitive effect, the licensing arrangement will typically not be challenged.[157]

The first step in the review process, the analysis of anticompetitive effects, calls for consideration of such things as market structure, foreclosure, and coordination.[158] The IP Guidelines point out that when a licensor and its licensees compete in technology or goods markets, a restraint in a licensing arrangement may increase the risk of coordinated pricing, output restrictions, or the acquisition or maintenance of monopoly power, and that the potential for competitive harm generally increases with the degree of concentration in, the difficulty of entry into, and the inelasticities of supply and demand in markets in which the licensor and licensees are in a horizontal relationship.[159] As such, market structure is an important consideration in this analysis.

The IP Guidelines are also concerned about the possibility that a vertical relationship between a licensor and its licensees may result in harm to competition if a restraint in the relationship forecloses competitors' access to, or increases their costs of obtaining,

cont.
relevant that the licensee's use of its technology was solely captive to its own production. Also, for the firms to be in a horizontal relationship, it is not necessary that the licensee actually use its technology prior to licensing technology from the licensor, provided that the licensee's technology is an economic alternative to the licensor's. Ibid.

[152] Example 10 of the IP Guidelines.

[153] Ibid.

[154] See IP Guidelines § 3.4, described in Chapter 4.

[155] Ibid. § 4.2.

[156] Ibid. § 4.4. The IP Guidelines point out that the DOJ's comparison of anticompetitive harms and procompetitive efficiencies is necessarily a qualitative one, and that the risk of anticompetitive effects in a particular case may be insignificant compared to the expected benefits, or vice versa. Ibid.

[157] Ibid. §§ 4.2 (citing *NCAA* v. *Board of Regents of the University of Oklahoma*, 468 U.S. 85 (1984)) and 4.4.

[158] Ibid. § 4.3.1.

[159] Cf. 1992 Horizontal Merger Guidelines, §§ 1.5, 3.

important inputs. For example, foreclosure may occur in a situation where a licensor has entered into a licensing arrangement with most of the established manufacturers in an industry which prevents those manufacturers from using any other technology.[160] Also, harm to competition from a restraint in a vertical licensing arrangement may occur if the restraint facilitates coordination to raise prices or reduce output in markets in which one of the parties participates, as might be case if the owners of competing technologies impose similar restraints on their licensees in order to make it easier to coordinate their pricing.[161]

Another situation in which a restraint might have an anticompetitive effect is when the benefits of a restraint in a licensing arrangement to the licensor or its licensees are derived primarily from reductions in competition that likely would have occurred absent the license rather than from the relationship of the restraint to any efficiency-producing objectives of the arrangement.[162] Factors such as a history of rivalry and a rapid pace of innovation are relevant to an analysis of the potential for harm to competition. The presence of these factors may indicate that licensors and licensees are less likely to successfully engage in coordinated behavior to raise prices or restrict output, and their absence may signal a greater likelihood of such behavior.[163]

Many of the procompetitive efficiencies which might follow from a licensing arrangement have already been described.[164] It is important to remember that the weight attached to a particular factor may vary depending on the market context. For example, a restraint that may be justified by the needs of a new entrant may not have a procompetitive efficiency justification in different market circumstances.[165] As to the determination of whether or not a restraint is reasonably necessary in order to achieve efficiencies, regulators will attempt to identify any practical and significantly less restrictive alternatives.[166] If it is clear the parties could have achieved similar efficiencies by means that are significantly less restrictive, then regulators will not give weight to the parties' efficiency claim; however, IP Guidelines indicate that regulators will not search for a theoretically least restrictive alternative that might, in hindsight, be easier to construct than in the practical prospective business situation faced by the parties.[167] The IP Guidelines also state that the effective duration of a restraint can be important in determining whether or not it is reasonably necessary to achieve the putative procompetitive effect.[168]

[160] IP Guidelines § 4.3.1 (noting also that "[T]he risk of foreclosing access or increasing competitors' costs is related to the fraction of the markets affected by the licensing restraint and to other characteristics of the input and output markets, such as concentration, difficulty of entry, and elasticities of supply and demand.")

[161] Ibid.

[162] Ibid. § 4.3.3.

[163] Ibid. § 4.3.4.

[164] See IP Guidelines § 2.3, described above.

[165] Ibid. § 4.4. Cf. *United States* v. *Jerrold Electronics Corp.*, 187 F. Supp. 545 (ED. Pa. 1960), *aff'd per curiam*, 365 U.S. 567 (1961).

[166] Ibid. § 4.4.

[167] Ibid. The evaluation of the reasonable necessity of a restraint to achieve procompetitive efficiencies may depend on the market context. Cf. *United States* v. *Jerrold Electronics Corp.*, 187 F. Supp. 545 (ED. Pa. 1960), *aff'd per curiam*, 365 U.S. 567 (1961).

[168] IP Guidelines § 4.4. The effective duration of a restraint may be dependent on a number of factors, including the option of the affected party to terminate the arrangement unilaterally and the presence of contract terms (e.g. unpaid balances on minimum purchase commitments) that encourage the licensee to renew a license arrangement. Ibid. The evaluation of the duration of the restraint may depend on the market context. Ibid. Cf. *United States* v. *Jerrold Electronics Corp.*, 187 F. Supp. 545 (ED. Pa. 1960), *aff'd per curiam*, 365 U.S. 567 (1961).

E. Antitrust "Safety Zone"

The IP Guidelines recognize an antitrust "safety zone" to inform parties of situations in which it is not likely that regulators will challenge the use of specified restraints in licensing agreements.[169] Specifically, absent extraordinary circumstances, regulators will not challenge a restraint placed in a licensing arrangement if (1) the restraint is not of a type which normally warrants condemnation under the *per se* rule and (2) the licensor and its licensees collectively account for no more than 20 percent of each relevant market affected by the restraint.[170] If there is no reliable market share data, regulators will not challenge a restraint if it is not facially anticompetitive and if there are four of more independently controlled technologies in addition to the technologies controlled by the parties to the arrangement that may be suitable for the licensed technology at a comparable cost to the user.

The IP Guidelines make it clear that the "safety zone" is designed to provide intellectual property owners with a degree of certainty, thereby encouraging the parties to enter into procompetitive licensing arrangements. By setting the boundaries of the "safety zone," it is not intend that parties involved in arrangements falling outside the safety zone be discouraged from adopting restrictions that are reasonably necessary to achieving efficiency-producing integration of economic activity.[171] The IP Guidelines explicitly specify that arrangements which fall outside the "safety zone" will continue to be analyzed based on the other considerations outlined therein.[172]

IV. The European Community: Technology Transfer Block Exemption Regulation

A. Introduction

The Patent Licensing Regulation was originally applicable until December 31, 1994, and the Commission was obligated to replace it, either by simple renewal or by developing some new form of group exemption which took into account its experience over the decade in which the original Regulation was in place. In anticipation of the need to replace this group exemption, the Commission published for comment a draft of a new Group Exemption for Technology Transfer Agreements (the "Proposed Technology Licensing Regulation"),[173] with the intent that it would replace not only the expiring Patent Licensing Regulation, but also the know-how licensing regulation (Regulation 556/89), which is not scheduled to expire until the end of 1999. By replacing both of these

[169] The "safety zone" does not apply to transactions that amount to mergers or acquisitions, which will be governed by the 1992 Merger Guidelines. IP Guidelines § 4.3.

[170] Ibid. As stated in section 1.41 of the 1992 Merger Guidelines, market shares for goods markets "can be expressed either in dollar terms through sales, shipments, or production, or in physical terms through measurement of sales, shipments, production, capacity, or reserves." The DOJ will include innovation market shares in its evaluation of whether a licensing arrangement falls within the safety zone only if the assets required to compete in research and development are specialized and identifiable. If not, the DOJ will confine its analysis to the goods and technology markets affected by the licensing arrangement. Ibid.

[171] Ibid.

[172] Ibid.

[173] 94/C 178/03.

regulations, the Commission apparently hoped to eliminate existing uncertainties about the treatment of mixed patent and know-how licenses.

While the Proposed Technology Licensing Regulation included a variety of important changes from the then-existing block exemptions, the most controversial feature of the proposal was the imposition of maximum market share limitations as one of the conditions for the availability of the exemption. This caused concern among many who argued that this radically reduced the benefit of the block exemption by creating uncertainty about its application. Moreover, because the EC is notorious for narrowly defining markets, even relatively small companies could be precluded from enjoying the benefits of the block exemption. Some even claimed that the Proposed Technology Licensing Regulation was a deliberate attempt on the part of the Commission to gain greater control over the entire process of technology transfer in the EC. After much discussion, the Commission finally adopted a new block exemption, which became effective as of April 1, 1996, which did not include market share tests as a condition, but which still made it clear that the Commission would continue to take the market share of the parties into consideration.

B. *Summary of Proposed Technology Licensing Regulation*

Territorial protections and market share conditions Article 1 of the Proposed Technology Licensing Regulation enumerated those obligations which, while potentially restrictive of competition, would nonetheless be exempted from Article 85(1) on the grounds that they generally contribute to the improvement of the production of goods and to the promotion of technical progress. Article 1 covered a variety of territorial protections, including the grant of exclusive licenses and restrictions on activities that may impact rights retained by the licensor, or granted to other licensees, in territories other than those in which the licensee was to be active. The exemptions in Article 1 were largely dependent upon the continuing validity of the licensed subject matter (i.e. patents and know-how).[174]

The Proposed Technology Licensing Regulation exempted obligations on the licensor not to license other undertakings to exploit the licensed technology in the licensed territory,[175] subject to the following conditions:

(i) In the case of pure patent licensing agreements, the licensed product must be protected by parallel patents in the licensee's territory.[176]

(ii) In the case of pure know-how licensing agreements, the period of the exemption could not exceed 10 years from the date when the licensed product was first put on the market in the EC by the licensor or one of his licensees.[177]

[174] In the case of pure know-how licensing agreements, the exemptions in Art. 1 would apply only where the parties identified in any appropriate form the initial know-how and any subsequent improvements to it, which became available to one party and were communicated to the other party pursuant to the terms of the agreement and for the purpose thereof, and only for as long as the know-how remained secret and substantial. Proposed Technology Licensing Regulation Art. 1, Para. 3. In the case of mixed patent and know-how licensing agreements, they qualified for the exemption in Art. 1 only for as long as the patents remained in force and provided the know-how was identified and for as long as it remained secret and substantial. Proposed Technology Licensing Regulation Art. 1, Para. 4.

[175] Proposed Technology Licensing Regulation Art. 1, para. 1.

[176] Proposed Technology Licensing Regulation Art. 1, para. 2.

[177] Proposed Technology Licensing Regulation Art. 1, para. 3.

(iii) In the case of mixed patent and know-how licensing agreement, the exemption would apply in member states in which the licensed technology was protected by (essential) patents[178] for as long as the licensed product or process was protected in those member states by such patents if the duration of such protection exceeded the period specified in point (ii) above.[179] In addition, the exemption of the obligation on the licensor not to grant other licenses would apply only provided:
   (a) that the products manufactured by the licensee which were capable of being improved or replaced by the contract products and other goods manufactured by him which were considered by users to be equivalent in view of their characteristics, price and intended use accounted for no more than 40 percent of the entire market in those products in the common market or a substantial part of it; and
   (b) that the licensee was not operating in an oligopolististic market, which, for purposes of the Regulation, was a market in which three undertakings or less held together a market share of more than 50 percent, or five undertakings or less held together a market share of more than two thirds and the licensee was one of the undertakings and held a market share of more than 10 percent.[180]

The Proposed Technology Licensing Regulation also exempted obligations (i) on the licensor not to exploit the licensed technology in the licensed territory himself, (ii) on the licensee not to exploit the licensed technology in territories within the common market which were reserved for the licensor, and (iii) on the licensee not to manufacture or use the licensed product, or use the licensed process, in territories within the common market which were licensed to other licensees, and (iv) on the licensee not to pursue an active policy of putting the licensed product on the market in the territories within the common market which were licensed to other licensees, and in particular not to engage in advertising specifically aimed at those territories or establish any branch or maintain any distribution depot there,[181] subject to the following conditions:

(i) In the case of pure patent licensing agreements, the licensed product must be protected by parallel patents in the protected party's territory(ies) (e.g. the licensee, the licensor, or other licensees).[182]
(ii) In the case of pure know-how licensing agreements, the period of the exemption could not exceed 10 years from the date when the licensed product was first put on the market in the EC by the licensor or one of his licensees.[183]

178 Proposed Technology Licensing Regulation Art. 10 defined "essential patents" as "patents which play an important role in the putting into effect of the conceded technology or which provide effective protection against the exploitation of the licensed technology by third parties, excluding in particular patents for mere details of construction and patents for improvements to a technology whose exploitation has already been the subject of an admissible period of territorial protection, where those improvements are of significantly less importance than the technology initially licensed or require no new expensive and risky investment."
179 Proposed Technology Licensing Regulation Art. 1, para. 4.
180 Proposed Technology Licensing Regulation Art. 1, para. 5.
181 Proposed Technology Licensing Regulation Art. 1, para. 1.
182 Proposed Technology Licensing Regulation Art. 1, para. 2.
183 Proposed Technology Licensing Regulation Art. 1, para. 3. This represented a change from the then-existing know-how block exemption where the ten years began to run as soon as the first license was signed for the territory. The change was intended to account for long lead times in developing products that may be

(iii) In the case of mixed patent and know-how licensing agreement, the exemption would apply in member states in which the licensed technology was protected by "essential" patents for as long as the licensed product or process was protected in those member states by such patents if the duration of such protection exceeded the period specified in point 2 above.[184]

(iv) The party that was to be protected by such obligations held a market share of no more than 20 percent.[185]

The Proposed Technology Licensing Regulation also exempted obligations on the licensee not to put the licensed product on the market in the territories licensed to other licensees within the common market,[186] subject to the following conditions:

(i) In the case of pure patent licensing agreements, the period of the exemption could not exceed five years from the date when the product was first put on the market within the common market by the licensor or one of his licensees, inasmuch and for as long as in these territories, the product was protected by parallel patents.[187]

(ii) In the case of pure know-how licensing agreements, the period of the exemption could not exceed five years from the date when the licensed product was first put on the market in the EC by the licensor or one of his licensees.[188]

(iii) In the case of mixed patent and know-how licensing agreement, the exemption could not exceed the five year period.[189]

(iv) The party that was to be protected by such obligations held a market share of no more than 20 percent.[190]

"White list" provisions The "white list" in the Proposed Technology Licensing Regulation permitted obligations on the licensee (i) not to divulge know-how, (ii) not to grant sub-licenses, (iii) not to exploit licensed know-how or patents after license termination, (iv) to grant back to the licensor a non-exclusive license to the licensor if the licensor is granting licenses under its improvements to the licensee, (v) to observe quality control standards, (vi) to assist the licensor in the enforcement of licensed intellectual property rights, (vii) to pay royalties after expiration, (vii) to remain within fields of use, (ix) to continue the grant-back license, and (x) to remit minimum royalties, produce minimum quantities, or carry out a minimum number of operations.[191]

"Black list" provisions The "black list" in the Proposed Technology Licensing Regulation was substantially shortened, but still restricted agreements that involved price-fixing of licensed products, restrictions on a party's research and development activities,

cont.
subject to such licenses; however, given that there was only a five year limit on passive sales bans, there still may be little time left in cases where the licensor takes several years to find licensees and to negotiate agreements in a succession of territories.

184 Proposed Technology Licensing Regulation Art. 1, para. 4.
185 Ibid., para. 6.
186 Ibid., para. 1.
187 Ibid., para. 2.
188 Ibid., para. 3.
189 Ibid., para. 4.
190 Ibid., para. 6.
191 See generally Proposed Technology Licensing Regulation Art. 2.

blockages of a licensee's or licensor's satisfaction of market needs, customer allocations, or full grant-backs of licensee improvements.[192]

"Grey clauses" and the opposition procedure As noted in the prior discussion of the prior block exemptions for patent and know-how licenses, inclusion of "grey clauses" (i.e. clauses that were not either specifically exempted or forbidden) in the agreement would, like a "black listed" clause, bring the entire agreement outside the block exemption and require that the parties consider notifying the agreement to the Commission. However, the Proposed Technology Licensing Regulation included a provision to the effect that inclusion of "grey clauses" would not affect the application of the block exemption to the remainder of the agreement, although the clause itself would risk fines and unenforceability unless notified. In what could be viewed as an exchange for this more lenient treatment of grey clauses, the Commission proposed to eliminate from the so-called "opposition procedure" for notifying agreements which contained grey clauses, which apparently was rarely used because it was thought that the relatively short time period would lead to cautious Commission opinions.

C. Industry Reactions to Proposed Group Exemption

General objections to market share limitations The Proposed Technology Transfer Block Exemption was originally scheduled to be in force upon the expiration of the Patent Licensing Regulation on December 31, 1994. However, a firestorm of protest from industry, albeit somewhat belated, about the proposed market share limits convinced the Commission that it had better hold a public hearing to air grievances about its proposal. A number of concerns were raised, including the following:

(i) The market share tests would lead to total lack of certainty, because the tests were vague and subjective, and not scientific. They involved investigation of, and judgment on, a number of inter-related legal, technical and economic questions. In particular, difficulties could be expected in identifying "equivalent" goods and services. Moreover, for parties to reach even an unreliable answer would be expensive, given the need to engage legal, technical and economic consultants.

(ii) The tests were continuous, not "one-off," in that they would apply not only at the time the license agreement was entered into, but also throughout the period of the license. This would have led to additional analytical concerns from year to year. It was believed that the problem would be particularly acute in this field, since market shares are especially fluid, as new developments are introduced, win acceptance, and jostle for position alongside competing products.

(iii) Successful innovations would actually be penalized, since they would typically take a larger share of the market, or even form a new market of their own. In such cases, the block exemption might not be available long enough to give the innovator the protection from competition that might be necessary to allow him to recover his sunk costs plus an adequate risk adjusted return on invest-

[192] See generally Proposed Technology Licensing Regulation Art. 3.

ment. In contrast, an agreement dealing with trivial, technical modifications would easily gain and keep a block exemption because the product in question will deservedly have a small share of the market.

(iv) It was argued that the notion of market share limitations was more in tune with the concept of "dominant position," and abuse thereof, under Article 86, rather than anticompetitive agreements governed by Article 85. These have always been separate questions, as *Tetra Pak* confirmed.[193] Yet the test proposed was stricter than the test of dominance, because it failed to take into account other significant factors, like barriers to entry.

(v) It was argued that the rigid market share test was crude, not selective, and would hinder the dissemination and exploitation of technology.[194] It was asserted that the Commission should rely upon its discretion to withdraw the benefits of the block exemption in appropriate circumstances, which would then allow it to discriminately review all of the surrounding circumstances relating to a given transaction.

(vi) The market share test would have a far greater adverse impact on small and medium enterprises, which have long been the subject of various supportive efforts by the Commission, than on large firms, since such enterprises (e.g. research institutes, the university departments, and start-up companies) are more likely to rely on the use of licensing arrangements due to their lack of the resources required for mass-manufacture or marketing of new products.[195] It was also suggested that the length of the exemption was too short for many smaller licensors, since they would need more time to ramp-up their commercialization program and locate licensees in various markets.

"Essential" patents Criticism was also leveled at the concept of "essential" patents. As noted above, the presence of essential patents along with unpatented know-how would lengthen the period of exemption beyond the ten years applicable to know-how licenses to cover the entire period of patent protection. Commentators argued that it is wholly impractical, if not impossible, to determine which patents are "essential."[196] They claimed that apart from introducing a new layer of uncertainty, the concept seemed to reflect an over-simplified idea that the technology involved in any technology transfer comes in a small tidy parcel, and that most licensing is of established, tested technology, where the picture is clearly laid out. It was rightly pointed out that most licenses cover a complex package of technology that is still under development, and that the parties typically have neither the time nor the desire to weigh up the value of individual elements, even if they could approximate the future value of a particular element of the licensed package.

Korah also indicated that he had great difficulty with the concept of an "essential patent."[197] He argued that if some of the technology is patented in both the licensed

[193] *Tetra Pak Rausing* v. *Commission* [1990] ECR II-309.

[194] Waite, "The Draft Technology Transfer Block Exemption," *European Intellectual Property Report* 7 (1994), 259, 260.

[195] Ibid.

[196] Ibid. at 261.

[197] Korah, "The Preliminary Draft of a New EC Group Exemption for Technology Licensing," *European Intellectual Property Report* 8 (1994), 262, 266–7.

territory and that to be protected, and it is worth using, then it should qualify for the exemption, particularly since the exemption would relate only to the licensed technology and products made thereby.[198] If, on the other hand, the patents turn out to be trivial or of little use for the particular purpose, then the impact on competition would be very small in any event, and there would be no need to endanger the entire exemption in such situations.[199]

"Ancillary" intellectual property rights The Proposed Technology Licensing Regulation provided that the exemption would apply only to licenses of patents and/or know-how, and Article 5(1)(4) of the draft Regulation provided that the block exemption would not be applicable in situations where the license also covers other intellectual property rights, unless such rights were "ancillary". This limitation was heavily criticized for the vagueness and uncertainties which it created, particularly since many licenses would necessarily include a full package of intellectual property rights for completion of the project, including copyrights and trademarks. As Korah pointed out:

> "A more grievous objection is that it is important that the parties can be advised firmly whether their transaction comes within a group exemption before they make commitments to investment. Deciding when complementary rights will be treated as ancillary confers a wide discretion on whoever applies the test, whether it be a national court or the Commission, and detracts from legal certainty."[200]

Early termination of know-how block exemption The decision to terminate the know-how exemption prior to its scheduled termination date was also criticized for the effect that it might have on investments previously made by firms in reliance on the exemption. As one commentator explained:

> "The loss of a safe harbour for firms that are subject to the cap of market share is particularly unfair for those who have invested on the assumption that the know-how regulation would continue until the end of 1999. It is repealed by Article 11 and the transitional provision in Article 9 allows such agreements to remain exempt only until the end of 1995. They may lose up to four years of territorial protection on the basis of which investments have already been made."[201]

Burdens of notification requirements In general, industry representatives indicated to the Commission's officials that, in spite of the officials' claims that businesspeople always know the market shares of their companies, markets are difficult to define and market shares are therefore difficult to determine. They argued that the market share limits would deprive the block exemption of the requisite certainty, and that the introduction of the limits would, in most cases, mean that applying for an individual exemption will be the only option.[202] It was predicted that the Commission would be inundated with additional notifications,[203] since no business could contemplate the burden

[198] Ibid.
[199] Ibid.
[200] Ibid.
[201] Ibid. at 264.
[202] Waite, supra note 194, at 260.

of investigating and regularly monitoring market shares in the ways required by the proposal to ensure that a block exemption was available at the start and continued during the life of the agreement.[204]

Critics claimed that the costs and effects of increased notification would be difficult to measure. Clearly, such a situation would lead to delays in implementing new projects. One key problem was that while the Commission had attempted to speed up its informal decisions with respect to the anticompetitive effects of joint ventures,[205] it was not likely that the Commission would be able to provide and publish more comfort letters regarding vertical agreements, such as technology license agreements.[206] Moreover, preparing a notification is a time-consuming and expensive exercise. It is an upfront cost, which may be totally wasted if the new product or process subsequently fails at a late stage of technical development or is superseded in the market. If innovators could not bear the time and expense of notifying their agreements, they would no longer license, and would either abandon the idea or sell it outright at a lower price than they might have obtained through licensing.[207]

Critics argued vigorously that the Proposed Technology Licensing Regulation was contrary to the Commission's objectives in promulgating block exemptions, which have been described as "firstly, to provide firms with legal certainty while at the same time avoiding the burdens and delays inherent in any individual notification and, secondly, to allow the Commission to concentrate on cases involving more serious restrictions of competition."[208] In the view of these critics, the purpose of the block exemptions would be frustrated if they were not simple and certain to apply, because parties would be forced to notify in cases of uncertainty or risk fines and unenforceability. Given the costs and delays of notification, parties would be less willing to make substantial investments relating to the licensed goods if there is a significant prospect of an unfavorable ruling many years down the road which would deprive the parties of their incentives to pursue the arrangement (e.g. loss of exclusivity). As a result, the market for new technology transfers and licenses would be adversely impacted, in spite of the Commission's recognition that the economic development of the EC depends on industry's "capacity to devise new technologies and disseminate them," and that therefore, "simplifying the way in which technology transfers are treated under the competition rules is an important ... measure."[209]

One method for reducing the adverse effects from enhanced requirements of notification is to improve the mechanics of the opposition procedure, which the Commission has indicated would be retained even though it was originally not included as part

[203] In fact, at one hearing, the President of Licensing Executives Society France, Mr Thierry Sueur, stated that on the basis of a survey of a cross-section of French companies, the Commission could expect an additional 900 notifications annually from France alone.

[204] The Commission argued that market share limitations are not new, since they have been incorporated into the existing block exemptions relating to research and development agreements and specialization agreements. However, critics respond by noting that those block exemptions have been of little practical use to firms in the marketplace precisely because of the existence of such market share limitations.

[205] Twenty-second Report on Competition Policy, 81.

[206] Korah, supra note 197, at 264.

[207] According to one commentator, innovators would receive less for the innovations by selling the technology as opposed to licensing. This is because a purchaser will pay a lower purchase price reflecting the inherent risk in the project while, in contrast, a licensee may be willing to agree to a formula of running royalties which allows the licensor to obtain higher rewards if the product proves to be a success. Waite, supra note 194, at 261.

[208] Twentieth Report on Competition Policy.

[209] Ibid.

of the original Proposed Technology Licensing Regulation. Industry representatives have suggested that the procedure, which has not been widely used under the Patent Licensing Regulation, should be shortened from the existing six months; however, the Commission has rejected this change on the understandable basis that member states must be given three months to intervene on notified grey clauses, making it difficult to complete the opposition procedure in less than six months. In addition, the Commission rejected the suggestion of applying the opposition procedure to any market share criteria ultimately included in the block exemption. On the other hand, the Commission did indicate that some form of opposition procedure would be reinstated in the final version of the block exemption, and sought feedback on the idea of introducing a new sort of short form notification system, whereby licensing parties falling above the limits would be required to notify their agreement to the Commission and it would thereby be subject to the block exemption unless the Commission chose to withdraw its application.

Following this initial round of comments, the Commission appeared willing to be a bit more flexible on various issues. For example, the Commission exhibited a certain willingness to consider points raised about "essential" patents and the length of the exemption period. The Commission also took all of the comments regarding the market share limits under consideration; however, at that time, it appeared that proposals to eliminate such limits entirely would not be successful, although it seemed that the Commission would consider applying the market share limits only to horizontal licenses or raising the percentages used in the market share limits.

D. Final Technology Licensing Regulation

Introduction On January 31, 1996, the Commission finally adopted Regulation 240/96, which became effective as of April 1, 1996 as the final form of the new Technology Transfer Block Exemption Regulation (the "Technology Licensing Regulation"). The actual Regulation was a surprising improvement over the prior drafts, especially because the Commission dropped the controversial and much-criticized market share tests. The new Regulation will be effective until March 31, 2006,[210] and provision was made for the protection of existing patent and know-how licenses which were in compliance with the block exemptions replaced by the Technology Licensing Regulation.[211]

Market share tests The most surprising part of the final Technology Licensing Regulation was the last minute withdrawal of the market share tests which had appeared in all the prior drafts as preconditions to the exemption of exclusivity and territorial restriction provisions. However, the market share tests were not dropped altogether. Instead, they reappeared in Article 7 of the Technology Licensing Regulation, which permits the Commission to withdraw the benefit of the exemption if it makes a finding that an exempted agreement nevertheless has certain effects which are incompatible with Article 85(3). Specifically, Article 7 of the new Regulation provides that the exemption may be withdrawn if, *inter alia*, the effect of the agreement is to prevent the licensed products from being exposed to effective competition from identical or "interchangeable or

[210]Technology Licensing Regulation, Art. 13, para. 1.
[211]Technology Licensing Regulation, Art. 11, para. 3.

substitutable" products,[212] "which may in particular occur where the licensee's market share exceeds 40 per cent."[213]

Recital (26) of the December 1995 draft of the proposed Regulation had provided, in explaining the situations where the Commission might exercise its discretion under Article 7, that "[i]n assessing the competition the Commission will pay special attention to cases where the licensee has, *at the time the contract is signed*, more than 40 percent of the whole market" (emphasis added). However, these words did not appear in the final version of the Technology Licensing Regulation, meaning that the Commission will likely be able to take into account the licensee's market share at any time in considering whether or not to withdraw the exemption. It remains to be seen how this will be enforced and, while the exemption can no longer be initially denied on the basis of market share, it does mean that parties will now need to consider difficult market share questions that previously only came up in Article 86 cases.

Starting point for exclusivity and territorial restrictions The relevant periods for the exemptions afforded for exclusivity and territorial restrictions now run from the date that the licensed product is first put on the market within the common market by one of the licensees.[214] This is more generous than in prior drafts where the period of the exemption was to be measured from the time the product was first put on the market by one of the licensees or by the licensor. The new starting point will reduce concerns that licensees might have in situations where it is anticipated that several years of preparation might be required before the licensed product is actually ready for sale into the market.

Use of licensor's trademarks and production limitations The Technology Licensing Regulation exempts an obligation on the licensee to use only the licensor's trademark or get up to distinguish the licensed product during the term of the agreement, provided that the licensee is not prevented from identifying himself as the manufacturer of the licensed products.[215] Where the agreement is a pure patent licensing agreement, the exemption of the obligations is granted only to the extent that and for as long as the licensed product is protected by parallel patents in the territories of the licensee.[216] Where the agreement is a pure know-how licensing agreement, the obligations are exempted for the lifetime of the agreement for as long as the know-how remains secret and substantial.[217]

The Technology Licensing Regulation exempts an obligation on the licensee to limit his production of the licensed product to the quantities he requires in manufacturing his own products and to sell the licensed product only as an integral part of or a replacement part for his own products or otherwise in connection with the sale of his

[212] Art. 10(7) of the Technology Licensing Regulation defines the "licensee's market share" to mean "the proportion which the products or services provided by the licensee, which are considered by users to be substitutable for licensed products in view of their characteristics, price and intended use, represent the entire market for the licensed products and all other substitutable goods and services in the common market or a substantial part of it."

[213] Technology Licensing Regulation, Art. 7(1).

[214] Technology Licensing Regulation, Art. 1, paras. 2 and 3.

[215] Ibid., para. 1(7).

[216] Ibid., para. 2.

[217] Ibid., para. 3.

own products, provided that such quantities are freely demonstrated by the licensee.[218] Where the agreement is a pure patent licensing agreement, the exemption of the obligations is granted only to the extent that and for as long as the licensed product is protected by parallel patents in the territories of the licensee.[219] Where the agreement is a pure know-how licensing agreement, the obligations are exempted for the lifetime of the agreement for as long as the know-how remains secret and substantial.[220]

Exemption for "white list" clauses Article 2 of the Technology Licensing Regulation contains a revised and extended version of the white list covering many ancillary provisions that rarely infringe Article 85(1), but which are exempted just in case.[221] The items enumerated in Article 2 include:

(i) an obligation on the licensee not to divulge the know-how communicated by the licensor; the licensee may be held to this obligation after the agreement has expired;[222]
(ii) an obligation on the licensee not to grant sublicenses or assign the license;[223]
(iii) an obligation on the licensee not to exploit the licensed know-how or patents after termination of the agreement in so far and as long as the know-how is still secret or the patents are still in force.[224]
(iv) an obligation on the licensee to grant to the licensor a licence in respect of his own improvements to or his new applications of the licensed technology, provided:
  (a) that, in the case of severable improvements, such a licence is not exclusive, so that the licensee is free to use his own improvements or to license them to third parties, in so far as that does not involve disclosure of the know-how communicated by the licensor that is still secret;
  (b) that the licensor undertakes to grant an exclusive or non-exclusive ence of his own improvements to the licensee, and when his right to use the licensee's improvements is of a longer duration than the licensee's right to use the licensed technology, the licensor agrees to pay appropriate royalties to the licensee or to allow him to continue exploiting the licensed technology.[225]
(v) an obligation on the licensee to observe minimum quality specifications for the licensed product or to procure goods or services from the licensor or from an undertaking designated by the licensor, in so far as such quality specifications. products or services are necessary for:
  (a) a technically satisfactory exploitation of the licensed technology; or
  (b) for ensuring that the product of the licensee conforms to the quality standards that are respected by the licensor and other licensees;

  and to allow the licensor to carry out related checks.[226]

[218] Ibid., para. 1(8).
[219] Ibid., para. 2.
[220] Ibid., para. 3.
[221] Technology Licensing Regulation Art. 2, para. 1.
[222] Ibid., para. 1(1).
[223] Ibid., para. 1(2).
[224] Ibid., para. 1(3).
[225] Ibid., para. 1(4).
[226] Ibid., para. 1(5).

(vi) obligations:
  (a) to inform the licensor of misappropriation of the know-how or of infringements of the licensed patents; or
  (b) to take or to assist the licensor in taking legal action against such misappropriation or infringements;[227]

(vii) an obligation on the licensee, in the event of the know-how becoming publicly known or the patents prematurely losing their validity other than by action of the licensor, to continue paying the royalties until the end of the agreement or the regular expiry of the patents, in the amounts, for the periods and according to the methods freely determined by the parties, without prejudice to the payment of any additional damages in the event of the know-how becoming publicly known or the patents losing their validity by the action of the licensee in breach of the agreement;[228]

(viii) an obligation on the licensee to restrict his exploitation of the licensed technology to one or more technical fields of application covered by the licensed technology or to one or more product markets;[229]

(ix) an obligation on the licensee to pay a minimum royalty or to produce a minimum quantity of the licensed product or to carry out a minimum number of operations exploiting the licensed technology;[230]

(x) an obligation on the licensor to grant the licensee any more favorable terms that the licensor may grant to another undertaking after the agreement is entered into;[231]

(xi) an obligation on the licensee to mark the licensed product with an indication of the licensor's name or of the licensed patent;[232]

(xii) an obligation on the licensee not to use the licensor's know-how to construct facilities for third parties: this is without prejudice to the right of the licensee to increase the capacity of his facilities or to set up additional facilities for his own use on normal commercial terms, including the payment of additional royalties;[233]

(xiii) an obligation on the licensee to supply only a limited quantity of the licensed product to a particular customer, where a know-how license was granted at that customer's request so that he might have a second supplier inside a licensed territory; this provision shall also apply where the customer is the licensee, and the license which was granted in order to provide a second source of supply provides that the customer is himself to manufacture the licensed products or to have them manufactured by a subcontractor;[234]

(xiv) a reservation by the licensor of the right to exercise the rights conferred by a patent to oppose the exploitation of the technology by the licensee outside the licensed territory;[235]

[227] Ibid., para. 1(6).
[228] Ibid., para. 1(7).
[229] Ibid., para. 1(8).
[230] Ibid., para. 1(9).
[231] Ibid., para. 1(10).
[232] Ibid., para. 1(11).
[233] Ibid., para. 1(12).
[234] Ibid., para. 1(13).
[235] Ibid., para. 1(14).

(xv) a reservation by the licensor of the right to terminate the agreement if the licensee contests the secrecy of the licensed know-how or challenges the validity of licensed patents within the common market belonging to the licensor or undertakings connected with him;[236]

(xvi) a reservation by the licensor of the right to terminate the licence agreement of a patent if the licensee raises the claim that such a patent is not necessary.[237]

"Black listed" clauses Article 3 of the Technology Licensing Regulation contains a shortened version of the "black list" covering those items which the Commission considers to be restrictive of competition.[238] The exemption offered under the Regulation will not apply where:

(i) One party is restricted in the determination of prices, components of prices or discounts for the licensed products.[239]

(ii) One party is restricted from competing within the common market with the other party, with undertakings connected with the other party or with other undertakings in respect of research and development, production, use or distribution of competing products without prejudice to an obligation on the licensee to use his best endeavours to exploit the licensed technology and without prejudice to the right of the licensor to terminate the exclusivity granted to the licensee and to stop licensing improvements to him when the licensee enters into such competition, and to require the licensee to prove that the licensed know-how is not being used for the production of products and the provision of services other than those licensed.[240]

(iii) One or both of the parties are required:
  (a) to refuse without any objectively justified reason to meet demand from users or resellers in their respective territories who would market products in other territories within the common market;
  (b) to make it difficult for users or resellers to obtain the products from other resellers within the common market, and in particular to exercise intellectual property rights or take measures so as to prevent users or resellers from obtaining outside, or from putting on the market in the licensed territory products which have been lawfully put on the market within the common market by the licensor or with his consent;

  or do so as a result of a concerted practice between them.[241]

(iv) The parties were already competing manufacturers before the grant of the licence and one party is restricted within the same technological field of use or within the same product market as to the customers he may serve, in particular by being prohibited from supplying certain classes of user, employing certain forms of distribution or, with the aim of sharing customers, using certain

[236] Ibid., para. 1(15).
[237] Ibid., para. 1(16).
[238] Proposed Technology Licensing Regulation Art. 3.
[239] Technology Licensing Regulation Art. 3, para. 1.
[240] Ibid., Art. 3, para. 2.
[241] Ibid., para. 3.

types of packaging for the products, save as provided in Article 1 (1)(7)[242] and Article 2 (1)(13)[243] of the Regulation.[244]

(v) The quantity of the licensed products one party may manufacture or sell or the number of operations exploiting the licensed technology he may carry out are subject to limitations, save as provided in Article 1 (1)(8)[245] and Article 2 (1)(13)[246] of the Regulation.[247]

(vi) The licensee is obliged to assign in whole or in part to the licensor rights to improvements to or new applications of the licensed technology.[248]

(vii) The licensor is required, albeit in separate agreements or through automatic prolongation of the initial duration of the agreement by the inclusion of any new improvements, for a period exceeding that referred to in Article 1(2) and (3) not to license other undertakings to exploit the same technology in the licensed territory; or a party is required for a period exceeding that referred to in Article 1(2) and (3) or Article 1(4) not to exploit the same technology in the territory of the other party or of other licensees.[249]

"Grey clauses" and the opposition procedure As promised, the opposition procedure has been reinstated in the final version of the Technology Licensing Regulation,[250] but the "partial exemption" alternative included in earlier drafts was not adopted. The opposition procedure has been made more attractive, in that the Commission now has only four, rather than six, months to state whether it has any objections to an agreement, and that the parties may not be required to submit a full-blown notification form in order to seek a decision under the procedure.[251] According to the Technology Li-

[242] Art. 1, para. 1(7) of the Technology Licensing Regulation exempts an obligation on the licensee to use only the licensor's trademark or get up to distinguish the licensed product during the term of the agreement, provided that the licensee is not prevented from identifying himself as the manufacturer of the licensed products.

[243] Art. 2, para. 1(13) of the Technology Licensing Regulation exempts an obligation on the licensee to supply only a limited quantity of the licensed product to a particular customer, where a know-how license was granted at that customer's request so that he might have a second supplier inside a licensed territory; this provision shall also apply where the customer is the licensee, and the license which was granted in order to provide a second source of supply provides that the customer is himself to manufacture the licensed products or to have them manufactured by a subcontractor.

[244] Tecchnology Licensing Regulation Art. 3, para. 4.

[245] Art. 1, para. 1(8) of the Technology Licensing Regulation exempts an obligation on the licensee to limit his production of the licensed product to the quantities he requires in manufacturing his own products and to sell the licensed product only as an integral part of or a replacement part for his own products or otherwise in connection with the sale of his own products, provided that such quantities are freely demonstrated by the licensee.

[246] Art. 2, para. 1(13) of the Technology Licensing Regulation exempts an obligation on the licensee to supply only a limited quantity of the licensed product to a particular customer, where a know-how license was granted at that customer's request so that he might have a second supplier inside a licensed territory; this provision shall also apply where the customer is the licensee, and the license which was granted in order to provide a second source of supply provides that the customer is himself to manufacture the licensed products or to have them manufactured by a subcontractor.

[247] Technology Licensing Regulation Art. 3, para. 5.

[248] Ibid., para. 6.

[249] Ibid., para. 7.

[250] Technology Licensing Regulation, Art. 4.

[251] Technology Licensing Regulation, Recital (25) ("The Commission may waive the requirement to supply specific information required in form A/B but which it does not deem necessary. The Commission will generally be content with communication of the text of the agreement and with an estimate, based on directly available data, of the market structure and of the licensee's market share.")

censing Regulation itself, use of the opposition procedure may be appropriate in situations where an agreement contains tie-in obligations that are not necessary for a technically satisfactory exploitation of the licensed technology nor necessary for maintaining quality standards;[252] and where an agreement obliges the licensee not to contest the "secrecy or substantiality" of the licensed know-how or the validity of the licensed patents.[253]

[252]Technology Licensing Regulation, Art. 4, para. 2(a).
[253]Ibid., para. 2(b).

11 Regulation of Common Licensing Practices in the United States and the EC

I. Introduction

While courts have generally upheld the validity of rights granted to inventors under the patent laws, there has long been substantial debate regarding the appropriate scope of the monopoly rights which should be granted to patent owners.[1] One area in which this debate has festered has been with regard to the enforcement of the antitrust or competition laws, with the issue being whether certain practices of patent owners with respect to the use of the patent rights should be subject to special administrative review and scrutiny. Among the areas of concern are attempts to use the exclusionary rights afforded by patents to restrict competition. Among those practices which have attracted the attention of the legislature, competition authorities and the courts are patent pools, regional or international divisions of marketing territories under exclusive licensing agreements and the imposition of various restrictions on the initiative and independence of licensees.[2]

In this chapter, we review how regulators in the United States and the EC have treated various common practices used in licensing agreements. Among the areas which will be covered are exclusive licensing, territorial and field of use restrictions, resale price restrictions, various quantitative and qualitative conditions on use (e.g. output restrictions, minimum production and output requirements), restrictions on competitive activities of licensees, tying and package licensing practices, grantback provisions, and other miscellaneous issues. Where appropriate, we provide an overview of some of the commercial practices and economic theories which are relevant to each issue.

II. Exclusive Licensing Arrangements

A. Introduction

The strategic utility of exclusive licenses stems, in part, from the fact that many firms lack the resources to fully exploit the results of their research and development activities

[1] See, e.g. Chapter 9 ("Patentee, Competition and the Law") in Dutton, *The Patent System and Inventive Activity During the Industrial Revolution 1750–1852* (1984).

[2] Cornish, *Intellectual Property: Patents, Copyright, Trade Marks and Allied Rights* (1989), at 14. The competition law of the European Community is based upon the general principles set out in the Treaty of Rome and must be read in tandem with the overall objective of facilitating the free movement of goods within the Common Market. As a result, attempts to use exclusive licensing agreements to divide up national markets in the Common Market have been substantially curtailed. Ibid. With regard to the regulatory interface between intellectual property and competition law in the United Kingdom and in the European Kingdom, see Whish, *Competition Law* (1989) 646–90. See also, Swann et al., *Competition in British Industry* (1974) 92–143 (theory of competition policy).

either in all geographic areas or in all potential fields of use. As such, they look to license these results, which generally take the form of patents and other intellectual property rights, to others who might be better placed to carry out the exploitation. In order for these relationships to be successful, however, the licensee presumably needs the incentives required for it to make the necessary investments (e.g. construction of manufacturing plants and development of distribution channels). One method for the creation of these incentives is for the licensee to be afforded the same degree of freedom from competition (i.e. exclusivity) that the licensor would presumably have had in the case where it had decided to go it alone.

B. Definitional Aspects

There are three general types of licensing categories which describe the license grant in terms of competitiveness. First, there are "exclusive" licenses, which provide the licensee with all of the rights held by the licensor with respect to one or more functional aspects of the technology rights. In effect, the licensor is precluded from competing with the licensee and must look to the anticipated stream of payments from the licensee as compensation for the costs associated with the development of the licensed technology. Exclusivity can be tailored to meet the goals and expectations of each of the parties. For example, exclusivity can be limited to a particular market or geographic area, such as when the licensee is given an exclusive license to make and sell products to retail outlets in a designated state or region, and other licensees are given non-exclusive rights to exploit other markets.

The second main type of license is a semi-exclusive, or sole, license. A sole license is similar to an exclusive license, although the licensor will "reserve" certain of the licensed rights for itself. These types of arrangements, which must be explicitly stated in the license, effectively create two "competitors," the licensor and the licensee, although it is most often the case that the license will include additional details as to how the parties are to utilize the right.[3] As implied by the title of the arrangement, the licensor cannot appoint any other licensees with rights that may conflict with those granted to the original licensee.

The third type of license is a non-exclusive license, which simply grants certain rights to the licensee and does not insulate the licensee from competition from the licensor or any other licensees created by the licensor. It may be possible for the licensor to grant a number of non-exclusive licenses covering essentially the same legal subject matter. If this is the case, the royalty rate which the licensor is able to obtain for the license will necessarily be lower than if it were granting an exclusive license. A non-exclusive license is often acquired by a licensee in order to insure that the licensor will not commence an infringement action against the licensee when a new product or application developed by the licensee arguably overlaps with the existing legal rights of the licensor.

[3] The term "exclusive" may have different meaning in other countries. For example, in the United States an exclusive license implies an agreement by the licensor not to license the rights to others and not to exploit the rights for its own account within the exclusively licensing territory. This is the meaning given to the term in the text. However, in France a grant of an exclusive license is actually what the text refers to as a "sole" license, since it normally does not prevent the licensor from exploiting a patent for personal benefit unless the license contains a specific restriction on the actions of the licensor.

C. Strategic Considerations in Exclusive Licensing

Introduction Obviously, the licensee perceives an exclusive license as being the most advantageous form of technology transfer, even though it can be anticipated that the "cost" of such a license will exceed the amount required for a non-exclusive license. An exclusive license provides the licensee with access to all of the results of the licensor's research and development work in the specific area, without the need to incur the substantial "up-front" costs associated therewith. Moreover, an exclusive grant allows the licensee to achieve its strategic objectives of being "first to market" with respect to the specified products.

For its part, a licensor will be understandably reticent to divest itself of control over the exploitation of the technology rights by granting an exclusive license, particularly when it has invested a significant amount of funds and time in developing the technology. However, in many cases, the grant of an exclusive license to a specific licensee may be advantageous to the licensor given the perceived skills and resources of the licensee in the areas of manufacturing and distribution. Accordingly, it is more than likely that any grant of an exclusive license will include a number of conditions calculated to insure that the licensee will perform in a manner that allows the licensor to realize the optimal benefits from the transferred technology.

Contractual restrictions on exclusivity One method that is often used to control the risks that a licensor might feel come with the grant of an exclusive license is to restrict the scope of the exclusive rights in some manner. For example, a licensor may restrict exclusivity by limiting it to that period of time which may be necessary for the licensee to achieve the desired advantage of being the first in the "market" with the product or technology covered by the license. This strategy may be appropriate when it is anticipated that the licensee will expend substantial amounts on promoting and distributing the product or technology. Once this initial investment has been recovered, or the licensee has had sufficient time to realize the benefits of its distribution strategies, the licensor will hopefully be able to enter the market directly or to obtain additional royalties by using other licensees.

When limitations are imposed on the duration of an exclusive license, a number of different measures can be utilized. For example, the restriction can be expressed in terms of a specific period of time, such as two or three years. On the other hand, exclusivity may terminate when the licensee has had the opportunity to recover all or a specified portion of its original investment as measured by the amount of revenues realized by the licensee from the sale of the product or technology or the amount of royalties which have been received by the licensor from the activities of the licensee.

Even when exclusivity is limited to a fixed period of time, the licensor should insist upon including provisions that the exclusivity will terminate prior to the end of the period in the event that the licensee fails to meet certain "minimum" goals with respect to sales or royalties. Minimum performance conditions will protect the licensor from the situation where the licensee may choose to abandon the rights prior to the end of the initial period set out in the license and allow the licensor to begin exploitation of the market itself or select another licensee to for that purpose before the competitive advantage disappears.

Another way to limit the effect of exclusive rights is to grant rights which are exclusive only as to specific products or geographic areas, but non-exclusive in all other respects. This type of arrangement makes sense when the licensee has particular strengths in the areas of exclusivity, such as a large and proven chain of distribution or a reliable reputation with respect to similar products. Similarly, exclusivity may be restricted to one or more markets within a specific geographic area, such as when a licensor retains the right to distribute products to "national accounts," while the licensee is granted an exclusive right to sell to all other potential customers.

Performance conditions for exclusivity In almost all cases, the ongoing exclusivity of a license grant will be conditioned upon continued satisfactory performance of the licensee in relation to specified measures included as part of the license agreement. While the conditions will vary depending upon the nature of the licensed rights and the objectives of each of the parties, the following types of tests might be considered.

Development of technology The licensee may be obligated to reduce the licensed rights to practice, or complete development of an improvement to the technology, by a certain date. A duty of this type insures that the licensee will actually expend the funds required to position the technology for sale into the designated market. Moreover, the licensor may be depending upon the right to acquire and use any improvements in its own internal efforts.

Establishment of distribution channel When the license pertains to a distribution relationship, the licensee may agree to establish a certain number of dealerships or agents by a certain date. A standard of this type evidences the willingness of the licensee to distribute the product; however, it does not guarantee that anticipated revenues from sales will be realized.

Development expenditures The licensee may be required to expend certain minimum amounts for further development or improvement of the licensed technology or for the promotion of the various products and services included in the license. This type of standard insures a certain minimum degree of commitment on the part of the licensee; however, it does not guarantee that the development will be successfully completed. Moreover, a duty to make specified expenditures may terminate once the licensed technology becomes obsolete.

Revenue and market share milestones Perhaps the most frequently used method for evaluating the performance of exclusive licensees is to require that the licensee meet specific revenue objectives or minimum royalties or, alternatively, capture a given minimum share of the target market for the licensed products. If a minimum royalty hurdle is used, the licensee should be prohibited from making payments other than out of product sales, since the object of the exclusivity is to penetrate the target market.

"Most favored licensee" clauses If the licensee obtains something less than an exclusive license as to all markets and applications, it may insist upon inclusion of a "most favored licensee," or "MFL," clause in the license agreement. This type of provision will require that the licensor extend to the licensee the benefits of any terms which may be included in subsequent license agreements covering the same subject matter which are more favorable to the licensee than those contained in the original license agreement. The most common item of concern to a licensee in seeking a MFL provision is the royalty rate charged to subsequent licensees. However, the scope of the MFL clause usually extends to all other material conditions set out in the license agreement.

MFL provisions are difficult to enforce and interpret for a variety of different reasons. First, the exact subject matter of the license, as well as the statutory protection available for the underlying technology in the area in which a new licensee resides, may be substantially different. Second, local restrictions and competitive pressures may differ in each area in which a licensing arrangement is created. Finally, certain of the benefits which may be granted in a subsequent license may be coupled with obligations for the new licensee which might be more onerous than those which were included in the original license. In such cases, the right of the licensee to claim the benefits under the MFL clause generally is conditioned upon acceptance of the new obligations.

Given some of the aforementioned problems experienced in interpreting MFL provisions, careful consideration should be given to drafting the MFL clause. It generally makes sense to limit the effect of the provisions to comparable licensing situations and terms. For example, the MFL clause might be narrowly drafted so as to apply only to the grant of licenses in the same country, rather than to licenses granted in other jurisdictions. Also, the MFL provisions might apply only to royalty rates based on volumes of revenues or products comparable to those included in the original license agreement. In any case, a procedure should be included in the license agreement to attempt to resolve any dispute over the enforcement of an MFL clause before the entire relationship erodes.

D. Regulatory Attitudes on Exclusive Licensing Arrangements

United States Exclusive licensing arrangements appear to present few analytical problems for regulators and courts in the United States. For example, the 1995 Antitrust Guidelines for the Licensing of Intellectual Property (the "IP Guidelines") stated that, as a general matter, exclusive licenses raise antitrust concerns only if the licensees themselves, or the licensor and its licensees, are actual or potential competitors in a relevant technology or goods market in the absence of the licensing arrangement.[4] In the case where the parties are not actual or potential competitors, the IP Guidelines recognize that exclusive licensing arrangements may actually promote competition in the marketplace, especially in those situations where the demand for the licensed product is uncertain and/or successful market penetration will require large promotional expenditures.[5]

The IP Guidelines are clear in stating that the DOJ will focus on the actual practice and its effects, not on the formal terms of the arrangement.[6] For example, a license denominated as non-exclusive may nonetheless give rise to the same concerns posed by formal exclusivity if it is structured in a manner which makes it costly for licensees to use competing technologies. On the other hand, the mere fact that the licensor chooses to deal with a single licensee, or elects to limit its activities to a single field of use or location, does not automatically raise anticompetitive concerns. Also, the final version of the IP Guidelines included a clarification that the same principles of antitrust analysis will be applied to certain

[4] IP Guidelines 4.1.2. Examples of exclusive licenses with possible competitive consequences include cross-licensing by parties collectively possessing market power, grantbacks, and acquisitions of intellectual property rights. The IP Guidelines are discussed in detail in Chapter 5.

[5] See, e.g. IP Guidelines Example 7. The IP Guidelines indicate that "non-exclusive" licensing arrangements generally do not present antitrust concerns, even if the parties are involved in a horizontal arrangement, although non-exclusive cross-licensing arrangements may reduce competition by levying royalties that increase competitors' costs or encourage price coordination. Ibid. § 4.1.2.

[6] Ibid.

exclusive forms of licensing arrangements as are applied to similar vertical types of restraints, such as exclusive territorial grants and exclusive dealing arrangements.

Exclusive licensing in the EC In contrast to the United States, exclusive licensing arrangements have been a major concern for the Commission and the ECJ for many years. In the years since the adoption of Regulation 2349/84 (the "Patent Licensing Regulation"),[7] the Commission generally followed the ECJ's decision in *Nungesser* v. *Commission*[8] by allowing licensors to be restricted from licensing other undertakings to exploit[9] the licensed invention in the licensed territory, provided that the licensed patents remained in force throughout the period of the exclusivity.[10] Licensors could also be prevented from exploiting the licensed invention themselves in the licensed territory for the same period,[11] and it was possible for exclusivity to be granted in all or a part of the common market, with or without areas that are outside the common market.[12]

However, while the Patent Licensing Regulation followed *Nungesser* in recognizing and permitting certain exclusive and sole licenses, the overall degree of protection from exploitation of the licensed products by other parties, including the licensor and other licensees, was still far from absolute. For example, under Article 3(11) of the Patent Licensing Regulation, neither of the parties could be required to refuse without any objectively justified reason to meet demand from users or resellers in their respective territories who would market products in other territories within the common market, nor could the parties be required to make it difficult for users or resellers to obtain the products from other resellers within the common market.[13] In addition, the Patent Licensing Regulation could not be used to save agreements which contained an obligation on either party to prevent parallel trading.

The Patent Licensing Regulation extended beyond the Commission's interpretation of *Nungesser* in other cases to apply to products that were not new[14] or which did not re-

[7] The Patent Licensing Regulation is discussed in detail in Chapter 5.

[8] *Breeders' rights: Maize Seed*, OJ 1978 L286/23 [1978] 3 CMLR 434 on appeal Case 258/78 *Nungesser* v. *Commission* [1982] ECR 2015 [1983] 1 CMLR 278. The Commission has continued to treat sole or exclusive technology licences of any importance as infringing Art. 85(1), but has usually exempted them individually. Korah, "The Preliminary Draft of a New EC Group Exemption for Technology Licensing," *European Intellectual Property Report* 7 (1994), 262.

[9] For purposes of the Patent Licensing Regulation, the term "exploitation" included all the means of exploitation contemplated under the patent laws, including the manufacture, use and sale of the patented product. See Recital (2).

[10] Art. 1(1)(1). Art. 9(3) provided that the licensor must, however, have the right to terminate the exclusivity within five years from the date of the agreement in the event the licensee failed to exploit the license. Territorial exclusivity could not be extended beyond the duration of the licensed patents, and the Commission stated that any exclusivity that might be granted for patent under an agreement is indissolubly linked to the existence and continued validity of such patents. Decision 85/410 *Velcro/Aplix*, OJ 1985 L233/22.

[11] Art. 1(1)(2).

[12] Art. 1(1)(1), Recitals (4) and (5).

[13] In particular, parties were precluded from exercising industrial or commercial property rights, or from taking any other measures, so as to prevent users or resellers from obtaining outside the licensed territory, or from putting on the market in the licensed territory, products which have been lawfully put on the market within the common market by the patentee or with his consent. Art. 3(11)(b).

[14] In one case, the Commission treated improvement patents subsequently added to a package license as not being new, although it seemed to think that patents it referred to as being basic, which were earlier added after the original license was granted, might escape the prohibition of Art. 85(1). See *Velcro/Aplix* OJ 1985, L233/22. The Patent Licensing Regulation did not contain any specific requirement that the licensed technology be "new," however, parties seeking an individual exemption might nonetheless have to establish that the technology is "new," notwithstanding the existence of a patent on the technology. See Korah, supra note 8, at 32.

quire a large amount of investment by a licensee.[15] However, commentators have noted shortcomings in the *Nungesser* decision, as well as in the Patent Licensing Regulation. For example, the ECJ did not address the question of the burden of proof with respect to whether or not an exclusive licensing arrangement infringes Article 85(1), and this is a matter that remains to be resolved in future cases.[16] Also, no attempt was made by the ECJ to determine how one might identify the degree of protection that might generate the optimal level of investment by the licensor and the licensee.[17] The ECJ, while recognizing that some protection is necessary in order to induce innovative activities, said little more other than that absolute territorial protection is excessive. But, it has been pointed out that licensees are adversely impacted by any form of parallel imports, regardless of their source (i.e. the licensor or other licensees), and that protection can only be effective if it reduces threats from all directions.[18] One commentator has suggested that this dilemma can be resolved by recognizing the existence of a rebuttable presumption that absolute territorial protection infringes Article 85(1), thereby allowing parties to present evidence regarding market characteristics, the level of investment, and other matters.[19]

As noted elsewhere, the market share conditions that were included as part of the initial draft of the proposed new Group Exemption for Technology Transfer Agreements (the "Proposed Technology Licensing Regulation")[20] with respect to the exemption of exclusive licensing arrangements generated a significant amount of controversy, and were ultimately dropped from the new block exemption (the "Technology Licensing Regulation").[21] Not surprisingly, the Technology Licensing Regulation will only permit obligations on the licensor not to license other undertakings to exploit the licensed technology in the licensed territory in pure patent licensing agreements if the licensed product is also protected by parallel patents in the licensee's territory.[22] Also, like the Patent Licensing Regulation, the Technology Licensing Regulation forbids the use of provisions under which one or both of the parties is required without any objectively justi-

[15] Ibid. at 39. See also Recital (11) of the Patent Licensing Regulation, where it was declared: "Exclusive licensing agreements, i.e. agreements in which the licensor undertakes not to exploit the 'licensed invention' ... in the licensed territory himself and not to grant further licenses there, are *not in themselves incompatible with Article 85(1)* where they are concerned with the introduction and protection of a new technology in the licensed territory, by reason of the scale of the research which has been undertaken and of the risk that is involved in manufacturing and marketing a product which is unfamiliar to users in the licensed territory at the time the agreement is made. ... [I]n so far as in other cases agreements of this kind may fall within the scope of Article 85(1), it is useful for the purposes of legal certainty to include them in Article 1."

[16] Ibid. at 5.

[17] Ibid.

[18] Ibid. at 6 (referring to Joliet, "Territorial and Exclusive Trademark Licensing under the EEC Law of Competition," [1984] 15 IIC 21, 35).

[19] Fox, "Maize Seed: a Comparative Comment," in B. Hawk (ed.), [1982] *Fordham Corporate Law Institute* 151.

[20] 94/C 178/03. The Proposed Technology Licensing Regulation is discussed in detail in Chapter 5.

[21] The Technology Licensing Regulation is discussed in detail in Chapter 5.

[22] Technology Licensing Regulation Art. 1, para. 2. In the case of a mixed patent and know-how licensing agreement, the exemption shall apply in member states in which the licensed technology is protected by necessary patents for as long as the licensed product or process is protected in those member states by such patents if the duration of such protection exceeds ten years from the date when the licensed product is first put on the market in the EC by one of the licensees. Technology Licensing Regulation Art. 1, para. 4. "Necessary patents" are "patents where a license under the patent is necessary for the putting into effect of the licensed technology in so far as, in the absence of such a licence, the realization of the licensed technology would not be possible or would be possible only to a lesser extent or in more difficult or costly conditions. Such patents must therefore be of technical, legal or economic interest to the licensee ..." Technology Licensing Regulation Art. 10(5).

fied reason to refuse to meet orders from users or resellers in their respective territories who would market products in other territories within the common market, nor may the parties be required to make it difficult for users or resellers to obtain the products from other resellers within the common market.[23]

III. Territorial and Field of Use Restrictions

A. Introduction

Definitional aspects While the licensor is certainly able to permit a licensee to exercise almost all the legal rights associated with the patent in question, it is typical for the license agreement to contain a variety of restrictions on the licensee's ability to lawfully use the licensed rights. Restrictions on a licensee can take a number of different forms; however, the two most common types of restrictions typically relate to the demographic or geographic markets in which the licensee is allowed to exploit the licensed subject matter, and the products which the licensee may produce and/or sell using the licensed subject matter. The first type of restriction generally is referred to as a "territorial" constraint, while the second type of restriction refers to the "field of use."

Territorial restrictions The licensor will often attempt to impose geographic and territorial restrictions upon the licensee's ability to make and/or distribute the licensed products outside of the region or countries specified in the license. Territorial restrictions include not only explicit limitations, but areas of primary responsibility, exclusive territories and restrictions on where products can be advertised. A restriction of this type is designed to allow the licensor to maintain control over the manner in which the licensed products are distributed and to permit the licensor to select the best mode for distribution in each geographic market. However, licensees will usually resist these limitations, since they generally sees the license as a means for expanding their own distribution capabilities outside of their traditional markets.

Territorial restrictions typically arise in an international license and the license agreement should clearly define each of the geographic areas, usually classified by national, community or continental boundaries, within which the licensee is to be permitted to use the licensed rights. However, the licensor may include geographic restrictions in domestic licenses, perhaps coupled with a product or application limitation. For example, a distributor with a strong network of retail outlets in the Western region of the United States may be given a license to sell the licensed products to the consumer market in the area, while other licensees are appointed to sell the licensed products to business accounts throughout the United States and through retailers outside of the Western area.

Field of use restrictions In many cases, the licensed rights may be used to manufacture a number of different "products," each of which may be used for one or more distinguishable applications. For example, in the computer and electronics area, the tech-

[23] Technology Licensing Regulation Art. 3(3). In particular, parties are precluded from exercising intellectual property rights, or from taking measures so as to prevent users or resellers from obtaining outside the licensed territory, or from putting on the market in the licensed territory, products which have been lawfully put on the market within the common market by the patentee or with his consent. Ibid. Art. 3(3)(b).

nology may be used to manufacture one or more products which might be suitable for sale into both the "consumer" market and into the "business" market. Similarly, in the pharmaceutical area, prescription and "over-the-counter" drugs form two distinct markets. Since a single licensee may not have the distribution skills for accessing more than one market, the licensor may wish to select two licensees whose activities will be limited to those applications best suited to their particular sales networks, thereby increasing the total anticipated stream of royalties from the use of the technology.

Further differentiation in defining licensed products may also be appropriate when the licensee lacks all of the technical capabilities to produce the full range of product embodiments of the licensed technology. For example, in the pharmaceutical area, the scope of the license may be limited to products for use in the treatment of a given disease that are to be administered by injection, while excluding from the scope of the license forms of ingestible products, such as syrups, capsules, caplets, and pills. These types of distinctions are generally driven by the particular assets, expertise and experiences of the prospective licensees, as well as the capabilities of the licensor itself.

In many situations, patents may have some distinguishable commercial use in more than one technical field or with different classes of users, such as a component which can be integrated into two or more end products to be offered to a different group of potential customers, and the patent owner may have sound business reasons for structuring his licensing and marketing program in a manner which recognizes and exploits these differences.[24] For example, a manufacturing licensor may want to reserve for itself a particular field of use or customer class, while recruiting other parties as licensees to service other potential areas in which the licensees might have greater experience or resources, thereby maximizing the amount of revenues available to the licensor. Also, the value of the invention may differ between particular areas of use, meaning that licenses may need to express different royalty rates in order to both allow the patent owner to optimize its financial return on the invention and permit the different licensee classes to pay only what the invention is worth to them. Finally, use of an exclusive license confined to a specific marketing area is a good way for a patent owner to induce the licensee to undertake the development risks while limiting the patentee's dependence on the resources of the licensee.

Another form of application restriction can be imposed with respect to end users. For example, in the software area, a license may restrict the use of the software to one or more specific types of computers using a particular form of operating system, possibly due to the limited functionality of the licensed software. Also, a software licensor may restrict the use of its licensed software to a single computer, or limit use to one computer at a time, in an attempt to increase the volume of sales.

[24] See generally Adamo, "Problems Connected With Acquisition, Licensing and Enforcement of Intellectual Property," *Alb. Law Review* 50, (1986), 475; Selinger, "Patent Licensing in the Afterglow of Sylvania: Practicalities of Life Under the Rule of Reason," *JPOS* 63, (1981), 353; Curley, "Patent Acquisitions and Restricted Licenses Under the Antitrust Laws," *Seton Hall Law Review* 8 (1977), 645, 658–60; Adelman & Juenger, "Patent Antitrust: Patent Dynamics and Field of Use Licensing," *New York University Law Review* 50, (1975), 273; Frost, "Restrictions on Fields of Use and Territories," *Antitrust Law Journal* 42 (1973), 633; Marquis, "Limitations on Patent License Restrictions: Some Observations," *Iowa Law Review* 58 (1972) 41, 65–71.

B. Regulatory Attitudes on Territorial Restrictions

General attitudes in the United States As a general matter, territorial restrictions in patent licenses have not generated a significant amount of controversy in the United States, at least to the extent that they were limited to the "first sale" of the patented product. For example, in *Ethyl Gasoline Corp.* v. *United States*,[25] the Supreme Court held that a patentee are allowed to "grant licenses to make, use or vend, restricted in point of space or time ... save only that ... he may not enlarge his monopoly."[26] In another case, the court noted the "fundamental rule" of patent law that "the owner of a patent may license another and prescribe territorial limitations."[27] Also, explicit language included in the Patent Act itself provides that the owner may "convey an exclusive right under ... his patents, to the whole or any specified part of the United States."[28]

General attitudes in the EC While the Commission also had originally endorsed, in its 1962 "Christmas Message," the use of territorial restrictions in licensing arrangements on grounds that they were consistent with the exercise of the licensor's otherwise exclusive rights relating to the exploitation of the invention,[29] attitudes have changed somewhat over the years. Nonetheless, the EC, in spite of its overriding objective of insuring free movement of goods within the common market, has tolerated some territorial restrictions in licensing agreements, although such restrictions are now deemed to be prima facie infringements of Article 85(1)[30] and both the Patent Licensing Regulation and the Technology Licensing Regulation include only limited exceptions for such restrictions, even when the restrictions apply only to sales in territories where other licensees or the licensor have rights under parallel patents. The relief that is available for such restrictions

[25] 309 U.S. 436 (1940).

[26] Ibid. at 456.

[27] *Brownell* v. *Ketcham Wire & Mfg. Co.*, 211 F.2d 121, 100 U.S.P.Q. 338 (9th Cir. 1954).

[28] Section 261 of the Patent Act (35 U.S.C. § 261) specifically states: "The applicant, patentee, or his assigns or legal representatives may in like manner [i.e. by a written instrument] grant and convey an exclusive right under his application for patent, or patents, to the whole or any specified part of the United States." While some commentators have argued that a review of the history of the statute dictates that it should only be applied to outright assignments, rather than mere licenses (see, e.g. Baxter, "Legal Restrictions on Exploitation of the Patent Monopoly: An Economic Analysis," *Yale Law Journal* 76 (1966) 267, 347–52; Gibbons, "Domestic Territorial Restrictions in Patent Transactions and the Antitrust Laws," *George Washington Law Review* 34 (1966), 893, 894–904; Turner, "The Patent System and Competitive Policy," *New York Uuniversity Law Review* 44 (1969), 450, 469–74; other commentators, as well as most courts, have reached the opposite conclusion (see, e.g. Selinger, "Patent Licensing in the Afterglow of Sylvania: Practicalities of Life Under the Rule of Reason," *JPOS* 63, (1981), 353; Marquis, "Limitations on Patent License Restrictions: Some Observations," *Iowa Law Review* 58 (1972) 41, 54–8; Wheeler, "A Reexamination of Antitrust Law and Exclusive Territorial Grants by Patentees," *University of Philadelphia Law Review* 199 (1971), 642, 646.

[29] The "Christmas Message" suggested that territorial restrictions would fall outside of Art. 85(1) all together. However, the ECJ's decision in *Nungesser* regarding which types of exclusive license would fall outside of Article 85(1) applied only to licenses that did "not affect the position of third parties such as ... licensees for other territories."

[30] See *AOIP/Beyrard*, OJ 1976 L6/8 [1976] 1 CMLR D14; *Velcro/Aplix*, OJ 1985 L233/22 [1989] 4 CMLR 157. See also Fourth Report on Competition Policy (1974), points 22 et seq.; Fifth Report (1975), point 11; *Zoller & Frohlich/Telemecanique*, Ninth Report on Competition Policy (1979), point 109; *Fondasol Techniques Louis Menard*, ibid., point 112; *Plant Royalty Bureau*, ibid., point 120. The Commission takes this position even where direct sales by one licensee into the territory of another would prima facie infringe the latter's patent rights since the Commission takes the view that Arts. 30 to 36 of the Treaty preclude the exercise of patent rights to prevent direct sales between licensees. See, e.g. submissions in Case 19/84 *Pharmon* v. *Hoechst* [1985] ECR 2281 [1985] 3 CMLR 775.

depends on several factors, including whether the restrictions apply in territories that are reserved for the licensor, the existence of parallel patents, the nature of the sales activities of the licensee, the duration of the restrictions, and the point when commercial use of the patented product commences.

Extraterritorial sales activities in the EC Provisions in a license agreement which restrict the licensee from making sales into territories reserved for the licensor would appear to fall outside of the ECJ's reasoning in *Nungesser*, since such restrictions are strictly a matter of contract between the parties and have no direct impact upon competition from parallel traders or licensees from other traders. As such, it may well be that such restrictions fall outside of Article 85(1), particularly when the licensor has patent protection in the territories which are covered by the restrictions.[31] If, however, Article 85(1) does apply, such a provision might well fall within the provisions of the applicable block exemption. The Patent Licensing Regulations allowed licensors to include provisions in its licenses that restricted licensees from selling in parts of the common market which the licensor had reserved for itself and in which the licensor had obtained patent protection.[32] The protection afforded by the Patent Licensing Regulation to licensors was substantial, and extended to both "active"[33] and "passive"[34] sales by a licensee, so long as parallel patents remained in force, although the exemption did not permit a restriction on the freedom of the licensee to meet, in his own territory, orders for goods which might be exported by third parties into the licensor's territory.[35] In addition, the Patent Licensing Regulation provided broad geographic protection for the licensor, as the territory reserved for the licensor included any place for which no license had yet been granted,[36] enabling the licensor to give a subsequent licensee a territory that had not been exploited by his other licensees.[37]

The protection afforded licensors under the Patent Licensing Regulation was actually broader than those in earlier drafts of the block exemption,[38] and were praised as particularly necessary for licensors wishing to exploit the patent on their own, since failure to provide such licensors with protection in their own territory might lead to both less

[31] It is, however, the Commission's position that any contractual restriction on a licensee's export activities, whether direct or indirect, will almost always infringe Art. 85(1) irrespective of the patent position in other territories. See, e.g. Fourth Report on Competition Policy (1974), points 24 and 25.

[32] Patent Licensing Regulation Art. 1(1)(3).

[33] An "active" policy of sales meant, in particular, engaging in advertising activities specifically aimed at territories which had been licensed to other licensees, or the establishment of a branch or distribution depot in those territories. Patent Licensing Regulation Art. 1(1)(5). Cf. Art. 2(2)(c) of Regulation 1983/83.

[34] See generally, Patent Licensing Regulation Recital (12). "Passive" sales are those as to which the licensee merely responds to unsolicited orders received.

[35] See Patent Licensing Regulation Art. 3(11).

[36] Recital (12) of the Patent Licensing Regulation defined "Territories reserved to the licensor" as "territories within the common market in which the licensor has patent protection and has not granted any licenses."

[37] The application of Art. 85(1) to this provision was not decided in *Nungesser* since the licensor did not produce seeds itself and thus did not require any protection of this type.

[38] For example, the draft block exemption published in 1979 provided that territorial sales restrictions were subject to four conditions: maximum turnover; the products had to be capable of being sold by dealers to ensure that parallel imports were possible; the product had to be made by the licensee; and the restrictions could not exceed the life of the patents originally licensed. OJ 1979, C58/12. The first two conditions did not appear in the final block exemption; the third condition was continued by Art. 1(2) of the block exemption, although the licensee can now have the products made by a sub-contractor; and the fourth condition has been relaxed to some extent. Korah, supra note 8, at 32.

exploitation therein, as well as to reduced licensing in other territories.[39] Nonetheless, the Proposed Technology Licensing Regulation, while continuing to exempt obligations on a licensee not to exploit the licensed technology in territories reserved for the licensor,[40] provided that, in the case of a pure patent license, the product is protected by parallel patents in the licensor's territory,[41] would deny the exemption in cases where the licensor holds a market share of more than 20 percent in the protected territory.[42] However, the market share exception for the exemption was dropped in the Technology Licensing Regulation, so that such provisions are still entitled to the same protection as was the case under the Patent Licensing Regulation.[43]

As to restrictions relating to a licensee's activities in territories which have been licensed to other licensees, the Patent Licensing Regulation exempted obligations on a licensee not to pursue active sales policies in such territories, so long as there were parallel patents in those other territories.[44] Also, licensees could be restricted from passive sale activities in territories licensed to other licensees within the common market for a period not exceeding five years from the date when the product is first put on the market within the common market by the licensor or one of his licensees, provided once again that there were parallel patents in the other licensed territories.[45] However, the Patent Licensing Regulation did not permit the use of restrictions on a licensee's ability to meet, in their own territory, orders for goods which may ultimately be exported out of that territory by third parties into territories which have been reserved for other licensees.[46]

The Technology Licensing Regulation also exempts obligations on the licensee not to pursue active or passive sales policies in the territories within the common market which are licensed to other licensees.[47] However, with regard to limits on active sales imposed in pure patent licensing agreement, the licensed product must be protected by parallel patents in the licensee's territory,[48] and restrictions on passive sales in pure patent licensing agreements cannot exceed five years from the date when the product is first put on the market within the common market by one of the licensees, inasmuch and for as long as, in these territories, the product is protected by parallel patents.[49] In addition, the Technology Licensing Regulation prohibits (i) restrictions on the ability of either party to refuse without any objectively justified reason to meet demand from users or resellers in their

[39] Ibid. at 36.

[40] Proposed Technology Licensing Regulation Art. 1, para. 1.

[41] Ibid. Art. 1, para. 2.

[42] Ibid. para. 6.

[43] See Technology Licensing Regulation Art. 1, para. 1(3) (exempting obligations on the licensee not to exploit the licensed technology in the territory of the licensor within the common market).

[44] Patent Licensing Regulation Art. 1(1)(5).

[45] Patent Licensing Regulation Art. 1(1)(6).

[46] See Art. 3(11)(a), which stated that the Patent Licensing Regulation would not apply if the parties "are required to refuse without any objectively justified reason to meet demand from users or resellers in their respective territories who would market products in other territories within the common market."

[47] Technology Licensing Regulation Art. 1, para. 1(5). Active sales activities include, but are not limited to, advertising specifically aimed at those territories or establishing a branch or maintaining a distribution depot in those territories.

[48] Technology Licensing Regulation Art. 1, para. 2. In the case of a mixed patent and know-how licensing agreement, the exemption shall apply in member states in which the licensed technology is protected by necessary patents for as long as the licensed product or process is protected in those member states by such patents if the duration of such protection exceeds ten years from the date when the licensed product is first put on the market in the EC by one of the licensees. Technology Licensing Regulation Art. 1, para. 4.

[49] Technology Licensing Regulation Art. 1, para. 2. In the case of mixed patent and know-how licensing agreement, the exemption cannot exceed the five year period. Technology Licensing Regulation Art. 1, para. 4.

respective territories who market products in other territories within the EC, or (ii) other requirements that make it difficult for users or resellers to obtain the products from other resellers within the common market.[50]

The degree of protection afforded by restriction on the ability of a licensee to engage in sales activities outside of its own territory turns, in part, upon the timing of the first date upon which the subject products are "put on the market" in the EC. This rule raises at least two important issues. The first relates to identifying the party whose act of "putting the product on the market" would trigger the five-year period. In the Patent Licensing Regulation, the period could be started by the act of either the licensor or one of his licensees, which meant that if a licensor put the product on the market and then suffered delays in finding licensees, the licensor would not be able to provide licensees with the full period of protection that might be necessary to induce the licensee to take on the costs of exploiting the licensed products. However, the Technology Licensing Regulation solved this problem by providing that the five-year period would begin only when the licensed product was first put on the market by one of the licensees.[51]

Another issue has to do with defining what constitutes "putting on the market." Neither the Patent Licensing Regulation or the Technology Licensing Regulation included any precise form of definition of how this date is to be determined, as it might have by reference to concepts of "novelty" in patent statutes. It has been suggested that since the restrictions are intended to allow the protected parties to earn a more than competitive profit for the investment in tooling up and developing a market, then the period should begin only at the time that the investment itself begins to pay off.[52] Under this view, limited sales for experimental purposes, test marketing activity, or clinical trials, should not start the period running, especially if the tests or trials lead to changes in the product. On the other hand, marketing of the product on a commercial scale, even in a relatively small geographic area, would be sufficient to start the five-year period running.

Extraterritorial manufacturing activities in the EC The Patent Licensing Regulation took an even more lenient approach to restrictions on a licensee's ability to manufacture or use an invention outside of its territory, and exempted such restrictions provided there were parallel patents in the protected territories.[53] The scope of the exemption was consistent with the Commission's historical willingness to be more lenient with manufacturing exclusivity than with sales exclusivity. For example, the *Davidson Rubber* decision allowed the parties to enter into an exclusive manufacturing license, although they were forced to remove exclusivity in the sales area.[54] Moreover, an argument can

[50] Technology Licensing Regulation Art. 3, para. 3 denies the block exemption in cases where one or both of the parties are required without any objectively justified reason:

"(a) to refuse to meet demand from users or resellers in their respective territories who would market products in other territories within the common market;

(b) to make it difficult for users or resellers to obtain the products from other resellers within the common market, and in particular to exercise intellectual property rights or take measures so as to prevent users or resellers from obtaining outside, or from putting on the market in the licensed territory products which have been lawfully put on the market within the common market by the licensor or with his consent.

or do so as a result of a concerted practice between them."

[51] See Technology Licensing Regulation Art. 1, paras. 1(2) and 1(3).

[52] Korah, supra note 8, at 42.

[53] Patent Licensing Regulation Art. 1(1)(4). The exemption also covers restrictions relating to the use outside the territory of "know-how" licensed with the patent(s).

[54] [1972] CMLR D52, D62. Also, Arts. 30 to 36 appear to be limited to sale activities, and thus would not operate to overturn restrictions on manufacturing and usage.

be made that Article 85(1) does not even apply to restrictions relating to manufacture and use if the licensor and those deriving title from the licensor have patent protection in all other territories in the common market, since in that case the provision would not restrict any act that could otherwise lawfully take place; however, Article 85(1) would appear to be applicable if a provision in the license agreement expressly or impliedly prevents manufacture or use in a territory where there is no patent protection,[55] and the Commission has held on at least one occasion that a license to manufacture that was limited to only one member state was an infringement of Article 85(1).[56]

The Technology Licensing Regulation treats manufacturing and use prohibitions in the same manner as obligations relating to active sales activities, and exempts obligations on the licensee not to manufacture or use the licensed product, or use the licensed process, in territories within the common market that are licensed to other licensees,[57] provided that the licensed product is protected by parallel patents in the licensee's territory.[58] Once again, in the case of a pure patent licensing agreement, the exemption will be granted for as long as the licensed product is protected by necessary parallel patents in the territories of the licensee.[59] In the case of a mixed patent and know-how licensing agreement, the exemption shall apply in member states in which the licensed technology is protected by necessary patents for the longer of (i) the period the licensed product is protected in those member states by such patent, or (ii) ten years from the date when the licensed product is first put on the market in the EC by one of the licensees.[60]

Other factors in analyzing territorial restrictions There are other situations when territorial restrictions will be closely scrutinized. One area of concern is when the license agreement actually amounts to a cartel arrangement among the participants, rather than a bona fide exploitation of the licensed technology.[61] Courts in the United States have also struck down territorial limitations which gave the licensee veto power over the appointment of additional licensees within the assigned territory on the grounds that such a right in favor of the licensee fell outside of the lawful protections otherwise afforded under the patent laws since it was granted "not to protect the patentee but for the benefit of the licensee."[62] The "licensee benefits" test is somewhat limited, however, in that it is often

[55] Case 193/83 *Windsurfing International* v. *Commission* [1986] ECR 611 [1986] 3 CMLR 489, §§ 82 et seq.; see also *Velcro/Aplix*, OJ 1985 L233/22 [1989] 4 CMLR 157, (use of equipment outside licensed territory).

[56] *Boussois/Interpane*, OJ 1987 L50/30 [1988] 4 CMLR 124. See also Case 193/83 *Windsurfing International* v. *Commission* [1986] ECR 611 [1986] 3 CMLR 489, where the Commission left open whether a license limited to a part of a member state infringed Article 85(1).

[57] Technology Licensing Regulation Art. 1, para. 1(4).

[58] Ibid., para. 2.

[59] Ibid., para. 2.

[60] Ibid., para. 4.

[61] See, e.g. *United States* v. *Crown Zellerbach Corp.*, 141 F.Supp. 118, 110 U.S.P.Q. 399 (N.D. Ill. 1956), and Patent Licensing Regulation Art. 3(10) (the block exemption was not available where licensee did not put the licensed product on the market in territories licensed to other licensees within the common market as a result of a concerted practice between the parties).

[62] *United States* v. *Crown Zellerbach Corp.*, 141 F.Supp. 118, 128, 110 USPQ 399 (N.D. Ill. 1956). The court explained that "[r]estraints upon competition for the benefit of the licensee are to be measured not by the standards of the patent laws, but by the general rules governing the validity of competitive restrictions attached to the conveyance of personal property." Ibid. at 127. See also *American Industrial Fastener* v. *Flushing Enterprises, Inc.*, 362 F.Supp. 32, 37, 179 U.S.P.Q. 722 (N.D. Ohio 1973) ("Section 261 is intended to benefit patentees, not licensees or their vendees"); *Mannington Mills, Inc.* v. *Congoleum Industries, Inc.*, 610 F.2d 1059, 203 U.S.P.Q. 81 (3d Cir. 1979) ("[w]here the license restriction results primarily in benefits for the licensees rather than the patentee," the restriction is "not immunized" from "antitrust scrutiny.").

difficult to determine if a provision is advantageous to the patentee or the licensee. Also, in many cases, the apparent benefit to the licensee may actually be necessary in order for the patentee to be able to commercially exploit the licensed invention or process.[63]

Territorial restrictions on resales As to attempts to impose restrictions on the sale of patented products following the "first sale," such restrictions will be analyzed under the antitrust laws, rather than the patent laws, in the United States. The important case in this area is the Supreme Court's decision in *Continental TV, Inc.* v. *GTE Sylvania*,[64] which actually related to territorial restrictions in a trademark license, rather than a patent license. In finding that such restrictions should be subject to analysis under the rule of reason, rather than being declared *per se* illegal, the Court specifically referred to a number of possible competitive justifications for such restraints, such as the special needs of new market entrants when attempting to induce distributors to carry their brands, and the possible needs of even established firms to induce point-of-sale promotional or servicing activities by retailers.[65]

While *Sylvania* did not include a specific list of the factors to be taken into account in rule of reason analysis of vertical restraints, subsequent cases have identified at least four major considerations that should go into a balancing test which weighs the pro-competitive and anticompetitive effects of a given restraint on both interbrand and intrabrand competition. The four considerations include the definition of the "market" in which the impacted competition occurs or is to occur, since a determination must be made as to whether the conduct of the defendant amounted to an "unreasonable" restriction on market competition;[66] the apparent purpose of the restraint and whether or not this purpose offers a sufficient business justification for any inhibition on the level of competition;[67] the market power of the licensor, with the likelihood that restraints would be upheld increasing in inverse relation to the total market share of the firms involved;[68] and

[63] See, e.g. *SCM Corp.* v. *Xerox Corp.*, 645 F.2d 1195,209 U.S.P.Q. 889 (2d Cir. 1981) (upholding an exclusive license covering a promising but commercially undeveloped invention, where the patentee had made earlier unsuccessful attempts to find licensees to develop the invention, and where the exclusivity served to induce the licensee to enter into the arrangement); *International Wood Processors* v. *Power Dry, Inc.*, 792 F.2d 416 (4th Cir. 1986) ("A court analyzing whether a particular patent-related practice uulawfully extends the limited [patent] monopoly should hesitate to apply antitrust sanctions to prohibit the practice if the effect of such would be to discourage innovation and investment by reducing potential rewards that otherwise would accrue to the patentee").

[64] 433 U.S. 36 (1977).

[65] Ibid. at 55.

[66] *Standard Oil Co.* v. *United States*, 221 U.S. 1, 65 (1911). See also *Broadcast Music, Inc.* v. *Columbia Broadcasting System, Inc.*, 441 U.S. 1, 19-20, 201 U.S.P.Q. 497 (1979); *National Society of Professional Engineers* v. *United States*, 435 U.S. 679, 690 (1978).

[67] See, e.g. *Martin B. Glauser Dodge Co.* v. *Chrysler Corp.*, 570 F.2d 72 (3d Cir. 1977), *cert. denied*, 436 U.S. 913 (1978); *Sandura Co.* v. *FTC*, 339 F.2d 847 (6th Cir. 1964); *Continental T.V., Inc.* v. *GTE Sylvania, Inc.*, 461 F. Supp. 1046 (N.D. Cal. 1978); *Sports Center, Inc.* v. *Biddell, Inc.*, 673 F.2d 780 (5th Cir. 1982); *Krehl* v. *Baskin-Robbins Ice Cream Co.*, 664 F.2d 1348 (9th Cir. 1982); *Clairol, Inc.* v. *Boston Discount Center of Berkley, Inc.*, 608 F.2d 1114, 204 U.S.P.Q. 89 (6th Cir. 1979).

[68] As to cases involving firms with small market shares, see, e.g. *Sports Center, Inc.* v. *Riddell, Inc.*, 673 F.2d 786 (5th Cir. 1982); *Muenster Butane, Inc.* v. *Stewart Co.*, 651 F.2d 292, 298 (5th Cir. 1982); *Continental T.V., Inc.* v. *GTE Sylvania, Inc.*, 461 F. Supp. 1046 (N.D. Cal. 1978) (only about 5 percent); *Amway Corp.* 93 F.T.C. 618 (FTC 1979) ("insignificant" share). As to cases involving firms with larger market shares, see, e.g. *Eiberger* v. *Sony Corp. of America*, 622 F.2d 1068 (2d Cir. 1980) (defendant's market share 12 percent after only four years in the market, and increasing rapidly); *Munters Corp.* v. *Burgess Indus., Inc.*, 1971-2 Trade Cas. (CCH) ô 62,149 (S.D.N.Y. 1978) (defendant's product had displaced virtually all other brands in the market). On the other hand, however, see *Cowley* v. *Braden Indus., Inc.*, 613 F.2d 751 (9th Cir.), *cert. denied*, 446 U.S. 965 (1980) (restraints upheld even though defendant's market share exceeded 70 percent).

the overall competitive condition in the relevant market (e.g. the amount of concentration, brand differentiation, active competition from other brands, barriers to entry).[69]

It would appear that many of the considerations which the Court recognized in *Sylvania* with respect to trademark licensing would also be applicable with respect to patent licensing.[70] For example, a new market entrant seeking licensees willing to incur the risks of developing and promoting the licensed invention may need to provide the licensees with protection against territorial raiding by dealers purchasing the goods in other neighboring territories. While such restrictions could not be supported merely by reference to the Patent Act due to the application of the exhaustion doctrine, the holding in *Sylvania* suggests that the appropriate antitrust law test for such restrictions is the rule of reason analysis, rather than *per se* illegality.

In the EC, regulation of resale restrictions can be understood in the context of Regulation 1983/83, which is the block exemption concerning exclusive distribution agreements.[71] While such agreements can appoint exclusive distributors for the whole of the EC, it is common for a manufacturer to develop a network of distributors with exclusive rights in less than all of the EC (i.e. one or more country(ies)). Among other things, the Regulation permits the use of restrictions on the distributor that would prevent the distributor from prospecting outside of its territory, including actively marketing, establishing a subsidiary, or maintaining a warehouse for the products outside of the contract territory.[72] However, the parties may not include provisions in their agreement which render it difficult for users or intermediaries to obtain contract goods from other dealers inside the EC, including through the exercise of intellectual property rights.[73]

C. Regulatory Attitudes on Field of Use and Customer Restrictions

General attitudes In general, both the United States and the EC have allowed patent owners to impose various "field of use" restriction in any patent license which they might grant, even though the licensor and the licensee might otherwise be in direct competition with each other, although the attitudes of regulators in the EC is significantly narrower than the policies followed in the United States. The basic rule in the United States was set out in *General Talking Pictures Corp.* v. *Western Electric Co.*,[74] which involved a scheme where one group of manufacturing licensees were granted patent licenses to practice patented electronic sound amplification devices in what was referred to as the "commercial field," an area that primarily encompassed motion picture equipment used in theaters, and another group of manufacturers received noncxclusive licenses cover-

[69] Holmes, *Intellectual Property and Antitrust Law*, Clark Boardman (1983), 32-11.

[70] See, e.g. *International Wood Processors* v. *Power Dry, Inc.*, 792 F.2d 416 (4th Cir. 1986) ("A court analyzing whether a particular patent.related practice unlawfully extends the limited [patent] monopoly should hesitate to apply antitrust sanctions to prohibit the practice if the effect of such would be to discourage innovation and investment by reducing potential rewards that otherwise would accrue to the patentee"). See generally Note, "Characterizing Horizontal Market-Division Agreements Under the Ancillary Restraints Doctrine," *Loyola University of Chicago Law Journal* 16 (1985), 775; Wheeler, supra note 4, at 651, 657–62; Posner, "Antitrust Policy and the Supreme Court: An Analysis of the Restricted Distribution, Horizontal Merger and Potential Competition Decisions," *Colorado Law Review* 75 (1975), 282.

[71] OJ L173/1 [1983]; amended by OJ L281/24 [1983].

[72] Regulation 1983/83, Art. 2(2)(c).

[73] Ibid., Art. 3(d).

[74] 305 U.S. 124, 39 U.S.P.Q. 329 (1938).

ing the "home field," which included radio receivers and amateur radio equipment. The arrangement was upheld as "reasonably within the reward which the patentee by the grant of the patent is entitled to secure."[75]

Another rationale for recognizing the legality of "field of use" restrictions was articulated in *United States* v. *Ciba Geigy Corp.*[76] In that case, the patent owner reserved for itself the bulk sales market for a patent drug, while granting a license to a competing manufacturer to produce the drug in finished form, subject to a provision that prevented the licensee from any bulk sales to other manufacturers. In upholding the restriction as a valid field of use restraint, the court noted that the restraint was within the legitimate scope of the exclusionary patent powers and that the license "opened up competition in an area" where the patentee "had the legal right to shut off all competition."[77] Similarly, in *United States* v. *Westinghouse Electric Corp.*,[78] the court upheld the patent owner's practice of granting licenses on a special "case-by-case, customer-by-customer" basis, arguing that the patent owner's rights established under *Ciba Geigy* to "shut off all competition" includes the "lesser right" to "restrict the exercise of the granted privilege," provided that the patentee does not attach a condition that "enlarges" the legitimate monopoly granted by the patent.[79]

In the EC, the general attitude is that field of use restrictions don't infringe Article 85(1).[80] However, the Commission remains concerned about the possibility that a field of use restriction might still be used as a device to divide markets and share customers, and the Patent Licensing Regulation took the position such restrictions were only acceptable in those cases where the restrictions related to different products belonging in different markets.[81] Also, while customer-based restrictions are quite common in the United States, the Patent Licensing Regulation did not allow parties to place restrictions on either party as to the customers he may serve, regardless of the ease or difficulty in identifying the members of the customer class in question.[82] In addition, the parties to a licensing agreement could not attempt to divide the market on the basis of customer types by employing different distribution methods or using certain packaging for products.[83]

Like its predecessor, the Technology Licensing Regulation permits the use of an obligation on the licensee to restrict his exploitation of the licensed technology to one or

[75] Ibid. The case actually involved the question of whether the licensor could maintain an infringement action against a party who, knowing of the field-of-use limitations in the license, purchased a patented amplifier from the licensee for commercial use. The Court found that the purchaser's actions constituted infringement. See also *Westinghouse Electric Corp.* v. *Bulldog Electric Products Co.*, 106 F. Supp. 819 (N.D.W.Va. 1952), *aff'd* 206 F.2d 574 (4th Cir.) (upholding restrictions on the use of patented electrical components); *Benger Laboratories Ltd.* v. *R.K. Laros Co.*, 209 F.Supp. 639, 648, (E.D. Pa. 1962), *aff'd per curiam* 317 F.2d 455 (3d Cir. 1962) (upholding two licenses to manufacture drugs, one for veterinary use, the other for human use).

[76] 508 F.Supp. 1118 (D.N.J. 1976).

[77] Ibid. at 1151.

[78] 471 F.Supp. 532 (N.D. Cal. 1978), *aff'd*, 648 F.2d 642 (9th Cir. 1981).

[79] Ibid. at 541–2.

[80] Patent Licensing Regulation Art. 2(1)(3). Also, Patent Licensing Regulation Art. 2(2) contained the necessary exemption for most cases.

[81] See Case 193/83 *Windsurfing International* v. *Commission* [1986] ECR 611 [1986] 3 CMLR 489.

[82] Patent Licensing Regulation Art. 3(7). See also *French State/Suralmo*, Ninth Report on Competition Policy (1979), point 114–15.

[83] Patent Licensing Regulation Art. 3(7) and Recital (23). A party could also not be prohibited from employing certain forms of distribution or, with the aim of sharing customers, using certain types of packaging for the products, subject to Arts. 1(1)(7) (obligations on licensees to use only the licensor's trademark) and 2(1)(3) (restrictions on the field(s) of use). Ibid.

more technical fields of application covered by the licensed technology or to one or more product markets.[84] However, in a significant change from prior practice, the Technology Licensing Regulation appears to permit the use of customer field of use restrictions provided that the parties are not competitors before the license agreement commences. Specifically, the prohibition against one party being restricted as to the customers he may serve within a given technological field of use or a product market applies only if the parties were already competing manufacturers before the grant of the licence in the specified field or market.[85]

Challenges to field of use restrictions A number of other cases have upheld field of use restrictions;[86] however, there are situations where such restrictions have been successfully challenged under the exhaustion doctrine. For example, in *Ansul Co.* v. *Uniroyal, Inc.*,[87] the court declared customer restrictions to be *per se* illegal under Section 1 of the Sherman Act. It is worth noting, however, that the ultimate holding in *Ansul* was dictated by the application of the then-prevailing Supreme Court rule that vertical territorial and customer restrictions were *per se* violations of the Sherman Act.[88] In light of the subsequent decision of the Court in *Sylvania*,[89] it is now likely that the distribution system would have been analyzed under the more flexible rule of reason analysis which

[84] Proposed Technology Licensing Regulation Art. 2, para. 1(8).

[85] Technology Licensing Regulation Art. 3, para. 4. If the prohibition does apply, the Technology Licensing Regulation makes it clear that the agreement cannot restrict the ability of a party to supply certain classes of users, employ certain forms of distribution or, with the aim of sharing customers, use certain types of packaging for the products, save as provided in Art. 1 (1) (7) and Art. 2 (1) (13) of the Regulation. Ibid. Art. 1 (1) (7) of the Technology Licensing Regulation exempts an obligation on the licensee to use only the licensor's trademark or get up to distinguish the licensed product during the term of the agreement, provided that the licensee is not prevented from identifying himself as the manufacturer of the licensed products. Art. 2 (1) (13) of the Technology Licensing Regulation exempts an obligation on the licensee to supply only a limited quantity of the licensed product to a particular customer, where a know-how license was granted at that customer's request so that he might have a second supplier inside a licensed territory; this provision shall also apply where the customer is the licensee, and the license which was granted in order to provide a second source of supply provides that the customer is himself to manufacture the licensed products or to have them manufactured by a subcontractor. This is a change from the provisions contained in the existing block exemptions.

[86] See, e.g. *Smith Int'l, Inc.* v. *Kennametal, Inc.*, 621 F.Supp. 79 (N.D. Ohio 1985) (field of use restrictions); *In re Yarn Processing Patent Validity Litigation*, 541 F.2d 1127, 192 U.S.P.Q. 241 (5th Cir. 1976), *cert. denied*, 433 U.S. 910 (1977) (customer restrictions); *Ciba-Geigy Corp.* v. *Bolar Pharmaceutical Co.*, 212 U.S.P.Q. 712 (E.D.N.Y. 1981) (field of use restrictions); *Bela Seating Co.* v. *Poleron Products, Inc.*, 438 F.2d 733, 168 U.S.P.Q. 548 (7th Cir.), *cert. denied* 403 U.S. 922 (1971) (design restrictions); *Armstrong* v. *Motorola, Inc.*, 374 F.2d 764, 152 U.S.P.Q. 535 (7th Cir.), *cert. denied*, 389 U.S. 830 (1967) (field of use restrictions); *Chemagro Corp.* v. *Universal Chemical Co.*, 244 F. Supp. 486, 146 U.S.P.Q. 466 (E.D. Tex 1965) (field of use); *Henger Laboratories, Ltd.* v. *R.K. Laros Co.*, 209 F. Supp. 639 (E.D. Pa. 1962), *aff'd*, 317 F.2d 455 (3d Cir.), *cert. denied*, 375 U.S. 833 (1963) (field of use restrictions); *Sperry Products, Inc.* v. *Aluminum Co. of America*, 171 F.Supp. 901, 120 U.S.P.Q. 362 (N.D. Ohio 1959), *rev'd in part*; 285 F.2d 911 (6th Cir. 1960), *cert. denied*, 368 U.S. 890 (1961) (field of use restrictions); *Eversharp, Inc.* v. *Fisher Pen Co.*, 204 F.Supp. 649, 132 U.S.P.Q. 423 (N.D. Ill. 1961) (field of use restrictions); *United States* v. *Birdsboro Steel Foundry and Machine Co.*, 139 F.Supp. 244, 108 U.S.P.Q. 428 (W.D. Pa. 1956) (customer restrictions); *Automatic Radio Mfg. Co.* v. *Hazeltine Research, Inc.*, 176 F.2d 799, 82 U.S.P.Q. 324 (1st Cir. 1949), *aff'd*, 339 U.S. 827 (1950) (field of use restrictions); *United States* v. *Consolidated Car Heating Co.*, 1950 Trade Cas. (CCH) §62,656 (S.D.N.Y. 1950) (field of use restrictions); *Turner Glass Corp.* v. *Hartford-Empire Co.*, 173 F.2d 49, 80 U.S.P.Q. 539 (7th Cir.), *cert. denied*, 338 U.S. 830 (1949) (field of use restrictions). See also Case 10 of 1988 International Guidelines (generally approving field of use licensing).

[87] 306 F.Supp. 541, 163 U.S.P.Q. 517 (S.D.N.Y. 1969), *rev'd in part on other grounds*, 448 F.2d 872 (2d Cir. 1971), *cert. denied*, 404 U.S. 1018 (1972).

[88] See *United States* v. *Arnold, Schwinn & Co.*, 388 U.S. 365 (1967).

[89] 433 U.S. 36 (1977).

would have taken into consideration the various practical and competitive reasons that the patent owner may have had for using nonprice vertical restraints,[90] such as the fact that the patent owner wanted to attract those dealers best able to educate farmers and otherwise provide point-of-sale services for the product, and felt that lower dealer profits on the product would discourage the type of dealer that it desired, leading it to adopt a distribution system designed to minimize dealer price competition.[91]

A "field of use" restriction has also been subjected to challenge on the basis that the restriction is for the benefit of the licensee, and not for the benefit of the licensor.[92] This type of analysis, however, is difficult to apply in practice, since it is often impossible to identify which of the licensor or the licensee is the fullest beneficiary of the effects of a particular provision. For example, there are often situations in which the patent owner owns a basic invention which may have promise in a number of different fields, but which also carries with it a high degree of development and promotional risk. In such cases, the use of one or more field restrictive licenses, while arguably being beneficial to the licensees, can also be an attractive strategy for the licensor to attract licensees on a field-by-field basis while reserving other fields for itself or other potential licensees. It may be that licensee-benefit analysis is most valuable as a means of uncovering circumstantial evidence that a licensing arrangement conceals illegal cartel activity between the parties.[93]

Field of use restrictions impacting unpatented items A number of cases have also dealt the legality of field of use restrictions in patent licenses covering patented machines or processes used to produce unpatented products, or unpatented products which contain key patented components. There are two lines of cases as to how field of use restrictions in these types of licenses should be treated for patent and antitrust law purposes. On the one hand, some courts have indicated that the exclusive rights embodied in a machine, process or component patent should be narrowly construed, and that the patentee's rights do not extend to unpatented items which do not fall within the specific claims of the patent.[94] On the other hand, a number of courts appear to have taken a more lib-

[90] See, e.g. *Business Electronics Corp.* v. *Sharp Electronics Corp.*, 108 S. Ct. 1515, 1524 (1988) ("a horizontal agreement to divide territories is *per se* illegal ... while GTE Sylvania held that a vertical agreement to do so is not"); *Valley Liquors, Inc.* v. *Renfleld Importers, Ltd.*, 822 F.2d 656 (7th Cir. 1987) (applying the rule of reason to uphold vertical non-price restraints imposed by a national distributor on its local dealers); *Assam Drug Co.* v. *Miller Brewing Co.*, 798 F.2d 311(8th Cir. 1986) (applying a "market power filter" under the rule of reason to summarily dismiss an antitrust case challenging vertical non-price restraints imposed by a manufacturer not shown to possess market power within a properly defined market).

[91] 306 F.Supp. 541, 546–7, 163 U.S.P.Q. 517 (S.D.N.Y. 1969), *rev'd in part on other grounds*, 448 F.2d 872 (2d Cir. 1971), *cert. denied*, 404 U.S. 1018 (1972). See also *Continental T.V., Inc.* v. *GTE Sylvania, Inc.*, 433 U.S. 36, 55 (1977), where the Supreme Court cited "services" justifications as one possible ground for upholding nonprice vertical restraints against antitrust charges of illegality.

[92] See, e.g. *Munter's Corp.* v. *Burgess Industries, Inc.*, 450 E Supp. 1195 (S.D.N.Y. 1977), a case in which a manufacturing licensor had imposed field of use restrictions with respect to a patented product and had agreed with the buyer not to sell the product to any of the buyer's competitors. On reargument following the *Sylvania* ruling, the court concluded that under a rule of reason analysis, the restriction would be unlawful (450 F. Supp. at 1205–11). The court relied heavily on the fact that the license restriction was for the benefit of the licensee.

[93] Thus, those cases that have employed the licensee-benefit line of reasoning have not relied upon this analytical ground alone. Rather, they have additionally invalidated the licensing arrangements as illegal conspiracies to allocate customers or other market areas under the guise of patent licenses.

[94] The effect of this narrow construction is that such restrictions are governed solely by antitrust law considerations, rather than patent law principles. See, e.g. *Mercoid Corp.* v. *Minneapolis-Honeywell Regulator Co.*, 320 U.S. 680, 60 U.S.P.Q. 30 (1944) (striking down price restraints on unpatented products produced containing patented parts); *Robintech, Inc.* v. *Chernidus Wavin, Ltd.*, 628 F.2d 142, 205 U.S.P.Q. 873 (D.C. Cir. 1980) (striking down territorial restraints on unpatented products produced with a patented

eral approach, and have emphasized patent law considerations as a basis for upholding field of use and other competitive restrictions on unpatented products which are produced using patented machines, processes or key components.[95] Some of these cases, notably *United States* v. *Studiengesellschaft Kohle, m.b.H.*,[96] appears to indicate that restrictions as to field of use in vertical arrangements,[97] even when they impact unpatented items, will not be struck down without an analysis of whether the particular restraint is really anti-competitive, with the extent of the patentee's effective exclusionary control under its patent being a relevant factor in the competitive reasonableness of the restraint imposed.

IV. Price Restrictions

A. Introduction

One area in which antitrust and competition laws have been applied in order to regulate licensing restrictions that arguably fall wholly within the scope of the licensed patent is price restrictive licensing, where the licensor imposes restrictions on the prices which the licensee may charge for the sale of licensed products to third parties. Price restrictions may often be a useful alternative to the licensor in cases where other common restraints which might be placed on a licensee (e.g. field of use, territorial, or customer) are either inappropriate or inadequate to protect the licensor's remaining interest in the licensed subject matter. For example, when a new product is in its infancy, and a licensor is unable to predict how his licensee's unit costs are likely to be affected by the volume of production as the product market develops, the licensor may want to impose price restrictions on the licensee in order to retain the flexibility to grant future licenses to larger competitors.

cont.
method); *Cummer-Graham Co.* v. *Straight Side Basket Corp.*, 142 F.2d 646, 61 U.S.P.Q. 395 (5th Cir.), *cert. denied*, 323 U.S. 726 (1944) (striking down price restraints on unpatented products produced using a patented machine).

95 See, e.g. *General Talking Pictures Corp.* v. *Western Electric Co.*, 305 U.S. 124, 39 U.S.P.Q. 329 (1938) (upholding field of use restrictions on unpatented products containing patented components); *United States* v. *Studiengesellschaft Kohle, m.b.H.*, 670 F.2d 1122, 212 U.S.P.Q. 889 (D.C. Cir. 1981) (upholding field of use restrictions on unpatented products produced using a particularly unique patented process); *In re Yarn Processing Patent Validity Litigation*, 541 F.2d 1127, 192 U.S.P.Q. 241 (5th Cir. 1976), *cert. denied*, 433 U.S. 910 (1977) (upholding customer restrictions in machine patent licenses designed to prevent contributory infringement of the patentee's process patents); *Ethyl Corp.* v. *Hercules Powder Co.*, 232 F.Supp. 453, 143 U.S.P.Q. (D. Del. 1964) (upholding field of use restraints on unpatented products produced using a particularly unique patented process); Turner *Glass Corp.* v. *Hartford-Empire Co.*, 173 F.2d 49, 80 U.S.P.Q. 539 (7th Cir.), *cert. denied*, 338 U.S. 830 (1949) (upholding use restrictions on the types of unpatented products produced by a patented machine); *Straight Side Basket Corp.* v. *Webster Basket Co.*, 82 F.2d 245 (2d Cir. 1936) (upholding price restraints on unpatented products produced using a patented machine and process); *Westinghouse Electric Corp.* v. *Bulldog Electric Products Co.*, 106 F.Supp. 819, 94 U.S.P.Q. 233 (N.D.W.Va. 1952), *aff'd*, 206 F.2d 574 (4th Cir.), *cert. denied*, 346 U.S. 909 (1953) (upholding use restrictions on the types of unpatented products produced using a key patented component); *Q-Tips, Inc.* v. *Johnson & Johnson*, 109 F.Supp. 657, 95 U.S.P.Q. 258 (D.N.J. 1951), *aff'd*, 206 F.2d 144 (3d Cir.), *cert. denied*, 347 U.S. 935 (1953) (upholding output restrictions on unpatented products produced using a patented machine). See also Antitrust Division, US Dept. of Justice, Antitrust Guide for International Operations, Case 11 (issued November 10, 1988) (generally approving territorial restrictions under process patent licenses).

96 670 F.2d 1122, 212 U.S.P.Q. 889 (D.C. Cir. 1981).

97 Obviously, field of use restrictions in a horizontal relationship may be struck down, given the general antipathy of the Supreme Court to cooperation among competitors.

B. Economic Arguments Regarding Price Restrictions

From a purely economic perspective, proponents of allowing price restrictions in licensing arrangements argue that such a provision will increase the likelihood that an inventor will be able to benefit from his intellectual property right and, as such, would be consistent with the overriding objectives of the patent laws. In addition, price restrictions have been praised as a means for promoting licensing, and technology transfer, by providing the licensor with adequate comfort that he will not be forced to compete in his own markets with a licensee that might enjoy cost-based advantages. Finally, it has been claimed that a patentee may be able to achieve efficiencies in the exploitation of his patent rights through price restrictive licensing in a manner that is analogous to efficiencies thought to arise from vertical price maintenance.[98]

Other advocates of price restrictions in the context of patent licenses base their views on the scope of the licensor's rights under the patent laws, and claim that it is unlikely that there would be any real objection since the only competition which is restrained is that which is created by the licensor's decision to grant the license in the first place. Since the underlying premise of the patent right is that the inventor is to enjoy complete discretion as to the use of the invention once the patent has been granted for a fixed period of time, there would appear to be little reason not to allow a patent owner to subdivide his monopoly by licensing another to manufacture a certain number of the units to be sold and to control the price for such units in the same manner as the patent owner would have been able to do had he simply chosen to manufacture those units on his own. Put another way, society is no worse off by the grant of the license containing the restrictions on pricing, since monopoly profits would have existed in any case, and allowing the license to go forward on such terms may ultimately be beneficial to society in that the licensee will develop the expertise to compete fully with the patent owner following the date that the patent expires.

Economists arguing against the use of price restrictions generally base their criticism on the fact that sharing of the monopoly profits by the licensor with the licensee discourages the licensees from engaging in activities that would otherwise be pro-competitive. For example, one commentator has explained the issues as follows:

> "[A] price restriction clause that is operative always has the primary consequence of dividing with the licensees, in proportion to their share of market sales, the monopoly profit potential of the end-product. Because it does so it has several secondary consequences. First, licensees have no incentive to challenge validity of the patent ... Second, the licensees have an incentive to suppress any unpatented product that competes with the patented product ... Third, licensee incentive to invent around the license patent is far less than if all monopoly profits were being absorbed by royalty obligations ...
>
> Since the anticompetitive conduct is inherent in the price restriction, use of such clauses should be flatly prohibited. Not only do they threaten monopolistic exploitation far beyond that commensurate with the value of the patentee's contribution, but they also undermine the objectives of patent policy by weakening incentives of licensees to invent around the licensed patent since it is a source of benefits rather than of royalty obligations."[99]

[98] See, e.g. Bowman, *Patent and Antitrust Law: A Legal and Economic Appraisal* (1974) 129–39. Bowman's analysis does not accord special significance to the fact that licensor–licensee relationships with respect to patents are often horizontal.

[99] Baxter, "Legal Restrictions on Exploitation of the Patent Monopoly: An Economic Analysis," *Yale Law Journal* 76 (1978) 267, 336–7.

C. Regulatory Analysis of Price Restrictions

General attitudes In general, agreements to fix, raise or otherwise stabilize prices are considered *per se* illegal under Section 1 of the Sherman Act in the United States, regardless of whether the agreements are horizontal (i.e. among firms at the same level of the market structure)[100] or vertical (i.e. between manufacturers or other suppliers and their customers fixing the customers' resale prices).[101] In order for resale price maintenance to occur, there must be an agreement to set the resale price at some level, although the price need not be set at a specific level.[102] The prohibitions on resale price maintenance even apply to an attempt to establish maximum resale prices or price ceilings.[103]

When price restraints appear in the context of a patent license, the rules as to the rights of the patent owner are far from clear.[104] In the United States, some of the cases in this area have recognized that patent licensors have a good deal of legal authority to restrict the prices established by firms that are involved in the production and distribution of the patented products subject to the license;[105] however, there are just as many cases which appear to hold just the opposite.[106] The wide disparity of judicial opinion in this area has caused licensors to avoid the use of any kind of price restrictions, given the risk that such provisions will ultimately become the subject of costly validity proceedings.[107]

The EC shares the same reticence regarding the use of pricing restrictions. An obligation on either the licensor or the licensee with regard to the prices at which that party may sell the patented products to third parties usually infringes Article 85(1). For example, in one case, a provision imposing a requirement on the licensee not to charge prices below a minimum amount specified by licensor was held to infringe Article 85(1) where there

[100] See, e.g. *United States* v. *Trenton Potteries Co.*, 273 U.S. 392 (1927); *Catalano, Inc.* v. *Target Sales, Inc.*, 446 U.S. 643 (1980); *Arizona* v. *Maricopa County Medical Society*, 102 S. Ct. 2466 (1982); but see *National Collegiate Athletic Ass'n* v. *Board of Regents of Univ. of Oklahoma* 104 S. Ct. 2948 (1984) (applying rule of reason, rather than *per se*, analysis to horizontal price restraints arising in an industry in which horizontal cooperation was deemed essential for the industry to function).

[101] See, e.g. *Dr. Miles Medical Co.* v. *John D. Parke & Sons*, 220 U.S. 373 (1911); *California Retail Liquor Dealers' Ass'n* v. *Midcal Aluminum, Inc.*, 445 U.S. 97 (1980); *Monsanto Co.* v. *Spray-Rite Service Corp.*, 104 S. Ct. 1464 (1984); *Business Electronics Corp.* v. *Sharp Electronics Corp.*, 108 S. Ct. 1515, 1519 (1988).

[102] *Business Electronics Corp.* v. *Sharp Electronics Corp.*, 485 U.S. 717 (1988).

[103] *Arizona* v. *Maricopa County Medical Society*, 457 U.S. 332 (1981).

[104] See generally Adarno, "Problems Connected With Acquisition, Licensing and Enforcement of Intellectual Property," *Alb. Law Review* 50 (1986), 475; Wetzel, "Did Line Material Really Rule Out Patent Price Fixing?" *Mercer Law Review* 26 (1975), 471; Marquis, "Limitations on Patent License Restrictions: Some Observations," *Iowa Law Review* 58 (1972), 41, 50–4 (1972); Curley, "Patent Acquisitions and Restricted Licenses Under the Antitrust Laws," *Seton Hall Law Review* 8 (1977) 645, 660–5; Austern, "Fish Traps, Indians, and Patents: The Antitrust Validity of Patent License Restrictions on Sales Price, Field of Use, Quantity, and Territory," *University of Pittsburgh Law Review* 28 (1966) 181, 184–8.

[105] See, e.g. *United States* v. *General Electric Co.*, 272 U.S. 476 (1926); *In re Yarn Processing Patent Validity Litigation*, 541 F.2d 1127, 1135, 192 U.S.P.Q. 241 (5th Cir. 1976) ("Normally, a patentee may license other parties to manufacture products incorporating his invention and fix the price at which the products are sold by the licensee.").

[106] See, e.g. *United States* v. *Line Material Co.*, 333 U.S. 287, 76 U.S.P.Q. 399 (1948); *Newburgh Moire Co., Inc.* v. *Superior Moire Co., Inc.*, 237 F.2d 283 (3d Cir. 1956); *Tinnerman Products, Inc.* v. *George K. Garrett Co.*, 185 F.Supp. 151, 126 USPQ 130 (E.D. Pa. 1960), *aff'd* 292 F.2d 137 (3d Cir.), *cert. denied*, 368 U.S. 833 (1961).

[107] See, e.g. Banner, "The Patent System and the Antitrust Laws: Stabilization of the Tensions?" *Antitrust Law Journal* 49 (1980), 41, 47; ("[T]he utilization of price fixing clauses in patent licenses ... has, I suspect, virtually disappeared. Few licensors, and fewer product lines, could support the overwhelming expense which could well be required to test the validity of such a clause").

is an appreciable effect on trade between member states.[108] The difficulties to the licensor are compounded by the fact that price restrictions cannot be saved by the block exemptions. Patent Licensing Regulation restrictions on a party with respect to "the determination of prices, components of prices or discounts for the licensed products",[109] and a similar prohibition is included in the Technology Licensing Regulation.[110]

***General Electric* case** The best-known case relating to price restrictions in patent licensing arrangement was the decision of the United States Supreme Court in *United States* v. *General Electric*.[111] In that case, the Court, in permitting the licensor to impose price restriction, recognized the right of the patent owner to impose restrictions reasonably designed to optimize the return on its patent, such as the right to "acquire profit" by controlling the "price at which the [patented] article is sold."[112] The Court relied almost solely upon the patent law for validation of the license, assuming that as long as the restriction affected only commerce within the scope of the patent, the restriction would normally be lawful.[113] The Court also based its ruling on an analysis of the effect the grant of the license might have on the competitive position of the licensor, pointing out that price restrictions might be necessary to induce the licensor to grant the license in the first place, since without them the licensor might be left unprotected against the possibility that the licensee would be able to undermine the licensor's expected profits from its own manufacturing and sales activities.[114]

The *General Electric* rule was subjected to a good deal of criticism and scrutiny, and its effect has been severely limited by subsequent cases. For example, the *General Electric* rule does not extend to patent license provisions which are simply devices to conceal what are actually horizontal price fixing cartels. In one case, the Supreme Court struck down a scheme established by two of the more dominant firms in an industry involving the creation of a patent holding company to grant licenses to more than 200 firms, comprising substantially all competitors in the industry.[115] Each of the prospective licensees was informed that all other firms were being offered licenses, and that the licenses, which provided that prices would be set in accordance with pricing schedules which were to be issued by the holding company, would serve to stabilize prices in the industry. The Court took note of the fact that the "purpose and result" of the licensing scheme was to "fix prices on the products for [the patentees] and their licensees."[116]

[108] Case 27/87 *Erauw-Jacquery* [1988] ECR 1919 [1988] 4 CMLR 576.
[109] Patent Licensing Regulation Art. 3(6) and Recital (23).
[110] Technology Licensing Regulation Art. 3, para. 1.
[111] 272 U.S. 476, 490 (1926).
[112] Ibid.
[113] In its decision, the Court quoted the following passage from *Bement* v. *National Harrow Company*, 186 U.S. 70, 91 (1902): "The very object of [the patent] laws is monopoly, and the rule is, with few exceptions, that any conditions which are not in their very nature illegal with regard to this kind of property, imposed by the patentee and agreed to by the licensee for the right to manufacture or use or sell the article, will be upheld by the courts. The fact that the conditions in the contracts keep up the monopoly or fix prices does not render them illegal."
[114] According to the Court, it seemed "entirely reasonable" in such a situation that the patentee "should say to the licensee, 'Yes, you may make and sell articles under my patent but not so as to destroy the profit that I wish to obtain by making them and selling them myself.'" *United States* v. *General Electric Co.*, 272 U.S. 476, 490 (1926). Commentators, taking note of this specific line of argument, have suggested that it is the basis for limiting the ruling in *General Electric* to cases where the patentee and licensee are in actual competition with one another regarding the patented technology.
[115] *United States* v. *New Wrinkle, Inc.*, 342 U.S. 371, 92 U.S.P.Q. 158 (1952).
[116] Ibid.

In *United States* v. *Gypsum Co.*,[117] the Supreme Court struck down as *per se* unlawful price restrictive patent licenses on an improved form of gypsum wallboard given to several competing licensees, concluding that such an arrangement was an unlawful cartel. Another decision that really undercut the viability of *General Electric* was *United States* v. *Line Material Inc.*,[118] in which the Court held *per se* unlawful a price-restrictive cross-license on two "blocking" patents which represented a major advance in the handling of electrical power, adding that such an arrangement was beyond the intended economic scope of the patent right.[119] Distaste for the use of price restrictions has even led to rejection of schemes to establish a maximum price, which is what the Court did when it struck down an arrangement among doctors to set a maximum price for services sold to particular insurers.[120]

In addition, attempts by patent licensors to regulate the resale, as opposed to "first sale" prices, of patent products have been overturned by the so-called "exhaustion of monopoly" doctrine. For example, the Supreme Court refused to apply the *General Electric* rule to price restrictions on the resale of a patent product, holding such restrictions to be *per se* unlawful violations of the Sherman Act.[121] The Court noted that "[s]ale of the [patented article] exhausts the monopoly in that article and the patentee may not thereafter, by virtue of his patent, control the use or the disposition of the article."[122] Having decided that patent rules were not applicable to the purchaser's resale practices, the Court went on review the restrictions under the normal antitrust rules[123] which, in general, dictate that resale price fixing arrangements are unlawful.

At this point, it appears the *General Electric* rule can be construed as valid only in the limited situation in which there is a single license and the price restriction is imposed by a licensor involved in manufacturing for its own protection,[124] and the IP Guidelines make it clear that the DOJ will continue to enforce the *per se* rule against resale price maintenance in the context of intellectual property licensing.[125] However, it would still appear to be possible for a manufacturer to announce the prices it wants its dealers to charge and refuse to sell to any dealers who do not adhere to the announced price, so long as the manufacturer does not attempt to coerce the dealer to adhere to particular prices.[126] In addition, suggested resale prices or price lists do not violate the antitrust laws. Finally, as discussed below, output restrictions can still be used in order to induce licensees

[117] 333 U.S. 364 (1948).

[118] 333 U.S. 287, 314 (1948).

[119] Ibid. at 319.

[120] *Arizona* v. *Maricopa County Medical Society*, 457 U.S. 332 (1981).

[121] *United States* v. *Univis Lens Co.*, 316 U.S. 241 (1942). See also, e.g. *Bauer & Cie* v. *O'Donnell*, 229 U.S. 1 (1913); *Ethyl Gasoline Corp.* v. *United States*, 309 U.S. 436, 44 U.S.P.Q. 614 (1940); *Edward Katzinger Co.* v. *Chicago Metallic Mfg. Co.*, 329 U.S. 394, 72 U.S.P.Q. 18 (1947); *MacGregor* v. *Westinghouse Electric & Mfg. Co.*, 329 U.S. 402, 72 U.S.P.Q. 21 (1947); *United States* v. *General Electric Co.*, 80 F. Supp. 989, 79 U.S.P.Q. 124 (S.D.N.Y. 1948).

[122] 316 U.S. at 250.

[123] See, e.g. *Adams* v. *Burke*, 84 U.S. 453, 456 (1873); *Hobbie* v. *Jennison*, 149 U.S. 355 (1893); *Keeler* v. *Standard Folding Bed Co.*, 157 U.S. 659 (1895); *Bauer & Cie* v. *O'Donnell*, 229 U.S. 1, 18 (1913); *United States* v. *Masonite Corp.*, 316 U.S. 265,53 U.S.P.Q. 396 (1942). See also *Atari Games Corp.* v. *Nintendo of America, Inc.*, 897 F.2d 1572, 14 U.S.P.Q. 2d 1034 (Fed. Cir. 1990) ("[A] patent owner may not take the property right granted by a patent and use it to extend his power in the marketplace improperly, i.e. beyond the limits of what Congress intended to give in the patent laws").

[124] See, e.g. *Newburgh Moire Co.* v. *Superior Moire Co.*, 237 F.2d 283, 291–4 (3d Cir. 1956); *Royal Indus. Inc.* v. *St. Regis Paper Co.*, 420 F.2d 449, 453–4 (9th Cir. 1969).

[125] IP Guidelines § 5.2.

[126] *United States* v. *Colgate & Co.*, 250 U.S. 300 (1919).

to maintain a specified pricing level, as can a unit royalty scheme which makes it economically infeasible for the licensee to sell the products at less than a certain price.

Economic analysis in price restriction cases While price restrictions do raise potential problems as devices for collusion among actual or potential competitors, it may well be argued that the legitimate justifications for the use of such restrictions in various circumstances are sufficient to dictate the use of rule of reason analysis, rather than simply dismissing such provisions without conducting any substantial economic analysis of the effects of the restriction. In general, the case law relating to price restrictions has developed without substantial attention of economic analysis, relying either on strict application of patent law principles, as was the case with the original *General Electric* decision, or on antitrust doctrine, which almost uniformly rejects any agreement with respect to price, even if some beneficial economic effects can be offered as justification.[127] For example, the Supreme Court's analysis in the *General Electric* case did not touch upon a number of factors which might otherwise have been reviewed had the case been decided under a rule of reason standard. For example, the Court did not address, at least in the formal record, why General Electric, which appeared to have a true monopoly in the relevant market, wanted to grant a license at all. A rule of reason review might well have asked whether the licensee may have developed a "second best" technology which General Electric sought to suppress through the licensing arrangement. Inquiries might also have been made as to whether or not General Electric would have granted the license without the restriction, perhaps by seeking the required protections in some way other than regulating the sales price.[128]

In addition, the Supreme Court's decision in *United States* v. *Line Material Inc.*[129] was reached without considering the economic effects of the licensing arrangement, including the need to promote licensing agreements where there are blocking patents. Significantly, the Court noted that "when parties join in agreement, as here, to maintain prices of their several products, that agreement, however advantageous it may be to stimulate the broader use of patents, is unlawful *per se* under the Sherman Act."[130] This reasoning is consistent with other cases which have invalidated actions by two or more parties which, if undertaken unilaterally, would be considered to be lawful,[131] and teaches

[127] See generally, *United States* v. *General Motors Corp.*, 384 U.S. 127, 148 (1966) ("protection of price competition from conspiratorial restraint is an object of special solicitude under the antitrust laws"); *United States* v. *Socony-Vacuum Co.*, 310 U.S. 150, 224–6, n59 (1940) (price fixing poses a "threat to the central nervous system of the economy").

[128] For example, Westinghouse's license also contained a graduated royalty such that whenever the Westinghouse market share exceeded 15 percent of the combined market shares of both General Electric and Westinghouse, Westinghouse's royalty obligation was increased by a factor of five. See 15 F.2d 715, 718. Arguably, this provisions, which was not challenged by the government and not mentioned by the Court, would have provided General Electric with most if not all of the same incentive to license that the price provision did.

[129] U.S. 287, 314 (1948).

[130] 333 U.S. at 314. The Court actually considered overruling *General Electric*; however, it split four to four in favor of overruling. In 1965, in *United States* v. *Huck Manufacturing Co.*, 382 U.S. 197 (1965), the Court again divided equally on the question.

[131] See, e.g. *Monsanto Co.* v. *Spray-Rite Service Corp.*, 104 S. Ct. 1464, 1469 (1984) ("there is a basic distinction [under the Sherman Act] between concerted and independent action"); *United States* v. *Gypsum Co.*, 333 U.S. 364, 401 (1948) ("it is familiar doctrine that lawful acts may become unlawful when taken in concert").

that once a patent owner decides to contract with others through a licensing arrangement and thereby avail himself of resources beyond those which he independently owns or controls, he no longer is able to rely on the argument that the practical effect of the arrangement is no different than what would have been the case if he had continued with independent exploitation of the patented invention.

Even when courts have attempted to conduct some review of the economic effects of a given price restriction, the results have been less than satisfactory. For example, a particularly alarming aspect of the Supreme Court's reasoning in the *Gypsum* was its emphasis on the fact that unpatented wallboard had disappeared from the marketplace as evidence that the arrangement was anticompetitive, particularly since any significant new invention is likely to lead to the disappearance of predecessor products or methods.[132] The Court also stressed that the arrangement would be unlawful because it was the licensees and not the licensor which actually benefitted from the price restrictions.[133] The Court did not offer any guidance as to how licensee benefits might be identified, and its suggestion that one criterion might be that the restrictive licenses had been accepted by licensees with knowledge that other licensees did likewise,[134] which is often the case, meant that numerous multi-licensee arrangements might well be caught within the scope of the *Gypsum* rule.[135]

[132] Also, the Court did not make clear whether a "purpose" test or an "effects" test was to be applied in this connection. If an "effects" test is to be applied, there parties may be placed at a significant risk that the arrangement will be overturned well after the agreement begins in those situations where it takes some amount of time for the market impact of the new invention to become apparent.

[133] Consideration of the benefits to the licensee, rather than the licensor, from licensing restrictions have been considered in several cases, both before and after *Gypsum*. See, e.g. *Ethyl Gasoline Corp.* v. *United States*, 309 U.S. 436, 458–9 (1943) ("The licensing conditions are thus not used as a means of stimulating the commercial development and financial returns of the patented invention which is licensed, but for the commercial development of the business of the [licensees] and the exploitation of a second patent monopoly not embraced in the first."); and *Mannington Mills* v. *Congoleum Industries Inc.*, 610 F.2d 1059, 1073 (3d Cir. 1979) ("Where [termination of a license occurs] after a number of nonexclusive licenses have been granted and in concert with competing licensee ... there is a greater risk that the restriction is designed not to reward the patent monopoly, but to increase the licensee's reward.")

[134] See *United States* v. *Gypsum Co.*, 333 U.S. at 387 ("industry-wide license agreements, entered into with knowledge on the part of the licensor and licensee as to the adherence of others ... [are] sufficient to establish a prima facie case of conspiracy.").

[135] The Court rejected the following finding by the trial court: "[S]ince a patent owner may lawfully divide his patent monopoly with a plurality of licensees, there will in the usual course be with each of such licensees the same negotiation and discussion, agreement upon terms, purpose to execute and carry out a license contract and to accomplish its normal results, and combined action in so doing, as in the case of a single licensee. And each licensee will be informed of and discuss with the licensor the terms and conditions of the proposed licenses; otherwise no more than a single license could be executed. A patent owner would not be able to license competing manufacturers upon different price terms; no one such would be willing to suffer competitive disadvantages; no one such would be willing to sign in the dark as to the terms to be extended to others; ordinarily, moreover, there will be discussion at large, i.e. within the trade, of the advantages and disadvantages of the licenses proposed by the patent owner. Each of a plurality of licensees will, moreover. have the same purpose to make a license and to secure its resulting advantages. The licensor and each licensee of such a plurality constitute a "combination" to effectuate the purposes of their license. Since a plurality of licenses is lawful, all of this must be lawful. Further, if in practical effect the licensor and the plurality of licensees are a "combination" to the same end, such a 'combination" is not stigmatized by the law provided in purpose and effect it does not secure to the patent owner more than the normal reward of a patent monopoly, nor to any of the licensees with whom that monopoly is divided more than the advantages which naturally result to a licensee, as well as to a licensor, from patent licensing. All of this necessarily follows from the General Electric case" (67 F.Supp. at 439–0).

V. Quantitative and Qualitative Conditions on Use

A. Introduction

Quantitative and qualitative conditions on the use of the licensed subject matter can take a variety of forms, each of which are designed to influence the level of investment that the licensee will, or must, make in exploiting the license. On the one hand, output restrictions are, as the name implies, limits on the quantity of products which can be manufactured and/or sold by the licensee. While such limitations may be advantageous to the licensor and its other licensees, they may impact the economics of the manufacturing process for the licensee, unless the level of royalties is set so as to permit the licensee to share in any increase in gross revenues which may result from the restrictions on output. On the other hand, minimum output requirements and "quality" standards impose obligations on the licensee to make some minimum investment in the licensed subject matter.

B. Strategic Considerations

Output restrictions Output restrictions control the quantity of the licensed products which can be manufactured and/or sold by the licensee. The controls can be accomplished through the use of direct output quotas,[136] or by including other terms (e.g. higher royalty rates once prescribed production levels had been reached)[137] which effectively inhibit the licensee from producing above a certain level.[138] Output restrictions are a viable alternative to price restrictions, and it has been observed that such restrictions are equally effective with price restrictions in generating monopoly profits, and that the anticompetitive effects of restrictions on output can be just as onerous as those cited with respect to price restrictions.[139]

Minimum required exploitation of licensed patents The ability and willingness of the licensee to exploit the licensed technology is a key factor in the success or failure of the business relationship, particularly when the licensee is to be given, for some period of time, the exclusive right to use the technology for a specific application or product or in a designated market. Since an exclusive license represents what amounts to a wholesale delegation of a commercial opportunity to the licensee, the licensor develops a keen dependence upon the capabilities of the licensee to successfully utilize the

[136] See, e.g. *Q-Tips, Inc.* v. *Johnson & Johnson*, 109 F. Supp. 657, 95 U.S.P.Q. 258 (D.N.J. 1951), *aff'd on other grounds*, 206 F.2d 144 (3d ir.), *cert. denied*, 347 U.S. 935 (1953).

[137] See, e.g. *United States* v. *E.I. du Pont de Nemours & Co.*, 118 F. Supp. 41, 99 U.S.P.Q. 462 (D. Del. 1953), *aff'd on other grounds*, 351 U.S. 377 (1956); *Rubber Tire Wheel Co.* v. *Milwaukee Rubber Works Co.*, 154 F. 358 (7th Cir. 1907).

[138] See, e.g. the royalty provisions imposed by the licensor in *United States* v. *General Electric*, 272 U.S. 476.

[139] See, e.g. Baxter, supra note 99, at 339 (1966): "A license restriction on licensee output – restriction to set the percentage of industry sales, or to a specific quantity or to a quota set from time to time by the patentee – should be treated just as a price restriction for the consequences are identical. Licensees share the benefit of restriction on industry output through resultant high prices; they have no incentive to market competitive products or to challenge validity of or to invent around that patent."

licensed technology and products. In the event that the licensee is unable, or unwilling, to adhere to its obligations under the arrangement, the licensor may suffer substantial economic losses and, in the worst case, the licensor will be unable to market and sell the new products in a timely fashion once the rights have reverted back to it.

At a minimum, any license should contain some sort of undertaking by the licensee to use its "best efforts" to exploit the licensed technology in the manner contemplated by the parties. Moreover, with exclusive licenses, it may be appropriate to provide that the licensee will lose its exclusive rights unless it meets one or more specific measures of performance. For example, as described above, the exclusivity may be conditioned upon whether or not the licensor is able to complete development of the technology, establish distribution channels for the licensed products or, at a minimum, expend minimum amounts on development and marketing. In other cases, the licensee may be denied exclusivity upon failure to meet or exceed certain revenue and market share milestones, even if the licensee has otherwise been diligently attempting to exploit the technology.

Minimum royalty provisions may also be used to provide the licensor with assurances that the licensing arrangement will allow it to recover a portion of the costs associated with developing the licensed technology. Such provisions obligate the licensee to remit a specified minimum royalty amount on a periodic basis during the license term, whether or not the royalties are actually earned under the standard specified in the agreement. Failure of the licensee to make the minimum royalty payments may lead to the termination of the agreement or an action for damages. Also, even if the licensee continues to make all the minimum payments, if any underlying performance milestones are not being achieved, the licensor may seek the right to terminate the license or, in the event that the license is exclusive, convert the license to a non-exclusive arrangement.

Minimum royalty payments may be particularly appropriate in those cases, such as in the pharmaceutical area, where regulatory requirements may create long delays in the commencement of sales of royalty-producing products. The licensor may require the licensee to make either minimum periodic payments during the development and testing of the new product, or to make benchmark payment whenever the licensee has passed certain objective milestones in the process of bringing the product to market. While the licensee may agree to such demands, it may argue for the right to credit amounts paid in the development stage against royalties which might become due once the product is sold in the market.

Quality standards Minimum requirements in the licensing agreement regarding the quality of products produced using the licensed subject matter can serve several purposes. First, quality standards insure that the licensee expends the money and time which may be required in order to perfect and monitor it manufacturing and distribution processes with respect to the licensed product. Presumably, this investment will provide sufficient incentive for the licensee to continue to vigorously market the licensed product, thereby increasing the return to the licensor in the form of royalties.

Second, as courts continue to apply statutory and common law principles of strict or products liability to manufacturers for claims for damages arising out of injuries suffered due to defects in any products manufactured using their licensed technology, it has become even more important for licensors to include one or more provisions in the licensing agreement designed to insure quality control in the use of the licensed technology. For example, it is now common to require licensees to obtain and maintain products liability insurance coverage with respect to products manufactured using the licensed technology.[140] In addition, in those cases where the licensor is concerned about the manner in which the licensee might be using the technology in the manufacturing process, the licensee agreement may provide for the licensee to indemnify the licensor for damages arising from the failure of the licensee to follow safety procedures which may be supplied by the licensor.

In some cases, particularly when a trademark license is involved, the licensor may take a more active involvement in the activities of the licensee, even though the licensor might well be exposing itself to a greater risk of a direct products liability claim. For example, the license agreement may set out detailed standards of quality control relating to the use of the technology for the manufacture of licensed products. The license agreement may also provide for technical assistance from licensor personnel in order to improve and maintain the quality of the manufacturing process. Finally, the licensor may retain the right to visit the licensee's facilities and laboratories and inspect the use of the licensed technology in connection with the maintenance of the specified quality standards.

C. Judicial and Regulatory Attitudes

Output restrictions While there has not been a significant number of cases on output restrictions, in general, courts in the United States have held that a patent owner has the right to impose conditions as to the maximum number of units that can be produced by its licensees,[141] and it appears that production limitations will be tested under the rule of reason and will be upheld as reasonably necessary for the protection of the business of the licensor[142] and/or within the permitted scope of the discretion of the licensor,[143]

[140]The licensor is often also required to obtain products liability insurance. In either case, the party required to carry that insurance must deliver to the other party a certificate to the effect that the insurance has been obtained and will be maintained.

[141]See, e.g. *Atari Games Corp.* v. *Nintendo of America, Inc.*, 897 F.2d 1572, 14 U.S.P.Q. 2d 1034 (Fed. Cir. 1990) (applying Ninth Circuit law to uphold the validity of output restrictions in patent licenses); *United States* v. *E.I. duPont de Nemours & Co.*, 118 F. Supp. 41, 224–6, 99 U.S.P.Q. 462 (D. Del. 1953), *aff'd on other grounds*, 351 U.S. 377 (1956) ("The cases are to the effect that the owner of a valid patent may by license restrict production of the licensee to a specified quantity."); *Q-Tips, Inc.* v. *Johnson & Johnson*, 109 F. Supp. 657, 95 U.S.P.Q. 258 (D.NJ. 1951), *aff'd on other grounds*, 206 F.2d 144 (3d Cir.), *cert. denied*, 347 U.S. 935 (1953); *United States* v. *Aluminum Co. of America*, 148 F.2d 416, 438, 65 U.S.P.Q. 6 (2d Cir. 1945); *Rubber Tire Wheel Co.* v. *Milwaukee Rubber Works Co.*, 154 F. 358 (7th Cir. 1907). But see *American Equip. Co.* v. *Tuthill Bldg. Material Co.*, 69 F.2d 406, 21 U.S.P.Q. 198 (7th Cir. 1934) (holding that a machine patent may not be used to set output restrictions on unpatented products used in connection with the machine).

[142] See, e.g. *United States* v. *General Electric*, 272 U.S. 476, in which the trial court noted: "[Another] provisions of the license [calls for] the exaction of 10 per cent royalty on all sales above 15 per cent of the [combined] sales of electric lamps having tungsten filaments, and only of 2 per cent royalty on a total business of 15 per cent or less. As to this provision, government counsel conceded on the hearing that it is only reasonably necessary to the protection of the business of the owner of the patent (15 F.2d 715, 718).

[143] See, e.g. *Q-Tips, Inc.* v. *Johnson & Johnson*, 109 F.Supp. 657, 660 (D.N.J. 1951), *aff'd on other grounds*, 206 F.2d 144 (3d Cir. 1953), where the district court approved a series of licenses to a number of competing manufacturers on a machine for the production of cotton applicators, where the license limited directly the number of applicators that could be produced. The court reasoned that the patent owner could have refused to grant any license at all, and therefore could control production. See also United *United States* v. *E. I. duPont de Nemours & Co.*, 118 F. Supp. 41, 226 (D.Del. 1953), *aff'd on other grounds*, 351 U.S. 377 (1956) ("owner of a valid patent may by license restrict production to a specified quantity").

unless the output restrictions are included in an agreement amongst two or more competitors and are found to be an unlawful horizonal cartel agreement.[144]

One area in which there is some controversy is the issue of whether an owner of a patent on a process or machine can impose restrictions on the number of unpatented articles that can be produced with the process or machine. Some cases have struck down such restrictions on the ground that they are beyond the patentee's limited rights under the specific claims of the patent.[145] However, recent cases have upheld such provisions and have apparently concluded that ownership of a process or machine patent includes the right to control the use thereof, including the right to foreclose the production of the unpatented articles by declining to grant any licenses.[146]

The EC takes a much stricter approach to restrictions on the maximum amount of products which can be manufactured or sold under a license. In the view of the Commission, output restrictions are violative of Article 85(1),[147] even though it could be argued that such limitations are merely one form of limited license. The Commission believes that such restrictions prevent the party accepting it from increasing his output, which may have effects similar to an export ban.[148] Output restrictions were forbidden in the Patent Licensing Regulation;[149] and are also largely prohibited in the Technology Licensing Regulation,[150] subject to two relatively narrow exceptions.[151]

[144] See, e.g. *Hartford-Empire Co.* v. *United States*, 323 U.S. 386 (1945).

[145] *American Equipment Co.* v. *Tuthill Building Material Co.*, 69 F.2d 406, 408, 201 U.S.P.Q. 198 (7th Cir. 1934). See also *United States* v. *General Electric Co.*, 82 F.Supp. 753, 814, 80 U.S.P.Q. 195 (D.N.J. 1949), *modified*, 115 F.Supp. 835 (1953). For arguments supporting this view, see Marquis, "Limitations on Patent License Restrictions: Some Observations," *Iowa Law Review* 58 (1972) 41, 52–3; Turner, "The Patent System and Competitive Policy," *New York University Law Review* 44 (1969) 450, 470.

[146] See, e.g. *Q-Tips, Inc.* v. *Johnson & Johnson*, 109 F.Supp. 657, 95 USPQ 258 (D.N.J. 1951), *aff'd on other grounds*, 206 F.2d 144 (3d Cir.), *cert. denied*, 347 U.S. 935 (1953). See also *United States* v. *Studiengesellschaft Kohle, m.b.H.*, 670 F.2d 1122, 212 U.S.P.Q. 889 (8th Cir. 1981) (patentee who controlled the only commercially feasible process for producing an unpatented drug was allowed to impose field of use restrictions on firms licensed to use the process in producing the drug). See also *Ethyl Corp.* v. *Hercules Powder Co.*, 232 F.Supp. 453, 143 U.S.P.Q. 57 (D.Del. 1964) (upholding field of use restrictions on unpatented products produced using a patented process, where the process was the only one commercially feasible for producing the products). See also Antitrust Division, US Dept. of Justice, Antitrust Guide for International Operations, Case 11 (issued November 10, 1988) (generally approving territorial restrictions under process patent licenses). For arguments favoring the use of output restrictions under even process patents, see Austern, "Fish Traps, Indians and Patents: The Antitrust Validity of Patent License Restrictions on Sales Price, Field of Use, Quantity, and Territory," *University of Pittsburgh Law Review* 28 (1966), 181, 190–1.

[147] See, e.g. Fourth Report on Competition Policy (1974), points 31–2; and *Breeders' rights: Maize Seed*, OJ 1978 L286/23 [1978] 3 CMLR 434 (obligation on licensee to produce only part of requirement and buy in the rest held to fall under Art. 85(1)).

[148] Ibid.

[149] Patent Licensing Regulation Art. 3(5) (forbidding restrictions on "the quantity of licensed products one party may manufacture or sell or the number of operations exploiting the licensed invention he may carry out are subject to limitations") and Recital (23).

[150] Technology Licensing Regulation Art. 3, para. 5.

[151] Art. 1, para. 1(8) of the Proposed Technology Licensing Regulation exempts an obligation on the licensee to limit his production of the licensed product to the quantities he requires in manufacturing his own products and to sell the licensed product only as an integral part of or a replacement part for his own products or otherwise in connection with the sale of his own products, provided that such quantities are freely demonstrated by the licensee. Art. 2, para. 1(13) of the Technology Licensing Regulation exempts an obligation on the licensee to supply only a limited quantity of the licensed product to a particular customer, where a know-how license was granted at that customer's request so that he might have a second supplier inside a licensed territory; this provision shall also apply where the customer is the licensee, and the license which was granted in order to provide a second source of supply provides that the customer is himself to manufacture the licensed products or to have them manufactured by a subcontractor.

Minimum required exploitation of licensed patents In large part, antitrust laws in the United States permit licensors to impose various requirements regarding the minimum exploitation of the licensed subject matter. As a general rule, there is no implied duty or obligation under a non-exclusive license in the United States which requires that a licensee must actively exploit the license; however, the grant of an exclusive license may create an implied obligation on the licensee to use its "best efforts" to exploit the licensed subject matter in "good faith." Express obligations included in the license agreement will typically be enforced, although a court is unlikely to require that a licensee take actions which extend beyond reasonable commercial exploitation undertaken in good faith. Accordingly, the licensee will not be required to take unreasonable steps, such as selling products at a substantial loss, to distribute products which prove to be unmarketable without fault on the part of the licensee.

Minimum royalty provisions are also generally accepted, and fall within the overriding principle that patent owners in the United States are entitled to lawfully negotiate royalties "as high as he can negotiate with the leverage of that monopoly,"[152] provided that the royalties result from arm's length bargaining between the parties, and are not a subterfuge for conspiratorial action among competitors. The wide latitude given to patent owners with respect to the amount of royalties is seen as part of the incentive which should be provided to patent owners in order to reward them for the development and public disclosure of their inventions.

Attitudes in the EC are similar to those in the United States in this area. For example, licensors may impose obligations on their licensees to produce a specified minimum quantity of the licensed products or complete a minimum number of operations exploiting the licensed invention (e.g. sales) without infringing Article 85(1), since such provisions are typically intended to ensure that the licensee is adequately exploiting the licensed subject matter.[153] Minimum royalty provisions also will generally fall outside of Article 85(1) as a means for ensuring that the licensed subject matter is being exploited.[154] In addition, obligations imposed on a licensee to use his "best endeavours" with respect to exploitation of the licensed technology will also generally be permitted under the Technology Licensing Regulation,[155] as they were under the Patent

[152] *Brulotte* v. *Thys Co.*, 379 U.S. 29, 33, 143 U.S.P.Q. 264 (1964). See also *Akzo N.V.* v. *International Trade Commission*, 808 F.2d 1471, 1 U.S.P.Q. 2d 1241 (Fed. Cir. 1986) (a royalty arrangement that varied the royalty amount by the "value" of the particular end use was not unreasonable); *Hennessy Indus., Inc.* v. *FMC Corp.*, 779 F.2d 402 (7th Cir. 1985); *In re Yarn Processing Patent Validity Litigation*, 541 F.2d 1127, 1140, 192 U.S.P.Q. 241 (5th Cir. 1976), *cert. denied*, 433 U.S. 910 (1977) ("Normally, a patent holder may charge whatever it likes for a license."); *LaSalle Street Press, Inc.*, v. *McCormick & Henderson, Inc.*, 445 F.2d 84, 95, 170 U.S.P.Q. 305 (7th Cir. 1971) ("Incidental to the monopoly grant of a patent are the rights in the patentee to exact royalties 'as high as he can negotiate with the leverage' of his monopoly."); *Carter-Wallace, Inc.* v. *United States*, 449 F.2d 1374, 1383, 171 U.S.P.Q.s 359 (Ct. Cl. 1971) ("Absent any overriding unlawful conduct, patentees can charge for their patented products and licenses whatever the market will bear"); *Warner-Jenkinson Co.* v. *Allied Chemical Corp.*, 477 F.Supp. 371, 206 U.S.P.Q. 837 (S.D.N.Y. 1979), *aff'd*, 633 F.2d 208 (2d Cir. 1980); *W.L. Gore & Associates* v. *Carlisle Corp.*, 381 F.Supp. 680, 183 U.S.P.Q. 459 (D. Del. 1974), *rev'd in part*, 529 F.2d 614 (3d Cir. 1976); *American Photocopy Equipment Co.* v. *Rovico, Inc.*, 257 F.Supp. 192, 199–200, 150 U.S.P.Q. 180 (N.D. Ill. 1966), *aff'd*. 384 F.2d 813 (7th Cir. 1967), *cert. denied*, 390 U.S. 945 (1968).

[153] Patent Licensing Regulation Art. 2(1)(2); First Report on Competition Policy (1971), point 75; Technology Licensing Regulation Art. 2, para. 1(9).

[154] Patent Licensing Regulation Art. 2(1)(2); Technology Licensing Regulation Art. 2, para. 1(9).

[155] Technology Licensing Regulation Art. 3(2).

Licensing Regulation,[156] provided that such obligations do not amount to a restriction on competition with the licensor or affiliates of the licensor with respect to research and development, production, use or distribution of competing products.[157]

Quality standards Quality standards generally arise in United States cases involving the licensing of trademarks, as well as patents. As a general rule, while vertical non-price restrictions adopted for the purpose of market domination and to eliminate price-cutting competition will not be permitted, trademark licensors may be allowed to use vertical restrictions that can be justified under rule of reason analysis as enhancing overall competition.[158] For example, courts have upheld restrictions implemented in order to attract better dealers to support a deteriorating distribution system,[159] to assure adequate quality control and proper use of the product,[160] to encourage dealers to engage in local advertising and promotional activities,[161] and to induce dealers to provide the facilities necessary for post-sale service and maintenance.[162]

In the EC, the Technology Licensing Regulation exempts obligations on the licensee to observe minimum quality specifications for the licensed product or to procure goods or services from the licensor or from an undertaking designated by the licensor, in so far as such quality specifications. products or services are necessary for: (i) a technically satisfactory exploitation of the licensed technology; or (ii) ensuring that the licensee's products conform to the quality standards that are respected by the licensor and other licensees.[163] This provision may be slightly broader than prior practice, which required that the parties must agree in advance on the quality standards on the basis of objectively verifiable criteria.[164] The Technology Licensing Regulation also allows the licensor to carry out related checks.[165]

[156] Patent Licensing Regulation Recital (21) and Art. 3(3). Recital (21) noted that the licensor has a "legitimate interest ... in having his patented invention exploited to the full and to this end to require the licensee to use his best endeavours to manufacture and market the licensed product."

[157] Technology Licensing Regulation Art. 3(2).

[158] *Continental T.V., Inc.* v. *GTE Sylvania, Inc.*, 433 U.S. 36 (1977).

[159] *Martin B. Glauser Dodge Co.* v. *Chrysler Corp.*, 570 F.2d 72 (3d Cir. 1977), *cert. denied*, 436 U.S. 913 (1978); *Sandura Co.* v. *FTC*, 339 F.2d 847 (6th Cir. 1964); *Continental T.V., Inc.* v. *GTE Sylvania, Inc.*, 461 F. Supp. 1046 (N.D. Cal. 1978).

[160] *Sports Center, Inc.* v. *Biddell, Inc.*, 673 F.2d 780 (5th Cir. 1982); *Krehl* v. *Baskin- Robbins Ice Cream Co.*, 664 F.2d 1348 (9th Cir. 1982); *Clairol, Inc.* v. *Boston Discount Center of Berkley, Inc.*, 608 F.2d 1114, 204 U.S.P.Q. 89 (6th Cir. 1979).

[161] *Muenster Butane, Inc.* v. *Stewart Co.*, 651 F.2d 292 (5th Cir. 1981); *Red Diamond Supply Inc.* v. *Liquid Carbonic Corp.*, 637 F.2d 1001 (5th Cir.), *cert. denied*, 102 S. Ct. 119 (1981).

[162] *Sports Center, Inc.* v. *Riddell, Inc.*, 673 F.2d 786 (5th Cir. 1982); *Martin B. Glauser Dodge Co.* v. *Chrysler Corp.*, 570 F.2d 72 (3d Cir. 1977), *cert. denied*, 436 U.S. 913 (1978).

[163] Technology Licensing Regulation Art. 2, para. 1(5).

[164] See, e.g. Case 193/83 *Windsurfing International* v. *Commission* [1986] ECR 611 [1986] 3 CMLR 489, § 46.

[165] Technology Licensing Regulation Art. 2, para. 1(5). In some cases, concerns that the licensor's need to audit and monitor the receipts of the licensee in order to verify the amount of royalties lead to anticompetitive sharing of market information have prompted regulators to suggest that it may be appropriate for any necessary inspection of the licensee's books to be made by an independent third party. See, e.g. Spitzer/Van Hool, Twelfth Report on Competition Policy (1982), point 86. See also Art. 2(2) of the Patent Licensing Regulation, which granted an exemption for provisions which allowed licensors to conduct some form of periodic review of the licensee's activities in order to ensure that the specifications are being satisfied.

VI. Restrictions on Competitive Activities of Licensee

A. Introduction

Since a licensing arrangement generally represents a strategic decision on the part of the licensor to rely, at least in part, on the competitive assets and resources of the licensee to secure an adequate return on the licensor's prior investment in developing the patented product, the licensor will typically seek comfort that the licensee will not engage in activities which might undermine its commitment to the relationship with the licensor. In many cases, the licensor will include restrictions on the licensee's ability to engage in competitive activities during the licensing relationship. Another method that often was used was to prohibit the licensee from challenging the legal validity of the patent rights covered by the license.

B. Strategic Considerations

Restrictions on competitive activities by licensee In some cases, patent licensors may wish to protect their investment in skills and abilities of the licensee by imposing a restriction on the licensee's ability to use or deal in products which are competitive with those covered by the license. In so doing, the licensor is effectively precluding the licensee from pursuing other alternatives in a specified field of interest, with the hope and intent that this will cause the licensee to invest greater effort in using and exploiting the licensed patents. As such, restrictions on competitive activities can be viewed as being similar to minimum obligations to exploit the licensed patents.

For its part, the licensee will be reluctant to agree to any substantial limitations on its activities, especially in cases where the licensing relationship is seen as an opportunity to reduce the time and expenditures required to absorb new forms of technology which will serve as a basic for product development efforts by the licensee in the future. Nonetheless, it is probably not unreasonable for the licensee to be asked to agree to certain restrictions. For example, the licensee will usually be prohibited from using the licensed technology to manufacture and sell new products which may compete with the licensed products. Also, the licensee will be restricted from acting as a manufacturer or distributor of competing products engineered by third parties. Finally, in appropriate circumstances, a specific prohibition on "reverse engineering" by the licensee is included.

Restrictions on licensee challenges to patent validity Provisions in a license agreement which restrict the licensee's ability to challenge the validity of the licensed patents are referred to as "no-challenge" clauses. In agreeing to such a provision, the licensee effectively compromises a right which is retained by all others, including competitors: the right to have the patent invalidated, or the scope of its permitted claims narrowed, in reexamination proceedings before the agency which originally issued the patent. In return, the licensee has obtained the right to practice the subject matter of the patent without fear of an infringement action, albeit at a price equal to the royalties and other consideration due under the license.

As described below, absolute restrictions on a licensee in this area will typically not be permitted in either the United States or in the EC. However, it is still possible for parties to include certain provisions relating to the consequences of a licensee's decision to challenge the licensed patents. From the licensor's perspective, it will certainly want to provide that the licensee's default in the payment of royalties, which would almost certainly occur if the licensee mounted a challenge, would allow the licensor to terminate the license. If this occurred, and the patent proved to be valid, the licensor would then have a cause of action against the former licensee for damages, since the use of the patent during the challenge would be infringement. In turn, the licensee may bargain for a provision that would allow it to cease making royalty payment once a challenge has been filed, provided that the challenge is made in good faith. If this arrangement is used, the licensor may require that the licensee deposit the royalties that would have otherwise been paid under the license agreement into escrow, thereby creating a financial disincentive for the licensee to engage in protracted litigation regarding the validity of the patent.

C. Judicial and Regulatory Attitudes

Restrictions on competitive activities of licensee Restrictions on competitive activities of the licensee are commonly referred to as "patent tie-outs," and the effect is similar to any form of exclusive dealing agreement.[166] Under an exclusive dealing arrangement, the dealer will agree to purchase certain products only from a specified manufacturer. A similar form of agreement arises when a dealer agrees to purchase all of its requirements of a product from a single manufacturer. These forms of exclusive dealing agreements raise concerns that competitors will not be able to gain access to a portion of the market for the products which are the subject of the arrangement.

In the United States, exclusive dealing arrangements will generally be analyzed under the rule of reason and will take into account a number of factors, including market share, duration, ease of entry to the relevant market, and the business justifications for the arrangement.[167] However, the treatment of "tie-out" arrangements in the context of a patent license depends on whether the restrictions are directed at infringing items or non-infringing items. In most cases, courts have upheld restrictions on a purchaser's ability to procure patented items from other sources, reasoning that such restrictions are a legitimate attempt by the patent owner to exercise its rights against patent infringement by third parties. On the other hand, courts have not looked kindly on attempts to bar a purchaser from buying competing products which do not infringe the licensed patent. In such situations, most courts have then applied what amounts to a standard of *per se* liability analogous to that imposed in patent tie-in cases, although a few courts have suggested that such arrangements should be subject to the more flexible analysis typically applied to various other forms of exclusive dealing arrangements.

[166] A patent "tie-out" can be distinguished from a "tie-in" arrangement by the fact that the former is generally limited to the product area in which the patent itself exists, while the effect of the later often is to impact competition outside of the scope of the licensed patent. For further information concerning the distinction between patent "tie-ins" and patent "tie-outs," see *Steiner Sales Co.* v. *Schwartz Sales Co.*, 98 F.2d 999, 1010–11, 38 U.S.P.Q. 553 (10th Cir. 1938) *cert. denied*, 305 U.S. 662 (1939); and *Baldwin-Lima-Hamilton Corp.* v. *Tatnall Measuring Systems Co.*, 169 F.Supp. 1, 26, 120 U.S.P.Q. 34 (E.D. Pa.), *aff'd*, 268 F.2d 395 (3d Cir. 1958), *cert. denied*, 361 U.S. 894 (1959).

[167] See, e.g. *Tampa Electric Co.* v. *Nashville Coal Co.*, 365 U.S. 320 (1961) (adopting a rule of "qualitative substantiality" in lieu of the prior test of "quantitative substantiality" adopted in *Standard Oil Co.* v. *United States*, 337 U.S. 293 (1949), which focused on market shares to measure the extent of market foreclosure); *Empire Volkswagen, Inc.* v. *World Wide Volkswagen Corp.*, 814 F.2d 90 (2d Cir.); *Great Escape, Inc.* v. *Union City Body Co.*, 791 F.2d 532 (7th Cir. 1986).

Baldwin-Lima-Hamilton Corp. v. *Tatnall Measuring Systems Co.*[168] provides an example of how most courts have handled restrictions directed primarily at infringing items. In that case, allegations were made that the patentee had violated the antitrust laws by licensing manufacturing firms to produce up to 10 percent of their requirements of the patented product in-house, with the remainder to be purchased exclusively from the patentee-licensor. The court declined to hold that the restrictions constituted patent misuse by the patent owner, noting that they were essentially "undertakings not to infringe the patents," protected by the patent laws and perfectly consistent with the concern of the antitrust laws with "legitimate competition." In holding that the licensee had, in fact, infringed the licensor's patent rights, the distinguished the various patent tying cases cited by the defendant when it made the following observation: "What the tying clause cases condemn is the extension of the monopoly of one patented product to create a monopoly or restraint of competition with respect to a second product not countenanced by the patent applicable to the first product ... Neither the rationale nor the doctrines involved in a tying clause case can be utilized to dispose of a case involving a requirements provision."[169]

In contrast, *United States Shoe Machinery Corp.* v. *United States*[170] illustrates the attitude of the courts when the patentee attempts to extend exclusive dealing arrangements to cover goods that are outside the scope of the subject patent. In that case, the owner of a patent covering certain shoe machinery had leased its machines under agreements that did not specifically preclude the use of competing machines, but which had the practical effect of imposing such an exclusive dealing requirement. For example, the lease agreements contained provisions which provided that the machines could not be used on shoes upon which other manufacturing operations had been performed using competitors' shoe machinery; and the leases were subject to cancellation by the patent owner in the event that the lessees failed to use certain of the defendant's machines exclusively. Also, the lessees were required to purchase certain supplies exclusively from the defendant; and the leased machines could be used only on shoes that had specified operations performed on them using other machines leased from the defendant. The Supreme Court held that the aggregate effect of these restrictions was an illegal exclusive dealing requirement.[171]

Compelling arguments can be made that strict *per se* liability is not appropriate in situations where the scope of the restriction is generally limited to the particular product area in which the patented item itself competes, particularly since many exclusive dealing arrangements often will generate desirable pro-competitive economic benefits. For example, *per se* treatment would seem to be unreasonable in the situation where exclusive dealing restrictions were imposed by a franchisor that conducted national advertising on behalf of its franchisee and provided warranty coverage for the licensed prod-

[168] 169 F.Supp. 1, 26, 120 U.S.P.Q. 34 (E.D. Pa.), *aff'd*, 268 F.2d 395 (3d Cir. 1958), *cert. denied*, 361 U.S. 894 (1959).

[169] 169 F.Supp. 1, 26, 120 U.S.P.Q. 34 (E.D. Pa.), *aff'd.* 268 F.2d 395 (3d Cir. 1958), *cert. denied*, 361 U.S. 894 (1959). See also *Steiner Sales Co.* v. *Schwartz Sales Co.*, 98 F.2d 999, 38 U.S.P.Q. 553 (10th Cir. 1938), *cert. denied*, 305 U.S. 662 (1939) (upholding an arrangement regarding a patented hand towel dispenser where the patent owner had licensed some firms to manufacture and other firms to use the dispenser, provided that the "user" firms purchased their requirements only from the manufacturing licensees); *Speed Shore Corp.* v. *Denda*, 605 F.2d 469, 203 U.S.P.Q. 807 (9th Cir. 1979); *Cardinal Films, Inc.* v. *Republic Pictures Corp.*, 148 F.Supp. 156, 112 U.S.P.Q. 292 (S.D.N.Y. 1957).

[170] 258 U.S. 451 (1922).

[171] Ibid. at 457–8.

ucts, particularly when the volume of total sales of the licensed products represented a relatively small portion of the overall market.[172] Other potential benefits to purchasers from these types of arrangements include assured supply, protection of the purchaser's capital investment, and enhanced long-term planning ability. From the seller's perspective, exclusive dealing may reduce selling expenses, produce other economies by smoothing out production and market fluctuations, and assure loyalty in product manufacture or distribution.

Some courts have begun to look at exclusive dealing arrangements involving patented items under a "rule of reason" standard,[173] recognizing that such arrangements are not so "clearly unreasonable" and without "redeeming virtue" as to warrant the *per se* label.[174] For example, in *Hensley Equipment Co.* v. *Esco Corp.*[175] the court refused to sanction as patent misuse a "policy statement" that the licensor had sent to its distributors informing them that they were not to carry competing products without its prior consent. The court found that the policy statement did not amount to exclusive dealing arrangement, since the evidence indicated that the patent owner had given its approval whenever requested;[176] however, the court went on to acknowledge the need to consider market conditions and competitive justifications for the apparent restrictions by noting that "[t]here was evidence that because of the limited market ... it is neither practical nor economical for a dealer to carry more than one manufacturers' line of products."[177]

Other cases have also taken a more flexible approach to analyzing exclusive dealing provisions in licensing situations. For example, one court upheld a licensing provision that converted the license from an exclusive to a non-exclusive license if the licensee carried competing products.[178] In another case, the court rejected a patent misuse defense where the defendant failed to prove anticompetitive "effects" resulting from the patent owner's requirement that its sole distributor in a designated geographic territory refrain from carrying competing merchandise.[179] As such, it appears that courts recognize

[172] *F.C. Russell Co.* v. *Consumers Insulation Co.*, 226 F.2d 373, 107 U.S.P.Q. 131 (3d Cir. 1955). For an example of a court taking into account market conditions in analyzing an exclusive dealing arrangement, see *United Airlines, Inc.* v. *Austin Travel Corp.* 867 F.2d 737 (2d Cir. 1988) (exclusive dealing claim failed where the market remained competitive with alternative sources of supply available).

[173] See, e.g. *United Airlines, Inc.* v. *Austin Travel Corp.*, 867 F.2d 737 (2d Cir. 1988) (exclusive dealing claim failed where market remained competitive with alternate sources of supply available); *Tampa Electric Co.* v. *Nashville Coal Co.*, 365 U.S. 320 (1961); *Empire Volkswagen, Inc.* v. *World Wide Volkswagen Corp.*, 814 F.2d 90 (2d Cir. 1987); *Great Escape, Inc.* v. *Union City Body Co.*, 791 F.2d 532 (7th Cir. 1986); *Perington Wholesale, Inc.* v. *Burger King Corp.*, 631 F.2d 1369 (10th Cir. 1979); *American Motor Inns, Inc.* v. *Holiday Inns, Inc.*, 521 F.2d 1230 (3d Cir. 1975).

[174] See, e.g. *Atlantic Richfield Co.* v. *USA Petroleum Co.*, 110 S. Ct. 1884 (1990) (a competitor – as opposed to a consumer or dealer – lacked standing to challenge a nonpredatory "vertical" maximum price fixing agreement between a manufacturer and its authorized dealers); *FTC* v. *Indiana Federation of Dentists*, 106 S. Ct. 2009 (1986); *Northwest Wholesale Stationers* v. *Pacific Stationery & Printing Co.*, 105 S. Ct. 2613 (1985); *National Collegiate Athletic Ass'n* v. *Board of Regents of Univ. of Oklahoma*, 104 S. Ct. 2948 (1984); *Broadcast Music, Inc.* v. *Columbia Broadcasting Sys., Inc.*, 441 U.S. 1, 201 U.S.P.Q. 497 (1979); *National Society of Professional Eng'rs* v. *United States*, 435 U.S. 679 (1978); *Continental T.V., Inc.* v. *GTE Sylvania, Inc.*, 433 U.S. 36 (1977).

[175] 383 F.2d 252 (5th Cir. 1967).

[176] 383 F.2d 252, 262 (5th Cir. 1967).

[177] Ibid. See also *Smith Machinery Co.* v. *Hesston Corp.*, 878 F.2d 1290 (10th Cir. 1989) (holding that a manufacturer's "full line" requirements imposed on its dealers was a form of exclusive dealing, governed by the rule of reason, rather than a form of tying subject to *per se* analysis).

[178] *Naxon Telesign Corp.* v. *Bunker Ramo Corp.*, 517 F. Supp. 804 (N.D. Ill. 1981), aff'd. 686 F.2d 1258 (7th Cir. 1982).

[179] *Swofford* v. *B&W, Inc.*, 251 F. Supp. 811, 149 U.S.P.Q. 32 (S.D. Tex. 1966), *aff'd*, 395 F.2d 362 (5th Cir.), *cert. denied*, 393 U.S. 935 (1968).

the such exclusive dealing requirements may well be a legitimate means of protecting the licensor's interests in situations where it has limited its own competitive options by granting an exclusive license.

The DOJ has, since the late 1980s, taken a more permissive attitude toward tie-outs, maintaining that such arrangements are generally not of competitive concern.[180] In the IP Guidelines, the DOJ points out that such restraints can be procompetitive in some instances;[181] however, the DOJ cautions that it is important to guard against the possibility that rivals may be denied access to a sufficient number of outlets to allow them to competitively exploit their technologies[182] and that such restraints can also lead to anticompetitive pricing practices. As a general matter, the DOJ is unlikely to challenge an exclusive dealing arrangement if several firms offer competing products, the relevant manufacturing and distribution markets are unconcentrated, and entry is easy.[183]

In contrast to the United States, the EC has generally been quite hostile to noncompetition covenants. Cases have held that noncompetition clauses under which a licensee agrees not to acquire, manufacture, use or sell competing goods, generally will not be permitted,[184] nor will restrictions on the ability of a licensee to carry out any independent research and development.[185] The Patent Licensing Regulation prohibited restrictions on the freedom of a party to compete with respect to research and development, manufacture, use or sales, beyond those listed in Article 1 of the Regulation, as well as restrictions on the ability of a party to involve himself in techniques other than those included in the license.[186] Although it has been argued that such prohibitions can provide the requisite incentive for the licensee to exploit the licensed products,[187] the Commission's attitude has been that such restrictions generally impede technical and economic progress.[188]

The Technology Licensing Regulation similarly prohibits restrictions on any party to a licensing agreement from competing within the common market with the other party, with undertakings connected with the other party or with other undertakings in respect

[180]Antitrust Division, U.S. Department of Justice, Antitrust Guide for International Operations, Case 10 (issued November 10, 1988).

[181]For example, a licensing arrangement that prevents the licensee from dealing in other technologies may encourage the licensee to develop and market the licensed technology or specialized applications of that technology. IP Guidelines § 4.1.2. See also IP Guidelines Example 8.

[182]Ibid. § 5.4.

[183]See IP Guidelines Example 7. As with exclusive licensing arrangements, the DOJ will examine the substance of the transaction to determine if there are material anticompetitive effects. For example, a royalty arrangement based on total sales of a licensee's product, regardless of whether it is made using the licensed technology, may increase the cost to a licensee of substituting alternative technologies, and thus may have effects similar to an exclusive dealing arrangement. Ibid. § 4.1.2. See also Complaint, *United States* v. *Microsoft, Inc.*, Civ. No. 94-1564 (D.D.C., filed July 15, 1994); Competitive Impact Statement, ibid. (filed July 27, 1994).

[184] See, e.g. *Breeders' Rights: Maize Seed*, OJ 1978 L286/23 [1978] 3 CMLR 434 (prohibition on dealing in competing products). Cf. *Delta Chemie/DDD*, OJ 1988 L309/34 [1989] 4 CMLR 535 (upholding obligations on a licensee not to manufacture or distribute similar products without the licensor's consent in a situation where such consent could not be withheld if sufficient certainty existed that the licensee would be able to honor its contractual obligations).

[185] *AOIP/Beyrard*, OJ 1976 L6/8 [1976] 1 CMLR D14 at D24.

[186] Patent Licensing Regulation Art. 3(3),(10) and (11) and Recitals (13), (15) and (21).

[187] See Korah [1979] JBL 291.

[188] Patent Licensing Regulation Art. 3(3), (10) and (11) and Recitals (13), (15) and (21). In Recital (21), the Commission stated that the block exemption should not apply to "restrictions on the freedom of one party to compete with the other and in particular to involve himself in techniques other than those licensed, since such restrictions impede technical and economic progress."

of research and development, production, use or distribution of competing products.[189] However, the Technology Licensing Regulation does not prohibit the use of obligations on a licensee to use his best endeavors to exploit the licensed technology, nor does it restrict the right of the licensor to terminate the exclusivity granted to the licensee and to stop licensing improvements to him when the licensee begins to compete with the licensor or the licensor's right to require that the licensee prove that any licensed know-how is not being used for the production of products and the provision of services other than those licensed.[190]

The block exemption for exclusive purchasing agreements is also illustrative of the Commission's attitudes in this area.[191] The Regulation applies where the reseller agrees with the supplier to purchase certain goods specified in the agreement for resale only from the supplier or from a connected business or from another source which the supplier has entrusted with the sale of its goods,[192] and the Regulation contemplates that resellers will be obligated to purchase all of its requirements as to specified goods from the supplier and not to manufacture or distribute goods that compete with the contract goods, and that suppliers will agree not to compete with the reseller in the reseller's principal sales area for the contract goods.[193] The Recitals to the Regulation enumerate several potential advantages to such arrangements, particularly improvements to the distribution system, and indicate that the block exemption is based on the assumption that the supplier will be lending significant support to the reseller as to promotion of the contract goods.[194]

The benefits of the block exemption for exclusive purchasing agreements are, however, explicitly limited. For example, Article 3 of the Regulation provides that the exemption will not apply in situations where manufacturers of identical goods or goods that are considered by users as equivalent in view of their characteristics, price and intended use either (i) enter into reciprocal exclusive purchasing agreements between themselves in respect of such goods or (ii) enter into a non-reciprocal exclusive purchasing agreement in respect of such goods, unless at least one of them has a total annual turnover of no more than 100 million ECU.[195] The exemption will also not be available for agreements cover more than one type of goods where the contract goods are neither by their nature nor according to commercial usage connected to each other.[196] In addition, the Regulation does not exempt agreements which are for an indefinite duration or for a period of more than five years.[197] Finally, while the objective of the exemption is to insure that the reseller concentrates his sales efforts on the contract goods, the supplier cannot impose restrictions on the reseller's choice of customers or ability to determine the prices and terms of sale for the contract goods.[198]

[189] Technology Licensing Regulation Art. 3(2).
[190] Technology Licensing Regulation Art. 3(2).
[191] Regulation 1984/83, OJ L173/5 [1983]; amended by OJ L281/24 [1983].
[192] Ibid., Art. 1.
[193] Ibid., Art. 2.
[194] Ibid., Recitals (5) and (6).
[195] Ibid., Arts. 3 (a) and (b) and Recital (10).
[196] Ibid., Art. 3 (c).
[197] Ibid., Art. 3 (d) and Recital (11).
[198] Ibid., Recital (8).

Korah has taken issue with the Commission's continuing reluctance to permit the use of non-competition clauses, arguing that "where the parties are not even potential competitors, the clause is unlikely to be anti-competitive and is justified by the need to encourage a sole or exclusive licensee to exploit the technology."[199] He concedes that it often is difficult to really ascertain whether the parties to a licensing agreement are, in fact, "potential" competitors, and notes that licensors will likely be satisfied with using a "best efforts" clause, which is permitted under the Technology Licensing Regulation while restraints on competition are not.[200]

Restrictions on licensee challenges to patent validity As a general matter, the United States and the EC both look with disfavor on provisions in a license agreement which restrict the licensee's ability to challenge the validity of the licensed patents. The primary concern is that by preventing a licensee from seeking release from the obligations imposed under the license by demonstrating that the underlying patent rights are invalid, the life of the patents may be extended artificially, leading to the possibility of windfall profits for the licensor in the form of the royalties that the licensee would be compelled to pay for the use of the otherwise invalid patents.

In the United States, it was said that a licensee was effectively "estopped" from challenging the validity of any patent under which he was licensed. However, in *Lear* v. *Adkins*,[210] the Supreme Court abolished the doctrine of licensee estoppel on antitrust grounds. The Court's decision was quite consistent with the then-prevailing attitude toward encouraging challenges to the validity of issued patents. Accordingly, since the *Lear* decision in 1969, attempts to enforce a "no challenge" clause will be considered to be patent misuse and, although there is no case law on the matter, it can also be argued that merely including such a provision in an agreement could amount to patent misuse in situations where the licensor does so with knowledge of the *Lear* doctrine.[202]

While there are certainly good public policy reasons for minimizing the possibility that invalid patents will be enforced, the existence of the *Lear* doctrine may often impact the balance of interests between the licensee and the licensor. For example, the licensee's ability to challenge the validity of the licensed patent at any time can create significant concerns for the licensor, particularly when the license is entered into at a time when the licensed invention has not yet been fully developed and there is no history of exploitation to look to in determining the appropriate royalty amount. In such situations, it might well be anticipated that the greater the success of the patent, the more likely

[199] Korah, supra note 8, at 266.

[200] Technology Licensing Regulation Art. 3(2). Also, the Technology Licensing Regulation retains the provision in the know-how licensing block exemption which permitted the licensor to agree not to supply further improvements once the licensee was using rival technology. Ibid.

[201] 395 U.S. 653, 670–1 (1969). A full discussion of *Lear* is beyond the scope of this chapter. Some useful references are McCarthy, "Unmuzzling the Patent Licensee: Chaos in the Wake of Lear v. Adkins," *George Washington Law Review* 45, (1977), 429; Dreyfuss, "Dethroning Lear: Licensee Estoppel and the Incentive to Innovate," *Valparaiso Law Review* 72, 677; Smith, "Foreclosure of Licensee Validity Challenges with Procedural Barriers: Federal Jurisdiction," *Idaho Law Review* 16 (1985), 177; Schlicher, "Judicial Regulation of Patent Licensing, Litigation and Settlement Policies Created in Lear v. Adkins," *AIPLA Papers* 3 (1985).

[202] If merely including a "no challenge" clause was found to be patent misuse, the licensor would be faced with the possibility that it would be unable to enforce the license (i.e. collect any royalties) during the period of misuse. This alone might well be sufficient incentive to comply with *Lear*. Cf. *Panther Pumps & Equipment Co.* v. *Hydrocraft, Inc.*, 468 F.2d 225, 231, 175 U.S.P.Q. 577 (7th Cir. 1972 ("[T]he cases limiting the manner in which a patentee may exploit his monopoly contain no suggestion that the existence of such a clause should itself be considered a 'misuse.'")

it is that the licensee might consider a challenge to the validity thereof rather than paying what might well become a significant royalty obligation, especially as the comparative costs of litigation diminish in relation to the amount of royalties. As such, *Lear* complicates the ability to allocate the cost and risk of development between licensor and licensee on a binding and predictable basis. Moreover, the licensee has enormous bargaining power in "renegotiating" the terms of the license, because the licensor faces the risk of permanent loss of his property value through collateral estoppel in a validity challenge.[203]

Lear also makes it difficult for patent owners to settle infringement litigation with any degree of certainty. For example, a plaintiff may be unwilling, after incurring significant legal fees in connection with the infringement litigation, to settle the dispute on terms which call for royalty payments from the alleged infringer if, immediately following the settlement, the reluctant licensee is able to rely on *Lear* to repudiate the agreement and challenge validity. There is certainly a danger this might happen if the defendant/licensee is seeking to delay an injunction or is attempting to bring the dispute before a more favorable court.

In the EC, a no-challenge clause may, depending on the legal and economic context, infringe Article 85(1).[204] For example, such a provision would have no restrictive effect where the license has been granted for no consideration, since the licensee would not be confronted with a competitive disadvantage as a result of having to pay royalties to the licensor. Similarly, such a provision would not be deemed a restriction where the license has been granted for consideration but relates to a technically outdated procedure not used by the licensee which is the party accepting the no-challenge obligation.[205] Beyond that, however, a no-challenge clause put the license agreement outside the protections of the Patent Licensing Regulation, unless the restriction took the form of a right in favor of the licensor to terminate the license in the event that the licensee elects to challenge the validity of the licensed patents.[206] The Technology Licensing Regulation removed no-challenge clauses from the black list and explicitly permits a licensor to terminate the license agreement in the event of a patent challenge by the licensee;[207] however, the Technology Licensing Regulation clearly suggests that inclusion of no-challenge clauses requires the use of the opposition procedure included in the Regulation.[208]

203 Under *Blonder-Tongue Labs., Inc.* v. *University of Illinois Found.*, 402 U.S. 313 (1971), a patent found invalid in one proceeding cannot be asserted in any subsequent proceeding.

203 Case 65/86 *Bayer* v. *Sullhofer* [1988] ECR 5249 [1990] 4 CMLR 182. Previously, there had been a general presumption that no-challenge clauses fell within Art. 85(1). See, e.g. Case 193/83 *Windsurfing International* v. *Commission* [1986] ECR 611 [1986] 3 CMLR 489 §§ 89–94.

204 Case 65/86 *Bayer* v. *Sullhofer* [1988] ECR 5249 [1990] 4 CMLR 182.

205 Patent Licensing Regulation Art. 3(1) and Recital (20). See, however, Case 193/83 *Windsurfing International* v. *Commission* [1986] ECR 611 [1986] 3 CMLR 489 (Art. 85(1) does apply to clauses which allow the licensor to terminate the license in the event the licensee commences any manufacturing activities outside of the licensed territory).

206 Technology Licensing Regulation Art. 2, para. 1(15).

207 Technology Licensing Regulation Art. 4, para. 2(b).

208 *Northern Pacific Ry. Co.* v. *United States*, 356 U.S. 1, 5-6 (1958). See also *United States* v. *Loew's, Inc.*, 371 U.S. 38, 44–5 (1962); *Foriner Enterprises, Inc.* v. *United States Steel Corp.*, 394 U.S. 495, 498 (1969); *Jefferson Parish Hospital District No. 2* v. *Hyde*, 104 S. Ct. 1551 (1984).

VII. Tying and Package Licensing Arrangements

A. Introduction

Tying and package license arrangements are two similar methods of conditioning the grant of a patent license on the purchase by the licensee of additional products or licenses from the licensor, even though the licensee may have no need or desire for the additional items. Tying has been defined as "an agreement by a party to sell one product but only on the condition that the buyer also purchases a different (or tied) product."[209] Tying is a concept that extends beyond the most obvious forms of tie-in arrangements, such as when the patent owner explicitly conditions a license of the patent on the purchase of a "tied" product or service by the prospective licensee,[210] to include the more subtle cases of "label licensing," which involves the sale of unpatented products with a label authorizing their use under a patent owned by the seller,[211] and "implied licensing," in which the patent owner simply sells the unpatented product for implied use under its patent.[212] A packaging arrangement occurs when the licensee is required to take a license under a group of patents, including the patent that the licensee was originally interested in acquiring.[213]

The practice of marketing products and services which are related to other products covered by a patent is quite common among prospective patent licensors. For example, a patent owner may want to sell component parts used in a patented product or enter into service contracts for maintenance of the product. In general, these dual marketing activities are not inherently illegal; and a number of economists have argued that, in appropriate situations, tying arrangements may produce some benefits (e.g. economies of production or sale, or protection of licensor's goodwill) and increase the overall level of economic output.[214] Similarly, package licensing is often necessary in order for the licensee to achieve the desired technical or business goals and objectives of the relationship, such as when a license to a number of patents must be granted in order for the licensee to produce one or more specific products,[215] each patent covers a separate

[209] See, e.g. *International Salt Co.* v. *United States*, 332 U.S. 392 (1948); *Morton Salt Co.* v. *G.S. Suppiger Co.*, 314 U.S. 488, 52 U.S.P.Q. 30 (1942).

[210] See, e.g. *Rex Chainbelt, Inc.* v. *Harco Products, Inc.*, 512 F.2d 993, 185 U.S.P.Q. 10 (9th Cir.), *cert. denied*, 423 U.S. 831 (1975); *Amul Co.* v. *Uniroyal, Inc.*, 306 F.Supp. 541, 163 U.S.P.Q. 517 (S.D.N.Y. 1969), *modified*, 448 F.2d 872 (2d Cir. 1971), *cert. denied*, 404 U.S. 1018 (1972).

[211] See, e.g. *Leitch Mfg. Co.* v. *Barber Co.*, 302 U.S. 458, 36 U.S.P.Q. 35 (1938); *Barber Asphalt Corp.* v. *La Fera Grecco Contracting Co.*, 116 F.2d 211, 47 U.S.P.Q. 367 (3d Cir. 1940).

[212] See generally Adamo, "Problems Connected With Acquisition, Licensing and Enforcement of Intellectual Property," *Alb. Law Review* 50 (1986), 475; Marquis, "Limitations on Patent License Restrictions: Some Observations," *Iowa Law Review* 58 (1972), 41, 85–9; Siekman, "Post Expiration Royalty Payments and Mandatory Package Licensing as Patent Misuses," *Vill. Law Review* 11 (1966), 382; Frost, "Tying Clauses and Package Licensing," *University of Pittsburg Law Review* 28 (1966), 207.

[213] See, e.g, Singer, *Antitrust Economics*, (1968), 187–95; Baxter, supra note 99, at 318 et seq.; Bowman, "Tying Arrangements and the Leverage Problem," *Yale Law Journal* 67 (1957), 19–36.

[214] See, e.g. *Zenith Radio Corp.* v. *Hazeltine Research, Inc.*, 395 U.S. 100, 161 U.S.P.Q. 577 (1969) (license of a number of patents generally related to the production of radios); *International Mfg. Co.* v. *Landon, Inc.*, 336 F.2d 723, 142 U.S.P.Q. 421 (9th Cir. 1964), *cert. denied*, 379 U.S. 988 (1965) (license of various patents used in the production of specific products).

[215] See, e.g. *Cutter Laboratories, Inc.* v. *Lyophile-Cryochem Corp.*, 179 F.2d 80, 84 U.S.P.Q. 54 (9th Cir. 1949); *Mobil Oil Corp.* v. *W.R. Grace & Co.*, 367 F.Supp. 207, 180 U.S.P.Q. 418 (D. Conn. 1973).

component of a larger product or process,[216] or lawful exploitation of one patent cannot proceed without a license to other "blocking" patents.[217] Also, package licensing has been approved as a reasonable means of efficiently monitoring the use of the licensed subject matter.[218]

However, tying and packaging arrangements may run afoul of antitrust and competition laws if the arrangement is, in fact, the product of coercion exercised by the licensor over its licensees.[219] The fundamental concern is that a patent owner may use a valuable and sought-after patented product as leverage to require licensees to accept, in the case of a tying arrangement, other products that are outside the scope of the patent or, in the case of a packaging arrangement, a package license that includes other less desirable patents.[220] In either situation, the danger exists that the patent owner can use the market power embodied in the patent right to force a licensee to pay more, through the purchase of additional items or royalty obligations on the additional patents, than may be warranted by the original patent rights. Moreover, the licensee in such situations may be foreclosed from pursuing other alternatives or internally developing their own substitute technology, thereby harming competition and reducing innovation.[221]

The tying analogy has been so popular that, in fact, for a number of years, courts expanded the tying concepts, sometimes unartfully, to track the idea of "extension of monopoly" by treating the protected article or work as the "tying" product and specified commercial activity that fell outside the scope of the intellectual property right as the "tied" product. For example, in *Mercoid Corp.* v. *Mid-Continent Inv. Co.*[222] and *Mercoid Corp.* v. *Minneapolis-Honeywell Regulator Corp.*,[223] the Supreme Court found, by analogy to tying, that the plaintiff's attempt to enforce its patent on a furnace system against seller in a line of commerce (i.e. stocker switches) not precisely covered by the claims of the patent constituted misuse.[224] Similarly, the Court's ruling in *Brulotte* v. *Thys Co.*[225] that

[216] See, e.g. *Carpet Seaming Tape Licensing Corp.* v. *Best Seam, Inc.*, 616 F.2d 1133, 1142 (9th Cir. 1980); *International Mfg. Co.* v. *Landon, Inc.*, 336 F.2d 723, 142 U.S.P.Q. 421 (9th Cir. 1964), *cert. denied.* 379 U.S. 988 (1965); *North American Philips Co.* v. *Stewart Engineering Co.*, 319 F.Supp. 335 (N.D. Cal. 1970).

[217] See, e.g. *Broadcast Music, Inc.* v. *Columbia Broadcasting System, Inc.*, 441 U.S. 1, 201 U.S.P.Q. 497 (1979) (approving the use of package licensing of music copyrights as a competitively reasonable means of efficiently monitoring copyright use). See also *Zenith Radio Corp.* v. *Hazeltine Research, Inc.*, 395 U.S. 100, 137–8, 161 U.S.P.Q. 577 (1969); *Automatic Radio Mfg. Co.* v. *Hazeltine Research, Inc.*, 339 U.S. 827, 834, 85 U.S.P.Q. 378 (1950); Antitrust Division, U.S. Department of Justice, Antitrust Guide for International Operations, Case 10 (issued November 10, 1988) (generally approving the use of patent package licensing).

[218] See Brinson, "Proof of Economic Power in a Sherman Act Tying Arrangement Case: Should Economic Power be Presumed When the Tying Product is Patented or Copyrighted?" *Louisiana Law Review* 48 (1987), 29; Note, "The Presumption of Economic Power for Patented and Copyrighted Products in Tying Arrangements," *Columbia Law Review* 85 (1985), 1140; Hornick, "The Per Se Rule in Tying Contexts: A Critical View," *Delaware Journal of Corporate Law* 10 (1985), 703; Strasser, "An Antitrust Policy for Tying Arrangements," *Emory Law Journal* 34 (1985), 253; Slawson, "A New Concept of Competition: Reanalyzing Tie-In Doctrine After Hyde," *Antitrust Bulletin* 30 (1985), 257.

[219] See e.g. *Rocform Corp.* v. *Acitelli-Standard Concrete Wall, Inc.*, 367 F.2d 678, 151 U.S.P.Q. 305 (6th Cir. 1966); *American Security Co.* v. *Shatterproof Glass Corp.*, 268 F.2d 769, 122 U.S.P.Q. 167 (3d Cir.), *cert. denied*, 361 U.S. 902 (1959).

[220] See Frost, "Tying Clauses and Package Licensing," *University of Pittsburg Law Review* 28 (1966), 207; Note, "Mandatory Package Licensing: A New Patent Misuse," *Virginia Law Review* 44 (1958), 727.

[221] 320 U.S. 661 (1944).

[222] 320 U.S. 680 (1944).

[223] 320 U.S. 661, 666 (1944).

[224] 379 U.S. 29 (1964).

[225] Specifically, the Court stated: "A patent empowers the owner to exact royalties as high as he can with the leverage of that patent. But to use that leverage to project those royalty payments beyond the life of the

a patent owner could not attempt to collect royalties which might accrue after the patent has expired was based, in part, on the Court's conclusion that an unlawful tying relationship existed between the valid term of the licensed patent, which was the "tying" product, and the post-expiration royalties, which were the "tied" product.[226] Unfortunately, such an approach ignored the fact that post-expiration royalties were not a separate economic entity.[227]

B. General Attitudes in the United States

As a general matter, tying arrangements have long been disfavored in the United States as amounting to unlawful "extensions of the monopoly."[228] For example, in *Northern Pacific Ry. Co.* v. *United States*, the Supreme Court said:

> "Indeed, 'tying agreements serve hardly any purpose beyond the suppression of competition.' ... They deny competitors free access to the market for the tied product, not because the party imposing the tying requirements has a better product or a lower price but because of his power or leverage in another market. At the same time buyers are forced to forego their free choice between competing products."[229]

Courts in the United States tend to treat tying arrangements as *per se* unlawful,[230] provided that (i) the tied item is a separate product or service from the tying

cont.
patent is analogous to an effort to enlarge the monopoly of the patent by tieing [sic] the sale or use of the patented article to the purchase or use of unpatented ones." Ibid. at 33.

226 See also *International Business Machines* v. *United States*, 298 U.S. 131 (1936) (tying unpatented cards to a patented machine violated Clayton Act); *Hazeltine Research, Inc.* v. *Zenith Radio Corp.* 388 F2d 25, 33–5 (7th Cir. 1967), *aff'd in part.* 395 U.S. 100 (1969) (requiring a licensee to take licenses under more than one patent violated Sherman Act); *Stearns* v. *Tinkor & Rasor*, 252 F.2d 589, 602–3 (9th Cir. 1957) (refusal to sell a patented component except as part of an assembly violated Clayton Act and was misuse).

227 The antipathy of the courts to tying arrangements is similar to that which had led to judicial criticism of requirements that licensees must manufacture only the patented device (see, e.g. *National Lockwasher Co.* v. *George R. Garrett Co.*, 137 F.2d 255 (3d Cir. 1943) or cannot use competing articles (see, e.g. *McCullogh* v. *Kammerer Corp.*, 166 F.2d 759 (9th Cir. 1948), and grantbacks (see, e.g. *Duplan Corp.* v. *Deering Milliken, Inc.* 444 F.Supp. 648, 700 (D.S.C. 1977), *aff'd* 594 F.2d 979 (4th Cir. 1980)).

228 356 U.S. 1, 6 (1958). See also *Jefferson Parish Hospital Dist. No. 2* v. *Hyde*, 104 5. Ct. 1551, 1560–1 (1984) (Court voted five to four to retain the use of *per se* rules to analyze tying).

229 The traditional tying situation was evidenced in *International Salt Co.* v. *United States*, 332 U.S. 392 (1947), where the patent owner leased patented machines for the utilization of salt products, giving to each lessee an implied license to use the machine, but only with salt obtained from the defendant. The Supreme Court found this practice to violate the Sherman Act, stating: "The patents confer no right to restrain use of or trade in unpatented salt. By contracting to foreclose this market for salt against competition, International has engaged in a restraint of trade for which its patents afford no immunity from the antitrust laws ... It is unreasonable, per se, to foreclose competitors from any substantial market. 322 U.S. at 395–6."

230 Courts in the United States appear to have had little difficulty in finding the requisite "tie." For example, in *Jefferson Parish Hospital District No. 2* v. *Hyde*, the Supreme Court indicated that the requirement of "separateness" would be satisfied upon a showing that it would be economically feasible to offer one item separately from the other. 104 S. Ct. 1551, 1561–5 (1984). In cases where the tied items consist of staple, separately marked commodities like salt, films, and finished electrical goods, the challenger generally has no problems with this element. See, e.g. *International Salt Co.* v. *United States*, 332 U.S. 392 (1948); *Leitch Mfg. Co.* v. *Barber Co.*, 302 U.S. 458, 36 U.S.P.Q. 35 (1938) (misuse case); *International Business Machines Corp.* v. *United States*, 298 U.S. 131, 37 U.S.P.Q. 224 (1936); *Motion Picture Patents Co.* v. *Universal Film Mfg. Co.*, 243 U.S. 502 (1917) (misuse case). However, some courts have recognized that the answer may not always be so clear in situations where "[a]rticles, though physically distinct, may be related through circumstances." *Dehydrating Process Co.* v. *A.O. Smith Corp.*, 292 F.2d 653, 655 (1st Cir.), *cert. denied*, 386 U.S. 931 (1961). For example, one court found that electric power cable sold by an electric

item;[231] (ii) an actual tie exists, meaning that the availability of the tying item is conditioned on taking the tied item;[232] (iii) the party imposing the tie has sufficient economic power in the market for the tying item to appreciably restrain competition in the market for the tied item; and (iv) the tie has affected a not insubstantial amount of commerce[233] in the tied item.[234] A number of courts have added a fifth element of proof, which is a requirement that the party imposing the purported tie have an economic interest in the tied as well as the tying item.[235] If, however, the seller does not have sufficient market power,[236] the arrangement will then be analyzed under the rule of reason[237] and may be permitted if there is a legitimate business justification, such as the needs of a new entrant.

Given that the tying party's economic power in the tying market is the key issue in determining the relevant standard of review, it is not surprising that this is the issue which

cont.
utility was part of the same overall electrical service marketed by the company, and thus rejected antitrust claims that the utility had illegally "tied" the availability of low priced cable to purchase of its electricity. *Washington Gas Light Co.* v. *Virginia Electric and Power Co.*, 438 F.2d 248 (4th Cir. 1971).

231 Since most cases in this area involve either sale or rental agreements, the condition that an "actual tie" exist is usually easily satisfied. See, e.g. *International Salt Co.* v. *United States*, 332 U.S. 392 (1947); *Morton Salt Co.* v. *G.S. Suppiger Co.*, 314 U.S. 488, 52 U.S.P.Q. 30 (1942) (misuse case); *International Business Machines Corp.* v. *United States*, 298 U.S. 131, 37 U.S.P.Q. 224 (1936); *Motion Picture Patents Co.* v. *Universal Film Mfg. Co.*, 243 U.S. 502 (1917) (misuse case); *Senza-Gel Corp.* v. *Seifihart*, 803 F.2d 661, 231 U.S.P.Q. 363 (Fed. Cir. 1986) (misuse case); *Duplan Corp.* v. *Deering Milliken, Inc.*, 444 F. Supp. 648, 197 U.S.P.Q. 342 (D.S.C. 1977), *aff'd* 594 F.2d 979 (4th Cir. 1979), *cert. denied* 444 U.S. 1015 (1980) (misuse found but antitrust liability not established, where plaintiffs failed to prove requisite injury); *Switzer Bros., Inc.* v. *Locklin*, 297 F.2d 39, 131 U.S.P.Q. 405 (7th Cir. 1961) (trademarks tied to patents); *United States* v. *American Linen Supply Co.*, 141 F. Supp. 105, 109 U.S.P.Q. 272 (N.D. Ill. 1956).

232 As to this requirement, courts have essentially required simply that the dollar amount involved not be *de minimis*, and have not assessed substantiality in percentage terms in relation to some defined market. *Fortner Enterprises, Inc.* v. *United States Steel Corp.*, 394 U.S. 495, 501 (1969); *Rex Chainbelt, Inc.* v. *Harco Products, Inc.*, 512 F.2d 993, 1003, 185 U.S.P.Q. 10 (9th Cir.), *cert. denied*, 423 U.S. 831 (1975). The requirement has been satisfied when the amount involved was $500,000 (*International Salt Co.* v. *United States*, 332 U.S. 392 (1947)); $200,000 (*Fortner Enterprises, Inc.* v. *United States Steel Corp.*, 394 U.S. 495 (1969)); and $61,000 (*United States* v. *Loew's, Inc.*, 371 U.S. 38, 49 (1962)). But see *Parts and Electric Motors, Inc.* v. *Sterling Electric, Inc.*, 826 F.2d 712 (7th Cir. 1987) (dictum that threatened market power in the tied product market should be an element of a *per se* tying claim).

233 *Northern Pacific Ry. Co.* v. *United States*, 356 U.S. 1, 6 (1958); *United States* v. *Loew's, Inc.*, 371 U.S. 38, 45 (1962); *Fortner Enters., Inc.* v. *United States Steel Corp.*, 394 U.S. 495, 498 (1969); *Jefferson Parish Hosp. Dist. No. 2* v. *Hyde*, 104 S. Ct. 1551 (1984).

234 See, e.g. *Sandburg Village Condominium Ass'n* v. *First Condominium Dev. Co.*, 758 F.2d 203 (7th Cir. 1985); *Keener* v. *Sizzler Family Steak Houses*, 597 F.2d 453 (5th Cir. 1979). But see *Gonzalez* v. *St. Margaret's Housing Dev. Fund Corp.*, 880 F.2d 1514 (2d Cir. 1989) (rejecting proof of the defendant's economic interest in the tied product as an element of a *per se* tying claim).

235 See *Jefferson Parish Hospital District* v. *Hyde*, 466 U.S. 2 (1984) (30 percent held insufficient for *per se* treatment).

236 Under this standard, the party challenging the arrangement must establish the competitive unreasonableness of the tying arrangement, "on the basis of a more thorough examination of the purposes and effects of the practices involved." *Fortner Enters., Inc.* v. *United States Steel Corp.*, 394 U.S. 495, 500 (1969). See also *United States Steel Corp.* v. *Fortner Enters., Inc.*, 429 U.S. 610, 612 n1 (1977); *Jefferson Parish Hosp. Dist. No. 2* v. *Hyde*, 104 S. Ct. 1551(1984); *Hand* v. *Central Transport, Inc.*, 779 F.2d 8 (6th Cir. 1985) (articulating on analytical framework for a "rule of reason" tying case).

237 See, e.g. *United States Steel Corp.* v. *Fortner Enterprises, Inc.*, 429 U.S. 610, 619 (1977); *United States* v. *Loew's, Inc.*, 371 U.S. 38, 45–6 (1962); *International Salt Co.* v. *United States*, 332 U.S. 392 (1947); *Senza-Gel Corp.* v. *Seiffhart*, 803 F.2d 661, 231 U.S.P.Q. 363 (Fed. Cir. 1986); *Rex Chainbelt, Inc.* v. *Harco Products, Inc.*, 512 F.2d 993, 1003, 185 U.S.P.Q. 10 (9th Cir.), *cert. denied*, 423 U.S. 31 (1975); *George R. Whitten, Jr., Inc.* v. *Paddock Pool Builders, Inc.*, 424 F.2d 25, 36 (1st Cir.), *cert. denied*, 400 U.S. 850 (1970); *Duplan Corp.* v. *Deering Milliken, Inc.*, 444 F. Supp. 648, 673, 197 U.S.P.Q. 342 (D.S.C. 1977), *aff'd*, 594 F.2d 979 (4th Cir. 1979), *cert. denied*, 444 U.S. 1015 (1980).

has raised the greatest controversy in this area in the United States. The traditional presumption was that the requisite power could be presumed from the use of a patent as the tying item, and most of the prior cases involving patent tie-ins have historically been quick to conclude that the existence of a patent, with its legalistic exclusionary rights, is sufficient to create a presumption of economic power.[238] As the Supreme Court explained in *United States Steel Corp.* v. *Fortner Enters., Inc.*,[239] economic power is somewhat analogous to monopoly power in that both involve an ability to "require" purchasers to "accept burdensome terms" (i.e. the purchase or rental of the tied item) that otherwise could "not be exacted in a completely competitive market."[240] In the Court's view, the possession of this leveraging power can be demonstrated in a number of ways other than by actual dominance, such as by the particularly desirable nature of the tying item itself, coupled with some economic or legal barrier that prevents competitors from offering the tying item on comparable terms.[241] The Court then concluded that use of a patent as the tying item creates a "presumption of economic power."[242]

However, in recent years, a number of courts have begun to abandon the presumption that patent rights automatically afford the patent owner economic power, and have undertaken the same difficult structural and market power inquiries found in many non-patent tying cases. For example, in *A.I. Root Co.* v. *Computer Dynamics, Inc.*,[243] a case which involved allegedly *per se* illegal tying restraints involving copyrighted computer software, the court held that the copyrighted nature of the software did not necessarily equate with economic power, and stated in dictum that the same would be true of a case involving a patented tying item. The court noted that "[m]ore often than not, a patent or copyright provides little, if any, market power."[244] Other courts have since reached similar conclusions,[245] and courts are also more willing to recognize that the presumption, if any, is not conclusive and that the trier of fact "must also take note of material facts in the record which may rebut that presumption." Therefore, the patentee can introduce evidence to the effect that its patent does not block competitors from freely producing competing products, thereby reducing the commercial significance of the patent and any resulting inference of market power from control over the patent.

The market-based approach suggested by *Root* achieved acceptance in revised statutory language in the Patent Code covering the situation where an alleged patent in-

[238] 429 U.S. 610, 620 (1977).

[239] Ibid.

[240] Ibid. at 620–1. See also *United States* v. *Loew's, Inc.*, 371 U.S. 38,45 (1962); *Jefferson Parish Hosp. Dist. No. 2* v. *Hyde*. 104 S. Ct. 1551, 1559 (1984) (the "special ability – usually called 'market power'" – possessed by the seller and enabling it to "force a purchaser to do something that he would not do in a competitive market").

[241] 429 U.S. 610, 619 (1977). See also *United States* v. *Loew's, Inc.*, 371 U.S. 38, 45–6 (1962) ("The requisite economic power is presumed when the tying product is patented or copyrighted").

[242] 1986-2 Trade Cas. (CCH) §67,363 (6th Cir. 1986).

[243] Ibid. at 61,930.

[244] See, e.g. *Mozart Co.* v. *Mercedes-Benz of N. America, Inc.*, 833 F.2d 1342 (9th Cir. 1987) (dictum that market power should not be presumed from the mere fact that an item is patented or copyrighted); *Nobel Scientific Indus., Inc.* v. *Beckman Instruments*, 670 F. Supp. 1313 (D. Md. 1986) (market power not inferred from the patented status of the tying item).

[245] See, e.g. *In re Data General Corp. Antitrust Litigation*, 490 F.Supp. 1089, 1112–13, 207 U.S.P.Q. 549 (N.D. Cal. 1980), *rev'd on other grounds*, 1984-1 Trade Cas. (CCH) §66,053 (9th Cir. 1984) (also taking note of the fact that "the presumption of economic power attendant upon patent or copyright protection of the tying product is not conclusive."); *H&G Affiliates, Inc.* v. *Knoll International, Inc.*, 1984-1 Trade Cas. (CCH) §66,047 (S.D.N.Y. 1984) (the fact that products in a manufacturer's "full line" of products were patented did not in itself establish economic power for purposes of a tying claim, where it was unclear precisely what significance, if any. the patented products had in the overall mix).

fringer attempts to assert patent tying practices as a "misuse defense."[246] Prior to the Patent Misuse Reform Act of 1988,[247] tying could be established as a misuse defense regardless of whether market power was shown in the relevant market. However, as a result of the amendments to the patent laws effected by the Act, a tying arrangement will no longer be considered patent misuse unless, in view of the circumstances, the patent owner has market power in the relevant market for the patent or patented product on which the license or sale is conditioned.[248] Congress enacted this provision in order to eliminate any vestiges of a *per se* inference of patent misuse from certain tying practices; therefore, it is now necessary to make an affirmative showing of market power in order for the misuse defense to prevail.

The intent of the drafters of the Patent Misuse Reform Act is clarified by a review of its legislative history. This history makes it clear that courts are now expected to "reasonably assess the potential for anticompetitive effect of a particular practice," and "to avoid the use of inflexible rules."[249] In particular, the determination of whether "market power" exists, courts would be called upon to evaluation "'market power' in the context of the patent, where a patent license is involved, the product and the market in which the tie-in occurs."[250] As described in a relevant portion of the legislative history:

> "The use of the term relevant market is designed to import into the court's analysis the idea that the scope of the product involved focuses the court's attention on the nature of the property right. If a patented product is unique because no practical substitute exists, the scope of the relevant market would be coextensive with the patent. In the situation where the product is sold in a marketplace context where there are substitute products, the scope of the market should resemble the typical antitrust analysis of relevant market."[251]

The use of the term "in view of the circumstances" also was designed to facilitate a more flexible analytical approach by the courts which took into account the potentially competitive or anticompetitive effects of the tie-in practice,[252] and the history suggested that courts might consider whether the tied product was a staple or a nonstaple.[253] The

[246] Pub. L. No. 100-703, § 202 (November 19, 1988).

[247] The relevant provision of Section 271(d) of the Patent Act (35 U.S.C. § 271(d)) reads as follows: "No patent owner otherwise entitled to relief for infringement or contributory infringement of a patent shall be denied relief or deemed guilty of misuse or illegal extension of the patent right by reason of his having done one or more of the following ... (5) conditioned the license of any rights to the patent or the sale of the patented product on the acquisition of a license to rights in another patent or purchase of a separate product, unless, in view of the circumstances, the patent owner has market power in the relevant market for the patent or patented product on which the license or sale is conditioned ..."

[248] Cong. Rec. HR 10,648 (October 20, 1988).

[249] Ibid.

[250] Ibid.

[251] Ibid.

[252] The language in the legislative history explained: "In the case of tying a patented product to a nonstaple the net effect of such arrangement may serve to expand the economic rights of the patent owner. This result, however, is generally appropriate because in most situations involving high technology the market for the nonstaple product would not exist but for the existence of the patented product. On the other hand, courts that apply a rule of reason analysis to the tie-in of a patented product involving a staple may evaluate it in a slightly different manner. The ability of a party with a patented product to require that the purchaser or the licensee of that product use a particular staple could have an anticompetitive effect. Thus, for cases involving the tie-in of staple products, the courts should be sensitive to the potential anticompetitive burden on commerce such a practice may have if the maker of a competing staple has its market substantially diminished as a result of the tie-in."

[253] Ibid.

history also makes it clear that a finding that market power exists does not necessarily mean that the practice should be automatically prohibited. The following passage is instructive:

> "It is also our intention to avoid the use of inflexible rules once a court has found that market power exists. There may be circumstances in which there is market power and a tie-in, but where a finding of misuse would be inappropriate. One example would be where the patent owner has a business justification for the licensing practice. In real world situations where the only practical way to meter output is to tie the sale of a patented product to the sale of another separate product, then such a practice would be legitimate, unless such a practice-on balance-has a generally anticompetitive effect."[254]

It remains to be seen to what extent this statutory rule – technically limited as it is to patent misuse as a defense in infringement actions – will ultimately carry over to the antitrust arena as well. Legislative proposals have periodically been introduced in Congress to extend a similar express "market power" requirement to antitrust claims of patent tying; however, no definite action has yet been taken.[255] One of the latest attempts in this regarding was a bill introduced in November 1995 which would provide that "[i]n any action in which the conduct of an owner, licensor, licensee, or other holder of an intellectual property right is alleged to be in violation of the antitrust laws in connection with the marketing or distribution of a product or service protected by such a right, such right shall not be presumed to define a market, to establish market power (including economic power and product uniqueness or distinctiveness), or to establish monopoly power."[256]

While legislative initiatives in this area have proceeded cautiously, there has recently been some indication that the courts have moved away from the expansive application of the "extension of monopoly" doctrine and tying principles in licensing cases. For example, *Dawson Chemical Co.* v. *Rohm & Haas Co.*[257] involved facts that were similar to those which were present in *Mercoid.* Rohm & Haas held a process patent which employed an unpatented chemical, propanil, as an herbicide. Rohm & Haas sold propanil and included with each sale an implied license to practice the patented process. Rohm & Haas refused to license the patent separately from the sale of propanil. The

[254] Ibid.

[255] See, e.g. HR 469 (introduced in the House on January 8, 1989) and S 270 (introduced in the Senate on March 13, 1989). These bills would have effectively eliminated any judicial presumption that intellectual property rights conveyed market power; however, the bills were defeated in the face of claims by opponents that the legislation would have created an antitrust exemption of somehow legalize conduct that may have otherwise been illegal, such as permitting tie-ins in case when a licensor actually has market power for reasons other than intellectual property rights.

[256] HR 2674 (introduced in the House on November 20, 1995 as the Intellectual Property Antitrust Protection Act of 1995). In his speech to the House, Representative Henry J. Hyde, the primary sponsor of the bill, explained: "Because of increasing competition and a burgeoning trade deficit, our policies and laws must enhance the position of American businesses in the global marketplace. This concern should be a top priority for this Congress. A logical place to start is to change rules that discourage the use and dissemination of existing technology and prevent the pursuit of promising avenues of research and development. Some of these rules arise from judicial decisions that erroneously create a tension between the antitrust laws and the intellectual property laws ... For too long, Mr. Speaker, court decisions have applied the erroneous presumption of market power thereby creating an unintended conflict between the antitrust laws and the intellectual property laws. Economists and legal scholars have criticized these decisions, and more importantly, these decisions have encouraged innovation to the detriment of the American economy."

[257] 448 U.S. 176 (1980).

defendant, Dawson, also sold propanil, which was useful only for infringing the patent. Rohm & Haas brought suit against Dawson for infringement, who defended by arguing that Rohm & Haas was guilty of misuse since it refused to license except as an incident to the sale of this unpatented product. A bare majority of the Supreme Court held that there was no misuse,[258] and explained that Sections 271(c) and 271(d) of the Patent Act, as they existed at that time, permitted the licensing of a process patent on terms which required the licensee to purchase from the patent owner an unpatented chemical needed for the process.[259]

The decision in *Dawson* also appeared to signal a growing sensitivity to the need to permit patent owners to attempt to recover their investments in innovation through the use of various schemes for commercialization. For example, the Court made it clear that the patent owner should have the right to refuse to license its patent, and that to treat refusals to license as misuse would essentially deprive the patent owner of the important right to be able to exploit the technology exclusively.[260] The Court also acknowledged that allowing the defendants in that case to prevail would significantly undermine the economic incentives for the first innovators that form the basis for the recognition of patents:

> "Under the construction of [principles of misuse] that petitioners advance, the rewards available to those willing to undergo the time, expense, and interim frustation of such practical research would provide at best a dubious incentive. Other could await the results of the testing and then jump on the profit bandwagon by demanding licenses to sell the chemical used in the newly developed process. Refusal to accede to such a demand, if accompanied by any attempt to profit from the invention through sale of the unpatented chemical, would risk forfeiture of any patent protection whatsoever on a finding of patent misuse. As a result, non-inventors would be almost assured of an opportunity to share in the spoils, even though they had contributed nothing to the discovery. The incentive to await the discoveries of others might well prove sweeter than the incentive to take the initiative oneself [448 U.S. at 221–2]."

In recent years, the DOJ has taken a more permissive attitude toward tie-ins, noting that they can facilitate risk sharing with respect to new technologies and allow the patentee to meter demand among licensees in order to better derive its true market value. The IP Guidelines recognize that, under some circumstances, tying can be efficiency-enhancing,[261] and the DOJ indicated that it would likely challenge a particular tying

[258] Four of the justices would have continued the misuse doctrine. As stated in the dissent: "[E]very licensed granted under this patent has been conditioned on the purchase of an unpatented product from the patentee. This is a classic case of patent misuse." Ibid. at 240.

[259] The Court noted that the revised language in Sections 271(c) and 271(d) of the Patent Act dictated a major change in the law as it existed following the decisions in the *Mercoid* cases. The relevant portions of Section 271(c) read as follows: "Whoever sells ... a material or apparatus for use in practicing a patented process, constituting a material part of the invention, knowing the same to be especially made ... for use in an infringement of such patent ... shall be liable as a contributory infringer." The Court found Dawson's actions to be contributory infringement within the scope of this language. The relevant portions of Section 271(d) which were read as protecting the acts of *Rohm & Haas* read as follows: "No patent owner ... shall be ... deemed guilty of misuse or illegal extension of the patent right by reason of his having done one or more of the following: (1) derived revenue from acts which if performed by another without his consent would constitute contributory infringement of the patent; (2) licensed or authorized another to perform acts which if performed without his consent would constitute contributory infringement of the patent ..."

[260] 448 U.S. at 215.

[261] See, e.g. *Jerrold Electronics Corp.* v. *Westcoast Broadcasting Co.*, 341 F.2d 653 (9th Cir.), *cert. denied*, 382 U.S. 817 (1965).

arrangement only if: (i) the seller has sufficient economic power in the market (i.e. "market power")[262] for the tying product to enable it to restrain trade in the market for the tied product, (ii) the arrangement has an adverse effect on competition in the relevant market for the tied product, and (iii) efficiency justifications for the arrangement do not outweigh the anticompetitive effect.[263] With regard to the practice of granting "package licenses," the IP Guidelines make it clear that such licenses may be efficiency enhancing, as might be the case when multiple licenses are need to practice any single item of intellectual property.[264]

C. General Attitudes in the EC

The EC's tolerance for tying arrangements is even more limited, and they will typically only be permitted in situations where it can be shown that the arrangement is indispensable for the exploitation of a patent. For example, under the Patent Licensing Regulation, licensees could be required to obtain goods (e.g. semi-finished products, raw materials, or components) or services from the licensor or a firm designated by the licensor, provided that the goods or services were necessary for the "technically satisfactory exploitation of the invention".[265] Otherwise, tying obligations generally were considered to fall within Article 85(1) and were not exempted under the Patent Licensing Regulation in situations where the licensee was required to accept further licenses he did not want, or to use patents or goods or services he did not want.[266] Similarly, a provision that licensed products be sold only in conjunction with other products, not covered by the patent, may also infringe Article 85(1)[267] and also was not exempted under the Patent Licensing Regulation.

The Technology Licensing Regulation does hold the promise of a bit more liberal attitude in this area. As with the Patent Licensing Regulation, the Regulation continues to permit the use of obligations on licensees to procure goods or services from the licensor when it is necessary for a technically satisfactory exploitation of the licensed technology or ensuring that the licensee's products conform to specified standards of quality.[268] However, in a departure from the prior regulations, the Technology Licensing Regulation eliminates tying arrangements that are not expressly permitted from the blacklist, leaving it for the Commission to decide on a case-by-case basis whether tying may be allowed. However, given the ECJ's clear precedent in the *Windsurfing* judgment finding that tying violates Article 85(1), it will be risky to include a tying provision in an agreement without notifying it to the Commission, and the Technology Licensing Regulation

[262] As noted above, the DOJ will not presume market power solely from the existence of a patent or other intellectual property right. The DOJ points out that this policy is consistent with the changes made in the Patent Misuse Reform Act in the patent law area.

[263] IP Guidelines § 5.3.

[264] Package licensing may be a form of tying arrangement, but only if the items licensed constitute "separate products" and the licensing of one product is used to force the acceptance of a license of another. Ibid. If a package license constitutes a tying arrangement, the DOJ will evaluate its competitive effects under the same principles it applies to other tying arrangements. Ibid.

[265] Patent Licensing Regulation Art. 2(1)(1).

[266] Patent Licensing Regulation Art. 3(9) and Recital (24).

[267] Case 193/83 *Windsurfing International* v. *Commission* [1986] ECR 611 [1986] 3 CMLR 489 (obligations to sell rigs only with approved sailboards and not to sell rigs separately).

[268] Technology Licensing Regulation Art. 2, para. 1(5).

clearly suggests that inclusion of such a provision requires the use of the opposition procedure included in the Regulation.[269]

VIII. Grant-back Provisions

A. Introduction

While the primary element of the technology package in any licensing relationship is the legal rights owned and developed by the licensor, in many cases the licensee will possess both the technical skills, and the economic motivation, to develop various improvements to the originally licensed technology which conducting its activities under the license. If it is anticipated that the licensee will develop improvements to the original technology, the license agreement should clearly describe the respective rights of both parties with respect to the ownership and use thereof, as well as any contemplated changes that the creation of such improvements might have upon the basic economic structure of the relationship.

The first issue which should be addressed is the ownership of the improvements. The answer to this question generally lies with the original intent of the parties. For example, the licensor may be willing to accept a lower royalty rate for the license of the original technology if it is contemplated that the licensee will be developing new improvements which will be owned by the licensor at the time that the development work is completed. However, if the improvements are of the type that could only be developed by the licensee in light of its particular skills and resources, or are of a type that are unexpected or unrelated to the original licensed subject matter, it may be appropriate for the licensee to be vested with full title to such improvements, perhaps with an appropriate adjustment to the amount of royalties for the original technology or, as described below, an appropriate agreement regarding use of the new technology by the licensor.

Once ownership issues have been resolved, consideration must be given to the rights of each party to use the new technology in similar or new applications or fields of activity. A number of different strategies might be used in this situation. For example, if the improvements are owned by the licensor, the licensee may either be required to assign all of its rights in the improvements to the licensor and have no right to use the improvements under the original license agreement or, as is more typical, the licensee may be given the right to use the improvements for purposes limited to the goals and objectives of the original licensing relationship.[270] If the improvements are owned by the licensee, it may retain all of the rights in the new discoveries, without any further obligation to the licensor;[271] however, the more common solution is to use a "grant-back" provision, which provides that the licensee will grant a license back to the licensor to use the improvements.[272]

[269] Technology Licensing Regulation Art. 4, para. 2(a).

[270] Another alternative in this situation gives the licensee a right of first offer on any license of the improvements which the licensor may intend to give to third parties.

[271] This may occur when the licensee has made a large lump-sum payment to the licensor at the time the license agreement was originally signed. In effect, the licensor is taking the position that it purchased the original technology in order to serve as a starting point for further research and development work.

[272] See generally Adamo, "Problems Connected With Acquisition, Licensing and Enforcement of Intellectual Property," *Albany Law Review* 50 (1986), 475; Schmalbeck, "The Validity of Grant-Back Clauses in Patent Licensing Agreements," *University of Chicago Law Review* 42 (1975), 733; Dunne, "Anti-Competitive Considerations of Patent Accumulation by Licensee Grant-Back Provisions," *Journal of*

In many ways, a grant-back provision amounts to an entirely new license or assignment, and the scope and competitive impact of such a provision can be as broad as the grant of the original license itself. For example, while the parties may limit the grant-back clause to a non-exclusive license that covers blocking improvement patents developed by the licensee,[273] a grant-back may also be structured to require outright assignments of all patented or unpatented technology acquired by the licensee in a specified general field.[274]

There are clearly a variety of procompetitive benefits of grant-back clauses. For example, grant-backs may actually serve to encourage licensing activities which might not otherwise have taken place, since without a grant-back clause the licensor may be hesitant about licensing its patent technology to firms that can then develop and exclusively retain improvement technology made possible by the licenses which might ultimately render the original technology obsolete or less marketable.[275] Moreover, grant-backs allow the patentee to continue making improvements in its own technology. Grant-back provisions also may allow the parties to share the risks and compensate the licensor for making possible further innovation based on or informed by the licensed technology. Finally, the use of grant-back clauses are seen as facilitating efficient bargaining for licenses.[276]

On the other hand, there is no doubt that grant-back provisions can have adverse effects on competition in a variety of circumstances.[277] For example, when the licensor holds a dominant market position, an analysis will be necessary to determine if the grant-back clause tends to further entrench the patentee's existing market dominance, in which case the arrangement will be subject to challenge as patent misuse.[278] Concerns may also

cont.
the Patent Officers Society 57 (1975), 124; Timberg, "Antitrust and Industrial Property: Tie-Ins and Grant-Backs," *Antitrust Law Journal* 42 (1973), 651; Comment, "Transparent Wrap: The Continuing Case for Grant-Back Agreements," *Cath. University Law Review* 17 (1967), 228; Chevigny, "Validity of Grant-Back Agreements Under the Antitrust Laws," *Fordham Law Review* 34 (1966), 569; Stedman, "Acquisition of Patents and Know-How by Grant, Fraud, Purchase and Grant-Back," *University of Pittsburgh Law Review* 28 (1966), 161, 168–72.

[273] See, e.g. *International Nickel Co., Inc.* v. *Ford Motor Co.*, 166 F.Supp. 551, 565, 119 U.S.P.Q. 72 (S.D.N.Y. 1958) ("A fair reading of the agreements discloses that the licensees are required to license back only patents which could not be exploited without risking infringement of ... [the licensor's] basic patent.").

[274] See, e.g. *Chandler* v. *Stern Dental Laboratory Co.*, 335 F.Supp. 580 (S.D. Tex. 1971) (patent misuse found where grant-back obligation encompassed all products in the general field, "whether within the scope of the ... [basic] patent or not.").

[275] *International Nickel Co., Inc.* v. *Ford Motor Co.*, 166 F.Supp. 551, 566, 119 U.S.P.Q. 72 (S.D.N.Y. 1958) ("The net effect may well be to increase rather than to decrease competition.") See also *Santa Fe-Pomeroy, Inc.* v. *P & Z Company, Inc.*, 569 F.2d 1084, 1101 n34, 167 U.S.P.Q. 449 (9th Cir. 1978) (the patentee is "entitled to some protection for its original investment" against improvements that would "commercially displace the latter."). See also Antitrust Division, U.S. Department of Justice, Antitrust Guide for International Operations, Case 11 (issued November 10, 1988) (generally approving the use of grant backs).

[276] See, e.g. 1988 International Guidelines Case 11 and IP Guidelines § 5.6.

[277] See generally Klitzke, "Patents and Section 7 of the Clayton Act: The Significance of Patents in Corporate Acquisitions," *Loyola University Law Journal* 12 (1981) 401, 415; Marquis, "Limitations on Patent License Restrictions: Some Observations," *Iowa Law Review* 58 (1972), 41, 77–81.

[278] See, e.g. *United States* v. *General Electric Co.*, 82 F. Supp. 753, 80 U.S.P.Q. 195 (D.N.J. 1949) (held that General Electric monopolized domestic incandescent lamp industry in part through the use of non-exclusive grant-back clauses which tended to "funnel" new industry innovations back through it); *Transparent-Wrap Machine Corp.* v. *Stokes & Smith Co.*, 329 U.S. 637, 646–7, 72 U.S.P.Q. 148 (1947) ("As patents are added to patents a whole industry may be regimented. The owner of a basic patent might thus perpetuate his control over an industry long after the basic patent expired."); *United States* v. *General Electric Co.*, 80 F. Supp. 989, 79 U.S.P.Q. 124 (S.D.N.Y. 1948); *United States* v. *Timken Roller Bearing Co.*, 83 F. Supp. 284, 81 U.S.P.Q. 28 (N.D. Ohio 1949), *aff'd*, 341 U.S. 595 (1951); *United States* v. *Aluminum Co. of America*, 91 F. Supp. 333, 410 (S.D.N.Y. 1950).

arise when there is evidence that the grant-back provisions are actually being used as a means of facilitating illegal cartel behavior, as might occur when new product developments are being funneled back through a central licensing entity. This situation arose in the case of *United States* v. *Associated Patents, Inc.*,[279] where a patent holding company was used to consolidate patents developed by machine tool manufacturers and to then use licenses under the patents to allocate fields of use among the participant firms. Since each of the licenses required the firms to assign all improvement patents they developed to the holding company, the licensees could not undermine the conspiracy by licensing outsiders. The court held that the arrangement, including the grant-back restrictions, was a violation of Section 1 of the Sherman Act.[280]

Another potential anticompetitive factor which must be taken into account is the adverse effect that the arrangement might have on the licensee's innovative activity. For example, a grant-back clause may dampen the licensee's incentive to engage in further research and development if the results thereof must be disclosed to the licensor. Such a result would not only improperly shield the licensor's own technology from competition by the licensee's, but will also reduce the overall flow of invention and innovation in the marketplace.[281] This problem was addressed in *Associated Patents*, where the court included the following language in its opinion: "[i]nvention and technological development have been discouraged by the limitations imposed on the members' rights to use and license improvement patents developed by them."[282] Other cases have also examined the effect of grant-back provisions on the licensee's incentive to innovate,[283] and the DOJ takes the position that the decision as to whether or not to challenge a grant-back will be based upon consideration of the extent to which, as compared with no license at all, a license which contains the grant-back clause may diminish total investment in research and development or lessen the degree of competition in innovation or technology markets.[284]

B. General Attitudes in the United States

Recognizing that grant-back provisions may have both procompetitive and anticompetitive effects, the Supreme Court and lower courts appear to have rejected the notion that grant-back clauses should be subject to *per se* illegality, choosing instead to employ a standard of overall reasonableness in evaluating grant-backs. For example, the case of *Transparent-Wrap Machine Corp.* v. *Stokes & Smith Co.*[285] involved an exclusive license

[279] 134 F.Supp. 74 (E.D. Mich. 1955), *cert. denied*, 350 U.S. 960 (1956).

[280] See also, e.g. *United States* v. *Timken Roller Bearing Co.*, 83 F. Supp. 284, 81 U.S.P.Q. 28 (N.D. Ohio 1949), *aff'd*, 341 U.S. 593 (1951) (market allocation and price fixing conspiracy); *United States* v. *General Electric Co.*, 80 F. Supp. 989, 79 U.S.P.Q. 124 (S.D.N.Y. 1948) (market allocation and price fixing conspiracy).

[281] See, e.g. IP Guidelines § 5.6.

[282] 134 F.Supp. 74 (E.D. Mich. 1955), *cert. denied*, 350 U.S. 960 (1956).

[283] See, e.g. *Transparent-Wrap Machine Corp.* v. *Stokes & Smith Co.*, 329 U.S. 637, 646, 72 U.S.P.Q. 148 (1947); *Blohm & Voss A.G.* v. *Prudential Grace Lines, Inc.*, 346 F. Supp. 1116, 174 U.S.P.Q. 484 (D. Md. 1972), rev'd on other grounds, 489 F.2d 231 (4th Cir. 1973), *cert. denied*, 419 U.S. 840 (1974); *International Nickel Co.* v. *Ford Motor Co.*, 166 F. Supp. 551. 119 U.S.P.Q. 72 (S.D.N.Y. 1958); *Old Dominion Box Co.* v. *Continental Can Co.*, 273 F. Supp. 550, 155 U.S.P.Q. 70 (S.D.N.Y. 1967), *aff'd*, 393 F.2d 321 (2d Cir. 1968).

[284] IP Guidelines § 5.6.

[285] 329 U.S. 637, 72 U.S.P.Q. 148 (1947).

that required the licensee to assign to the patentee any improvement patents it acquired relevant to the licensed machine. While the Second Circuit had held the grant-back provision to be *per se* invalid, analogizing it to coercive tying arrangements,[286] the Supreme Court rejected the tie-in analogy, noting that while tying restrictions improperly extend the legal patent monopoly to encompass unpatented items, grant-back provisions merely are "one legalized monopoly to acquire another legalized monopoly."[287] The Court went on to point out that grant-back provisions were merely another form of lawful "consideration" which might be tendered by a licensee for the right to use the patentee's licensed invention.[288] The Court did not deny that grant-back provisions might, under certain circumstances, be employed in a manner which might amount to patent misuse or an antitrust violation;[289] however, it concluded that the mere fact of the "inclusion in the license" of a grantback clause was "not" in and of itself "per se illegal and unenforceable."[290]

A number of factors are considered in assessing the reasonableness of a grant-back arrangement. For example, if the arrangement calls for the licensor to be given exclusive rights in improvements to the original licensed technology, it is more likely that a successful challenge will be made[291] than if the grant-back is non-exclusive, thereby leaving the new technology reasonably accessible to the licensee and others.[292] Also, concern regarding the scope of the grant-back clause increases as one moves from provisions limited to non-exclusive rights in improvement patents, which are generally upheld as reasonable in the face of antitrust or patent misuse challenges,[293] to exclusive rights in all of the licensee's patented or unpatented inventions in the general field.[294] In addi-

[286] *Stokes & Smith Co.* v. *Transparent-Wrap Machine Corp.*, 156 F.2d 198, 201, 69 U.S.P.Q. 278 (2d Cir. 1946), *rev'd*, 329 U.S. 637 (1947).

[287] *Transparent-Wrap Machine Corp.* v. *Stokes & Smith Co.*, 329 U.S. 637, 644, 72 U.S.P.Q. 148 (1947).

[288] Ibid.

[289] Ibid.

[290] Ibid. On remand, the legality of the particular grant-back clause in issue was upheld. *Stokes & Smith Co.* v. *Transparent-Wrap Machine Co.*, 161 F.2d 565, 73 U.S.P.Q. 200 (2d Cir.), *cert. denied*, 331 U.S. 837 (1947).

[291] See, e.g. *Chandler* v. *Stern Dental Laboratory Co.*, 335 F. Supp. 580 (S.D. Tex. 1971) (licensee "coerced" into paying royalties on its own improvements); *United States* v. *Associated Patents, Inc.*, 134 F. Supp. 74, 106 U.S.P.Q. 129 (E.D. Mich. 1955), *cert. denied*, 350 U.S. 960 (1956) (conspirators given veto power over further licenses); *United States* v. *General Electric Co.*, 80 F. Supp. 989, 79 U.S.P.Q. 124 (S.D.N.Y. 1948) (most of the grant-backs were exclusive).

[292] See, e.g. *Hull* v. *Brunswick Corp.*, 704 F.2d 1195 (10th Cir. 1983); *Santa Fe-Pomeroy, Inc.* v. *P & Z Company, Inc.*, 569 F.2d 1084, 167 U.S.P.Q. 449 (9th Cir. 1978); *Robintech, Inc.* v. *Chemidus Wavin, Ltd.*, 450 F. Supp. 817, 197 U.S.P.Q. 657 (D.D.C. 1978), *aff'd*, 628 F.2d 142 (D.C. Cir. 1980); *Blohm & Voss A.G.* v. *Prudential Grace Lines, Inc.*, 346 F. Supp. 1116, 179 U.S.P.Q. 484 (D. Md. 1972) *rev'd on other grounds*, 489 F.2d 231 (4th Cir. 1973), *cert. denied*, 419 U.S. 840 (1974); *Old Dominion Box Co.* v. *Continental Can Co.*, 273 F. Supp. 550. 155 U.S.P.Q. 70 (S.D.N.Y. 1967), *aff'd*, 393 F.2d 321 (2d Cir. 1968); *Zajicek* v. *Koolvent Metal Awning Corp. of Am.*, 283 F.2d 127 (9th Cir. 1960), *cert. denied*, 365 U.S. 859 (1961). See also IP Guidelines § 5.6 (a nonexclusive grant-back provision is less likely to be anticompetitive as compared with an exclusive grantback, since a nonexclusive provision does not restrain the licensee from licensing improvements to the technology to other parties).

[293] See, e.g. *Hull* v. *Brunswick Corp.*, 704 F.2d 1195 (10th Cir. 1983); *Santa Fe-Pomeroy, Inc.* v. *P & Z Company, Inc.*, 569 F.2d 1084, 167 U.S.P.Q. 449 (9th Cir. 1978); *Barr Rubber Products Co.* v. *Sun Rubber Co.*, 277 F. Supp. 484, 156 U.S.P.Q. 374 (S.D.N.Y. 1967), *aff'd*, 425 F.2d 1114 (2d Cir.), *cert. denied*, 400 U.S. 878 (1970); *Old Dominion Box Co.* v. *Continental Can Co.*, 273 F. Supp. 550, 155 U.S.P.Q. 70 (S.D.N.Y. 1967), *aff'd*, 393 F.2d 321 (2d Cir. 1968); *International Nickel Co., Inc.* v. *Ford Motor Co.*, 166 F. Supp. 551, 119 U.S.P.Q. 72 (S.D.N.Y. 1958).

[294] See, e.g. *Duplan Corp.* v. *Deering Milliken, Inc.*, 444 F. Supp. 648, 197 U.S.P.Q. 342 (D.S.C. 1977), *aff'd* 594 F.2d 979 (4th Cir. 1979), *cert. denied*, 444 U.S. 1015 (1980) (the "scope" of the grant-back obligation "extended far beyond" that of the "basic patent," resulting in an impermissible extension of the patentee's

tion, grant-back restrictions with a relatively short term are less likely to give rise to anticompetitive effects or to discourage the innovative activities of the licensee,[295] while grant-backs that extend far into the future can raise significant concerns and are especially susceptible to allegations of antitrust or patent misuse.[296] Finally, the analysis will take into account whether the parties are competitors and, if so, their respective market power.

C. General Attitudes in the EC

The EC permits the use of grant-back provisions only in relatively narrow situations. If, for example, the licensee's obligation is nothing more than an agreement to grant-back to the licensor a non-exclusive license on any such improvements or enhancements, and a reciprocal obligation has been imposed on the licensor to license to the licensee the benefits (i.e. any improvements or enhancements) of the licensor's experience with the licensed subject matter following the date of the initial license, then Article 85(1) will normally not be infringed. Under the Patent Licensing Regulation, both parties could agree to exchange experience gained in exploiting the licensed invention, and they could also agree to grant to each other licenses for improvements and new applications; however, any such exchanges or licenses needed to be non-exclusive.[297] On the other hand, if the licensee was required to assign ownership of the improvements or enhancements exclusively to the licensor[298] or grant co-ownership rights to the licensor,[299] Article 85(1) would come into play, and such a provision was not be eligible for an exemption under the Patent Licensing Regulation.[300]

The Technology Licensing Regulation includes provisions relating to grant-back provisions which are quite similar to those set out in the Patent Licensing Regulation. The basic rule is that the parties may agree that the licensee will be obligated to grant the licensor a licence in respect of the licensee's own improvements to, or new applications of, the technology that is the subject of the licence;[301] however, such a provision may only be included if, in the case of severable improvements, the licence is non-exclusive,[302] and the licensor must also agree to grant a licence of his own improvements to the licencee.[303]

cont.
tee's lawful patent monopoly). See also *Chandler* v. *Stern Dental Laboratory Co.*, 335 F. Supp. 580 (S.D. Tex. 1971) (misuse found where grant-back required licensee to assign over patented and unpatented inventions it developed in the general field, even though not related to the basic licensed patents).

295 See, e.g. *Santa Fe-Pomeroy, Inc.* v. *P & Z Company, Inc.*, 569 F.2d 1084, 1101, 167 U.S.P.Q. 449 (9th Cir. 1978); *Old Dominion Box Co.* v. *Continental Can Co.*, 273 F. Supp. 550, 155 U.S.P.Q. 70 (S.D.N.Y. 1967), *aff'd*, 393 F.2d 321 (2d Cir. 1968); *Stokes & Smith Co.* v. *Transparent-Wrap Machine Co.*, 161 F.2d 565, 73 U.S.P.Q. 200 (2d Cir.), *cert. denied*, 331 U.S. 837 (1947).

296 See, e.g. *United States* v. *Associated Patents, Inc.*, 134 F. Supp. 74, 106 U.S.P.Q. 129 (E.D. Mich. 1955), *cert. denied*, 350 U.S. 960 (1956); *United States* v. *Timken Roller Bearing Co.*, 83 F. Supp. 284, 81 U.S.P.Q. 28 (N.D. Ohio 1949), *aff'd.* 341 U.S 599 (1951); *United States* v. *General Electric Co.*, 82 F. Supp. 753, 80 U.S.P.Q. 195 (D.N.J. 1949).

297 Patent Licensing Regulation Art. 2(1)(9).

298 See, e.g. *Raymond/Nagoya*, OJ 1972 L143/39 [1972] CMLR D45; *Velcro/Aplix*, OJ 1985 L233/22 [1989] 4 CMLR 157.

299 Nodet-Gougis/Lamazou, Tenth Report on Competition Policy (1980), point 127.

300 Patent Licensing Regulation Art. 3(8) and Recital (24).

301 Technology Licensing Regulation Art. 2, para. 4.

302 Ibid.

303 In addition, if the licensor's right to use the licensee's improvements is of a longer duration than the licensee's right to use the licensed technology, the licensor must agree to pay appropriate royalties to the licensee or to allow him to continue exploiting the licensed technology. Ibid.

On the other hand, the Technology Licensing Regulation does not permit obligations on the licensee to assign in whole or in part to the licensor the rights to improvements to or new applications of the licensed technology.[304]

IX. Royalty Obligations

A. Base for Computation of Royalties

In some cases, a patent licensor may attempt to base royalty payments on measures which are not reasonably related to the licensee's sale of patented products. For example, a "total sales" royalty provision bases the license fee on total sales of indicated types of products by the licensee, whether or not the specified products actually incorporate any technology covered by the license. While the DOJ formerly took issue with such royalty practices, it now takes the view that it is not concerned with "the basis upon which license royalties are measured" and that such systems of payment may have the beneficial effect of allowing the patentee to meter or extract the true value of the patent.[305] Total sales royalties have also been upheld in the face of antitrust challenges and claims of patent misuse so long as it can be shown that the arrangements are truly voluntary and not the result of coercion by the patentee.[306] For example, in *Automatic Radio Mfg. Co.* v. *Hazeltine Research, Inc.*,[307] the Supreme Court upheld a total sales royalty provisions in the absence of patentee coercion, noting that "[s]ound business judgment" arguably favored such a royalty system as "the most convenient method" of "fixing the business value" of the patent use privileges granted by the licensee.[308] However, the attitude may well be different if the patentee and licensee are horizontal competitors, in which case the practice would probably be *per se* illegal.

Similarly, royalty formulas which are based on the price of other products into which the patented item is incorporation will typically not cause concerns under Article 85(1) in the EC, at least in those cases where the complexity of the production process makes it difficult to identify the number or value of the patented items, nor is there any problem if there is no separate independent demand for the patented items. Also, so long as the royalties payable on the whole product are no higher than could otherwise be justified if the royalty were based solely on the component, then Article 85(1) would not be infringed.[309] On the other hand, Article 85(1) may apply in situations where the royalty is charged on sale prices that include unpatented items if the effect of the royalty is to discourage the licensee from meeting any separate demand for the unpatented items in question.[310]

[304] Technology Licensing Regulation Art. 3, para. 6.

[305] See 1988 International Guidelines Case 10.

[306] See generally Note, "Total Sales Royalties Under the Patent Misuse Doctrine: A Critique of Zenith," *Michigan Law Review* 76 (1978), 1144; Wallace, "Proper Use of the Patent Misuse Doctrine," *Mercer Law Review* 26 (1975), 813.

[307] 339 U.S. 827, 85 U.S.P.Q. 378 (1950).

[308] Ibid. at 834. See also *Zenith Radio Corp.* v. *Hazeltine Research, Inc.*, 395 U.S. 100, 161 U.S.P.Q. 577 (1969); *United States* v. *Westinghouse Electric Corp.*, 471 F.Supp. 532 (N.D. Cal. 1978), *aff'd*, 648 F.2d 642 (9th Cir. 1981).

[309] Case 193/83 *Windsurfing International* v. *Commission* [1986] ECR 611 [1986] 3 CMLR 489.

[310] Ibid.

B. Package and Hybrid Licenses

There are two special situations in which the licensing arrangement may obligate the licensee to pay royalties relating to licensed patents even after the patent may have expired. In the first situation, referred to as "package licensing," the licensee receives the right to use a "package" of two or more patents, but is typically obligated to pay the licensor the same royalty amount throughout the entire license term,[311] even though particular patents may expire prior to the end of the term of the license and, presumably, significantly reduce the overall value of the licensed technology.[312] The second situation is the hybrid license, which includes a license for both an issued or pending patent and a license for some other form of intellectual property right, such as trade secrets, that is not coextensive with the licensed patent, and which obligates the licensee to pay the same royalty for so long as the license remains in effect, even if the patents expire before the appurtenant intellectual property rights terminate (e.g. the trade secrets are publicly disclosed).

The key concern in both situations from the perspective of antitrust and competition law regulators is whether allowing the licensor to continue to collect the same amount of royalties after one or more of the included patents expires amounts to an unlawful extension of the patent monopoly beyond its intended scope.[313] It can be argued that package licensing arrangements, like total sale royalty provisions, should not be struck down provided they are voluntarily entered into for the "convenience of the parties."[314] Moreover, such provisions often are more in accord with commercial realities than an inflexible rule that the entire value of the licensed patents must be recovered over an arbitrary period corresponding to the term of the patent. For example, in those industries where a substantial development period is needed before profits can be realized, or where there is a real need to defer royalties to maximize the capital available to the licensee during the early years of commercialization, both parties might prefer to spread the payments over a longer period.

Until 1964, courts in the United States tended to honor the intent of the parties as set out in their licensing contract. However, in 1964, the Supreme Court, in *Brulotte* v. *Thys Company*,[315] used an antitrust rationale analogous to misuse in ruling that agreements that provided for payment of royalties beyond the life of an issued patent were unenforceable *per se*, reasoning that allowing royalties to be paid under those circumstances effectively sanctioned royalty payments for inventions that were properly in the public domain.[316] The Court stated that use of post-expiration royalties was an improper extension

[311] See, e.g. *Duplan Corp.* v. *Deering Milliken, Inc.*, 444 F.Supp. 648, 197 U.S.P.Q. 342 (D.S.C. 1977), *aff'd* 594 F.2d 979 (4th Cir. 1979), *cert. denied*, 444 U.S. 1015 (1980).

[312] See generally Note, "Standard Antitrust Analysis and the Doctrine of Patent Misuse: A Unification Under the Rule of Reason," *University of Pittsburgh Law Review* 46 (1984), 209; Roberts, "Antitrust and Patent Misuse Considerations of Royalty Rate Differences Among Patent Licensees," *Journal of the Patent Officers Society* 65 (1983), 344; Fisher, "The Misuse Doctrine and Post-Expiration-Discriminatory and Exorbitant Patent Royalties," *Indiana Law Journal* 43 (1967-68), 106; Wallace, "Proper Use of the Patent Misuse Doctrine," *Mercer Law Review* 26 (1975), 813; Note, "Post Expiration Royalty Payments and Mandatory Package Licensing as Patent Misuses," *Villanova Law Review* 11 (1966), 382.

[313] Similar issues arise in licensing agreements covering a number of patents which provide for the payment of the same royalty until the last patent covered by the agreement expires. This would mean that the licensee would presumably be paying the same amount for the use of inventions which, upon earlier expiration of their patent, had entered the public domain.

[314] See, e.g. *Zenith Radio Corp.* v. *Hazeltine Research, Inc.*, 395 U.S. 100, 138, 161 U.S.P.Q. 577 (1969).

[315] 379 U.S. 29, 143 U.S.P.Q. 264 (1964).

[316] Ibid.

of the temporal scope of the patent monopoly.[317] The license involved in the case was limited to patents and the royalty rate, which was based on the licensee's use of a patented machine, was the same before and after expiration of the patent.

Subsequent cases have attempted to refine the basic rules set out by the Supreme Court and reconcile them with the emerging view that free access to inventions should be placed above the private contractual intent of the parties, which was evidenced by the 1969 ruling that licensees were not estopped from challenging patent validity.[318] In general, lower courts have upheld last-to-expire royalty clauses that have been voluntarily accepted,[319] and struck down similar clauses where the evidence establishes that the patent owner had engaged in coercive activities, as might be the case when there is no reduction in the license fee even when the patent or patents which clearly have the greatest value expire prior to the end of the royalty term.[320] A similar approach appears to have been taken with hybrid agreements. For example, in a case involving a license of a pending patent application as well as trade secrets, the Supreme Court upheld the enforceability of the licensee's obligation to pay royalties even if the patent was not issued. The Court noted that the required royalty payments were substantially reduced and reasoned that because of the reduction it was unlikely that the agreement was obtained using the leverage of the patent.[321]

On the other hand, there are situations where the courts will not permit post-expiration royalties, even though there has not been an explicit act of coercion. For example, in one situation, the district court, while appearing to accept the proposition that royalties could extend beyond the life of a patent if the reduction was meaningful, nonetheless struck down a hybrid license which provided for a reduction in the minimum royalty rate if a pending patent never issued because the licensee had always paid well above the minimum royalty rate and, therefore, a reduction in the minimum rate did not provide the licensee with sufficient incentive to challenge the patent.[322]

[317] Ibid. at 32.

[318] *Lear, Inc.* v. *Adkins*, 395 U.S. 653 (1969).

[319] *Hull* v. *Brunswick Corp.*, 704 F.2d 1195 (10th Cir. 1983) (patent misuse case); *Cohn* v. *Compax Corp.*, 87 A.D.2d 364, 451 N.Y.S.2d 171 (App. Div. 1982); *Beckman Instruments, Inc.* v. *Technical Development Corp.*, 433 F.2d 55, 67 U.S.P.Q. 10 (7th Cir. 1970), *cert. denied*, 401 U.S. 976 (1971) (patent misuse case); *Well Surveys, Inc.* v. *Perfo-Log, Inc.*, 396 F.2d 15, 158 U.S.P.Q. 119 (10th Cir. 1968), *cert. denied*, 393 U.S. 951 (1969) (patent misuse case); *Hensley Equipment Co.* v. *Esco Corp.*, 383 F.2d 252, 155 U.S.P.Q. 183 (5th Cir. 1967) (patent misuse and antitrust case); *McCullough Tool Co.*, v. *Well Surveys, Inc.*, 343 F.2d 381, 145 U.S.P.Q. 6 (10th Cir. 1965), *cert. denied*, 383 U.S. 933 (1966) (patent misuse case); *Mobil Oil Corp.* v. *W.R. Grace & Co.*, 367 F.Supp. 207, 180 U.S.P.Q. 418 (D. Conn. 1973); *Cohn* v. *Compax Corp.* 1982-2 Trade Cas. (CCH) §64,800 (N.Y. S.Ct. 1982) (patent misuse case).

[320] *Rocform Corp.* v. *Acitelli-Standard Concrete Wall, Inc.*, 367 F.2d 678, 151 U.S.P.Q. 305 (6th Cir. 1966) (patent misuse case); *American Security Company* v. *Shatterproof Class Corp.*, 268 F.2d 769 (3d Cir.), *cert. denied*, 361 U.S. 902 (1959) (patent misuse case); *Duplan Corp.* v. *Deering Milliken, Inc.*, 444 F.Supp. 648, 197 U.S.P.Q. 342 (D.S.C. 1977), *aff'd* 594 F.2d 979 (4th Cir. 1979), *cert. denied*, 444 U.S. 1015 (1980) (patent misuse and antitrust case). Cf. *Boggild* v. *Kenner*, 853 F.2d 465, 7 USPQ 2d 1642 (6th Cir. 1988) (a post- expiration royalty formula that failed to reduce the royalty upon expiration of the licensed patent constituted patent misuse even though the license also covered trade secrets); *Pitney Bowes, Inc.* v. *Mestre*, 701 F.2d 1365 (11th Cir. 1983) (striking down a post-expiration royalty formula that combined patents with trade secrets and continued to charge a non-reduced royalty following expiration of the patents and continued use of the trade secrets).

[321] *Aronson* v. *Quick Point Pencil Co.*, 440 U.S. 257, 201 U.S.P.Q. 1 (1979).

[322] *Timely Products, Inc.* v. *Costanzo*, 465 F Supp 91, 201 U.S.P.Q. 567 (D Conn 1979). See also *Meehan* v. *PPG Industries, Inc.*, where the Seventh Circuit held that the royalty provision of an agreement negotiated before any patent application was filed, but expressly contemplating such a filing, was *per se* unlawful because the provision did not distinguish between pre- and post-expiration royalties. 802 F 2d 881 (7th Cir 1986).

In the EC, the Commission has often stated its belief that Article 85(1) will be infringed in situations where the licensee is required to pay royalties with respect to products which are manufactured without making any use of the licensor's patents.[323] The Commission believes such a provision will have the effect of strengthening what is an otherwise objectionable prohibition on competition. Moreover, it is considered likely that such a clause would have the effect of "burdening manufacturing costs without any economic justification, thereby weakening the competitive position of the licensee."[324] In addition, in the words of the Commission, such a provision restricts competition "because there is a danger of the licensee's production, as compared with that of competitors, having to bear costs for which the licensee is not compensated through the advantages conferred by exploitation of the patent."[325]

The exemption in the Patent Licensing Regulation was not available if the licensing agreement contained an obligation on the part of the licensee to continue to pay royalties after all the licensed patents have expired, or after licensed know-how has entered the public domain other than by the fault of the licensee;[326] however, such provisions were permitted if they really were an arrangement for spreading payments for the previous use of the licensed invention prior to the expiration of protection.[327] The Technology Licensing Regulation permits obligations on a licensee to continue paying royalties over a period that goes beyond the duration of the licensed patents in order to facilitate payment.[328] In addition, at least one court has held that an obligation to pay a royalty throughout the term of the licensing arrangement, whether or not the licensed subject matter was still eligible for legal protection, would not infringe Article 85(1) if the licensee has the right to terminate the agreement.[329] In rendering the decision, the court recognized the possibility that the parties may sometimes set the duration of the payment obligation by reference to factors that are unrelated to the legal term of protection, including an economic assessment of the commercial value of the possibilities of exploitation granted under the licensing arrangement.

X. Miscellaneous Licensing Issues

A. Cross-Licensing and Pooling Arrangements

Patent pooling, or cross-licensing, involves the exchange of license rights between two or more patent owners, which not only resolves only conflict which may otherwise exist

[323] *AOIP/Beyrard*, OJ 1976 L6/8 [1976] 1 CMLR D14 at D24; *Preflex/Lipski*, Tenth Report on Competition Policy (1980), point 126 (patents); *Spitzer/van Hool*, Twelfth Report, point 86 (know-how); *Neilson-Hordell/Richmark*, Twelfth Report (1982), points 88 et seq. (copyright); *UARCO*, Fourteenth Report (1984), point 93 (patents and know-how).

[324] *AOIP/Beyrard*, OJ 1976 L6/8 [1976] 1 CMLR D14 at D24.

[325] *Windsurfing International*, OJ 1983 L229/1 [1984] 1 CMLR 1.

[326] Patent Licensing Regulation Art. 3(4) (prohibiting "charging royalties on non-patented products, or on know-how that is within the public domain"). See also Recital (22).

[327] Ibid. See also *Boussois/Interpane*, OJ 1987 L50/30 [1988] 4 CMLR 124.

[328] Technology Licensing Regulation Art. 2, para. 7(b).

[329] *Ottung* v. *Klee & Weilbach A/S & Anor* (Case 320/87), May 12, 1989, [1989] ECR 1177; [1990] 2 CEC 674. Case involved a license agreement under which the licensee was obligated to pay royalties in respect to two devices, one being patented after the agreement was entered into and the other being complementary to the first, and the agreement was entered into prior to the time that the first patent was issued. The court noted that if there was no right in favor of the licensee to terminate the license, then the arrangement would infringe Art. 85(1) as a restriction on the licensee's post-termination freedom of action.

among the pooled patents but also provides each of the patent owners with a means for sharing information and access to proprietary technologies. Patent pooling arrangements may well violate the antitrust laws, particularly when they result in an accumulation of monopoly power in the relevant market,[330] or if they include mutual territorial restrictions,[331] price-fixing clauses,[332] or field of use restrictions.[333] Since patent pools can often be used for legitimate purposes, such as resolving conflicting claims in patent litigation, they are generally revised under the rule of reason, with courts focusing upon the purpose of the pool, the nature of the rights exchanged, the market power of the parties to the arrangement, and clauses which restrict competition.

The IP Guidelines specifically acknowledge that cross-licensing and pooling arrangement may promote economic welfare by integrating complementary technologies,[334] reducing transaction costs, clearing blocking positions,[335] avoiding the costs of infringement litigation, and enhancing the dissemination of technology.[336] On the other hand, such arrangements can also have anticompetitive effect in certain circumstances. For example, when these arrangements are a mechanism to accomplish price fixing, or market or customer allocation, they can lead to a significant lessening of competition.[337] Pooling arrangements also may be anticompetitive in situations where participation therein deters or discourages participants from engaging in research and development, thus re-

[330] *United States* v. *Krasnov*, 143 F. Supp. 184 (E.D. Pa. 1956), *aff'd per curiam*, 355 U.S. 5 (1957) (patent pool between parties with monopoly power in which they agreed to allocate customers, maintain prices established by the licensor, to jointly determine infringement litigation strategy, and to license others only with mutual consent violated Sections 1 and 2 of the Sherman Act); *United States* v. *National Lead Co.*, 63 F. Supp. 513, 531–2 (S.D.N.Y. 1945), *aff'd*, 332 U.S. 319 (1947) (agreement between two largest competitors in titanium oxide industry to grant the other non-exclusive licenses in all patents, past, present, and future, and the right to inspect annually the other's technology and plants violated the antitrust laws).

[331] See, e.g. *United States* v. *Holophane Co.*, 119 F. Supp. 114, 117–19 (S.D. Ohio 1954), *aff'd per curiam*, 352 U.S. 903 (1956) (division of territories between cross-licenses violated antitrust laws); *United States* v. *Imperial Chemical Industries, Ltd.*, 100 F. Supp. 504, 519–31 (S.D.N.Y. 1951).

[332] *United States* v. *Line Material Co.*, 333 U.S. 287, 305–15 (1948).

[333] See, e.g. *United States* v. *Associated Patents, Inc.*, 134 F.Supp. 74, 82–3 (E.D.Mich. 1955), *aff'd mem. sub nom*, *Mac Investment Co.* v. *United States*, 350 U.S. 960 (1956) (patent pool dividing fields of use amongst members violated the antitrust laws); cf. *United States* v. *Birdsboro Steel Foundry & Mach. Co.*, 139 F. Supp. 244, 260 (W.D. Pa. 1956) (cross-license agreement dividing fields of use permitted).

[334] Example 10 in the IP Guidelines builds on the fact situation in Example 9, discussed above, by analyzing the case where two manufacturers assign to a separate jointly-owned joint venture only patents are blocking. In the example, the DOJ states that pooling of patents would not raise competitive problems in the relevant technology market if the pool involved complementary patents and enabled licensing of a package whose value exceeded the sum of its component patents.

[335] In Example 10 in the IP Guidelines, discussed above, the DOJ notes that "[u]nlike the previous example, the joint assignment of patent rights to the wholly owned corporation in this example does not adversely affect competition in the licensed technology among entities that would have been actual or likely potential competitors in the absence of the licensing arrangement. Moreover, the licensing arrangement is likely to have procompetitive benefits in the use of the technology. Because the manufacturers' patents are blocking, the manufacturers are not in a horizontal relationship with respect to those patents. None of the patents can be used without the right to a patent owned by the other firm, so the patents are not substitutable."

[336] IP Guidelines § 5.5.

[337] See *United States* v. *New Wrinkle, Inc.*, 342 U.S. 371 (1952) (price fixing); *United States* v. *United States Gypsum Co.*, 333 U.S. 364 (1948) (customer allocation). In Example 10 of the IP Guidelines, which involves the joint assignment of patent rights to a jointly-owned joint venture, the DOJ states that in the absence of evidence suggesting that the assignment is also contributing to coordinated pricing of the firms' final products, it would be unlikely to challenge the arrangement.

tarding innovation.[338] Such an arrangement is more likely to cause competitive problems where it includes a large fraction of the potential participants in research and development.

The IP Guidelines cover certain additional aspects of cross-licensing and pooling arrangements. For example, the DOJ notes that the joint marketing of pooled intellectual property rights, with collective price setting or coordinated output restrictions, may violate Section 1 of the Sherman Act.[339] As to settlements of pending infringement litigation, the IP Guidelines note that while such settlements can reduce the costs to the parties, the DOJ will nonetheless consider challenging cross-licensing arrangements between horizontal competitors entered into as part of a settlement if the effect thereof is to diminish rivalry that would otherwise have occurred.[340] The IP Guidelines also confirm that pooling arrangements need not necessarily be open to all parties that might wish to join; however, the DOJ points out that such arrangements may be harmful to competition when the participants collectively possess market power.[341]

In the EC, horizontal license agreements between actual or potential competitors, including patent pools and cross licenses, may violate Article 85(1). As a general rule, the Technology Licensing Regulation does not cover agreements between members of a patent or know-how pool relating to the pooled technologies, or to reciprocal licenses of competing products;[342] however, an exemption may be available for such agreements and licences provided that they don't contain any territorial restrictions.[343]

B. *"Bad Faith" Enforcement of Patent Rights*

As a general rule, the mere act of commencing legal proceedings in order to enforce a patent is not, in and of itself, a violation of the antitrust laws even if the suit is intended to exclude a competitor from the market for the goods covered by the subject patent. However, There are certain situations where filing, or threatening to file, an infringement suit may lead to antitrust problems. For example, in *Walker Process Equipment, Inc.* v. *Food Machinery & Chemical Corp.*,[344] the Supreme Court held that an antitrust remedy for attempted monopolization under Section 2 of the Sherman Act may be available to a party

[338] The IP Guidelines note that "[a] pooling arrangement that requires members to grant licenses to each other for current and future technology at minimal cost may reduce the incentives of its members to engage in research and development because members of the pool have to share their successful research and development each of the members can free ride on the accomplishments of other pool members." See IP Guidelines § 5.5. See also generally *United States* v. *Automobile Manufacturers Association*, 307 F. Supp. 617 (C.D. Cal 1969), *modified sub nom. United States* v. *Motor Vehicle Manufacturers Association*, 1982-3 Trade Cas. (CCH) § 65,088 (C.D. Cal. 1982); *United States* v. *Manufacturers Aircraft Association*, 1976-1 Trade Cas. (CCH) § 60,810 (S.D.N.Y. 1975).

[339] Compare *NCAA* v. *Board of Regents of the University of Oklahoma*, 468 U.S. 85 (1984) (output restriction on college football broadcasting held unlawful because it was not reasonably related to any purported justification) with *Broadcast Music, Inc.* v. *CBS*, 441 U.S. 1(1979) (blanket license for music copyrights upheld because the cooperative price was found necessary to the creation of a new product).

[340] The DOJ states that "[i]n the absence of offsetting efficiencies, such settlements may be challenged as unlawful restraints of trade." Cf. *United States* v. *Singer Manufacturing Co.*, 374 U.S. 174 (1963) (cross-license agreement was part of broader combination to exclude competitors).

[341] Cf. *Northwest Wholesale Stationers, Inc.* v. *Pacific Stationery & Printing Co.*, 472 U.S. 284 (1985) (exclusion of a competitor from a purchasing cooperative not unlawful absent a showing of market power).

[342] Technology Licensing Regulation Art. 5, para. 1.

[343] Technology Licensing Regulation Art. 5, para. 2(2).

[344] 382 U.S. 172 (1965).

accused of infringing a patent which was procured by fraud on the Patent and Trademark Office;[345] however, while claims under the *Walker* doctrine have been numerous, they have not generally been successful.[346] In another case, the Ninth Circuit allowed treble damages to be assessed against a patent owner who attempted to enforce a patent which the owner knew was invalid, although noting that sanctions in such situations would only be appropriate upon presentation of "clear and convincing" evidence,[347] and the IP Guidelines state that the DOJ will challenge attempts to enforce intellectual property rights which are invalid under the antitrust laws.[348] Finally, according to at least one court, a bad faith threat to sue on intellectual property rights may also constitute a violation of Section 2 of the Sherman Act.[349]

C. Acquisitions of Intellectual Property Rights

Antitrust and competition law regulators have dealt with the competitive effects of an acquisition of intellectual property rights, which can occur as part of an asset acquisition or in the course of acquisition in which one of the assets transferred is the rights of the transferring party under a licensing arrangement. The IP Guidelines reiterate that the legality of transactions which result in an actual or effective acquisition of intellectual property rights will be analyzed under section 7 of the Clayton Act and Sections 1 and 2 of the Sherman Act.[350] The IP Guidelines state that intellectual property acquisitions will be analyzed in the same manner as any other acquisition of assets.[351] In cases where the transaction amounts to a horizontal acquisition, the DOJ will apply the analysis

[345] In addition to fraud, a plaintiff must also show a restraint in a relevant market and some causal connection to his injury. The Supreme Court in *Walker* rejected the notion that a patent, in and of itself, gives rise to a completed monopolization offense, remanding the case for a determination of "the exclusionary power of the illegal patent claims in terms of the relevant market" (382 U.S. at 177). Moreover, the monopoly power of the defendant must arise directly from the claims themselves.

[346] The only reported recoveries appear to be *Keamey & Trecker Corp.* v. *Cincinnati Milacron, Inc.*, 562 F.2d 365 (6th Cir. 1977) and *Kearney & Trecker Corp.* v. *Gidding & Lewis, Inc.*, 452 F.2d 579 (7th Cir. 1971), both involving a patent owner who surreptitiously employed a former examiner with a conflict of interest to obtain a reissued patent broader than the original. The patent suit was based on the original claims, which were not tainted by the fraud, but damages assessed as the cost of defending the patent suit were nevertheless awarded to the defendants. See also *Kimberly Clark Corporation* v. *Johnson & Johnson*, 745 F.2d 1437, 1454, 223 U.S.P.Q. 603 (Fed. Cir. 1984) ("'fraud in the PTO' has been overplayed, is appearing in nearly every patent suit, and is cluttering up the patent system."); *Argus Chemical Corp.* v. *Fibre Glass-Evercoat Co.*, 812 F.2d 1381 (Fed. Cir. 1987) (enforcement of a patent obtained by mere inequitable conduct before the Patent and Trademark Office cannot be the basis of a section 2 claim, because inequitable conduct does not involve knowing and willful patent fraud).

[347] *Handgards, Inc.* v. *Ethicon, Inc.*, 743 F.2d 1282, 1288–9 (9th Cir. 1984), *cert. denied*, 469 U.S. 1190 (1985) (patents). See also *Professional Real Estate Investors, Inc.* v. *Columbia Pictures Industries, Inc.*, 113 S. Ct. 1920, 1928 (1993); *Handgards, Inc.* v. *Ethicon, Inc.*, 601 F.2d 986, 992–96 (9th Cir. 1979), *cert. denied*, 444 U.S. 1025 (1980) (patents); *CVD, Inc.* v. *Raytheon Co.*, 769 F.2d 842 (1st Cir. 1985) (trade secrets).

[348] IP Guidelines § 6.

[349] *CVD, Inc.* v. *Raytheon Co.*, 769 F.2d 342 (1st Cir. 1985). Finding of bad faith was based on findings that all the details of the purported trade secrets, other than those which were either obvious or insignificant, could be found in published government reports or in scientific journal articles that had been prepared by the trade secret owner's employees, and that the owner did not follow its own procedures for protecting the information.

[350] IP Guidelines § 5.7, citing *SCM Corp.* v. *Xerox Corp.*, 645 F.2d 1195, 1210 (2d Cir. 1981) (patents); *United States* v. *Columbia Pictures Corp.*, 189 F. Supp. 153, 183 (S.D.N.Y. 1960) (copyrights).

[351] Ibid.

contained as part of the 1992 Horizontal Merger Guidelines in order to evaluate the effects of the acquisition of intellectual property in affected technology, innovation, and goods markets.[352] In particular, the DOJ intends to take into account any integrative efficiencies that could not reasonably be achieved in the absence of the acquisition, and will also consider any anticompetitive effects resulting from the acquisition, such as the lessening of competition among existing technologies or goods or the lessening of competition to develop new technologies.[353]

The acquisition of intellectual property rights in the EC may raise concerns under Article 86 of the Treaty of Rome in cases where the acquisition amounts to an abuse of a dominant position. The ECJ considered this question in *Tetra Pak* v. *Commission*.[354] In that case, Tetra Pak acquired a United States company and, in so doing, acquired an exclusive license for a technique for manufacturing and filing sterile cartons. While the actual terms of the license brought it within the Patent Licensing Regulation, the Commission nonetheless issued a decision condemning Tetra Pak for its acquisition of the license on the basis that it constituted an abuse of a dominant position which infringed Article 86 of the Treaty. The ECJ upheld the decision, stating that the grant of an exemption under Article 85(3), whether individually or by virtue of the patent licensing block exemption, did not provide protection against the possible effects of Article 86.[355]

[352] See Example 10 of the IP Guidelines, where the DOJ stated that a horizontal relationship which amounted to an acquisition of intellectual property rights would be evaluated in the same way as any acquisition of the assets of a likely potential competitor, using the methodology described in the Merger Guidelines. In particular, the DOJ indicated in Example 10 that it would consider the impact of the acquisition on market concentration, other factors that might affect the likelihood that competition will be affected by the acquisition, and possible efficiency defenses.

[353] The degree to which competition will be lessened will also be considered. For example, in analyzing a license agreement in the context of an acquisition, the DOJ will take into account the fact that a license is non-exclusive, the exclusivity is temporary, or that the acquisition is otherwise structured to allow the parties freedom to compete independently in related products. IP Guidelines § 5.7.

[354] Case T-51/89 *Tetra Pak* v. *Commission* [1990] II ECR 309 [1991] 4 CMLR 334.

[355] Ibid.

12 Principles for Review of Joint Ventures under United States Antitrust and EC Competition Laws

I. Introduction

As described above, there appears to be little doubt that joint ventures and other forms of collaborative arrangements are often an extremely attractive strategy for businesses to reduce the costs of expanding current capacity, entering new markets, or developing new products.[1] For example, joint venturers may be able to realize transactional efficiencies, and share the costs and risks of the endeavor. In addition, joint research and development activities may lead to quicker results, or joint production may lead to reduced costs. In each case, the joint venture or other collaborative relationship actually has the effect of promoting competition and/or enhancing consumer welfare, particularly when the result is new competition by companies that may not have entered the market on their own, or an increase in the productive capacity in the marketplace.

However, while collaboration can, and often does, have procompetitive effects, joint ventures also raise a number of special concerns from an antitrust and competition law perspective.[2] For example, if the joint venturers would have entered the market on their own, without the formation of the joint venture, then the joint venture might serve to eliminate actual competition or potential competition that otherwise would have existed between them. There are also always concerns that the cooperation between the parties in the context of the joint venture may lead to "spill over" collusion or collateral restraints which are unrelated to the main purpose of the joint venture. These dangers are heightened when the joint venturers are traditional competitors. Finally, joint venture agreements often include various restrictions on the freedom of the parties to conduct their own research and development activities or to engage in other activities that might otherwise allow the party to gain a competitive advantage over the other party.[3]

As a general matter, antitrust and competition law regulators in the United States and the EC recognize that joint ventures and other collaborative arrangements have a number of advantages which can, and must, be balanced against any of the anticompetitive effects from the collaboration, and regulators and economists in both regions have acknowledged the importance of the competitive byproducts of collaborations.[4] This is

[1] See Chapter 5.

[2] See Chapter 5.

[3] The restrictions included in the agreements generally flow from the need of the parties to impose conditions which are designed to insure that each gets a fair share of the results, as well as the benefits of any exclusive rights which may be created during the joint work.

[4] See, e.g. Bellamy and Child, *Common Market Law of Competition* 4th edn., London (1993) § 5-002. The increasing competitive importance of the joint venture form in the EC in recent years can be traced by a review of the Commission's own annual reports on competition since the middle of the last decade. See, e.g.

particularly true when the collaboration occurs one or more steps (e.g. research and development) away from the market for the final products. For example, based on recent cases and statutory language, research and development and production joint ventures will be granted relatively relaxed treatment in the United States. On the other hand, marketing joint ventures are likely to raise antitrust problems, whether or not the joint venture is involved in research and development and production activities, because they can often fail to add any productive capacity to the marketplace and present horizontal competitors with an attractive vehicle to coordinate pricing and marketing decisions.

In this chapter, we review the general principles which guide antitrust and competition law review of joint ventures in the United States and the EC. While, as we have seen with respect to licensing arrangements, the procedural aspects of the review process are substantially different in the United States and the EC, one can also identify certain similarities. For example, a common focus of regulatory inquiry is whether or not there are any actual or potential anticompetitive effects from the creation and operation of the joint venture. In addition, regulators are concerned that the collaboration between the joint ventures may "spillover" into areas outside of the activities of the joint venture, thereby distorting competition in markets that are unrelated to the purpose of the joint venture. Finally, regulators will closely scrutinize the effect of collateral, or "ancillary," restraints imposed on the parties as part of the joint venture agreement.

II. Principles for Review of Joint Ventures under United States Antitrust Law

A. Introduction

As a general matter, a legitimate joint venture will be analyzed under the rule of reason in the United States. Accordingly, a preliminary issue is whether or not the arrangement is, in fact, a joint venture, or is merely some other form of business arrangement which the parties have labeled a joint venture, but which is actually a *per se* illegal arrangement between competitors to fix prices or allocate geographic markets.[5] In such situations, the courts and regulators will closely scrutinize the purpose and structure of the arrangement, and will deny rule of reason treatment to relationships which lack the necessary integration and risk sharing associated with joint ventures.

Assuming that the arrangement is a joint venture, a number of different methods for analyzing the relationship from an antitrust context have been put forward. For example, the Department of Justice ("DOJ") has stated that:

> "The antitrust inquiry into the legality of a particular joint venture generally involves three major issues. The first is whether the creation of the joint venture itself unreasonably restrains competition. The second is whether the joint venture has any unreasonable collateral restraints that must be struck down even if the venture is allowed. The third is whether the joint venture is in essence a 'bottleneck monopoly' which is so

cont.
the section generally entitled "The Development of Concentration, Competition and Competitiveness," which usually appears as Part Four of the Commission's Annual Reports on Competition Policy.

[5] *Timken Co.* v. *United States*, 341 U.S. 593 (1951).

> important to those in the business that it must be opened to all on reasonable and nondiscriminatory terms."[6]

Commentators have generally emphasized the same basic areas of inquiry, usually making it clear that the competitive impact of a joint venture should be analyzed not only in the market for the product(s) or service(s) covered by the joint venture, but also in markets outside of the scope of the joint venture in which the venturers are actual or potential competitors. Another factor that must be taken into account is the possibility that the joint venturers' cooperation in the activities of the joint venture will "spill-over" into areas unrelated to the joint venture, with a resulting adverse effect on competition in the unrelated areas.

B. Analysis of Actual and Potential Competitive Effects

Introduction As a general matter, concerns about the competitive effect of a proposed joint venture will be greatest in situations where both parents were already in the market competing horizontally, or would have entered independently but for the decision to establish a joint venture. This follows from the fact that a new competitor (i.e. the joint venture) would not be added to the marketplace as a result of the joint venture; rather, the joint venture arrangement would merely substitute a single competitor for what otherwise already was, or would most likely have been, two separate business units. In addition, that fact that both of the parents were already in the market independently or willing to attempt entry independently suggests that collaboration was not necessary in order to achieve some of the commonly cited benefits from joint venturing, such as overcoming barriers to entry or achieving levels of minimum efficient scale that would not be possible from independent action.

In these so-called "both-in" situations, defendants of the joint venture would appear to have a significant burden to overcome in justifying the arrangement. For example, in order for the joint venture to survive a challenge, a showing would need to be made that the efficiencies from the joint venture would outweigh the elimination of actual or potential competition between the parents, perhaps by facilitating entry or expansion at an earlier date than what would otherwise be possible if the parties had continued to act independently. In fact, at least one economist has argued, at least at some point, that a finding that such a joint venture arrangement was legitimate would be so rare that a *per se* rule of illegality might be justified in such situations.[7]

However, the law in this area has evolved in the direction of rule of reason analysis.[8] There appear to be a number of good reasons for this conclusion. First, the particular joint venture may be so limited in time and scope that anticompetitive effects are modest. Second, unlike a merger situation, the joint ventures do not disappear from market, but rather may remain in the market as competitors in other areas of their businesses. Third, the independence of the joint venture parents makes it possible to impose remedies at a later date in the event that the joint venture arrangement turns out to be anticompetitive,

[6] See the DOJ's 1988 Antitrust Guide for International Operations (hereinafter, the "International Guidelines").

[7] See, e.g. Pitofsky, "Joint Ventures Under the Antitrust Laws: Some Reflections on the Significance of Penn-Olin," *Harvard Law Review* 82 (1969), 1007, 1042–5.

[8] *Virginia Excelsior Mills* v. *FTC*, 256 F.2d 538 (4th Cir. 1958); see *United States* v. *Minnesota Mining & Mfg. Co.*, 92 F. Supp. 947 (D. Mass. 1950).

which means that a more lenient attitude might be taken at the outset of the arrangement when there is greater uncertainty regarding the potential impact of the collaboration. Finally, courts have tended to be more solicitous of business arrangements that achieve substantial efficiencies, and have thus been willing to consider and weigh procompetitive effects in such situations.[9]

On the other hand, where neither of the joint venturers are in the market or likely entrants into the market, it is difficult to see why the joint venture should be treated other than legal *per se*. Even in situations where substantial market power will result from the arrangement, it makes more sense to attempt to allow outsiders access to the benefits of membership, assuming they cannot organize a comparable joint venture on their own, than to prevent the organization of a joint venture which, by definition, adds a new competitor and productive capacity where none otherwise would have existed.[10]

The more difficult joint venture case involves the situation, treated by the Supreme Court in the famous *United States* v. *Penn-Olin Chemical Co.*[11] case, is the "one-in, one-out" situation where one parent is in the market or is a highly likely entrant and the other parent is not likely in the near future to enter on its own. In most cases, it is clear the contribution of the joint venturer currently on the sidelines may be substantial, and may generate efficiencies that would not otherwise be possible. However, opponents would argue that these efficiencies must be balanced against the loss of potential competition stemming from the fact that the nonentering parent's decision to join the joint venture means that it likely will never enter the market independently, nor will it continue to impact current market participants by the fact that it might be considered to be a potential entrant.

The *Penn-Olin* case In the *Penn-Olin* case,[12] the Supreme Court adopted a two-stage analysis of potential competition in looking at the legality of joint ventures that has been followed and refined in a number of subsequent cases. The fact situation at issue was the formation of a joint venture by Penn Salt Company and Olin-Mathieson Company to produce sodium chlorate, which is a white crystalline chemical used principally in bleaching pulp paper. Each company owned half of the stock of the new company, and each had an equal representation in the management. Penn Salt was to have the primary responsibility for operating the joint venture, while Olin would be responsible for the joint venture's sales activities.

The DOJ brought suit against the participants on the theory that the joint venture violated § 1 of the Sherman Act and § 7 of the Clayton Act. Specifically, the DOJ argued that (i) in the absence of the joint venture, both Penn Salt and Olin probably would have entered the relevant market; and (ii) even if both parties would not have entered, one of the two parent companies would have entered, leaving the other as a potential entrant. In either case, the DOJ believed that the consequence of the joint venture would be to

[9] See, e.g. *Yamaha Motor Co.* v. *FTC*, 657 F.2d 971, 977–80 (8th Cir. 1981) (*Yamaha-Brunswick*), *cert. denied*, 456 U.S. 915 (1982) (weighing possible procompetitive effects after finding Yamaha was a potential market entrant).

[10] *Associated Press* v. *United States*, 326 U.S. 1 (1945) (joint venture of newspapers to gather news worldwide not challenged, though other arrangements restricting new entrants found illegal); see also *United States* v. *FCC*, 652 F.2d 72 (D.C. Cir. 1980) (en banc) (*Satellite Business Systems*) (upholding FCC order requiring admission of third partner before approving communications joint venture).

[11] 378 U.S. 158 (1964).

[12] Ibid.

substantially lessen competition. The district court concluded that only if both participants would have entered the market would § 7 of the Clayton Act be violated.

The Supreme Court, however, said that the lower court had erred in not considering the possibility that one participant would have entered the market while the other remained on the sidelines as a potential competitor poised to enter the market, with the theory being that this possibility would create a more competitive atmosphere than would the joint venture. The Court remanded the case back to the lower court to determine whether (i) either company would have entered the market independently; and (ii) there would have been a lessening of competition by the formation and operation of the joint venture.

The Supreme Court's mandate to the district court in this particular case defines the two-stage analysis which must now be followed in assessing a joint venture. First, an attempt must be made to determine if either participant in the joint venture have entered the market by itself. If the answer is no, there is no need for further inquiry, because formation of the joint venture in that situation could not possibly foreclose or lessen the level of competition and, in fact, the joint venture is actually procompetitive in that it creates a new competitor where before there would not have been new competitors. If, however, one or both of the companies are found to be potential entrants into the market, then it is necessary to make the difficult assessment of whether the joint venture will result in a substantial lessening of competition.

The first stage of analysis mandated by *Penn-Olin* is known as the "potential competition doctrine," and actually focuses on whether or not there is a reasonable probability that either both parents would have individually entered the joint venture market but for the formation of the joint venture, or that one of the parents would have entered the venture market but for the joint venture and the other would have remained a significant potential competitor. In determining whether there is a reasonable probability of individual entry, consideration will be given to whether a parent had sufficient resources to enter the market,[13] the know-how and technical expertise to enter the market,[14] motivation to enter the market, including the likely profitability of entry,[15] and a long-sustained and strong interest in the market.[16] In addition, the ease of entry into the market will be considered.[17]

On the related issue of determining whether or not one of the joint venturers would have remained a significant potential competitor in the market in the absence of the joint venture, a number of case have identified a variety of factors that might be considered. For example, consideration should be given to the attractiveness of the market in which the joint venture is to be active,[18] the nearness of the alleged potential entrant to the joint venture's market,[19] the venturer's ownership or control of the resources required to enter the market,[20] and ease of entry into the joint venture's market.[21]

Assuming that one can demonstrate the requisite level of probability of individual entry, it is then necessary to analyze the actual and potential competitive effects of the new joint venture. This analysis requires a balancing of the relative procompetitive and anticompetitive effects of the transaction. If the venturers have previously been com-

[13] Ibid. at 175.
[14] Ibid.
[15] *United States* v. *FCC*, 652 F.2d 72, 100 (D.C. Cir. 1980).
[16] *United States* v. *Penn-Olin Chemical Co.*, 378 U.S. 158, 175.
[17] *United States* v. *FCC*, 652 F.2d 72, 100 (D.C. Cir. 1980).
[18] *Northern Natural Gas Co.* v. *Federal Power Commission*, 399 F.2d 953, 964 (D.C. Cir. 1968).
[19] *United States* v. *El Paso Natural Gas Co.*, 376 U.S. 651, 660 (1964).
[20] Ibid.

petitors or potential competitors in the same line of commerce, it may be necessary to evaluate a wide variety of factors. This requires an economic analysis of the companies involved, including their size and relationship, as well as consideration of the length of time, scope of activities and other elements of the venture. The scope of the inquiry is suggested by the *Penn-Olin* case itself, in which the Court indicated that the following considerations might need to be explored by the trial court:

> "the number and power of the competitors in the relevant market; the background of their growth; the power of the joint venturers; the relationship of their lines of commerce; the competition existing between them and the power of each in dealing with the competitors of the other; the setting in which the joint venture was created; the reasons and necessities for its existence; the joint venture's line of commerce and the relationship thereof to that of its parents; the adaptability of its line of commerce to noncompetitive practices; the potential power of the joint venture in the relevant market; an appraisal of what the competition in the relevant market would have been if one of the joint venturers had entered it alone instead of through [the joint venture]; the effect, in the event of this occurrence, of the other joint venturer's potential competition; and such other factors as might indicate potential risk to competition in the relevant market."

Penn-Olin itself did not illustrate how the various factors cited by the Court might be evaluated, given that the district court determined that neither of the companies would have entered the market on its own, and thus did not have to reach the second stage of the analysis. The district court appeared to rely almost completely on subjective factors in reaching its decision, notably the testimony of the presidents of the companies that they would not have entered the market absent the joint venture. On appeal, the DOJ argued that the testimony of the corporation as to its intentions was less valid than what it referred to as objective economic evidence, which presumably would include such things as the company's technical know-how to produce the product, its financial capability, and the incentive each company would have to enter the industry because of the profit margins. The Court split four-to-four on the question of subjective versus objective intent, left the lower court's decision standing, and failed to provide definitive guidance as to the type of evidence that might be used in determining whether one would have entered the market.

FTC* v. *Procter & Gamble *FTC* v. *Procter & Gamble*,[22] which was actually a merger case, supplemented the factors to be considered in assessing potential competition that were laid out in *Penn-Olin*. The case arose from the FTC's challenge of Procter & Gamble's proposed acquisition of Clorox. Procter and Clorox accounted for 65 percent of all sales of household cleaning products, although neither was an actual competitor at the time of the proposed acquisition. Procter produced household detergents, soaps, and cleansers, and Clorox manufactured household liquid bleach. The Supreme Court considered two issues. First, would the merger reduce future competition by increasing barriers to entry? Second, would Procter enter the bleach market if there were no merger? The Court concluded that the merger would likely lead to further concentration in an already concentrated market. The Court also noted that Procter's presence on the "edge" of the bleach market influenced the pricing decisions of the other bleach manufacturers

[21] *United States* v. *FCC*, 652 F.2d 101.
[22] 386 U.S. 568 (1967).

resulting in lower prices, and that Procter was the most likely independent entrant into that market. The Court concluded that Procter's subjective intentions on whether or not it would independently enter the bleach market were irrelevant, and relied on objective factors such as the market share and competitive behavior in the bleach market.

Procter & Gamble is significant in that it focused on the analysis on probabilities and requires some attempt to predict the merger's possible effect on competition rather than focusing on the actual effects. It also adds three important elements to the potential competition doctrine. First, a determination must be made as to whether into a new market through an acquisition, as Procter appeared to be doing, increase the barriers to entry by discouraging smaller firms from competing. Second, and this was apparently a key factor in the actual case, an analysis must be made to determine if the acquisition would eliminate a potential entrant which previously exerted a procompetitive influence on the relevant market. Finally, the result will, to some extent, depend upon the number of identifiable potential entrants, with the adverse impact of the acquisition on competition obviously being greater if there are only one or two potential entrants than if there are a number of potential competitors.

United States* v. *Falstaff Brewing Corporation The Supreme Court built on *Procter & Gamble* in Justice Marshall's concurring opinion in *United States* v. *Falstaff Brewing Corporation*.[23] In that case, the government sought to stop a merger between Falstaff and Naragansett Brewing Corporation, a small regional brewer. Falstaff had never marketed in the same region, and it was attempting through this merger to extend its market geographically for an existing product. The issue for the Court was whether Falstaff exerted any procompetitive influence on the market prior to the merger. Falstaff argued that it was not a potential entrant under the *Penn-Olin* rationale because it would not have been profitable for it to enter the market on its own, perhaps through internal expansion by increasing its manufacturing capability or distribution and sales outlets.

The Supreme Court rejected this attempt to limit the notion of potential competition, and Justice Marshall described three types of potential entrants which might lessen or eliminate competition thereby violating § 7 of the Clayton Act:

(i) A "dominant entrant" who has overpowering resources and who could drive out marginal companies and discourage other entrants is prohibited under § 7 from acquiring a company in the relevant market.

(ii) A "perceived potential entrant" may violate § 7 if it is perceived by present competitors as a potential competitor who is "waiting in the wings" resulting in lower prices and profit margins.

(iii) An "actual potential entrant" would be prohibited from entering an oligopolistic market if the acquiring company could be expected in the future to enter the relevant market *de novo* or through a "toehold" acquisition of a company with only a small share of the market.

The case was remanded to determine whether Falstaff had exerted a procompetitive influence before the merger by being "in the wings" or on the fringe of the market.

The *Falstaff* test admittedly raises significant issues regarding proof, since it is based on speculation regarding the future conduct and competitive impact of a company that is currently outside the market and intends to remain there. The doctrine has not been expressly sanctioned by the Supreme Court, and one circuit court noted the difficulties surrounding that speculative nature of proof as follows:

"Even if the likelihood of a firm's entry is a probability, as distinguished from an 'ephemeral possibility,' see Brown Shoe Co. v. United States, 370 U.S. 294 at 323, 82 S. Ct. 1502, at 1522, 8 L. Ed. 2d 510 (1962), its potential entry does not promote existing competition, since at most it may become a competitor in future. Moreover, entry by acquisition may have a procompetitive effect on the market, such as where the outsider acquires a failing firm and thus strengthens the acquired firm's market position."[24]

United States* v. *Marine Bancorporation In *United States* v. *Marine Bancorporation*,[25] the Supreme Court, while not validating the "actual" potential entrant doctrine, did set out two preconditions that must be present in order for it to be applicable. The case involved an attempt by Marine Banc, a nationally chartered bank in northwest Washington state, to enter the heavily concentrated eastern Washington state banking market by having one of its subsidiaries acquire a state-chartered bank in that area. While the government argued that the merger should not be allowed on the grounds that Marine Banc could have entered the market independently and that it exerted procompetitive effects on the eastern region by being on the fringe of the market, the Court concluded that § 7 of the Clayton Act was not violated because the government had failed to demonstrate that future competition would actually be lessened.

The Court explained that in order to find an "actual" potential entrant, a showing must be made that the acquiring company, Marine Banc in this case, could be expected to enter the market through "available feasible means," and that the alternative means would have ultimately produced deconcentration of that market or other procompetitive effects on the market. The Court also provided guidance as to the "perceived" potential entrant doctrine, which is concerned with the present effect that the acquiring outside company has on the present competition in the market. The Court explained that § 7 of the Clayton Act could be violated by potential entrance of a company into a highly concentrated (i.e. oligopolistic) market, provided that two elements are present: (i) the acquiring company is perceived by existing competitors as a potential independent entrant; and (ii) the acquiring company has had a "tempering impact" on the competitive conduct of the existing companies.

Other circuit court cases Several circuit court cases have attempted to clarify the doctrine of "potential competition" in order to deal with certain of the difficulties in actually applying the doctrine. In *FTC* v. *Atlantic Richfield Co.*,[26] the Fourth Circuit addressed the issue of the amount of proof required to show that a party to a proposed merger would have entered the applicable market independently, and concluded that there must be a clear showing, more than speculative market predictions, that the acquiring company would have entered the market *de novo* for it to be an actual potential entrant.

In *United States* v. *Siemens Corporation*,[27] the Second Circuit addressed the question of the standard of proof for showing whether an acquiring party in a merger was either an "actual" potential or a "perceived" potential entrant. The case involved Siemen's attempt to merge with Searle Diagnostics ("Searle"). Searle had been the pioneer in the

[23] 410 U.S. 526 (1973).

[24] See *United States* v. *Siemens Corp.*, 621 F.2d at 504 (2d Cir. 1980) (quoting *BOC International Ltd.*, 557 F.2d at 28.).

[25] 418 U.S. 602 (1974).

[26] 549 F.2d 289 (4th Cir. 1977).

[27] 621 F.2d 499 (2d Cir. 1980).

nuclear medical equipment market. The market had grown rapidly during the 1960s, but had leveled off due to increased government regulation of hospital costs and increased price competition. At the time of the merger, market share was concentrated among four companies, of which Searle was one. Searle was experiencing both loss of market share and loss of millions of dollars in profit. Siemens, which manufactures and sells X-ray equipment and computer tomography equipment throughout the world, had attempted several times to enter the nuclear medical equipment market by both de novo and a toehold acquisition. Apparently, it concluded that it was too late to become a new entrant, since it would not be able to recover its investment. Accordingly, it sought to enter the market through a merger with Searle.

The Antitrust Division challenged the merger because it would eliminate a potential competitor, Siemens. However, the Second Circuit, held that the government failed to prove that Siemens was either an "actual" potential or a "perceived" potential entrant. The court stated that the "actual" potential entrant doctrine required proof that (i) there was an oligopolistic market; and (ii) the acquiring company would have entered the market either de novo or through a toehold acquisition. The court noted that competitive behavior in a "relatively new, volatile and competitive" market could counter evidence of high concentration and high barriers to entry. "The bare market concentration ratio or percentages may not here 'accurately depict the economic characteristics' of the market."[28] The court went on to note that while Siemens had the financial resources to enter and the incentive to enter the nuclear medical equipment market in order to offer a full line of products, the court concluded that it would have entered the market only through acquisition and not *de novo*.

The court gave a number of reasons for concluding that acquisition was the only route available for Siemens to enter the market. First, there was a lack of evidence that Siemens had the technological competence to develop its own nuclear medical equipment or to apply its capability in other types of equipment to nuclear medical equipment. The court made it clear that where the development of an entirely new product is involved, the court must "scrutinize the technological feasibility of that expansion." Also, the court noted that there would be a long lead time in developing the technology, which was a strong factor that would likely inhibit new entry by Siemens or other firms. Second, the state of the market and limited profit prospects for new entrants militated against a de novo entry, given that the period of rapid growth had passed and existing competitors were losing money. Finally, there was no evidence of any toehold purchases that were available to Siemens.

The Second Circuit also rejected the notion that Siemens was a "perceived" potential entrant, explaining that "[a] claim of 'perceived' potential entry will not be upheld in the absence of evidence that present competitors have altered or tempered their conduct as a result of the acquiring firm's presence." The court stated that direct evidence was required, and that although there was testimony that some companies viewed Siemens as a likely entrant, this alone was not sufficient to show that the pricing or competitive conduct of those companies was influenced. Moreover, in a market that is fairly competitive, it is less likely that the existing companies would be influenced by the threat of a new entry on the fringe.

The government also failed to show that the elimination of Siemens as an influencing fringe company would make an appreciable competitive difference. This is usually

[28] Quoting *United States* v. *Black & Decker Mfg. Co.*, 430 F. Supp. 729, 748–55 (D. Md. 1976).

established where only one or two actual or perceived potential entrants are waiting. There was testimony from one of the existing competitors that there were at least six potential entrants. Therefore, the elimination of Siemens did not appear to pose a significant threat to competition.

The Second Circuit also opined that subjective testimony, if corroborated by objective evidence, may receive some weight. However, the testimony in the case was in the form of memoranda from lower echelon Siemens employees which indicated an intent to enter the market de novo. The court refused to give this much weight, explaining that "The opinions of lower level employees without management responsibility, absent some indication that the senior management has seriously considered and endorsed those views, simply do not constitute evidence of a corporate intent or commitment to enter a business de novo." Similarly, testimony from employees of existing companies was viewed with skepticism, especially in the situation where the proposed merger would likely infuse some competitive strength into a currently failing market participant (i.e. Searle).

General guidelines Since *Penn-Olin*, the courts have developed a variety of hurdles that challengers to a "one-in, out-out" joint venture must overcome. First, a showing must be made that the market in which the joint venture would operate is not competitive, so that an "in the wings" entrant could have some significant effects on levels of competition. This generally requires proof of high levels of concentration and of substantial barriers to entry. Second, the plaintiff must introduce credible evidence that the other competitors in the market to be served by the joint venture perceive the nonentering parent as a likely potential entrant. Next, the nonentering parent must be one of a few, probably two or three, most likely potential entrants so that the loss of in the wings potential competition was a substantial loss. Finally, the challengers must show that the "in the wings" entrant tempers anticompetitive conduct in the joint venture market.[29]

In *United States* v. *FCC*,[30] the D.C. Circuit found that a joint venture between IBM and Comsat to provide domestic satellite communications services would not lessen competition because it was unlikely, given the significant barriers to entry and limited potential profitability, that either firm would independently enter the market. In addition, neither firm was perceived by the present competitors as a potential competitor in light of the significant barriers to entry. On the other hand, a joint venture between Brunswick, the second largest seller of outboard motors in the United States, and Yamaha was found to have a substantial likelihood of lessening competition by eliminating the possibility of Yamaha entering the American outboard motor market where Yamaha had the feasible means to enter the market, had demonstrated a repeated interest in entering the market, and such entry would have a substantial procompetitive effect on a market dominated by four companies.[31]

C. Analysis of Collateral Restraints in Joint Ventures

The next legal hurdle to obtain clearance for a new joint venture involves a review of any restrictions, referred to as "collateral restraints" or "ancillary restraints," which may be placed on the activities of the participants, remembering that for antitrust purposes

[29] See *Tenneco, Inc.* v. *FTC*, 689 F.2d 346, 355 (2d Cir. 1982).
[30] 652 F.2d 72, 100–2.
[31] *Yamaha Motor Co., Ltd.* v. *Federal Trade Commission*, 657 F.2d 971 (8th Cir. 1981).

the joint venture itself is treated as an independent entity separate from its parents. A number of restrictions may be imposed on the parties in the course of negotiating the transaction relating to the businesses already conducted by the joint venture parties, as well as to the uses of the technology or products which might be developed and/or manufactured by the joint venture. For example, restrictions may be placed on pricing, territory and customers, and there may also be limitations on the kinds of products each company will make or sell.[32] Other common obligations include a duty to exchange the results of prior research in the area of the joint research effort, a duty to avoid disclosure to outside parties prior to securing patents, or a division of particular aspects of the research task among the participants. In the majority of cases, it is the actual or potential anticompetitive effect of these proposed restrictions, rather than the existence of the joint venture itself, which will cause the most antitrust difficulties.

Antitrust concerns with respect to collateral agreements arise because some of the agreement, if entered into outside of a lawful joint venture, would normally be *per se* illegal. Examples would include agreement which depress, stabilize or fix prices or allocate markets among the venturers. However, the courts have developed the "ancillary restraints" doctrine, which permits the joint venturers to enter into these "ancillary restraints," even if the restraints would have been unlawful in other contexts and the parties are or have been horizontal competitors, if: (i) the purpose of the underlying contract is lawful; (ii) the restriction on competition is closely related to the main objective of the contract; and (iii) the restriction on competition is reasonably limited in time and scope, taking into account the nature of the underlying contract.

Courts typically test ancillary restraints related in some manner to the venture's main objective under a "reasonable necessity" standard. In general, it is permissible to have some restrictions which are only incidental to the conduct of the joint venture, and the Sherman Act will be violated only if the restraints are found to be unreasonable. The first issue is to determine whether or not the restriction is ancillary to carrying out the lawful primary purpose of the venture. Restrictions which seem inappropriate-that is, not reasonably related-to the asserted efficiencies defendants claimed they were attempting to produce, will generally be successfully challenged.[33] Next, an analysis must be conducted to determine if the scope or duration of the restriction is only as broad as is necessary to support the primary purpose. Finally, an assessment should be made as to whether or not the restriction is otherwise reasonable under the circumstances, including a review of whether there is a less restrictive provision that would accomplish the desired result.

In the joint venture context, the ancillary restraints doctrine will frequently operate to justify agreements involving the venturers which would be questionable in other contexts, since a joint venture, by its very nature, requires a high degree of cooperation and non-competition between the ventures with respect to the activities of the venture in order for it to realize some of the efficiencies often attributed to the joint venture structure. For example, if the joint venture will develop new technology, the parties must reach an agreement about such matters as who will be entitled to exploit the technology and for what purposes, and whether outsiders may be licensed. Licensing restrictions of some

[32] The restrictions are often set out in the various collateral agreements that the joint venture will enter into with one or more of the venturers relating to licensing the technology developed by the joint venture or distributing the products which are manufactured by the joint venture.

[33] See, e.g. *Arizona* v. *Maricopa Medical Society*, 457 U.S. 332 (1982). To the same effect, see *United States* v. *Realty Multi-List, Inc.*, 629 F.2d 1351, 1381–87 (5th Cir. 1980) (membership criteria create restraints not justified by competitive needs).

type will usually be permitted; however, grant-backs and similar provisions which may result in pooling of patents among the joint venturers should be avoided. The doctrine even justifies "fixing" of the prices at which the products of the joint venture will be sold in some cases, if the joint venture was formed for the purpose of jointly marketing the product.[34]

One common example of the type of ancillary restraint which typically would be upheld arises when one party sells all of the assets of his business, including the "goodwill" associated with the business, to another party and the seller agreements not to compete with the buyer for a limited period of time in the area which was previously served by the business when it was owned by the seller. The type of non-competition agreement ordinarily would be upheld as necessary to protect the value of the assets and the associated goodwill to the buyer. However, since the doctrine is intended only to situations where the restraints are reasonably limited to the legitimate purposes of the underlying agreement, non-competition agreements which extend to lines of business other than those engaged in by seller prior to the transaction, or to geographic areas outside of where the business may have operated in the past, will generally not be permitted.

A non-competition agreement between joint venture partners was upheld in *Polk Bros., Inc.* v. *Forest City Enterprises, Inc.*[35] In that case, the joint venture parents, a dealer in large appliances and a dealer in building materials, constructed a single store to house both retailers, and agreed not to sell a restricted list of competing products. In upholding the non-competition agreement, the court found that the agreement made the retailers' cooperation possible and, thus, was ancillary to an output enhancing joint venture.

Another example of a legal ancillary restraint might be an agreement between parents, contributing valuable technology to a joint venture, that they would not try to expand their own operations or develop new markets alone or with other parents without giving the other parent an option to participate. If either parent declines to participate, the other would be free to proceed. Such arrangements would, outside of a lawful joint venture, normally be regarded as *per se* illegal attempts to divide markets. However, in the context of a joint venture, the parties might argue that neither was willing to contribute valuable technology without being assured that the other parent would not use that technology to develop markets outside the scope of the partnership.

The classic ancillary restraint case is *Timken Roller Bearing Co.* v. *United States*,[36] which involved the situation where three companies, US Timken, British Timken, and French Timken, entered into various agreements, including a joint venture. The agreements allocated trade territories, fixed prices on products sold by one company in another's territory, and contained other restrictions. The companies were charged under § 1 of the Sherman Act with cooperating to protect each other by eliminating competition and engaging in other cartel-like activities. While Timken tried to justify the restrictions as merely "ancillary" to the otherwise lawful joint venture, the Supreme Court disagreed, saying that agreements that permit "an aggregation of trade restraints such as those existing in this case are illegal agreements which suppress competition cannot be justified by calling them a joint venture."

[34] At one time, the Assistant Attorney General for the DOJ Antitrust Division stated that the Department "will not challenge joint venture restrictions ... if the restrictions are reasonably tailored to bring about significant efficiencies – such as large-scale economies in marketing."

[35] 776 F.2d 185.

[36] 341 U.S. 593 (1951).

United States v. *Sealy, Inc.*[37] involved a group of small mattress manufacturers, all of which were Sealy licensees, who adopted a common trademark and allocated exclusive territories among themselves to manufacture and sell mattresses under the Sealy trademark. The licensees, who accounted for about 20 percent of mattress sales in the United States, did not impose territorial restrictions on their sales of non-Sealy mattresses. The Supreme Court held the territorial allocation to be part of a horizontal scheme to fix prices and found it unlawful. However, if the Court had focused not so much on the form of the restriction, but rather on its effect, it might well have reached a different conclusion. In this case, it is likely that the allocation of territories provided an incentive for each licensee to promote and develop the Sealy trademark in its territory. Each licensee would be constrained from exercising market power within its territory both by non-Sealy manufacturers and other licensees selling mattresses but not under the Sealy label, which meant that there was only negligible harm to competition. Viewed in this way, the practice in Sealy is equivalent to a cost reduction (the trademark lowered the cost of providing information to consumers) that lowers price and increases output.

The ancillary restraints in *Citizens Publishing Co.* v. *United States*[38] were struck down as *per se* illegal in that they amounted to price-fixing, market control, and profit pooling provisions. In a private suit, *Association of Retail Travel Agents, Ltd.* v. *Air Transport Association of America*,[39] the court struck down a joint venture based on an ancillary agreements which involved setting and controlling the terms on which airlines deal with travel agents, imposed uniform payment schedules and bond-posting requirements, and required that persons buying a travel agency must assume its obligations in order to be accredited. While the defendant argued that the restrictions were necessary and contributed to the demonstrated efficiencies of the joint venture's operations, the court responded that the antitrust laws protect competition, not efficiencies. Finally, in *Engine Specialties, Inc.* v. *Bombardier Ltd.*,[40] which involved a joint venture between Bombardier, the world's largest manufacturer of snowmobiles, and Agrati, an Italian manufacturer of minibikes, to manufacture and market minibikes in the United States, the court found that restrictions in the joint venture which limited the ability of Bombardier to manufacture minibikes outside the United States and the ability of Agrati to sell or manufacture minibikes within the United States amounted to a *per se* violation of § 1 of the Sherman Act.

D. "Bottleneck" or "Essential Facilities" Concerns

In situations where participation in a joint venture will confer a "significant competitive advantage,"[41] and the venture itself hold a substantial position in the market, it may well be necessary for the courts to mandate that competitors must be allowed to participate in the joint venture or otherwise obtain the advantages of membership on reasonable and nondiscriminatory terms. This rule, sometimes referred to as the "bottleneck" or "essential facility" rule, originated with joint ventures that controlled an essential physical facility, such as a railway terminal, the concept has been much more broadly ap-

[37] 388 U.S. 350 (1967).

[38] 394 U.S. 131 (1969).

[39] 1987-1 Trade Cas. (CCH) § 67,449 (D.D.C. 1987).

[40] 605 F.2d 1 (1st Cir. 1979), *cert. denied*, 446 U.S. 983 (1980).

[41] A significant competitive advantage does not mean that membership in the joint venture is "indispensable," but only that lack of opportunity to participate would seriously diminish the company's ability to compete. Associated Press, 326 U.S. 1.

plied to cover other strategic assets, such as patents and other forms of intellectual property. In those cases where a "bottleneck" monopoly has been found, the remedy has been to require the parties not to refuse to deal with nonmembers, or to admit members to the group by extending patent licenses or access to patent pools at reasonable rates. It would appear that the rule is most likely to be applied in those situations where economies of scale are high and the venturers control a high market share.

Silver v. *New York Stock Exchange*[42] involved the refusal of the NYSE to provide two nonmember, over-the-counter securities brokers with access to its wire service. The Supreme Court held that the NYSE and its members had committed a *per se* violation of § 1 of the Sherman Act because it denied the plaintiffs access to a "valuable business service which they needed in order to compete effectively as broker-dealers in the over-the-counter securities market."[43] Similarly, the First Circuit has held that a group of wholesalers must rent space in a commercially unique building to competitors on a nondiscriminatory basis.[44]

On the other hand, if a joint venture does not possess market power or control exclusive access to an element essential to effective competition, any denial of access to a competitor will be judged under the rule of reason.[45]

One can certainly imagine a number of counterarguments against any requirement of equal access, particularly when the joint ventures have taken on significant risks to establish and complete the project. If the rule were carried to an extreme, the incentives for risk taking and innovative efforts will certainly diminish if advantages of the undertaking must be made available to nonparticipants. Certainly, the access requirement might be imposed to take the anticompetitive "edge" off what would otherwise be an illegal combination, thereby allowing what would other be an illegal joint venture to proceed and produce useful social and economic benefits. However, as a general rule, courts will apply the remedy sparingly, taking into account at least two important limitations.

The first condition to requiring equal access is a showing that there are no reasonable alternative sources for the joint venture advantages. For example, if "outsiders" can reasonably put together a joint venture of their own with comparable advantages, the original joint venture probably need not take in unwanted additional members.[46] In practice, it means that if the joint venture itself includes parents accounting for no more than 30 or 40 percent of a market, a comparable joint venture usually can be organized by the outsiders. Even where the original joint venture includes parents representing a higher proportion of the market, access may not be required if the minimum efficient scale

[42] 373 U.S. 341 (1963).

[43] Ibid. at 347.

[44] *Gamco, Inc.* v. *Providence Fruit & Produce Bldg., Inc.*, 194 F.2d 484 (1st Cir.), *cert. denied*, 344 U.S. 817 (1952). See also *United States* v. *Terminal R.R. Ass'n*, 224 U.S. 383 (1912).

[45] In *Northwest Wholesale Stationers, Inc.* v. *Pacific Stationery & Printing Co.*, 472 U.S. 284 (1984), the Court held that the *per se* rule should not be applied to a wholesale buying cooperative that expelled a member without procedural safeguards because the cooperative did not possess market power. The Court distinguished Silver on the ground that the N.Y. Stock Exchange held a "dominant position" in the market. The Court remanded the case for a rule of reason analysis. Cf. *United States* v. *Realty Multi-List, Inc.*, 629 F.2d 1351 (5th Cir. 1980) (rules restricting access to realty listing service with market power analyzed under the rule of reason).

[46] Antitrust Division, U.S. Department of Justice, Antitrust Guide Concerning Research Joint Ventures (1980); cf. *United States* v. *Terminal R.R. Ass'n*, 224 U.S. 383, 405–6 (1912) (finding of violation based on fact that physical conditions prevent nonowners of terminal facility from developing own terminal).

of the joint venture is low enough that comparable advantages can be achieved by a smaller undertaking.

The second condition that must be considered stems from the fact that the courts have not required "free" access, but only that membership in the joint venture must be made available on fair and nondiscriminatory terms.[47] It seems clear that if the access problem is solved by equity ownership in the joint venture, the latecomers can be required to pay their fair share of total investment. If access to the advantages of membership is more appropriate, reasonable royalties or fees can be charged. Moreover, if the early investment was a risky undertaking, a premium over the price of original investment should be justified and, if the assets of the joint venture have appreciated since its initial organization, latecomers should properly be required to pay a share of the appreciated price.

E. "Spill-Over" Effects of Joint Venture Collaboration

Another area of potential concern with respect to joint ventures, particularly when the venturers are actual or potential competitors in other markets, is that the collaboration between the venturers will "spill-over" into other areas and reduce the degree of competition which might otherwise exist between the venturers. The possibility of "spill-over" collusion was addressed in *United States* v. *Minnesota Mining & Mfg. Co.*[48] The case related to a relatively small export association which was organized in 1929 under the Webb–Pomerene Act of 1918. by nine domestic manufacturers of coated abrasives. At the time that the joint venture was formed, the economic activity of the venturers accounted for 86 percent of the total dollar value of United States exports of coated abrasives. By 1948, the share of the exporters not in the company rose to almost 40 percent, and the share of the export group declined, correspondingly, to 60 percent. While the court held the joint venture activity protected by Webb–Pomerene, it objected to two extensions of the export joint venture: the creation of joint manufacturing activities in foreign countries, and the cross-licensing of foreign patents required in order to execute the foreign joint manufacturing operations.

The defendants argued that the additional step of forming joint manufacturing subsidiaries abroad was necessary in order "to preserve and expand their foreign markets which were disappearing in the face of foreign countries' tariffs, quotas, import controls, dollar shortages, foreign exchange restrictions, local preference campaigns and like nationalistic measures."[49] However, the court objected to the fact that the "the intimate association of the principal American producers in day-to-day manufacturing operations"[50] might result in spill-over collusion in connection with their business activities in the United States. The court appeared to place form over substance, and ignored the fact that four of the nine original members of the export association had dropped out years

[47] See *Terminal R.R. Ass'n*, 224 U.S. at 411 (admission of nonmembers on terms providing a "plane of equality in respect of benefits and burdens" required); *Silver* v. *New York Stock Exch.*, 373 U.S. 341 (1963); *Gamco Inc.* v. *Providence Fruit & Produce Bldg.*, 194 F.2d 484 (1st Cir.), *cert. denied*, 344 U.S. 817 (1952).

[48] 92 F. Supp. 947 (D. Mass. 1950).

[49] Ibid. at 958.

[50] The Court stated that "[t]he intimate association of the principal American producers in day-to-day manufacturing operations, their exchange of patent licenses and industrial knowhow, and their common experience in marketing and fixing prices may inevitably reduce their zeal for competition inter se in the American market." Ibid. at 963.

before the suit was brought. The court also did not seriously consider that the uncertainties and political risks of the 1930s and 1940s made it prudent to seek to share the risks of joint manufacturing activity in France, Germany, and England, where the main overseas operations were taking place. Moreover, these jointly owned foreign subsidiaries also faced competition from the domestic producers in those foreign countries, accordingly it made sense to use local manufacturing as a means for reducing the costs of supplying the local markets.

The court may also have erroneously concluded that the collaboration in foreign markets necessarily amount to price-fixing and/or other anticompetitive collusion in the domestic market. There was no evidence to support a conclusion that there had been any diminution in competition in the domestic markets. In fact, in 1947, the abrasive products market consisted of 236 companies with a four-firm concentration ratio of 49 percent, and by 1954 there were 290 companies with a four-firm concentration ratio of 50 percent. Accordingly, domestic industry collusion, with or without the export agreement, was virtually impossible.

In determining whether there is a significant risk of "spill-over" collusion with respect to a specific joint venture, reference is often made to the governance structure agreed to by the parties. For example, in cases where the board of directors of a venture is comprised of an equal number of representatives from the joint venturers, there will be concerns regarding the ability of either of the joint venturers to block the actions of the joint venture entity and the increased ease by which the parent companies could collude.[51] Accordingly, it is generally recommended that the venturers seek independent members of the board of directors and independent viable support staff.[52] Also, joint venturers are encouraged to include requirements that certain types of competitively sensitive business information (e.g. pricing, cost allocations, market division) be aggregated or disclosed only to neutral third parties, as well as insuring that meetings which involve representatives of the venturers should be monitored by knowledgeable counsel and that accurate records of the discussions at such meetings be maintained.[53] In no case, obviously, should the venturers share information about their independent businesses which would have a tendency to reduce or eliminate competition between the parties.

III. Principles for Review of Joint Ventures under EC Competition Law

A. Introduction

The Commission has attempted to state its overall policy objectives with respect to the regulation of joint research and development activities. In its Thirteenth Report on Competition Policy, the Commission stated that it intended "to seek the best possible balance between on the one hand a reinforcement of the competitivity of European industry and on the other hand the maintenance of workable competition."[54] The Fifteenth Report on Competition Policy included the Commission's promise "to maintain work-

[51] See, e.g. *Northern Natural Gas Co.* v. *Federal Power Commission*, 399 F.2d 953, 967, 972 (D.C. Cir. 1968).

[52] If possible, the joint venture should also avoid interlocking directorates and shared officers or facilities.

[53] International Guidelines § 3.43.

[54] Thirteenth Report on Competition Policy (1983), point 42.

able competition and to ensure that the technical progress resulting from the research does not merely serve to produce monopoly profits."[55] The foregoing indicates that the Commission realizes collaboration on research and development need not be totally free from deleterious effects on competition in order to be beneficial, provided that the arrangement does not destroy workable competition in the industry.

In large part, the Commission has concluded that restrictions included in pro-competitive agreements which would not occurred in the absence of such restrictions generally will not be considered to be restrictive of competition. Also, if none of the parties are likely to have entered the market independently, the fact that they will not be able to compete with the joint venture does not mean that the arrangement is anticompetitive. However, the Commission has had much difficulty in balancing pro-competitive effects and potential restrictions on competition in other situations. For example, agreements which restrict some forms of competition in order to encourage other types of competition have generally fallen under Article 85(1), even though it is clear that the market has become more competitive as a result of the agreement.[56]

The case law in this area indicates the development of a list of potential anticompetitive effects against which is used to assess each particular joint venture. For example, one of the most direct anticompetitive effects of a joint venture is the diminution, if not outright elimination, of actual or potential competition between the various parties to the joint venture. In addition, concerns will arise that the collaboration in the joint venture area will ultimately "spillover" into areas outside of the joint venture, thereby adversely effecting competition in other markets. Also, joint ventures may foreclose possibilities for other parties to cooperate with the partners, a situation which is of particular concern when the industry is already concentrated. Finally, the Commission is concerned about what has been referred to as the "network effect," which essentially is a concern regarding the possibility that a party may seek to divide the single EC market by using two or more parallel joint ventures, all of which are under common control, but formed in order to act separately in particular geographic markets.[57]

The Commission has generally approached joint ventures by first finding that restrictions on conduct of the parties which have important result in the market constitute an infringement of Article 85(1), and then going on to determine whether or not it is appropriate to grant an exemption under Article 85(3).[58] This procedure has tended to increase the contracting risks to the parties, particularly since the Commission often imposes various conditions or obligations on the parties at the time that the exemption is proposed. In many situations, the Commission's view of particular joint ventures and contractual provisions is much clearer in its annual reports on competition policy then it is in any given case or series of cases.[59]

[55] Fifteenth Report on Competition Policy (1985), point 284.

[56] Korah, *R&D and the EEC Competition Rules: Regulation 418/85*, ESC Publishing, (1986), 11.

[57] See, e.g. Bellamy and Child, supra note 4, at § 5-003. See also ibid. at § 5-005 ("The Commission's principal underlying aims in relation to joint ventures are to ensure that the affected Community markets remain sufficiently competitive and are not compartmentalised."). In its Fifteenth Report on Competition Policy (1985), at point 26, the Commission noted that while joint ventures may produce a number of desirable competitive effects, they may also be anti-competitive to the extent that lead to "market sharing, raising of barriers to entry, and the intensification of market power."

[58] Korah, supra note 56, at 2.

[59] A further problem is that the Commission sometimes does not grant an actual exemption, but rather issues a so-called "Rovin" letter, which merely states that in its view the agreement merits an exemption, however, the Commission is closing its file without granting one. See Rovin, OJ 1983, C295/6 and 7; [1984] 1 CMLR 128 and 87. Such a letter raises the important implication that the arrangement does fall within Art.

B. Overview of the Analytic Framework

Assuming that the joint venture or cooperative arrangement falls under the general rubric of Article 85, as opposed to being a concentrative joint venture which is to be analyzed under the EC's merger controls, the Commission appears to engage in a two-stage analysis of the arrangement. First, the Commission will attempt to determine whether or not Article 85(1) is applicable to the joint venture itself and, if so, whether Article 85(3) is available to exempt the formation and actual operation of the joint venture. Thereafter, the Commission will examine each of the particular restrictions that the participants might include as part of the joint venture agreements, including, for example, shareholders' agreements or intellectual property licenses,[60] under Article 85(1) and, if necessary, Article 85(3).

If the research and development collaboration is limited to the stage prior to industrial application, it may fall completely outside of Article 85(1).[61] In those cases, however, where the collaboration extends into the area of joint exploitation of the results of the research (e.g. joint production, distribution or licensing), consideration will be given to the restrictions that the joint venture might have upon actual or potential competition between the parties and whether it is likely the collaboration will put third parties at a significant competitive disadvantage (i.e. whether the collaboration will appreciably foreclose others from participating in the market or from outlets and sources of the products covered by the arrangement).[62] Also, even if there is no appreciable adverse effect on competition and/or danger of any significant foreclosure of competition by third parties, Article 85(1) may nonetheless apply if it is believed that there will be a material effect on competition as a result of spillovers or the existence of a network effect.

As a general rule, there appears to be a "hierarchy of acceptability," with respect to the Commission's attitudes with respect to cooperation which extends downstream from what are referred to as "pure research and development joint ventures" to include joint production and then joint marketing. It would appear that joint research would generally be permissible even between partners having strong market positions.[63] However, the Commission tends to be hostile toward ventures extending to joint production, and even more

cont.
85(1), thereby making it difficult for any national court within the EC to independently enforce the agreement, without providing the parties with some degree of certainty from an exemption which can only be granted by the Commission itself. However, the Commission also has a process for a "short-form" exemption, which takes about a year following the date that the agreement is notified. See, e.g. *BP/Kellogg* OJ 1985, L369/6; [1986] 2 CMLR 619. Short-form exemptions require less translation. Parties may now choose between a "Rovin" letter, which can be sent in a few months, or a short-form exemption. In either case, the transaction can proceed much more quickly than when a full-blown exemption analysis is necessary.

[60] Bellamy and Child, supra note 4, at § 5-026.

[61] See, e.g. the discussion of the Commission's 1968 Notice on Co- operation Between Enterprises, OJ 1968 C75/3 as corrected OJ 1968 C93/14, below. However, as noted therein, the Notice is now of limited application to most collaborations.

[62] With regard to the so-called "foreclosure effect" in the area of research and development joint ventures, see, e.g. *BP/Kellogg*, OJ 1986 L369/6 [1986] 2 CMLR 619; *Continental/Michelin*, OJ 1988 L305/33 [1989] 4 CMLR 920; *KSB/Goulds/Lowara/ITT*, OJ 1991 L19/25 [1992] 5 CMLR 55; and *Elopak/Metal Box-Olin*, OJ 1990 L209/15 [1991] 4 CMLR 832. In the specialization area, see, e.g. *VW/MAN*, OJ 1983 L376/11 [1984] 1 CMLR 621 (exclusive dealing arrangements between the parties precluded component supply by sub-contractors).

[63] See *Continental/Michelin*, OJ 1988 L305/33 [1989] 4 CMLR 920, citing *Henkel/Colgate*, OJ 1972 L14/14) ("It is beyond dispute that joint research is permissible even between partners having strong market positions").

hostile to cooperation which runs all the way through to joint marketing.[64] Nonetheless, it seems that the Commission has become less inclined to view "R&D-to-selling" collaborations as being intrinsically objectionable,[65] and this change in approach has been evidenced not only in the case law,[66] but also in those provisions of the group exemption for research and development arrangement which allow joint selling in certain circumstances when the parties do not have market control.[67]

At least one commentator has suggested that the analytical approach which the Commission takes with respect to a production joint venture depends on the type of joint venture, the function which the joint venture is designed to achieve, and the character or "flavor" of the joint venture. The range of joint venture types is quite broad and extends from arrangements which are actually close to be "cartels," and which are greeted with hostility by the Commission, to "structural" or "full function" joint ventures that often are so close to full concentrations that they actually amount to a merger, and therefore are analyzed in terms of the impact of the joint venture on the structure and dynamics of the marketplace rather than the influence that the joint venture has on the market behavior of the parents.[68]

While production joint ventures are characterized as such because they have, as at least one of their immediate and ultimate functions, the production of goods or services, it is quite often the case, as with other joint ventures, that the activities of the enterprise also include upstream (e.g. research and development) and/or downstream activities, such as marketing and sales.[69] In fact, as the production joint venture extends the functions in which it is involved, it begins to approach the "structural" or "full function" which triggers merger-type concerns in the eyes of the Commission. In large part, function ultimately determines the type of joint venture for analytical purposes.[70]

[64] The Commission is not disposed to allow joint ventures dedicated to the joint distribution and sale of products which are unrelated to prior collaborative research activity. The Commission has equated these joint distribution and sales arrangements to "classic horizontal cartels ... [and] will therefore in principle assess sales JVs negatively." Commission's Joint Venture Notice at § 60.

[65] It is often possible to gain an exemption for collaboration with respect to sales when it is considered to be inseparable from, or necessary to attain the full benefits of, joint development and production previously conducted by the parties. Bellamy and Child, supra note 4, at § 5-051. Also, if the market share of the joint venture does not exceed 10 percent, the Commission will typically presume that the effects on competition created by the joint sales activity will fall within acceptable bounds. Ibid.

[66] See, e.g. *KSB/Goulds/Lowara/ITT*, OJ 1991 L19/25 [1992] 5 CMLR 55 (R&D and production JV between major competitors exempted for five years subject to reporting requirement); *Alcatel Espace/ANT*, OJ 1990 L32/19 [1991] 4 CMLR 208 (R&D, production and selling JV in telecommunications sector exempted for ten years); *Bayer/Hoechst*, XXth Report on Competition Policy (1990) (R&D, production and marketing JV for new AIDS drug dealt with by comfort letter). See also *Elopak/Metal Box-Olin*, OJ 1990 L209/15 [1991] 4 CMLR 832 and *Konsortium* ECR 900, OJ 1990 L228/31 [1992] 4 CMLR 54 (both R&D, production and sale JVs), which were negatively cleared, but involved non-competitors.

[67] See Art. 3a of Regulation 418/85, as amended by Regulation 151/93 (extending the group exemption for research and development agreements to cover joint agreements that cover selling in circumstances where the aggregate relevant market share of the parties does not exceed 10 percent in the EC or any substantial part of the EC). Regulation 418/85 is described in greater detail below.

[68] Bellamy and Child, supra note 4, at § 5-020. See also Faull, "Joint Ventures under the EEC Competition Rules," (1984) ECLR 358.

[69] Bellamy and Child, supra note 4, at § 5-021. However, in most cases, where the joint venture involves actual or potential competitors and extends to marketing and sales, the Commission will only grant its approval upon a demonstration of a special need for cooperation beyond just research and production, and will often require various agreements and undertakings from the parties to insure that the cooperation is contained. Ibid.

[70] Ibid. at § 5-022.

The character or "flavor" of a joint venture goes to its primary strategic goals and objectives, regardless of the functions which may be conducted by the enterprise. For example, a joint venture may have as its ultimate aim production rationalization, as might be the case when the participants are involved in an industry which is in the midst of overall restructuring.[71] In other cases, the production aspects of the joint venture may be secondary to the desire of the participants to fully develop a new product. Other production joint ventures are primarily designed for the purpose of transferring technology or combining resources in order to build new plants and thereby increase existing production capacity.[72]

Although the ECJ has not often been called upon to address the issue of whether or not a joint venture falls within Article 85(1), it has analyzed the need for protection from competition in other contexts, such as exclusive distribution agreements and licensing arrangements. In such cases, the ECJ has taken into account the level of investment made by the parties and the practices in the particular industry and has held that the grant of exclusive rights designed to afford protection from other dealers, licensees, or the licensor itself do not infringe the competition rules if the necessary investment would not have made in original innovation or by the dealer or licensee without such protections.[73] In one case, the burden of proof with regard to demonstrating that competition was restricted rather than enhanced by the arrangement was placed on the plaintiff.[74]

In general, the period of the exemption will vary depending upon the type of arrangement[75] and the particular industry involved; however, in most cases, a production joint venture will receive an exemption which is long enough for the parties to have a reasonable opportunity to achieve the anticipated benefits from the specific collaboration.[76] For example, with respect to one production joint venture, the Commission set the exemption period for twelve years, reasoning that this was the amount of time necessary for each of the parties to be able "to rely on the enforceability of the agreements and to obtain a satisfactory return on their capital."[77] In another case, the Commission granted a renewal for twenty years after an initial exemption period of just nine years on the basis of evidence that the degree of competition in the industry made it more difficult for the joint venture to achieve its objectives in the shorter period.[78] In other cases, the exemption period has been set by reference to the life cycle of the joint venture product[79] and the amount of time required in order for commercial production and sales to commence.[80] However, the Commission will not hesitate to set an exemption period which

[71] See, e.g. *Enichem/ICI*, OJ 1988 L50/18 [1989] 4 CMLR 54; *Bayer/BP Chemicals*, OJ 1988 L150/35 [1989] 4 CMLR 24; and *EMC/DSM*, Eighteenth Report, point 58 (all pertaining to restructuring of the EC's thermoplastics and petrochemicals industries).

[72] See, e.g. *Rockwell/Iveco*, OJ 1983 L224/19 [1983] 3 CMLR 709.

[73] See, e.g. *La Technique Minière* v. *Maschinenbau Ulm* [1966] E.C.R. 235, [1966] CMLR (exclusive distribution agreement); *Nungesser* v. *Commission* [1982] ECR 2015, [1983] 1 CMLR 278 (licensing); and *Coditel (2)*, [1982] ECR. 3381, [1983] 1 CMLR 49 (copyright licensing).

[74] *Coditel (2)*, [1982] ECR 3381, [1983] 1 CMLR 49.

[75] See, e.g. *Eirpage*, OJ 1991 L306/22 [1991] 4 CMLR 233, § 23 (service joint ventures might justify a longer exemption period than production joint ventures based on the need for continuity in the relationship between the joint venture parents and third parties).

[76] See Sixth Report on Competition Policy (1976), point 59.

[77] *Olivetti/Canon*, OJ 1988 L52/51, § 60.

[78] *De Laval/Stork*, OJ 1977 L215/11 [1977] 2 CMLR D69 and OJ 1988 L59/32 [1988] 4 CMLR 714.

[79] See, e.g. *Ford/Volkswagen*, OJ 1993 L20/14 (ten years equated to life cycle of the joint venture product).

[80] See, e.g. *GEC/Weir*, OJ 1977 L327/26 [1978] 1 CMLR D42 (exemption granted for twelve years and commercial production and sale only likely in last three and a half years of such period).

is shorter than the contractual term of the joint venture if it believes a longer term is unwarranted in the situation.[81]

C. Application of Article 85(1) to Cooperative Arrangements

Introduction In general, Article 85(1) will apply to a joint venture in those situations in which the existence of the joint venture itself results in an appreciable restriction or distortion of competition in the whole or any substantial part of the EC which may affect trade between the member states. There are a number of key issues which the Commission must consider in this regard:

(i) Does the joint venture restrict actual or potential competition between the parents?
(ii) Does the joint venture restrict actual or potential competition between any of the parents and the joint venture itself?[82]
(iii) Does the joint venture result in the foreclosure of competitive opportunities for any third party?
(iv) Does the joint venture carry a risk of "spillover"?
(v) Is the joint venture part of a network of parallel interlocking joint ventures?
(vi) Are the competitive effects of the joint venture "appreciable" in a particular market?
(vii) Does the joint venture result in the isolation of parts of the common market?[83]

Competition between joint venture parents The first consideration in analyzing a joint venture is whether or not the parties are actual or potential competitors, an issue which generally is a question of fact. The key factors which the Commission believes might be considered in making an evaluation in this area have been enumerated as follows.

Contributions to the joint venture Does each parent company have sufficient financial resources to carry out the planned investment? Does each parent company have sufficient managerial qualifications to run the joint venture? Does each parent company have access to the necessary input products?

Production of the joint venture Does each parent know the production technique? Does each parent make the upstream or downstream products himself and does it have access to the necessary production facilities?

Sales by the joint venture Is actual or potential demand such as to enable each parent company to manufacture the product on its own? Does each parent company have access to the distribution channels needed to sell the product manufactured by the joint venture?

[81] See, e.g. *KSB/Goulds/Lowara/ITT*, OJ 1991 L19/25 [1992] 5 CMLR 55.
[82] In some cases, the first two issues are considered together (i.e. does the joint venture restrict actual or potential competition among the parties, including the new joint venture).
[83] This listed factors are derived, in part, from Bellamy and Child, supra note 4, at § 5-027.

Risk factors Can each parent company on its own bear the technical and financial risks associated with the production operations of the joint venture?

Access to the relevant market What is the relevant geographic and product market? What are the barriers to entry into that market? Is each parent company capable of entering that market on its own? Can each parent overcome existing barriers within a reasonable time and without undue effort or cost?[84]

The aforementioned factors have been neatly combined into the following fundamental test with regard to competition: whether the parties themselves have, or could have gain access to, all of the financial, technical and commercial resources which may be required in order for them to independently enter the joint venture's market within a reasonable timetable and without undertaking a level of technical and/or financial risk which is unreasonable for the party in question.[85] In general, the Commission has focused on a "medium term" view of what constitutes a reasonable timetable for any party to a joint venture to independently commercialize the products to be produced by the joint venture;[86] however, in appropriate cases, a shorter timeframe has been used by the Commission.[87]

When the joint venture is primarily formed for the purpose of conducting research and development, the threshold inquiry is whether there would have been a reasonable likelihood that, in the absence of the joint venture, that the joint venture parents would have independently engaged in the particular research and development project.[88] As a general rule, the likelihood of independent research and development work is a function of the degree of activity that the parents have in the same or adjacent markets (i.e. are the parents actual or potential competitors downstream of the development activities). However, the Commission may also consider other factors which might make it difficult for any one parent to independently conduct the development program, such as the development timetable imposed under the terms of any contract which the proposed joint venture intended to contest, the scale of the expenditures involved in conducting the development program (including the risk that the contract actually awarded would not be sufficiently large to allow the parties to fully amortize their research expenditures), and the difficulties involved in assembling a pool of qualified engineers and other personnel to finish the project.[89]

[84] Commission's Joint Venture Notice, §§ 18–20. See also the Commission's Thirteenth Report on Competition Policy (1983), point 55, which listed four of the factors referred to in the Joint Venture Notice (i.e. inputs, production, sales, and risk factors), and went on to state: "In applying these objective criteria the Commission will be guided by what in its opinion could reasonably be expected of the participating firms in the light of all relevant economic factors. Clearly, the weight given to the various criteria in assessing potential competition will vary from case to case."

[85] Bellamy and Child, supra note 4, at § 5-033.

[86] See Sixth Report on Competition Policy (1976), point 57.

[87] See, e.g. *Konsortium* ECR 900, OJ 1990 L228/31 [1992] 4 CMLR 54 (short-term view required due to the fact that the joint venture was formed in order to bid for tenders which were due under a very tight timetable).

[88] See, e.g. Case 56/65 *Technique Minière* v. *Maschinenbau Ulm* [1966] ECR 235; Case 31/80 *L'Oreal* v. *De Nieuwe* [1980] ECR 3775.

[89] See, e.g. *Konsortium* ECR 900, OJ 1990 L228/31 [1992] 4 CMLR 54 (holding that based on the factors referred to in the text that none of the parties could have been expected to carry out the development work on their own and therefore they could not be considered to be potential competitors). It is not necessary, however, for the parties to be considered "non-competitors" in order for the Commission to allow the cooperation. See, e.g. *KSB/Goulds/Lowara/ITT*, OJ 1991 L19/25 [1992] 5 CMLR 55, where the

The Commission has gone on record as realizing that its analysis of the competitive effects of a prospective joint venture should be made "in the most realistic way possible,"[90] and that it should attempt to evaluate whether each parent alone would have the ability to fulfill the functional activities delegated to the joint venture and whether or not the parent effectively restricts, or even forfeits, it ability to conduct such activities on its own by participating in the joint venture.[91] However, in general, the Commission has taken an expansive view with respect to finding that the parties to a joint venture are at least potential competitors so as to bring the joint venture within the scope of Article 85(1),[92] and, in the view of at least one commentator, it has continued to find potential competition in a number of situations based on somewhat unsatisfactory analysis.[93]

In one case, the fact that the parties both had the necessary technology was sufficient to make them potential competitors for purposes of Article 85(1).[94] In another situation, parties which had never manufactured the joint venture products were nonetheless found to be competitors because their resources and expertise might have led them to extend their range of products to include the joint venture products, even though there was evidence to the effect that, but for the joint venture, one party would have withdrawn from the market and the product probably never would have been developed.[95] Potential for competition was also found when one party failed to have any of the technology contributed to the joint venture by the other due to the fact that, in the opinion of the Commission, the non-technology party could have acquired the technology and made all the investments required to compete with the party that had contributed the technology to the joint venture.[96] Finally, potential competition has even been based on the fact that the parties agreed to enter into non-competition covenants as part of the arrangement.[97]

The issue of whether a research and development joint venture might be anti-competitive is typically of the greatest importance when the parties are competitors. However, even when the joint venture is between non-competitors, Article 85(1) will nonetheless apply if it entails the likelihood of an appreciable restriction or distortion of competition.[98] For example, Article 85(1) may apply to a fairly common type of research

cont.
Commission concluded that although all four parties to the joint venture were competitors, the cooperation was indispensable. The Commission relied, in part, on assurances from two of the parties that they together would not have pursued the research project.

90 Thirteenth Report on Competition Policy (1983), point 55.

91 See, e.g. Bellamy and Child, supra note 4, at § 5-032.

92 Ibid. at § 5-031.

93 Ibid. at § 5-034. This criticism has been particularly pointed in those cases where the parties are already are involved in the same industry, since the Commission often takes the position that the parties are *ipso facto* potential competitors in those situations. See, e.g. *Clima-Chappee*, OJ 1969 L195/1 [1970] CMLR (sufficient financial and technical resources to develop competing products); and *VW/MAN*, OJ 1983 L376/11 [1984] 1 CMLR 621 (both of the parties potentially capable of expanding into the range covered by the agreement). However, even the critics have noted that there have been some recent decisions by the Commission which appear to include "a more robust, economic analysis" on the question of potential competition. Ibid.

94 *KEWA*, OJ 1976 L51/15 [1976] 2 CMLR D15.

95 *Vacuum Interrupters (No. 1)*, OJ 1977 L48/32 [1977] 1 CMLR D67. See also *Sopelem/Vickers (No. 1)*, OJ 1978 L70/47 [1978] 2 CMLR 146; *Beecham/Parke Davis*, OJ 1979 L70/11 [1979] 2 CMLR 157.

96 *Cekacan*, OJ 1990 L299/64 [1992] 4 CMLR 406.

97 See, e.g. *Iveco/Ford*, OJ 1988 L230/39 [1989] 4 CMLR 40. But cf. e.g. *Konsortium* ECR 900, OJ 1990 L228/31 [1992] 4 CMLR 54 (parents found not to be potential competitors regardless of the fact that they entered into non- competition agreements).

98 See Commission's JV Notice, § 33.

and development arrangement under which to previously non-competing parties agree to cooperate with respect to further development of a product created by one of the parties which might have other applications in a field which is of interest to the other party, since the collaboration might have the effect of foreclosing third parties from entering the field or may result in an anti-competitive spillover or network effect.[99]

Competitive analysis becomes even more difficult if one assumes that there are actually two markets in which firms might compete: the market for innovation, which involves the creation of what is essentially a new solution to a particular need of the end user, and the market for the products to be improved or replaced by the results of the development work.[100] White has argued that in those cases where the parties have complimentary skills and resources they are unlikely to restrict competition in the marketplace for innovation; however, if the joint venture also includes marketing and distribution activities, significant restrictions on competition may arise as the new products then begin to compete with the products to be improved or replaced.[101] While this may be true in many cases, it may not be true in those situations where the innovation would not have taken place but for the expectation of joint marketing.

As to the market for the products which are to be improved or replaced, it is sometimes argued that if the parties already are involved in competition in that market, they may be reluctant to exploit the innovations arising out of the development work. Also, the parties making the existing products are the ones most likely to compete with the innovative products. In summary, since research and development may well be one of the few forms of remaining competition in oligopolistic markets, collaboration between firms at that stage may well diminish a significant element of competition.[102]

If it is found that the parties are competitors either in the market for products or innovation, then there are a number of other factors which might be considered.[103] First, the combined market share of the parties should be taken into account, with the assumption being that joint ventures involving firms which have relatively small market shares will typically not have any significant impact on competition. Second, barriers to entry may well be an important factor, particularly in those cases where the market is more concentrated. It is interesting to note that the Commission has expanded the notion of barriers to entry beyond that which may be used in the United States to include advantages, such as plants[104] or commercial networks,[105] which new firms could acquire simply be making the same expenditures as the incumbent firms. This approach has been criticized since they do not confer upon the holder any power to engage in monopolistic pricing activities, nor do they present any material difficulties for firms interested in participating in the long-term market for innovation. Finally, few concerns regarding re-

[99] See, e.g. *EMI/Jungheinrich*, Seventh Report on Competition Policy (1977) points 119–20 [1978] 1 CMLR 398; *BP/Kellogg*, OJ 1985 L369/6 [1986] 2 CMLR 619. See also *ACEC/Berliet*, OJ 1968 L201/7 [1968] CMLR D35 (agreement between supplier of transmissions and manufacturer of buses to develop a transmission suitable for a bus with reciprocal exclusive dealing obligations accepted between the parties).

[100] See, e.g. White, *Research and Development Joint Ventures under EEC Competition Law* (1985) 16 IIC 663, 672.

[101] Ibid. at 684.

[102] Korah, supra note 56, at 6.

[103] Ibid.

[104] See, e.g. *Michelin* v. *Commission* (322/81) [1983] ECR 3461; [1985] 1 CMLR 282 (although the Court noted that the cross elasticity of supply to the existence of a dominant position was a relevant factor in assessing the market, it ignored the fact that a factory for producing heavy tires could actually be built by competitors on the grounds that the construction would take time).

[105] *Hoffman-La Roche* v. *Commission (Vitamins)* (85/76) [1979] ECR 461; [1979] 3 CMLR 211.

strictions on competition arise when the research is basic, not closely related to any particular product, and the results can be obtained on reasonable terms by any firm in the market, even in those cases where the collaboration involves virtually all the firms in the industry.[106]

Competition between the joint venture and the parents Even if the parties themselves are not actual or potential competitors in the field of joint venture operations, a separate issue may arise as to whether the existence of the joint venture restricts competition between the joint venture and any of the parties. As a general matter, the joint venture will include specific contractual restrictions on the ability of the parties to engage in the field of the joint venture during the term of the joint venture. Moreover, there need not be a specific covenant or restriction in order for the Commission to find that the formation of the joint venture, in and of itself, is sufficient to conclude that the parties will no longer be actual or potential competitors with the joint venture in the joint venture's field of operations.[107] However, this "presumption" can be rebutted by a showing that under the circumstances the parties to a joint venture will, in spite of their interest in the joint venture, continue to compete with the joint venture.[108]

Concerns regarding restrictions on the potential level of competition may exist in those situations where the joint venture is operating in the same market as one or more of the parties, or if the joint venture is operating in upstream, downstream or adjacent markets. As a general matter, it would seem that any restriction on competition in the joint venture agreement (e.g. a non-competition agreement from the party which is the potential competitor) might be characterized as either an exempt ancillary restriction or as an indispensable restriction.[109] However, if the potential coordination creates a danger of market-sharing or price collusion, then the Commission will likely takes steps to insure that the joint venture is insulated from contact with, and potential pressure from, its parents.[110]

Another potential competitive problem in this area stems from the fact that the parties, by entering into the joint venture, are effectively surrendering their ability to make independent decisions with regard to the field in which the joint venture is operating. For example, in most cases, decisions regarding such things as investments, production levels, and pricing with respect to the joint venture will generally require the joint consent of the parties. While the earlier Commission decisions characterized these require-

[106] See *Eurogypsum* OJ 1968, L57/9; [1968] CMLR D1.

[107] See, e.g. *GEC/Weir*, OJ 1977 L327/26 [1978] 1 CMLR D42 ("Even in the absence of express provisions, the creation of a joint venture generally has a notable effect on the conduct of parent parties who have a significant holding in the joint venture. Within the field of the joint venture and in related fields such parties are likely to coordinate their conduct and be influenced in what would otherwise have been their independent decisions and activities. Where the parent parties are actual or potential competitors their participation in a joint venture is accordingly likely to impair free competition between them, regardless of explicit restrictive provisions to that effect.")

[108] See, e.g. *Bayer/BP Chemicals*, OJ 1988 L150/35 [1989] 4 CMLR 24 (finding that joint venture did compete with 50 percent parent); and Cases 142 and 156/84 *BAT* and *Reynolds* v. *Commission* [1987] ECR 4487 (ECJ upholding the Commission's findings that a 24.9 percent interest in a competitor did not necessarily create any restriction on competition between the two companies).

[109] See, e.g. *Elopak/Metal Box-Odin*, OJ 1990 L209/15; and *Mitchell Cotts/Sofiltra* OJ 1987 L41/31 [1988] 4 CMLR 111, and the discussion below of "ancillary" and "indispensable" restrictions.

[110] See, e.g. *Montedison/Hercules-Himont*, Seventeenth Report on Competition Policy (1987) and *duPont/Merck* IP(91) 381.

ments as restrictions on competition in the joint venture's end-markets,[111] it has been suggested that provisions of this type are better viewed merely as a means for exercising the type of joint control over the joint venture which is implicit in its structure, and that any competitive restrictions arise primarily from the formation of the joint venture itself and not from the fact that the parties must act in concert in order to manage its operations.[112]

Foreclosure of third parties The "foreclosure" effect of a joint venture refers to the possibility that the formation of the joint venture may alter the structure of the market to such an extent that it becomes much more difficult for third parties (i.e. firms not parties to the joint venture) to continue to compete, particularly in those situations where the market was highly concentrated before the joint venture was formed.[113] For example, it is likely that the parties to a joint venture will become less willing to cooperate with third parties in the same field.[114] Also, the combined resources of the parties in the joint venture may give it a competitive advantage[115] or economies of scale[116] which cannot be overcome by competitors, either acting alone or together.[117] In addition, certain restrictions included as part of the joint venture arrangement, such as the grant of exclusive intellectual property licenses to the joint venture by the parties, will compound the difficulties which confront potential competitors in entering the market.[118]

At least one commentator has argued that the "foreclosure" effect should be analyzed in a manner which takes into account the procompetitive effects of the joint venture. For example, in those cases where existing assets are acquired, such as when the parties acquire all the existing patents which protect the known methods for satisfying a particular demand, then it would seem that the joint venture would cause concerns, since outsiders are foreclosed from the existing market and the potential sources of supply are substantially reduced. However, with regard to the products of the joint venture activity, it might be argued that the joint venture does not foreclose competition, assuming that the products would not have been created without the joint venture, since one source of the new product is better and far more efficient than having no sources at all. In many ways, foreclosure is actually another version of inherent restrictions on competition and

[111] See, e.g. *De Laval/Stork (No. 1)*, OJ 1977 L215/11 [1977] 2 CMLR D69; *Rockwell/Iveco*, OJ 1983 L224/19 [1983] 3 CMLR 709; *Alcatel Espace/ANT*, OJ 1990 L32/19 [1991] 4 CMLR 208.

[112] Bellamy and Child, supra note 4, at § 5-038 n64.

[113] See, with respect to "foreclosure," the Commission's Joint Venture Notice, §§ 23–5. The existence and degree of "foreclosure" is a question of fact which needs to be decided in each case, although the Commission appears to be attempting to develop some general guidelines as it struggles with the concept of articulating what is an "appreciable effect" for purposes of invoking Art. 85(1). Bellamy and Child, supra note 4, at § 5-040.

[114] See, e.g. *GEC/Weir*, OJ 1977 L327/26 [1978] 1 CMLR D42.

[115] See, e.g. *Eurosport*, OJ 1991 L63/32 [1991] 4 CMLR 228 (joint venture was to have substantial advantages with respect to access to broadcast rights and bank of sports programming).

[116] See, e.g. *KSB/Goulds/Lowara/ITT*, OJ 1991 L19/25 [1992] 5 CMLR 55.

[117] The Commission has even recognized that potential competitors may actually be "psychologically" deterred from competing with a joint venture involving parties with substantial competitive resources. See, e.g. *Eirpage*, OJ 1991 L306/22 [1991] 4 CMLR 233, § 12 ("the fact that potential competitors are faced by a [joint venture] between the national telecommunications company and a subsidiary of one of the world leaders in mobile telecommunications ... may have a deterrent effect on potential market entrants and thus further restrict competition").

[118] See, e.g. *Elopak/Metal Box-Odin*, OJ 1990 L209/15; *Eurosport*, OJ 1991 L63/32 [1991] 4 CMLR 228; and *Astra*, OJ 1993 L20/23. Also, in *Optical Fibres*, OJ 1986 L236/30, the Commission noted the absence of foreclosure effects due to the fact that the joint venture would be selling to third parties on terms which were nondiscriminatory and could obtain supplies from third parties.

is only likely to occur if more than one of the parties could have produced the products independently.[119]

"Spillover" effects The "spillover" effect[120] refers to the possibility that the collaboration between the parties as part of the joint venture may diminish the competitive which might have otherwise existed between the parties in other areas. While "spillover" is a danger whether or not the parties are actual or potential competitions in the areas in which the joint venture will be operating, the risk of "spillover" is particularly high in those cases where the parties remain active in the same market as the joint venture or are competitors in the markets which are upstream, downstream, or adjacent to the joint venture's market. It is not necessary that the parties actively cooperate with each other or co-ordinate their activities; rather, "spillover" can occur whenever there appears to be a lessening of competitive tension between the parties.[121] However, theoretically the potential spillover must be capable of having an "appreciable" effect on competition in order for Article 85(1) to come into play.[122]

In one case, the supply of intermediate products by the joint venture to its parents created a risk that competition between the parents in the downstream supply of finished products utilizing the intermediate products would be restricted.[123] This is a particular danger in those situations where the intermediate products are the key cost item and key factor in the purchasing decisions of the end users.[124] When the competition between the parties is only outside of the joint venture field, the cases have focused on whether the cooperation between the parties within the joint venture would create other "opportunities or strong inducements" for anticompetitive activities in these "overlapping markets."[125] For example, the Commission has previously expressed concerns regarding "spillover" when the personnel of the parties assigned to the joint venture retained an interest in other aspects of the activities of the parties, some of which were overlapping[126], and has sometimes required additional disclosures of interlocking shareholdings or cross-directorships in situations where the joint venture products might be used in other markets[127]. In extreme cases, the parties may actually be required to abandon certain downstream activities as a condition for approval if the Commission believes that the danger of "spillover" is too high.[128]

[119] Korah, supra note 56, at 8.

[120] The "spillover" effect is sometimes referred to as the "group" effect.

[121] Bellamy and Child, supra note 4, at § 5-041.

[122] See, e.g. *GEC/Siemens-Plessey*, OJ 1990 C239/2. See also the original Art. 19(3) Notice in *P&G/Finaf*, OJ 1992 C3/2, IP(92)496 (the "spillover" effect was quantified by reference to the market shares of the parties in certain "adjacent" markets).

[123] See *Vacuum Interrupters (No. 1)*, OJ 1977 L48/32 [1977] 1 CMLR D67.

[124] *KSB/Goulds/Lowara/ITT*, OJ 1991 L19/25 [1992] 5 CMLR 55. See also *Ford/Volkswagen*, OJ 1993 L20/14 (joint venture to develop and make vehicles to which parents would add options/design features and market separately under their respective trade marks).

[125] See, e.g. *WANO Schwarzpulver*, OJ 1978 L322/26 [1979] 1 CMLR 403 (joint venture might give "opportunities and strong inducements" for market sharing between parties in products of mutual interest which fell outside immediate scope of the joint venture).

[126] *GEC/Weir*, OJ 1977 L327/26 [1978] 1 CMLR D42 (parties competed in other product markets; senior staff seconded to the joint venture retained interests in other aspects of the business activities of the parties).

[127] See, e.g. *Bayer/Gist-Brocades*, OJ 1976 L30/13 [1976] 1 CMLR D98 (reporting required in relation to any products capable of being manufactured from raw penicillin).

[128] See, e.g. *Montedison/Hercules (Himont)*, Seventeenth Report on Competition Policy (1987), point 69.

The Commission's concerns regarding "spillover" effects extend to joint ventures primarily concerned with research and development work, rather than manufacturing. For example, care will be taken to insure that collaboration on one research and development project does not lead to more wide-ranging technical collaboration, thereby blunting the competition between the parties in the research and development area.[129] Also, the Commission has been concerned that collaboration in research and development may diminish the level of downstream competition between the parties with respect to production and sales. As such, it is likely that the Commission might seek to impose various containment mechanisms on the activities between the parties, such as product differentiation and tight controls on the information exchanges between the parties.[130]

"Network" effects The Commission has concerns that the goal of a single EC market will be endangered when a series of commonly controlled joint ventures are established with local partners in a number of different countries, since this network of joint ventures could be used a means for the controlling party to coordinate the behavior of the joint ventures in a manner which might eventually lead to market sharing and collusion among the joint ventures.[131] While the "network" effect is now becoming part of the analytic framework which the Commission uses in order to determine whether a joint venture falls within Article 85(1), the issue obviously will arise in a limited number of cases, and is generally dealt with by securing various undertakings and agreements from the parties which are designed to prevent the effects from occurring. When there is a danger of the "network" effect, the Commission's strong aversion to territorial divisions of the EC market makes it likely that it will take a fairly hard line with respect to what is appreciable and will conclude that Article 85(1) is infringed in almost all cases other than those which can be sorted out under the general *de minimis* test that applies to all joint ventures.[132]

"Non-appreciable" competitive effects Even if a joint venture does appear to have an actual or potentially restrictive or distortive effect on competition, it may nonetheless escape Article 85(1) if the effect in the particular market is not "appreciable." The Commission's recognition of this additional standard is apparently part of its recent efforts to attempt to take a more realistic and fact-based economic approach to assessing the competitive effects of joint ventures.[133] The notion of "appreciability" first appeared in a letter written by the Commission in 1990 to a party complaining about the potential effect of a pending joint venture.[134] In that letter the Commission rejected the complainant's arguments that the joint venture would infringe Article 85(1) by referring to many of the structural considerations which had theretofore been reserved for analysis of concentrations under the Merger Regulation, including the small market shares of the parties, the presence of a sufficient number of competitors, geographic and product complementarity, barriers to entry, and countervailing monopsony power. This approach was noteworthy in that it was taken at the stage at which a decision was being made regarding the applicability of Article 85(1), rather than as part of the Article 85(3) balancing process.

[129] See, e.g. *Ford/Volkswagen*, OJ 1993 L20/14 (a joint venture for the development and production of a car raised concerns regarding potential spillover into upstream vehicle development know-how in adjacent markets).

[130] Ibid.

[131] See, e.g. *Optical Fibres*, OJ 1986 L236/30.

[132] Bellamy and Child, supra note 4, at § 5-043.

The Commission's Joint Venture Notice also takes the position that Article 85(1) should not apply to a joint venture unless the competitive effect of the venture is appreciable. Among the factors which the Commission will consider as important in determining whether or not there is an appreciable effect upon competition are the following:[135]

(i) the market shares of the parties and the joint venture, the structure of the relevant market and the degree of concentration in the sector concerned;
(ii) the economic and financial strength of the parties and any commercial or technical advantage they have over competitors;
(iii) the market proximity of the joint venture's activities;
(iv) whether the fields of the joint venture and the parties are identical or interdependent;
(v) the scale and significance of the joint venture's activities in relation to those of the parents;
(vi) the extent to which the arrangements between the undertakings concerned are restrictive;
(vii) the extent to which market access by third parties is restricted.

Isolation of parts of Common Market It has been noted that the Commission has special concerns regarding the possibility that a joint venture will result in the isolation of national markets within the EC. In some cases, the objective of integrating the Common Market, which Article 2 of the Treaty describes as one means of achieving a higher standard of living in the EC, has become an independent objective which is sometimes pursued even if it results in the loss of efficiencies and creates greater costs and risks for the parties. This may well occur when the right to appropriate the results of the joint venture is allocated to the parties in accordance with geographic (i.e. national) territories, rather than on an accounting basis. For example, all or a portion of a geographic territory may be more difficult or expensive to exploit than another due to such things as higher costs of transport. Similarly, a party may be able to obtain a greater portion of the results if it is allowed to sell into areas or sales channels which are easier to exploit than some others, such as major cities or supermarket chains.[136]

D. Exemption of Joint Ventures Under Article 85(3)

Introduction If Article 85(1) is applicable to a joint venture and/or any of the restrictions which are a part thereof, then an exemption may be sought. In some cases, the group

[133] Ibid. at § 5-029.

[134] The letter was sent to Plessey, which had registered complaints regarding agreements between GEC and Siemens relating to their proposed acquisition of Plessey and the manner in which its various businesses would be operated following the acquisition. The Commission rejected the complaints holding, in part, that the agreements were non-infringing. See *GEC/Siemens–Plessey*, OJ 1990 C239/2. See also the report of the United Kingdom Monopolies and Mergers Commission on the original proposed merger, which had been published on April 21, 1989.

[135] See Commission's Joint Venture Notice, § 26.

[136] Korah, supra note 56, at 8.

exemptions for research and development and specialization arrangements may be used. In other cases, the parties may need an individual exemption under Article 85(3) by following the notification procedures. It has been suggested that the general economic criteria that will be used by the Commission in analyzing a joint venture under Article 85(3) can be summarized as follows:[137]

(i) The joint venture must provide certain benefits to third parties, such as improvements in production, technology or distribution.
(ii) The formation of the joint venture must not allow the parties to substantially eliminate competition in the market(s) in which the joint venture is to be active. Among the factors to be considered would be the market shares of the parties and the degree of concentration in the market prior to the formation of the joint venture.
(iii) The restrictions on competition which are included as part of the joint venture must be no greater than those which are indispensable to attaining the identifiable economic benefits, and the economic benefits from the joint venture must not have been obtainable through some less restrictive form of collaboration (i.e. the joint venture form and the restrictions associated therewith must be indispensable to the attainment of the benefits).[138]

Within the general analytic framework, there will be a need for a good deal of balancing, particularly between the potential economic benefits of the collaboration and the effect which it may have upon competition. For example, it is likely that a new joint venture among parties which are already major competitors involved in a highly concentrative market will not be approved, regardless of the short-term economic benefits which might flow from the joint venture.[139] Similarly, when the joint venture includes joint research and development or allows the parties to rationalize their existing production activities, a determination must be made as to whether the "benefits" from the proposed arrangement will accrue largely to the parties to the joint venture, perhaps reducing the overall economic benefits to others (e.g. consumers) through reduction in competition, or whether the collaboration will eventually serve to enhance competition through the diffusion of new technologies or the removal of unprofitable production capacity in the market.[140]

Economic benefits of the joint venture As noted above, there are a number of potential economic benefits which can be achieved through the use of a joint venture structure. For analytical purposes, however, the key areas for review appear to be whether the joint venture will result in any improvements in the competitive structure of the relevant mar-

[137] Bellamy and Child, supra note 4, at § 5-045.

[138] *Bayer/Gist-Brocades*, OJ 1976 L30/13 [1976] 1 CMLR D98; *Olivetti/Canon*, OJ 1988 L52/51 [1989] 4 CMLR 940, § 56; *Continental/Michelin*, OJ 1988 L305/33 [1989] 4 CMLR 920 § 25.

[139] See, e.g. *WANO Schwarzpulver*, OJ 1978 L322/26 [1979] 1 CMLR 403; *Henkel/Colgate*, OJ 1972 L14/14, Eighth Report on Competition Policy (1978), points 89 to 91.

[140] See Commission's Joint Venture Notice, § 55, stating that production rationalization will only succeed as a justification for a joint venture "if they serve to overcome a structural crisis, to accelerate the removal of unprofitable production capacity from the market and thereby to reestablish competition in the medium term."

ket, whether the joint venture relates to the development of advanced technology products, and whether the joint venture will lead to the creation of new industry capacity.

Improvements in competitive structure An important potential benefit from a joint venture is that the collaboration will, either immediately or in the long term, be a catalyst for procompetitive changes in the structure of the relevant product or geographic market. For example, the Commission will look favorably on joint ventures which facilitate the entry of new competitors into the market, as might be the case when it can be shown that the parties have either abandoned their own efforts to enter the market or that they would cease their current activity in the market but for the opportunity to participate in the joint venture.[141] Improvements in the competitive structure may also flow from joint ventures which allow the parties to penetrate a market more speedily and efficiently,[142] or to otherwise achieve a level of competitiveness which could not have been attained if the parties had only acted independently.[143] It is important to remember, however, that the improvements in competitive structure will only be recognized if the Commission is convinced that the parties will be able to continue to compete independently after the termination of the joint venture.[144]

Advanced technology products Joint ventures which are formed for the primary purpose of enabling the parties to combine their complementary expertise and resources to overcome what are often substantial technical difficulties and financial risks associated with developing new advanced technology products will be given favorable treatment, even in the case when the parties are relatively large companies.[145] Among the factors which the Commission has considered are whether or not the joint venture permits resolution of development problems more "effectively, economically and quickly" than if the parties had continued to work on the project independently.[146] The Commission will also be disposed to favorably view the joint venture if the product to be developed is of major economic importance or represents a breakthrough,[147] is a successful solution to a problem that has been particularly sought by customers,[148] is

[141] See, e.g. *Vacuum Interrupters (Nos. 1 and 2)* OJ 1977 L48/32 [1977] 1 CMLR D67; OJ 1980 L383/1 [1981] 2 CMLR 217; *Beecham/Parke- Davis*, OJ 1979 L70/11 [1979] 2 CMLR 157.

[142] See, e.g. *De Laval/Stork*, OJ 1988 L59/32 [1988] 4 CMLR 714; *GEC/Weir*, OJ 1977 L327/26 [1978] 1 CMLR D42; and *Eirpage*, OJ 1991 L306/22 [1991] 4 CMLR 233 (parents could not have achieved the objectives of the joint venture independently on the same "rapid timescale").

[143] See, e.g. *KEWA*, OJ 1976 L51/15 [1976] 2 CMLR D15.

[144] See, e.g. *KEWA*, OJ 1976 L51/15 [1976] 2 CMLR D15; and *United Reprocessors*, OJ 1976 L51/7 [1976] 2 CMLR D1. At issue in these cases was insuring that the formation and operation of the joint ventures was consistent with the long-term objective of nurturing a new industry to the point where it would eventually be competitive. See also *Eirpage*, OJ 1991 L306/22 [1991] 4 CMLR 233, where the Commission required that the parties delete a restriction in the joint venture agreement which prevented competition for a period of three years following termination of the joint venture.

[145] See Seventh Report on Competition Policy (1977), point 150. See also, e.g. *Optical Fibres*, OJ 1986 L236/30; *Ford/Volkswagen*, OJ 1993 L20/14; *KSB/Goulds/Lowara/ITT*, OJ 1991 L19/25 [1992] 5 CMLR 55.

[146] See, e.g. *GEC/Weir*, OJ 1977 L327/26 [1978] 1 CMLR D42; *Vacuum Interrupters (Nos. 1 and 2)* OJ 1977 L48/32 [1977] 1 CMLR D67; OJ 1980 L383/1 [1981] 2 CMLR 217.

[147] See, e.g. *Beecham/Parke Davis*, Eighth Report on Competition Policy (1978), point 93 (the fact that the new product to be developed by the joint venture would pharmacologically and therapeutically different from all known medicines was described as a "conclusive element" in granting an exemption under Art. 85(3)).

[148] See, e.g. *GEC/Weir*, OJ 1977 L327/26 [1978] 1 CMLR D42.

urgently required for economic or social reasons,[149] is likely to strengthen the EC industry against import competition, or will facilitate the transfer of important foreign technology to an EC party.[150]

Creation of new industry capacity In most cases, the Commission will favorably view any joint venture formed for the purpose of constructing and operating new production facilities, or enlarging existing facilities, if the parties can adequately demonstrate that the joint activities of the parties will avoid the construction of excess capacity and result in economies of scale, thereby permitting the parties to reduce the unit costs of production and achieve profitability.[151] However, the Commission will also want to be sure that the benefits achieved by the parties to the joint venture does not harm the position of third parties[152] and may impose conditions on the joint venture, such as safeguards to guarantee that the facility's output will not be unduly restricted.[153]

Substantial elimination of competition One of the conditions to the grant of an exemption under Article 85(3) is a finding that the proposed collaboration will not "substantially eliminate"[154] or "appreciably reduce"[155] competition in the relevant market. The analysis of whether or not competition is sufficiently reduced so as to preclude an exemption will be based on the particular facts of the case, including the benefits expected from the collaboration.[156] However, the Commission has put forward some general non-binding guidelines that tend to serve as the basis for rebuttable presumptions. For example, in the case of a production joint venture that does not extend downstream to actual sales and distribution activities, it is usually presumed that there will be no substantial decrease in competition unless the market share of the joint venture exceeds 20 percent.[157] But, if the joint venture does include production and sales, then this de minimis safe harbor is reduced to a 10 percent market share.[158] In those cases where the market share exceeds the amounts set forth in the guidelines, then it can be anticipated that the Commission will undertake a "careful examination" of the case,[159] resulting not only in the possibility that the exemption will be denied, but also in a significant delay in the review of the arrangement in relation to joint ventures which fall under the threshold amounts.

When the activities of the joint venture are extensive enough so that the joint venture can be characterized as "structural" or "full-function," then the test for the effect that it may have on competition tends to shift toward that which is used in the context of the Merger Regulation for concentrative joint ventures: does the formation of the joint ven-

[149] Ibid.

[150] See, e.g. *Bayer/BP Chemicals*, OJ 1988 L150/35 [1989] 4 CMLR 24, § 27; *Olivetti/Canon*, OJ 1988 L52/51 [1989] 4 CMLR 940, § 56.

[151] Among the cases involving the construction and operation of production facilities are *KEWA*, OJ 1976 L51/15 [1976] 2 CMLR D15; *Rockwell/Iveco*, OJ 1983 L224/19 [1983] 3 CMLR 709; *Optical Fibres*, OJ 1986 L236/30; and *Bayer/BP Chemicals*, OJ 1988 L150/35 [1989] 4 CMLR 24.

[152] See, e.g. *Optical Fibres*, OJ 1986 L236/30 (joint venture required to sell on non-discriminatory terms to third parties).

[153] Ibid.

[154] Bellamy and Child, supra note 4, at § 5-046.

[155] See Sixth Report on Competition Policy (1976), point 59.

[156] Bellamy and Child, supra note 4, at § 5-046.

[157] Commission's Joint Venture Notice, §§ 63–4. This threshold has been incorporated into the block exemptions for research and development and specialization agreements.

[158] Ibid.

[159] Ibid.

ture result in the attainment or the strengthening of a dominant position.[160] In general, the Commission is not likely to grant an exemption for a structural joint venture which will have such an effect.[161]

Indispensability As a general matter, the Commission is inclined to readily accept that the joint venture is indispensable to the attainment of the anticipated benefits if it is shown that the parties would not have incurred the costs and risks of developing the new product on their own.[162] Similarly, the Commission will approve a joint venture if it believes that the desired objectives, such as market entry, could not have been obtained more quickly or efficiently by using any less close form of cooperation, such as a license agreement or a specialization agreement.[163] Finally, if technology transfer is considered to be a key element of the arrangement, a joint venture is often seen to be the best, if not only, way to insure that all the required technical information and expertise will be completely and effectively transferred from the developer to the other parties to the arrangement.[164]

E. Analysis of Restrictions in the Joint Venture Agreement

General analytical framework Whether or not the joint venture itself infringes Article 85(1), the Commission will also examine whether the restrictions included in the joint venture agreements run afoul of Article 85(1). The analytical framework used by the Commission depends on whether the joint venture itself is infringing. In the case where the joint venture itself does not infringe Article 85(1), a restriction which contained in the joint venture agreement will, ipso facto, also be considered to be non-infringing provided that the restriction can be considered to be "ancillary" to the joint venture.[165] If, however, a restriction cannot be regarded as ancillary to a non-infringing joint venture, then it must be examined under Article 85(1).[166] On the other hand, if the joint venture itself has been found to infringe Article 85(1), then the so-called "ancillary restraints" doctrine will not apply and it will then be necessary to determine whether any specific restriction falls with Article 85(1) and, if so, whether that restriction qualifies for an exemption under Article 85(3).

Determining whether restrictions are "ancillary" In order for a restriction to be considered ancillary, it must be one which "cannot be disassociated from the creation

[160] See, e.g. *PRB/Shell*, Seventeenth Report (1987), point 74; *P&G/Finaf*, IP(92)250; *Fiat/Hitachi*, OJ 1993 L20/10 at § 26.

[161] See, e.g. Commission's Joint Venture Notice, § 64 (in evaluating "full-function" joint ventures "the Commission has to take special care ... that no position of market power will be created or strengthened ..."). See also ibid. at § 58 ("This category [unexemptable JVs] includes JVs which ... achieve, consolidate or strengthen a dominant position.")

[162] Bellamy and Child, supra note 4, at § 5-052.

[163] See, e.g. *De Laval/Stork*, OJ 1977 L215/11 [1977] 2 CMLR D69 and OJ 1988 L59/32 [1988] 4 CMLR 714; *GEC/Weir*, OJ 1977 L327/26 [1978] 1 CMLR D42.

[164] See, e.g. *Olivetti/Canon*, OJ 1988 L52/51 [1989] 4 CMLR 940 ("The major involvement of the parties inherent in a manufacturing joint venture permits a permanent and intense flow of technology."); *Optical Fibres*, OJ 1986 L236/30.

[165] OJ 1990 L209/15 [1991] 4 CMLR 832.

[166] Bellamy and Child, supra note 4, at § 5-028.

of [a non-infringing joint venture] without undermining its existence and purpose."[167] For example, an obligation to obtain only from the other party products that may have been developed under the agreement has been held not result in any infringement of Article 85(1),[168] as has a restriction on a joint venture from making or dealing in competing products.[169] On the other hand, if one party is obligated to supply certain intermediate products only to the other party, or not to license such products to third parties, it may infringe Article 85(1).[170] The Commission also has taken issue with the grant of unduly broad exclusive licenses to the joint venture by one of the parties.[171]

Elopak/Metal Box-Olin[172] was one of the major cases in which the Commission attempted to define those provisions which it considered as ancillary restraints which would not infringe Article 85(1). On the specific facts of that case, the following provisions were considered to be ancillary restraints:

(i) Exclusive know-how licenses from each party to the joint venture covering the relevant technology, provided that the "exclusivity" is strictly limited to the specific products to be produced by the joint venture and to a restricted field-of-use.[173]

(ii) Non-exclusive licenses from the joint venture back to each of the parties covering improvements which (i) the joint venture does not wish to exploit itself (ii) are only to be used outside of the field in which the joint venture is active.

(iii) First options in favor of parties with respect to rights to obtain licenses from joint venture for exploitation of technology in geographic areas other than those in which the joint venture is active.

(iv) Each party granted the freedom to compete with the joint venture with respect to research, production, and sales, provided that the party does not utilize technology that was either contributed by the other party or developed by the joint venture.

(v) If one party purchases the other party's interest in the joint venture, non-exclusive cross-licensing agreements would take effect in order to insure that the parties had equal access to the joint venture's technology, provided that (a) the joint venture's use of any technology that was originally contributed by the departing party would be limited to the joint venture's field-of-activity and (b) neither party would be allowed, for a period of five years, to use the other's know-how or any improvements developed by the joint venture with a competitor of the other party.

(vi) Neither party would be allowed to dispose of its interest in the joint venture without the consent of the other.

[167] *Elopak/Metal Box-Odin*, OJ 1990 L209/15 [1991] 4 CMLR 832, § 36. The rule is restated in the Commission's Joint Venture Notice at §§ 65–76.

[168] See, e.g. *EMI/Jungheinrich*, Seventh Report on Competition Policy (1977) points 119–20 [1978] 1 CMLR 398.

[169] *Mitchell Cotts/Sofiltra*, OJ 1987 L41/31 [1988] 4 CMLR 111.

[170] See, e.g. *BP Kellogg*, OJ 1985 L369/6 [1986] 2 CMLR 619. See also *ACEC/Berliet*, OJ 1968 L201/7 [1968] CMLR D35.

[171] *Mitchell Cotts/Sofiltra*, OJ 1987 L41/31 [1988] 4 CMLR 111 (holding that exclusive know-how license granted to the joint venture by one of the parties covering the United Kingdom extended beyond what was reasonably necessary for the purpose of the joint venture).

[172] OJ 1990 L209/15 [1991] 4 CMLR 832.

[173] Ibid. at § 30 (the scope of the joint venture's exclusivity was "very narrowly defined to include only the highly specific product in question").

(vii) If one party purchases the other party's interest in the joint venture, it would not be allowed to sell the joint venture to a third party for a period of five years without first offering it on the same terms to the other party.

In addition, the Commission noted approvingly that there were no restriction placed on the joint venture concerning pricing, quantities, or its geographic area of operations, implying that any such restrictions might not be considered as ancillary to the joint venture. However, the Commission has held in other situations that geographic restrictions included in an exclusive license to the joint venture may, at least during the start-up phase of the joint venture, be permissible.[174]

Another recent case dealing with the ancillary restraints doctrine was *Konsortium ECR 900*,[175] where the Commission found that the following provisions should, at least in that case, be considered as ancillary:

(i) Neither party would be allowed to compete with the joint venture in the joint venture's product and geographic markets; however, outside of the geographic market of the joint venture, the parties would be permitted to market such parts of the joint venture's product as to which the party was involved in developing.
(ii) Parties were allowed to pool technical documentation regarding joint developments in the course of the joint venture; however, pooling was not permitted as to any unilateral developments.
(iii) For a period extending up to eight months prior to the termination of the joint venture, the parties were not allowed to use any pooled technical information to make the joint venture product for sale in the geographic market served by the joint venture; however, following the termination of the joint venture, each party would have a non-exclusive license to use the information in all geographic markets.
(iv) Following termination of the joint venture, the parties could not sub-license any jointly developed technical information without (a) the consent of the other parties to the joint venture and (b) agreeing to equally share any licensing fees obtained from the sub-license.
(v) A party that breached its obligations under the joint venture contract would forfeit all rights to use any technical information acquired during the course of the joint venture.[176]

While the aforementioned cases are useful in determining whether a specific restriction will be considered as ancillary, and further guidance can be obtained through a review of some of the Commission's decisions under the Merger Regulation,[177] the ques-

[174] *Mitchell Cotts/Sofiltra*, OJ 1987 L41/31 [1988] 4 CMLR 111 (holding that exclusive know-how license granted to the joint venture by one of the parties covering the United Kingdom extended beyond what was reasonably necessary for the purpose of the joint venture, and therefore violated Art. 85(1), but then granting an exemption for eight years). See also Commission's Joint Venture Notice, § 76.

[175] OJ 1990 L228/31 [1992] 4 CMLR 54.

[176] Ibid. at § 34 ("competition not based on performance is not protected by Article 85(1).")

[177] The principles which have been developed by the Commission in its decisions regarding ancillary restrictions in this area should approximate the Commission's analysis as to whether or not a restriction is "indispensable," which is one of the criteria for an exemption under Art. 85(3).

tion of whether or not a particular restriction is ancillary ultimately depends on the specific facts of the case. There are, however, certain restrictions which usually will not qualify for any sort of exemption. For example, restrictions which might be placed on the joint venture's output, prices, and customer bases will generally not be exempted, and restrictions placed on the geographic markets which the joint venture may serve are also not likely to be permitted, particularly in those cases where the restrictions have the effect of dividing the EC market.[178]

The doctrine of ancillary restraints has become a useful analytical tool in evaluating the competitive effects of joint venture arrangements. However, the doctrine does not involve a balancing of pro- and anticompetitive effects, as would be the case if regulators had chosen to use a rule of reason approach to reviewing restrictions in the joint venture agreement.[179] The key consideration remains whether or not the main transaction (i.e. the formation of the joint venture) is found to be anticompetitive, since a finding of non-infringement at that level will render any necessary, or "ancillary," restraints included in the arrangement non-infringing.[180]

Restrictions falling within Article 85(1) If the joint venture itself infringes Article 85(1), then restrictions included in the joint venture agreements must be analyzed under Article 85(1), as must "non-ancillary" restraints in the agreements for a joint venture that does not infringe Article 85(1). One example of a restriction that would be violative of Article 85(1) is an exclusive license of technology by the parties to the joint venture, since the resultant restrictions on the ability of the parties to grant licenses to third parties impairs their pre-existing commercial freedom of action and also prevents them from pursuing other competitive opportunities.[181] Another restriction that would infringe Article 85(1) is when the use of the technology developed by the joint venture is limited to the parties,[182] or when third parties are suddenly unable to obtain certain products directly from the parties due to covenants in the joint venture agreement which preclude independent competition by the parties.[183] Also, if the parties are obligated to obtain all of their requirements from the joint venture, then third parties would no longer be able to compete for the business of the parties,[184] and if the joint venture is required to obtain all of its requirements of raw materials or inputs from the parties, then third

[178] Bellamy and Child, supra note 4, at § 5-060.

[179] Korah, supra note 56, at 9 n4.

[180] Ibid.

[181] See, e.g. *Continental/Michelin*, OJ 1988 L305/33 [1989] 4 CMLR 920 ("Through the exclusive commitment to [the JV] the parties lose their freedom to grant licences independently to third parties"); *KSB/Goulds/Lowara/ITT*, OJ 1991 L19/25 [1992] 5 CMLR 55 ("The fact that third parties are prevented from having access to the process technology is a restriction of competition"). However, amendments were made to the patent and know-how license exemptions to include licenses granted in the context of a joint venture. Reg. 151/93, OJ 1993 L21/8.

[182] See, e.g. *Continental/Michelin*, OJ 1988 L305/33 [1989] 4 CMLR 920 (Commission required that technology developed by the joint venture would be offered on reasonable terms to competitors); *Olivetti/Canon*, OJ 1988 L52/51 [1989] 4 CMLR 940 (provision which required consent of both parties in order to license third parties to sell joint venture products amended upon the request of the Commission so that joint consent would only be required in those cases where one party wanted to license the sale of products which were based essentially on the technology of the other party).

[183] See, e.g. *De Laval/Stork (No. 1)*, OJ 1977 L215/11 [1977] 2 CMLR D69; *GEC/Weir*, OJ 1977 L327/26 [1978] 1 CMLR D42; *Alcatel/ANT*, OJ 1990 L32/19 [1991] 4 CMLR 208.

[184] *WANO Schwarzpulver*, OJ 1978 L322/26 [1979] 1 CMLR 403; *Vacuum Interrupters (No. 2)*, OJ 1980 L383/1 [1981] 2 CMLR 217.

parties will be unable to compete for the opportunity to sell such materials or inputs to the joint venture.[185]

Exemption of restrictions under Article 85(3) In general, the Commission has tended to find that most of the common types of restrictions which are contained in joint venture agreements (e.g. non-competition agreements) are, in fact, subject to Article 85(1). In such cases, however, it is often possible to obtain an exemption for the restriction under Article 85(3) if it can be shown that the specific restriction is indispensable to the achievement of the lawful objectives of the collaboration.[186] Among the many restrictions generally included in joint venture agreements which are usually allowed by the Commission are non-competition obligations imposed on the parties for the life of the joint venture;[187] duties on the parties to purchase only from the joint venture or obligations on the joint venture to source only from the parties,[188] at least for some limited period of time following the start-up phase of the joint venture[189] or for so long as the arrangements remain competitive;[90] requirements and procedures with respect to joint decision making and control,[191] although the Commission often has acted to reduce the powers of one or more of the parties and/or to bolster the autonomy of the joint venture in certain instances;[192] and, at least in some cases, the grant of exclusive intellectual property rights to the joint venture.[193] However, even though the Commission may grant exemptions for these various restrictions, it has also taken steps to ensure that once the joint venture has terminated, the parties are free to compete "with no restrictions as to time, territory or of any other kind."[194]

185 Ibid.

186 Bellamy and Child, supra note 4, at § 5-056.

187 The Commission has upheld non-competition covenants during the term of the joint venture in a number of cases, such as *GEC/Weir*, OJ 1977 L327/26 [1978] 1 CMLR D42; *Amersham Buchler*, OJ 1982 L314/34 [1983] 1 CMLR 619. On the other hand, the Commission has generally not been willing to approve even short-term post-joint venture non-competition covenants. See, e.g. *Eirpage*, OJ 1991 L306/22 [1991] 4 CMLR 233 (requiring deletion of three-year post-joint venture non-competition covenant).

188 See, e.g. *Vacuum Interrupters (No. 1)*, OJ 1977 L48/32 [1977] 1 CMLR D67, and *(No. 2)*, OJ 1980 L383/1 [1981] 2 CMLR 217.

189 *Rockwell/Iveco*, OJ 1983 L224/19 [1983] 3 CMLR 709 (eight years from first production).

190 *Olivetti/Canon*, OJ 1988 L52/51 [1989] 4 CMLR 940.

191 See, e.g. *Rockwell/Iveco*, OJ 1983 L224/19 [1983] 3 CMLR 709.

192 See, e.g. *Optical Fibres*, OJ 1986 L236/30 (reduction of Corning's voting rights and veto powers due to fears regarding the possible "network" effect); *duPont/Merck*, IP(91)381 (bolstering of joint venture's decision-making autonomy to address concerns about possible coordination between the parties and between the parties and the joint venture).

193 However, as noted in Chapters 10 and 11, the Commission will often attempt to limit the effects of an exclusive license, even forcing the parties to modify the arrangements to make them non-exclusive. See, e.g. Continental/Michelin, OJ 1988 L305/33 [1989] 4 CMLR 920 ("Through the exclusive commitment to [the JV] the parties lose their freedom to grant licences independently to third parties"); *KSB/Goulds/Lowara/ITT*, OJ 1991 L19/25 [1992] 5 CMLR 55 ("The fact that third parties are prevented from having access to the process technology is a restriction of competition").

194 See, e.g. *De Laval/Stork (No. 1)*, OJ 1977 L215/11 [1977] 2 CMLR D69. However, with regard to non-competition restrictions on the parties following termination of the joint venture, it is considered likely that the Commission will begin to follow certain of its decisions in the context of the Merger Regulation and allow short-term non-competition restrictions on the grounds that they are relatively benign and also generally should be considered as ancillary to the entire arrangement.

13 Development of Judicial and Regulatory Policies in the United States and the EC Relating to Joint Ventures

I. Development of Antitrust Law Policy in the United States

A. Introduction

The development of antitrust policies in the United States regarding joint ventures can be traced through a historical review of some of the seminal cases in this area. Joint venture cases earlier in this century paid little attention to the factors which now occupy the debate among economists and regulators, such as efficiencies. In fact, the main concern of the courts appeared to be whether a joint venture or similar arrangement was actually a device for effectuating a monopoly, fixing prices or otherwise imposing an unreasonable restraint on trade. For example, in one case, the Supreme Court upheld the creation of a joint venture licensing concern among various oil refiners for the purpose of entering into cross-licensing agreement in order to resolve competing claims of infringement on the basis that the arrangement did not restrict competition.[1] On the other hand, restrictions in the by-laws of a major news service which prevented the sale of news to non-member competitors were deemed to be agreements in restraint of trade, and thus illegal.[2] Since the early 1960s, however, the joint venture cases have exhibited an increasingly sophisticated level of economic analysis.

B. United States *v.* Columbia Pictures

United States v. *Columbia Pictures*[3] (*Screen Gems*) involved a combination of existing resources. Columbia Pictures had been engaged for some time in licensing its library of films for television exhibition through Screen Gems, a wholly-owned subsidiary. Universal also had a library of feature films but no distribution organization. Columbia and Universal entered into an arrangement whereby Screen Gems would distribute both Universal and Columbia films for television viewing, an arrangement that had an effect on price because of a provision that Universal films might not be licensed to television exhibitors on terms less favorable than Columbia films of comparable quality.[4] The court

[1] *Standard Oil Co.* v. *United States*, 283 U.S. 163 (1931) (action under §§ 1 and 2 of the Sherman Act against Standard Oil Company of Indiana, the Texas Company, Standard Oil of New Jersey and others).

[2] *Associated Press* v. *United States*, 326 U.S. 1 (1945) (rejecting argument that AP's news was not "indispensable").

[3] 189 F. Supp. 153 (S.D.N.Y. 1960).

found the arrangement legal after an extended rule of reason analysis, apparently concluding that joint use of an existing distribution operation with excess and unused capacity would be more efficient than for each of the film companies to distribute its film library separately.[5]

The reasoning behind the *Screen Gems* decision has been criticized for its failure to make any estimate of the magnitude of the alleged efficiencies, nor to balance such efficiencies against potential anticompetitive effects. One possible result of such an analysis would have been a finding that many of the same efficiencies may have been achieved in a way which was less anticompetitive way. For example, Universal could have used the existing distribution system of other companies which, unlike Columbia, were not horizontal competitors, or could have looked to a significant number of distribution agents who did not have their own film library or represent a company that did have a film library. If these alternatives had been considered, then it might have reasonably been concluded that the integration of the existing resources of the participants in Screen Gems, while very real, had almost no procompetitive value.

C. Broadcast Music, Inc. *v.* Columbia Broadcasting System

Courts have tended to view favorably joint ventures which create a new product or market which, presumably, would not exist in the absence of the cooperative arrangement. In those types of situations, any horizontal restraints included in the agreement will generally be analyzed under the rule of reason, and will be valid so long as they are reasonably necessary for marketing the new product. One case that illustrates the recognition of new products as a procompetitive efficiency is *Broadcast Music, Inc. (BMI)* v. *Columbia Broadcasting System, Inc.*,[6] which involved the operations of BMI and the American Society of Composers, Authors, and Publishers ("ASCAP"). Both BMI and ASCAP were associations of composers, lyricists, and publishers of music that acted as a clearing house between its members and those who wish to perform copyrighted works from their repertory. Some 20,000 authors and composers had granted nonexclusive rights to ASCAP to license their musical compositions for a fee, and another 20,000 granted similar rights to BMI. Licensees, such as a radio or television station, typically were required to take a full license to an artist's entire repertoire, with fees based on the licensee's advertising revenues and no limitation on the number of times each song is performed.[7] Because BMI and ASCAP acquired only nonexclusive rights to its members' compositions, each member still retains the right to license performance rights for his compositions, and users could still choose between a license from onc of the associations or a direct license from the copyright holder of songs that he wishes to perform.

The transactional efficiencies from the BMI and ASCAP arrangements were substantial. For example, from the point of view of sellers, the cost of negotiating and monitoring contracts was greatly reduced since, in effect, both BMI and ASCAP were able to substitute a single transaction for what was potentially thousands of contracts. As to the users, the fact that almost all musical compositions were held in one of the association's libraries meant that the license in effect provided an assurance of authorization that would not be possible without gaining the collaboration of the thousands of members of each society.

[4] Ibid. at 179–80.
[5] Ibid. at 178–81.

CBS argued that the blanket license was an agreement among competitors to fix prices for performance rights, presumably relying on the fact that the BMI and ASCAP each had several of the characteristics which are commonly associated with a cartel. For example, BMI and ASCAP were both common sales agencies, in that they acquired nonexclusive performance rights from its members and licensed them to users. Also, BMI and ASCAP members pooled their performance revenues and then distributed them using an elaborate formula that takes account of both the number and types of plays. These procedures were similar to those which are commonly employed among cartel members to discourage cheating. Finally, BMI and ASCAP engaged in price discrimination, which means price differences unrelated to cost differences, in the manner in which it set licensing fees by reference to revenues of the licensees. Price discrimination normally seen as a sign of market power. Since cartels exercise market power, the ability to price discriminate might signal the presence of a cartel.

The Supreme Court rejected the plaintiff's characterization of the arrangement as a cartel, and proceeded to balance the procompetitive and anticompetitive effects of the arrangement. The Court found that, even assuming some restraining effect on price, the "integration of sales, monitoring and enforcement against unauthorized copyright use,"[8] and the efficiencies that resulted, created a new product that could not otherwise exist but for the joint arrangement. The Court then went on to conclude that the magnitude of procompetitive effect properly should counterbalance the limited price and market restrictions involved in the ASCAP–BMI licensing system.[9] It is also worth noting that, at least in theory, BMI and ASCAP actually had a positive effect on output (i.e. as measured by the performance or plays of copyrighted music) due to the reduced transaction costs, and that the licensing fees charged by the associations were regulated by the fact that potential licensees could still seek licenses directly from the artists.

D. United States *v.* Columbia Pictures Industries, Inc.

Creation of a new product is an essential condition to the avoidance of a finding of *per se* illegality with respect to any horizontal restraints in a joint venture. The case of United States v. Columbia Pictures Industries, Inc.[10] involved a joint venture among four major motion picture producers and Getty Oil to exhibit feature films on pay television. The court found that the pricing formula that the joint venture used to pay its members for contributed films fixed prices in violation of the Sherman Act, and that the venturers' agreement to withhold designated films from third parties for a nine-month period was actually a group boycott that also violated § 1 of the Sherman Act. The court was not sympathetic to the defendants' argument that the joint venture created a new product, thereby warranting rule of reason analysis, noting that it found "completely unwarranted ... [the

[6] 41 U.S. 1, 19 (1979).

[7] Ibid. at 5.

[8] 441 U.S. 1, 20 (1979).

[9] To the same effect, see *National Bancard Corp. (NaBANCO)* v. *VISA U.S.A., Inc.*, 779 F.2d 592 (11th Cir. 1986), where the court sustained under a rule of reason a discount arrangement between various classes of banks within the VISA credit card system. The court found that the arrangement could not exist but for the price arrangement between categories of banks and concluded, consistent with the reasoning of this part, that the arrangement should be sustained because "the restraint at issue potentially could create an efficiency enhancing integration to which the restraint is ancillary." Ibid. at 599.

[10] 507 F. Supp. 412, 426–30 (S.D.N.Y. 1980).

assertion] that Premiere creates a totally new product in the pay-television market consisting of a package of movies and their transmission and promotion."[11] Moreover, the court explained that the horizontal restraints in the joint venture would, in any event, fail under any rule of reason analysis because they were not necessary to market the purported joint venture product, would increase the price of films licensed to pay television, would eliminate competition in the network program service market, and would put competing pay television stations out of business.[12]

E. The Brunswick–Yamaha *Joint Venture Case*

One of the most illustrative cases in the joint venture arena involved Brunswick and Yamaha, who formed a joint venture for the purpose of manufacturing outboard motors jointly in Japan and marketing them in different parts of the world. This case, *Yamaha Motor Co.* v. *FTC*,[13] called for analysis in a number of different areas, including the loss of potential competition stemming from the joint venture, high industry concentration, and review of the effect of several restrictive marketing agreements. The parties entered into three agreements that were found illegal by the Federal Trade Commission. On appeal, the FTC's findings were affirmed by the Eighth Circuit Court of Appeals.

Brunswick had begun manufacturing outboard motors in 1961 when it acquired its Mercury Marine Division, and was selling Mercury motors in the United States, Canada, Australia, Europe, and Japan. Before entering the joint venture, Brunswick had considered adding a second line of motors, to be called Mariner. For its part, Yamaha had entered the outboard motors market in 1969 by acquiring 60 percent of Sanshin Kohyo Co. Ltd., and was selling these motors under the Yamaha brand name throughout the world, but not in the United States. Prior to the joint venture, Yamaha had twice attempted to enter the United States market. First, it tried a low horsepower motor, but it proved too costly relative to substitutes and was of the wrong technical design for the United States market. It has next tried selling through Sears, but again its engine was oo costly for Sears customers.

On November 21, 1972, Brunswick and Yamaha entered into a joint venture in which Brunswick acquired 38 percent of Sanshin and Yamaha's share fell to 38 percent. Others held the remaining 24 percent. Sanshin was to continue to manufacture motors and sell them to Yamaha which, in turn, would resell some of them to Brunswick for distribution by Brunswick under its Mariner brand name. Motors not resold to Brunswick would be distributed by Yamaha in its traditional markets under the Yamaha label.[14] Thus, an identical motor was to be sold as a Yamaha and as a Mariner. Sanshin's board was to have eleven directors, six from Yamaha and five from Brunswick, and decisions on budgets and expansion of product lines required seven votes of approval. The joint venture was scheduled to last for an initial term of ten years, with automatic three-year extensions, but either party could terminate by giving three years written notice.

The arrangement also included several collateral marketing and production agreements. First, Brunswick was given exclusive rights to sell Sanshin-produced motors in the United

[11] Ibid. at 430.
[12] Ibid. at 432.
[13] 657 F.2d 971 (8th Cir. 1981), *cert. denied*, 456 U.S. 915 (1982).
[14] Ibid. at 974 (Brunswick owned both the Mercury and Mariner outboard motor labels, and intended to market joint venture products under the Mariner name).

States, Canada, Mexico, Australia and New Zealand, and Yamaha was given exclusive rights to Japan. The parties would be free to compete in the rest of the world – Europe, South America and Africa – with one reservation: they agreed not to compete for each others' dealers in these market areas. Second, Yamaha was not allowed to manufacture directly or indirectly the same or a similar motor to that made by Sanshin, nor could it resell any motors purchased from other suppliers outside of Japan. Finally, in exchange for technical information not relating to outboard motor products, Brunswick was prohibited from manufacturing any existing Yamaha products, such as motorcycles or boats, except for snowmobiles.

The first key issue considered by the FTC was the effect that the arrangement would have on competitive conditions in the market. The relevant economic market was determined to be the United States, in which four firms dominated the market with 98.6 percent of dollar sales in 1973: Outboard Marine Corporation ("OMC"), Brunswick, Chrysler and Eska. The top two firms, OMC and Brunswick, had 85 percent of total sales, with Brunswick's share in 1973 being around 25 percent while OMC had 60 percent. While industry sales were growing, concentration was increasing, which three of the eight firms which were in the market in 1965 having exited the market by 1973.

The FTC's administrative law judge found that, even though Yamaha was eliminated as a potential entrant, the net effect of the joint venture was procompetitive, since it would add a new engine to the market. The judge also found that Yamaha was not considering additional attempts at entry into the United States, so that distribution in the United States would not be any different than it had been prior to the arrangement. However, the full Commission reversed the judge, finding that Yamaha was a potential entrant and that, as such, the joint venture would have the effect of reducing competition from Yamaha's "waiting in the wings." The FTC also rejected the argument that the Mariner engine represented a new entrant into the United States market, reasoning that Brunswick would not allow Mariner to compete with its Mercury motor line.[15]

The FTC's decision was upheld on appeal. The court took a strict structuralist approach, analogous to the approach that has traditionally been applied to horizontal mergers, and found that Yamaha would have entered the market on its own and that in an oligopolistic market five firms are more competitive than four.[16] The court did concede that there may have been some combination of technology and perhaps marketing know-how; however, all in all, the overall level of integration was relatively slight and the anticompetitive effect fairly clear; hence, the arrangement was found to be in violation of the antitrust laws.[17] Also, since Yamaha was to be denied the opportunity to market non-joint venture products in the United States, it appeared that one of the main purposes of the joint venture, at least from Brunswick's perspective, may have been to eliminate Yamaha as an immediate or eventual entrant into the United States market.

The FTC and the court also examined whether each of the collateral arrangements with respect to marketing were reasonably related to the operation of the joint venture and no broader than necessary. In each case, they found that the restrictions were beyond the range

[15] "[I]f the parent companies are in competition, or might compete absent the joint venture, it may be assumed that neither will compete with the progeny in its line of commerce." Ibid. at 980 (quoting *United States* v. *Penn-Olin Co.* 378 U.S. 158 (1964)).

[16] Ibid. at 97–9.

[17] Ibid. at 979–80.

of legitimate commercial concerns.[18] As to the exclusive marketing arrangements, the FTC and the court found that while they might well be ancillary and lawful with respect to a worldwide manufacturing and marketing joint venture, the exclusion of Yamaha from selling non-joint venture outboard motors in the United States was beyond that which was necessary to the operation of the joint venture, particularly since Brunswick was not excluded from the Japanese market. Also, the agreement between the parents not compete for each other's dealers in "non-exclusive markets" was found not to be necessary to further the legitimate purposes of the joint venture. Finally, any agreement regarding the use of technology which was unrelated to the type exchanged as a part of the joint venture was found to go beyond the reasonable needs of the parties.

The decisions in *Yamaha–Brunswick* have been subject to criticism. For example, with regard to the determination of the relevant market, it has been argued that if Yamaha actually was a potential entrant into the United States, then supply elasticity was high and the market that should have been considered extended beyond the United States.[19] If this was the case, then there were probably a number of firms, other than Yamaha, that also would be likely entrants, and the concentration ratios calculated as part of the court's analysis were clearly overstated.

One might also take issue with the conclusions reached regarding the impact of the new Sanshin-Mariner motor on the United States market. While it is certainly true that Yamaha had previously experienced difficulties in entering the United States market with the Sanshin technology independently, this does not mean that entry through the joint venture arrangement would not have had a significant procompetitive effect on the market. For example, if the Sanshin-Mariner motor was a superior motor, and indications pointed to Yamaha as a technological leader, then its entry would have benefited consumers and pressured rivals to improve their motors.[20] In addition, the argument that Brunswick could prevent competition between Mercury and Mariner is without substance, since there would be a high cross-price elasticity of demand across the comparable products.[21]

It could also be contended that the find that the various collateral agreements were anticompetitive at variance with the *Sylvania* decision[22] and the economic efficiency consequences of avoiding free-riding in partnerships.[23] It is generally agreed that exclusive territories can lead to greater competition by avoiding free-riding by some dealers on the selling efforts of other dealers and, thus, provide incentives for dealers to invest in special services for consumers. The agreement of the parties in connection with the joint venture not to compete for dealers, thereby avoiding the disruption of the existing dealer networks of the parents, is also consistent with enhanced interbrand competition. Finally, it has been argued that the restriction on Yamaha not to produce its own motor or resell that of another is simply a provision preventing a partner from competing with the partnership and diluting its interbrand competitive efforts.[24]

[18] Ibid.

[19] Weston and Ornstein, "Efficiency Considerations in Joint Ventures," in *Collaboration Among Competitors* (1990) 875, 879.

[20] Ibid.

[21] Ibid.

[22] *Continental TV., Inc.* v. *GTE Sylvania Inc.*, 433 U.S. 36 (1977).

[23] Weston and Ornstein, supra note 19, at 880.

[24] Ibid. at 880.

F. Northrop Corporation *v.* McDonnell Douglas Corporation

A teaming arrangement is typically a limited production joint venture involving two or more private contractors who decide to pool their financial and technical resources to complete a single, discrete project. As a general matter, bona fide teaming agreement will not be subject to antitrust challenge on the grounds that, far from restricting competition, "there would be no competition but for the agreements."[25] One example of a teaming arrangement is an agreement between Northrop Corporation and McDonnell Douglas Corp. to develop the new F-18 aircraft. Under the terms of the agreement between the venturers, McDonnell Douglas was to act as the prime contractor for the development of carrier-based aircraft, and Northrop was to be the prime contractor with the Army for the development of land-based aircraft. The Ninth Circuit rejected the argument that the clause was *per se* unlawful, since the clause did not foreclose head-to-head competition in the international market between the two variants of the joint F-18 aircraft.[26] The Ninth Circuit did note, however, that an agreement that restricted the parties' right to develop and market either land-based or carrier-based aircraft other than the jointly developed F-18 may be *per se* unlawful.[27]

G. National Collegiate Athletic Association *v.* Board of Regents of The University of Oklahoma

National Collegiate Athletic Association ("NCAA") v. *Board of Regents of the University of Oklahoma*[28] is an example of a joint arrangement which was found to be illegal. The NCAA is a joint agency which establishes rules for athletic competition among college athletes and sponsoring tournaments. In addition, the NCAA acted as the bargaining agent between member colleges and television networks, negotiating the total amount that the networks would pay for college football telecasts and the maximum number of television appearances for each school. Within these guidelines, individual schools negotiated their own fees with the various networks. In the course of acting as the bargaining agent, the NCAA sought to limit the number of telecasts in order, in its view, to moderate the impact of television on the actual attendance at the games.

The NCAA's bargaining activities were challenged by one of the member colleges on the ground that by limiting output (i.e. the number of appearances a school could make on television), it was artificially creating high prices.[29] The Supreme Court upheld the challenge to the NCAA's activities, pointing out that the NCAA's role as bargaining agent achieved no efficiencies since separate bargaining between the colleges and networks would still have to occur; it simply assured that the bargaining would take place in a restricted (that is, noncompetitive) market.[30] Moreover, the NCAA "joint venture," at least in its dealings with the TV networks, lacked sufficient efficiency generating char-

[25] *Northrop Corp.* v. *McDonnell Douglas Corp.*, 705 F.2d 1030, 1053 (9th Cir.), *cert. denied*, 464 U.S. 849 (1983).

[26] Ibid. at 1052.

[27] Ibid. at 1053 n29.

[28] 468 U.S. 85 (1984).

[29] Ibid. at 94–8.

[30] Ibid. at 113–15.

acteristics to overcome the obvious restraints on price and output it imposed in its negotiations with the networks.[31]

H. Arizona *v.* Maricopa Medical Society

Arizona v. *Maricopa Medical Society*[32] was another case, like NCAA, where a joint arrangement was found to be illegal due to the absence of real integration or significant efficiencies. In Maricopa, competing doctors agreed through their medical societies to set a maximum fee they would claim in full payment for certain types of health services provided under health care plans.[33] The Supreme Court struck down the arrangement as a naked cartel and therefore illegal *per se*, hinting that the purported goal of setting fee ceilings was really a device to set minimum prices.[34] The defendants sought rule of reason treatment based on the claim that the fee schedules were analogous to the kind of blanket licenses upheld in *Broadcast Music, Inc.* v. *Columbia Broadcasting System.*[35] However, the Court was able to distinguish BMI on the basis that the BMI arrangement generated unusual levels of efficiency and also permitted the sale of a distinguishable new product. The Court also distinguished the medical fee plan from the licensing scheme in BMI by noting the absence in Maricopa of efficiencies derived from integration:

> "The foundations are not analogous to partnerships or other joint arrangements in which persons who would otherwise be competitors pool their capital and share the risks of loss as well as the opportunities for profit. The agreement under attack is an agreement among hundreds of competing doctors concerning the price at which each will offer his own services to a substantial number of consumers ... If a clinic offered complete medical coverage for a flat fee, the cooperating doctors would have the type of partnership arrangement in which a price-fixing agreement among the doctors would be perfectly proper. But the fee arrangements disclosed by the record in this case are among independent competing entrepreneurs."[36]

I. The General Motors–Toyota *Joint Venture Case*

One important case in which regulators explicitly attempted to make a comparative analysis of joint venture efficiencies and possible anticompetitive effects of a joint venture was the joint venture between General Motors ("GM") and Toyota.[37] This case involved a joint venture between GM and Toyota under which the parties agreed to jointly manufacture a subcompact automobile in the United States, with the auto being marketed independently by each of the parents. The FTC originally found that the proposed venture would result in a violation of § 7 of the Clayton Act; however, after extensive negotia-

[31] Another way of looking at NCAA is to conclude that the arrangements with the networks were collateral rather than ancillary to the formation and operation of the joint venture. There are joint venture cases in which the issue of absence of efficiency and the issue of "ancillary" are two ways of looking at roughly the same question.

[32] 457 U.S. 332 (1982).

[33] Ibid. at 339–42.

[34] Ibid. at 348.

[35] 441 U.S. 1 (1979).

[36] 457 U.S. at 356–7.

[37] *General Motors Corp. and Toyota Motor Corp.*, FTC File No. 821-0159, Statement of Chairman Miller and Commissioners Douglas and Calvani, 48 Fed. Reg. 57,314 (1985).

tions, the parties entered into a consent decree[38] which allowed a majority of the commissioners to conclude:

> "The Commission has provisionally determined that the proposed Fremont venture – as outlined in the Memorandum of Understanding and as restricted by the consent agreement – creates a substantial likelihood of producing significant procompetitive benefits to the American public. The Commission has concluded that these procompetitive benefits to the American consumer would outweigh any anticompetitive risks, *provided* that the scope of the venture is restricted. Accordingly, the Commission has provisionally accepted a consent agreement that provides the appropriate safeguards."[39]

The consent decree cited with approval by the commissioners included limitations on the duration of the joint venture (i.e. twelve years); limitations on the joint venture's annual output (i.e. 250,000 cars per year); prohibitions on any exchange or discussion of pricing, costs, sales or marketing information relating to non-joint venture automobiles, or marketing information on the automobile produced by the joint venture; and restrictions on any exchange or discussion of changes to the design of the joint venture automobile, sales or production forecasts of the joint venture automobile, or costs of GM or Toyota products supplied to the joint venture, except as necessary to accomplish the goals of the joint venture.[40] The parties were also required to maintain complete files and records of all communications between and among GM, Toyota, and the joint venture concerning the joint venture product, and make such files available to the FTC.[41]

For the majority, the key efficiency consideration was the opportunity which the joint venture presented for determining if Japanese manufacturing methods could be imported into the United States. Specifically, the commissioners noted that, "To the extent the Fremont venture can demonstrate successfully that the Japanese system can work in America, the Commission finds that this will lead to the development of a more efficient, more competitive US automobile industry."[42] The majority dismissed concerns about possible anticompetitive effects stemming from the possibility of collusion between the venturers, arguing that this was not possible in light of their findings regarding the various characteristics of the automobile industry, including high levels of product differentiation, cyclical demand, and unequal cost structures among competitors.[43] Moreover, since other firms were ready to augment output if GM or Toyota individually or jointly attempt to control supply, the two firms did not have monopoly power in the marketplace.[44]

[38] *In the Matter of General Motors Corporation et al.*, 103 F.T.C. 374 (1984).

[39] 48 Fed. Reg. 57,316 (emphasis in the original).

[40] *In the Matter of General Motors Corporation, et al.*, 103 F.T.C. 374, 383–6 (1984).

[41] Ibid.

[42] 48 Fed. Reg. at 57,315.

[43] In the majority's words: "Given the characteristics of the automobile market, it appears that the joint venture will not result in an increased risk of tacit or explicit collusion. Automobiles are highly differentiated in terms of quality, performance, appearance, and durability. The demand for cars shows cyclical swings. Because firms have unequal cost structures, each firm has a different profit-maximizing price. These and other characteristics make a collusive agreement difficult to establish and maintain." Ibid. at 57,316.

[44] Reliance was also placed on the fact that the consent agreement reached with respect to the joint venture was designed to further reduce any possibility of collusion. See *General Motors Corp.*, and *Toyota Motor Corp.*, Proposed Consent Agreement, 48 Fed. Reg. 57,246 (1983).

The *GM–Toyota* case was instructive for a number of reasons, not the least of which was that there are least five aspects of the arrangement which differentiated it from some of the earlier joint ventures which were found to be illegal (i.e. 3M and Yamaha). These factors including the narrow scope of the joint venture, the limited duration and small intersection of the activities of the two firms in the joint venture, the absence of collateral restraints, the limitations on the exchange of information beyond what is required by the operation of the joint venture itself, and the low level of concentration in the relevant market (i.e. subcompacts). The case is also notable in its focus on balancing the procompetitive and anticompetitive effects of the joint venture, including the intangible benefits of gaining access to the innovative methods of the foreign joint venture partner.

J. Subsequent Developments

The FTC's approval of the *GM–Toyota* joint venture was the culmination of a trend in antitrust policy that had begun at least ten years before[45] and which had been foreshadowed by cases dealing with mergers, monopolies,[46] and distribution.[47] In each of these areas, consideration was given to comparing efficiencies against possible anticompetitive effects. After that case was decided, the Congress took further action to reduce the potential antitrust hazards to research and development arrangements by passing the National Cooperative Research Act of 1984, which was subsequently amended in 1993 by the National Cooperative Research and Production Act of 1993[48] to extend certain protections to production joint ventures.

While the primary targets for these legislative acts were small- to medium-sized firms, regulators were now becoming less reluctant to approve appropriate joint ventures between parties that were clearly industry leaders. For example, in 1991, Apple Computer and IBM formed a joint venture, Taligent, to develop the next generation of operating systems. While Apple and IBM were direct competitors in the sale of operating systems and hardware, the joint venture was formed with the objective of producing a revolutionary, object-oriented operating system that neither company then produced. The FTC did not seek a consent decree, and ultimately approved the joint venture, on the basis that the arrangement was an output-enhancing venture[49] that created a new product and was structured so as to avoid both unnecessary collateral restraints and spill-over effects. For example, the joint venture was structured as an independent company, and the joint venture's board was structured to minimize both the risk that either company could control the joint venture, and the risk that competitively sensitive information would be exchanged by the venture parents. Moreover, the operating system was to be openly licensed to the joint venture parents and to third persons, which meant that the parents would compete head-to-head both against each other, as well as against the joint venture.

[45] The delineation of this evolution has been cogently developed in a series of publications by Dr. Betty Bock. See, e.g. Bock, "Rediscovering Economic Realism in Defining Competition," *RECORD*, June 1974, at 6; "Notes on the Future of Competition," *RECORD*, March 1975, at 28. See also the publications of the Conference Board Antitrust Conferences, particularly for the years 1979 through 1983.

[46] *Telex Corp.* v. *IBM*, 510 F.2d 894 (10th Cir.), *cert. dismissed*, 423 U.S. 802 (1975).

[47] 433 U.S. 36 (1977).

[48] 15 U.S.C. §§ 4301–5.

[49] The object-oriented operating system to be developed by the joint venture greatly reduced the time to market of software applications and, thus, increased the output of software vendors.

II. Development of Competition Law Policy in the EC

A. Early Commission Attitudes Regarding R&D Collaborations

Introduction The first public pronouncements by the Commission as to its attitude regarding joint research and development arrangements came in 1968, when the Commission handed down its decision in the *Eurogypsum*[50] case, and then followed that decision six months later with the publication of its Notice on Co-operation Between Enterprises.[51] Both the *Eurogypsum* case and the Notice dealt with a relatively simple, and limited, group of arrangements in which the parties to the collaboration essentially limited there activities to an exchange of experiences and information without any restrictions on their ability to conduct their own research and development activities.[52] The Notice now has only limited application[53] and, in fact, the *Eurogypsum* collaboration is probably the only relevant case of pure research and development where the Commission has held that Article 85(1) does not apply.

The *Eurogypsum* case *Eurogypsum* involved a trade association that was formed for the purpose of promoting development of the plaster and gypsum industry by conducting research on scientific problems and providing information, through lectures and publications, to the individual members and the general public. A majority of the undertakings within the Common Market involved in the manufacture of plaster, gypsum, or their constructional derivatives, belonged to the association, either directly or through a national trade association. The Commission appeared to have little difficulty in granting negative clearance to the rules of the association, given the open nature of the constitution of the association and the fact that no restrictions were placed upon the use of the results of its research by the members.[54] Clearly, members of the association retained the freedom to continue to carry on their own research and development activities, and any benefits from participating in the association supplemented, rather than replaced, their own development work.[55]

1968 Notice on Co-operation Between Enterprises The Notice was intended to identify certain types of cooperative arrangements relative to research and development which, in the Commission's view, would not infringe Article 85(1), even in those cases where the participants were actual or potential competitors. The Notice defined several different forms of collaboration which might be eligible for the exclusion from Article 85(1); however, the Notice contained a number of broad exceptions which, taken together, tended to significantly limit the scope and utility of the comfort provided by the Notice.

The Notice began by stating that Article 85(1) would not apply to cooperative agreements which were strictly limited to the stage prior to industrial application and having as their sole object joint implementation of research and development projects; joint placing of research and development contracts; sharing out of research and development

[50] [1968] CMLR D1.

[51] OJ 1968 C75/3 as corrected OJ 1968 C93/14.

[52] See Goyder, *EC Competition Law* 2nd ed, Oxford: Clarendon Press (1993) 184.

[53] The notice is endorsed in a qualified fashion in the Commission's JV Notice, which states the Commission's current views in this area.

[54] See Goyder, supra note 52, at 184 (noting that not only were all members free to use the results of the research work, but that such results were often made available to the public).

[55] Ibid.

projects among the participants; or the mere exchange of results and experience in the field of research. However, the Notice itself, as well as subsequent Commission decisions, limited its effect as follows:

(i) The exclusion from Article 85(1) may not be appropriate if the participants agreed to refrain[56] from research work on their own in some or all of the fields covered by the agreement.[57]
(ii) The parties were limited in their ability to accept restrictions as to the exploitation of the products of the joint development work.
(iii) The parties needed to have access to the results of the joint development work at least in proportion to the amount of their participation in the work.
(iv) The grant of licenses to third parties was expressly or tacitly excluded from the effect of the Notice.[58]
(v) Based on subsequent Commission decisions, agreements between significant competitors (e.g. parties having a combined market share in excess of 20 percent) would fall within Article 85(1), even if the collaboration was limited in scope to that described in the Notice.[59]
(vi) The Notice provided no protection in situations where the joint applied research program overlapped with other joint activities directed toward industrial application of the results.[60]

B. Case Law Developments from the Early 1970s to Mid-1980s

Introduction While the Commission's 1968 Notice was often cited for the proposition that the Commission would take a fairly liberal view of cooperative arrangements which extend only to the stage immediately prior to industrial application, as a practical matter the case law which developed from the early 1970s was to reflect a relatively cautious approach by the Commission in these situations.[61] In one case, the Commis-

[56] The Commission's interpretations of the circumstances in which the parties may be considered to have "agreed" to refrain from independent research have been very wide, meaning that more agreements would fall outside the protections of the Notice. See, e.g. *Henkel/Colgate*, OJ 1972 L14/14 and *Beecham/Parke-Davis*, OJ 1979 L70/11 [1979] 2 CMLR 157.

[57] See also Recital 2 of the group exemption covering research and development arrangements, which was finally adopted in 1985, stating: "(2) As stated in the Commission's 1968 notice concerning agreements, decisions and concerted practices in the field of cooperation between enterprises, agreements on the joint execution of research work or the joint development of the results of the research, up to but not including the stage of industrial application, generally do not fall within the scope of Article 85(1) of the Treaty. In certain circumstances, however, such as where the parties agree not to carry out other research and development in the same field, thereby foregoing the opportunity of gaining competitive advantages over the other parties, such agreements may fall within Article 85(1) and should therefore not be excluded from this Regulation."

[58] The Notice actually contained a proviso that the need to obtain the consent of the other parties to grant licenses does not infringe Article 85(1). However, this is no longer the law after *Beecham/Parke-Davis*, OJ 1979 L70/11 [1979] 2 CMLR 157.

[59] *Continental/Michelin*, OJ 1988 L305/33, [1989] 4 CMLR 920 (Article 85(1) applied to joint venture between parties who were competitors with a combined market share of over 20 percent, even though joint venture was confined to research and development and cooperation short of joint selling).

[60] See, e.g. *Beecham/Parke-Davis*, OJ 1979 L70/11 [1979] 2 CMLR 157.

[61] Other cases in this area include *MAN/SAVIEM* JO 1972, L31/29 [1974] 2 CMLR D123; *Bayer/Gist- Brocades* OJ 1976, L30/13 [1976] 1 CMLR D98; *GEC/Weir Pump* OJ 1977, L327/26 [1978] 1 CMLR D42; *Beecham/Parke Davis* OJ 1979, L70/11 [1979] 2 CMLR 157.

sion expressed concerns that upstream collaboration short of the stage of industrial application among large competitors in an oligopolistic market could have sufficiently adverse competitive consequences to bring the arrangement within Article 85(1), even though the parties were not restrained from their own independent research and development activities.[62] Apparently, the Commission feared that the oligopolistic nature of the market heightened the importance of competition at the research and development stage and that collaboration between the parties with respect to that activity would dampen the competitive tension which otherwise existed between them with respect to downstream activities, even though the collaboration itself did not extend that far.[63]

The *Henkel-Colgate* case In the early 1970s the Commission reviewed a proposed joint venture between two leaders in the European detergent market, Colgate, which held approximately 10 percent of the market, and Henkel, which had a roughly 30 percent market share.[64] The parties had proposed a new joint venture which was to be limited solely to research and development activities, and which would not extend to distribution and marketing. The Commission noted that competition in the detergent industry was centered largely in the ability of firms to improve the technical quality of their products, meaning that research and development capabilities were very important factors in the industry. The Commission also noted that the joint venture would, as a practical matter, remove almost all incentive for Colgate and Henkel to conduct their own research activities, given that each party was bound to encourage the joint venture to exploit any success progress which it might make on its own. As such, the joint venture would improve the joint competitiveness of the parties, while eliminating any competition between them in the area.

While it is not surprising that the Commission had concerns regarding a collaboration between firms of the size and strength of Colgate and Henkel, the Commission nonetheless granted an exemption for the research and development activities based on a finding that the collaboration might enable technical and economic progress and that the structure of the joint venture contained no restrictions on the independent activities that were not necessary in order for the objectives of the arrangement to be achieved.[65] The Commission noted that the research results would be available to both parties upon payment of a modest royalty and on a non-exclusive basis.[66] Although the joint venture was prohibited from granting licenses to third parties without the consent of both par-

[62] *Henkel/Colgate* JO 1972 L14/14; Commission's First Report on Competition Policy, p. 47 ("the prohibition of Article 85(1) could apply to joint research by two large enterprises with world-wide markets and with a very strong position on the Community markets which are characterized by their oligopolistic structure and great technical homogeneity of products offered. as well as by particularly intensive and costly advertising, which makes penetration of the market very difficult for other manufacturers who wish to compete").

[63] See also Recital 3 in the published draft of the group exemption for research and development arrangement which stated, in relevant part, that "an appreciable restriction of competition will often occur where the parties are large undertakings with significant market shares, where competition in research and development plays an important role in the competitive performance of undertakings, [and] where the research and development cooperation is particularly long in duration or extensive in scope." This language was not included in the version of the Regulation which was ultimately adopted.

[64] See JO [1972] L14/14. In three of the member states, the combined market share of Colgate and Henkel was over 50 percent; however, they did have major competitors (e.g. Unilever and Procter and Gamble). See Goyder, supra note 52, at 185.

[65] Ibid.

[66] Ibid. (the parties would be required to pay a royalty not to exceed 2 percent of the net price of the product).

ticipants, this was permitted on the basis that such a limitation was an "inherent consequence" of establishing the joint venture for the benefit of the firms.[67]

While the Commission authorized the research and development collaboration, it sought to take steps to ensure that the parties did not extend their arrangement to other activities, such as market-sharing arrangements or agreements regarding production and sales policies. This was achieved by imposing requirements that the parties provide the DG IV with certain information on their policies with respect to licensing the results of the joint research, as well as certain information regarding changes in the shareholdings of the joint venture or in other connections between them.[68]

Vacuum Interrupters *Vacuum Interrupters*[69] involved the formation of a new joint venture company by AEI Ltd. and Reyrolle Parsons Ltd. in order to conduct research and development on a "circuit breaker," which was a key component for heavy electrical equipment. Both venturers were manufacturers of heavy switchgear and earlier development work had proceeded quite slowly and with only limited success. Further work would continue to be expensive and it appeared unlikely that either party would have sufficient inducement to continue unless they were able to pool their resources with the other party.[70] Under the proposed terms of the joint venture arrangement, the joint venture company would only manufacture the circuit breakers and not the switchgear into which they would be installed.[71] The joint venture company would be free to manufacture and sell the circuit breakers to any manufacturer and would handle all orders, whether from a parent company or otherwise, on a confidential and independent basis.[72] For their part, the parents agreed not to develop, design, manufacture, or sell circuit breakers and to use their reasonable efforts to purchase all their requirements for the circuit breakers from the joint venture.[73]

The Commission found that, even though AEI and Reyrolle were not actually competing in the market for circuit breakers, Article 85(1) would apply to the joint venture since both AEI and Reyrolle were potential competitors and the entry of both parties into the market separately would not only have resulted in direct competition between them, it would also have had other effects on the market for circuit breakers in both the United Kingdom, where the parties were located, and throughout the EC.[74] The Commission arrived at this decision even though there were actually no manufacturers of the circuit breaker in the market at that time. The Commission then went on to grant an

[67] Ibid. It is interesting to note that the parties chose not to commence the venture at the time that the original exemption was granted and that seven years later, when the parties once again sought an exemption from the Commission, the application was denied on the basis that the requirement of mutual consent for any license to third parties had too great an effect on competition between the parties. Ibid. (citing 8th Annual Report, 77–8).

[68] Ibid. at 185–6.

[69] [1977] 1 CMLR D67.

[70] See Goyder, supra note 52, at 186.

[71] Ibid. at 186–7. The switchgear itself is actually ordered on an individual basis. Under the arrangement, either AEI or Reyrolle Parsons would give the specifications for the circuit breakers to the joint venture company who would then design and construct the circuit breakers for ultimate installation by the specifying party in the switchgear to be delivered to the end user. Ibid. at 187.

[72] Ibid.

[73] Ibid.

[74] Ibid. For example, the pattern of exports and imports between the United Kingdom and the rest of the Common Market would be affected, and there also would have been a substantial reduction in the possibility that Common Market undertakings would have entered the market for circuit breakers in the United Kingdom. Ibid.

exemption for the joint venture based on the belief that the technical progress from the joint venture, in the form of a more durable and efficient low power circuit breaker which could be manufactured at a reasonable cost, would provide a fair share of the benefits to the end users of the electricity.[75]

De Laval/Stork *De Laval/Stork* involved a joint venture which covered not only research and development, but which also extended to cover joint production and distribution.[76] The Commission had little trouble finding that the arrangement violated Article 85(1), given that parties were competitors,[77] which had operated independently of each other in the Common Market, and that the production and marketing of the joint venture products would no longer be done independently. The Commission observed that neither party was likely to compete with its joint subsidiary; moreover, Stork was contractually prohibited from competing with de Laval in some non-member states. Accordingly, it was held that the effect of the agreement was to eliminate competition between the parties, since the parties had deprived themselves of the opportunity for autonomous costing and pricing.[78]

Although the Commission rarely grants exemptions for joint ventures which are designed to include not only research, but to also cover marketing, it did so in this case. However, in so doing, it required that each party should receive any licenses relating to patents or know-how licensed to the joint venture and indispensable for the use of such patents or know-how, at royalty rates which were no greater than the minimum charged to any third party, even after the three years originally agreed to in the joint venture agreement. This ruling by the Commission raised several concerns in this particular context. First, given that de Laval apparently had provided about 85 percent of the basic designs that were to be used in the joint venture's development activity, the obligation of equal access may have upset the commercial balance between the parties. Second, the Commission took over six years to render its decision. During that time, the relative bargaining power of the parties may have changed substantially, and the Commission's conditions may have led one of the parties to insist upon renegotiation of the entire arrangement as part of its decision as to whether or not to consent to the Commission's requirements.

De Laval/Stork illustrates the potential impact of the Commission's attitudes and procedures on the commercial actions of firms in the marketplace. Clearly, while the conditions that were eventually imposed by the Commission may well have had the effect of increasing competition by making it easier for each party to effectively compete after the joint venture itself had terminated, they may also have had an adverse effect on the feasibility of the transaction to the extent that they modified the commercial balance the parties had been able to independently achieve in the original contract. In the long run, if potential joint venturers have concerns regarding the attitude of the Commission with respect to specified terms and conditions, it may lead them to seek other structure for a collaborative effort, some of which may be less efficient than the joint venture form.

[75] Ibid.

[76] OJ 1977, L215/11; [1977] 2 CMLR D69.

[77] It is interesting to note that it was not clear that Stork was capable of entering the marketing independently, since its technical capabilities were based, at least in part, on an earlier license from de Laval, which was not considered to be part of the same transaction.

[78] There were also a variety of specific restrictions in the joint venture agreement, including the grant of a blocking vote to each of the parties on the board of directors of the joint venture.

GEC/Weir *GEC/Weir*[79] involved the creation of a joint committee in which GEC and Weir Group were to hold an equal interest and which was formed to develop a very specialized product to be used in fast nuclear reactors.[80] Both parties were major UK industrial groups, with a number of overlapping business activities, including the development and manufacture of pumps for fast reactors. Although GEC previously had built a sodium circulator which had been used as part of a pilot fast reactor, it had concluded that there was a good deal of further development work which must be completed on the product. Based, in part, on the requests of the only major UK customers for the new product, GEC agreed to work with Weir on the development work, largely because Weir could supply the needed supply pump technology.[81]

Under the terms of the various collaboration agreements entered into in connection with the joint committee, the parties would effectively exercise joint control over all of their activities with respect to the new product, including such things as planning, financing, research, development, production and sale. The development and production work, which would require a wide variety of costly and highly specialized skills, was to be allocated between the parties, rather than done jointly, and the agreements described those areas for which each party was to have primary responsibility. Both parties agreed to grant non-exclusive licenses to the committee covering all the technical information that would be required in order to complete the development work, and the information was freely shared between the factories and facilities of each of the parties.[82]

The parties would not independently compete in the joint committee's field of activity during the period the committee was in existence, except in certain limited cases where either a customer declined to contract with the joint venture or the other party did not wish to accept a specific contract with a third party.[83] However, both companies were to have the right to freely used the information from the venture, including the right to sub-license others, following termination of the venture, subject to an obligation to pay royalties for such use.[84]

The Commission had no difficulty in concluding that the restrictions placed on competition by the contractual arrangements were sufficient to bring the committee within Article 85(1). It seemed clear that each of the parties was abandoning its right to freedom of action with respect to business activities relating to sodium circulators. For example, not only had the parties explicitly agreed not to compete with the joint venture, the fact that they were also exchanging all technical information meant that neither could obtain a technological advantage over the other. In addition, the fact that the parties were involved in overlapping activities and that senior personnel from both parties would be involved in such activities and with the joint venture meant that it was likely that competition between them would be impaired in other areas. Finally, third parties were adversely effected by the existence of the arrangement since they could no longer participate with either of the parties with respect to activities relating to the new product, nor would they be able to have more than a single supplier of the new products.[85]

[79] [1978] 1 CMLR D42.

[80] See Goyder, supra note 52, at 187. The product was a sodium circulator which is required to maintain circulation of the coolant for the high-powered density cores of fast nuclear reactors. Proper design of the product required a high degree of accuracy and reliability. Ibid.

[81] Bellamy and Child, *Common Market Law of Competition* 4th edn, London (1993) § 5-071.

[82] See Goyder, supra note 52, at 188.

[83] Bellamy and Child, supra note 81, at § 5-071.

[84] See Goyder, supra note 52, at 188.

[85] Bellamy and Child, supra note 81, at § 5- 072.

The Commission then went on to grant an exemption for the joint committee under Article 85(3). The decision was based on a variety of procompetitive factors which the Commission found in the analyzing the arrangement. First, the committee combined the complementary skills of the parties in a manner which permitted them to overcome the substantial risks and potential development problems associated with commercializing the new product. Second, it would not have been possible for either of the parties to independently to complete the planned work as quickly, effectively or economically as could be done as part of the joint committee. Third, although the parties could not compete with each other during the term of the arrangement, following the completion of the collaboration, both of the parties would have substantially greater technical knowledge and, presumably, a much improved capacity for future competition.[86] Fourth, even though consumers may suffer some short-term loss in potential alternative suppliers, they would eventually benefit from the cost savings generated by the joint committee and the long-term competitive strengths of the parties. Finally, as to the non-competition provisions, the Commission concluded that such restrictions were required in order to convince the parties to dedicate their resources to the project.[87]

The duration of the exemption granted by the Commission under Article 85(3) was twelve years, a period which was set with reference to the fact that the parties would only begin to receive the benefits from the actual sale of the products in the last three and one-half years of that period.[88] The Commission also required that the parties continue to supply it with information regarding developments in the industry (i.e. the United Kingdom nuclear power industry) which affected the terms of the collaboration between them.[89]

Beecham/Parke Davis The attitude of the Commission toward collaboration in an industry which is driven by innovation competition, particularly at the research and development stage, was illustrated by its decision in *Beecham/Parke Davis*.[90] The case involved a proposed agreement between Beecham and Parke Davis for joint research into drugs that would prevent the impairment of blood circulation.[91] At the time that the research agreement was proposed, there were no drugs available on the market to treat the condition.[92] The development work probably would not have been undertaken by either company individually without the assistance of the other. The difficulties and risks of the development program could be seen by the fact that the time-scale for the work was believed to be at least a decade, with testing alone occupying a significant period of time.[93]

The arrangement involved the joint planning of individual research and the exchange of results. The parties agreed to share the costs of the research work equally and were entitled to receive sole patents covering their own developments. The parties would agree to cross-license patents on a non-exclusive royalty basis, and neither party would be restricted from conducting its own individual research. If the joint research was success-

[86] This was particularly true in the case of Weir, which had the basic supply pump technology, but had not theretofore been involved in the production of the new product. See Bellamy and Child, supra note 81, at § 5-073.

[87] See Goyder, supra note 52, at 188.

[88] Bellamy and Child, supra note 81, at § 5-073.

[89] See Goyder, supra note 52, at 188.

[90] OJ 1979 L70/11 [1979] 2 CMLR 157.

[91] Ibid.

[92] See Goyder, supra note 52, at 188.

[93] Ibid.

ful, a joint development program focusing on marketing the new product(s) was contemplated, although that program was also expected to take a significant period of time, at least five years by some estimates. Each party would have the right to individually carry out the marketing, including the ability to grant licenses to third parties without the consent of the other party (except in the case of any Japanese patents); however, the parties agreed to share any information regarding improvements in the manufacturing process for a period of ten years following the date of initial marketing and to cross-license such improvements to the other party.[94]

The Commission found that the proposed joint arrangement was covered by Article 85(1), even though the parties had previously decided not to conduct any further independent research in the area in which the joint venture was to operate. Although there was no express prohibition on independent research, it was clear that the joint planning and regular exchange of research results contemplated by the arrangement would discourage any individual research work. As a result, neither party would be able to obtain any competitive advantage with respect to innovation, which was considered to be a key element of competition in this particular industry. Also, the agreement between the parties regarding information exchange after marketing began took the arrangement outside of the Notice on Co-operation Agreements, which was limited to development joint ventures.

Even though the joint agreement fell within Article 85(1), the Commission granted an exemption for ten years, noting specifically that a conclusive element in its deliberations was the fact that the research program relating to a new product that was considered to be "pharmacologically and therapeutically different from all known medicine," which meant that the agreement did not relate merely to improving production of existing products but actually was intended to create a wholly new product.[95] It was acknowledged that regardless of the combined market shares of the parties and their tremendous resources, such an undertaking had a substantially better chance of succeeding if carried out on a joint basis[96] and, in fact, the product probably would not have been developed but for the joint venture. The provisions regarding information exchange were considered to be indispensable to the aims of the venture.

The Commission also approved of the fact that both parties were free to use and license the research results on their own; however, it did request certain modifications in their original licensing arrangements. For example, the Commission asked the parties to delete any obligation to remit royalties for cross-licenses on the ground that the royalty burden might present a considerable disincentive for the parties to actually compete with one another.[97] The Commission also required that the parties remove an original restriction on independent sub-licensing in France. Finally, the Commission objected to a provision which would have extended the agreement every time the product was

[94] Ibid.

[95] Ibid.

[96] Ibid.

[97] The Commission condemned the royalty provisions as "likely to create a considerable disincentive for the parties to compete with one another, particularly where marketing raises difficulties, since substantial returns through royalties would have been obtained without production or marketing expenditure." § 43(c) of the Decision. The case involved parties which were to make substantially equal contributions to the joint venture. The Commission did recognize that its position regarding royalty provisions might be different in situations where the parties had not made equal contributions.

to be marketed in a new country and declared that the exemption was to run only for a determinate period.[98]

Rockwell/Iveco In *Rockwell/Iveco*,[99] the Commission was called upon to consider the issue of whether the parties to the proposed joint venture were potential competitors in the market. Rockwell possessed the technology required to make single reduction axles for heavy trucks. Iveco, which was part of the Fiat group, was involved in the production of heavy trucks; however, it lacked the capabilities to produce the new axle technology. In almost all cases, truck manufacturers made their own axles; therefore, it was not surprising that Iveco would find it attractive to form a new joint venture with Rockwell under which axles would be made using Rockwell's technology in one of Iveco's empty factories. The joint venture would not only allow Iveco to make a truck that was more competitive, the parties could also foresee that the new joint venture might be able to create a new market for axles that were produced by an entity other than a firm which was wholly owned by a truck manufacturer.

It would appear that the joint venture was pro-competitive, since Iveco lacked the technology and would not have entered the market on its own. However, the Commission found that the joint venture restricted competition and, as such, was violative of Article 85(1). The Commission explained that both parties were potential competitors and that they each chosen to forego the opportunity to independently engage in the manufacture of the new axles. Moreover, the parties could not enter the downstream market for the axles on their own, since the joint venture was to have the right to distribute other axles made by the venturers; Rockwell had agreed not to grant licenses to third parties in the territory to be worked by the joint venture; and Iveco agreed to take its requirements of certain kinds of axles from the joint venture. It was only when the Commission turned to an analysis of the joint venture under Article 85(3) that it actually made a realistic assessment of the market, finding that an exemption was warranted in light of the fact that the joint venture ultimately made the market more competitive.

Carbon Gas Technology *Carbon Gas Technology*[100] was a case involving a joint venture among three companies for the development of a process that would permit conversion of coal into gas, a process that promised to have significant economic consequences in the EC, given its dependence on imported oil. Each of the parties to the joint venture brought a specific set of complimentary skills and assets to the project, which was to consist of research and development and construction of a demonstration plant costing approximately DM400-500 million.[101] The parties agreed to various restrictions on their own activities. For example, they agreed not to compete with the joint venture, which essentially meant that they would not compete with one another, during the joint development program or for a period of five years following withdrawal from the joint ven-

[98] Since the date of first marketing could not be known in advance, the Commission chose a fixed period for the exemption (twelve years).

[99] OJ 1983, L224/19 [1983] 3 CMLR 709.

[100] [1984] 2 CMLR 275

[101] See Goyder, supra note 52, at 189. PCV, a German company, provided specialized skills in the basic process technology and the manufacture of the equipment which would be necessary for the conversion; Deutsche Babcock would provide the expertise required to construct the large-scale plants and facilities in which the conversion equipment would be housed; and Deutsche BP had experience in refining oil with a technology that was similar to that which would be employed in the attempted gasification of the coal. Ibid.

ture.[102] They also agreed that if they withdrew from the joint venture, they would not disclose for a period of five years any know-how that had belonged to the joint venture, whether the know-how originated with one of the parent companies or from the activities of the joint venture itself.[103]

The Commission found that Article 85(1) would apply to the joint venture, since at least some of the parties could have been able to complete the project independently. The Commission found, therefore, that the parties were potential competitors who had, by entering into the joint venture, accepted significant restrictions on their abilities to compete with one another. However, an individual exemption was granted under Article 85(3) since it was reasonable to assume that the cooperation between the parties, each of which had achieved significant size in their own specific markets, would result in considerable efficiencies and savings of time and money. It also seemed plausible, given that the results of research work would produce a new product that would actively compete with alternative fuel sources, that consumers would be able to share in the benefits from the research.

While the Commission exempted the joint venture, it did so only after imposing conditions designed to insure that the venture did not ultimately become a vehicle for the parties to establish a market-sharing arrangement, which was considered to be possible in light of the individual strength and contacts of the parties in their specialized areas.[104] Specifically, the parties were made to submit all licensing agreements between the joint venture and its parents to DG IV for prior approval so that the Commission could be sure that the parties were not making anticompetitive arrangements for the sale of the equipment or facilities in certain markets in the Common Market.[105]

C. The Block Exemption and Subsequent Cases

Introduction The biggest development with respect to joint ventures during the middle of the decade was the final adoption as of March 1, 1985 of the long delayed group exemption for research and development arrangements – Regulation 418/85[106] – which covered joint venture agreements relating to research and development and certain activities with respect to the actual exploitation of the results of such research and development work. The block exemption is discussed in greater detail in the following chapters; however, since the exemption was adopted, the Commission has continued to expound upon the boundaries of the permissible scope and activities of collaborative ventures in various decisions relating to requests made by thc parties for individual exemptions. The decisions in these cases, as well as the Commission's previous opinions, provide some guidance as to the Commission's analysis in this area; however, it appears that Regulation 418/85 may well serve as the model for individual exemption decisions, at least with regard to the definition and interpretation of certain basic concepts.[107]

[102] The parties other than BP expressly agreed not to compete with the joint venture. The Commission held that BP was de facto not likely to compete with the joint venture even though it had not agreed to be bound by any express restriction.

[103] See Goyder, supra note 52, at 189.

[104] Ibid.

[105] Ibid.

[106] Regulation 418/85 was subsequently amended as of April 1, 1993 by Regulation 151/93 and expires on December 31, 1997.

[107] See, e.g. *Continental/Michelin*, OJ 1988 L305/33 [1989] 4 CMLR 920 § 23 ("the individual exemption should also directly base themselves on [Reg. 418/85] at least as far as the definition of basic concepts is concerned").

Cases relating to the "ancillary restraints" doctrine *Mitchell Cotts/Sofiltra*[108] involved a joint venture between Mitchell Cotts, which was a United Kingdom-based manufacturer of air filtration devices, and Sofiltra, which produced certain high efficiency pleated glass fibre paper filters that were included as part of the devices manufactured by Mitchell Cotts. Before the establishment of the joint venture, the filters, which were a key technical component and major cost of the air filtration devices, had to be imported. The proposed joint venture was to facilitate the transfer of the necessary technology to Mitchell Cotts in the United Kingdom to allow the final products to be locally manufactured. Sofiltra was to grant an exclusive know-how license to the joint venture and the joint venture was to have the right to manufacture and sell the finished products in the United Kingdom. The joint venture itself was prohibited from manufacturing or dealing in any competitive products, and the exclusive license from Sofiltra included terms which prevented Sofiltra from manufacturing in the United Kingdom or granting licenses to others to manufacture in the United Kingdom and prohibited sales by the joint venture outside of the United Kingdom, presumably into markets where Sofiltra was still able to operate freely. Sofiltra would have a 25 percent interest in the joint venture; however, by virtue of certain block powers in the joint venture agreement, the parties effectively shared control of the venture.

In determining whether Article 85(1) would apply to the proposed joint venture, the Commission concluded that the parties were not, on the facts, actual or potential competitors at the manufacturing level, since it was highly unlikely that Mitchell Cotts would ever independently develop the production technology which was being supplied by Sofiltra. The Commission, which took the entire EC as the relevant geographic market, also found that there was no appreciable risk of foreclosure to third parties from the joint venture in light of the relatively small market shares of the parties in the EC. Accordingly, the Commission was able to determine that Article 85(1) was not infringed by the joint venture with respect to manufacturing.

The Commission then considered some of the restrictions included as part of the joint venture arrangement. In large part, the Commission accepted that the restriction on the joint venture's ability to manufacture or deal in competitive products was essential to the success of the joint venture and did not appreciably impact competition, particularly in light of the high capital costs that would be incurred in commencing manufacturing. However, there were some problems stemming from the fact that Sofiltra, but for the concessions made under the terms of the exclusive license granted to the joint venture, would be at least a potential competitor to the joint venture. The Commission did believe that Article 85(1) was infringed by the combined effect of the terms of Sofiltra's exclusive license, and the Commission stated that these restrictions could not, particularly in light of the fact that they were unlimited in duration, be regarded as "strictly necessary for the creation and proper operation" of the joint venture, which was the standard that must be satisfied in order for the restriction to be considered "ancillary," and thus fall outside of Article 85(1).

The Commission then proceeded to exempt the exclusive license under Article 85(3), noting that the various restrictions contained in the license were indispensable "since without them the parties would not be prepared to make the necessary effort and commitments." The Commission noted that the capital costs associated with establishing the new manufacturing facility would be significant, and it appeared that the Commission was willing to accept the fact that an exclusive license which included a territorial re-

[108] OJ 1987 L41/31 [1988] 4 CMLR 111.

striction could be accepted as ancillary, at least during the start-up phase of the joint venture.[109]

Elopak/Metal Box-Odin[110] was another important case with respect to the Commission's policies in analyzing potential competition and ancillary restrictions. Odin was a 50/50 joint venture company established by Elopak and Metal Box. The joint venture began as an effort to conduct research and development relating to a new kind of container and the development of the technology and machinery that would be necessary to fill such containers. The container to be developed did not exist at the time that the joint venture was to commence. If the development efforts were successful, it was anticipated that the joint venture, which was to be of indefinite duration, would have the responsibility for producing and selling the containers and the filling machines throughout the EC.

The parties each agreed to grant the joint venture an exclusive license to exploit their relevant existing intellectual property rights worldwide within the joint venture's anticipated field of activity. In turn, the joint venture was prohibited from exploiting those rights outside of its field of activity. There were also a number of other restrictions included as part of the joint venture agreements. The parties agreed that upon termination of the joint venture, each party would have access to the relevant technology, including the intellectual property rights originally contributed by the other party, in order for them to be able to compete independently in the joint venture's field of activity, although the parties were restricted for a certain period of time from exploiting the technology in collaboration with a third party competitor.[111]

The Commission approached the question of whether the joint venture infringed Article 85(1) by analyzing four key issues. The first question was whether or not the joint venture restricted the level of actual or potential competition between the parties. The joint venture's product would be a paperboard-based carton with a separate metal lid, and could be used for aseptic filling with shelf-stable UHT-treated particulate foods such as soups and baby foods. Metal Box had a number of existing packaging products, but no paperboard cartons. Elopak was a manufacturer of paperboard cartons, but lacked the know-how required for producing either the cartons or the filling machines for aseptic filling. The parties believed that the new product, once developed, would compete with some of the products produced by Metal Box (i.e. metal cans), but not with many other currently produced products. However, it was more likely that the new products would eventually become their own distinct market.[112]

The Commission analyzed the particular structural and dynamic market characteristics and concluded that, in light of the costs of developing the new product, the parties would not have entered the market independently. Among the specific characteristics which the Commission took into account were the existing technical skills and capabilities of the parties, the commercial risks associated with attempting to persuade consumers to put aside their aversion to the innovative nature of the product, and the fact that the parties would need to be able to provide extensive customer support at the time

[109] Ibid. at § 23. See also *Elopak/Metal Box-Odin*, OJ 1990 L209/15 [1991] 4 CMLR 832 and *Konsortium* ECR 900, OJ 1990 L228/31 [1992] 4 CMLR 54, which the Commission found that exclusive licenses to a joint venture without territorial restrictions could properly be regarded as ancillary.

[110] OJ 1990 L209/15 [1991] 4 CMLR 832.

[111] The technology developed by the joint venture, as opposed to technology licensed to the joint venture by the parties, was not subject to any post-termination field of use restrictions; however, collaboration with third party competitors was not to occur for a period of five years following termination of the joint venture.

[112] Bellamy and Child, supra note 81, at §5-064.

that the new product was introduced in order to insure that it would be accepted. The Commission also felt that it was unlikely that either party would be willing, or able, to make the large investment necessary for it to acquire or develop the technology possessed by the other party. In light of the foregoing, the Commission found that it was unlikely that either party would be able to independently enter the market in the "short term," and that the parties were therefore not competitors in the particular field in which the joint venture was active.[113]

The question of whether the joint venture would provide competition for one of the parties was a bit more difficult, since it was possible that the joint venture's product would, at least indirectly, compete with Metal Box's existing metal cans products. However, the Commission did not believe that this situation would bring the joint venture within Article 85(1), particular since there were substantial disincentives for Metal Box to attempt to coordinate its commercial behavior with the joint venture, given that Elopak had the unilateral right to dissolve the joint venture and thereby trigger an extensive network of cross-licensing rights which would deprive Metal Box of any competitive advantage.[114]

The Commission next examined whether the joint venture would have the effect of foreclosing competition from third parties. It concluded that even though the various packaging markets for cans, jars and brick cartons were generally quite concentrated, there were nonetheless a number of other parties throughout the EC which had the capability to develop or acquire the technology necessary to compete with the joint venture's products.[115] Finally, as to the potential "spillovers" from the joint venture, the Commission found that the parties were not actual or potential competitors outside of the joint venture field, since Metal Box had no products in the specific area (i.e. the manufacture of cartons for fresh or pasteurized liquids) in which Elopak was active.[116]

After concluding that Article 85(1) did not apply to the formation of the joint venture itself, the Commission went on to examine the specific restrictions included in the joint venture agreements. Upon analysis, the Commission concluded that each of the restrictions also fell outside of Article 85(1), since they were restrictions which were ancillary to the formation of a non-infringing joint venture. Among the permitted provisions were exclusive know-how licenses from each party to the joint venture covering the relevant technology, provided that the "exclusivity" is strictly limited to the specific products to be produced by the joint venture and to a restricted field-of-use; non-exclusive licenses from the joint venture back to each of the parties covering improvements which (i) the joint venture does not wish to exploit itself (ii) are only to be used outside of the field in which the joint venture is active; first options in favor of parties with respect to rights to obtain licenses from joint venture for exploitation of technology in geographic areas other than those in which the joint venture is active; and restrictions on the ability of a party to dispose of its interest in the joint venture without the consent of the other.

The "network effect": *Optical Fibres* The *Optical Fibres* case pertained to three joint ventures involving Corning, a United States glass manufacturer which had developed new optical fibre technology.[117] The joint ventures were with major cablemakers in the United Kingdom (50/50 with BICC), France (40/60 with COFO) and Germany (50/50) to pro-

[113] Ibid. at § 5-066.
[114] Ibid. at § 5-067.
[115] Ibid. at § 5-068.
[116] Ibid. at § 5-069.
[117] OJ 1986 L236/30.

duce optical fibres for use in telecommunications. Cablemakers were the main purchasers of optical fibres. The Commission made a variety of findings regarding the joint ventures and the actual relationships between them. First, each joint venture was involved in substantially the same business. Second, the common parent, Corning, owned a substantial interest in each of the joint ventures and participated in their management. Finally, each joint venture was materially dependent upon the technology to be provided by Corning.

The Commission believed that there were no restrictions on competition between Corning and the other parents, since they were not actual or potential competitors. Also, the Commission found that no competitive opportunities were foreclosed to third parties as a result of the joint ventures. However, in spite of the foregoing, the Commission did conclude that Article 85(1) was infringed by virtue of the relationship that existed between the joint ventures themselves. Specifically, the Commission held that Corning was in a position to influence and coordinate the various activities of each of the joint ventures in a market which already was oligopolistic, and this possibility of influence was sufficient to bring the arrangements within Article 85(1).

The Commission went on to grant an exemption to each of the joint ventures, noting that the chosen structure provided a number of benefits. For example, optical fibres were a new high technology product that had significant advantages over the traditional cables, and the Commission found that the joint ventures enabled this technology to be rapidly transferred to other companies within the EC and also facilitated the production of a more advanced product. The Commission also noted that the transfer of the relevant technology, which was a key step in the exploitation of the new product, could be more expeditiously achieved by Corning's active involvement in the joint ventures then by Corning establishing its own plants and/or granting licenses to third parties, including the partners in the joint ventures.

The period of the exemption was set at fifteen years; however, the Commission did require that a number of changes be made to the joint venture arrangements before the exemption was granted in order to reduce the potential dangers of Corning's management and voting rights. For example, amendments were required to: (i) reduce Corning's voting rights, management participation and rights of veto; (ii) provide each of the joint ventures with a non-exclusive sales license and the right to make active sales into the territories of the other joint ventures; (iii) provide that the joint ventures could make passive sales into the territories of other Corning licensees and vice-versa; (iv) allow each parent to require that the relevant joint venture's capacity be expanded, thereby preventing Corning from artificially limiting the output of the joint venture; (v) provide each joint venture partner other than Corning with the right to purchase optical fibres from independent suppliers; and (vi) obligate each joint venture to sell to all users on non-discriminatory terms, thereby allowing third parties to compete with the joint venture. Also, each of the joint ventures were restricted from engaging in the exchange of competitive information with respect to prices, costs, sales, production, and marketing plans.

Adverse competitive effects: *Eurosport* Eurosport was a 50/50 joint venture between Sky Television and the Eurosport Consortium, which consisted of certain members of the European Broadcasting Union ("EBU"). The joint venture was formed for the purpose of operating a sports-dedicated trans-national satellite television channel that could be received in twenty-two European countries, including eight members of the EC. Sky had experience in operating and marketing commercial satellite television channels and had assumed responsibility for financing and marketing the new channel. The Eurosport Consortium, through its access to programs by virtue of their membership in the Eurovision system, was able to provide the channel with a large pool of otherwise unused broadcast coverage of sporting events that were likely to be of interest to viewers in a number of EC member countries.[118] The joint venture was notified following complaints by Screensport, a competitor of Eurosport, that it violated Article 85(1).

The Commission concluded that the joint venture did indeed infringe Article 85(1).[119] The Commission found that the parents of the joint venture had been potential competitors in the field of sports dedicated satellite television broadcasting and that, in fact, the joint venture was motivated, at least in part, by the desire of both parents to effectively "neutralize" the possibility of competition from the other parent, especially since both of the parents had previously been attempting or contemplating there own sports dedicated programming channel.[120] It went on to state that the formation of the joint venture effectively precluded further competition between the parents. In addition, the Commission cited the foreclosure effect that the joint venture had in the markets for (i) the broadcast of sports programming; (ii) the acquisition of sports programming rights; and (iii) the marketing of cable distribution rights. However, the Commission made no separate assessment in its decision with respect to the appreciability of the competitive restrictions or the foreclosure effect.

The Commission refused to grant an exemption to the joint venture under Article 85(3). While the parties attempted to claim that the joint venture would improve production and distribution of services, the Commission found that any such benefits were clearly outweighed by the distortion of competition in a new and developing industry and the prejudicial effects on potential new entrants that were caused by funneling the pool of sports programming through the joint venture. The Commission, noting that Sky had announced that it would continue to provide extensive sports programming, came to the conclusion that failing to exempt the joint venture would not materially reduce the benefits to consumers, since they still would have the coverage provided by Sky and Screensport to choose from. Finally, the Commission noted that the alliance was not "indispensable" to the establishment of a dedicated sports channel with transnational dimensions[121] and that, in fact, a collaboration between the Eurosport Consortium that was backed by the ECU and the most likely competitor was "excessive."

[118] Eurovision was a system for the free pooling of coverage (i.e. basic video and international sound feed) of events, such as sporting events, which might be of interest to viewers in EC member countries other than the one in which the event originated. It was possible for the EBU to acquire, albeit in the face of competition from independent broadcasts, the exclusive right to various sports events and exchange the signal among the EBU members for broadcast using the Eurovision system; however, it appeared that only about 15 percent of the sports material available for broadcast using the Eurovision system was actually broadcast by EBU members, meaning that the remainder was available for use by the new joint venture.

[119] OJ 1991 L63/32 [1991] 4 CMLR 228.

[120] Apparently, the Eurosport Consortium members had previously attempted, unsuccessfully, to establish a satellite channel, and it was generally known that Sky was considering a launch of a sports dedicated channel as an addition to its family of existing channels. The fact that the Eurosport Consortium had not been able to independently succeed with the project was not deemed to be sufficient by the Commission to preclude it from being a potential competitor.

[121] The Commission also objected to specific restrictions included in the joint venture agreement as not being indispensable, such as the privileged position accorded to Eurosport with regard to a 48 hour window of exclusivity.

14 Regulation of Research and Development and Production Joint Ventures in the United States and the EC

I. Regulation of Research and Development Joint Ventures

A. Introduction

While innovation is often a solitary process engaged in by the lone inventor, it is now increasingly common for research and development activities to be carried out as collaborative joint ventures. As noted earlier, collaborations can spread the risks and costs of invention, reduce unnecessary duplication in research and development expenditures without diminishing the output of new knowledge, and facilitate the dissemination of knowledge without undermining its appropriability. Moreover, in both the United States and the EC, research and development joint ventures have been promoted as effective solutions to address competitive threats to domestic high-technology industries.[1]

Collaborations in this area can take a number of different forms.[2] Research and development collaborations, regardless of the legal form in which they are carried out, generally include a variety of restrictions and obligations that will apply to the activities of the parties with respect to the subject matter of the research and development program. In large part, the restrictions imposed on the parties impact their ability to freely compete in the business area in which the collaboration is to occur. Such restrictions can be characterized in two ways. First, the parties may be restricted in their ability to engage in independent research activities outside of the specific joint venture. Second, the parties may be restricted in the manner in which they are allowed to exploit the results of the joint work of the venture.

The analysis of restrictions on competition in the context of research and development joint ventures is similar to that which must occur when the joint venture relates to downstream activities, such as production and sales; however, since research and development is generally materially removed from the market place for the products of the development work, it may be assumed that the potential for an appreciable effect on competition will be much less than in other types of joint ventures which involve activities closer to the market. In the last decade or so, both the United States and the EC have

[1] Among the programs were ESPRIT (European Strategic Programme for R&D in Information Technologies), RACE (Research in Advanced Telecommunication Technology for Europe Programme), BRITE (Basic Research in Industrial Technology for Europe Programme), EUREKA (European Research Coordination Agency).

[2] See Chapters 7 and 8 regarding the various forms of strategic business relationships which might involve research and development activities.

taken steps to clarify how research and development joint ventures will be treated under the relevant antitrust or competition laws through, in the United States, the promulgation of a specific statute on cooperative research and development, and in the EC, the adoption of a block exemption relating to such joint ventures.

B. United States: The National Cooperative Research and Production Act

In the United States, the most important recent legal initiatives relating to cooperative arrangements focusing on "R&D" activities and early-stage commercialization have been the National Cooperative Research Act of 1984 ("NCRA"), and the amendments thereto included in the National Cooperative Research and Production Act of 1993 ("NCRPA").[3] The NCRA was enacted to address certain concerns about the risks which parties to various research and development joint ventures faced under the antitrust laws as they existed prior to the time the NCRA was enacted. Specifically, the NCRA clarified that joint research and development ventures that were engaged in activities which were limited to basic research would be analyzed under the rule of reason standard,[4] that a successful plaintiff in any antitrust action which involved a covered venture could only recover actual damages, and not treble damages, provided that the parties to the venture had made certain disclosures to the DOJ and the FTC,[5] and that persons who made any unreasonable challenges to a venture would be subjected to specified penalties. The disclosures to the DOJ and the FTC, which included a general description of the activities of the joint venture and the identity of the parties to the joint venture, were not mandatory; however, they are required in order for the venture to claim the single-damage protection.

The NCRA was strictly limited in the scope of its protection, covering only ventures that were only involved in pure research activities, such as theoretical analysis, development or testing of basic engineering techniques, and the extension of scientific or technical findings into practical application "for experimental and demonstration purposes." The NCRA could not be used by joint ventures involved in production or marketing of products which might be developed by the venture. In addition, the NCRA did not protect exchanges of information concerning costs, profitability or sales by the parties to a joint venture, even when the venture itself was formed only to conduct the covered basic research activities.

The functional scope of the NCRA was disappointing to those who advocated a more liberalized attitude toward downstream collaboration aimed at commercializing the results of the initial research and development. Teece and Jorde originally argued that the NCRA should be extended to include joint manufacturing and production of innovative products and processes (i.e. commercialization), claiming that such joint activities were often necessary to provide the partners with the type of feedback information which the

[3] 15 U.S.C. §§ 4301–5.

[4] 15 U.S.C. § 4302 makes it clear that in any state or federal antitrust action "the conduct of any person in making or performing a contract to carry out a joint research and development venture shall not be deemed illegal per se," but shall be evaluated under the rule of reason.

[5] 15 U.S.C. § 4304 provides that in any state or federal antitrust action, recovery is limited to actual damages sustained, plus interest and costs including a reasonable attorney's fee; provided that the following two conditions set out in 15 U.S.C. § 4303 are satisfied: (i) the damages claimed result from conduct within the scope of the notification required by the NCRA, and (ii) the claim is filed after such notification becomes effective.

simultaneous model of innovation holds is required in order to assist in any further innovation and product development.[6] Jorde and Teece also point out that firms may lack incentives to even engage in early stage, joint development activities in those cases where later stage, stand-alone commercialization of the results of the development effort will be too expensive for the firms to carry out on their own.[7]

Expansion of the functional coverage of the NCRA was certainly not universally popular among economists. Brodley, for example, pointed out that while it is true that the application and commercialization of research is vital to innovation, it does not automatically follow that greater amounts of collaboration in production and marketing should be pursued as a policy objective, particularly since there are smaller problems of appropriability and "free riding" with these functions.[8] Brodley acknowledged that innovation will generally require informational feedback from users in the marketplace[9]; however, he points out that such feedback can be obtained by the parties in the course of their separate sales activities, and does not necessarily require joint selling arrangements.[10] Brodley also argued that any allowance for greater collaboration in production and marketing should require an affirmative demonstration of material economies of scale or scope.[11]

Nonetheless, several years after the NCRA was enacted, there was a movement in Congress to extend its scope to include some production activities. Extensive hearings were held in both houses of Congress, and eventually the NCRPA was passed, with the effect of extending the protections originally included as part of the NCRA to certain production activities, while maintaining the same mechanism and general purpose of the prior statute. Under the NCRPA, the activities which separate business firms may collaborate to engage in are specified as follows:

(i) theoretical analysis, experimentation, or systematic study of phenomena or observable facts;
(ii) the development or testing of basic engineering techniques;
(iii) the extension of investigative findings or theory of a scientific or technical nature into practical application for experimental and demonstration purposes, including the experimental production and testing of models, prototypes, equipment, materials, and processes;
(iv) the production of a product, process, or service;
(v) the testing in connection with the production of a product, process or service by such venture;

[6] Jorde and Teece, "Innovation and Cooperation: Implications for Competition and Antitrust," in *Collaboration Among Competitors* (1990) (hereinafter *Collaboration*) 887, 900. They believed that the NCRA implicitly accepted the serial model of innovation, not the simultaneous model of innovation. Ibid.

[7] Ibid. at 906.

[8] Brodley, "Antitrust Law and Innovation Cooperation," *in Collaboration*, supra note 3, at 913. See also Shapiro and Willig, "On the Antitrust Treatment of Production Joint Ventures," in *Collaboration*, supra note 3, at 938.

[9] Brodley, supra note 5, at 920 (citing von Hippel, *The Sources of Innovation*, New York: Oxford University Press (1988), for the proposition that users themselves are often a very good source of innovation, especially when they are better able to appropriate the rents from innovation than the manufacturers).

[10] Ibid. (noting that manufacturers can obtain the desired feedback through the formation of user groups, as is the case with computer software).

[11] Ibid. at 913. In fact, Brodley believed that protection for a joint production collaboration should not be granted without a showing of a particular need, although the standard might be eased in situations where the parties have small combined market shares. Ibid. at 920.

(vi) collection, exchange, and analysis of research or production information;
(vii) any combination of the foregoing activities;
(ix) establishment and operation of facilities for conducting such venture; and
(x) prosecution of applications for patents and granting of licenses for results of such venture.[12]

As with the NCRA, there are significant limitations on the activities which can be carried under the umbrella of the NCRPA. Specifically, the following activities are excluded from the scope of the NCRPA:

(i) exchanging information among competitors relating to costs, sales, profitability, prices, marketing, or distribution of any product, process, or service that is not reasonably required to conduct the research and development that is the purpose of such venture;
(ii) entering into any agreement or engaging in any other conduct restricting, requiring, or otherwise involving the production or marketing by any person who is a party to such venture of any product, process, or service other than the production or marketing of proprietary information developed through such venture, such as patents and trade secrets; or
(iii) entering into any agreement or engaging in any other conduct to restrict or require the sale, licensing, or sharing of inventions or developments not developed through such venture, or to restrict or require participation by such party in other research and development activities, that is not reasonably ;required to prevent misappropriation of proprietary information contributed by any person who is a party to such venture or of the results of such venture.[13]

Under the voluntary notification system, companies disclose certain basic information in exchange for exemption from trebling of damages in civil suits. The DOJ stresses that the adequacy of the information is the key to the protection from trebled damages. Any party to a joint venture may file with the FTC and the Attorney General a written notification disclosing the identities of all parties to the venture and the nature and objectives of the ventures. The filings should be simultaneous and must occur not later than ninety days after entering into a written agreement concerning the joint venture. The enforcement agencies will publish a notice in the Federal Register concerning the proposed joint venture thirty days after receipt of the information.

The reduction of damages protection for production joint ventures included in the NCRPA will only apply if the principal production facilities are located in the United States or its territories, and all parties are United States entities or a "foreign person from a country whose law accords antitrust treatment no less favorable to United States persons" who participate in joint ventures in that country.[14] A country that is a party to an international agreement with the United States that provides for national treatment would satisfy this requirement, including treaties of Friendship, Commerce and Navigation; Bilateral Investment Treaties; and Free Trade Agreements (such as the North American Free Trade Agreement).[15]

[12] 15 U.S.C. § 4301 (a)(6).
[13] 15 U.S.C. § 4302(b).
[14] 15 U.S.C. § 4306.
[15] H.R. Rep. No. 94, 103d Cong., 1st Sess. 20, reprinted in 1993 U.S.C.C.A.N. 176, 193.

C. EC's Group Exemption for R&D Arrangements

Introduction The seminal regulatory declaration of the Commission with respect to research and development arrangements is the block exemption on the subject. Although the Commission had the authority, by virtue of Regulation 2821/71, to promulgate a group exemption which covered research and development arrangements, a variety of legal and political considerations delayed the Commission's efforts until early 1984, when an initial draft of a research and development group exemption was circulated for comment.[16] The consultation process which followed resulted in substantial changes to the exemption, including a significant lengthening of the original draft, and the final exemption was adopted as Regulation 418/85 effective from March 1, 1985.[17]

As with the other group exemptions, Regulation 418/85 is preceded by various recitals that are designed to set out the underlying economic and public policy considerations relating to the scope of the exemption. The terms of the exemption itself are then laid out in various articles. Article 1(1) of Regulation 418/85 exempts specified agreements relating to research and development and certain activities with respect to the actual exploitation of the results of such research and development work. The terms "joint research and development of products and processes" and "exploitation of the results" are essential to determining the application of the exemption, and are defined in Articles 1(2) and (3) of Regulation 418/85. Article 2 explains that the exemption will only apply on the conditions specified therein, and those conditions are more fully described in the recitals. Article 3 sets out certain conditions relating to the market share of the parties and the duration of the agreement. These conditions, in effect, deny the exemption in situations where the parties have a significant market share, and provide for the termination of the exemption within five years after the products are first put on the market if the parties have achieved a significant market share at that time.

The remainder of Regulation 418/85 generally follows the format of many other block exemptions which have been promulgated by the exemption. For example, Article 4 enumerates provisions which are exempted even though, in the Commission's view, they do have the effect of restricting competition. Article 5 includes a list of other obligations which are permissible, and has been included to provide specific guidance to the parties, rather than because the Commission believes they will lead to restrictions on competition. In fact, in the Commission's view, the provisions in Article 5 rarely have an impact on competition. On the other hand, provisions included in Article 6, the so-called "black list," are considered to be so restrictive on competition that their existence will deny the exemption to the parties. Article 7 sets out a "fast-track" opposition procedure for agreements that do not run afoul of Article 6, yet which include provisions that are not specifically permitted by other Articles of the exemption. Finally, Regulation 418/85 has other miscellaneous provisions relating to confidentiality, connected undertakings, termination of the exemption by decision of the Commission, and agreements notified to the Commission before Regulation 418/85 came into force.

Although Regulation 418/85 is complex, it appears from a review of the various recitals included therein that, in general, the Commission will not interfere with cooperative arrangements involving basic research, where the market is not oligopolistic or highly

[16] Goyder, *EC Competition Law* 2d edn, Oxford: Clarendon Press (1993) 189–90.

[17] Regulation 418/85 was amended as of April 1, 1993 by Regulation 151/93 and expires on December 31, 1997. References herein to "Art." and "Recitals" refer to Articles and Recitals included in Regulation 418/85 unless specified otherwise in the text.

concentrated, where the cooperation does not extend to industrial applications, or where not more than one of the parties could have independently conducted the activities which are to be carried out by the joint venture; provided that any restrictions included in the arrangement are not broader than those which might be considered necessary to insure the viability of the transaction, and provided further that other alternatives remain for research and development activities.

In many ways, Regulation 418/85 is consistent with the Commission's established desire to promote cooperative agreements which facilitate the introduction of new technology in the EC. For example, in its First Annual Report on Competition Policy, the Commission indicated that it would be sympathetic to joint research and development arrangements, including ventures which covered utilization of the results, with the most favored treatment being reserved for agreements between small and medium-sized firms which related only to research and development, as such agreements did not generally present any significant danger to competition.[18] This approach was originally embodied in Article 4(2)(iii)(b) of Regulation 2822/71 (which dealt both with specialization and research and development agreements), which provided that all joint research and development agreements would not have to be notified provided that the objects did not extend beyond the research and development stage.[19]

Unfortunately, many of the rules in Regulation 418/85 are difficult to apply in practice. For example, while the group exemption will only be available in situations where the parties meet specified "market share" conditions, it is often impossible to compute a market share for new products due to the problems with establishing the correct market definition. Moreover, even when the market share is ascertainable, it will often be exceeded by the parties since, presumably, the innovative nature of the new products means that they have some initial advantage over any potential competitors. Also, the group exemption provides for a five-year time limit on exclusive territorial protection, which is often seen as too short by the parties, particularly for innovative products with slow sales growth.[20] Accordingly, it is often necessary to notify these types of agreements.[21]

Scope of arrangements falling within exemption Regulation 418/85 applies to three general types of arrangements:[22]

(i) joint research and development of the products or processes without subsequent joint exploitation of the results;
(ii) joint research and development of the products or processes with joint exploitation[23] of the results;

[18] See Goyder, supra note 16, at 183 (citing 1st Annual Report at 45). In the same Report, however, the Commission did note that competition problems could arise with respect to the exploitation of the results of the joint research and development effort. Ibid.

[19] Ibid.

[20] No time limit for the exemption was included in the original draft of the block exemption. Ibid. at 190.

[21] Ibid. at 183 ("only a minority of R and D agreements can ... meet the stringent standards of the group exemption, especially if joint exploitation of the results is required, and there are, therefore, a majority which nevertheless still require individual exemption").

[22] Regulation 418/85 Art. 1. The extensive definition of rcscarch and agreements was not part of the original draft of the block exemption. Ibid. at 191. The block exemption applies to decisions of associations as well as to agreements. Regulation 418/85 Art. 12.

[23] "Joint" exploitation is considered to take place: (i) where the exploitation is carried on by a joint team or is jointly entrusted to a third party or is allocated between the parties by way of specialization in research,

(iii) joint exploitation of the results of research and development carried out as the result of earlier research and development agreements between the same undertakings.

For purposes of Regulation 418/85 the term "research and development" includes "the acquisition of technical knowledge and the carrying out of theoretical analysis, systematic study of experimentation, including experimental production, technical testing of products or processes, the establishment of the necessary facilities and the obtaining of intellectual property rights for the results."[24] "Exploitation of the results"[25] includes manufacturing of the "contract product,"[26] as well as licensing or assignment of the intellectual property rights required in order to manufacture the product to third parties. However, there are situations where manufacturing activities are considered to be part of the research and development process, rather than exploitation, such as the manufacture of prototypes. Also, joint clinical testing by pharmaceutical companies in order to comply with national safety rules is considered to be research and development, even when the drug was developed outside of the EC and outside of the agreement.[27]

In establishing the scope of the exemption, the Commission recognized that it is likely that arrangements will extend beyond basic research to include some additional aspects of exploiting the results of the research and development effort.[28] Recital 3 of Regulation 418/85 focuses upon the potential anticompetitive effects that might be expected in those situations, specifically noting that: "Agreements providing for both joint research and development and joint exploitation of the results may fall within Article 85(1) because the parties jointly determine how the products developed are manufactured or the processes developed are applied or how related intellectual property rights or know-how are exploited."[29] Nonetheless, under the Regulation, the parties are allowed to engage in collaborative manufacturing arrangements relating to the actual results of the research and development activities;[30] however, joint sales activities are not within the scope of the exemption.[31]

cont.
development or production; or (ii) the parties collaborate in any way in the assignment or licensing of intellectual property rights or know-how resulting from the joint research and development. Regulation 418/85 Art. 1(3).

[24] Regulation 418/85 Art. 1(2)(a).

[25] Regulation 418/85 Art. 1(2)(d).

[26] The term "contract products" means "products or services arising out of the research and development or manufactured or provided [sic] applying the contract processes"; Regulation 418/85 Art. 1(2)(b). "Contract processes" means processes arising out of the research and development. Ibid. In situations where the parties have brought technical knowledge that they owned prior to the joint venture to the collaboration and such knowledge has become a necessary component of the exploitation of the results of the joint development effort, then such pre-existing knowledge is considered to be part of the contract products and processes. Accordingly, such knowledge could then be jointly exploited within the block exemption if it was decisive in the production or processing. See Continental/Michelin, OJ 1988 L305/33 [1989] 4 CMLR 920.

[27] Korah, *R&D and the EEC Competition Rules: Regulation 418/85*, ESC Publishing (1986), 22–3.

[28] In Recital 7 of Regulation 418/85 itself, it is noted that "joint exploitation of the results can be considered as the natural consequence of joint R&D."

[29] In the view of some commentators, the potential application of the language in Recital 3 may be excessively broad, at least in situations where it can be shown that not more than one of the parties could have actually completed the development work independently. Korah, supra note 27, at 14–15.

[30] Art. 1(2)(d) and 3. Permitted manufacturing activities include the grant of manufacturing licenses to third parties.

[31] Arts. 2(e) and (f).

Article 2: preconditions for application of exemption Article 2 of Regulation 418/85 contains a number of preconditions which will be applicable to the framework of any joint venture that seeks to be covered by the block exemption. These preconditions in the case of a research and development agreement up to the stage of industrial application are: (i) that the joint research and development work is carried out within the clear framework of a program that defines the object of the work and the fields in which it is to be carried out;[32] (ii) that all parties have access to results of the work;[33] and (iii) that each party is free to exploit the results of the work, and any pre-existing technical knowledge or know-how, independently.[34] In situations where the joint venture extends to exploitation, the following additional preconditions apply: (iv) that the joint exploitation of the research and development relates only to the results which are protected by intellectual property rights or which constitute know-how which substantially contributes to technical or economic progress and the results are decisive for the manufacture of the contract products or the application of the contract processes; (v) that no joint undertaking or third party responsible for manufacture of the contract products is required to supply them only to the joint venture parties; and (f) that any undertakings required to manufacture by way of specialization in production are required to fulfill orders for supplies from all parties in the venture.[35]

Defined research program requirement The Commission's direction that research and development work must be undertaken within a defined program stems from a concern that the collaboration between the parties in one field might easily extend to other areas. However, while the exemption will only apply when the objectives and field of the research and development work have been defined in advance, it is not really necessary that the parties impose excessive limitations on the scope of their collaboration.[36] For example, the parties may agree to include in the research program not only an initial set of potential products, but also work on any new problems which might arise during the initial research work. Similarly, the parties might agree in advance that if the initial research is unsuccessful, they will jointly explore other potential solutions as specified in the agreement.

Access to results of research requirement The requirement that all of the parties have access to the results of the research and development work is intended to further the Commission's objectives that the collaboration must result in an increase in the dissemination of technical knowledge between the parties, a reduction in the duplication of research and development work, and the stimulation of new advances through the exchange of complementary technical knowledge.[37] The desire to encourage the exchange and dissemination of technical knowledge appears also in a requirement that the parties must pool the background knowledge necessary for the exploitation of the new

[32] Regulation 418/85 Art. 2(a). For an example of where the research program failed to meet the requirements of specificity set out in the exemption, see GEC/ANT/Telettra/SAT, OJ 1988 C180/3. See also Recital 4 of the Regulation, wherein it is asserted that the benefits of promoting technical and economic progress can be attained only where the research and development program and its objectives are defined and the results are pooled.

[33] Regulation 418/85 Art. 2(b).

[34] Regulation 418/85 Art. 2(c). See Regulation 418/85 Art. 1(2)(c) for a definition of "technical knowledge."

[35] Regulation 418/85 Art. 2(c).

[36] Korah, supra note 27, at 26.

[37] See, e.g. Recital 4 of Regulation 418/85.

results, which means that the parties must be willing to permit the other parties to sublicense their background technology to third parties to the extent necessary to allow the other parties to independently exploit the products of the research and development work. Since it is likely that a party may object to such a provision, this requirement may cause the arrangement to fall outside of the block exemption.[38]

The "access" requirement can present a number of interesting problems, particularly when the contributions of the parties to the research work are not equal and when the results of the work are not equally valuable to the participants. For example, if one of the parties has contributed to the technology required in order for the joint venture to proceed, but has no particular need for the results of the joint venture's development work, then it may insist that it be compensated on a basis which takes into account that it will not be independently exploiting the results. Article 5(1)(f) of Regulation 418/85 does allow for the parties to accept obligations to pay royalties for unequal exploitation or contribution;[39] however, since it is difficult to determine the ultimate value of research results before the development work has been completed, the parties may not be able to reach a satisfactory agreement regarding the amounts to be paid to a party which will not also be exploiting the results.

Right to exploitation requirement When the agreement covers joint exploitation of the research and development, the aforementioned requirement as to independent exploitation obviously does not apply; however, the scope of the joint exploitation must be limited only to intellectual property rights or know-how that substantially contributes to technical or economic progress and the results of the joint research project must be decisive for the manufacture of the contract products or the application of the contract processes.[40] In addition, if any undertakings are to specialize in the manufacture of some of the products, they must agree to meet orders from all the parties to the agreement.[41]

Significance of results requirement Another method for limiting the potential anticompetitive effects of collaboration is the requirement that Regulation 418/85 will only be available when the research and development work is an essential element for the manufacture of the contract products or the application of the contract processes and produces results that are important enough to qualify for some form of protection under the intellectual property laws.[42] Therefore, parties cannot rely on Regulation 418/85 to cover agreements, such as a patent license, in which research and development is not one of the main purposes of the collaboration. It is important to note that the parties

[38] Korah, supra note 27, at 26–7.

[39] The Commission's Notice on Cooperation, OJ 1968 C75, only required that the parties must be able to share in the result to the extent of their participation in order to avoid the application of Art. 85(1) (§ II,3).

[40] Regulation 418/85 Art. 2(d). Note that Art. 85(3) uses what appears to be a less exacting standard of "a mere contribution to the promotion of technical or economic progress." See, e.g. *SB/Goulds/Lowara/ITT*, OJ 1991 L19/25, § 26.

[41] Regulation 418/85 Art. 2(f). This requirement is designed to provide some assurance that the exemption is not used in situations where the agreement is not intended as a disguise for arrangements in which research and development is actually a small or ancillary factor. See also Regulation 418/85 Recital (7).

[42] See Art. 2(d) of the Regulation, which limits the exemption to agreement where "the joint exploitation relates only to results which are protected by intellectual property rights or constitute know-how which substantially contributes to technical or economic progress and that the results are decisive for the manufacture of the contract products or the application of the contract processes." See also Recital 7 of the Regulation, which states that the exemption should be limited to agreements that substantially contribute to technical or economic progress.

may not be able to determine whether the conditions for the exemption will actually be satisfied at the time that the agreement is first negotiated, since the research and development work, which by its very nature is risky and uncertain, has not been completed.[43]

Third party manufacturing activities As noted above, the exemption in Regulation 418/85 does not extend to joint sales, since the exemption will only be available if "any joint undertaking or third party charged with manufacture of the contract products is required to supply them only to the parties."[44] This condition is designed primarily to insure that consumers will be able to purchase the products from more than one independent supplier and that the parties will be able to continue to compete in downstream markets. However, if the parties decide to take most of their profits in the joint venture, as might be the case when the transfer price is set at a level that approximates the ultimate sales price to consumers, then the amount of price competition between the parties will be significantly reduced. The objective of multiple suppliers is further advanced by the condition that "undertakings charged with manufacture by way of specialization in production are required to fulfill orders for supplies from all the parties."[45]

Articles 4-6 of Regulation 418/85 Articles 4 and 5 of Regulation 418/85 contain the clauses which are permitted (the "white list"),[46] while Article 6 is designed to insure that the restriction are not too wide by listing provisions which cannot be included in order for the exemption to apply (the "black list"). The exemption will be lost if restrictions other than those included on the white list are also included in the agreement, even though such other restrictions are not specifically excluded under Article 6 of Regulation 418/85.[47] The clauses described in Article 4 generally would fall within the scope of Article 85(1); however, the clauses in Article 5 typically would fall outside of Article 85(1), but are included as part of Regulation 418/85 in order to provide the parties with certainty and definitive guidance.[48]

Article 4: provisions which infringe Article 85(1) and permitted under the group exemption The provisions listed in Article 4 of Regulation 418/85 are admittedly restrictions on competition; however, they are exempted because they concentrate the research activities of the parties in order to improve their chances of success, and to facilitate the introduction of new products and services onto the market.[49] Article 4(1) includes various restrictions on competition imposed on the parties which will be exempted, while Article 4(2) makes it clear that the exemption in Article 4(1) will also apply where in a particular agreement the parties undertake obligations of the types referred to in paragraph 1 but with a more limited scope than is permitted by that paragraph.

Specific provisions permitted under Article 4(1) include obligations:

[43] It might be argued that if the technology produced from the research and development work is not "important," then the parties will probably not have gained any competitive advantage and the availability of the exemption become a moot point.

[44] Art. 2(e) of Regulation 418/85.

[45] Art. 2(f) of Regulation 418/85.

[46] Both Arts. 4 and 5 apply to restrictions of narrower scope than those actually listed therein. See Regulation 415/85 Arts. 4(2) and 5(2).

[47] See, e.g. *BP/Kellogg*, OJ 1985 L369/6 [1986] 2 CMLR 619 § 15(b).

[48] Regulation 415/85 Recital (11) and Art. 5(2).

[49] Recital 6 of Regulation 418/85, which also states that "These restrictions are generally necessary to secure the desire benefits for the parties and consumers."

(i) not to carry out on an independent basis research and development in the same field or a closely related field during the joint venture program;[50]
(ii) not to enter into agreements with third parties for research and development in such fields during the joint venture program;[51]
(iii) an obligation to procure contract products exclusively from specified sources;[52]
(iv) an obligation not to manufacture contract products or apply the processes in territories reserved for other parties (a form of territorial protection);[53]
(v) an obligation to restrict the manufacture or application of the products or processes to specific fields of use, unless the parties in the joint venture are already competitors at the time it is signed;[54]
(vi) an obligation not for five years at the time when the contract products are first marketed within the Common Market to follow an active policy of marketing in territories reserved for the other parties; so long as users of the products can obtain them from other sources and the parties do not themselves obstruct the entry of parallel imports or make it otherwise difficult for supplies to be obtained;[55]
(vii) an obligation to grant exclusive distribution rights to one of the parties or to a joint entity or to a third party subject in any of those cases to a prohibition on distributing competing products of a third party[56]; and
(viii) a mutual obligation to communicate experience or improvements to each other an to grant non-exclusive licenses with respect thereto.[57]

Article 5: provisions which do not infringe Article 85(1) and permitted under the group exemption Article 5 of Regulation 418/85 permits the inclusion of the following provisions:

(i) an obligation to communicate the patented or non-patented technical knowledge re quired to carry out the research program or to exploit its results;[58]
(ii) an obligation not to use know-how received from another party except to carry out the research and development program and to exploit its results;[59]
(iii) an obligation to maintain intellectual property rights and take actions against infringers;[60]
(iv) an obligation to maintain the confidentiality of know-how following termination of the arrangement;[61]

[50] Ibid., Art. 4(1)(a).
[51] Ibid., Art. 4(1)(b).
[52] Ibid., Art. 4(1)(c).
[53] Ibid., Art. 4(1)(d).
[54] Ibid., Art. 4(1)(e).
[55] Ibid., Art. 4(1)(f).
[56] Ibid., Art. 4(1)(fa) and (fb), inserted by Regulation 151/93.
[57] Ibid., Art. 4(1)(g).
[58] Ibid., Art. 5(1)(a).
[59] Ibid., Art. 5(1)(b).
[60] Ibid., Arts. 5(1)(c) and (e). Art. 5(1)(e) exempts "an obligation: (i) to inform other parties of infringements of their intellectual property rights, (ii) to take legal action against infringers, and (iii) to assist in any such legal action or share with the other parties in the cost thereof."
[61] Regulation 418/85, Art. 5(1)(d).

(v) an obligation to pay royalties or render services to other parties as a means of compensation for unequal contributions to the joint research and development or unequal exploitation of the results;[62]
(vi) an obligation to share royalties received from third parties with other parties;[63] and
(vii) an obligation to supply other parties with minimum quantities of contract products and to observe minimum standards of quality.[64]

Article 6: provisions which infringe Article 85(1) and prohibited under the group exemption Article 6 of Regulation 418/85 sets out the "black list" of clauses which may not appear in the agreement if the exemption is to apply. The following clauses, many of which the Commission required to be deleted in evaluating earlier individual requests for exemption, are prohibited:

(i) where the founders of the joint venture are restricted in their own freedom to carry out research and development either independently or with third parties in separate fields of activity during the joint venture program or after completion of the research program in the same field;[65]
(ii) where after completion of the research and development program the parties are restricted from challenging the validity of intellectual property rights which protect the results of that research and development within the Common Market;[66]
(iii) where there are restrictions relating to the quantity of the contract products that the participants may sell or manufacture;[67]
(iv) where the parties are restricted in their free determination of prices or discounts within the contract products;[68]
(v) when the parties are restricted as to the customers they serve, except so far as It is required by field of use restriction;[69]
(vi) when the parties are unable to pursue an active sales policy in territories within the Common Market reserved for other parties after the end of the initial five-year period;[70]
(vii) where the parties are prohibited from allowing third parties to manufacture contract products or apply contract processes when joint manufacture is outside the scope of the agreement;[71] and
(viii) when the parties are required to refuse "without any objectively justified reason" to meet orders from sources within their respective territories that wish to market the products in other territories within the Common Market, or are required to make it difficult for customers (users or dealers) to obtain the con-

62 Ibid., Art. 5(1)(f).
63 Ibid., Art. 5(1)(g).
64 Ibid., Art. 5(1)(h).
65 Ibid., Art. 6(a).
66 Ibid., Art. 6(b).
67 Ibid., Art. 6(c).
68 Ibid., Art. 6(d).
69 Ibid., Art. 6(e).
70 Ibid., Art. 6(f).
71 Ibid., Art. 6(g).

tract products from other dealers and to obstruct the free movement of products within the Common Market which have been lawfully marketed by another party.[72]

Article 3: market share conditions Article 3 of Regulation 418/85 restricts the application of the block exemption to situations where the market shares of the participants fall below certain ceilings. Since the results of the research and development work are unknown at the time that the original agreement is entered into, the Commission turns to the market for the "products capable of being improved or replaced by the contract products" in order to determine the competitive effect of the proposed collaboration. The question of whether the parties are competitors in this downstream market will impact not only the availability of the block exemption, but also, as discussed below, the duration of the exemption if it can be utilized initially by the parties. While the applicable test is whether or not the parties are competitors in the research and development "market," the existence of downstream competition is an indication of competition at the innovation stage. However, it is important to separately analyze the various conditions for innovation, since it may be that the innovation could not have been independently developed by any of the parties and owes its existence only to the ability of the parties to actually combine complementary skills and other resources at the research and development stage.[73]

The exemption will not be available where the parties, at the time of the agreement, are competing manufacturers of products "capable of being improved or replaced by the contract" and their combined market shares of such products exceed 20 percent in the EC or a substantial part thereof;[74] moreover, if the distribution of the products concerned is carried on by one of the contracting parties, a joint venture, or a commonly designated third party, the above market share limit amount is reduced to 10 percent.[75] Manufacturers with higher market shares may still apply for individual exemptions and, in deciding such cases, the Commission is required to take an expansive view of market definition by taking special account of "world competition and the particular circumstances prevailing in the manufacture of high technology products."[76]

Several of the Recitals address the issue of insuring that there are more than one source for innovation within the EC in any economic sector, as well as the Commission's concern that the research and development activities do not lead to market shares which restrict the ability of others to compete with the parties to the joint venture.[77] The Commission realized that there would be substantial difficulties in developing rules that would be applicable to all industries and eventually devised a scheme based on the aggregate market share of the parties which would insure that there would be enough room for com-

[72] Ibid., Art. 6(h).

[73] Korah, supra note 27, at 31–2.

[74] Regulation 418/85 Art. 3(2). The original draft of the block exemption provided that the exemption would only be available if the aggregate annual turnover of the undertakings did not exceed 500 million ECU, and if not more than one of the three actually or potentially leading undertakings in the sector to which the research and development relates at the conclusion of the agreement were a party to the agreement. Goyder, supra note 16, at 190.

[75] Regulation 415/85 Art. 3a, which was inserted by Regulation 151/93.

[76] Regulation 418/85, Recital 10. See, e.g. the review of the *KSB/Goulds/Lowara/ITT* joint venture by the Commission in OJ 1991 L19/25 [1992] 5 CMLR 55, where an individual exemption was granted after it became apparent that the market share restrictions in Regulation 418/85 would take the joint venture outside of the block exemption.

[77] See Recitals 8–10 of Regulation 418/85.

petition in research and development within the EC.[78] However, if the joint venture does not fall within Regulation 418/85's "safe harbor," the parties may still seek an individual exemption.

In determining whether the exemption is available, the Commission will analyze actual, not potential, competition between the parties. In the view of some commentators, in order for the parties to be actual competitors, they must be involved in manufacturing products which are close substitutes and which are sold in the same geographic markets. This narrow view of competition is considered to be appropriate given that the exemption is designed to induce the parties to undertake the risks associated with innovation.[79] However, there have been some cases where the Commission appears to have treated parties which are selling in different territories as competitors, even though the parties have a nil market share in the territory where the other parties are active.[80]

Interesting issues may arise in situations where the goods to be replaced or improved are not homogenous, since it is difficult to define the relevant product market. While there is no precise definition of a "market" in Regulation 418/85, it does indicate that consideration must be given to other products which users may believe to be equivalent due to their characteristics, price and intended use.[81] Accordingly, products that may be substitutes on the demand side of the market should be included in defining the total market. It has been noted that this concept may be more concrete than the standard test of cross elasticity of demand used by economists;[82] however, it is still dependent upon an assessment of how far the price of the products to be improved or replaced would have to rise before users will switch to substitutes and how long it will take for the switch to occur.[83] The size of the market will also be effected by whether the time required for the construction of new productive facilities is taken into account.[84]

Another difficult issue raised by the language of the exemption is determining what constitutes a "substantial part" of the common market for purposes of determining whether or not the exemption is available. This concept appears in a number of the block exemptions; however, it is difficult to arrive at a precise method for delineating a substantial part of the market. In the first instance, it is clear that the answer will depend upon the particular product at issue.[85] Beyond that, the Commission and the ECJ have not been very helpful, with one case construing the same phrase as it occurs in Article 86 by noting that it depends on "the pattern and volume of production and consumption of the said product as well as the habits and economic opportunities of vendors and purchasers ...".[86]

Some guidance on this issue might come from the Commission's objective, as stated in Recital 9, to maintain an effective level of competition during joint exploitation. In light

[78] Korah, supra note 27, at 31.

[79] Ibid. at 32–3.

[80] See, e.g. *Siemens/Fanuk*, OJ 1985 L376/29.

[81] Art. 3(3) of the Regulation. See also Regulations 1983 and 1984/83, Arts. 3(a) and (b), guideline 21, OJ 1983 C355/7 as corrected in OJ 1984 C101/2.

[82] Cross elasticity refers to the extent to which consumers would switch their demand in response to a percentage rise in the price of one product, the other(s) remaining constant.

[83] Korah, supra note 27, at 34–5.

[84] The market will be wider if time is permitted for the construction of new supply facilities. Ibid. at 35. The Commission often appears to be taking a shorter term view in market definition. See, e.g. *Michelin* v. *Commission* [1983] ECR 3461; [1985] 1 CMLR 282, ground 41 (in analyzing the supply side of the market the Court did not take into account the possibility that new supply facilities could be built since construction would take too much time).

[85] Korah, supra note 27, at 35.

[86] *Sugar Cartel* [1975] ECR 1663; [1976] 1 CMLR 295, ground 371.

of this objective, it has been argued that the Commission should take into account the ease with which other products could enter the part of the common market where the parties have their plants and the extent to which the output from those plants is exported to other member states.[87] For example, if a certain product is particularly valuable in relation to the cost of transport, it would appear to be likely that the products will be able to move freely through the EC and that there would be no need to focus upon a smaller part of the market.[88] One commentator has suggested that the following test set forth in an unpublished draft of a revision to the Commission's notice on minor agreements might be useful in this context:

> "12. The relevant geographic market coincides with the whole of the common market where intermediaries and users can obtain throughout the Community under comparable conditions at least one of the products which are included in the relevant product market. Thus small or medium sized undertakings which do not operate throughout the Community but face competition from undertakings which do also benefit from the larger definition of the relevant geographic market."[89]

Duration of the exemption under Regulation 418/85 If the block exemption is available, the duration thereof in a particular situation will depend upon whether the parties are downstream competitors with respect to the manufacture of the products falling within the scope of the cooperative arrangement. If, on the one hand, the parties are not downstream competitors, then Regulation 415/85 will provide an exemption which extends to the end of the research program and, if the results are going to be jointly exploited, for a period of five years thereafter.[90] In those cases where the parties are downstream competitors, the block exemption will continue for the same period (i.e. length of the research program plus five years for exploitation) only if the combined production of the parties is less than 20 percent at the time of the initial agreement.[91] The five year period was thought to be long enough for market shares to stabilize and provide the parties with sufficient incentive for the innovation.

Once the initial exploitation period of five years ends, the exemption may be extended indefinitely so long as the continued production of the parties of the contract products and any other substitutable products does not exceed 20 percent of the total market for such products, not the products that were to be replaced or improved, in the Common Market or a substantial portion thereof.[92] However, it will still be possible to use the exemption even when the aforementioned thresholds are exceeded if the excess is not more than one-tenth during any two consecutive years.[93] Once the relevant market share limitation has been exceeded, the block exemption will cease to apply six months after the end of the financial year in which the limitation was exceeded.[94]

[87] Korah, supra note 27, at 35.

[88] Ibid. at 35–6.

[89] Ibid.

[90] Regulation 418/85, Art. 3(1). Experimental production is included in the definition of research and development and it is likely that sales to customers for experimental trials will not be deemed to be "putting on the market" for purpose of beginning the five year period. Similarly, the supply of drugs for clinical trials should not begin the period until they are available for general use; however, test marketing within the EC will probably begin the five year period.

[91] Regulation 418/85, Art. 3(2).

[92] Regulation 418/85, Art. 3(3). If the contract products are components which are used "in-house," the 20 percent test applies to the market for the finished products.

[93] Regulation 415/85 Art. 4, which was inserted by Regulation 151/93.

[94] Regulation 418/85, Art. 3(5).

It is important to note that the Commission does have the authority to terminate the exemption prior to the end of the five year period if "the contract products are not subject in the whole or a substantial part of the common market to effective competition from identical products or products considered by users as equivalent in view of their characteristics, price and intended use."[95] This means that if the parties develop a new innovative product which is so successful that it overruns the market, then the Commission may step in to reduce their monopoly profits by allowing others to share in the results made possible by the effort of the parties. The concern in this area is similar to that which applies to compulsory licensing; that the parties will be reluctant to invest in innovation if they believe that the exemption will not extend for the full period and, as a result, the EC will not be able to compete in technology markets.

Article 7: opposition procedures An opposition procedure is included in Article 7 of Regulation 418/85 for agreements which contain restrictions which are not covered by Articles 4 or 5, provided that the agreement also does not contain restrictions prohibited by Article 6. In such cases, the agreement can be notified to the Commission and will be deemed covered by the block exemption unless within six months from notification the Commission formally opposes the exemption and notifies the parties that any exemption would have to be on an individual basis.[96] It is also still possible for a member state to request the Commission oppose proposed restrictions; however, such a request must now be justified on the basis of considerations relating to the competition rules of the Treaty rather than on an open-ended basis.

D. Antitrust Analysis of R&D Joint Ventures in the United States

Research and development joint ventures, even before the adoption of the NCRA in 1984, did not cause the same degree of antitrust concerns as might apply to production joint ventures. According to Congressional testimony at the time that the NCRA was being considered, the DOJ never challenged a pure research venture without ancillary restraints, even if the joint venture involved firms with considerably high market shares.[97] Moreover, the aura of legality assigned to research joint ventures by the NCRA's requirement of rule-of-reason analysis has somewhat precluded the generation of the same amount of case law in this area as in that of production joint ventures. As noted by then Assistant Attorney General James F. Rill in a speech before the 1989 Annual Meeting of the Section of Antitrust Law, American Bar Association:

> "[C]oncerns regarding the [unreasonable] application of the antitrust laws to joint research and development were dealt with successfully a few years ago. The National Cooperative Research Act of 1984 clarified the application of the rule of reason to joint R&D ventures in conspicuous public fashion, sending messages simultaneously to po-

[95] Art. 10(d) of the Regulation.

[96] While the opposition procedure has not been heavily utilized, several important Commission decisions have arisen out of Commission review following an Art. 7 opposition. See, e.g. *Elopak/Metal Box-Odin*, OJ 1990 L209/15 [1991] 4 CMLR 832 and *KSB/Goulds/Lowara/ITT*, OJ 1991 L19/25 [1992] 5 CMLR 55.

[97] The DOJ has even left unchallenged a joint venture that involved the entire industry. See Centrifugal Pump Industry, DOJ Business Review Letter (June 5, 1985) (proposed research collaboration included all the competing US producers in the centrifugal pump industry).

tential joint venturers, to possible challengers, and to the courts. The messages were clear:
– Congress sees substantial procompetitive potential in joint research and development potential that would enhance US international competitiveness;
– The risk of anticompetitive harm from joint R&D has been lowered substantially by international competition and globalization of markets; and
– The courts should take particular care to avoid overly strict antitrust approaches to joint research and development ventures."

The NCRPA has made it clear that research and development joint ventures are to be analyzed under the rule of reason, which requires that courts realistically analyze the competitive effects of any challenged joint research and development venture. If a joint research program has no anticompetitive effects, or if any such effects are outweighed by its procompetitive effects, then it should not be deemed to violate the antitrust laws.[98] Section 3 of the NCRPA identifies "effects on competition in properly defined, relevant research and development markets" as an important, but not exclusive, focus of this analysis.[99] This language recognizes that, in many industries, particularly those that are based on rapidly evolving technology, remaining competitive in innovation and being the first to develop new processes and products may be crucial to success. Thus, under the rule of reason, courts will consider whether any challenged joint venture could reduce research and development competition, and thus deter innovation.

One of the few cases dealing with research and development joint ventures is Pump Research and Development Committee.[100] The DOJ did not challenge a research and development joint venture in the centrifugal pump industry that arguably was overinclusive, given that it included all the domestic pump manufacturers. The DOJ noted that total projected expenditures by the venture would not be extremely large, and that competitive research would still come from overseas manufacturers. Another case, Polyisocyanurate Insulation Manufacturers Association ("PIMA")[101] involved a research and development joint venture to investigate and evaluate substitutes for chiorofluorocarbon foam blowing agents ("CFCs"). Although the venture members accounted for 80 percent of the total domestic sales of CFC's, the DOJ did not express concerns regarding the possibility that the venturers might restrain the development of the technology, noting that PIMA members would not have an incentive to deliberately delay the production of CFC substitutes because government restrictions on the use of CFCs was likely no matter what action PIMA took.

While there is an absence of case law with respect to analysis of research and development joint ventures, there are other sources for determining the likely attitude of the DOJ toward a particular arrangement. One particularly helpful guide has been the DOJ's 1988 Antitrust Enforcement Guidelines for International Operations (the "International

[98] The rule of reason condemns only research and development joint ventures that on balance are anticompetitive. If no anticompetitive effects are established, a venture does not violate the antitrust laws, regardless of whether or not it creates any demonstrable efficiencies.

[99] The NCRA prohibited courts from condemning a research and development joint venture on antitrust grounds unless the joint venture was proven to have anticompetitive effects in a properly defined, relevant market that outweigh the joint venture's procompetitive benefits. 15 U.S.C. § 4303; H.R. Conf. Rep. No. 1044, 98th Cong. 2d Sess. 11–12, reprinted in 1984 U.S.C.C.A.N. 3105, 3135–6 (hereinafter "NCRA House Conf. Rep.").

[100] B.R.L. 85-12.

[101] B.R.L. 89-2.

Guidelines"). The International Guidelines obviously address joint ventures in an international context; however, much of the discussion of joint research ventures focuses on the legislative history of the NCRPA. Accordingly, it is likely that the DOJ would analyze the procompetitive effects of any research and development joint venture in a manner similar to that described in Section 3.4 of the International Guidelines.[102]

The International Guidelines call for a four-step analysis, the first three of which involves a determination of whether the joint venture would have any anticompetitive effect in the joint venture market,[103] whether the joint venture would likely have any spillover effects in markets where the ventures are actual or potential competitors,[104] and the likely competitive effect of any non-price vertical restraints imposed in connection with the joint venture.[105] If the DOJ's analysis under the first three steps raises significant anticompetitive risks, it will consider whether any procompetitive efficiencies achieved by the venture would outweigh the risk of anticompetitive harm.[106]

The International Guidelines may, in appropriate cases, be read in light of the DOJ's 1992 Merger Guidelines, even though the Merger Guidelines do not include a specific reference to the antitrust analysis of joint ventures. For example, the Merger Guidelines may be useful in situations where the analysis of a particular joint venture requires some definition of the market, determination of market shares, or the identification of barriers to entry. However, the DOJ itself has made it clear that joint venture analysis involves issues which clearly differentiate it from mergers, where the review often centers on factors such as market share and concentration. In that regard, the DOJ stated that "[b]ecause mergers generally do not carry with them the same promise for new operating capacity as many production joint ventures, the market concentration threshold at which anticompetitive concerns arise generally should be greater for production joint ventures than for mergers between the parties to the joint venture."

In order to properly assess the effect a joint venture might have on competition in the research and development area, it is first necessary to attempt to identify the "relevant research and development" market, which includes firms that have the ability and incentive, either alone or in collaboration with one another, to undertake research and development comparable to that of the proposed joint venture. Whether a firm has the requisite "ability" and "incentive" is an issue to be determined on the basis of objective standards, taking into account the firm's business objectives, facilities, technologies, and other available assets. It is not necessary that a firm be an actual competitor at the production or marketing stage. Market shares in existing markets or in projected future markets can be relevant, but will not be determinative of a firm's ability and incentive to compete in a relevant research and development market. In addition, because there normally are no geographic barriers to the free movement of information, foreign research and development competitors in most cases would also be significant competitive factors in the research and development market, even if they could not sell products incorporating the technology in the United States.

102 This assumes, of course, that the joint venture represents a legitimate integration of efforts to produce valuable research and development. An research and development "joint venture" that was not formed to pursue research and development, but was formed merely to disguise a naked agreement to fix prices or allocate territories or customers for the sale of some product or service would be viewed as a sham, and would probably be prosecuted as a criminal violation of the Sherman Act.

103 International Guidelines § 3.42.

104 International Guidelines § 3.43.

105 International Guidelines § 3.44.

106 International Guidelines § 3.45.

After a relevant market is determined, the next step is to analyze the effect that the proposed joint venture will have on the level of competition in that market. Since a joint venture cannot be successfully challenged unless there are some anticompetitive effects, the first issue is to determine whether any such effects are present. As a general matter, potential harm to competition is most likely to occur when a joint venture is overly inclusive, in that it includes a greater percentage of the potential market than it should under the circumstances. The concern in this regard is that an overly inclusive collaboration will ultimately reduce the number of separate competing research and development projects, thereby increasing the costs to society of mistakes in research and development strategy because there will be fewer other businesses pursuing different and potentially successful paths. In addition, it has been claimed that larger collaborations, involving a number of participants, will decrease the incentives to innovate since the benefits of succeeding and the costs of failing in research and development competition will be reduced.

There are a number of factors which might be considered in determining whether a given research and development joint venture is overly inclusive, including the relative capability and incentive of the firms that are in and out of the venture. Unless the facts in a particular case indicate otherwise, the DOJ typically regards all possible comparable research efforts in the market to be of equal competitive significance. As a general matter, and consistent with the legislative history of the NCRA, the DOJ believes that an anticompetitive effect in an research and development market is unlikely where there are at least four other comparable research and development efforts underway or where there is a substantial potential for such efforts by firms or groups of firms included in the market.[107] It follows from the foregoing that a decision to exclude a potential competitor from a joint venture generally will not, in and of itself, be deemed to be anticompetitive, assuming the excluded firm is able mount, either alone or in collaboration with others, a competing effort; however, when the joint venture in question is the only means for efficient pursuit of the research goals, exclusion may be anticompetitive.

Another way to look at the potential anticompetitive effect of a prospective joint venture is to assess whether or not the joint venture will be able to exercise market power. In many cases, if the joint venture or the parents cannot exercise market power, the joint venture will be allowed to go forward, even in a concentrated market.[108] Once again, the relevant considerations with respect to market power include such things as the assets and resources of those firms not included in the joint venture to undertake research and development comparable to that performed by the subject joint venture, or to develop products and services which ultimately will be competitive with those to be developed by the joint venture.

Another possible anticompetitive risk of a joint research and development venture would be that the participants would in some manner agree to slow the pace of innovation, or the exploitation of the fruits of the venture. This is particularly true in situations where the joint venture includes a large portion of the competitors in properly defined relevant markets for the goods or services to be produced with the products of the re-

[107] In enacting the NCRA, Congress indicated that the existence of four or more comparable research and development ventures would generally be sufficient to insulate an research and development joint venture from antitrust condemnation. See NCRA House Conf. Rep., at 10, 1984 U.S.C.C.A.N. at 3134–5. The HHI of a market with five comparably situated firms would be 2000. This

[108] See, e.g. *Sewell Plastic, Inc.* v. *Coca-Cola*, 720 F. Supp. 1196, 1213–4 (W.D.N.C. 1989), *aff'd*, 912 F.2d 462, *cert. denied*, 111 S. Ct. 1019 (1991).

search and development. In such cases, regulators are concerned about any anticompetitive "spill-over" effects, such as collusion among those competitors with respect to the price or output of goods or services sold outside the joint venture, or with respect to other strategic business decisions. The risk of collusion is typically a function of how closely related the research and development being carried out is to the production and marketing of the product or service in question, with the risk of collusion least significant when the research and development program involves basic research which is far removed from current price and output decisions.

In those cases where some anticompetitive effects from the proposed joint venture have been demonstrated, it is then necessary to consider any procompetitive effects from the joint venture, and to determine whether the procompetitive effects will outweigh the anticompetitive effects. For example, a joint venture is more likely to be approved when the cost of the venture relative to a single firm's budgetary limits is quite high, or it is apparent that the joint venture will produce substantial economies of scale. Conversely, if there are potential anticompetitive effects from the collaboration, and it can be reasonable demonstrated that the activities which are to be conducted by the joint venture could be performed by a single firm without significant budgetary hardship and without reasonable efficiency, then any procompetitive benefits from the joint venture may be outweighed by its anticompetitive effects.

Obviously, there are a variety of procompetitive effects that might be realized from a research and development joint venture. One benefit that joint research and development ventures can capture is economies of scale. Also, combining the complementary abilities of different competitors may lead to synergies that further reduce the cost of R&D. However, while there are a number of recognized procompetitive benefits from a research and development joint venture, the Congress resisted any attempt to actually enumerate the benefits that might be included in the rule of reason weighing process. For example, the conferees agreed to drop from the bill Senate language that expressly instructed a court to consider the "effects in promoting competition through innovation or enhancement of efficiency," thereby resolving the concerns of House conferees that a more detailed list might be read as an exclusive list, thereby precluding consideration of other competitive factors.

E. Competition Law Analysis of R&D Joint Ventures in the EC

Introduction The Commission's general attitude regarding research and development joint ventures apart from Regulation 418/85 has been explained as follows:

> "Pure research and development JVs which do not fulfil the conditions for group [block] exemption under Regulation (EEC) No. 418/85 can still in general be viewed positively. This type of co-operation normally offers important economic benefits without adversely affecting competition. That is also the case where the parents entrust the JV with the further task of granting licenses to third parties. If the JV also takes on the manufacture of the jointly researched and developed product, the assessment for the purpose of exemption must include the principles which apply to production JVs. JVs which are responsible for R&D, licensing, production and distribution are full function JVs and must analysed accordingly."[109]

[109] Commission's JV Notice, § 59. The JV Notice also states at § 20 that: "The economic pressure towards co-operation at the R&D stage does not normally eliminate the possibility of competition between the participating undertakings at the production and distribution stages."

Restrictions on competitive research and development activities Article 85(1) will apply in situations where the parties effectively preclude themselves from establishing a competitive advantage over one another with respect to conducting research and development, either in the field covered by the joint venture or in some other area.[110] Certainly, Article 85(1) would be triggered by an express agreement not to carry out independent research. Moreover, even in the absence of a formal agreement, if the effect of the contract or arrangement of the parties is that de facto the parties will not conduct research independently of one another, than Article 85(1) will also apply.[111] For example, if the parties cooperate with each other research and development work and/or exchange information regarding research results, it is likely the parties will not have the incentives to engage in independent research that they might have had in the absence of any contact with the other party, nor would they have the opportunity to seize any competitive advantage which might exist from sole ownership and knowledge of the results of the research work.[112]

Under Regulation 418/85, the parties may be restricted from carrying out on an independent basis research and development in the same field or a closely related field during the joint venture program,[113] as well as from entering into agreements with third parties for research and development in such fields during the joint venture program.[114] However, the parties cannot be prevented from carrying out research and development, either independently or with third parties, in separate fields of activity during the joint venture program, nor can they be restricted from conducting research and development activities in the same field after completion of the research program.[115]

Restrictions on competitive research and development are considered to be necessary in order to prevent less than all of the parties from appropriating the results of the development work to the detriment of the others, a situation which would substantially increase the risks of the collaboration to all of the parties. In fact, if a party was able to appropriate the initial results of the research and development work and develop them further, the party may ultimately be able to obtain exclusive rights to an enhanced or improved product which would effectively make the products arising from the original work obsolete, thereby totally undercutting the utility of the joint venture to the other parties.

A key concept is how one should define whether or not a research field is "closely connected" to the activities of the joint venture. One commentator has suggested that

[110] See, e.g. *Henkel/Colgate*, JO 1972 L14/14; *Beecham/Parke Davis*, OJ 1979 L70/11 [1979] 2 CMLR 157; and *Vacuum Interrupters* (No. 1), OJ 1977 L48/32 [1977] 1 CMLR D67.

[111] See, e.g. *Continental/Michelin*, OJ 1988 L305/33 [1989] 4 CMLR 920 (despite the freedom of the parties as a matter of formal contract to pursue independent research and development in competition with the joint venture, the Commission found that in practice the parties did not engage in independent work and, in fact, forgoing independent work was part of the understanding that existed between the parties); and *Olivetti/Canon*, OJ 1988 L52/51 (the parties had no interest in devoting resources to independent development work due to joint development work of the venture).

[112] See, e.g. *Bayer/BP Chemicals*, OJ 1988 L150/35 [1989] 4 CMLR 24 § 22 ("Continual co-operation between competitors at the R&D level is bound to have a direct impact on competition between them.") These concerns have led to some of the attempts to contain the exchange of information at the point of upstream cooperation. See *Ford/Volkswagen*, OJ 1993 L20/14, and *Continental/Michelin*, OJ 1988 L305/33 [1989] 4 CMLR 920. In addition, the Commission has often indicated concerns that the reduced incentives might actually spillover to downstream activities, like production and sales. See, e.g. *Henkel/Colgate*, JO 1972 L14/14; *Beecham/Parke Davis*, OJ 1979 L70/11 [1979] 2 CMLR 157; and *Bayer/Gist-Brocades*, OJ 1976 L30/13 [1976] 1 CMLR D98.

[113] Regulation 418/85, Art. 4(1)(a).

[114] Regulation 418/85, Art. 4(1)(b).

[115] Regulation 418/85, Art. 6(a).

the closeness of the field should be assessed by reference to the probability that the results of the research and development work conducted in the course of the joint venture will be appropriate by one of the joint venturers to the detriment of the others.[116] However, it generally will be quite difficult to identify all the possible connections between various fields, particularly since one can find connections among all the ways of improving the manufacture of some product or its substitutes, all methods for performing some function, and the ways one can exploit some new invention or utilize some other asset.[117]

There are several ways in which the parties might attempt to deal with regulatory uncertainties in this area. For example, since the prohibition in Regulation 418/85 applies to unconnected fields, it may be wise for the parties to utilize the opposition procedure in order to obtain comfort regarding the validity of restrictions which may appear to be in less closely or remotely related fields.[118] Another method would be for the parties to deliberately structure the initial research and development program as broad as possible, thereby increasing the breadth of the fields in which restrictions might be enforced. As noted above, Regulation 418/85 does not regulate the breadth of research and development programs, instead it only requires that the program be clearly defined. On the other hand, if a party is concerned about restrictions that may become overly broad, it may insist that the research program be quite narrowly defined and that duration of the program be relatively short. If following this initial stage, the parties find that research in new fields is required, they can then enter into a new agreement.[119]

Restrictions on exploitation of research results Article 85(1) will be applicable to a joint venture or other collaboration in which the parties accept, either expressly or by implication, limitations or restrictions on their right and ability to independently exploit the results of the joint research work.[120] Restrictions of this type may take a variety of different forms, including direct restrictions on production or marketing,[121] which are particularly offensive in situations where the effect thereof is to impede the free flow of the products derived from the research work throughout the EC;[122] an agreement to engage in joint production or joint marketing of the products, which has the effect of reducing incentives to invest in independent activities;[123] and restrictions on their ability to independently grant licenses to third parties with respect to the exploitation of the research results.[124] Article 85(1) will also apply where the parties are obligated to pay royalties

[116] White, "Research and Development Joint Ventures under EEC Competition Law," (1985) 16 IIC 663, 689.

[117] Korah, supra note 27, at 41.

[118] Ibid.

[119] However, other parties may be concerned about the possibility that the party will elect not to enter into a new agreement and to proceed independently with the new development work. This possibility may impede the flow of research and the exchange of information in the initial stage, thereby threatening the success of the preliminary work. Ibid.

[120] See, e.g. Regulation 418/85, Arts. 4 and 6.

[121] *MAN/SAVIEM*, OJ 1972 L31/29 [1974] 2 CMLR D123 (restrictions placed on producing or marketing products derived from the joint research work in which the other party specialized).

[122] See, e.g. *Rank/Sopelem*, OJ 1975 L29/20 [1975] 1 CMLR D72 (export restrictions); and *Beecham/Parke-Davis*, OJ 1979 L70/11 [1979] 2 CMLR 157 (agreement involved an exclusion of France and the Commission stated that an exemption would be granted only "if the results of such joint research can be used by both parties freely and independently without any territorial or other restriction on production or marketing within the common market"). See also Regulation 418/85 Art. 4(1)(c) to (f) and Art. 6(c) to (h).

[123] See, e.g. *Vacuum Interrupters (No. 1)*, OJ 1977 L48/32 [1977] 1 CMLR D67, and *(No. 2)*, OJ 1980 L383/1 [1981] 2 CMLR 217; *GEC/Weir*, OJ 1977 L327/26 [1978] 1 CMLR D42. See also *Alcatel Espace/ANT*, OJ 1990 L32/19 [1991] 4 CMLR 208, which involved a joint venture for research and

to one another with respect to their use of their research results, unless the obligations can be justified by reference to unequal contributions of the parties;[125] when the parties do not have access to the research results in proportion to the level of their participation;[126] and when restrictions are imposed on the ability of the parties to freely use the results following the termination of the joint venture.[127]

As a general rule, the parties may not accept any restriction regarding the customers that they serve.[128] However, if the parties are not competitors[129] at the time that the joint venture agreement is executed, they may include therein an obligation to restrict the manufacture or application of the products or processes to specific technical fields of use.[130] As is often the case with the various provisions relating to restrictions, one of the fundamental issues is defining the technical field of use. One commentator has taken the position that a technical field of use is separate only if it relates to separate markets, basing that conclusion on the fact that an exemption will not be available if the parties already are competitors (i.e. active in the same market) at the time that the agreement is

cont.
development, production and sales in the telecommunications section. While the venture was exempted for ten years, the Commission noted that while the agreement allowed independent marketing, the structure of the joint venture meant that joint marketing would occur whenever possible. But cf. De Laval/Stork *(No. 1)*, OJ 1977 L215/11 [1977] 2 CMLR D69, §40, where the Commission found in relation to a joint venture which covered research and development, production and sale that "Even the fact that research is being undertaken jointly will not prevent either company from continuing to use the results separately."

[124]See, e.g. *Beecham/Parke-Davis*, OJ 1979 L70/11 [1979] 2 CMLR 157 §42. See also *Continental/Michelin*, OJ 1988 L305/33, [1989] 4 CMLR 920 § 15, in which the Commission found that grants of third party licenses required consent of both parties to a joint venture when the contract provided that the joint venture was to grant a license on request of one party and following consultation with the other party. The Commission did obtain a statement from both parties that "they will offer licences on reasonable terms to all interested competitors."

[125]See, e.g. *Beecham/Parke Davis*, OJ 1979 L70/11 [1979] 2 CMLR 157. See also Art. 5(1)(f) and 5(1)(g) of Reg. 418/85. *A fortiori* cross royalty provisions will also be condemned if they are related to marketing territories allocated to the participants in the joint venture. *Re Research and Development* [1971] CMLR D31.

[126] Regulation 418/85 Art. 2(b) and (c).

[127] See, e.g. *De Laval/Stork*, OJ 1977 L215/11 [1977] 2 CMLR D69 (post-termination restrictions infringed Art. 85(1): upon termination both parties must have access to technology of joint venture and freedom to exploit throughout the Common Market; royalties permitted but to be no higher than those payable by third parties). See also *Beecham/Parke Davis*, OJ 1979 L70/11 [1979] 2 CMLR 157 § 42 (obligations after termination to pay 75 percent of other parties' costs if exploiting products emanating from that party based on joint work held to infringe Art. 85(1) but exempted under Art. 85(3); and *Carbon Gas Technologie*, OJ 1983 L376/17 [1984] 2 CMLR 275 (obligation on withdrawing shareholder not to use joint know-how for five years infringed Art. 85(1) but exempted under Art. 85(3)). However, a post-termination obligation to preserve the confidentiality of know-how does not infringe Art. 85(1). See also Art. 5(1)(d) of Regultion 418/85. But cf. Art. 6(b) of Regulation 418/85 (a no-challenge clause following termination of the program precludes exemption).

[128] Regulation 418/85, Art. 6(e). See Art. 4(1)(e) (permitting the imposition of restriction on manufacturing to one or more technical fields of use, provided that the parties are not competitors).

[129] The parties must not be competitors "within the meaning of Art. 3" of the Regulation.

[130] Regulation 418/85, Art. 4(1)(e). The Commission has held that Art. 4(1)(e) does not cover a prohibition in the context of a joint venture between non-competitors on the grant of licenses in the other party's field of use to competitors of the latter, even though such a prohibition is recognized as ancillary to the matters which Art. 4(1)(e) is intended to cover; however, the Commission has accepted that such a provision was not prohibited and granted an exemption to the agreement under the block exemption's "short-form" notification procedures. Seventeenth Report on Competition Policy (1987), point 31(c). Note that field of use restrictions are included in Art. 4 rather than 5 of the Regulation, thereby implying that such restrictions fall within Art. 85(1), while such restrictions are on the "white list" of the block exemption for patent licensing. However, this may be appropriate in light of the fact that Art. 4(1)(e) is much broader in application, in that it extends to situations when the results are not protected by patents.

executed.[131] However, it is clear that this test will be difficult to administer, since it is almost always impossible to clearly demarcate two or more markets.[132]

Regulation 418/85 does permit some degree of territorial protection, although less than is permitted under the exclusive distribution, patent licensing and specialization regulations. The parties are permitted to include an obligation not to manufacture the contract products or apply the processes in territories which are reserved for other parties.[133] In addition, for five years from the time when the contract products are first marketed within the Common Market, the parties can be restricted from engaging in any active policy of marketing in territories which are reserved for other parties;[134] provided that the users of the products can obtain them from other sources and the parties do not themselves obstruct the entry of parallel imports or make it difficult for supplies to be obtained.[135]

Territorial protection can extend only to manufacturing[136] activities and to active sales. Parties cannot restrict passive sales activities, nor can they be required to refuse "without any objectively justified reason" to meet orders from sources (e.g. users or dealers) which are established within their respective territories that wish to market the products in other territories within the Common Market, or to make it difficult for customers to obtain the contract products from other dealers and to obstruct the free movement of products within the Common Market which have been lawfully marketed by another party.[137]

Regulation 418/85 reflects the Commission's reluctance to run a risk that the parties will divide the Common Market through the use of territorial restrictions on sales.[138] The Commission does concede that such provisions may, in some cases, be required in order for the parties to engage in collaborative innovation as part of the joint venture; however, the Commission has tempered the effects of such provisions by include time limitations and through the use of the requirement that the contract products must be also be available elsewhere. It worth noting that the prohibitions on parties from making it difficult for customers to obtain any of the goods from other sources effectively overrides any intellectual property rights which the party might other have relating to the contract products.[139]

By limiting the duration of the restrictions to five years following the date that the contract products are first put on the market anywhere in the EC, the Commission appears

[131] White, supra note 116, at 692.

[132] See, e.g. *Windsurfing*, OJ 1983, L229/1 [1984] 1 CMLR 1, paragraph 74 (sailboards for use on water were a single technical field of use); *France/Suralmo*, Ninth Report on Competition Policy 72 (military and civilian applications were a single field of use).

[133] Regulation 418/85, Art. 4(1)(d).

[134] "Active marketing" includes, without limitation, engaging in advertising specifically aimed at such territories or the establishment of any branch or the maintenance of any distribution depot for the products in the territories.

[135] Regulation 418/85, Art. 4(1)(fc), inserted by Regulation 151/93. In turn, Art. 6(f) of Regulation prohibits a provision which restricts the parties from pursuing an active sales policy in territories within the EC reserved for other parties following the date which is five years from when the products were first marketed in the EC. See *Quantel International (Continuum)/Quantel SA*, OJ 1992 L235/9; *BBC Brown Boveri*, OJ 1988 L301/68 [1989] 4 CMLR 610 § 21. Note that the restrictions against active sales may terminate before the exemption for the entire joint venture expires.

[136] It is worth noting that the protection against competitive manufacturing in a particular territory can actually be quite broad, particularly if the products cannot be transported very easily. The manufacturing protection is available even in cases where the process is not protected by patents. Korah, supra note 27, at 45.

[137] Regulation 418/85, Art. 6(h).

[138] See, e.g. *Re Davidson Rubber* JO 1972 L143/31; [1972] CMLR D52 and Regulation 2349/84, Art. 1(1).

[139] Korah, supra note 27, at 45.

to be attempting to provide an incentive to the parties to introduce the products throughout the EC as quickly as possible. Although the parties might be able to increase their competitive advantage through a strategy of manufacturing and selling the products outside of the EC before entering the EC market, the five-year period may be triggered by a disposition of the products in the EC by a buyer which purchased them outside of the EC.[140]

Restrictions on independent exploitation It is not surprising that several of the "black listed" items in Article 6 of Regulation 418/85 are intended to insure that the parties are free to engage in independent competition following the completion of the joint research and development, particularly in those cases where the parties have decided against joint exploitation of the results of the development work. For example, Regulation 418/85 cannot be relied upon in situations where there are restrictions relating to the quantity of the contract products that the parties may sell or manufacture or the number of operations employing the contract process which the parties may carry out;[141] where the parties are restricted in their ability to independently set prices or discounts for the contract products;[142] or where the parties are prohibited from allowing third parties to manufacture contract products or apply contract processes when joint manufacture is outside the scope of the agreement.[143] These limitations supplement the prohibitions on certain field of use restrictions (i.e. customers) which have been discussed above.

There are several practical issues raised by the items in Article 6 relating to independent exploitation. First, while the parties cannot be prohibited from licensing others to manufacture contract products when joint manufacturing is not taking place, there is no mechanism for insuring that licensees will be able to obtain access to any background technology which is owned by a party other than the licensor and which is necessary in order for the licensee to effectively exploit the new technology. Another issue relates to determining what constitutes "joint manufacture" for purposes of the exemption. For example, if the parties have developed more than one product during the course of the research work and jointly decide to commercialize one product before they decide how to exploit the other products, does this amount to a prohibition with respect to such other products that falls within Article 6? At least one commentator has argued that any joint decision regarding the order of exploitation should not prevent the exemption from applying, lest it reduce the incentive to engage in joint collaboration on more than one potential product at the same time.[144]

Exclusive purchasing obligations In cases of joint exploitation the parties may agree to an obligation to procure the contract products exclusively from other parties, joint

[140] The five-year period begins when the results are first marketed in the EC, whether or not by the parties. In contrast, a similar provision in the patent licensing block exemption states that the period does not begin until a party has marketed any of the products covered by the license.

[141] Regulation 418/85, Art. 6(c).

[142] Regulation 418/85, Art. 6(d). See also *Alcatel Espace/ANT*, OJ 1990 L32/19 [1991] 4 CMLR 208, § 17.

[143] Regulation 418/85, Art. 6(g), as substituted by Regulation 151/93. The term "joint manufacture" includes the allocation of manufacturing responsibilities among the parties by way of specialisation (Art. 1(3)) and the agreement to procure exclusively from a supplier jointly charged by the parties with manufacture (Art. 4(1)(c)). See *KSB/Goulds/Lowara/ITT*, OJ 1991 L19/25 [1992] 5 CMLR 55 § 22.

[144] Korah, supra note 27, at 55. Korah uses the example of joint collaboration which produces two drugs which are capable of performing the same function, yet which are different with respect to potential side effects and other matters. Korah notes that the parties might decide to proceed with exploiting one drug first, which means expensive clinical trials and other activities, and not to license or otherwise exploit the other

organizations or undertakings or third parties, jointly charged with their manufacture.[145] This obligation is the reciprocal of the requirement in Article 2(e) of Regulation 418/85 that any joint organization or outsider charged with manufacturing the contract products must sell those products only to the parties. The Commission believes that an exclusive purchasing obligation may well be required in order for the parties to invest in the joint research and development and exploitation. It is not necessary to permit the parties to purchase from other parties if they are able to purchase the contract products more cheaply;[146] however, the obligation can only extend to purchases from one of the parties or a person "charged with their manufacture," therefore it is not possible to gain an exemption for purchases from licensees of the parties or the joint venture.[147]

Feed and grantback obligations The parties are permitted to allow for the exchange of technical information gained during the course of exploiting the results of the research and development work. Specifically, Article 4(1)(g) of Regulation 418/85 provides that an exemption will be available for "an obligation on the parties to communicate to each other any experience they might gain in exploiting the results and to grant each other non-exclusive licenses for inventions relating to improvements or new applications." The exemption appears as part of Article 4, rather than Article 5, because of the Commission's concern that such post-development information exchanges may have an adverse competitive effect, particularly if the market in which the parties are operating is highly concentrated. Although the obligations must be reciprocal in order to fall under the terms of the exemption, there is no requirement that the licensing terms be identical. Accordingly, each party may agree to grant licenses to the other parties to manufacture only in its own territory.

Article 4(1)(g) of Regulation 418/85 has at least two interesting effects when interpreted in relation to certain of the other provisions of the exemption. First, the exchange of information relating to any new applications or improvements actually broadens the collaboration beyond the scope of the original research and development work. Second, although the parties cannot be restricted from doing their own independent research and development work following the termination of the initial joint development program, the existence of cross-licensing obligations tends to undermine the freedom of the parties to freely exploit the results of the development work, at least to the extent that the sharing of information makes it more difficult for a party to obtain and maintain a competitive technological advantage over the other parties.

Case law Several recent cases have dealt with pre-industrial development of new products. For example, one case involved a proposed joint venture between the two largest tire suppliers in the oligopolistic EC market, Continental and Michelin. Both companies had been independently working on the development of a design for a new tire that

cont.
drug until it is clear whether or not the first drug will be marketable. Korah argues that this joint exploitation of a portion of the results of the research work should constitute "joint manufacture" for purposes of the block exemption.

[145] Regulation 418/85, Art. 4(1)(c). This extends the benefits of the block exemption to certain research-based joint ventures which include production.

[146] The block exemption for specialization agreements (Regulation 417/85) provides that exclusive purchasing obligations will only be exempted if the parties are free to purchase from third parties if the products can be purchased more cheaply.

[147] Korah, supra note 27, at 42 n1.

could be driven for considerable distances even when it was flat. If such a tire could be successfully developed and adopted within the automobile industry, it would require changes in the standards for auto construction. Apparently, Michelin's work had been somewhat unsuccessful; however, Continental was making progress on the new necessary new design. Realizing that Michelin's prior work might be useful to it in furthering the pre-industrial design work, Continental approached Michelin regarding the possibility of their working together to complete the design. Also, the parties were to work together on the commercial launch of the new tire and, through a newly-created joint venture, they would jointly license and administer the intellectual property associated with the new tire. The parties would not, however, be involved in joint commercial manufacturing or joint selling activities, although they did agree to give each other prior notice of their intention to market the new product.

The Commission found that the arrangement fell within the scope of Article 85(1).[148] The Commission identified three major restrictions in making its determination. First, it noted that even though there was no contractual restriction on the parties from engaging in independent research and development on the new type of tire, the arrangement effectively deprived the parties of the incentive to engage in such activities. Second, there was the restriction on the ability of either party to freely grant any licenses to third parties without the consent of the other party. Finally, the restrictions on the marketing activities of each of the parties which flowed from their mutual obligations to give the other party prior notice of their intention to market and also to coordinate the commercial launch of the new tires vis-a- vis the industry and the relevant authorities.

The Commission granted an exemption to the cooperative arrangement and, in so doing, apparently took into consideration each of the following:

(i) The scope of the collaboration was narrowly defined and, despite the cooperation between the parties at the pre-industrial stage, room still remained for a substantial amount of product differentiation of a more than merely cosmetic nature at the marketing stage, where the parties would still be able to compete with each other and with any licensees.[149]

(ii) The parties had stated their intention to also offer licenses covering the new technology to any interested competitors.

(iii) The parties' arguments that the industry's aversion to single sourcing would mean that any new product design that would require changes in the standards of new auto construction would not gain broad acceptance unless there were two or more sources for the new product, which the Commission believed would be the case in light of the aforementioned representations of the parties regarding the availability of licenses to third parties.

The Commission decided to grant three individual exemptions covering those three distinct aspects of the collaboration which it could identify. First, an exemption of ten years was granted for the parties to conduct joint research and development. Second, the parties were to have two years for joint commercial development of the new product. Finally, an exemption of twenty years was provided for the activities of the joint ven-

148 OJ 1988 L305/33 [1989] 4 CMLR 920.

149 Product differentiation was also mentioned as a factor in the Commission's analysis in *Ford/Volkswagen*, OJ 1993 L20/14.

ture (i.e. licensing and the administration of the jointly owned intellectual property). All of the exemptions were based on the assumption that the product would first be marketed no later than January 1, 1989.

Another case arose in the early 1980s when British Petroleum sought to form a new joint venture with MW Kellogg, a large United States company, for the development of designs for the construction of ammonia plants using catalysts for the production of ammonia that had been developed by BP. The venture was attractive to BP in that it had no experience in designing, constructing or operating a large-scale plant, and Kellogg specialized in the development of such plants. It was agreed that both parties would continue to own any know-how that it may make available for use in the joint venture and would jointly own any new know-how which might be created during the course of the joint venture. In addition, each party would be granted the irrevocable world-wide non-exclusive right to use and sub-license any joint patents from the venture. If the development work was successful, Kellogg would exploit the process by licensing it to the operators of the plants and remit a percentage of the license fee to BP. It was anticipated that the development work would take about six years.

A variety of ancillary restraints were placed on each of the parties. For example, BP was restricted from making disclosures of information relating to the catalyst to third parties during the term of the venture and could not supply catalysts to any other person without Kellogg's consent during that time. BP also agreed to use its best efforts to have Kellogg be the design contractor if the first commercial plant using the new process was built by BP. For its part, Kellogg agreed not to carry out any other research work on competitive catalysts, either independently or as part of a collaboration with others, and not to commercialize any new ammonia process that was likely to be more commercially attractive than BP's process without first informing BP. Kellogg also was not allowed to bid for the construction of ammonia plants other than by offering a jointly developed process unless it first informed BP. BP was given the option to terminate the joint venture in the event that Kellogg was not actively developing the process.

The Commission found that the joint venture fell with Article 85(1) since the parties' freedom of commercial decision was to be materially restricted by the various mutual obligations, such as Kellogg's inability to develop or sell competing processes and BP's undertaking not to supply the catalyst to third parties. However, the Commission approved the joint venture[150], emphasizing that it represented a considerable technical advance, and that success would provide substantial consumer benefits in the form of energy savings and lower ammonia prices. The Commission believed that it was unlikely that the parties would have otherwise competed with one another in the design of process plants using the new form of catalyst. The Commission also felt that the ancillary restraints imposed on Kellogg were necessary in order to allow BP to protect its investment in the required research over the term of the joint venture. In addition, the obligations on Kellogg were reasonably designed to insure that Kellogg remained sufficiently committed to the project in order to give it an acceptable chance of success and also allowed BP to obtain the information necessary in order for it to be able to intelligently exercise its termination rights. It has also been noted that the Commission may have been disposed to approve the joint venture because of the positive effect that it might have on enhancing

[150] [1986] 2 CMLR 619. The block exemption was not available to the parties due, in part, to certain of the restrictions placed on the ability of the parties to exploit patents and processes that had not been jointly developed by the parties.

the EC competitiveness in the international marketplace,[151] as well as the fact that there would be significant competition from other parties.

II. Antitrust Analysis of Production Joint Ventures

A. United States Antitrust Laws

Production joint ventures involving traditional competitors can generally be segregated into three categories for purposes of the United States antitrust laws. First, there are joint ventures between competitors to produce a commodity or service that both parents currently produce. In such cases, the joint venture is analyzed as a horizontal merger under § 7 of the Clayton Act.[152] Second, there are joint ventures between competitors to produce a commodity or service that only one parent currently produces. These usually are analyzed as a potential competition merger under § 7 of the Clayton Act.[153] Finally, there are joint ventures between competitors to produce a commodity or service that neither parent currently produces, which ventures are also analyzed as a potential competition merger.[154] In addition, any form of production joint venture that utilizes agreements ancillary to the purpose of the joint venture will be analyzed under § 1 of the Sherman Act.[155]

For years, counsel considered analysis of a production joint venture to be markedly different from that of a research joint venture. This was largely due to the fact that the DOJ released, in 1980, its Antitrust Guide Concerning Research Joint Ventures which had served as the basis of any antitrust analysis of a research joint venture. Since there were no comparable guidelines for production joint ventures, it was assumed that reference should be made to merger-type analysis, which tended to change with each new enforcement agency administration. However, with the passage of the NCRPA, it now appears that the DOJ's analysis of both research and production joint ventures will, relying on the legislative history of the NCRPA, be merger-type.[156] Moreover, the legislative history makes it clear that both joint venture types will be subject to rule-of-reason analysis by the courts and/or the reviewing federal agency.[157]

151 Goyder, supra note 16, at 195.

152 *United States* v. *Avouch, Inc.*, 704 F. Supp. 1409 (W.D. Mich. 1989) (joint venture between two competing manufacturers of automatic tampers that would create a firm controlling 70 percent of the automatic tamper market would violate §7 of the Clayton Act).

153 *United States* v. *Penn-Olin Co.* 378 U.S. 158 (1964).

154 Ibid.

155 See *Polk Brothers, Inc.* v. *Forest City Enterprises, Inc.*, 776 F.2d 185 (7th Cir. 1985).

156 See statement by Charles F. Rule, former Assistant Attorney General, in Rule, "The Administration's Views on Joint Ventures," *Antitrust Law Journal* 54, (1985), 1121: "I don't know how many people are still looking at the 1980 Guidelines instead of the legislative history of the NCRA, but I can tell you our current analysis is based on the NCRA legislative history I don't even think we have any copies of the 1980 Guidelines left in the Antitrust Division."

157 The legislative history itself states: "The legislative history of the 1984 act described in some detail the proper application of the rule-of-reason standard by the courts and the Federal agencies in analyzing the competitive effects of a challenged research and development joint venture. As was noted in the 1984 legislative history, the rule of reason condemns only those joint ventures that, on balance, are anticompetitive; if anticompetitive effects are established, the court must weigh them against any demonstrated procompetitive effects in determining whether an antitrust violation has occurred.

Although that discussion related specifically to research and development joint ventures, the underlying framework is the same for production joint ventures and should serve as guidance to any court or

In addition to the concerns regarding the competitive effects of a proposed production joint venture, regulators will closely evaluate the purported efficiencies realized from the particular arrangement. Realizing the difficulties that often may be encountered in identifying and measuring efficiencies, it is useful to develop some method for weighing procompetitive effects from collaboration. One device that might be used is to analyze the level of integration of resources. Full integration, and hence the highest level of likely efficiencies, usually occurs where two companies agree to create a completely new facility for the manufacture and sale of a product. For example, in *Penn-Olin*, Pennsalt, which manufactured and sold sodium chlorate on the West Coast, and Olin Mathieson, which was a purchaser and reseller of the product, joined forces to construct a new facility in the southeastern United States.[158] Differing processes and expertise were contributed by each parent along with substantial amounts of capital to create a wholly new source of supply.

United States v. *Standard Oil*[159] was another case involving the procompetitive joint use of existing assets. *Standard Oil* involved a joint arrangement among four patentees to interchange blocking patents. Although the arrangement called for an equal division of royalties, which the government argued amounted to price fixing, the Supreme Court nevertheless sustained the arrangement, relying on findings that the patents embodied legitimate conflicting claims and that the interchange permitted an efficient process which could not otherwise be achieved without danger of litigation.[160] Moreover, the integration of technology was substantial and, in this case, led directly to a procompetitive effect.

On the other hand, in cases where the joint undertaking involved little or no integration, no real improvement in the functioning of the market, and no increase in firm or group efficiency, the anticompetitive effects have been enough to lead to a finding of illegality. For example, in *United States* v. *Socony Vacuum* Oil Co.,[161] a group of major oil companies agreed to a buying program in which each agreed to stabilize market prices by purchasing "excess" oil from independent refiners. Unlike other types of cooperative arrangements, the agreement among the major oil companies was totally at arm's length, introduced no efficiencies, such as joint warehousing or marketing, into the market,[162] and involved no integration of the existing resources of the parties. Since the necessary effect of drawing "excess" oil off the market was to affect price and there were no apparent efficiencies from the arrangement, the Supreme Court characterized it as a naked cartel and opted for a rule of *per se* illegality.[163]

United States v. *Topco Associates*[164] involved a cooperative joint buying arrangement among various grocery chains who were actual or potential competitors. A term of the arrangement was that each co-op member might sell Topco brands in an exclusive marketing territory and that no other member could compete in that territory with

cont.
Federal agency examining the effects on competition of a production joint venture. It should be noted, however, that each joint venture must be analyzed on its own merits and that the underlying facts of a production joint venture may be much different from the underlying facts of a research and development joint venture. To the extent two joint ventures differ, the competitive analysis which a court or Federal Agency undertakes, will, and should, be different." S. Rep. No. 51, 103d Cong., 1st Sess., pt. V. (1993).

158 378 U.S. 158, 160 (1960).

159 283 U.S. 163 (1931).

160 Ibid. at 170–1.

161 310 U.S. 150 (1940).

162 Ibid. at 177–91 (discussing formation of joint buying programs).

163 Ibid. at 219–24.

164 405 U.S. 596 (1972).

Topco-labeled products without the approval of the member enjoying the exclusive territory.[165] The Supreme Court rejected the notion that the arrangement should be subject to rule of reason analysis, noting that there was no integration of manufacturing, processing, or warehousing facilities. The Court went on to characterize the agreement as a *per se* illegal market division arrangement.[166] Moreover, there is evidence to indicate that whatever efficiencies may have been achieved in the joint buying program, they probably could have been achieved without using territorial restrictions.[167]

B. EC Competition Laws

Research and development collaboration extending to joint production among major competitors was the subject of a 1992 case involving KSB, Lowara, Goulds Pumps and ITT Corporation.[168] Each of the companies was a significant competitor in its principal field of activity. KSB was the largest manufacturer of water pumps in the world; Lowara was a key manufacturer of water pumps in Italy and a subsidiary of Goulds, the third largest manufacturer of water pumps in the world; and ITT, an American conglomerate, had a number of subsidiaries which were active in some aspect of the relevant product markets in the EC. In 1985, KSB and Lowara began to work together on the development of wet end components for water pumps which were made of chrome nickel stainless steel rather than the cast iron that had customarily been used. The parties believed that the new material would have significant advantages over cast iron which they expected would make it competitive in a variety of different applications.

In 1987, Goulds and ITT joined the other two parties in formulating an arrangement to conduct coordinated research and development on the new materials technology. In addition, the parties agreed that all manufacturing responsibilities would be sub-contracted to Lowara, which would manufacture the new products exclusively for the joint venture parents who would then use the products as part of the pumps that they each manufactured on their own. Each of the parties would retain the sole rights to any patents or other intellectual property rights arising out of their own development work. The parties did not make any provision for third party licensing. The parties also agreed that they would all maintain the joint venture technology in confidence for a period of five years following the cessation of the joint venture. As might be expected, the proposed joint venture drew several complaints from competitors of the joint venture parents, who claimed that the arrangement was not joint research and development, but rather was really intended as a scheme to jointly exploit the new technology in manner calculated to eliminate competition from the producers of cast-iron components.[169]

[165] Ibid. at 601–2.

[166] Ibid. at 608.

[167] For example, the same efficiencies could have been achieved if there had been both a Brown's Topco line and a Black's Topco line in those cities where Topco members operated. Also, concerns that original Topco franchisees might have had about the possibility of "free-riding" by later franchisees could have been alleviated through primary responsibility or profit pass-over clauses. These less restrictive alternatives were authorized by the Supreme Court when it eventually affirmed the entry of a final decree against Topco. See *United States* v. *Topco Assocs.*, 1973-2 Trade Cas. (CCH) § 74,728 (N.D. Ill. 1973), *aff'd*, 414 U.S. 801 (1973) (per curiam).

[168] OJ 1991 L19/25 [1992] 5 CMLR.

[169] In fact, five competitors filed adverse comments with the Commission regarding the proposed arrangement. One competitor even included an application to the Commission to compel Lowara to supply third parties *pari passu* with the joint venture parents.

Not surprisingly, the Commission found that the joint venture parents were actual competitors in the production of the cast iron components and that they must also be regarded as having been at least potential competitors, given their own financial strengths and technological capabilities, with respect to the product that was to be produced by the joint venture. Moreover, the Commission believed that competition among the parents would be possible due to the fact that the basic technology may well have been obtained from sources other than Lowara and that the parties were probably large enough to realize any required economies of scale as to the production of the new products without resorting to a joint venture structure.

In the Commission's view, the joint venture placed substantial restrictions on the parties' commercial freedom of action and also would result in a reduction in supplier choice, given that the new product was a key product in the end product (i.e. the pump) and the parties' own brand products would thereafter be homogenous as to that particular component. The foregoing restrictions were even more problematic in light of the market shares of the parties, both absolutely and in relation to competitors, in the conventional pumps. Moreover, if the customers actually were to prefer the new pumps, it could be expected that the already strong position of the joint venture parents would be significantly enhanced. Finally, as reflected in the complaints of competitors excluded from the joint venture, the Commission felt that the arrangement would result in the foreclosure of competitive opportunities, given the existence of the exclusive manufacturing license to Lowara, the absence of any provisions for third party licensing, and the obligations with respect to maintenance of confidentiality following the term of the joint venture.

For a variety of reasons, Regulation 418/85 did not apply to the proposed joint venture;[170] however, the Commission was inclined to exempt the arrangement under Article 85(3) for a period of five years, which appeared to be about the period required in order for the actual effects of the joint venture on competition to become more apparent.[171] In granting the exemption, the Commission made a few interesting findings. First, the Commission had no problem in finding the requisite contribution to be made by the joint venture to the improvement of production, technical and economic progress and consumer benefit. Second, while the Commission had found that the joint venture parents were certain potential competitors, it also believed that the cooperation among all four parents was an indispensable condition to completion of the project, since the financial risks associated with the development project would make it unlikely that KSB and Lowara would have pursued it on their own without the assistance of the other parties. Finally, it appeared that the Commission felt that it was still too early to tell if the joint venture would actually eliminate competition in the field, since it could not yet be known if customers would prefer the new pumps over the conventional pumps which competitors would continue to manufacture and market.

While the Commission did exempt the joint venture, it was still concerned that if the new products were successful then there would be material structural market changes which would be caused, in large part, by the inability of third parties to gain access to

[170] Although the Commission did make a finding that the arrangement pertained to joint research and development as well to joint exploitation of the new product (i.e. further research and development work was necessary at the time that Goulds and ITT became involved in the project), the joint venture fell outside of the research and development block exemption because of the fact that the combined EC market share of the parties with respect to water pumps exceeded the 20 percent ceiling in the Regulation. See Art. 3(2) of the Regulation.

[171] Five years was also the exemption period set out in the research and development block exemption. See Art. 3(1) of the Regulation.

the technology for the new products. The Commission attempted to deal with this potential foreclosure effect by requiring that the parties provide the Commission with periodic reports regarding the structure of the market and market penetration.

In another case, Ford and Volkswagen ("VW"), two major competitors in the EC automobile industry, formed a 50/50 joint venture in order to develop, engineer and manufacture at new, purpose-built plant in Portugal a type of car which was known as an MPV, which stood for "multi-purpose vehicle." The total investment was estimated to be about $3 billion. It was anticipated that the life cycle of the MPV would be about 10 years, and the proposed joint venture was to extend for that period. VW would assume the primary responsibilities for developing the new product, while Ford would contributes its expertise in the areas of manufacturing and plant engineering. Both parties would supply engines and transmissions to the joint venture; however, all the other necessary components would be externally sourced. The joint venture would produce differentiated vehicles for both parties under their respective trademarks. The vehicles made for each of the parties would be differentiated with respect to exterior design, optional extras, product range and engines. For example, all VWs would have VW engines. The parties would buy vehicles from the joint venture at "cost plus," and would then do their own independent marketing and sales activities. Each of the parties agreed to accept a fixed minimum quantity of vehicles and parts from the joint venture which equalled, in each case, 50 percent of the anticipated output of the joint venture.[172]

In analyzing the proposed joint venture under Article 85(1), the Commission assumed that the MPVs would be part of an existing market which had one manufacturer with an EC market share of over 50 percent,[173] while the next major competitor had approximately 16 percent of the market. The current market was relatively slow, but fast-growing; however, it was believed that ultimate size of the market was not going to be very large.[174] The Commission believed that entry into the market was quite difficult, given that the development and productions costs were quite high and that the minimum efficient scale for production actually exceeded the total size of the market at the time that the joint venture was to be formed.[175]

It seemed clear that both Ford and VW could, in principle, produce their own versions of the MPV without assistance from the other, in spite of the costs and difficulties associated with the development work. As such, given that the development of new products was a key element of competition in car manufacturing, the decision of the parties to collaborate rather than to go their separate ways must necessarily result in a restriction upon competition between them. Any problems associated with the joint venture were further compounded by the real possibility that the cooperation between the parties, including the extensive sharing of technical information, would extend outside the joint venture into adjacent markets, such as those for light vans and estate cars. As such, it was apparent that the joint venture would infringe Article 85(1).[176]

[172] Bellamy and Child, *Common Market Law of Competition* 4th ed, London (1993) § 5-074.

[173] The Renault Espace had a market share of over 50 percent at the time the new joint venture was announced.

[174] In 1992, the total market was about 100,000 units. It was estimated that by 1995, the total market would grow to about 350,000 units.

[175] It was believed that the minimum efficient scale of production was approximately 110,000 units per year, which compared to a total annual market in 1992 of 100,000 units. Bellamy and Child, supra note 172, at § 5-075.

[176] Ibid.

The Commission then went on to grant an exemption for the joint venture under Article 85(3) for a period of thirteen years. The Commission found a number of positive effects from the proposed joint venture. First, the joint venture would result in the entry of two new MPVs into the market, each of which would be technically and environmentally superior to existing products due, in part, to the marriage of the engineering and technical capabilities of each of the parties. Second, the Commission noted favorably that the joint venture would result in the construction of a new state-of-the art manufacturing plant in the EC which would incorporate the best elements of the Japanese Kanban system for "just-in-time" components delivery. Third, the joint venture represented the single largest investment ever made in Portugal, creating more than 15,000 direct and indirect jobs and thereby furthering one of the basic goals of the Treaty of Rome to reduce regional disparities in economic capability.[177] Fourth, the Commission concluded that consumers would ultimately benefit from the increased competition in the MPV market, given that the parties would actually be competing with one another at the distribution level,[178] and that the joint venture would eventually lead to a more balanced market structure.[179] Fifth, the Commission found that while the independent resources of each of the parties were certainly great, it was still unlikely that either one of them could achieve the break-even production sales volume on their own.

The Commission's decision to grant the exemption, which was challenged by the manufacturer of the Espace that was marketed and distributed by Renault,[180] was made subject to several conditions designed to address the possibility of certain anticompetitive effects from the collaboration between the parties. First, in order to limit the possibility of spillovers from the cooperation in the course of the joint venture, the parties were compelled to establish certain procedures and safeguards regarding the types of information that could be exchanged, and the persons to whom and the basis upon which it could be disclosed. Second, in order to insure that there would be sufficient product differentiation to permit competition at the distribution level, Ford was prevented from using VW engines in more than 25 percent of its vehicles over any three-year period.[181] Third, the risk of market sharing was to be reduced by requiring that each party make its vehicles available throughout the EC. Fourth, the parties were required to implement various procedures for effective separation in the event that the exemption was not renewed. Finally, the Commission set out a system of reporting requirements designed to allow it to monitor the progress of competition in the MPV market during the term of the joint venture.[182]

[177] While the creation of new jobs and the enhancement of regional productivity was a desirable side-effect of the proposed joint venture, the Commission explicitly noted that "This would not be enough to make an exemption possible unless the conditions of Art. 85(3) were fulfilled ..." (§ 36).

[178] Although the parties would both be purchasing their vehicles from the joint venture under a "cost plus" formula, the Commission accepted that the product differentiation between the vehicles manufactured by the joint venture for the parties and their differing profit margins would leave "sufficient scope for a certain degree of competition" between them. The Commission found that it was likely that the parties would still compete with each other in order to maximize their individual sales profits and to reinforce overall loyalties to their respective trademarks (§ 38).

[179] As noted [above], the Renault Espace held more than half of the market at the time that the joint venture was organized.

[180] Matra, the manufacturer of the Renault Espace, filed a formal complaint against the Commission's decision to exempt the joint venture. See Case T-17/93.

[181] While all standard Fords would have Ford engines, the top-of-the-range and diesel Fords would use VW engines.

[182] Bellamy and Child, supra note 172, at § 5-077.

15 Conclusions and Areas for Further Study

I. Introduction

This work has touched upon a variety of different areas, including patent rights, innovation, licensing, and regulatory policies. As we have seen, each of these subjects is relevant the complex issues surrounding the motivations and incentives for individuals and firms to make the real investments necessary for dynamic technological progress and economic growth. In this chapter, we conclude with a survey of the key issues which must be addressed in the areas of patents and intellectual property rights, and antitrust and competition law and policy. We also examine some alternative schemes which have been suggested for protecting exclusivity in new inventions and innovations. Thereafter, we look at some of the possible lessons that this work might hold for developing countries seeking to create their own intellectual property and competition law regimes. Finally, we suggest areas which might benefit from further study and consideration.

II. Patents and Intellectual Property Rights

A. Introduction

The earlier discussion of some of the major economic justifications of the patent system illustrates a number of the difficulties confronting economists, judges and lawmakers with respect to formulating a coherent system of laws and regulations regarding the products of inventive activity.[1] In spite of the long history of the patent system, a number of high-level studies of costs and benefits associated with patents in developed countries have failed to produce solid testimonials regarding the utility of the patent system. For example, a Canadian report in 1976 advised that "... the country should give serious consideration to the possibility of abandoning ... the patent system in any form."[2] Another analysis conducted in Australia in 1982 found that "there is little room for doubt that the benefit-cost ratio of the patent system is negative, or at the very best, in balance."[3] Major studies in the United States and United Kingdom have reached similar conclusions.[4]

[1] On the general policy debate regarding patents, see Meinhardt, *Inventions, Patents and Monopoly* (1950), 31–9.

[2] See Working Paper on Patent Law Revision, Department of Consumer and Corporate Affairs (1972), 92.

[3] Mandeville, Lamberton and Bishop, *Economic Effects of the Australian Patent System*, Report to the Industrial Property Advisory Committee (1982).

[4] With regard to the United States, see Machlup, *An Economic Review of the Patent System* (Subcommittee on Patents, Trademarks, and Copyrights of the Senate Committee on the Judiciary, Study No. 15, 85th Congress 2nd Session) (1985) (noting that if no patent system existed it would be a mistake to establish one; however, since one is in place it may as well be maintained). With regard to the United Kingdom, see

Whether or not one is willing to conclude that the patent system should be dismantled based, at least in part, on the mixed inferences which might be drawn from an analysis of each of the alternative theories for justifying the use of patents, it appears that there are a wide variety of additional areas which might be profitably investigated. Among those suggested by the discussion set out above are the following:

(i) The proper role of the state, and its power to make and enforce laws and regulations, in the marketplace for new ideas and knowledge. Specifically, one must ask whether the "public goods" character of new ideas and knowledge, as well as the positive externalities flowing from the creation of such ideas and knowledge, dictates the need for state intervention or whether it can be left to the market to develop private mechanisms for protecting the products of creative activity such that a sufficient incentive exists for inventors and innovators to take on the risks associated with those activities.

(ii) The social costs associated with recognizing a right of exclusivity as a means of providing an incentive for specified activities. Concerns in this area include the dangers of creating monopolies in the hands of the patent owner, the allocative inefficiencies of not allowing all consumers to gain immediate access to the invention, and the viability of alternative incentives which might have the effect of minimizing the negative effects of monopoly and underconsumption of newly-created resource.

(iii) The role of patents, and patent-like rights, as a tool for public policy. Among the issues in this area are assessing whether current standards for granting patent rights are consistent with the incentives required for innovative activities, and studying the impact that technological development might have on other areas of concern to public policymakers (e.g. employment, education, resource allocations, etc.).

B. The Role of the State in the Marketplace for New Ideas

Introduction The recognition of patents and other forms of intellectual property rights represents an affirmative decision by the state to intervene in the marketplace for the creation and use of new ideas. The state provides inventors, innovators and potential competitors with a set of laws and regulations with respect to registering the products of creative activity, determining the obligations of the public in relation to the owner of the rights, and the enforcement of such rights in the event that one or more persons fail in their duties with respect to the use of the object of the intellectual property right. The affirmative actions of the state in this area are not without controversy and signal a broader concern about the proper balance between market forces on the one hand and the use of the policy tools which are at the disposal of the government.

Market failure and state intervention Government intervention often occurs in those cases where there has been some sort of failure in the market which results in competi-

cont.
Taylor, "Do We Still Need a Patent System?," CIPA (1972–3) 292, 306 (concluding that the interest of the United Kingdom in preserving the existing international patent system was no more than "modest but perceptible").

tive imperfections.[5] Monopoly is a classic example of a market failure and commonly is treated by the promulgation and enforcement of antitrust and competition laws. Other examples of potential market failure include the existence of public goods, externalities, or the presence of high transaction costs which make it difficult for the market to function efficiently.[6] Whenever one of these conditions occurs, it would appear that the system of private property rights is not performing properly and, in many cases, some economists will advocate that the state take whatever action is required in to order to make the necessary corrections in the marketplace.

Knowledge, after it is created, is a classic example of a "public good."[7] As opposed to a private good, a public good is non-exclusive and non-rivalrous.[8] Public goods are non-exclusive because once the good is available to the public, it is difficult to prevent anyone from using it.[9] A public good is non-rivalrous since the use of the good by one person does not reduce the quantity of the good which is available for use by others.[10] Since it is often difficult, if not impossible, for any private party to appropriate a "public good," economists often advocate that having the state provide the good is the only or the least costly way of producing the good.[11] This strategy is particularly compelling in the case of the objects of intellectual property protection, given the various arguments which are often made against any restrictions on the use of new ideas and knowledge.

As noted below, "state provision" of knowledge, which may be accomplished through a system of rewards or through some form of direct government financing scheme, may not feasible or efficient. However, that does not necessarily mean that the government has no alternative but to create a system of intellectual property rights in order to provide an incentive for invention and innovation. It has been suggested that the "public good" character of new ideas and information might simply be viewed as one of the "costs" to the producer which must be borne as part of developing new products and services embodying the ideas and information.[12] If viewed from this perspective, the issue becomes whether or not the marketplace, on its own, can develop its own mechanisms for controlling these costs and prevent others from simply "free riding" on the efforts of the original inventor or innovator.[13]

[5] With regard to "market failure," see, e.g. Samuelson, "The Pure Theory of Public Expenditure," *Review of Economics and Statistics* 36 (1954), 387–9; Bator, "The Anatomy of Market Failure," *Quarterly Journal of Economics* 72 (1958), 351–79; Barrett, *The Theory of Microeconomic Policy* (1974).

[6] Mackaay, "Economic Incentives in Markets for Information and Innovation," *Harvard Journal of Law and Public Policy* 13 (1990), 867, 880.

[7] The defense and criminal justice systems are also generally offered as examples of goods which are "pure" public goods. Ibid. at 881.

[8] Ibid. at 880.

[9] Ibid.

[10] Ibid.

[11] Ibid. at 883. Some public goods, such as the "environment," are characterized as "free goods." If there is no incentive to limit the use of the free good, it will be overconsumed, with the result being that it no longer is available for use in future periods (e.g. pollution of the environment through use in the present period diminishes it value in future periods). The problem has been referred to as the "tragedy of the commons." Hardin, "The Tragedy of the Commons," *Science* 162 (1968), 1243, 1244. See also generally Hardin and Baden (eds), *Managing the Commons* (1977). New ideas and information are not susceptible to "overuse" in the same manner as free goods; however, the concern remains about non-compensated use of the ideas and information in the present period diminishing the incentive to produce new ideas and information for consumption in future periods.

[12] Palmer, "Intellectual Property Rights: A Non-Posnerian Law and Economics Approach," *Hamline Law Review* 12 (1989), 261, 273.

[13] "Free-riding," the ability of others to use new ideas and information without compensating the creator, plays a prominent role in the debate regarding intellectual property rights. For example, consequentialists argue that new ideas will underproduced unless free-riding is eliminated by the creating of a system for

Some economists have taken issue with the basic premise that new ideas and knowledge cannot, by their very nature, be made the subject of private efforts to achieve and maintain a modicum of exclusivity with respect thereto such that some form of market can be maintained.[14] They point to the example of the manner in which broadcast signals, originally a public goods technology, were converted into the current quasi-privatized radio spectrum through the development of radio advertising.[15] Other areas where private good technologies have been able to compete with public good technologies include video cameras and players in the private area and television signals in the public area; burglar and fire alarms in the private area and the services of police and fire departments in the public area; and books and computers in the private area and schools in the public area.[16]

However, while many supporters of intellectual property rights are willing to concede that market participants will act, often in very ingenious ways, to convert what are seemingly "public goods" into private property rights, they also point out that the problems extend beyond privatization. For example, merely creating a private property right in a new idea or body of knowledge does not guarantee that the idea or knowledge will be effectively utilized. In order for this to occur, it may be necessary to develop a legal right which has the essential elements typically associated with classical property rights: a clear right to exclude, an incentive for investment, and transferability. Only when the right has these characteristics can there be a market in which bargaining can occur leading to transaction which result in movement of the ideas and knowledge to their highest valued uses.

New ideas and knowledge also raise issues with respect to "externalities." An externality occurs whenever one person's use of his own "property" disturbs another's use of his own property ("negative" externalities) or, more rarely, confers some sort of benefit upon the other person ("positive" externality) without the first person, in deciding upon such use, taking into account the disturbance or benefit which might be conferred upon others.[17] As is the case with public goods, some economists have argued that externalities must be corrected by government intervention, which often takes the form of new taxes imposed with the intent of internalizing externalities. However, other economists respond that the concerned parties will themselves be able to internalize the exter-

cont.
registering the utility of the idea to others by compelling them to pay for its use. Nance, "Forward: Owning Ideas," *Harvard Journal of Law and Public Policy* 13 (1990), 757, 765. The advocates of moral rights take issue with the ability of free-riders to copy and use the ideas created by others without their consent. Ibid. However, it is important to note that anti-consequentialists take a different view of the free-riding problem. Rather than assuming the illegitimacy of free-riding, anti-consequentialists, who tend to generally take a very narrow view of property rights, refer to the common law example of free-riding simpliciter which, they note, in the absence of fraud, trespass, consumer deception, and the like, is generally not an actionable wrong. See generally Chafee, "Unfair Competition," *Harvard Law Review* 53 (1940), 1289. Therefore, anti-consequentialists appear to have little concern regarding the free-riding problem, concluding that it either is not wrongful or not sufficiently wrongful to justify legal redress since the use of information which is "nonrivalrous" is not a morally compelling reason to employ the coercive powers of the state. Nance, supra at 772.

14 Mackaay, supra note 6, at 893.

15 See Nance, supra note 13, at 765.

16 Ibid. at 922. See also Coase, "The Lighthouse in Economics," *Journal of Law and Economics* 17 (1974), 157 (describing how English lighthouses were actually operated by private initiative); and Stroup and Baden, *National Resources: Bureaucratic Myths and Environmental Management* (1983) (describing efforts to privatize the protection of the environment).

17 Mackaay, supra note 6, at 881.

nalities without direct government action, provided that the law provides the parties with guidance as to what rights are owned by which party.[18]

New ideas and knowledge create positive externalities for large numbers of people; however, these beneficiaries generally lack any incentive to pay for them. Since the beneficiaries figure that they are able to receive the benefits for free, the aggregate amount that people will pay for the collective good is understated and a danger exists that the good will either not be produced or will be "underproduced" relative to the true needs of the various beneficiaries. If one assumes that new ideas and knowledge are goods which have significant social value, the public policy issue is the same as that raised with respect to public goods: can the market independently develop the means for reducing externalities to the point where adequate resources will be devoted to creating new ideas and knowledge or will it be necessary for the state to intervene in some manner.

It is apparent that private actors have been successful in developing mechanisms for controlling some of the externalities that are associated with new ideas and information. For example, physical or programmed copy protection has already been developed for software information, and television signals can be protected through cable or scrambling.[19] Another method that is used is to vary the medium used to convey the information, such as occurs with hardback books, which are released first and have a higher price, and paperback books, which are released much later and have a lower price.[20] A prospective purchaser may make her own choice with regard to the timing and pricing of the information in accordance with her own individual tastes and preferences.

Support can be gather for the proposition that inventors and innovators can develop their own mechanisms for reducing the costs to them associated with free riding, as well as capturing some of the value associated with the positive externalities which might be available to others as a result of the activities of the inventor or innovator. If this is true, an argument might well be made that state intervention, in the form of intellectual property rights, is not necessary in order for the market for new ideas and knowledge to function effectively. However, unless one is able to estimate the real transactional costs associated with privately developing these mechanisms,[21] it is difficult to know whether or not such a strategy will produce the socially optimal level of invention and innovation. In any event, the proper balance between intervention by the state and private market mechanisms must constantly be monitored in analyzing any incentive-based exclusivity system.

C. Patents, Monopolies and Allocative Inefficiencies

Background Some commentators attempt to make a distinction between property rights, which they consider to be legitimate rights or entitlements to exclusivity which can be protected by recourse to the coercive power of the state, and privileges, which are illegitimate rights and entitlements that are also granted by the state and which are, in the

[18] See, e.g. Coase, "The Problem of Social Cost," *Journal of Law and Economics* 3 (1960), 1.

[19] Mackaay, supra note 6, at 893.

[20] Ibid. at 894.

[21] Arguing that public goods will ultimately be privatized in some fashion by private entrepreneurs, some economists advocate that it is more efficient to create exclusive rights (e.g. patents) directly at the outset. By so doing, the high transaction costs associated with privatizing the commons can be avoided, and the rights can be transferred and exchanged in the market until the achieve their highest valued use.

minds of some critics, unjustifiable institutions. In large part, property rights, especially with respect to tangible goods, are considered to be a justifiable and paradigmatic institution. It has been said that property rights and contracts, as enforced by the power of the state, are a "prelude to almost any economic relationship."[22] Under this view, the state involvement "[does not] inhibit but rather make[s] possible competition (as a process)."[23] On the other hand, patents are sometimes characterized as illegitimate state-created privileges which are to be severely restricted and regulated.

The notion of patent rights as illegitimate privileges can be traced back to their historic origins and has survived to find a place in the language used by the United States Supreme Court with respect to the powers given to Congress under the Patent Clause. For example, the Court has stated that "[E]very patent is the grant of a privilege of exacting tolls from the public,"[24] and often refers to patents as "monopolies."[25] Critics of the patent system have been quick to point out that patents may provide the owner of the patent with the opportunity to behave as a monopolist, thereby reducing consumer welfare to a point which cannot be justified as a means for providing an incentive for the patent owner to engage in inventive or innovative activities.[26]

Monopolistic behavior can, in theory, have a number of significant disadvantages. First, monopoly pricing prevents certain consumers from purchasing the monopolist's goods and services. The presumption is that these consumers would have purchased the goods and services if they were offered at the price which would have been established in a competitive market and that the monopolist's pricing policies causes such consumers to purchase other goods which were of less utility to them. As a result, there is a "misallocation of resources."[27] Second, objections are often made to the monopoly profits which monopolists do receive from those consumers who are willing to purchase the goods and services of the monopolist at the monopoly price.[28] Third, it is argued that the monopolist is in a position to control factors regarding the goods and services which extend beyond the price and that such control may have deleterious effects for consumers.[29] Finally, it is speculated that monopolists lack the incentive to reduce costs that generally drives prices down in competitive markets.[30]

Commentators have expressed concerns about the ability of patent owners to establish prices for their patented goods which are far in excess of the cost of production and any additional amounts which might be necessary in order for the patent owner to recover the costs of developing the product. Monopoly prices may make it difficult, if not impossible, for consumers to purchase the goods, thereby reducing or eliminating the utility of the inventive activity which the patent system is intended to support. One area in

[22] O'Driscoll and Rizzo, *The Economics of Time and Ignorance* (1985) 155.

[23] Ibid.

[24] *Great Atlantic & Pacific Tea Co.* v. *Supermarket Equip. Corp.*, 340 U.S. 147, 154 (1950). See also *Graham* v. *John Deere Co.*, 383 U.S. 1, 5–6 (1966) ("Congress ... may not ... enlarge the patent monopoly without regard to the innovation, advancement or social benefit gained thereby ...").

[25] Ibid.

[26] Obviously, the incentive-to-innovate and prospect theories described in Chapter 2 are based, to a large extent, on the patent right providing the patent owner with "monopoly-type" powers which can be used to achieve the goals specified by each of the theories. Reference should be made to the discussions in Chapter 2 regarding whether or not monopoly conditions actually enhance innovation or facilitate the desired level of control over subsequent research and development activities.

[27] Cornish, *Intellectual Property: Patents, Copyright, Trade Marks and Allied Rights* (1989), at 16.

[28] Ibid.

[29] Ibid.

[30] Ibid.

which this problem has been particularly acute is with respect to the development of new pharmaceuticals which are often priced at levels which exceed the means available to the patient population. Patent owners may also be able to influence the kinds of service supplied, continuity of supply, the number of different versions, and the amount of research and development into future products and services.[31]

Patent rights as market monopolies Attempts to characterize patents as monopolies have been rejected on a variety of different grounds. First of all, it has been pointed out that the rights granted to a patent owner relate to an invention that, presumably, did not exist before the act of invention and, as such, can be distinguished from the restraints on vested liberties which occur when monopolistic barriers arise with respect to the sale or exploitation of products, services or skills which were already in existence.[32] As one commentator wrote almost two hundred years ago with respect to patents: "Who could reasonably complain about a merely apparent privilege? It neither harms nor hinders any branch of industry that was previously known. The costs are paid only by those who do not mind paying them: their wants ... are not less fully satisfied than before."[33]

Other commentators point out that patents are just one of a number of resources which can be used to gain a competitive advantage in the marketplace.[34] Commentators also point out that every form of property right has a degree of exclusivity which may impact the market for the subject matter of the property right.[35] For example, the owner of a large tract of land in a desirable location should be able to exercise a good deal of market control, as can a popular entertainer, author, or other service provider.[36] The key issue in analyzing any exclusive right is the effect, if any, that the existence of the right will have upon competition.[37] The answer to that question depends upon a number of different factors, such as the availability of substitutes for the goods and services which are the subject of the exclusivity and, perhaps even more importantly, the manner of competition in the marketplace.

It has often been argued that a patent does not, in and of itself, lead to creation of monopoly conditions in the marketplace. It has been pointed out that one must always

[31] Ibid.

[32] Blackstone's *Commentaries* contained the following definition of "monopoly": " a license or privilege allowed by the King, for the sole buying and selling, making, working or using of anything whatsoever; whereby the subject in general is restrained from that liberty of manufacturing or trading which he had before." In addition to monopoly rights, the government may create various privileges which effectively limit the ability of presumptively qualified competitors to gain access to the market. Occupational and professional licensing requirements are examples of these privileges. See Mackaay, supra note 6, at 890.

[33] J. Say, *Traité d'économie politique* 1st ed, 1803), 263.

[34] Kitch, "Patents: Monopolies or Property Rights?," *Research in Law and Economics* 8 (1986), 31, 38 ("A patent can have value like any input that gives a firm a comparative advantage over its competitors, but that does not mean that the owner of the patent owns a 'monopoly'"). Moreover, patents are only one of the legal tools which are used in competition. In many cases, firms will rely on trade secrets for protection, rather than patents. The fact that patents are not always chosen by their intended beneficiaries is sometimes cited as evidence that the alleged monopoly power associated therewith is not as strong as some have argued. Mackaay, supra note 6, at 905.

[35] See, e.g. Easterbrook, "Intellectual Property is Still Property," *Harvard Journal of Law and Public Policy* 13 (1980), 108, 109 ("Patents are not monopolies ... Patents give a right to exclude, just as the law of trespass does with real property. Intellectual property is intangible, but the right to exclude is no different in principle from General Motors' right to exclude Ford from using its assembly line, or an apple grower's right to its own crop. A patent may create a monopoly – just as an auto manufacturer may own all of the auto production facilities – but property and monopoly usually differ.")

[36] Meiners and Staaf, supra note 15, at 921.

[37] Carter, "Owning What Doesn't Exist," *Harvard Journal of Law and Public Policy* (1990) 99, 104.

conceive of a market in terms of all the goods that consumers will treat as substitutes for one another and it is difficult to assess how consumers will react to any changes in pricing policies.[38] In addition, market domination largely depends upon innovative activities which follow the initial invention and which are calculated to commercialize the new invention in the relevant markets.[39] As such, delays in this process may allow others to "catch up" and will serve to reduce the present value of the investment return which the inventor will ultimately be able to achieve.[40] Finally, it is likely that the new invention itself, while providing the initial patent holder with some limited degree of protection, will cause potential competitors to scrutinize the market opportunities associated with the underlying technology, which will lead to substitutes and improvements which reduce the competitive advantage of the initial patentee,[41] as well as the profits available to her.[42]

Patents have long raised questions of monopoly. While, in theory, a patent holder has the potential to extract monopolist prices from consumers for some period of time, it does not appear that most patents create that sort of opportunity. Most of the inventions which qualify for patent protection do not carry with them the ability to exercise monopolistic control in the market and, in fact, it has been argued that only a relative handful of patented inventions are even suitable for exploitation.[43] This all suggests that the concerns regarding monopoly in the patent area may be misplaced. Moreover, to the extent

[38] Cornish, supra note 27, at 17. The ability of a patent holder to act as a monopolist will depend upon the extent to which the new invention fulfills a demand from consumers that was not previously met at all and the degree to which it is cheaper or leads to more efficient operations in comparison with the substitutes which may have preceded it. As a general matter, true monopolist behavior is only possible for a small proportion of patent holders. Ibid. at 81. See also Bainbridge, *Intellectual Property* (1992), at 246 (the optimal pricing practice for the patent owner depends upon the relationship between volume of sales and profit margin and the patent owner may actually make more money by reducing the price to a level where sales volume increases). Any new invention faces some degree of market risk since consumers have previously managed without the invention and may continue to do so by refusing to pay high prices. If this is the case, the patent holder may be forced to reduce her prices down to a level which approximates the prices charged for products which might be perceived as substitutes for the new invention. Ibid.

[39] See, e.g. Aubrey, "A Justification of the Patent System," in Phillips (ed.), *Patents in Perspective* (1985) at 1 ("there must be very few companies who would claim to have dominated an industry by the use of patents alone. To become master of an industry demands technical, commercial and entrepreneurial flair, resources, determination and courage. No patentee could ever promise that his patent ensures unlimited freedom to use the patented invention.").

[40] Cornish, supra note 27, at 81. Among the difficulties which might be encountered are: the unpredictability of the investment and time needed to arrive at a viable product; the costs associated with manufacturing and distributing the new product; and even competition from within the organization as to the investment projects which the firm should pursue. Ibid.

[41] Ibid. at p. 82. Obviously, the scope of the initial patent will have a bearing upon the research activities of competitors and their ability to generate substitutes and improvements by "inventing around" the initial patent. See also Meiners and Staaf, supra note 15, at 921; and Mackaay, supra note 6, at 888 (noting that the high returns in the secured market will provide the potential rewards that are necessary for the development of a new parallel market and for substitute products).

[42] Carter, supra note 37, at 104 (arguing that the pressures in the market arising from the availability of substitutes and, near the end of the patent term, outright copies, will usually prevent the patent owner from being able to extract monopoly rents for her products). See also White, Jeffs, Jacob, Cornish and Vitoria, *Encyclopedia of United Kingdom and European Patent Law* (1977), at § 1–105 (noting that competitors may be able to benefit from the experiences of the pioneer manufacturer with regard to the underlying technology and the development of the market without incurring many of the costs borne by the manufacturer).

[43] See Staines, "Some Thoughts on the 1983 Green Paper," in Phillips, supra note 39, at 103 (citing data from Taylor and Silberston, *The Economic Impact of the Patent System: A Study of the British Experience* (1973), at 320, to the effect that between 1948 and 1971 the NRDC estimated that only 0.5 percent of the inventions submitted to it from private sources, including individuals and small firms, were acceptable for exploitation).

that monopoly profits flow only in those cases where the monopolists markets a good or service which has substantial social utility, it also provides some indication of the inadequacy of the patent rights as an incentive for the creation of net consumer welfare.

Allocative efficiency problems of patent rights The fact that patent rights place significant restrictions on the use of the patented invention following the date of creation is a source of continuing debate. Opponents of patent rights remind us that the knowledge embodied in these inventions, once in existence, could be used by everyone without depriving anyone else of the right to use them.[44] Accordingly, they claim that the various restrictions which patent rights place upon the use of the patented invention may well result in a sub-optimal use of the invention, since consumers may either be forced to pay higher "monopoly-type" prices for the invention or, in extreme cases, may be unable to use the invention at all if the patent holder chooses not to make it available for sale.[45]

Supporters of the patent system are not unmindful of the paradox of attempting to provide economic incentives for engaging in inventive activities by creating rights which tend to restrict the diffusion of new information during the short-term. One writer has explained the problem as follows:

> "A patent is a device to prevent the diffusion of new methods before the original investor has recovered profit adequate to induce the requisite investment. The justification of the patent system is that by slowing down the diffusion of technical progress it ensures that there will be more progress to diffuse ... Since it is rooted in a contradiction, there can be no such thing as an ideally beneficial patent system, and it is bound to produce negative results in particular instances, impeding progress unnecessarily even if its general effect is favorable on balance."[46]

Various commentators have argued that concerns regarding the short-term allocation problems associated with patent rights should be placed in the proper context. They point out that the social gains from inventions and innovations encouraged by patent rights will always exceed the private gains of patent holders, since the private gains are only part of the total social gain.[47] For example, if a patent owner has an effective monopoly

[44] Hettinger, "Justifying Intellectual Property," *Philosophy and Public Affairs* 18 (1989) 31, 35 (noting that the subject matter of intellectual property rights, the so-called "intellectual objects," are nonexclusive: they can be at many places at once and are not consumed by their use).

[45] Since ideas are, in many ways, public goods, the risk of the type of over-grazing often cited as the basis for private property does not apply. Instead, the real potential tragedy is that there will be under-grazing because of the exclusion induced by patent rights. See Meiners and Staaf, supra note 15, at 919.

[46] Quoted in Nelkin, *Science as Intellectual Property* (1984), 15. See also Hettinger, supra note 44, at 36 (noting that the restrictions embodied in the patent right restrict the free flow and use of ideas, which retard the advance of technological innovation and human knowledge).

[47] A group led by Edwin Mansfield conducted case studies of seventeen industrial innovations and found the median estimated social rate of return to investment in research and development to be 56 percent, while the median private rate of return was about 25 percent. In about 30 percent of the cases studied, the private rate of return was so low that with the benefit of hindsight no firm would have actually invested in the innovation; however, in such cases the social rate of return was so high that the investment was worthwhile from society's perspective. It also appeared that the gap between the social and the private rate of return was greatest when the innovation was more important or in those cases where the innovation could be imitated by competitors at a small cost. Mansfield et al., "Social and Private Rates of Return from Industrial Innovations," *Quarterly Journal of Economics* 91 (1977), 221, 233–7. See also O'Brien, "Patents: An Economist's View," in Phillips, supra note 39, at 34 (noting that while there are some welfare losses once a new invention has been created due to the patent "monopoly," the savings and cost reductions resulting from

on a product due to the lack of substitutes, she can only appropriate the total social benefit if she is able to charge each consumer the maximum price that the particular consumer would be willing to pay for the product. Since it is impractical to enter into negotiations with each consumer, the patent owner will set a single price which would apply to the entire market. Consumers who would be willing to pay more than the established price thus receive a net benefit equal to the difference between the price that they would be willing to pay and the price actually charged by the patent owner.[48]

> "Advocates of patent rights also point out that commentators who base their criticisms of the patent system on the fact that new ideas should be made available to everyone as common property are ignoring the fact that new inventions are generally the result of purposeful inventive activity involving resources which are subject to redirection toward other uses in future periods.[49] Those in favor of patent rights point out that there is a clear element of scarcity in inventions which includes not only the opportunity costs of inventors inventing or firms spending money on research and development, but also the opportunities that inventors may choose to forego if there is no guarantee of exclusivity which will allow them to recover their investment."[50]

D. Alternative Reward Systems for Inventive Activities

Introduction In spite of the evidence to the effect that the potential monopolistic effects of patents have been overstated, the fear that patented products will be "under-consumed" due to the monopolistic pricing practices of the patent owner has led to a number of suggestions that the exclusivity provided by the patent system following the point of invention should be replaced by alternative rewards for successful inventive activity.[51] For example, it has been suggested that an inventor might be employed in a prestigious position on a temporary or permanent basis.[52] Other rewards might take the form of stipends, research grants, commissions, prizes, or other honors.[53] Artisans may also receive voluntary contributions from persons who attend their performances or be compensated through the sale of individual copies of their works.[54]

cont.
the availability of the new invention generally increases that measure of economic welfare which economists refer to as consumer surplus). See also Scherer, *Industrial Market Structure and Economic Performance* (1970), at 383. However, others have pointed out that some of the conclusions regarding the utility of new inventions have been reached on the assumption of perfect knowledge, an assumption that is inappropriate to the uncertain world of the inventor and the innovator. O'Brien, above, at 32–3.

48 As to consumers who are not able to purchase the new product due to the high price charged by the patent holder, if the patent is required in order to induce the required investment in research and development, the consumer is no worse off from the grant of the patent. Schwartzman, *Innovation in the Pharmaceutical Industry* (1976), at 165. Therefore, the issue becomes whether or not the patent provided the incentive that is required in order for the new invention to be created.

49 Meiners and Staaf, supra note 15, at 920.

50 Ibid. Critics of the patent system have argued that knowledge and ideas, once created, are "non-scarce" and, as such, cannot be the subject of traditional property rights. They note that intellectual property rights lead to an artificial scarcity in knowledge and ideas which is unjustifiable.

51 Mackaay, supra note 6, at 872.

52 Ibid.

53 Ibid.

54 Ibid.

The theory behind the use of each of these alternative forms of incentive is that researchers would be able recover the costs of their research and development from sources other than the monopoly prices which they might charge to consumers if granted exclusive rights to sell the products under the patent laws. In turn, the new inventions, including the know-how therein, would be available to all who might wish to use them at a cost which essentially is established by reference to the marginal cost of production and not the research and development costs associated therewith. As a result, it is argued that the overall social welfare associated with the introduction of the new invention would be increased as a result of relying upon these alternative incentives.

While there is some appeal to the use of incentive systems which do not rely on the grant of exclusive rights, it is clear that the use of such alternatives has not necessarily created the desired result. There has been substantial criticism regarding the efficacy of prizes, rewards and substantial financing of research and development activities by the state, and it is now fairly clear, at least in the United States, that the preexisting levels of financing for basic research will cannot be sustained in the years to come. Also, as noted above, supporters of the patent system claim that the grant of a patent rarely, if ever, provides the patentee with monopoly control in the marketplace, and that private firms are generally better situated to identify those areas where scarce research and development resources should best be applied to address the desires of consumers.

Prize systems One alternative to the patent system in attempting to stimulate inventive activity is the use of a prize system which would award cash to deserving inventions based upon their utility to the economy, with monetary awards being limited to inventions which are deemed to be truly significant. It is argued that the possibility of a prize would provide the requisite incentive, while the misallocations of the patent monopoly would be avoided. The innovating firm would not only receive the prize, but presumably would also be able to rely upon the technological lead as a further means for recouping the costs of development and earning a return. It might be possible to structure such a system so as to provide awards not only to inventors, but to other firms who are willing to make the investment required in order to bring the inventions to commercial fruition. Arguably, the costs of administering such a system would be lower than those associated with running a patent system in which most of the patents become obsolete.[55]

A prize system is subject to criticism on a number of different grounds. First, it may be quite difficult to trace the actual inventor and, in many cases, there are really a number of inventors involved in the creation of a new product or process. As such, it may well be necessary to retain some portion of the current patent system as it relates to proof of inventorship and novelty. Second, the patent system has the advantage of being impersonal in its application, while a prize system may be subject to higher levels of corruption and special-interest rent seeking. Third, awards can become somewhat arbitrary and it may be problematic to develop standards regarding the award of prizes which can be followed by inventors. Fourth, since a prize system is likely to be subject to budgetary constraints, in many cases the size of the award will bear no relation to the costs

[55] For an example of the arguments put forward in favor of a prize or reward system, see Polanyi, "Patent Reform," *Review of Economic Studies* 11–12 (1943–5), 61–76. See also Cohen, Awards to Inventors (The Holdsworth Club of the University of Birmingham: 1960) (describing the system for providing rewards to inventors for inventions used by the Crown during World Wars I and II pursuant to provisions in the applicable Patents Acts permitting the Crown to use patented inventions in certain circumstances).

of development or the value of the invention.[56] Finally, the rewards provided by the patent system are largely determined by the market and, while the market is sometimes imperfect, it is generally better than results produced by committee.[57]

Government subsidies for research and development Direct government subsidies for research and development is another alternative to the patent system. Government funding may be provided in a number of different forms, including the use of direct research grants to public institutions and private firms and the establishment of research consortium. Alternatively, aid may be provided indirectly through tax incentives designed to increase the level of research and development spending by private firms in specified areas. It is argued that the government would be able to identify national needs and apply economic resources in a manner which conforms with those needs. In addition, the availability of government subsidies in certain areas of interest would result in a shift in the emphasis of research planning from private economic objectives to projects which presumably would enhance the general public welfare.

As noted previously, in the United States there is already a significant amount of government involvement in research activity, a phenomenon which actually has been cited as proof that the patent system has not been sufficient to generate the optimal level of investment in the production of new knowledge. However, there are still a number of commentators who are skeptical of involving the government in widespread funding and coordination of research and development projects. One main concern is the effect that further funding might have on the efficiency of research and development. An argument can be made that providing funds and grants to private firms results in a loss of the discipline and rigor which normally must be present when a firm weighs the risks of taking on a given research project as opposed to alternative uses of the funds. Other express skepticism about the efficacy of the government in any role other than basic research projects.[58]

There may also be other significant difficulties associated with government financed research and development. One study has concluded that increasing the use of government subsidies may have an adverse effect of private productivity growth, given that the study found that privately financed research expenditures have a significantly larger effect on private productivity than federally financed research.[59] Also, while it is claimed that government subsidization of research could be achieved without excessive amounts of control over the type of research, thereby leaving the recipients with the discretion to decide which new projects would be most beneficial,[60] it is hard to believe that the government would not have some degree of involvement in attempt to select the areas of research, an intervention which might have unanticipated effects on the competitive environment that exists among firms.[61] Finally, it is unclear what effect eliminating patents and relying on government financing might have upon the contributions that have been made to invention by individuals and small firms.

[56] See O'Brien, supra note 47, at 38, for a criticism of the arguments made by Polanyi.

[57] Ibid. at 41.

[58] Ibid. at 38 (citing the development of the Concorde aircraft as evidence that it may not be wise to have the government involved in innovation, as opposed to basic research).

[59] See Z. Griliches, "Productivity, R&D, and Basic Research at the Firm Level in the 1970s," *American Economic Review* 76 (1986), 141–54.

[60] Hettinger, supra note 44, at 49.

[61] It has been pointed out that modern patent systems offer a "standard formula" which delineates the protection which will be available for all inventions which the market decides to generate while a system of research grants and awards will tend to concentrate resources in areas chosen by the group or persons responsible for distributing the funds. Cornish, supra note 27, at 80.

It has been noted that proposals in the United States for modifying the antitrust laws to facilitate innovation have not authorized any significant funding, despite the fact that it has been estimated that between 50 and 70 percent of the funding for Japanese and European innovation collaboration is provided by government sources.[62] The foregoing has led some commentators to question the basic premise underlying many of the proposals that simply amending the antitrust statutes will be sufficient to achieve any significant advances in innovation.[63] In contrast, other legislation has been proposed that would provide some form of federal financial support for joint research and development, either between two or more private firms or between private firms and government-sponsored agencies, without granting any antitrust immunity.[64]

E. Patents and Public Policy

Laws and regulations are generally designed to further one or more public policy objectives. With respect to patents, the various economic theories described above tend to justify their use as necessary in order to facilitate invention and innovation. Some commentators, notably those who also favor the use of the private market in shaping rights or claims with respect to the use of knowledge and the other subjects of intellectual property law, have questioned whether invention and innovation are proper goals of public policy activity. In one case, it was argued that "the rate of innovative activity is a rather poor candidate for governmental overriding of people's revealed preferences in the production of goods and services, at least when compared to more basic issues in production related to health, safety, and even education."[65] It has also been claimed that to the extent that new ideas and inventions are removed from the discipline of the market by the creation of intellectual property rights, society loses the benefits of the market's ability to register the opportunity costs associated with the ideas and innovation, thereby making it more difficult to articulate a meaningful criterion of "optimality in government subsidy."[66]

In spite of the aforementioned arguments, there would appear to be little doubt regarding the fact that technological change has been an important factor in the rate of economic growth.[67] For example, one study estimated that approximately 80 percent of the growth in nonfarm output per worker in the United States during the first part of this century was attributable to technological change rather than increased capital inten-

[62] See Brodley, "Antitrust Law and Innovation Cooperation," in *Collaboration Among Competitors* (1990) (hereinafter *Collaboration*) 887, 918 (citing Ouichi, "The Logic of Research and Development," *California Management Review* Spring (1988), 30, 9–33; and *New York Times*, "Europeans United to Compete with Japan and US," August 21, 1989, 1).

[63] Ibid.

[64] Ibid. (citing Renewable Energy and Energy Efficiency Technology Competitiveness Act of 1989, Senate Bill 488, 101st Cong., 1st Sess., 1989, which provided for 50 percent governmental funding for joint ventures between the Department of Energy and private groups to demonstrate commercial application of nonnuclear energy technologies, such as solar and wind; and Senate Bill 1191, 101st Cong., 1st Sess., 1989 and Senate Bill 488, 101st Cong., 1st Sess., 1989, each of which provided federal funding for joint research and development, usually tied to specific types of projects).

[65] Nance, supra note 13, at 765–71.

[66] Ibid.

[67] See, e.g. Abramovitz, "Resource and Output Trends in the US Since 1870," *American Economic Review* 46 (1956), 5 (American Economic Association Papers and Proceedings).

sity.[68] Other economists have determined that new advances in scientific and technological knowledge, as well as expenditures on research and development, have had a substantial impact on increases in output per worker, rates of growth in productivity, and economic growth in general in the United States during various periods of time in the post-Depression era.[69]

Economists have also been able to present a good deal of empirical evidence that levels of inventions, which serve as at least one indicator of the rate of technological change, have been positively related to economic stimuli. One analyst, using patent statistics as an indicator of the level of invention, has found a strong and positive correlation between inventive activity and the level of investment in capital goods, with the peaks and troughs in invention preceding rather than following peaks and troughs in the level of capital goods investment, and has therefore concluded that inventions occur in response to rising demand in an industry.[70] Another study demonstrated that the timing of expenditures relating to research and development and innovation are responsive to the expectations of profits on the results of the project.[71]

It is also important to remember that advances in technical knowledge are a function not only of the new knowledge created by inventive activity, but also the rate at which the knowledge is actually discovered. As such, if, by virtue of the patent system's bias toward disclosure, the rate of discovery of inventions which would have otherwise been created is improved, then availability of patent protection may well have provided benefits to society. It is likely that these advantages will be most apparent with respect to inventions which, even in their earliest stages, are not easily protected through alternative means.

One of the difficult public policy issues associated with the patent system, or any other state-created mechanism for obtaining and maintaining a competitive advantage through exclusivity, is quantifying the desired public policy objectives and devising a scheme for measuring the net output from the system. As the complexity and sophistication of technology and new knowledge advances and changes, the role of a patent has now become quite different. In fact, the output of the patent system now depends upon a broad range of factors, such as the importance of various non-patent competitive factors, including trade secrets and other forms of intellectual property, and the risks and investment that may be associated with moving from the inventive step for patent purposes to the actual stage of commercial application.

While economists have attempted to study and theorize about the economic effects of patents and patent scope decisions, current law mandates that patents are to be granted and interpreted solely on the basis of scientific and technical standards. One area that war-

[68] Solow, "Technological Change and the Aggregate Production Function," *Review of Economics and Statistics* 39 (1957), 312, 320.

[69] See Denison, *The Sources of Economic Growth in the United States and the Alternatives Before Us* (1962) 271–2 (estimating that 36 percent of the increase in output per worker in the United States between 1929 and 1957 was attributable to the advances made in the levels of scientific and technological knowledge); Denison, *Accounting for United States Economic Growth, 1929–69* (1974) 128 (estimating that 27 percent of the economic growth in the United States between 1929 and 1969 was attributable to advances in knowledge); Scherer, "Inter-Industry Technology Flows and Productivity Growth," *Review of Economics and Statistics* 64 (1982), 627 (estimating that in the post-war era R&D has added to the rate of growth by about one percentage point per year, or about half of the annual rate of growth in productivity).

[70] See generally Schmookler, *Invention and Economic Growth* (1966).

[71] See Mansfield et al., *Research and Innovation in the Modern Corporation* (1971) (also showing that firms tend to invest in research and development project which have a high probability of success and profitability).

rants further study is whether the patent system can be improved as a public policy tool by modifying the criteria for state-recognition of private rights in the products of creative activity to explicitly consider the various economic factors that are referred in each of the various theories regarding the patent system. For example, should patent scope decisions be based not on technical standards, but on some assessment of the competitive factors that will impact the innovator's ability to achieve some target rate of return on investment. Some of the alternatives to the patent system which have been proposed are based, in large part, on economic standards, and raise challenging questions as to the appropriate public policy in this area.

III. Antitrust and Competition Law Policies

A. Introduction

Antitrust and competition law policy raise many of same issues which must be considered in optimizing the patent system. One obvious area of concern is the effect that the use of the patent right, including the refusal to directly exploit the patented invention or grant licenses to others to allow them to exploit the invention, has on competitive conditions in the marketplace. In addition, antitrust and competition law will apply to the efforts of firms to reduce the risks associated with investing in invention and innovation through collaborations with other firms possessing complementary resources. Finally, the decisions of antitrust regulators will necessarily be effected by many of the same factors which influence the debate regarding the use of the patent system, including the desire to encourage innovation and the need to take into account the specific technological factors which influence competitive conditions in a particular market or industry sector.

In this section we consider a variety of interesting issues relating to antitrust and competition law policies as they relate to patent rights, patent licensing, and business collaborations in the pursuit of new technologies. In turn, we will examine innovation and antitrust/competition law policies, the regulatory treatment of license agreements, the use of market share-based standards for antitrust/competition law analysis and the problems of market definition in innovation-driven industries, regulatory treatment of research and development arrangements, and antitrust and competition law policies relating to small firms. We then discuss some of the areas in which regulatory policies in the United States and the EC diverge, and then go on to conclude with a summary of additional issues which will impact the substance and development of regulatory policies relating to innovation-based competition in the years to come.

B. Innovation and Antitrust/Competition Policies

Introduction We have seen that one of the constantly reoccurring issues in the area of antitrust and competition law with respect to the regulation of patent licensing practices is achieving the optimal level of economic efficiency. On the one hand, antitrust and competition law is generally concerned with the maintenance of competitive conditions in the marketplace and, as such, will seek to regulate commercial practices which might have the effect of restricting the rapid diffusion of the benefits associated with innovative activities throughout the marketplace. These concerns are often referred to as "static," or "allocative," efficiencies, and are based on the premise that the allocation of benefits and costs between producers and consumers depends upon the presence or absence of monopolistic conditions in product markets. On the other hand, the recognition of patent rights, which clearly have the potential to impair the diffusion of knowledge thought to be necessary for static efficiency,[72] is viewed as necessary for the pursuit of another important form of economic efficiency, which is generally referred to as "dynamic," or long-run, efficiency.

It has often been said that the interface between the intellectual property laws, particularly patent rights, and the antitrust and competition laws, is the situs for continuing policy tradeoffs between static economic efficiencies, in the form of competitive markets, and dynamic economic efficiencies. There is certainly some truth to this; however, there is merit to conceptualizing decisions regarding various aspects of patent rights (e.g. patent scope) and decisions regarding the treatment of patent licenses under the antitrust and competition laws as part of a single continuum, so that the actions of regulators, legislators, and judges in these areas can be coordinated so as to pursue consistent economic goals and objectives. In order to accomplish this, one must first understand how antitrust and competition policies actually impact economic efficiencies. Then, decisions must be made regarding the socially desirable mix of static and dynamic efficiencies.

Antitrust/competition policy and economic efficiency One of the most exhaustive surveys of the effect of antitrust and competition laws on economic efficiency has been undertaken by Scherer.[73] He suggested that economic efficiency was, as suggested above, multifaceted, and distinguished three types of efficiencies that should be considered. In addition to dynamic efficiencies, Scherer indicated that two other types of efficiencies should be considered: allocative and productive efficiency, sometimes called "X-efficiency."[74] The former is a function of relative allocation of benefits between producers and consumers, while the later is primarily concerned with whether or not the available resources of the producers, and the economy as a whole, are being efficiently applied in the production and distribution of a given good.

Citing several sources, Scherer points out that if the antitrust and competition laws were concerned solely with allocative efficiency, then there relevance would be quite limited, since the potential gains and losses are generally so small in relation to the total sales or value added measures in the industry. For example, one attempt to measure the overall "deadweight" loss attributable to monopolies estimated the value at approximately one-tenth of one percent of the gross national product in the chosen year of measurement.[75] Scherer concedes that monopolistic conditions also raise wealth redistribution considerations, particularly where prices remain above costs; however, he believes that, in general, wealth redistribution is not considered to be a concern of antitrust regulators.[76]

[72] This occurs where the legal rights provided to patent owners to exclude others from using the patent invention increase prices in the market or otherwise restrict others from practicing the innovation until expiration of the patent rights.

[73] Scherer, "Antitrust, Efficiency, and Progress," in *Collaboration*, supra note 62, 33.

[74] The term "X-efficiency" was developed by Harvey Leibenstein. See Leibenstein, "Allocative Efficiency vs. 'X-Efficiency'," *American Economic Review* 56 (1966), 392.

[75] Harberger, "Monopoly and Resource Allocation," *American Economic Review* 44 (1954), 77, 84 (cited in Scherer, supra note 73, at 34. See also Marshall, *Principles of Economics*, 447 n1 (1890).

[76] Scherer, supra note 73, at 34–5. Cf. DOJ's 1984 Merger Guidelines § 1.0, 49 Fed. Reg. 26,827 (June 29, 1984), reprinted in 2 Trade Regulation Reports (CCH) § 4490 (1984) (stating that anticompetitive mergers transfer wealth and misallocate resources).

On the other hand, several arguments have been posited regarding the importance of using the antitrust and competition laws to reduce monopolistic conditions as a means of enhancing X-efficiency. For example, if monopoly producers, freed of the need to control costs by their ability to raise price, allowed costs to rise through inefficient operations, then it can be said that monopoly has been socially wasteful, at least to the extent the monopolist has dissipated the economy's scarce resources.[77] Another school of thought in this area is the so-called "theory of rent seeking," which argues that if restrictive practices cause incumbent producers to realize excess "rents," outsiders will incur legal, advertising, bribery, product-duplication, and excess-capacity costs in order to win a share of the rents. Since it can also be expected that the incumbents will then spend additional resources to defend their position, the net result of the rivalry created by the initial monopoly may be a significant waste of the economic resources of the firms involved.[78]

The most interesting results from Scherer's work relate to the relationship between market structure and the attainment of long-term, dynamic efficiencies. As described earlier, the pursuit of dynamic efficiencies, sometimes at the expense of allocative and productive efficiencies, is the foundation of the works of Schumpeter and others, and is the key element of the incentive-to-innovate theory in the patent law area. In effect, the underlying premise is that antitrust and competition law policies should be formulated in a manner that assumes that allocative efficiencies are far less important with respect to economic well-being than whether innovators are able to seize the opportunity to reduce costs through technical progress. In 1942, Schumpeter wrote:

> "The fundamental impulse that sets and keeps the capitalist engine in motion comes from the new consumers goods, the new methods of production or transportation, the new markets, the new forms of industrial organization that capitalist enterprise creates ... A system – any system, economic or other – that at every given point of time fully utilizes its possibilities to the best advantage may yet in the long run be inferior to a system that does so at no given point in time, because the latter's failure to do so may be a condition for the level or speed of long-run performance."[79]

Schumpeter then went on to argue that monopolistic conditions were better suited to producing rapid technological progress than "competitive" market structures, which had long been the preferred method for pursuing allocative efficiencies.[80] If this were true, then there was a tradeoff between allocative efficiencies, which were to be realized from relatively vigilant application of the antitrust laws, and long-run technological progress, which in Schumpeter's view would come largely from a relaxed approach to the dangers of monopolistic conditions.

[77] This analysis is similar to Adam Smith's observation that "monopoly ... is a great enemy of good management." See A. Smith, *An Inquiry into the Nature and Causes of the Wealth of Nations* (E. Cannon ed. 1937) (5th ed. 1789), 147.

[78] With regard to "rent-seeking," see, e.g. Tullock, "The Welfare Costs of Tariffs, Monopolies and Theft," *W. Economics Journal* 5 (1967), 224; Krueger, "The Political Economy of the Rent-Seeking Society," *American Economic Review* 64 (1974), 291; Posner, "The Social Costs of Money and Regulation," *Journal of Political Economy* 8 (1975), 807 (applying theory of rent-seeking to measuring social costs of monopoly).

[79] Schumpeter, *Capitalism, Socialism and Democracy* (1942), 83.

[80] Ibid. at 106.

However, Scherer's review of the large number of studies that have been made regarding the Schumpeterian postulates has highlighted a variety of flaws in Schumpeter's arguments.[81] In addition, these studies have themselves provided a basis for a several alternative explanations regarding the factors that might drive innovative activities in the marketplace. As a result, it may well be possible to make more informed choices regarding the antitrust and competition law policies which might be best suited to achieve any overriding objectives with respect to innovation and market structure.

One key finding is that market structure may actually be less important than demand and supply conditions as determinants of the level of profit-oriented innovative activity.[82] On the demand side, the rate of innovative activity will increase in relation to the depth of the anticipated stream of appropriable benefits, regardless of whether the innovation is carried out under monopolistic or competitive conditions.[83] It appears to be true that more competition actually stimulates and accelerates innovation, although when competition becomes so intense that any given rival can anticipate appropriating only a small share of the benefits from the innovation, more competition retards innovation.[84] On the supply side (i.e. the relevant knowledge base), competition appears to be more advantageous in those situations where the pace of the general advance of knowledge is rapid; however, if the relevant knowledge base advances fairly slowly, monopolies or even tightly-knit oligopolies appear, up to a certain level of concentration, likely to innovate more rapidly than fragmented industry structures.[85]

One of the basic assumptions of the Schumpeterian theories is that only large business enterprises are able to achieve the scale of effort, and bear with equanimity the risks, required to achieve important technological advances.[86] However, beyond a modest size threshold, there is no evidence that large companies are more progressive than medium-sized firms in the sense of supporting research and development, making patented inventions, and originating significant technological innovations and, in fact, the relationship between size and innovative input/output is roughly proportional.[87] As a general rule, dominant sellers lack strong incentives to be the first mover with respect to the development of new products, unless their dominant position in existing markets is seriously challenged by smaller (or new entrant) rivals, although there is some limited evidence that increases in company profitability lead to lagged increases in the level of research and development activity.[88] In turn, the research and development efforts of smaller,

[81] The results of a number of these studies have been summarized in Scherer, supra note 73, at 45–9. See also Baldwin and Scott, *Market Structure and Technological Change* (1987).

[82] Scherer, supra note 73, at 47.

[83] See, e.g. Scherer, *Innovation and Growth: Schumpeterian Perspectives* (1984), 79–80, 87.

[84] Ibid. at 127.

[85] Scherer, supra note 47, at 375–6. See also Levin, Cohen, and Mowery, "R&D Appropriability, Opportunity, and Market Structure: New Evidence on Some Schumpeterian Hypotheses," *American Economic Review* 75 (1985), 20 (R&D/sales ratios rise with concentration at modest concentration levels, especially in industries with a knowledge base that advances only slowly, until concentration ratios reach the 50 to 60 percent range, after which a declining pattern may set in).

[86] Schumpeter, supra note 79, at 89.

[87] Scherer, supra note 83, at 87.

[88] See, e.g. Branch, "Research and Development Activity and Profitability: A Distributed Lag Analysis," *Journal of Political Economy* 82 (1974), 999 (finding, in six of seven industries studied, possible or probable linkage between increased profits and increased R&D). It is not yet clear, however, whether the observed relationships provide support for the Schumpeterian "monopoly profits plowed back into innovation" hypothesis, or for a more complex scenario in which demand- or supply-side shocks simultaneously increase profitability and the attractiveness of intensified research and development. See, e.g. Pakes, "On Patents, R&D, and the Stock Market Rate of Return," *Journal of Political Economy* 93 (1985), 390, 406–7 (finding that events causing market reevaluation of firms correlate with unexpected changes in R&D).

potential entrants may turn upon their perception of the anticipated response by the dominant seller. For example, potential rivals may be dissuaded from innovation if they foresee that the dominant seller will respond with a "fast second" strategy or actually overtake the rival,[89] however, it does appear that relatively small firms and "outsiders" do originate a disproportionate fraction of the most radical innovations.[90]

Schumpeterians also argue that a monopoly position affords the security required to insure the appropriation of a sufficient amount of financial rewards after successful innovation to make the investment in research and development worthwhile.[91] This argument is based on certain implicit assumptions regarding the costs and risks associated with research development. However, studies indicate that the costs of "typical" industrial research and development projects are often quite modest, as are the risks of technical failure.[92] Clearly, the rate at which firms pursue the advantages of innovation is a function of their ability to appropriate the resulting social benefits,[93] but it may well be that a number of projects can produce an "acceptable" return, in relation to the costs and risks involved, even in cases where the innovator is not a monopolist, provided that the number of other competitors is not so great that the remaining benefits available to a firm considering duplicative research are not so meager that the additional research costs are not warranted.[94]

Schumpeter also suggested that the profits that might be derived from monopoly positions are an ideal source of funds to support continued industrial research and innovation.[95] This may be so; however, the decision by monopolists to reinvest their profits in innovation depends on a variety of factors, including the strategies of actual or potential competitors. Moreover, it is evident that financial markets have developed rapidly since the time that Schumpeter first put forth his theories, and that there are now a number of other capital sources for innovation, many of which are readily available to firms that are operating in oligopolistic markets. Also, the growth of venture capital activity in the United States, but much less so in other nations, shows that an internal flow of prior profits is by no means necessary to the funding of technological innovation.[96]

In summary, these studies have led Scherer to conclude:

> "Although much remains to be learned, these theoretical and empirical findings have definite implications for antitrust policy. Perhaps most important, they compel abandonment of the romantic but naive Schumpeterian belief that giant firms organized into highly concentrated oligopolies are essential to maintain the most vigorous pace of technological progress. There may be isolated instances in which the Schumpeterian view is correct, but they should be treated as such. More commonly, loosely structured oli-

[89] See, e.g. Scherer, supra note 83, at 104–5; Baldwin and Childs, "The Fast Second and Rivalry in Research and Development," *Southern Economics Journal* 36 (1969).

[90] See, e.g. Jewkes, Sawers and Stillerman, *The Sources of Invention* 2d edn. (1968) 203–9; US National Science Board, Nationall Science Foundation, *Science Indicators: The 1985 Report* (1985) 82–5.

[91] Schumpeter, supra note 79, at 88, 103.

[92] Scherer, supra note 47, at 415–18.

[93] Baldwin and Scott, *Market Structure and Technological Change* (1987) 19-20, 29-32, 40–54; Barzel, "Optimal Timing of Innovations," *Review of Economics and Statistics* (1968), 348.

[94] Ibid. It has also been suggested that if the number of research and development projects rises to bring the "market for innovations" to a zero-expected-profit equilibrium, the case for high seller concentration, and in extreme cases monopoly, is strengthened. Ibid. at 19–20; Dasgupta and Maskin, "The Simple Economics of Research Portfolios," *Economics Journal* 97 (1987), 581.

[95] Schumpeter, supra note 79, at 90.

[96] Scherer, supra note 73, at 48.

> gopolies are likely to be at least as progressive as industries dominated by one firm or a few, and relatively small technology-oriented enterprises often prove to be more dynamic innovators than the corporate giants. Above all, it is important to keep entry open so that challengers with new ideas can force the pace of innovation."[97]

Optimal objectives of antitrust/competition policy The preoccupation of economists and jurists regarding the rationalization of static and dynamic efficiencies has continued for many years. As noted earlier, courts and regulators often evidenced concern regarding potential misuse of patent rights, and in so doing tended to make decisions which emphasized their desire to achieve allocative efficiencies. Admittedly, many of these concerns were based, at least in part, on a fundamental skepticism of the legitimacy of intellectual property rights; nonetheless, the consequences of these decisions was to limit the restrictions that patent owners could place on their licensees in their efforts to recover the costs of innovation. Moreover, while Scherer did not seriously consider the issue, there are those that believe that allocation of wealth between consumers and producers that follows from the relationship between price and costs is actually an important concern in developing a proper antitrust/competition policy.[98]

On the other hand, a number of economists have argued that conflicts between static efficiencies and long-term benefits from collaboration should, in most cases, be resolved in favor of the long-term. As one set of commentators has said, "... where the two desiderata are unavoidably at loggerheads, it is generally our predisposition ... to favour the long-run goals over those of the short run. The reason is clear, and rests on the powers of compounding. If an increment in growth, even a small one, can be sustained indefinitely, it must ultimately overcome any static loss that its generation required."[99] Another commentator, citing the importance of innovation to economic growth and welfare, has stated that the central focus of antitrust policy ought to be on fostering "progressiveness," including, among other things, innovation in the form of new ways of competing in products, marketing and service.[100] He then goes on to argue that progressiveness should be given greater weight whenever there is any tradeoff between progressiveness and allocative efficiencies in antitrust analysis and policy.[101]

It would appear that any attempt to refocus the patent system on innovative activities also would require that antitrust and competition law policies be adapted in a manner which anticipates and tolerates, if not encourages, the post-invention relationships which might be necessary in order for innovation to properly occur,[102] even if such relationships

[97] Ibid. at 49.

[98] Ibid. at 34–5. Cf. DOJ's 1984 Merger Guidelines § 1.0, 49 Fed. Reg. 26,827 (June 29, 1984), reprinted in 2 Trade Regulation Reports (CCH) § 4490 (1984).

[99] Ordover and Baumol, "Antitrust Policy and High-Technology Industries," in *Collaboration*, supra note 62, 949, 978.

[100] See "Innovation, Rivalry and Competitive Advantage," Interview with Professor Michael E. Porter, *Antitrust* 5 (Spring 1991).

[101] Ibid. Allocative efficiencies are analyzed through an assessment of "price/cost" margins. One of the key general policy issues in this area is whether, and to what extent, the economic values of welfare and efficiency should be explicitly invoked in formulating exclusionary schemes for the purpose of providing incentives. The incentive-to-innovate and prospect theories each rely, to some extent, upon an assumption that a market monopoly is somehow more efficient for the achievement of certain social welfare goals.

[102] The Orphan Drug Act enacted in the United States, and briefly referred to below, is an example of how exclusive rights might be crafted in a manner that covers not only initial inventive activities, but also post-invention market development. On the general subject of innovation and antitrust/competition policies, see Jorde and Teece, *Antitrust, Innovation and Competitiveness* (1992), at 47–70 (discussing innovation,

sometimes heighten the danger of some short-term allocative inefficiencies. For example, an inventor may not be able to recover the cost of development without entering into one or more horizontal and/or vertical arrangements with other firms for the services and assets needed to bring the new invention to market. If antitrust and competition laws are too restrictive with regard to such arrangements, inventors may well abandon new projects before they even begin. Small firms, in particular, may be reluctant to remain involved in inventive activities, thereby depriving the market of a proven source of new products and ideas.

There is strong empirical evidence supporting the view of economists that strategic business relationships at one or more of the stages of innovation can produce substantial advantages for the participants and, it is argued, for consumers by virtue of the accelerated pace of technological advance and new product introduction. Assuming this is true, then regulatory policy should be crafted in a manner which encourages innovation, and which recognizes the consumer welfare must be analyzed over longer time periods than those generally associated with static analyses. Moreover, as Scherer's work seems to indicate that, in most instances, antitrust and competition law policies have only a modest impact on allocative efficiency, then what is left to regulators but to concentrate on dynamic efficiencies, as well as upon ensuring that collaborations do not "step over the line," and produce anticompetitive effects which, on balance, override the benefits from collaborative relationships.

Another factor which is likely to drive the direction of antitrust and competition law policies is the relatively rapid rate of technological progress in many industries, especially those industries where future economic growth is anticipated to be the greatest. In situations where technological progress is rapid, it is more likely that static analyses of market dominance will be far less relevant than they would be in industries where market position is acquired primarily through such "standard" means as predatory pricing or advertising.[103] Leadership in these technology-driven industries requires "skill, foresight, and industry," which tend to be reflected in continuous expenditures on new technologies and products. It has been argued that these expenditures on acquiring and maintaining market leadership have effects which go beyond the particular entrepreneur to further the broader social interest, thereby dictating a preference for encouraging the pursuit of dynamic economic efficiencies.[104]

There are several other reasons why static analyses may be misleading with respect to technology-driven industries.[105] For example, in technology-driven industries, product-

cont.
cooperation and antitrust in the context of amendments to the National Cooperative Research Act of 1984 (15 U.S.C. §§ 4301–5)). For a discussion of Jorde & Teece's proposals for amending the National Cooperative Research Act, see Jorde and Teece, *High Technology Law Journal* 4:1 (1989), 62–80 (discussing proposed National Cooperative Research and Commercialization Act).

103 See, e.g. D. Hay and J. Vickers, "The Economics of Market Dominance: Introduction," in Hay, and Vickers, *The Economics of Market Dominance*, Oxford, Basil Blackwell, (1987), 1–60. Ordover and Saloner, G., "Predation, Monopolization and Antitrust," in R. Schmalensee and R.D. Willig (eds.), *Handbook of Industrial Organization*, Amsterdam, North Holland (1988).

104 See Fudenberg and Tirole, "Understanding Rent Dissipation: On the Use of Game Theory in Industrial Organization," *American Economic Review Papers and Proceedings* 77 (1987), 176–83 for examples of models in which investments in the creation of market power are not socially wasteful.

105 Many scholars and policymakers question whether workable antitrust rules can ever be developed to account for the realities of technological competition. See, e.g. Easterbrook, "The Limits of Antitrust," *Texas Law Review*, 63 (1984), 1–40.

specific assets (i.e. assets most directly related to current products)[106] endow firms with only transitory market power, and rapidly changing technical standards will reduce the value of such assets as a foundation for future rents.[107] Also, while coordinated use of market-specific assets, such as reputation, distribution channels and service networks, may generate significant future rents, it does not necessarily follow that incentives for future investment in research and development will be significantly reduced in such industries.[108] As such, in order to provide continued incentives for such firms to enter additional races, it may be efficient to permit them to obtain prices which are above their non-sunk costs during the short window of market opportunity to allow them to recoup their sunk costs of research and development.[109] This approach is obviously not without controversy, particularly in situations where the first innovator set prices at levels which, while assuring a healthy level of profits, limit the diffusion of the products into the marketplace.

Second, while econometric and statistical studies have generally not been supportive of Schumpeter's view that monopoly stimulates rather than retards technological progress[110], it is not necessarily the case that diffusion of market power is necessary to achieve the kind of sustained investment in research and development that leads to product and process innovations. For example, it may be that firms with a dominant market share will be better able to appropriate the results of its innovation and protect them against imitation. If this is so, regulatory intervention that diminishes the market power of dominant firms may weaken incentives for future innovation activity without necessarily providing any offsetting gains in the current level of market efficiency.[111]

Finally, it can be argued that static analyses of market dominance at the downstream (product) level are inappropriate in technology-driven industries, since in such industries the real locus of interest should be on the upstream markets for research and development.[112] Ordover and Baumol elaborated on this view as follows:

> "[I]n high-technology industries, the relevant lines of inquiry would appear to be upon the following questions: (a) Does the dominant firm in the product market also currently dominate the R&D stage? If it does, how likely is it that such dominance will persist? (b) Does product-market dominance serve to perpetuate dominance in the R&D stage? (c) Given that lowering entry barriers into the R&D stage tends to improve short-run and long-run efficiency, are there significant entry barriers into the relevant R&D stage, and can those barriers be raised strategically by the dominant firm? And, (d), given the answers to questions (a)-(c), what trade-offs will antitrust intervention entail if incremental improvements in the current diffusion of the existing stock of knowledge alter the incentives for future investments in R&D?"[113]

[106] See Farrell and Shapiro, "Horizontal Mergers: An Equilibrium Analysis," Working Paper, Princeton University, May 1988, with regard to the three different types of assets (i.e. product-specific, market-specific, and research and development) that are brought into coordinated use by a merger.

[107] Ordover and Baumol, supra note 99, at 976.

[108] Ibid. at 977. For example, since the market-specific assets will retain their value, the merged firm may be less concerned about cannibalization of profits from introduction of innovative new products. Also, a merged firm with a substantial share of market-specific assets may be more capable of converting their investment in research and development into downstream market share. Ibid.

[109] Ibid. at 952.

[110] See, e.g. Baldwin and Scott, supra note 93.

[111] Ordover and Baumol, supra note 99, at 952.

[112] Ibid.

[113] Ibid.

Regulators are recognizing that innovation does necessarily require, in many instances, ongoing collaboration between firms with complementary assets along the chain of activities necessary for full commercial development of a new product concept. Also, courts and regulators in the United States have recently taken a more relaxed view about vertical restraints, apparently signaling that conflicts in this area are now more likely to be resolved in favor of long-run dynamic considerations.[114] In addition, the Schumpeterian view continues to be important in the review of mergers in industries where technological progress is the driving force of competition,[115] and regulators often follow suggestions of economists that mergers in high-technology industries, in which technologies and products are short-lived, should raise fewer antitrust concerns than would similar mergers in industries which have entered into a more mature and stable phase.[116]

C. Regulatory Treatment of Licensing Arrangements

Introduction While a licensing arrangement may merely be a mechanism for transferring technology and its related legal rights without a significant degree of ongoing interaction between the parties, a license often serves as the basic building block for a variety of collaborative business arrangements between firms. As discussed earlier, the licensing of innovations creates the potential for social gains from the improved *ex post* diffusion of existing innovations, thereby mitigating some of the concerns regarding the social costs of recognizing intellectual property rights.[117] However, licensing also raises the risky possibility that the process will be abused, and turned into a means for effectuating collusion at the upstream and downstream levels.[118]

There are a variety of issues relating to antitrust and competition law regulation of licensing arrangement which should be considered in greater detail. Among them are the quest for the proper subject matter of the patent right, which has served as a guiding principle for the review of patent licenses in both the United States and the EC; the impact of licensing agreements on upstream and downstream markets, a question which is extremely relevant in considering the overall competitive impact of the relationship between the licensor and the licensee; the virtue of distinguishing between vertical and horizontal relationships in setting the standards for antitrust and competition law review of licensing arrangements; the transactional costs associated with licensing arrangements; the effect of appropriability conditions in the relevant industry on regulatory attitudes

[114] Ibid. See also Baumol and Ordover, "Antitrust: Source of Dynamic and Static Inefficiencies?," Working Paper, C.V. Starr Center for Applied Economics, New York University (July 1988). These changing attitudes have not been limited to collaborations involving firms with limited market share; in fact, regulators realize that many important projects can only be accomplished by joint ventures between firms which, taken together, have a large share of existing markets. Nonetheless, collaboration among market leaders is permitted, if not encouraged, in cases where the objective is considered to be important for other reasons (e.g. national competitiveness, reduction of import dependence, job development, etc.).

[115] See, e.g. Ordover and Willig, "Antitrust for High-Technology Industries: Assessing Research Joint Ventures and Mergers," *Journal of Law and Economics* 28 (1985), 311–33; J.K. Reiniganum, "The Timing of Innovation: Research, Development and Diffusion," in R. Schmalensee and R.D. Willig (eds.), *Handbook of Industrial Organization*, Amsterdam: North Holland (1988).

[116] Ordover and Baumol, supra note 99, at 977.

[117] Licensing actually presents the inventor with a difficult decision, since the inventor must balance of the advantages of obtaining the incremental revenues from licensing against the potential disadvantage of the increased likelihood of future competition from licensees who might benefit in their own R&D from having working access to the licensed patent. Ibid. at 963.

[118] Ibid. at 961.

relative to licensing arrangements in the industry; and the social welfare benefits from licensing agreements.

Subject matter of the patent right Cases and regulations in both the United States and the EC have long focused on assessing the lawful boundaries of the subject matter of the patent right in the context of licensing arrangements. One school of thought suggests that patent licensing practices, however restrictive they might be, are lawful to the extent they allow the patent owner to capture the monopoly profits that are part of the superiority of the patented process or product over those of its competitors.[119] Under this view, many of the prohibited patent licensing practices, such as tying arrangements and attempts to impose price restraints on licensees, are not unlawful extensions of the patent monopoly, but rather are legitimate methods of fully capturing the profits that are rightfully attached to the patent grant. Others have argued that any advantage which might be obtained through the exercise of the patent monopoly should be treated no differently than any other competitive strategy that may be pursued by firms, and that eventually any monopoly position which may be achieved by the patent holder will fall to the efforts of competitors.[120] In either case, the lesson for antitrust policy is not to attempt to intervene substantially in the marketplace, lest intervention upset the normal competitive evolution.[121]

On the other hand, critics of extending the patent monopoly, many of whom have spent a good deal of time defining and enforcing the patent-antitrust doctrines, argue that restrictive patent practices cannot be justified simply because they permit a patent owner to capture the full monopoly profit.[122] They maintain that certain antitrust limitations on the exercise of patent rights would not discourage, or even significantly reduce the rewards of, inventive activity.[123] Moreover, they argue that patent licensing regulatory policy should not be directed by the proposition that "society should feel a moral obligation to compel all who use an idea to pay monetary tribute to its creator," but rather by simply insuring that society is able to provide an inventor with an economic return from his invention commensurate with its social value by preventing the practice of free-riding on economically valuable creations.[124]

On balance, it has often proved difficult to delineate the legal boundaries of the patent right with any great certainty, and it may well be that courts and regulators are best advised to concentrate on those areas where there clearly are appreciable economic costs associated with the patent holder's exercise of her exclusive rights.[125] Among the areas of potential concern are efforts to extend the patent to areas that provide no appreciable benefit to consumers; the creation of large patent portfolios that may be used to block the research efforts of others; the practice of granting licenses only to firms strong enough to contest the original patent, with the resultant concentration of market power in a few

[119] See, e.g. Bowman, *Patent and Antitrust Law* (1973). (Bowman analyzes vertical agreements and reaches the conclusion that they do not allow a monopolist to capture any more than the full value of the patent monopoly.)

[120] See Brenner, *Rivalry: In Business, Science, Among Nations* (1987).

[121] See Mackaay, supra note 6, at 888.

[122] Turner, "The Patent System and Competitive Policy," *New York University Law Review* 44 (1990), 450.

[123] Ibid.

[124] Ibid. at 457.

[125] See, e.g. O'Brien, supra note 47, at 41 ("there is ... a strong case to be made for the use of competition policy to monitor and where necessary modify the enjoyment of patent rights ...").

firms; and, finally, the use of the patent position to harass smaller firms and discourage them from attempting to enter the market.[126]

Impact of licensing on upstream/downstream markets One of strongest arguments advanced for liberalizing antitrust and competition law treatment of licensing agreements is that to the extent that private firms already have inadequate incentives to engage in ex post dissemination of the results of their research, policies which might interfere with dissemination may only exacerbate the problem.[127] Advocates of a "relaxed" rule-of-reason approach toward licensing arrangements maintain that regulators should focus their analysis on the effects that a licensing agreement will have on actual and potential competition in the upstream and downstream markets, and that agreements that do not significantly reduce the likelihood of future competition in the upstream market beyond the level implied by the patenting itself should be approved without much additional scrutiny.[128]

Applying this approach, in situations where the licensor is not vertically integrated into the downstream market, as might be the case with a licensor in a joint venture strictly limited to research and development, it may be prudent to allow the licensor to exploit its patent rights without any restrictions, provided the licensor does not impose terms that substantially reduce the likelihood of present and future independent competition from licensees in the upstream research and development and licensing market.[129] Moreover, while it is important to consider imposing some limitations on restrictive licensing terms, it should be recognized that some such restrictions may be necessary to reduce returns from imitation and to prevent unauthorized leakage of licensed technology.[130] On the other hand, if the licensor is vertically integrated into the downstream market, its licensing strategies should be closely reviewed to insure that the licensor is not using the license to facilitate downstream collusion.[131]

Horizontal/vertical nature of licensing arrangements One of the key elements in the analysis of any licensing arrangement is assessing whether the parties stand in either a "horizontal" or "vertical" relationship to one another as to the activity which is the subject of the license. The importance of this distinction has been explicitly recognized as part of the 1995 Antitrust Guidelines for the Licensing of Intellectual Property (the "IP Guidelines") in the United States. Moreover, as discussed in greater detail below, one of the main criticisms of EC regulation of patent licensing arrangements is that the Commission has been too quick to view such licences as horizontal relationships and ignore the procompetitive benefits which might flow from collaboration.

Regulators typically have little problem recognizing a horizontal relationship in situations where the participants are already competing with one another in the relevant market. The far more difficult question is determining whether a licensee should be treated

[126] Ibid. at 35 and 36. O'Brien also notes differences between the way that the antitrust and competition regulators in the United States and the United Kingdom have reacted to anti-competitive uses of patents. Ibid. at 39 and 40. See also Whish, *Competition Law* (1989), at 262–7 (block exemptions under EC competition laws relating to patent licensing, know-how licensing and cooperative research and development arrangements) and Van Hise, *Concentration and Control* (1973).

[127] See, e.g. Ordover and Baumol, supra note 99, at 962.

[128] Ibid. at 966–7.

[129] Ibid. at 965.

[130] Ibid.

[131] Ibid.

as a "likely potential competitor." Initially, the DOJ suggested that the analysis of this question should be made by reference to the standards set out in the 1992 Horizontal Merger Guidelines. However, the problem with this approach is that the Merger Guidelines' standard for "likely" entry entails a complex, difficult-to-apply test that requires a predictive assessment about the prospect of new market entry, after and in light of a proposed merger's expected changes in market structure and related market conditions. The test set out in the Merger Guidelines test depends on the assumption that there exists a well-defined market, characterized by accessible, meaningful price information; however, as noted below, many licensing transactions are unlikely to occur in situations conforming to the paradigm. Accordingly, the use of the Merger Guidelines test would likely have led to undesirable uncertainty as to whether the prospective licensee following the execution of the agreement might be deemed to have been a "likely potential competitor" even though they were not competitors at the time they negotiated the license arrangement.

It is clearly understandable that regulators might want to treat a licensor and licensee as in a horizontal relationship in some cases where, as of the time of licensing, they are not yet "actual" competitors. However, rather than attempting to apply the Merger Guidelines standards, another alternative is to focus on evidence to the effect that, in the absence of the particular license, the parties would have actually competed in the relevant market. This approach is consistent with the type of standard followed by the United States Supreme Court in a recent case where two firms, expecting to become competitors of each other, instead entered into a license agreement granting exclusive rights to one of them in Georgia and to the other for the rest of the country.[132] The Supreme Court held the license to be *per se* illegal.[133] In fact, the final form of the IP Guidelines largely adopted such an approach, indicating that "[a] firm will be treated as a likely potential competitor if there is evidence that entry by that firm is reasonably probable in the absence of the licensing arrangement."[134]

Assuming that a decision can be made as to whether a particular licensing arrangement is "horizontal" or "vertical," the next issue is whether there are any circumstances under which a truly vertical arrangement should be challenged. Korah, for one, advocates a very liberal regulatory attitude with respect to restrictions in a licensing arrangement where the licensor and the licensee would not have been competitors in the absence of the license. In fact, he has argued that even the reduced black list in the new block exemption for technology licenses in the EC (the "Technology Licensing Regulation") should "be abrogated where the agreement is vertical."[135] The IP Guidelines are generally consistent with this approach, noting specifically that "in the vast majority of cases, restraints in intellectual property licensing arrangements are evaluated under the 'rule of reason."[136] Moreover, the IP Guidelines go on to address several licensing situations by differentiating between vertical and horizontal relationships and, in many cases, develop rules that for use in the licensing area that are more consistent with those which are normally applied to vertical relationships outside of the licensing area.

[132] *Palmer* v. *BRG of Georgia, Inc.*, 499 U.S. 46 (1990).

[133] Ibid. at 49–50 ("Such agreements are anticompetitive regardless of whether the parties split a market within which both do business or whether they merely reserve one market for one and another for the other.").

[134] IP Guidelines § 3.1 (n. 14).

[135] Korah, "The Preliminary Draft of a New EC Group Exemption for Technology Licensing," *European Intellectual Property Report* 7 (1994), 262, 266. See Chapters 5 and 6.

First of all, the IP Guidelines provide that an exclusive license may raise antitrust concerns only if the licensees themselves, or the licensor and its licensees, are involved in a horizontal relationship.[137] This final language differs from the initial draft of the IP Guidelines, where the DOJ noted that, "generally, such a grant of exclusivity may raise antitrust concerns only if the licensees themselves, or the licensor and its licensees, are actual or potential competitors in a relevant technology or goods market in the absence of the licensing arrangement."[138] The position taken in the IP Guidelines is consistent with the relatively permissive treatment afforded other types of vertically-granted exclusivity under generally applicable vertical restraints law.[139] As noted in the IP Guidelines, the areas where exclusive licensing may lead to antitrust concerns include cross-licensing by parties which collectively possess market power, grantbacks, and acquisitions of intellectual property rights.[140]

As to nonexclusive licensing, the IP Guidelines make it clear that nonexclusive intellectual property licenses which do not contain competitive restraint on either party generally do not present anticompetitive concerns, even in situations where the parties are in a horizontal relationship, because that type of license normally does not diminish competition which would have otherwise occurred in its absence.[141] This position is consistent with, and even extends beyond, judicial precedent or regulatory action in the United States, where there has never been any action taken against a genuinely nonexclusive licensing arrangement between parties in a purely vertical relationship, particularly where no market power existed.

Second, the IP Guidelines discuss licenses that involve "exclusive dealing" as "when a license prevents or restrains the licensee from licensing, selling, distributing, or using competing technologies"[142]. The IP Guidelines note some of the general concerns regarding potential anticompetitive effects of such restraints, including the fact that "[s]uch restraints may anticompetitively foreclose access to, or increase competitors' costs of obtaining, important inputs, or facilitate coordination to raise price or reduce output ...".[143] However, some of the procompetitive effects of such restraints are also noted, such as the fact that they "may encourage the licensee to develop and market the licensed technology or specialized applications of that technology."[144] On balance, the IP Guidelines recognize that restraints of this type in a licensing arrangement should be treated no more strictly than product manufacturers' extraction of comparable commitments from their distributors or resellers, which have been given relatively permissive treatment in the case law.[145] The IP Guidelines also acknowledge that "the fact that intellectual property may, in some cases, be misappropriated more easily than other forms of property may justify the use of some restrictions that might be anticompetitive in other contexts."[146]

136 IP Guidelines § 3.4.
137 Ibid. § 4.1.2.
138 Draft IP Guidelines § 4.1.2.
139 See, e.g. *Continental T.V., Inc.* v. *GTE Sylvania Inc.*, 433 U.S. 36 (1977) (manufacturer's grant of territorial or exclusivity to and among their distributors or resellers).
140 IP Guidelines § 4.1.2.
141 Ibid.
142 Ibid.
143 Ibid.
144 Ibid.
145 See, e.g. *Tampa Electric Co.* v. *Nashville Coal Co.*, 365 U.S. 320 (1961).
146 IP Guidelines § 4.1.2.

The IP Guidelines also express concerns that a restraint in a vertical license may harm competition if it "facilitates coordination among entities in a horizontal relationship to raise prices or reduce output in a relevant market."[147] It then offers examples involving owners of competing technologies imposing similar restraints on their licensees or competing licensees benefitting from common license restraints, and notes that "[t]he risk of anticompetitive coordination is increased when the relevant markets are concentrated and difficult to enter."[148] However, the IP Guidelines recognize that the prevalence of a restraint, rather than being inherently suspect, often signifies nothing more than that such a restraint is widely considered to be an efficient way of doing business in that industry.[149]

Finally, the IP Guidelines discuss the situation where the licensor and licensee, while in a vertical relationship with regard to the licensed subject matter, are otherwise involved in a horizontal relationship in another relevant market.[150] The IP Guidelines point out that harm to competition from a restraint in such a licensing relationship may occur if it forecloses access to important inputs or facilitates coordination to raise price or restrict output. However, the IP Guidelines also indicate that "[a] licensing arrangement does not foreclose competition merely because some or all of the potential licensees in an industry choose to use the licensed technology to the exclusion of other technologies. [E]xclusive use may be an efficient consequence of the licensed technology having the lowest cost or highest value."[151]

Managing the transactional costs of licensing It has also been argued that licensors should be given sufficient latitude in drafting their contract terms to overcome the high transactional costs and potential risks of market failure which often are present in licensing arrangements.[152] For example, transactional difficulties may arise from (i) the presence of asymmetric information about the value of the patent, (ii) the fact that the value of the license diminishes if part of the relevant information is transferred by the licensor to the licensee in order to convince the latter about the value of the patent, (iii) the fact that once the information covered by the license is transmitted, the licensee's ability to imitate is increased and the value of the knowledge asset to the seller is reduced, and (iv) the possible inability of the licensor to monitor ex post the intensity of exploitation of the patent.[153]

Appropriability conditions Another area of interest with respect to antitrust policies and some of the factors referred to in the general debate regarding patent policy concerns the general appropriability conditions which might exist in the relevant market. As noted above, the efficacy of the patent incentive varies substantially among industries. It has been pointed out that these finding might have very important consequences for antitrust policies regarding the review of patent licensing provisions. For example, if patent protection were considered to be inherently strong and imitation costs were found to be quite high in a particular market, then licensing restraints with substantial anticompeti-

[147] Ibid. at § 4.1.1.
[148] Ibid.
[149] Ibid.
[150] Ibid.
[151] Ibid.
[152] Ordover and Baumol, supra note 99, 966.
[153] C. Shapiro, "Patent Licensing and R&D Rivalry," *American Economic Review, Papers and Proceedings* 75 (1985) 25, 26–7.

tive consequences should be viewed with disfavor. On the other hand, in markets where patent protection is relatively weak and no other appropriation mechanism has proven to be particularly effective, it may be acceptable to take a more permissive view of restrictive licensing clauses.[154]

Social welfare benefits from licensing arrangements Other arguments advanced in support of reducing antitrust and competition law scrutiny of licensing arrangements are (i) to the extent that price and non-price vertical restraints improve static efficiency, when those restraints are applied to the sale of information they are even more likely to be conducive to social efficiency; (ii) to the extent that revenues from licensing add to the innovator's profit, they enhance the incentives for entrepreneurs to invest in the production of information and knowledge; and (iii) to the extent that licensing arrangements improves dissemination and adds to the effective stock of knowledge, it can reduce wasteful duplication of research and development efforts and can redirect research and development funds towards less imitative pursuits.[155]

D. Market Share Standards and Measurement

Market share-based regulatory "safety zones" The simultaneous evolution of the IP Guidelines and the Technology Licensing Regulation have, for the first time in the licensing context, focused attention on the notion of a market share-based exemption, or "safe harbor," from the rigors of antitrust or competition law challenge. For its part, the IP Guidelines include a "safety zone," which defines certain transactions by reference to the market shares of the parties which, if they fall within the definition, will not be challenged by the regulators absent extraordinary circumstances.[156] Initial drafts of what eventually became the Technology Licensing Regulation provided that the exemption would not be available if certain market share limitations were exceeded and, although these conditions were abandoned in the final form of the Regulation, they remain as explicit reasons for the Commission to consider withdrawing the exemption during the term of the relationship.[157]

There have been several previous attempts to include market share tests in regulatory guidelines. For example, the 20 percent limit included in the IP Guidelines was apparently derived by reference to the market share criterion incorporated into the safety zone for "Physician Network Joint Ventures" in the 1992 DOJ/FTC "Antitrust Enforcement Policy Statement in the Health Care Area." The EC has also included market share standards as part of its block exemption for research and development arrangements, which will only continue to be available to the participants after an initial start up period (i.e. five years after the products are first put on the market) if their combined market shares do not exceed a specified maximum amount. Moreover, economists have often suggested that a market share-based safe harbor should be adopted to protect cer-

[154] Levin, "A New Look at the Patent System," *AEA Papers and Proceedings* 76 2 (1986) 201.

[155] Ordover and Baumol, supra note 99, at 962.

[156] IP Guidelines § 4.3.

[157] See Chapters 5 and 6 for a discussion of the market share conditions included in the Technology Licensing Regulation and the drafts thereof.

tain cooperative agreements among smaller firms from possible liability under the antitrust laws.[158]

Market share-based safe harbors have been subject to criticism on a number of counts. For example, Shapiro and Willig have expressed concerns about the use of some form of market-power based safe harbor for joint ventures, arguing that the very definition of the relevant market usually requires a complete and complex analysis that is better left to expert agencies working within the context of a rule of reason approach to the problem.[159] They point out that a flexible approach under rule of reason analysis, rather than a rigid definition of market power, is the best way to deal with some of the subtleties of analysis, advances in knowledge, and the general accumulation of knowledge from reviewing a number of different arrangements.[160]

Another potential problem is the difficulty in determining what the relevant markets will be for the purpose of computing the market share. As described below, both the United States and the EC have needed to attempt to develop various guidelines with respect to market definition; however, this is always a difficult exercise, particularly in the case of technology-driven products where the product life cycle is relatively short and there is a strong possibility of rapid turnover in market leadership. The situation in the EC is further complicated due to the fact that there may be multiple geographic markets to consider, despite the efforts to develop and maintain a single market within the EC.

Even in those cases where there is some sympathy for articulating market share-based standards, numerous suggestions have been made for developing the most appropriate tests. For example, it may well be that a more liberal market share limit is warranted in situations involving purely vertical relationships. Such an approach would actually be consistent with a number of the "general principles" set out in the IP Guidelines, such as (1) intellectual property should be treated as "essentially comparable to any other form of property"; (2) "the [a]gencies do not presume that intellectual property creates market power in the antitrust context"; and (3) intellectual property licensing "is generally procompetitive".[161] While setting the appropriate standard may require further study, there is precedent for the adoption of a 35 percent threshold for potential market power concerns in this context.[162]

"Technology" and "innovation" markets Commentators have also stressed the need to specifically tailor the definition of the "market" for antitrust review of cooperative arrangements to the context in which innovation is apt to occur. This means that markets should not be defined in terms of products unless commercialization is included in the scope of the cooperative arrangement, rather the focus should be primarily on the mar-

[158] See, e.g. Jorde and Teece, "Innovation and Cooperation: Implications for Competition and Antitrust," in *Collaboration*, supra note 62, 887, 903 (safe harbor should be applicable to interfirm agreements that involve less than 20 to 25 percent of the relevant market).

[159] Shapiro and Willig, "On The Antitrust Treatment of Production Joint Ventures," in *Collaboration*, supra note 62, 929, 945.

[160] Ibid.

[161] IP Guidelines § 2.0.

[162] See, e.g. *Jefferson Parish Hospital District No. 2* v. *United States*, 466 U.S. 2, 26–27 (1984); *Times-Picayune Publishing Co.* v. *United States*, 345 U.S. 594, 611–13 (1953): DOJ/FTC Horizontal Merger Guidelines § 211 (35 percent threshold for unilateral market power in merger analysis); DOJ/FTC Enforcement Policy Statements in the Health Care Area, Part 5A (35 percent threshold for market power in joint purchasing arrangements). It should be noted that the policy statement on physician network ventures, on which the 20 percent standard in the IP Guidelines was based, relates only to horizontal collaborations.

ket for technology and know-how.[163] This is the approach taken in the IP Guidelines, with the emphasis on "technology" and "innovation" markets. However, recognition of these markets does create a number of analytical issues. For example, Bordley has argued that one cannot rely on traditional numerical measures of market power in this area, since they ignore the indeterminacy of markets for research and future technology.[164] Moreover, Bordley notes the difficulties which must be overcome in identifying all the potential innovators, a process which, at the very least, would be quite speculative.[165] In addition, the global nature of innovation dictates that the relevant market reach worldwide.[166]

The development of the discussion of "technology" and "innovation" markets in the IP Guidelines is instructive of the difficult issues in this area. The Draft IP Guidelines adopted the approach of the 1992 Merger Guidelines as the starting point for defining relevant markets, measuring market shares and analyzing competitive effects.[167] The 1992 Merger Guidelines define markets in terms of demand and supply responses to a "small but significant and nontransitory" increase in the price of the relevant product or service. This price-based approach to market definition presumes that there is a functioning market with a statistically significant number of transactions at an ascertainable market price. While these standards might well be appropriate for most goods markets, they are, for a number of reasons, less than serviceable when applied to technology and innovation markets.[168]

One important problem is that very few technology markets actually function in the same manner as a goods market. On the contrary, most technology markets are characterized by a very small number of transactions,[169] and ascertaining the "price" at which these transactions occur is virtually impossible for two reasons. First, as a practical matter, the data are very difficult to obtain: it is not collected or published, nor do companies tend to keep it internally in any easily retrievable form. Second, even if transactional information could be obtained, ascertaining the actual "price" of a license will in

[163] See, Jorde and Teece, supra note 158, at 903.

[164] See Brodley, supra note 62, at 921. Difficulties in defining research markets are exacerbated by the need to attempt to understand the extent of appropriability in the underlying technology. See, Jorde and Teece, supra note 158, 903.

[165] See Brodley, supra note 62, at 921.

[166] See, Jorde and Teece, supra note 158, at 903. Others have criticized the notion that the relevant market for technology should always be assumed to be global. They note that while research markets are generally worldwide, it should not bc forgotten that the markets for the products embodying or produced by the technology developed during the research efforts may be much narrower for a variety of reasons, including international trade restrictions, differing national standards, local subsidies, disadvantageous exchange rates, and government procurement policies. See, e.g. Brodley, supra note 162, at 912 (also noting that much of the defense portion of the United States high technology market is effectively limited to domestic suppliers).

[167] See Draft IP Guidelines §§ 3.2 (Markets affected by licensing arrangements), 3.2.3 (market definition and market shares); § 4.3.1 (horizontal effects); § 5.7 (acquisitions).

[168] The Draft IP Guidelines explicitly recognized the limitations in using the quantitative approach of the 1992 Merger Guidelines with either technology or innovation markets. For example, in discussing technology markets, reference was made to "practical problems in gathering appropriate data to determine 'prices' for the technology and its substitute processes" as an impediment to delineating a relevant market and the possibility that "market shares or other indicia of market power may not be readily available." See Draft IP Guidelines § 3.2.1 and Example 2.

[169] See, e.g. R. Caves, H. Crookell, and J.P. Killing, "The Imperfect Market for Technology Licenses," *Oxford Bulletin of Economics and Statistics* 45 (1983), 249, 250–1. In fact, the dearth of transactions for which economic activity which is essentially developmental has led some courts to conclude that such activity does amount to a line of commerce at all. See, e.g. *American Medicorp.* v. *Humana, Inc.*, 445 F.Supp. 589, 598 (E.D. Pa. 1977); *Babcock & Wilcox Co.* v. *United Technologies Corp.*, 435 F.Supp. 1249, 1275 (N.D. Ohio 1977).

most cases be a very complex task. Most licenses are complex transactions, involving reciprocal rights and obligations, and are seldom comparable, even where they involve the same type of intellectual property. Nor is the "price" easily determined. Royalties frequently require complex calculations based on uncertain levels of future sales. For many transactions, particularly cross-licenses, there may be no royalty payable at all. These difficulties make it virtually impossible to define technology markets or calculate market shares according to any methodology contained in the 1992 Merger Guidelines.

Even if prices could be readily determined and markets identified, it is not clear that the tools used to assess the competitive effects of a merger analysis are appropriate to intellectual property transactions. The Merger Guidelines, including the HHI concentration standards, reflect two concerns: (1) the creation or enhancement of single-firm market power; and (2) the likelihood of coordinated behavior.[170] In contrast, the principal concern in a technology transfer is not with collusion. Most technology is inherently heterogeneous; its value differs widely from one licensee to another; the complexity and secrecy of intellectual property transactions preclude detection of deviations; and the mechanics of punishing deviations are not readily apparent.[171] Since tacit coordination is inherently improbable in technology markets, employing the standards of the 1992 Merger Guidelines, including the HHI-based market share standards, appears inappropriate.

In light of these difficulties, it is questionable whether the concept of a separate technology market will prove be a valuable analytical tool. It would appear that most technology transactions would be more comfortably analyzed with reference to the downstream market for goods or services resulting from use of the technology. Indeed, the IP Guidelines appear to recognize that in most situations market share in the downstream market is the most logical proxy for assessing market power in the upstream technology market.[172] Also, when multiple technologies are alleged to be in the same relevant market, the IP Guidelines suggest comparing the relative efficiency of each technology – a task that almost necessarily relies on an analysis of the downstream market.[173]

The notion of "innovation markets" may be even more problematical than "technology markets." In order to define an innovation market, an effort must be made to attempt to identify all those companies which have the "capability" to carry out research and development activities in a given area. Once that has been done, a determination must be made as to whether any scarcity of that capability might be the basis for challenging a proposed merger or other transaction. According to Gilbert, the competitors in an innovation market include all those firms which "possess the assets or characteristics that are necessary to engage in the development of particular new or improved products or processes."[174] He asserts that a transaction should be the scrutinized in situations where it affects a large fraction of the assets that could be directed toward innovation.[175]

Gilbert points out that the analysis that was included in Case 6 of the 1988 International Guidelines, where the DOJ applied the concept of the "relevant R&D market" to the analysis of a research and development joint venture, is similar to the innovation

[170] 1992 Merger Guidelines, §§ 0.1, 2.1, 2.2.

[171] Ibid. §§ 2.11, 2.12.

[172] See also *United States* v. *General Motors Co.*, Civ. No. 93-530 (filed Nov. 16, 1993) (DOJ analyzed merger both in the context of a technology market and the goods market of heavy duty transmissions).

[173] See IP Guidelines § 3.2.2, n22.

[174] Gilbert, "An Interview with Richard Gilbert," *Competition: The Journal of the Antitrust and Trade Regulation Law Section of the State Bar of California* 4 (1994) 1, 2.

[175] Ibid.

market approach in the IP Guidelines.[176] Research and development markets were also relevant in several actions recently brought by the DOJ and the FTC, including the complaint filed by the FTC against the acquisition of Genentech by Roche Holding, where it was claimed that one of the relevant product markets was research and development related to vitamin C and therapeutics for treatment of human growth deficiency, AIDS and HIV), and the complaints which were filed by the DOJ in U.S. v. Automobile Manufacturers Association (competition in research and development of motor vehicle air pollution control equipment) and in *United States* v. *Manufacturers Aircraft Association* (competition in research and development of airplane technology).[177]

Unlike technology markets, innovation markets may not be as easily analyzed by focusing exclusively on the downstream market for goods. Nevertheless, the practical problems in applying the 1992 Merger Guidelines are similar for both kinds of markets. For example, in innovation markets there often will be no transactions from which to apply the price sensitivity test of the Merger Guidelines and no convincing basis for measuring market shares. The IP Guidelines suggest as alternative measures "shares of those identifiable assets or characteristics upon which innovation depends [or] shares of research and development expenditures."[178] However, there does not appear to be any empirical basis for assuming a correlation between the size of the budget for research and development or of "identifiable assets" and superior ability to innovate.[179]

It has been suggested that "innovation markets," while certainly not entirely new in theory, won't really enhance antitrust analysis beyond that which can be accomplished with existing tools.[180] For example, it has been pointed out that the 1992 Merger Guidelines already provide a method for the analysis of competitive conditions in markets where shares are not currently measurable in terms of existing goods and services (i.e. potential competition), and also address the analysis of barriers to entry.[181] It has also been argued that it is not likely that there will be a significant variation between the participants in an existing innovation-driven goods market and the "market" for continuing investment in the specific type of innovation, since the firms in the goods market are unlikely to forego the sunk costs of establishing themselves in the goods market without some form of ongoing competitive response to new innovations.[182] Moreover, there has been substantial skeptism about whether it is really possible for one or more companies to actually monopolize the capability for conducting research and development in any area.[183]

[176] Ibid.

[177] Ibid. at 2–11.

[178] IP Guidelines § 3.2.3.

[179] See, e.g. Scherer, *Innovation and Growth* (1984), 222–38 (no direct relation between corporate size or size of research and output of patented inventions).

[180] Taylor, "Counterpoint: Will the Guidelines Work?," *Competition: The Journal of the Antitrust and Trade Regulation Law Section of the State Bar of California* 4 (1994) 1, 12.

[181] Ibid.

[182] Ibid. His concern is that the lack of a "capacity to innovate" might be used as a basis for eliminating existing firms from the market, thereby reducing the size of the market and leading to challenges of transactions that would otherwise not be anticompetitive. His fears are amplified by the fact that the innovation capacities of various firms are to be ascertained, at least in part, from data to be provided by competitors of the company under scrutiny. Ibid. at 13.

[183] Ibid. at 12. He points out that even in those situations where only a handful of companies appear to be working on a given research and development project, analysis can be accomplished through such traditional tools as "potential competition" and "barriers to entry." Ibid.

Another important obstacle to defining and using "innovation markets" is that many of the elements necessary for the pursuit of innovation (e.g. human resources) are highly mobile, thereby making it difficult to take a static view of the "capacity to innovate." As Taylor rightly points out:

> " the capacity to innovate is virtually uncontainable, even by companies that try to do so. The last twenty-five years in Northern California (and other areas where start-up companies have proliferated) provides dramatic proof of the tremendous mobility of expertise and experience, which are the basic ingredients of the capacity to innovate. When scientists in the computer and pharmaceutical industries can leave their employers and start entire new industries around the truly complex technologies that those companies work with, it (sic) inappropriate to take a static view of the 'capacity to innovate' in any industry."[184]

Taylor and others have suggested that the DOJ must attempt to state more specifically the various factors that are likely to be used for market definition and the measurement of market power in connection with "innovation markets."[185] Specifically, the DOJ must determine whether it is even possible to make a quantitative assessment of market boundaries using research and development as a basis for market definition. One possible criteria associates innovation with "cost reduction," and suggests that the outer limit of an innovation market might be a five percent "SSNLC," which is a "small but significant lowering of cost."[186] However, on balance, critics of "innovation markets" maintain that the concept is just too speculative to be useful, particularly since there is almost no way of knowing whether a particular company has the "capacity to innovate" due to the great secrecy typically associated with any research and development project and the fact that the existence of innovative activity is not really evident until some transaction for value actually takes place.[187]

Gilbert responds that concerns regarding the use and application of innovation markets are misplaced, since he anticipates that the innovation market will only be relevant in those situations where the "capacity for research and development is scarce and can be associated with identifiable specialized assets or characteristics of specific firms."[188] He explains that innovation markets will be used only when it appears that there are only a few firms capable of innovation, and the likely innovators are easy to count.[189] In one case, the DOJ was able to rely on evidence to the effect that it was unlikely that new competitors would enter into the market for the design and production of improved automatic transmissions in the near-term in order to conclude that the innovation market was limited to the established producers.[190] Gilbert recognizes, however, that in many other situations, innovation can come from a number of different sources, including companies which currently are not active in the affected markets, and that if this is the case then it is less likely that a single transaction will have an adverse impact on innovative competition.[191]

[184] Ibid. at 13.
[185] Ibid.
[186] Ibid.
[187] Ibid.
[188] Gilbert, supra note 174, 11.
[189] Ibid.
[190] Ibid.
[191] Ibid.

It is likely that regulators will look to the National Cooperative Research Act of 1984 ("NCRA") for guidance regarding the concept of "innovation markets," since Congress expressed concern which considering that legislation that "over-inclusive" research and development efforts might stifle independent, competing efforts.[192] These concerns are similar to those which appear to have inspired the interest in possible "innovation markets." The NCRA and its legislative history emphasize several points to which the DOJ should be sensitive in analyzing any possible "innovation market": (i) any such market would be international in scope, and foreign sources are as important as domestic ones;[193] (ii) as long as the potential exists for "several other – perhaps four" – comparable research efforts, "anticompetitive effects are unlikely";[194] (iii) four is not a magic number, and economies of scale or other factors may dictate even fewer alternatives without raising undue competitive concerns[195] and (4) even if there is a possible over-concentration of research, its effects must be balanced against "all relevant factors affecting competition" in all of the markets involved.[196]

Further work remains to be done on the definition and application of innovation markets in antitrust analysis. For example, as part of its initiative on "The Changing Nature of Competition,"[197] the FTC has proposed additional studies of the following types of questions:

(i) Are there other circumstances, beyond those in the IP Guidelines, where it is appropriate to use innovation markets, as opposed to traditional goods and services markets, in competition analysis?

(ii) What improvements can be made in defining innovation markets, as well as in identifying the evidence which will be required by regulators as part of the process of definition?

(iii) Which of the "innovation models" should be used in analyzing innovation-driven industries? What is the effect, if any, on market definition if innovation occurs according to a "learning curve" or "leapfrog" model? For example, if innovation occurs according to a "leapfrog" model, does this mean that the market shares of a previous generation of products are no longer a good proxy for expected market shares in the next generation of products?

(iv) To the extent that innovation produces changes in product qualities, is there any way to measure changes in product attributes in order to assess product substitutability? What are the relative roles of price and attribute competition for market definition?

(v) What factors are relevant to market power – e.g. research and development expenditures; possession of intellectual property assets; first mover advantage; installed base; existence of/ownership of a standard; network externalities, etc.? Are there different barriers to entry in innovation markets?

[192] See Conference Report, BNA Antitrust and Trade Regulation Report 47 at 590; S. Rep. No. 427, 98th Cong., 2nd Sess. 948 (1984).

[193] 15 U.S.C. § 43O2; Conf. Report, BNA Antitrust and Trade Regulation Report 47 at 590.

[194] Ibid.

[195] Ibid. at 590–1.

[196] Ibid.

[197] See Federal Trade Commission, "Hearings on the Changing Nature of Competition" (Preliminary Version – June 29, 1995) (the "FTC Competition Agenda").

(vi) What are the risks to innovation, if any, that should lead regulators to err in one direction or another in defining innovation markets or what constitutes market power therein?

(vii) do firms decide to innovate, and what observable factors are relevant to determine whether a firm will enter an innovation market – e.g. access to capital; access to intellectual property assets; research and development capabilities; ability to collaborate or coordinate with other firms; producers of complementary products or producers with vertical relationships?

(viii) Are vertical mergers of greater concern in innovation markets because they may involve the most likely new innovator or the integration of two dominant upstream and downstream firms?[198]

E. Regulatory Treatment of Research and Development Agreements

While licensing arrangements obviously can, and often are, formed outside of a formal joint venture structure,[199] it is also quite common for licensing to occur within the boundaries of a larger collaborative arrangement formed to pursue one or more of the functions within the innovation process. Since many of these joint ventures concerns themselves primarily with the research and development required to develop new products which ultimately would be eligible for patent protection, then it is important to consider regulatory attitudes toward such joint ventures as part of any general discussion on the relationship between innovation and antitrust and competition law policies.

A number of theoretical models have been developed with respect to coordinated investment in research and development. In general, these models appear to lead to the conclusion that while research and development joint ventures are driven by a profit motive, on balance they improve social welfare.[200] Accordingly, it has been argued that antitrust treatment of such arrangements should utilize rule-of-reason principles, and that as long as the co-venturers involved face a reasonable threat from outside firms in the activities of the joint venture and in their ability to elevate prices after the innovation, the joint venture should be allowed to proceed.[201] Moreover, restraints on the research and development and the marketing strategies of venture participants, such as limits on independent research, should not automatically be condemned, but rather should be evaluated in light of their role in facilitating cooperation and generating adequate (expected) returns to the research and development investments undertaken by the venture.[202]

Liberalized treatment of research and development joint ventures in the United States was literally codified in the NCRA. While the NCRA mandated the use of rule of rea-

[198] Adapted from the FTC Competition Agenda, 3–5.

[199] See Chapter 6.

[200] The models demonstrate that research and development joint ventures reduce duplication and enhance spillovers, thereby reducing the social and private costs of generating a given amount of effective knowledge, and improve ex post dissemination through research and development output sharing arrangements, especially where effective licensing of knowledge is difficult to accomplish. Ordover and Baumol, supra note 99, 973. Examples of the theoretical models include Grossman and Shapiro, "Research Joint Ventures: An Antitrust Analysis," *Journal of Law, Economics and Organization* 2 (1986), 315–37; Katz, "An Analysis of Cooperative Research and Development," *Rand Journal of Economics* 17 (1986), 527–43; and Ordover and Willig, "Antitrust for High-Technology Industries: Assessing Research Joint Ventures and Mergers," *Journal of Law and Economics*, 28 (1985), 311–33.

[201] Ordover and Baumol, supra note 99, 969–70.

[202] Ibid.

son analysis for research and development joint ventures, it did not attempt to specify the factors which should be considered in applying such an analysis. Teece and Jorde have criticized this omission,[203] pointing out that merely requiring consideration of "all relevant factors affecting competition," did not afford enough importance to any of those special characteristics of the innovation process which exist in a rapidly changing industry.[204] They argue that the antitrust laws, specifically the NCRA, should be amended to make it clear that courts and regulators should take account of the appropriability regime, the pace of technological change, the diversity of sources of new technology, the need to access complementary assets and technologies, and the need for "cheek-by-jowl" cooperation to manage innovation simultaneously rather than serially.[205]

A number of commentators have been very critical of Teece and Jorde's proposals. For example, Brodley opposes any attempt to codify antitrust standards, as well as some of the standards that Teece and Jorde wish to have implemented.[206] Brodley points that existing antitrust legislation does not include a set of detailed formulations, but rather depends upon broad concepts, such as "injury to competition" and "monopolization," which can evolve and change gradually as experience is gained with specific types of transactions.[207] He argues that this approach is to be preferred over the use of detailed statutory rules, which Bordley believes must either be arbitrary or simply contribute to further complication of the decisional processes utilized in this area, particularly since each new statutory concept generally becomes a source of argument and uncertainty regarding the actual intent of the Congress in making the modification.[208]

Congress has already rejected several proposals to codify elements of the rule of reason analysis. For example, when the NCRA was originally being considered, Congress decided not to include detailed language to amplify the meaning of the rule of reason, fearing that the inclusion of such language would simply lead to arguments that relevant elements not enumerated in the statute need not be considered.[209] Brodley also argues that if standards are actually codified, it becomes extremely difficult if they cause interpretive problems, which is quite likely to occur in light of the significant levels of uncertainty in the area of economic and legal analysis of innovative collaboration. As such, Bordley and others conclude that the best approach to providing certainty and advance guidance to private firms is to rely on enforcement agency guide-

[203] See, Jorde and Teece, supra note 158, 900.

[204] Ibid.

[205] Ibid. at 903. While Teece and Jorde hope that the courts will independently choose to apply each of the aforementioned factors whenever they are called upon to make a rule of reason analysis, they have made it clear that they have a preference for legislation mandating changes in the analysis as the best means for rapidly achieving a package of substantive and procedural changes, thereby assuring certainty and predictability in the area. Ibid. at 904.

[206] Brodley, supra note 62, at 918. For example, Brodley asserts that Teece and Jorde's suggestions include obscure concepts such as "whether the innovation ... will compete with other technologies in a preparadigmatic stage of competition for particular products and processes." Ibid.

[207] Ibid. at 919. Brodley also points out that the antitrust laws have changed dramatically in recent years to take into account the growing complexity and diversity of business transactions, thereby suggesting that the use of broad concepts has proven to be quite adequate for regulatory purposes. Ibid. n 6. See also Kwoka and White (eds), *The Antitrust Revolution*, Glenview: Scott, Foresman and Company (1989), 46–79.

[208] Ibid.

[209] Ibid. (citing Conference Report, National Cooperative Research Act of 1984, 98th Cong., 2d Sess., House of Representatives, Rep. 98-1044 (1984)) It is also interesting to note that the Congress has similarly rejected proposals to codify the Department of Justice's Merger Guidelines. Ibid. n 8 (citing S. 2160, 99th Cong., 2d Sess., 132 Cong. Rec. S2280, 1986).

lines and judicial formulations which can be easily modified to take into account any changes in circumstance over time.[210]

While it is therefore unlikely that specific standards for rule of reason analysis will be implemented by statute in the immediate future, there are nonetheless definite opinions from economists regarding the areas upon which the analysis of these research and development joint ventures should be focused. For example, economists emphasize the importance of analyzing such arrangements for their effects on present and future competition, both at the upstream level (research and development) and downstream in the product markets.[211] With respect to upstream effects, the key issue is the effect that the collaboration may have upon the current and future production of new knowledge.[212] It can anticipated that the potential adverse effect that a joint venture might have upon the production and adoption of new technology will be strongly related to the joint venture's share of the assets that are critical to the production, dissemination, and adoption of new knowledge;[213] however, even if the venture controls a large portion of assets, its ability to repress technological progress may be weak if entry into the relevant upstream market is not very difficult, as might the case in the early stages of the evolution of the industry, when technological opportunities are still wide open.[214]

With respect to the potentially anticompetitive effects of research and development joint ventures in downstream markets, it would seem that the less vertically integrated downstream is the venture, the less likely it is to be used to facilitate collusion in the product market. The potential dangers stemming from downstream collaboration were, in part, reflected in the initial decision to limit the scope of the NCRA only to research and to downstream commercial activity "reasonably required" in order to complete the research, which was narrowly confined to marketing intellectual property rights which may be developed in the joint research and development program.[215]

Even in those cases where the parties to the original research and development program do not actually collaborate in downstream commercialization of the results of the program, it should be recognized that the joint venturers may need to impose some restraints on ex post competition in situations where there are concerns that unbridled downstream competition will dissipate the anticipated rents from successful development efforts.[216] For example, it may be prudent for the parties to agree on the territories within which each will be allowed to commercialize the results of the research program without competition from the other participants, since some degree of exclusivity may well be necessary in order for each party to recoup its shares of the costs of the original development program.

Another factor which economists often cite as relevant to analyzing a research and development joint venture is the level of research appropriability in the particular industry or field in which the participants are active. In one interesting study of the effect of the NCRA, Scott examined all the registrations under the NCRA that occurred over an 18 month period and found that collaboration was taking place not in industries where research appropriability was likely to be low, but rather in industries where concentration was high, productivity growth was strong, and where the firms already were involved

[210] Ibid.
[211] Ordover and Baumol, supra note 99, at 974.
[212] Ibid.
[213] Ibid.
[214] Ibid.
[215] See, Jorde and Teece, supra note 158, at 900.
[216] Ordover and Baumol, supra note 99, at 974.

in diversified research and development activities so that they would be able to exploit any research and development spillovers across other industries.[217] In light of these results, it has been argued that the collaboration fostered by the NCRA may have actually caused substantial losses with respect to the stimulation of competitiveness without achieving more than some minimal amount of appropriability gains.[218]

Research and development joint ventures should also be analyzed for the effects on excluded firms, since such firms will be placed at a relative disadvantage in downstream markets in situations where the joint venture increases effective research and development to the co-venturers relative to firms outside of the joint venture.[219] In some cases, the disadvantage may be so great that excluded competitors actually may be forced to exit from the downstream market, reducing competition there. However, excluding a subset of competitors from a research and development joint venture is not in and of itself anticompetitive, especially if the joint venture encompasses less than an overwhelming share (e.g. 60 percent) of the pertinent assets.[220]

F. Antitrust/Competition Policies and Small Firms

Another area where patents and competition policy overlap is the relationships between individual inventors and small firms on the one hand and larger firms with greater research resources on the other hand. As noted elsewhere, individual inventors and small organizations have historically generated a significant number of important technical advances; however, it is often difficult for these inventors to collect the non-patent resources required to actually exploit these new inventions. In turn, larger firms, while usually possessing many of the strategic and physical assets necessary for commercializing new products, will often look outside their organizations for new products, a phenomenon which is consistent with Von Hippel's finding[221] that the sources of invention and innovation with respect to a given product are often dissimilar.

The pharmaceutical industry, which is clearly one of the areas where patents remain important and innovative research is a key competitive element, provides a good example of the growing importance of collaborative business relationships. Larger firms, with an interest in acquiring access to the products of research in the relatively new area of biotechnology, are entering into a variety of relationships with smaller research-based firms. For example, smaller firms may contract to do research work for the larger firms, with the larger firm providing its extensive distribution skills to reduce the time and expense of bringing the new products to market, and the smaller firm providing technical expertise that falls outside of the in-house skills of the larger firm. These partnerships are an attempt to manage the substantial risks associated with product development in this area and, as such, have goals similar to those which might be addressed by some of the proposed modifications to the patent system.

The importance of individual inventors and smaller firms in developing new inventions has led to various proposals designed to assist them in exploiting those inven-

[217] See Scott, "Diversification versus Cooperation in R&D Investment," *Managerial Decision Economics* 9 (1988), 173–86 (cited in Brodley, supra note 62, at 915.

[218] Ibid.

[219] Ordover and Baumol, supra note 99, at 975.

[220] Ibid.

[221] Von Hippel, *The Sources of Innovation* (1988), at 11–42.

tions, primarily by enhancing their bargaining position in relation to larger firms possessing complementary assets seeking to contract with the smaller firms to commercialize the inventions. The prospect of modifying the patent system to facilitate cooperation among smaller and larger firms means that antitrust and competition laws must also be drawn and interpreted in a consistent manner if the modifications in the patent system are to have an appreciable impact on the activities of market participants.

Some steps have already been taken in the United States to shape antitrust policies in a manner which facilitates the types of collaborative arrangements between smaller firms and other firms at different stages of the innovative process (e.g. large-scale manufacturing and distribution) that allow smaller firms to bring their products to market. The adoption of the NCRA, and then the National Cooperative Research and Production Act ("NCRPA") several years later, represented an attempt by regulators and legislators to encourage cooperation between firms at the research and development and manufacturing stages by easing the level of antitrust scrutiny.[222] Moreover, inclusion of the "safe harbor" in the IP Guidelines, while perhaps not as broad as some economists and practitioners would like, was also an attempt to insulate smaller firms from the costs and delays of undertaking complex economic analyses as a pre-condition to collaborative arrangements with similarly sized firms.

There is also a sense among policymakers that more might be done for smaller firms. The FTC has indicated an interest in studying the question of whether current antitrust and consumer protection regimes operate so as not to impair unnecessarily the ability of small businesses to compete in a changing global, innovation-based economy. Among the specific issues and questions to be considered are the following:

(i) the particular competitive circumstances or regulatory barriers that impair the ability of small firms to compete in a global environment? How should antitrust enforcers define small firms in this context?
(ii) What are the competitive effects of the current antitrust treatment of small businesses?
(iii) What is the competitive effect of current safe harbors (e.g. absence of Hart–Scott–Rodino Act filing requirements, safe harbor for mergers of competitors with low market share) on small businesses? Should there be a greater consideration of the use of "safe harbors" for small business transactions? In what circumstances? How should small business be defined for this purpose? Should the definition of small business – or the safe harbor – always be keyed to market power?
(iv) Are there any significant antitrust impediments to beneficial strategic alliances or joint ventures (short of joint marketing) for small businesses?
(v) Do small businesses have advantages in terms of flexibility on account of their small scale?[223]

[222] The NCRPA attempts to minimize some of the antitrust concerns presented by the United States antitrust laws in the context of joint research and development ventures and joint production ventures. Regulations regarding cooperative activities also impact the debate regarding the inefficiencies associated with "duplicative" research efforts. In addition, cooperative research activities influence how new information actually is published and disclosed in the marketplace. See also Blair and Kaserman, *Law and Economics of Vertical Integration and Control* (1983); and Whish, *Competition Law* (1989), at 262–7 (block exemptions for patent licensing, know-how licensing and research and development contracts) and 575–600 (block exemptions for exclusive distribution arrangements).

[223] See *FTC Competition Agenda*, supra note 197, at 10.

G. Lessons from Comparative Analysis of United States and EC Regulation of Licensing Arrangements

Introduction Comparative analysis is an important element of this work, and serves as a useful guide for navigating a variety of complex and difficult policy issues. Comparison need not necessarily identify the "best practices," instead it generally provides a solid basis for determining those factors which may be relevant in forging regulatory policies and legal precedents. The United States and the EC were chosen for comparison because they have the most experience in dealing with antitrust and competition law regulation of licensing practices; however, it seems clear that experience has, for a variety of reasons, not led regulators in each area to the same conclusions on a number of issues.

There are a number of possible explanations for what are quite often material variations in regulatory policy between the United States and the EC. One overriding factor is that the United States generally has an integrated, federal, system of patents and antitrust laws and regulations,[224] while the EC is still an evolving economic union with separate national laws pertaining to patents and competition. As we have seen, perhaps the most difficult issue for EC competition law regulators has been reconciling the existence of national patent rights with the overriding EC Treaty objectives of reducing and eliminating barriers to trade among the member states. Clearly, some progress has been made in harmonizing the content of national patent and competition laws; however, attempts to supplant those laws with a universal EC-wide legal regime have moved slowly, meaning that actual or potential conflicts between EC and national laws will remain, at least for the foreseeable future.

Another factor to be borne in mind in comparing the antitrust and competition laws of the United States and the EC is the relative levels of demographic diversity within each of the regions. Certainly, there is a good deal of variety in the cultural and economic background in the various regions of the United States; however, it does not begin to approach the continuing nationalism within the EC. In spite of the single market objective, the EC will, for some time, likely remain a bundle of different markets demarcated by historical experience, location, culture, and language. This means that EC companies will continue to need the flexibility to develop, through there licensing practices, production and distribution strategies which are tailored to portions of the EC market. This will likely cause difficulties for EC competition law regulators which are not currently significant in the United States.[225]

Another important factor, given that exploitation of the patent right and patent licensing is most significant in the technology area, is the differences between the United States and the EC with respect to the mobility of the human resources and the capital

[224] In general patent cases in the United States are heard in the federal courts; however, there are instances where cases which involve patent licenses and assignments are heard in state courts. Advocates of a truly federal patent law system in the United States have argued that steps need to be taken to ensure that all patent cases are heard in federal courts in order to guard against variant decisions by state judges lacking experience in the subject matter. See, e.g. Henry, "State Courts Hearing Patent Cases: A Cry for Help to the Federal Circuit," *ABA Section of Intellectual Property Law Newsletter* 14 (1996), 3 (suggesting that the Court of Appeals for the Federal Circuit reverse previous decisions and hold that all cases involving a patent matter must be heard in a federal court).

[225] Among the areas of concern will be exclusive territorial licenses and field of use and customer restrictions. See below.

necessary for the effective commercialization of new ideas and inventions. Technology licensing in the United States is done in an environment where scientists and engineers can move relatively freely, the laws relating to the raising of capital (i.e. securities laws) are largely federal in nature, and the differences in state laws with respect to labor and business entities (e.g. corporation and partnership laws) are rarely significant. In contrast, innovation in the EC will likely berelatively more complicated until agreement is reached relating to free movement of labor and capital across national borders.

Beyond these macro-level differences between the United States and the EC, we can identify significant variations with respect to the general attitudes in each region with respect to regulation of licensing arrangements, including the procedures followed in each region and the analytical tools and assumptions applied to each relationship. In addition, one can examine how each region tends to deal with the most often used licensing practices, such as exclusive licenses and restrictions relating to the territories and technical areas in which the licensed subject matter can be used. Finally, it is useful to identify some of the other areas where disparate policy approaches might, directly or indirectly, impact the relationship between patent rights and innovation in each region.

Differences in general attitudes and practices regarding antitrust and competition laws Our study has identified several areas in which the divergent general attitudes and practices of regulators in the United States and the EC may have led to different views as to the legality of certain licensing practices. Among these areas are the characterization of licensing agreements as horizontal or vertical arrangements, the use of rule of reason analysis, the measurement and use of market shares in assessing the competitive impact of licensing agreements, durational limitations on permitted licensing practices, and institutional differences between the agencies and personnel responsible for interpreting and enforcing the relevant laws and regulations.

Characterization of licensing agreements One of the threshold factors influencing regulatory attitudes toward any form of cooperative arrangement is whether the relationship between the parties is horizontal or vertical. In the United States, the IP Guidelines clearly state that most licensing arrangements will be treated as vertical, although the relationship may be treated as horizontal in cases where the licensor and its licensees, or the licensees, would have been actual or likely potential competitors in a relevant market in the absence of the license.[226] In general, regulators in the United States tend to look at the position of each of the parties before the licensing relationship is consummated, and risky investment have been made, in determining whether or not the relationship is vertical.

In contrast, the Commission, at least in the eyes of one commentator, has usually tended to treat licensing arrangements as horizontal once each of the parties have begun to sell the licensed products.[227] By doing this, the Commission ignores the position of the parties in the event that the license had never been granted, thereby minimizing the weight of the procompetitive effects of the licensing agreement and undermining any incentives the parties might otherwise have to make *ex ante* investments.[228] In addition, this tendency to look at licensing arrangements as horizontal relationships has often raised un-

[226] IP Guidelines § 3.3.

[227] Korah, supra note 135, at 265. See also Venit, "In the Wake of Windsurfing: patent licensing in the common market," Fordham Corporate Law Institute (1986 ed. Barry Hawk) Chapter 22, 522.

[228] Ibid.

founded fears that the arrangement might turn into a cartel where the licensor and each of the licensees would conspire to fix prices or allocate customers or markets. As a result, the Commission, has generally prohibited a number of collateral restrictions, such as price and customers restrictions and maximum quantity limitations, without serious analysis of the commercial justifications for the use of such practices.

On balance, the approach taken in the IP Guidelines would appear to be more desirable. While certainly there is a danger that a purely vertical licensing arrangement may, over time, take on aspects of an unlawful horizontal cartel, the primary focus of antitrust and competition regulators at the time the relationship begins should be on whether the licensing arrangement may reduce or eliminate competition that would otherwise have occurred. If the parties to the agreement only become "competitors," in the sense that they are both able to practice the licensed subject matter, as a result of the license, then the arrangement should generally be allowed to proceed, since it will typically result in a dissemination of technology which might not have occurred in the absence of the license. Safeguards should, however, still be retained by ensuring that regulators work with a developed and focused definition of "horizontal" relationships. Moreover, the parties should also be given explicit guidance on the limited circumstances when licensing restraints in vertical agreements may raise significant antitrust concerns.

Rule of reason analysis As we have seen, regulatory attitudes regarding regulation of licenses have changed quite dramatically in the last two decades in the United States. Presently, the DOJ and most courts are of the view that the owners of intellectual property rights should be afforded the maximum degree of freedom in the commercialization of inventions and new ideas, and have abandoned the mechanical application of *per se* rules of illegality in favor of rule of reason analysis that takes into account the actual competitive impact of a particular restriction and the realities of the commercial setting in which the parties to the license are operating. In addition, the historical presumption in the United States that possession of a patent automatically confers the ability to control a market has eroded in the face of evidence that market power is often a function of other factors (e.g. quality, service, and speed of innovation). As a result, the vast majority of intellectual property licensing arrangements in the United States are now evaluated under the rule of reason.[229]

On the other hand, past practice in the EC, as well as it latest efforts in formulating the new Technology Licensing Regulation, has been sadly lacking in its appreciation of commercial realities and the dynamics of innovation. In the words of one leading commentator:

> "Industry in the EEC has suffered greatly from the lack of economic analysis in decisions under Article 85(1) and from its wide application by the Commission. The adoption of block exemptions has not satisfactorily solved the need for those investing in sunk costs to appropriate the benefit of their investment – to reap where they have sown. Large firms with manufacturing capacity seldom need to license their technology. The licensors most likely to suffer from the limitation of territorial protection they can grant to licensees are small firms who do not perceive themselves as able to ex-

[229] See IP Guidelines § 3.4. This change in attitude has been so significant that a number of courts in the United States no longer declare restrictions which reach beyond the licensed patents to be *per se* illegal (e.g. field of use and other competitive restrictions on unpatented products which are produced using patented machines, processes or key components will not be struck down in vertical arrangements without an analysis of whether the particular restraint is really anticompetitive).

> ploit the whole of the common market directly, such as Universities, research institutes and so on."[230]

It has been suggested from time to time that there may not be any need for a block exemption for technology licensing, since either the market share of the parties would be small, in which case the agreement could not be anticompetitive, or it is large, in which case the Commission should be considering an individual exemption.[231] If this were to occur, than the Commission and the national courts might finally begin to rely on the type of rule of reason analysis that forms the foundation of law and policy in the United States, and which the ECJ has often done in reviewing various commercial transactions.[232] However, legitimate concerns have been raised about the Commission's ability and willingness to abandon its historically formalistic approach, under which almost any important restriction on freedom included as part of a licensing arrangement is found to be contrary to Article 85(1) and therefore requires an exemption, and to begin to rely upon a realistic analysis of the relevant markets, including markets defined by reference to innovation-based assets.[233]

Market share tests The measurement and use of market shares with respect to innovative products has been the area of greatest debate and controversy on both sides of the ocean over the last few years. The IP Guidelines recognize, for the first time in the licensing context, an antitrust "safety zone" to inform parties of situations in which it is not likely that regulators will challenge the use of specified restraints in licensing agreements.[234] The IP Guidelines make it clear that the "safety zone" is designed to provide intellectual property owners with a degree of certainty, thereby encouraging the parties to enter into procompetitive licensing arrangements, and that it is not the intent that parties involved in arrangements falling outside the safety zone be discouraged from adopting restrictions reasonably necessary to achieving an efficiency-producing integration of economic activity.[235]

In contrast, the Commission gave long and careful consideration to conditioning the exemption for exclusive licenses in the Technology Licensing Regulation on the market share of the licensee not exceeding 40 percent of the market for the licensed products in the common market or a substantial part of it.[236] Ultimately, this condition was dropped

[230] Korah, supra note 135, at 268.

[231] Ibid. at 263.

[232] Ibid. at 267 (citing various cases (e.g. *Nungesser* v. *Commission*, Case 258/78, [1982] ECR 2015, *Louis Ereau Jacquery* v. *La Hesbignonne*, Case 27/87, [1988] ECR 1919) where the ECJ has held that where there is a legitimate transaction, a restriction on conduct that is necessary to make it viable does not infringe Article 85(1)).

[233] Korah is also concerned that judges in the national courts tend to be generalists unskilled in economics and market analysis, and will accordingly follow the Commission's approach rather than engage in rule of reason or market-based analysis. Ibid. In contrast, the Court of Appeals for the Federal Circuit was established in the United States to develop the requisite expertise to deal with economic aspects of patents, and a good deal of work has been done by regulators in the United States to develop more sophisticated analytical tools to deal with innovation-based goods and markets. See below.

[234] Specifically, absent extraordinary circumstances, regulators will not challenge a restraint placed in a licensing arrangement if (i) the restraint is not of a type which normally warrants condemnation under the *per se* rule and (ii) the licensor and its licensees collectively account for no more than twenty percent (20 percent) of each relevant market affected by the restraint. IP Guidelines § 4.3.

[235] The IP Guidelines explicitly specify that arrangements which fall outside the "safety zone" will continue to be analyzed based on the other considerations outlined therein (e.g. rule of reason analysis). Ibid.

[236] Proposed Technology Licensing Regulation Art. 1, para. 5.

in the final form of the Technology Licensing Regulation, although the Commission explicitly stated that a licensee market share in excess of 40 percent may serve as a basis for it to withdraw the exemption under Article 7(1) of the Technology Licensing Regulation if the effect of the agreement is to prevent the licensed products from being exposed to effective competition in the licensed territory.

While the Commission has retreated from its earlier positions regarding markets and market shares, a number of issues still merit discussion. One of the most important, of course, is the consequence of misapplying the market share standards. In the United States, the "safety zone" effectively saves parties to a licensing arrangement falling within it the time and expense of conducting a full blown antitrust and economic analysis of the licensing arrangement. If they are wrong, however, and their market shares are too high for the "safety zone" to apply, the only effect of their mistake would be that the rule of reason will be applied to judge any anticompetitive restrictions in the agreement (i.e. they are not automatically *per se* offenses).

On the other hand, if the market share conditions had remained in the Technology Licensing Regulation, and the parties relying on the Regulation mistakenly misjudged their market shares such that the block exemption was not available, then almost any restriction of competition in the agreement would have become essentially illegal *per se*, and would be unenforceable and subject to fines.[237] Such dire consequences would most likely have led prospective EC licensors to either abort the transaction or go through the time and expense of notifying the agreement to the Commission. In either cases, legitimate fears were raised about the overall effect that the introduction of market share factors might have on technology transfer activity in the EC.[238]

Another issue, related to the vertical/horizontal debate referred to above, is the point in time where the market share test is applied. The IP Guidelines note that "the status of the licensing arrangement with respect to the safety zone may change over time ... [a] determination by the Agencies that a restraint in a licensing arrangement qualifies for inclusion in the safety zone is based on the factual circumstances prevailing at the time of the conduct at issue."[239] The IP Guidelines indicate that the "conduct at issue" may be the transaction giving rise to the restraint or the subsequent implementation of the restraint.[240] Accordingly, the DOJ will typically make a single determination regarding the application of the "safety zone" at the time the agreement is first consummated, realizing that market share conditions may change thereafter, perhaps as a result of the success of the relationship itself.

In contrast, the market share test in both the Proposed Technology Licensing Regulation and in the Technology Licensing Regulation could be applied at any time during the term of the licensing agreement.[241] As noted above, the uncertainty stemming from

237 In certain cases, unenforceability of the restriction might be dangerous to the licensee than any fines, as might be the case where the licensee is suddenly denied the protection of an exclusivity clause and exposed to competition from the licensor and/or other licensees before the licensee has an opportunity to recover its initial investment in the licensed product.

238 While presumably the Commission's decision not to include the market share test as an explicit condition to the availability of the Technology Licensing Regulation will reduce the concerns of companies, one must wait and see how the Commission will actually choose to exercise its discretion under Article 7 of the Technology Licensing Regulation. Not surprisingly, Art. 7(1) of the Technology Licensing Regulation does not indicate that rule of reason principles will be applied by the Commission in deciding to exercise its discretion.

239 IP Guidelines § 4.3.

240 Ibid. at n32.

241 See Recital (11) of the Proposed Technology Licensing Regulation.

this provision was of great concern to potential licensing partners. In particular, companies argued that innovations that were truly important would be penalized, since they typically took a larger share of the market. Since such advantages also often carried the greatest level of risk and cost, the fear was that investment in innovation would be retarded, and scarce funding would be diverted into trivial, technical modifications that would not approach the market share ceiling. The approach of the Commission is consistent with its characterization of licensing arrangements as horizontal relationships, and it still remains to be seen whether the Commission uses its discretion under Article 7(1) of the Technology Licensing Regulation to withdraw the exemption in cases where the market share of the licensee rises above 40 percent during the term of the agreement.[242]

It is also interesting to compare how the United States and the EC approach the difficult questions of market identification and measurement. As to goods markets, the IP Guidelines adopted the approach of the 1992 Merger Guidelines as the starting point for defining relevant markets, measuring market shares and analyzing competitive effects.[243] The 1992 Merger Guidelines define markets in terms of demand and supply responses to a "small but significant and nontransitory" increase in the price of the relevant product or service. In turn, the Technology Licensing Regulation appears to define a market by reference to specified products and other products considered by users to be substitutable therefor in view of their characteristics, price and intended use.[244] Courts in the United States have also analyzed markets by reference to "product substitutes," as we have seen in Supreme Court decisions holding that patent holders have "no market power in any relevant sense if there are close substitutes for the patent product."[245]

While one can dissect the differences that might exist between the definitions of goods markets in the United States and the EC, it is useful to take a closer look at some other factors which the regulators have introduced into the general area. For example, in the United States, regulators have often broadened the scope of a particular goods market by recognizing that some consideration must be given not only to existing goods, but also to the effect of "potential competition."[246] In addition, the IP Guidelines recognize that there may not be just one market that is relevant in evaluating the effects of a license agreement, and that it may be necessary to consider both downstream and upstream markets as well.[247] Finally, the IP Guidelines break new ground in introducing "technology" and "innovation" markets as tools for analyzing the competitive effects of licensing arrangements.

[242] The Commission attempted to diffuse some of the criticism regarding the market share conditions in the Proposed Technology Licensing Regulation by reminding the critics that market share limitations have been incorporated into the block exemptions for research and development and specialization agreements. However, the critics responded that this was the very reason why each of those exemptions were rarely used.

[243] IP Guidelines § 3.2.1.

[244] Technology Licensing Regulation Art. 10(8). Also, the terms "competing manufacturers" or manufacturers of "competing products" mean manufacturers who sell products which, in view of their characteristics, price and intended use, are considered by users to be substitutable for the licensed products. Art. 10(16).

[245] *Jefferson Parish Hospital District No. 2* v. *Hyde*, 466 U.S. 2, 37 n7 (1984).

[246] See, e.g. IP Guidelines § 3.2 ("[L]icensing arrangements raise concerns under the antitrust laws if they are likely to affect adversely the prices, quantities, qualities, or varieties of goods and services either currently or potentially available.")

[247] IP Guidelines § 3.2.1 ("A restraint in a licensing arrangement may have competitive effects in markets for final or intermediate goods made using the intellectual property, or it may have effects upstream, in markets for goods that are used as inputs, along with the intellectual property, to the production of other goods.").

In contrast to the United States, the EC has generally been unable to definitively articulate how it might determine whether particular products are in or outside of a given market.[248] This has been particularly problematic in light of the fact that the Commission might look at the market share of a licensee at any point in time during the license agreement, given that it is not now possible for the parties to reasonably predict, at the time the license agreement is first entered into, what future markets might eventually be identified. Moreover, since the market share test would be based not on the share in the licensed territory, but on the "entire market in those products in the common market," computation of the relevant market share may be based on arbitrary decisions regarding which other areas outside of the licensed territory should be included for purposes of determining whether the limits are exceeded.[249] The problem is particularly acute in situations where the technology travels more quickly than the actual products, as is often the case.

Term of exemption While the IP Guidelines are typically not concerned with the term of the licensing arrangement, the EC is only willing to provide an exemption for a finite term. For example, licensees may only be restricted from putting the licensed product on the market in the territories licensed to other licensees in the common market for a period not exceeding five years from the date when the licensed product is first put on the market in the common market by one of the licensees,[250] and exclusivity provisions in pure know-how licensing agreements may not exceed ten years from the date when the licensed product is first put on the market within the common market by one of the licensees.[251] While the Commission is concerned that allowing restrictions to run for an indeterminate term may unduly impede the flow of goods, imposing arbitrary limitations of this type necessarily deprives the parties of the ability to structure the arrangement in a manner which makes the most the most economic sense. The problem is exacerbated when the permitted duration turns on vague definitional concepts like "necessary patents"[252] or "substantial know-how."[253]

[248] See Korah, supra note 135, at 265.

[249] Ibid.

[250] Technology Licensing Regulation Art. 1, para. 2.

[251] Ibid., para. 3.

[252] The Technology Licensing Regulation provides that the exemption for mixed licenses will apply in member states in which the licensed technology is protected by "necessary patents" for as long as the licensed product or process is protected in those member states by such patents if the duration of such protection exceeds ten years from the date when the licensed product is first put on the market in the EC by one of the licensees. Technology Licensing Regulation Art. 1, para. 4. "Necessary patents" are "patents where a license under the patent is necessary for the putting into effect of the licensed technology in so far as, in the absence of such a license, the realization of the licensed technology would not be possible or would be possible only to a lesser extent or in more difficult or costly conditions. Such patents must therefore be of technical, legal or economic interest to the licensee ..." Technology Licensing Regulation Art. 10(5).

[253] Certain obligations on the licensee (i.e. obligations to use only the licensor's trademark to distinguish the licensed product, or to limit the licensee's production of the licensed product to his own requirements) are exempted only for so long as the know-how in the licensed subject matter remains "secret and substantial." Technology Licensing Regulation Art. 1, para. 3. "Substantial" means that the know-how includes information which must be useful (i.e. can reasonably be expected at the date of conclusion of the agreement to be capable of improving the competitive position of the licensee, for example by helping him to enter a new market or giving him an advantage in competition with other manufacturers or providers of services who do not have access to the licensed secret know-how or other comparable secret know-how. Technology

Institutional differences There are some interesting procedural and institutional differences between regulatory practice in the United States and the EC. As noted above, the EC, like Japan, has what is essentially a pre-transaction review system, in that licensing arrangements must normally be vetted by the Commission for compliance with Article 85 before the parties enter into the formal agreement. In fact, the block exemptions, such as the Technology Licensing Regulation, are really intended to be guidelines for circumventing the pre-transactional review that might otherwise be required. In contrast, the United States does not normally regulate licensing arrangements before they begin, although the IP Guidelines are intended to give parties some idea of the provisions and practices which may ultimately catch the attention of regulators. The practical effect of these procedural differences is that the parties to an EC licensing agreement may be faced with substantial costs and possible delays in the event that notification is required or even advisable due to uncertainties in the application of the block exemption.

Another interesting institutional difference between the United States and the EC is the manner in which the regulatory agencies are staffed. For example, in the EC, the competition laws are administered by a permanent, elite group of career civil servants within the Commission, and the Commission historically has exercised both a regulatory and an enforcement role. As a general rule, these regulatory specialists serve largely free from political influence after their appointment. In contrast, DOJ policymakers are often political appointees, and regulators within the DOJ often have little experience in anything other than enforcement proceedings. The effect of these differences is difficult to measure. It may well be that the very strength of the Commission structure (i.e. lack of turnover and immunity from political influence) makes it unlikely that it will deviate from historical practices, thereby explaining the reluctance to fully investigate some of the analytical tools that have evolved so quickly in the United States as successive administrations have brought in experts schooled in some of the new theories of law and economics.[254]

Comparison of regulation of licensing practices

Exclusive licenses Exclusive licensing arrangements appear to be present few analytical problems for regulators and courts in the United States. As a general matter, exclusive licenses raise antitrust concerns only if the licensees themselves, or the licensor and its licensees, are actual or potential competitors in a relevant technology or goods market in the absence of the licensing arrangement.[255] In the case where the parties are not actual or potential competitors, the IP Guidelines recognize that exclusive licensing arrangements may actually promote competition in the marketplace, especially in those situations where the demand for the licensed product is uncertain and/or successful market penetration will require large promotional expenditures.[256] In contrast, while the EC has recognized that exclusivity may be necessary in order to provide the requisite incen-

cont.
Licensing Regulation Art. 10(3). Note that this definition is analogous to the "competitive advantage" condition for protection as a trade secret in the United States.

254 However, neither the DOJ's policymakers or the Commission's competition regulators are linked to broader industrial planning initiatives in the same way that competition regulators in Japan have been linked to other high-ranking trade and industry ministries.

255 IP Guidelines § 4.1.2.

256 Ibid.

tive for innovative activities, it has not yet afforded licensees the type of absolute territorial protection that can be obtained in the United States. This approach has often been criticized, and one commentator has suggested that a balance might be struck between the Commission's concerns about free movement of goods on the one hand and commercial realities on the other hand by recognizing the existence of a rebuttable presumption that absolute territorial protection infringes Article 85(1), and then allowing parties to present evidence regarding market characteristics, the level of investment, and other matters in support of an exemption.[257]

Territorial restrictions As a general rule, territorial restrictions in patent licenses have not generated a significant amount of controversy in the United States, at least to the extent that they were limited to "first sales" of the patented product. The EC, in spite of its overriding objective of insuring free movement of goods within the common market, has also tolerated territorial restrictions in licensing agreements; however, the relief that is available for such restrictions depends on several factors, including whether the restrictions apply in territories that are reserved for the licensor, the existence of parallel patents, the nature of the sales activities of the licensee, the duration of the restrictions, and the point when commercial use of the patented product commences. One of the most problematical aspects of EC regulatory practice is that territorial restrictions are only exempted under the Technology Licensing Regulation for a period not exceeding five years from the date when the product is first put on the market within the common market by one of the licensees. This arbitrary period may cause significant problems in cases where the licensee does not have a sufficient amount of time to recover its investment before other licensees begin selling into the territory.

Field of use restrictions In general, both the United States and the EC have allowed patent owners to impose various "field of use" restrictions in any patent license which they might grant. In the United States, such arrangements have been upheld as reasonably within the reward which the patentee by the grant of the patent is entitled to secure. In the EC, the general attitude is that field of use restrictions don't infringe Article 85(1), and the Technology Licensing Regulation permits the use of an obligation on the licensee to restrict his exploitation of the licensed technology to one or more technical fields of application covered by the licensed technology or to one or more product markets. Moreover, in a significant change from prior practice, the Technology Licensing Regulation appears to permit the use of customer field of use restrictions provided that the parties are not competitors before the license agreement commences.

Non-competition clauses In the United States, exclusive dealing arrangements will generally be analyzed under the rule of reason and will take into account a number of factors, including market share, duration, ease of entry to the relevant market, and the business justifications for the arrangement. However, the EC has typically been hostile to noncompetition covenants. Although it has been argued that such prohibitions can often provide the requisite incentive for licensees to exploit the licensed prod-

[257] See Fox, "Maize Seed: a Comparative Comment," in B. Hawk (ed.), [1982] Fordham Corporate Law Institute 151.

ucts,[258] the Commission's attitude has generally been that such restrictions impede technical and economic progress. Cases have held that noncompetition clauses under which a licensee agrees not to acquire, manufacture, use or sell competing goods, generally will not be permitted, nor will restrictions on the ability of a licensee to carry out any independent research and development.

Tie-ins In recent years, the DOJ has taken a more permissive attitude toward tie-ins, noting that they can facilitate risk sharing with respect to new technologies and allow the patentee to meter demand among licensees in order to better derive its true market value, and the IP Guidelines recognize that, under some circumstances, tying can actually be efficiency-enhancing.[259] The EC had historically limited the use of tie-ins to those situations where the goods or services at issue were necessary for technically satisfactory exploitation of the licensed technology or for ensuring the licensee's products conformed to specified standards of quality, and all other forms of tie-ins were blacklisted in the previous block exemption for patent licenses (the "Patent Licensing Regulation"). However, tie-ins were removed from the blacklist as part of the new Technology Licensing Regulation, although provisions which fall outside of those described above will most certainly need to be notified to the Commission under the opposition procedure in the new Regulation.[260]

Grant-backs The Supreme Court and the lower courts in the United States now apply rule of reason analysis in evaluating the effects of grant-back provisions. Among the factors which might be taken into consideration are the scope of the rights which are to be given to the licensor (i.e. exclusive or non-exclusive), the scope of the technology covered by the grant-back provision (i.e. improvements to the original licensed technology versus rights in all of the licensee technology in the relevant field), the duration of the grant-back obligations, and whether or not the parties are competitors and, if so, their relative market power. In contrast, the EC permits the use of grant-back provisions only in relatively narrow situations. The basic rule is that the parties may agree that the licensee will be obligated to grant the licensor a license in respect of the licensee's own improvements to, or new applications of, the technology that is the subject of the license;[261] however, such a provision may only be included if, in the case of severable improvements, the license is non-exclusive, and the licensor must also agree togrant a license of his own improvements to the licensee.[262] On the other hand, the Technology Licensing Regulation does not permit obligations on the licensee to assign in whole or in part to the licensor the rights to improvements to or new applications of the licensed technology.[263]

Cross-licensing and pooling arrangements The IP Guidelines include explicit recognition of the fact that cross-licensing and pooling arrangements may provide procompetitive benefits, such as integration of complimentary technologies, reduction of various transaction costs, and avoidance of costly litigation relating to alleged patent

[258] See, e.g. Korah [1979] JBL 291.

[259] See, e.g. *Jerrold Electronics Corp.* v. *Westcoast Broadcasting Co.*, 341 F.2d 653 (9th Cir.), *cert. denied*, 382 U.S. 817 (1965).

[260] Technology Licensing Regulation Art. 4, para. 2(a).

[261] Technology Licensing Regulation Art. 2, para. 4.

[262] Ibid.

[263] Technology Licensing Regulation Art. 3, para. 6.

infringement.[264] Accordingly, such arrangements will be permitted in the absence of a showing that they are actually mechanisms for naked price fixing or market division, that the participants collectively possess market power and by excluding other firms are preventing the excluded firms from effectively competing in the relevant market, or that the pooling arrangement deters or discourages participants from engaging in research and development.[265] On the other hand, the Technology Licensing Regulation only applies to agreements between two parties,[266] and does not apply to agreements between members of a patent or know-how pool for the pooled technologies.[267] Accordingly, many traditional pooling arrangements in the EC may need to be restructured in order to be covered by the terms of the new Regulation.

Related areas for further comparative study Adequate comparison of the regulatory attitudes of the United States and the EC toward patents, competition, and innovation should not be limited to antitrust and competition law policies. There are a number of other related areas suggested by the review of the economic aspects of patent rights which should also be examined. Among the more important are the following:

(i) Are the material differences in the scope of the patent rights available in each region? For example, are the differences with respect to the eligible subject matter for patents, the breadth of patent claims, and/or the availability of compulsory licenses?
(ii) Are the discernable differences in the objectives (e.g. invention, innovation) of the patent system in the two regions? While some differences may surface in the decisions made by the patent granting authority, it is more likely that a coherent policy will emerge from court decisions.
(iii) What role does the state play in supporting and financing innovative activities in the two regions, especially basic research and development?
(iv) Do the differences in the level of integration between the United States and the EC lead to significant differences in the level of allocative inefficiency or "monopolist's costs" (e.g. variations in labor costs between national markets in the EC)?
(v) Are there significant differences between the regions as to judicial attitudes and expertise with respect to interpretation of patent claims and/or enforcement of patent rights?
(vi) Are there differences in the industrial mix between the regions which might impact the development of antitrust and competition law policies relating to innovation? For example, is one region more dependent on products from an industry in which innovation or technological progress proceeds at a discernably different pace?
(vii) What are the preferred strategies for private financing of innovation in the two regions? For example, is the venture capital and technology-based securities market in the United States more sophisticated than in the EC?
(viii) Can we determine the relative preferences in each region with respect to the "tradeoff" between static and dynamic efficiencies?

[264] IP Guidelines § 5.5.
[265] Ibid.
[266] Technology Licensing Regulation Art. 1, para. 1.
[267] Technology Licensing Regulation Art. 5, para. 1(1).

(ix) Are there discernable differences between the regulatory attitudes in each region with respect to collaborative relationships and joint ventures in high technology industries?

H. Innovation-Based Competition and Regulatory Policy: Other Issues and Considerations

The most striking development in the environment for antitrust and competition policy in recent years has been the rapid move toward global and innovation-based competition. Innovation-driven industries have become leaders in the creation of new products, markets and jobs around the world, and the push of technological progress and development in those industries has challenged many of the traditional tools of analysis commonly applied by regulators and jurists in the antitrust/competition law arena. Some of the key issues in this area have already been discussed, including the tradeoffs between allocative and dynamic economic efficiencies in innovation-driven industries, the problems in defining markets in industries where a good deal of the new product competition is technology-based, and the special requirements of small firms. There are, however, other issues that may warrant further study.

First, there is a need to carefully consider the impact of changes in the nature of competition caused by the increase in innovation-driven product development. For example, it may well be important to consider how innovation has led to changes in industry structure, and the relative importance of innovation in relation to other environmental factors, such as deregulation, loss of government funds and subsidies, global competition, and cost-containment pressures. In addition, changes in industry and market structures demand a reassessment of market definition, as we have already seen, as well as refinement of innovation-related procompetitive (e.g. efficiencies) and anticompetitive effects of various collaborative arrangements. A further complication is the fact that the effects of innovation will likely vary among industries, as will the strategic behaviors of firms in a given industry, raising the question of whether it is prudent, if not practical, to develop industry-specific competition rules.

Second, the studies of innovation highlight the need for further analysis of the efficiencies which might be present in any proposed collaborative arrangement. Consideration should be given to identifying the relevant efficiencies (e.g. production, distribution, managerial, innovation, cost of capital, etc.) and formulating methods for their measurement. Also, decisions must be made as to when other antitrust concerns (e.g. concentration) will outweigh efficiencies, and should there by some level of market power which cannot be exceeded regardless of efficiencies. In addition, certain effects of attaining efficiencies might be assessed, such as whether the efficiencies are passed on to the consumer, whether the efficiencies occur domestically or abroad, or will the efficiencies lead to reductions in force. Finally, if efficiencies are to be given still greater weight in antitrust and competition law analysis, does that warrant any changes in current procedures for reviewing collaborations (e.g. standard of proof or "conditional" clearances by regulators and courts pending demonstration of efficiencies)?

One final area which merits further study both in the United State and in the EC is the continuing role of subsidies and prizes from the government for inventive activities, as well as how state-funded basic research can be effectively transferred to the private sector for use in commercial activities. While the evidence seems to be clear that government subsidies and prizes are, at best, inadequate policy tools, particularly in an era of diminishing public financial resources, it remains to be seen how the United States and EC will be able effectively integrate the disparate innovation assets (e.g. basic research capabilities, clinical investigation, scientific education and training, and engineering) now contained in government-operated and -funded research facilities, universities, hospitals, foundations, and the private sector. As progress is made in this area, one can expect further changes in antitrust/competition law policies, as well as in intellectual property rights.

IV. Alternative Means for Protecting Exclusivity

A. Introduction

While the patent system has a long history, and some have raised interesting arguments that this longevity alone may be sufficient justification for retaining the system, some benefits might well be gained by seriously examining various alternative means for establishing and protecting exclusivity in order to incentivize investments in new products and ideas. The study of the foundations and justifications for use of a patent system can be used as a means for identifying the variables which must be accounted for in devising such a system: the type of activities to be covered by the exclusive rights, the conditions for the grant of exclusivity, the method for measuring the period of exclusivity, the conditions which might be imposed in order for the holder to continue to exercise the rights during the period of exclusivity, and the manner in which the exclusive rights will be enforced against rivals.

B. Innovation Warrants and Patents

One interesting alternative has been advanced by William Kingston, who joins advocates of the incentive-to-innovate and prospect theories in arguing that legal protection with respect to new products should focus not on the initial inventive process, but rather upon encouraging investment in the innovation required after the initial research and development work.[268] Specifically, he calls for the creation of "innovation warrants," which would protect investments in commercializing products with a new feature, or combination of features, which cannot actually be obtained in the ordinary course of trade for the type of goods in question at the time that the warrant is issued.[269] The holder of an innovation warrant would have exclusive rights to exploit the product until the profit realized from sales of the product reached a given "multiple" of the holder's original

[268] See Kingston, "Innovation Warrants," in Phillips, supra note 39, at 213 ("rewarding the inventor is comparatively unimportant, because the human urge to keep on finding new things is apparently inextinguishable. The person who needs to be rewarded is the investor"). See also Kingston, *The Political Economy of Innovation* (1984), at 68–80. Kingston's proposals are similar to the "innovation patent" first advocated by Dr Hermann Kronz in *Mitteilungen der Deutschen Patentanwalte*, Munich, 3 (1976) and 7/8 (1983). Kronz's description of the innovation patent can also be found in "Patent Protection for Innovations: A Model," EIPR (1983), 178 and 206.

[269] Kingston, in Phillips supra note 269, at 216. Eligibility for protection would be based strictly upon commercial criteria, rather than novelty or non-obviousness, as is the case with the current patent system. Ibid.

investment.[270] In Kingston's view, it makes better sense to focus upon money as the means for monitoring the period of exclusivity, since money is the scarce resource being risked in the innovation process.[271]

While the innovation warrant has its roots in the arguments made in favor of the incentive-to-innovate theory, it is based on the notion that the current scope of the patent system is hardly adequate to provide the desired result. Recognition of innovation warrants would require that the focus of protection shift from the initial invention to the entire bundle of technology, including information which currently is protected under trade secret law. Innovation warrants would facilitate the disclosure of know-how and trade secrets, thereby substantially enhancing the information value of the "protection" system. Kingston also argues that a new system of innovation warrants would be more effective than any of the other government programs which have been adopted over the years as a means for increasing investment in innovation.[272] As such, it would no longer be necessary for the government to attempt to intervene in the marketplace with rewards, prizes, tax benefits or funding schemes.[273]

The concept of an innovation warrant or patent has certainly generated a good deal of criticism. For example, critics express concerns regarding the administration of such a review program. They note the difficulties which may arise in attempting to review and process information regarding proposed development programs for thousands of inventions in a given year, a good number of which will probably overlap to some extent.[274] In addition, it is likely that such a system will be accompanied by calls from entrepreneurs for modifications during the term of the exclusivity owing to unforeseen changes in the environment in which business and competitive decisions are made and executed. For some, linking the patent right to financing and development of the underlying invention is similar to the "sale of monopolies" which even the Tudors and Stuarts would not tolerate.[275] Finally, it has been observed that terminating exclusivity after the innovator has been able to earn a specified return makes little sense since it is only when an "abnormal" return is anticipated will the innovator be willing to endure the risks and uncertainties of development.[276]

[270] Ibid. at 215. Kingston would use of a series of multiples to take account of the different amounts of risk that might be involved in different types of innovation. For example, "basic research" has the highest risk of failure and the longest time to produce returns, together with the greatest difficulty in being protected; therefore, a large multiple would be used to cope with both the time and the risk elements. On the other hand, originative, or applied, inventive research would have a lower multiple than basic research work, as would small improvements with what is referred to as "intensive" invention. Ibid. at 215–16.

[271] Ibid. at 213. Kingston points out that there are now appropriate and accurate measures of "money" which can be used to monitor revenues and profits and competitors will also assist in this process. Ibid. at 215. Kingston draws analogies from the protection of "investment" provided by the innovation warrant and the protection of other forms of investment available under the law of property. See Kingston, in *Political Economy* supra note 269, at 74.

[272] Ibid. at 80.

[273] Attempts by the government to provide incentives for innovation might include direct funding of research and development programs, tax benefits tied to research and development activities, and rewards or prizes to firms involved in innovative product development in specified areas. See above for a discussion of the efficacy of some of the government programs.

[274] Aubrey, A Justification of the Patent System, in Phillips, supra note 39, at 4.

[275] Ibid.

[276] O'Brien, supra note 47, at 39.

C. Patent Protection for Results of Applied Research

Another alternative to the current patent system is expanding the scope of protection to include the results of applied research on new inventions with regard to their industrial applicability in so far as its scope is foreseeable over a specified period of time, such as three to five years.[277] This scheme recognizes the crucial importance of further development to effectively commercialize any new invention; however, it does not extend patent protection beyond the period necessary to test and refine the product. The "final" set of patents would cover the specific products and applications developed during the research and development period. Inventors and their assignees would have an incentive to undertake lengthy research programs and examine their results at an early stage.[278] In turn, it is anticipated that the level of disclosure generated by the patent system would improve, since the information included in the specifications would cover an earlier stage of development than do present day patents.[279]

D. Patent System Incentives for Cooperative Research

A variety of other ideas have been put forward regarding the structure of a system which might be used to encourage invention and investment in innovation. One economist has argued that it may be possible to diffuse the waste which might be caused by rivalries among inventive firms by granting patents on more favorable terms to licensed, semi-public, non-monopolistic trade associations, than to non-cooperating patentees.[280] Advocates of this type of system claim that it would make unlimited funds available for research, optimize patent pooling, eliminate secrecy, invention-suppression and duplication of research; and encourage small firms. However, critics argue that the sharing of information would not create the monopoly conditions required for innovation, which appears to call for some form of exclusivity. Also, it is unlikely that such a structure would allow for the selection of a "champion" to exercise the requisite degree of control over the innovation process.[281]

E. Assistance for Individual Inventors and Small Firms

Some commentators have suggested programs designed to assist specified groups of inventors in getting their inventions into the marketplace, thereby allowing them to receive the required return on their investment. Fuscone, noting that the role of innovation is being accorded greater importance in national economies, has put forth a menu of proposals designed to assist individual inventors and small businesses in exploiting new patented inventions.[282] Buckley, concerned about the fact that patents are not prod-

[277] See Eisenschitz, "The Value of Patent Information," in Phillips, supra note 39, at 51 and 52.

[278] Ibid.

[279] Ibid.

[280] See, e.g. Gilfillan, *Invention and the Patent System* (1964).

[281] Kingston, in Phillips supra note 269, at 209–10.

[282] Fuscone, "The Exploitation of Individual Invention," in Phillips, supra note 39, at 81–7. Among his suggestions are: establishment of a national office to coordinate relevant services to inventors and to small businesses; enhance facilities available to inventors to permit them to make or obtain working models/ prototypes of their inventions which can be made available to other firms interested in exploiting the invention; development of a system of "innovation exploitation brokers" to facilitate private exploitation of

ucts and that patents do not protect product concepts, has developed an "inventor's contract," which would encourage innovation by allowing inventors to approach potential manufacturers or investors without incurring the substantial expense and delays typically associated with attempting to have large firms review and assess new product proposals.[283] While neither of these proposals amount to an effective change in the protection directly provided by the patent system, they do shift the emphasis from invention to innovation and concentrate on reducing the transaction costs of commercialization which are so overwhelming to individual inventors and small firms.

It is important to note that some commentators has been critical of focusing upon the size of the firm involved in the inventive activity as opposed to other factors, such as the field of business or the characteristics of the technology involved.[284] As noted above, technical advance proceeds at different rates in various industries and empirical studies have shown that patents and other forms of intellectual property rights play a different role as one moves from pharmaceuticals to consumer electronics. Therefore, these commentators caution against making changes that are intended to assist smaller firms without taking into account the effect that such changes might have in industries which have very different technological characteristics.[285]

F. Industry-Specific Schemes for Creating Exclusivity

One of the most striking conclusions of the many empirical studies which have been conducted regarding patents and research and development activities is that there may well be substantial variations between industries which might be relevant in attempting to make appropriate public policy decisions. With respect to patents, Merges and Nelson have forcefully argued that invention and advancement of technology varies between industries and that defining the proper scope of patents in any particular industry depends largely upon two characteristics: the relationship between technical advances in the industry and the extent to which the firms in the industry license technologies to one another.[286]

Most of the analyses of the effects of the patent system on invention and innovation assume, at least implicitly, that the rate of technical advance is similar in every industry. However, Nelson has argued that this assumption is mistaken and that, in fact, one can identify at least four different generic models of technical advance.[287] The first is called "discrete invention." The second model applies to industries in which technologies are "cumulative." Chemical technologies exhibit their own unique characteristics. Finally, spe-

cont.
inventions; lengthening the period of "commercialization"; and development of standard license and option contracts which could be used as a means to simplify negotiations between inventors and potential exploiters. Ibid. at 86 and 87.

283 Buckley, "A Creative Alternative to the Private Inventor's Patent," in Phillips, supra note 39, at 88–99. Inventor's contracts would focus upon the potential exploitation of ideas and would operate separately from patents. Inventor's contracts would be a combined licensing and confidentiality agreement, would have standard terms and be used universally, and both the inventor and the manufacturer would agree to the terms before the inventor makes any disclosures to the manufacturer. Ibid. at 94–8.

284 See, e.g. Staines, "Some Thoughts on the 1983 Green Paper," in Phillips, supra note 39, at 101–3.

285 Ibid. at 106–8.

286 See Merges and Nelson, "On the Complex Economics of Patent Scope," *Columbia Law Review* 90 (1990), 839, 843.

287 Ibid. at 880.

cial consideration must be given to "science-based" technologies where the rate of technological advance is driven, or at least materially affected, by developments which occur outside of the industry, such as basic research discoveries in university laboratories.

As noted elsewhere, Merges and Nelson are have a particular interest in analyzing the proper scope of the patent grant. In that regard, they argue that each model of technical advance has its own special requirements as to the optimal policy in setting the breadth of allowable patent claims.[288] Moreover, they note that the analysis becomes even more complicated by the fact that one or more of the models might apply to an industry at a given point in time and appropriate characterization of an industry may, in many cases, reflect a distinctive mix of the various models.[289] Therefore, it may well be that the orthodox criterion used as a means for awarding patent protection should be supplemented by some further measure of technical advance determined by reference to the competitive characteristics of relevant industries.

Industry-specific schemes for granting exclusivity might also be suggested by the empirical studies showing that patents are of varying importance in different industries. The pharmaceutical industry is an area where patents have proven to be both necessary for research and development work and a source of controversy regarding undue monopolistic effects. The United States has experimented with the mechanism for exclusivity in the pharmaceutical industry by enacting the Orphan Drug Act, which was intended to encouraging investment in new pharmaceuticals which allegedly do not have a large enough market to justify the investment of funds necessary to bring the drug to market.[290] The Orphan Drug Act provides firms with a limited form of "market-based" exclusivity to in order to allow them to exploit the "orphan drugs." However, the Act has been subjected to extensive criticism, particularly with regard to determining the size of the market for purposes of determining orphan drug status and the prices at which firms often sell the drugs which are covered by the scheme.[291]

G. New Forms of Intellectual Property Rights

New technologies continue to present difficult challenges and problems for the existing system of intellectual property rights. It is already evident that patent protection alone may not be sufficient for an inventor or innovator to appropriate the maximum value of a new product or process. In fact, unless the patented invention is one of those relatively rare occurrences in which it leads to the creation of an entirely new market, it is likely that competition based on intellectual property rights will soon focus on a number of factors which fall outside the patent system: know-how, both technical and business, which might be protected under trade secret laws; design elements and other forms of expression, which fall under the umbrella of the copyright laws; and symbols and marks created for marketing purposes, which are protectable as a matter of trademark law.

[288] Ibid. at 880.

[289] Ibid. at 880.

[290] Orphan Drug Act, Pub. L. No. 97-414, 96 Stat. 2049 (1983) (codified as amended at 21 U.S.C. §§ 360aa–360ee (1988). The legislative history of the Orphan Act is thoroughly traced in Richardson, "The Orphan Drug Tax Credit: An Inadequate Response to an Ill-Defined Problem," *American Journal of Tax Policy* 6 (1987), 135, 138–68.

[291] See, e.g. Rin-Laures and Janofsky, "Recent Developments Concerning the Orphan Drug Act," *Harvard Journal of Law and Technology* 4 (1991), 269; and Thomas, "Re-Assessing the Orphan Drug Act," *Columbia Journal of Law and Socil Problems* 23 (1990), 413.

It would appear likely that new forms of intellectual property rights may soon appear in response to concerns from innovators as to how to protect their substantial investments in such areas as biotechnology, information technology, and software. These changes may take the form of modifications in the current standards used to secure intellectual property rights, "hybrid" rights which include elements of current patent, copyright and trade secret law to protect all of the elements of complex technology-based products, or the extension of current forms of intellectual protection into new areas (e.g. software patents). As these rights begin to develop, one might actually see a shift in emphasis away from rewarding the initial creative step, as is the case with the current patent system, toward recognition of a network of rights covering a more advanced stage of concept or product development.

H. Certification Procedure for Antitrust Exemption

The exclusionary rights recognized under the intellectual property laws are often not sufficient to provide the incentives required to fully commercialize and exploit the subject matter of the right. In many cases, true innovation is only possible when the inventor or developer is able to share the risks involved in innovative activities by entering into various collaborations with other firms. However, such collaborations often run the risk of challenge under applicable antitrust and competition laws, since they may have the effect of reducing competition in the markets in which the collaboration is taking place. In order for business collaborations to play a proper role in facilitating innovation, there may well be a need for some form of exclusivity that is analogous to that which is recognized in the intellectual property arena which allows innovative business collaborations to proceed for the time necessary to achieve the returns required to induce new investment, even if the near-term effects of the collaboration place it outside of acceptable norms of allocative economic efficiencies.

One of the most interesting areas of recent debate in the United States has been the notion of establishing a regulatory procedure that would allow firms that are involved in a joint venture or other form of collaborative arrangement that meets certain characteristics to apply for a certification by a governmental agency which would provide the arrangement with substantial protections from possible antitrust liability.[292] The premise for these proposal is that firms may be inhibited from entering into collaborative arrangements due to fears that the relationship will ultimately be declared illegal under the antitrust laws, even though the arrangement holds the promise of substantial long-term economic benefits. Interest in such a certification procedure has been particularly acute with respect to potential alliances between firms with relatively high market shares, since such arrangements would normally be most vulnerable to findings of *per se* illegality.

Teece and Jorde have suggested that an administrative procedure be established which would permit evaluation and certification of those cooperative arrangements among firms with higher market shares in situations where the dynamic efficiency gains are most likely

[292] See Jorde and Teece, supra note 158, at 903. Jorde and Teece's ideas regarding certification were included in House Bill 1024 (Boucher-Campbell), 101st Cong., 1st Sess., 1989, which would have provided a federal procedure for advance approval of innovation collaboration in research and production (but not marketing), and would also institute a registration system similar to other proposals. Brodley, supra note 62, at 917 n4.

and the level of rivalry is most robust.[293] In order to receive this type of protection, the cooperating firms would be required to make substantial disclosures and subject their business plans to a high degree of scrutiny.[294] Antitrust immunity, which would be available for up to seventeen years,[295] would be granted prior to commencement of the joint venture's operations.[296] The certification approach has been included in new bills which have been introduced in the Congress over the last few years;[297] however, it has not yet gained sufficient support outside of the limited area of proposed export activities which fall under the Export Company Trading Act.[298]

Opponents of the certification procedures have made a variety of arguments. For example, the point out that while the advance approval granted under the certification procedure does, in fact, provide the participants with assurances and presumably facilitates the venture through the reduction of legal risks, it also reduces the ability of the regulating agency to take corrective measures as the actual effects of the joint venture become apparent during future years.[299] They note that while it is true that advance rulings are made by regulators in the case of mergers, the criteria for reviewing a merger rest primarily on a variety of structural factors, like market share and barriers to entry, which are suitable for advance determination, while the analysis in the joint venture context, which generally requires a complex set of tradeoffs, is far more difficult and uncertain.[300] Also, with respect to mergers, advance decisions are necessary, as it is usually not possible to secure effective judicial relief after the transaction is completed. However, the same problem does not exist with respect to judicial intervention with joint ventures, since the participants retain their separate business identities.[301]

Criticisms have been made of the fact that information submitted to the regulatory agencies would not be subject to public disclosure, thereby impeding the ability of outside firms which might be adversely affected by the proposed collaboration from effectively challenging the proposal.[302] Brodley also is uncomfortable with the standard of judicial review with respect to decisions made by the regulatory agency. He notes that while there is a provision which calls for review of an agency's decision to certify a proposed collaboration, he is critical of the fact that the court would be required to uphold the findings of the agency if it was supported by substantial evidence, even if the court finds the opposing evidence to be stronger.[303] Brodley considers some of the aforementioned institutional differences between the United States and the EC to be quite

[293] See, Jorde and Teece, supra note 158, at 903–4. Under Teece and Jorde's proposal, certification would be available to joint ventures even in situations where the participants collectively have a sufficient market share to dominate the market. Jorde and Teece, "Innovation, Cooperation and Antitrust: Balancing Competition and Cooperation," *High Technology Law Journal* 4 (1989), 1–113.

[294] See, Jorde and Teece, supra note 158, at 904. Firms could choose between the registration procedures, which provides relief from the possibility of treble damages, and the new certification procedure. Ibid. at 903–4.

[295] See Brodley, supra note 62, at 922 (citing Jorde and Teece, supra note 158.

[296] Ibid.

[297] See Jorde and Teece, supra note 158 (citing legislation introduced in 1989 by Congressmen Boucher and Campbell (H.R. 1024)).

[298] See 15 U.S.C. 4301.

[299] See Brodley, supra note 62, at 922.

[300] Ibid.

[301] Ibid.

[302] Ibid.

[303] Ibid.

significant, and doubts if United States regulators, whose experience is generally limited to enforcement, can develop the skills required to administer a certification system.[304]

Shapiro and Willig have also criticized the use of a certification procedure.[305] They point out that any grant of immunity from antitrust liability would only be prudent if the government is willing, and able, to continue to monitor each of the activities of the joint venture on an ongoing basis in order to be sure that the parties are staying within the bounds of the original collaboration plan and have not introduced any further restrictions on their own activities or the activities of the joint venture. In addition, since the antitrust immunity would deprive other private parties from any incentive to monitor and review the activities of the joint venture, it is the government that would constantly be in the position of renegotiating the terms upon which the exemption was granted as the activities of the joint venture expand, change and evolve over time.[306] It is unlikely that the United States would be able, under existing circumstances, to implement a procedure for the ongoing review of the economic effects of cooperative arrangements after they are in place,[307] given that regulatory personnel are subject to rapid personnel turnover, and are more susceptible to the vagaries of political changes than their counterparts in Japan and the EC.[308]

While proposals such as that advanced by Teece and Jorde have encountered a good deal of criticism, even its opponents concede that the importance of innovation to continued economic progress and growth mitigates in favor of encouraging, rather than simply disallowing, cooperative production experiments involving two or more parties which collectively have large market shares.[309] One alternative suggestion would create a legal framework that would permit experimental production collaboration under carefully controlled conditions, with the ultimate goal of testing whether the collaborating group became more innovative or efficient than non-collaborating firms within the industry or in comparable firms outside the industry.[310]

Under this alternative scheme, experimental joint production collaborations would not be given complete antitrust immunity, rather they would be covered by the registration procedures of the NCRPA, meaning that the collaborators would still be subject to single damages if the collaboration is eventually found to reduce competition without an increase in efficiency.[311] Second, in order to insure that the production activities are related to the larger innovative process, protection would only be available for the production of products and processes which either arise out of or involve some prior collaboration at the research and development stage.[312] Third, protection would be conditioned upon a specific administrative finding by an appropriate regulatory body that the collaboration: (i) relates directly to prior research and development activities; (ii) involves the production or commercialization of a new product or process of production; (iii) includes substantial participation by United States firms; and (iv) appears

[304] Ibid.

[305] Shapiro and Willig, "On The Antitrust Treatment of Production Joint Ventures," in *Collaboration*, supra note 62, 920, 944.

[306] Ibid. (also noting that private parties would lack the information necessary to determine whether the joint venture's activities fell within the scope of the original exemption).

[307] See Brodley, supra note 62, at 922.

[308] Ibid.

[309] Ibid. at 924.

[310] Ibid. See also Brodley, "The Economic Goals of Antitrust: Efficiency, Consumer Welfare and Technological Progress," *New York University Law Review* 62 (1987), 1020–53.

[311] Ibid. at 925.

[312] Ibid.

likely to lead to significant efficiencies in a field of competitive importance to the United States.[313]

The alternative scheme would also create a mechanism for a post-registration examination of the actual conduct and economic performance of the joint venture after some specified period of operation, such as five years.[314] The actual experience of the joint venture would be examined in order to determine whether the anticipated benefits forecast at the outset of the collaboration have been achieved and how long the joint venture remains necessary in order for those benefits to continue to be achieved. In addition, the enforcement agency would attempt to determine whether there have been competitive losses which outweigh the actual benefits from the collaborative venture.[315] If, as a result of this examination and review, the regulatory agency discovers a problem with the arrangement, the agency may withdraw or modify its earlier findings. If the agency's prior findings are withdrawn, the venture would remain protected for past activities, but would not be protected for any future activities.[316] If the agency's prior findings are modified, subsequent protection would only extend to conduct which fell within the scope of the modified findings.[317]

The participants in permitted joint production collaborations would also be required to prepare and file periodic reports with the regulatory agency in order to monitor their activities and to build a bank of research information which can be used to evaluate the efficiencies and possible detrimental effects of the collaborations.[318] The information would serve as the basis for a report which would be issued after some reasonable period of time, such as ten years, by the enforcement agency or some other appropriate government department regarding the results of the experiment. Based on the results described in the report, the issuing agency or department would make some sort of recommendation regarding termination or extension of the experiment.[319]

V. Lessons for Developing Economies

A. Introduction

Most of the conclusions described above, as well the various areas suggested for further study, focus on issues which are most relevant to developed countries such as the United States and the EC. However, the topics that have been discussed in the course of this work may well have some broader application with respect to the evolving legal and economic environment which exists in that larger part of the world often referred to as the emerging or developing areas. In this section, we offer some thoughts on three

[313] Ibid. The findings of the administrative agency would not only allow the collaboration to be registered under the NCRPA, it would also be admissible as evidence in any future antitrust proceeding as to the joint venture. Ibid.

[314] Ibid. As to EC and German antitrust law practices, see Hawke, *United States, Common Market and International Antitrust: A Comparative Guide*, 2, New York: Harcourt Brace (1986) and Brodley, supra note 309.

[315] See Brodley, supra note 62, at 925. While the determination of costs and benefits is extremely difficult, Brodley believes that it becomes at least more feasible after the joint venture has been in operation for some period of time. Ibid.

[316] Ibid.

[317] Ibid. at 926.

[318] Ibid.

[319] Ibid.

important areas: the future direction of intellectual property laws around the world, the relevance of competition and technology transfer laws, and growing phenemonen of regional economic and political groupings and associations.

B. Intellectual Property Laws

Background As discussed at the beginning of this work, intellectual property laws are national laws which extend only to the borders of a single country. This means that an innovator will not be able to exercise exclusive rights with regard to an invention or other item eligible for intellectual property protection in more than one national market unless the laws of other countries allow the innovator to apply for and receive some form of exclusivity in that country. If the conditions for protection, and/or the terms of protection, differ from country-to-country, then there is some danger that the free flow of technology around the world may be impaired, since variations in intellectual property rights will create de facto barriers to trade.

In the second chapter we explored some of the possible economic justifications for creating and recognizing intellectual property rights, particularly patent rights. While it is clear that the discussion was relevant for countries and regions, such as the United States and the EC, which have both the financial and technical resources to consider investments in innovation, developing countries face a series of different problems in this area. Primarily, they tend to lack the entrepreneurial base that is necessary for indigenous inventive activities. As such, they often lack the incentive to create private legal rights in new technologies, particularly if it means expending the resources of the state in enforcing rights which will generally be granted to foreigners, rather than local firms. In fact, it is often in their best interests to permit, or require, rapid diffusion of new technologies throughout the local economy. The problem with this, however, is that technology creaters and owners in foreign markets will likely be unwilling to share their technologies without some form of compensation. This dilemma generally lies at the heart of much of the current debate about international intellectual property laws.

Harmonizing international intellectual property laws In the world today, intellectual property laws are far from uniform. There are several international conventions, treaties, and agreements relating to intellectual property rights.[320] In general, these treaties attempt to harmonize the laws of the various member states and reconcile procedural differences among the many national legal systems around the world. For example, the International Convention for the Protection of Industrial Property, originally enacted in Paris in 1883 (the "Paris Convention"),[321] is a comprehensive treaty administered by the World Intellectual Property Organization ("WIPO"), a specialized agency of the United Nations. More than 100 countries are currently members of the Paris Convention, which provides for equal treatment toward nationals among member states with respect to patent

[320] There are also a variety of regional conventions and treaties that have been structured to harmonize the various practice and filing procedures which must be adhered to in situations where a person or firm is seeking simultaneous statutory protection in more than one country. For example, see Appendix A for a description of the European Patent Convention (the "EPC"), which came into force on October 7, 1977 and is designed to assist persons wishing to acquire patent protection in more than one member state of the EC.

[321] 25 Stat. 1372, T.S. No. 379; revised in 1890, 32 Stat. 1936, T.S. No. 411; revised in 1911, 38 Stat. 1645, T.S. No. 579; revised in 1925, 47 Stat. 1789, T.S. No. 834; revised in 1934, 53 Stat. 1748, T.S. No. 941; revised in 1958, 13 U.S.T. 1 T.I.A.S. No. 4391; revised in 1967, 212 U.S.T. 1583, T.I.A.S. No. 6923.

rights, and establishes procedures to ease the process of filing multiple patent applications in different countries.[322] In the copyright area, the Berne Convention for the Protection of Literary and Artistic Works (the "Berne Convention"), which was established in 1886 and is also administered by the WIPO, ensures that all authors of works published in member states are treated in the same way as nationals would be under the national laws of the member state with respect to copyright protection.[323] The Madrid Agreement Concerning the International Registration of Marks, which is a special union within the Paris Convention, was formed to facilitate establishment of a centralized system of international registration and protection of marks.[324]

While the various treaties serve valuable practical purposes, particularly for multinational firms interested in easing the burdens of obtaining some form of protection in a number of markets, the treaties do not unify the national laws of the member states. For example, the equal treatment provisions included in the Paris Convention do not guarantee harmonization of all the substantive provisions of the various national laws, and many countries believe that true harmonization can occur only after a comprehensive multilateral treaty covering the protection and patentability of inventions is completed.

One of the most comprehensive efforts to harmonize international patent law was undertaken by the WIPO Committee of Experts on the Harmonization of Certain Provisions in Laws for the Protection of Inventions ("WIPO Committee"). The WIPO Committee originally intended to submit a final draft of the treaty to a Diplomatic Conference for approval in 1991; however, it is now unclear when, if ever, a treaty agreement can be reached within the WIPO. For example, significant differences surfaced among the various participants as to adoption of "first-to-file" procedures, publication of pending patent applications, the scope of items which could be patented, and compulsory working requirements in the territory in which the patent has been granted.

The WIPO discussion also dramatically illustrated another difficult problem with respect to harmonization of international intellectual property laws: the fundamental schism that exists between developed and developing countries regarding the benefits and perceived dangers of property rights in technologies and related items.[325] On the one

[322] Specifically, the Paris Convention provides that any patent application in a member nation may be filed in any other member nation within twelve months of the first filing date with the same priority as if it had been filed in each member nation on its first filing date. The terms of the Paris Convention have been supplemented by the Patent Cooperative Treaty ("PCT") which came into force in 1978 amongst some of the members of the Convention, andpermits an applicant to make a single filing in one nation of an international application. See 28 U.S.T. 7645, T.I.A.S. No 8733.

[323] The Berne Convention covers "literary and artistic works [that] shall include every production in the literary, scientific, and artistic domain whatever may be the mode or form of its expression." While the Berne Convention does not purport of regulate all of the standards for copyright protection that might be imposed within a member country, it does require that members adhere to certain minimum requirements in formulating their own national copyright laws, including copyright duration of the author plus 50 years and the recognition of the exclusive rights of translation, reproduction, performance, broadcasting, adaptation and arrangement.

[324] The Madrid Agreement entered into force in 1892, and has been revised several times since then. The most notable revisions were completed at Nice in 1957 and at Stockholm in 1967. For various reasons the goals of the Madrid Agreement have not been achieved, and only thirty-three countries were members of the Madrid Agreement as of January 1, 1993.

[325] For example, the "Group of 77," a group of developing countries, introduced two new articles for discussion in late 1990. The articles "would impose an obligation on the owner of a patent to make a complete and clear disclosure, including best mode of his invention, to provide information regarding corresponding foreign applications, to work the invention in the territory of the patent, to pay periodic fees, and to refrain from anticompetitive practices in respect to licenses and assignments." William J. Brunet, Eighth Session (Second Part) of Committee of Experts on the Harmonization of Certain Provisions in Laws

hand, an inventor in a developed country will seek strong intellectual property protection to prevent firms in developing countries from "free-riding" on his work and to establish additional markets through which to recover the costs of development. On the other hand, governments in many poor and developing countries are reluctant to provide any strong degree of protection to foreign inventors and firms, since protection of this sort may work as a disincentive to local innovators to build their own research and marketing capabilities and, perhaps more importantly, allow foreign firms to exercise undue control over the availability and affordability of the protected items.

Intellectual property rights and economic development Some advocates of enhanced intellectual property right protection in developing countries base their arguments on what they perceive to be the positive relationship between economic development and a strong system of intellectual property rights. The relationship of intellectual property protection to economic development has been studied in great detail. For example, using multiple regression analysis, one researcher claimed that the level of economic development is closely correlated with the level of intellectual property protection.[326] Specifically, this researcher concluded that many developing countries that failed to implement intellectual property protection systems experienced a correspondingly lower level of economic development, as well as a slower evolution in the size and complexity of their local markets.[327] These findings suggested that further reforms to the existing intellectual property laws would be a precondition to continued growth.

Other studies have focused on a specific subset of developing countries, such as Argentina, Brazil, Indonesia, South Korea, Mexico, the Philippines, Taiwan, and Turkey.[328] These countries, because of the size of their local economies or their scientific and technological capabilities, have either already experienced rapid economic growth or can be expected to enter the front lines of the global economy in the near future.[329] Researchers have claimed that intellectual property protection in these countries is "out of phase" with their actual level of economic development. Moreover, several of these countries were found to be active participants in the patent systems of developed countries.[330] These findings suggested that their failure to implement a stronger domestic patent system flows

cont.
for the Protection of Inventions, *Patent Trademark Copyright Newsletter* 9 (Sec. of Pat. Trademark and Copyright L., ABA, Chicago, Ill.), Winter 1990, at 15.

326 Richard T. Rapp and Richard P. Rozek, "Benefits and Costs of Intellectual Property Protection in Developing Countries," *Journal of World Trade* 24 1990) (Oct. 1990), 75. The researcher developed an "index" of patent protection based upon conformity of each nation's patent laws to the minimum standards proposed in the Guidelines for Standards for the Protection and Enforcement of Patents of the US Chamber of Commerce Intellectual Property Task Force. The index ranked the level of patent pro tection on a scale of zero to five. Zero is assigned to a nation having no patent protection law, and five corresponds to nations whose laws are fully consistent with the minimum standards. Ibid. at 79. It should be noted that the scores were based solely on the laws as written and did not account for actual enforcement experience.

327 Ibid. at 81. The study determined that Indonesia, New Guinea, Kuwait, Turkey, Bulgaria, Algeria, Iraq, Poland, Yugoslavia, East Germany, Brunei, and most of Central and South America had weaker patent regimes than would be expected, given their stage of economic development. However, India, Brazil, and Argentina, despite the size and complexity of their economies, were also among the weak protection regimes. Ibid.

328 J. Davidson Frame, "National Commitment to Intellectual Property Protection: An Empirical Investigation" ,*Journal of Law and Technology* 2 (1987) 209, 212.

329 Ibid. at 215.

330 Ibid. at 216.

not from a lack of appreciation for the value of such a system, but from a reluctance to incur the associated costs or from the belief that "free-riding" is the preferred strategy.[331]

Assuming the existence of a local science and technology base, it can be argued that intellectual property rights in a developing country might well serve as the same type of incentive for innovation as presumably exists in developed countries. If, however, the utility of an incentive for indigenous innovation is problematic, as is often the case in developing countries, then there must be other reasons that developing countries might want to consider an intellectual property rights regime. Some of these alternative arguments include the following:

(i) The availability of intellectual property protection increases the flow of new products into the developing country, thereby increasing the welfare of the population.
(ii) Enhanced intellectual property rights protection will increase the rate of inbound investment and technology transfer from foreign firms.
(iii) A strong intellectual property law regime improves the knowledge base concerning technical development due to increased imports, licensing activities, and patent application filings in the local market.
(iv) If enforcement regimes are improved, welfare gains are realized through a reduction of agency costs attributable to surveillance, verification, compliance, and enforcement.

Little doubt exists that developing countries are quite anxious to reduce what they perceive as a critical technological gap between developed and developing countries. However, many of these countries fear that patent protection for new products and technologies will merely enable large multinational corporations to secure global monopolies[332] and charge exorbitant prices for their goods. Also, lacking the scientific and financial infrastructure necessary to create patent-induced innovations, developing countries are far more interested in technology transfer than in encouraging domestic innovation.[333] It is not surprising, therefore, that developing countries have initially had little or no interest in creating a system that impedes their own ability to "appropriate" new technologies and products developed by foreign innovators. Another fundamental problem is that many developing countries do not necessarily share the same cultural attitudes regarding the nature of private rights to own and use various types of tangible and intangible property.[334]

331 Rapp and Rozek, supra note 327, at 86.

332 See Dembo, Dias and Morehouse, "Technology to Aid the Poor: Constraints to Access Resulting from the Privatization: The Case of Biotechnology," in *The International Context of Rural Poverty in the Third World* (1986), 103, 124–5. Also, the Director General of the Chinese Patent Office once stated that developed nations have a greater interest in protecting their own patent rights than in the economic growth of less fortunate countries. "Asia-Pacific Countries Agree to Cooperate on Patent Issues," *World Intellectual Property Report BNA* 4 (1990), 53.

333 Robert P. Mcrges, "Battle of Lateralisrns: Intellectual Property and Trade," *B U Interntional Law Journal* 8 (1990), 239, 244 n9 (citing *Science and Technology: Lessons for Development Policy* 353 (R. Evenson and G. Ranis (eds), 1990)) (adopting current international intellectual property conventions might be globally optimal for developing countries, but not in the best interests of individual developing countries, since the best regime for them is one that facilitates the transfer of technology).

334 For example, in some countries, certain forms of intellectual property (e.g. pharmaceuticals) are viewed as pure public goods, and some cultures are genuinely hostile to any notion that knowledge is a private capital good, which is the fundamental premise to the intellectual property systems of the industrialized economies.

Multilateral and bilateral initiatives regarding intellectual property rights In spite of their fundamental discomfort with intellectual property rights, developing countries have been forced to begin adopting some rudimentary form of intellectual property law. This movement has been driven by multilateral negotiations on a variety of trade-related issues (e.g. the Uruguay Round of the negotiations under the General Agreement on Tariffs and Trade ("GATT")),[335] as well as well-publicized concerns of many developed countries regarding various inadequacies in the content and enforcement of intellectual property laws in the developing countries, including no preliminary or final injunctive relief, lack of seizure and impoundment relief, lack of exclusion of infringing imports, inadequate civil remedies (i.e. monetary damages), and inadequate fines or other criminal penalties.

Intellectual property rights was a novel topic for discussion within the GATT and, in fact, the mere inclusion of the topic proved to be a source of great discord between the participating developed and developing nations. For developed countries, GATT was viewed as preferable to the WIPO for the resolution of intellectual property issues because it offered a dispute resolution mechanism. Moreover, since the United States believed that most intellectual property disputes involved trade, it argued that the GATT was the proper body to discuss these issues.[336] For their part, the developing countries push the WIPO, where they are numerically superior to the developed countries, as the lead organization in intellectual property matters.[337]

The negotiations on trade-related aspects of intellectual property rights ("TRIPs") were devoted to developing an agreement which would supplement, rather than override, the terms of any international convention or multilateral agreement that might be reached under the auspices of the WIPO. It was hoped that the agreement would provide minimum standards for copyright and related rights trademarks, geographical indications, industrial designs, patents, protection of undisclosed information (i.e. trade secrets), and the control of anti-competitive practices in contractual licenses. In addition, the agreement would cover the enforcement of intellectual property rights, the acquisition and maintenance of intellectual property rights, and prevention and settlement of disputes.

As with other aspects of the Uruguay Round, the TRIPs negotiations proved to be long and difficult; however, agreement on TRIPs was finally reached in April 1994. Among other things, patents would be available without discrimination as to the place of invention, the field of technology, and whether the product is imported or produced in the local market. In addition, the class of inventions for which patent protection could be denied was limited, although countries could still exclude inventions involving diagnostic, therapeutic, or surgical methods for the treatment of humans or animals. A number of conditions were imposed on the grant of compulsory patent licenses, and the minimum duration of a patent was set at twenty years from filing.

[335] The Uruguay Round was described by Arthur Dunkel, the Director-General of the GATT during that period, as "an effort to reorder the basis for economic relationships [among] countries." Harry B. Ensley, "Intellectual Property Rights in the GATT," *New Matter* 15 (1990), 1.

[336] Ibid.

[337] For example, a number of the developing countries, including India, Brazil and Mexico, quickly registered their concerns about the TRIPs discussions and actively encouraged the WIPO to develop its own draft treaty on resolution of intellectual property-related disputes. Chile even submitted a proposal to the TRIPs negotiations calling for a two-tiered mechanism under which WIPO would rule on the intellectual property segments of a specific dispute and then, if the parties agreed, the trade-related aspects of the dispute would be arbitrated within the GATT. However, this proposal was not well-received at the GATT. *World Intellectual Property Report BNA 4* (1990), 61.

The TRIPs agreement also contained various understandings relating to copyrights and trademarks. For example, it was agreed that computer programs must be treated as literary works for purposes of copyright protection, that the normal duration of copyright protection must be at least fifty years from the death of the author, except for photographs and works of applied art, and that fair use provisions and similar limitations on the powers of the copyright owner must be confined to certain special cases which do not conflict with the normal exploitation of a work and do not unreasonably prejudice the legitimate interests of the owner. With regard to trademarks, the TRIPs agreement provided that actual use of a mark would not be required as a condition for making application for registration, and that it would not be permissible for countries to discriminate in granting trademark protection based on the nature of the goods and services.

In addition to the multilateral negotiations relating to international intellectual property rights, various countries, notably the United States, have attempted to address perceived weaknesses in intellectual property protection in a number of countries through bilateral negotiations. This trend is hardly surprising, in that the share of United States exports that rely heavily on intellectual property protection (e.g. chemicals, pharmaceuticals, computers, software, sound recordings, books, movies, and scientific equipment) more than doubled since the end of World War II to reach a level that exceeded one-quarter of total United States exports by the end of the 1980s. The United States included authority to initiate bilateral negotiations relating to intellectual property matters in various domestic trade statutes and regulations.[338]

The bilateral activities of the United States were spearheaded through the office of the United States Trade Representative ("USTR"). On May 25, 1989, the USTR, with input from the Interagency Trade Policy Staff Committee, the Patent and Trademark Office, and the Copyright Office, released the results of its initial survey of intellectual property laws and market access issues in various countries.[339] That review[340] as well as subsequent annual surveys and updates[341] focuses on whether countries meet certain minimum standards of adequate protection for intellectual property rights.[342]

In 1989, the USTR identified no "Priority Foreign Countries" (PFCs), even though it had concluded that none of the surveyed countries fully satisfied the standards advocated by the United States during the course of multilateral discussions, such as the TRIPs negotiations.[343] Instead, the USTR implemented a new, non-statutory procedure which

338 The first statutory provision was included as an amendment to Section 301 in the reciprocity bill enacted as part of the Trade and Tariff Act of 1984. Trade and Tariff Act of 1984, Pub. L. No. 98-573, 98 Stat. 3002. See also Portion of US Trade Representative's Fact Sheet on 1988 Trade Act's "Special 301" Provision on Intellectual Property, Released 25 May, 1989, *World Intellectual Proprty Report BNA* 3 (1989), 162. (hereinafter "Special 301" Provision).

339 Ibid. at 162–3.

340 The review was conducted under the "Special 301" provisions of The Omnibus Trade and Competitiveness Act of 1988 (1988 Trade Act). Trade and Tariff Act of 1984, 19 U.S.C. § 2241(a)(1)(A)(i). A description of the "Special 301" procedures is included in Operation of the Trade Agreements Program, 41st Rep. 1989, ITC Pub. No. 2317, 61 (September 1990).

341 Office of thc US Trade Representative, National Trade Estimate Report on Foreign Trade Barriers 1 (1992) [hereinafter 1992 Barriers Report]. See also Office of the US Trade Representative, National Trade Estimate Report on Foreign Trade Barriers 1 (1991).

342 See "Special 301 " Provision, supra note 339, at 162. Section 182 of the Trade Act of 1974, as amended, which was added to the 1974 Trade Act in 1988, requires that the USTR identify those "Priority Foreign Countries" (PFCS) which, *inter alia*, "deny adequate and effective protection" to United States IP or deny "fair and equitable market access" to US citizens who rely on such protection. IIPA Submission to US Trade Representative on "Priority Foreign Countries," *World Intellectual Property Report BNA* 4 (1990), 84.

343 See "Special 301 " Provision, supra note 339, at 162.

designated twenty-five countries for special attention based on intellectual property practices or market barriers that concerned the USTR.[344] Of those twenty-five countries, eight were placed on the Priority Watch List ("PWL") and the remaining seventeen were placed on the Watch List ("WL").[345] The USTR developed accelerated action plans for the PWL countries designed to cure deficiencies over the 150 days following the announcement.[346] WL countries would simply be the subject of increased efforts to resolve any problems over the following year.[347]

The original PWL included Brazil, India, South Korea, Mexico, the People's Republic of China ("PRC"), Saudi Arabia, Taiwan, and Thailand.[348] The USTR expressed several concerns regarding these countries: (i) the need to improve and provide adequate patent protection;[349] (ii) the need to provide greater copyright protection, including the need to enhance enforcement efforts against the piracy and counterfeiting of software and sound recordings from the United States;[350] (iii) the need to improve protection of foreign trade and service marks;[351] (iv) the need to eliminate various market barriers which inhibited the ability of United States film manufacturers to penetrate foreign markets;[352] and (v) the need to engage these countries in constructive multilateral intellectual property negotiations, such as the TRIPs discussions.[353]

In subsequent years, the USTR has taken a more activist approach. For example, on April 26, 1991, the USTR designated India, Thailand, and the PRC as Priority Foreign Countries.[354] Each country had been on the Priority Wait List since 1989, and the USTR opined that the practices of these countries had been "egregious," resulting in an adverse impact on United States industry.[355] The USTR also stressed that no country had made "significant progress," either bilaterally or multilaterally, to address these practices.[356] The USTR criticized all three countries for their unwillingness to provide ade-

[344] Ibid.

[345] Ibid. Countries originally placed on the WL included Argentina, Canada, Chile, Colombia, Egypt, Greece, Indonesia, Italy, Japan, Malaysia, Pakistan, the Philippines, Portugal, Spain, Turkey, Venezuela, and Yugoslavia. Ibid.

[346] The USTR was to review each trading partner on the PWL to determine whether it should identify the trading partner as a PFC. The USTR could have designated a country as a PFC at any time during the period of review. It also could have initiated a "Special 301" action if the country failed to continue negotiations in good faith or ceased to make satisfactory progress on the specified improvements in the IP area. The USTR was to complete this initial review by November 1, 1989. Ibid. at 163.

[347] Ibid.

[348] Ibid.

[349] Inadequate patent protection was found to be particularly troubling in Brazil, the People's Republic of China (PRC), India, Mexico, and Thailand. Ibid.

[350] Brazil, India, Korea, the PRC, Saudi Arabia, Taiwan, and Thailand were all cited for inadequate copyright protection. Ibid. The PRC and Saudi Arabia were asked to enact copyright laws, while Taiwan was cited for its failure to fulfill its obligations under existing bilateral copyright agreements. Ibid.

[351] The USTR noted that both India and Thailand failed to protect foreign trade and service marks. Ibid.

[352] Brazil and India both maintained such market barriers. Ibid.

[353] The USTR particularly sought to involve Brazil, India, Korea, Mexico, and Thailand in the ongoing negotiations. Ibid.

[354] 56 Fed. Reg. 20,060 (1991). In addition, the USTR included Australia, Brazil, and the European Community (EC) on the Priority Watch List. Ibid. See also Foreign Trade: China, India, and Thailand Named in First "Priority Foreign County" List, *Patent Trademark and Copyright Journal* 42 (1991), 7, 7–8 (discussing priority watch list and secondary watch list countries). The USTR also placed twenty-three other countries on a secondary watch list for monitoring including: Argentina, Canada, Chile, Colombia, Cyprus, Germany, Greece, Hungary, Indonesia, Italy, Japan, Korea, the Philippines, Saudi Arabia, Spain, Turkey, and Yugoslavia. Ibid. at 8.

[355] 56 Fed. Reg. 20,060 (1991).

[356] Ibid.

quate levels of patent protection-particularly with respect to pharmaceutical and chemical products.[357] In addition, in December of 1991, the USTR made a finding that Thailand's copyright enforcement procedures were unacceptable.[358] Since then, other Asian countries, such as Taiwan and Japan, have been the subject of major initiatives by the USTR, often resulting in significant changes in intellectual property laws and practices in both the Asian country in question and in the United States.

The direction of further changes in international intellectual property laws It seems fairly evident that broader protection of intellectual property rights does not make sense for all countries, nor does the United States and the other developed countries have the right to impose its conventions upon the rest of the world.[359] The evidence which has been put forth regarding the relationship between intellectual property rights and economic development is unconvincing and, in fact, the recent rise of Japan and other "technology importers" appears to support the utility of alternative strategies, at least while the country is in a relatively early stage of development. Moreover, the developing countries are rightly skeptical of the motives of the United States and other developed countries, given the focus of bilateral efforts on countries which appear to have large domestic markets and an industrial capacity capable of competing with United States firms.

Any global agreement or understanding with respect to protection of intellectual property rights must address the "knowledge gap" between North and South. It should not be forgotten that even though almost all of the developing countries have adopted at least some rudimentary form of patent law, the technological disparity which has existed between industrialized and developing countries has continued to increase. For example, in the late 1970s, 90 percent of all the technologists and scientists worked in the developed countries[360] and 90 percent of their activities were concentrated in work for use in the industrialized world, thereby illustrating that the direction of research and development is largely determined by the ultimate commercial potential of the activities. Moreover, newer technologies, such as in the agricultural area, are threatening to displace Third World products and services, thereby exacerbating the relative poverty in those areas.[361]

[357] The USTR found patent laws in India deficient due to an inadequate level of patent protection, including overly broad compulsory licensing requirements, too short a term of protection, and specific failures to enforce foreign pharmaceutical patent rights.

[358] 1992 Trade Barriers Report, supra note 342, at 243.

[359] "Companies in rich countries often imply that all such disputes over intellectual property are a straightforward matter of piracy or theft ... However, the developing countries are more concerned about patent protection for technologically sophisticated goods such as drugs and chemical fertilizers ... In many cases, matching rich-county standards of patent protection would make such goods a lot more expensive to third-world buyers ... [I]n their dealings with the third world, companies regard the conventions agreed at home as self-evidently correct for everywhere else. It is not at all obvious that the developing countries are obliged, either morally of for the sake of sound economics, to meet the rich countries' demands." "Something Old, Something New," Special Survey on World Trade, *The Economist*, September 22, 1990, at 34, 35.

[360] P. Nanyenya-Takirambudde, Technology Transfer And International Law 5 (1980) (citing J. Tinbergen, A. Dolman and Q. Van Ettinger, *Reshaping The International Order: A Report To The Club of Rome* (1976), 19–20).

[361] For example, biotechnology has produced an economical industrial sweetener made from corn displacing natural sugar in the world market. See Dias, Dembo and Morehouse, "Product Displacement: Biotechnology's Impacts on Developing Countries," in *The International Context of Rural Poverty In The Third World* (1986), 129, 132. Further displacement is promised by aspartame, an artificial sweetener. Ibid. at 133. These technological advancements have damaged the economies of sugar-exporting countries who can no longer control the price and quantity of their exports. Ibid. at 132–33.

All countries must begin to work together on adapting existing intellectual property systems to address the disparate positions of each of the participants and the uncertainties created by new technologies with almost unknown potential. Any agenda on these matters might well include the following:

(i) Developed countries must look for ways to share their knowledge and assist developing countries in building their own technological infrastructure. Until this is done, intellectual property rights will have little or no value to developing countries.
(ii) Developed countries should consider providing funds to developing countries for research activities on products which would have commercial potential in the domestic market in which the research is being conducted. Once again, developing countries have little stake in a global technological community which fails to address their own unique needs.
(iii) Efforts should be made to encourage scientists and technology managers from developing countries to remain in those countries to build the industrial infrastructure necessary for local exploitation of the new technologies.
(iv) Developed countries should address some of their own concerns about return on investment for funds expended on new products by reviewing the domestic regulatory hurdles associated with the approval and introduction of new products.[362] Similarly, as noted above, the type of protection to be afforded to new technologies, such as biotechnology, must be decided upon.

For their part, developing countries must be willing to consider and implement various changes which will reduce the concerns that foreign firms might have in participating in the development of local technology transfer programs. One of the most important and difficult tasks in this area for developing countries is improving the legal infrastructure for commercial activities. Economic transitions and reforms in developing countries generally include a number of changes in the legal, regulatory, and judicial environment in which enterprises exist and commercial transactions take place. Specifically, there is a need to develop "rules of the game" for economic activity and provide procedures to ensure these activities are carried out efficiently and effectively.

While the elements of modern commercial and economic laws have been around for some period of time, they nonetheless remain quite new for many developing countries. While most developing countries are attempting to reform and modernize their economic laws, there is often a lack of understanding of need to make such reforms, and the adverse consequences which might flow from the failure to push forward with the development of a comprehensive system of commercial laws and institutions. Advocates of legal and regulatory reform in developing countries argue that it is necessary not only to increase equity in the society and raise productivity, but also to reduce various costs which adversely impact the rate of growth of economic activity in particular

[362] For example, regulatory hurdles in the United States often delay the introduction of new products for a number of years. The timeline for product introduction, and thus the internal rate of return on investment, can be effected by changes in clinical trials and other testing requirements, provided that material risks are not created for the ultimate consumers of the products. Since substantial differences exist among the developed countries regarding clinical trials requirements, it is clear that there is some room for change in this area.

markets. Among the costs which are of greatest concern are transactional costs, costs of avoiding the formal legal system, and capital costs.

Transaction costs In order to understand what transaction costs are let us go back to the origin of the phrase. In his 1960 article, "The Problem of Social Cost," Coase first used the phrase "the costs of market transactions." These became known as "transaction costs." According to Coase, what he had in mind was:

> "In order to carry out a market transaction it is necessary to discover who it is that one wishes to deal with, to inform people that one wishes to deal and on what terms, to conduct negotiations leading up to a bargain, to draw up the contract, and to undertake the inspection needed to make sure that the terms of the contract are being observed. These operations are extremely costly, sufficiently costly at any rate to prevent many transactions that would be carried out in a world in which the pricing system worked without cost."[363]

Transaction costs accrue throughout the process of negotiating a particular commercial transaction. For example, parties incur search and information costs as they begin to seek appropriate business partners and information regarding their resources and capabilities. Once the potential partner has been selected, parties are faced with bargaining and decision costs, including the direct costs associated with drafting all required legal documents for the transaction and, if necessary, forming new enterprises to participate in the transaction. Then, after the contract has been completed, the parties are faced with the costs of monitoring performance under the contract, including the possibility that adjustments will be required in cases where the transactions begin to stray from their initial objectives. Finally, there may be significant enforcement and award costs in the event that material disputes arise between the parties.

Transaction costs are a major impediment to efficiency, even in developed countries. North has estimated that almost one-half of United States national income is spent on transaction costs. Even more disheartening, this is a significant increase over a century ago.[364] Not surprising, transaction costs are an even greater impediment in emerging economies. Once again, it is North who found:

> "When we compare the cost of transacting in a Third World country with that in an advanced industrial cconomy, the costs per exchange in the former are much greater – sometimes no exchange occurs because costs are so high. The institutional structure in the Third World lacks the formal structure (and enforcement) that underpins efficient markets."[365]

Transaction costs can quickly become excessive, a problem which is exacerbated when the number and ideological diversity of the economic actors increases. As parties

[363] R.H. Coase, *The Firm, The Market, and the Law* Chicago: The University of Chicago Press (1988), 114.

[364] North wrote: "Wallis and North ... measuring the size of transaction costs that go through the market (such as costs associated with banking, insurance, finance, wholesale, and retail trade; or, in terms of occupations, with lawyers, accountants, etc.) in the U.S. economy found that more than 45 percent of national income was devoted to transacting and, moreover, that his percentage had increased from approximately 25 percent a century earlier. Thus the resources of the economy consumed in transacting are of considerable magnitude and growing." North, *Institutions,* 28.

[365] Ibid. at 67.

with widely divergent interests (e.g. entrepreneurs in developing countries and foreign investors from developed countries) being to engage in negotiations regarding economic contracts, it becomes impossible for them to rely on the use of informal agreement and enforcement mechanisms. Such parties need an objective framework within which to organize their transactions, or the transaction costs will be so high that the transactions will be so expensive that they will not be attempted.

Avoidance costs When transaction costs are excessive, access to the formal legal system is limited, or the risks of using the formal legal system remain too high due to inadequacies in the content and/or enforcement of laws and regulations, then parties may instead turn to strategies which avoid the formal legal system. For example, rather than applying for the licenses and permits which are nominally required as a condition of using the formal legal system, the parties may simply pursue the transaction without any formal legal approvals. While this may permit them to avoid some of the formal legal costs, it does expose them to another set of costs, including the need to pay protection money, bribe local officials, or downsize the scope of their activities in order to avoid detection. While activities conducted within this informal system may provide productivity gains, it is likely that those gains will be less than those that might have been achieved if the business had been conducted in the formal legal sector.

Capital costs An inadequate formal legal system can have an adverse impact on the cost of capital and the availability of credit for financing new transactions. This is due to the fact that the cost of capital is closely linked to the legal certainty and predictability implicit in a country's formal legal system. Capital costs reflect the relative security of a future claim with respect to a current transfer of resources. For example, in a commercial loan transaction, the lender is asked to transfer capital to the borrower in the present in exchange for a future right to receive the capital back along with some additional compensation for the loss of use of the capital during the time that the loan was outstanding. If the formal legal system does not include reliable enforcement mechanisms which increase the lender's confidence regarding the return of capital, then the cost of capital (i.e. the interest rate, collateral coverage) will be much higher in order to compensate the lender for the increased risks flowing from inadequacies in the legal system.

The problem is exacerbated in the case of developing countries, since many of the borrowers are new enterprises with no established track records, and many of the projects carry their own independent levels of business risk, such as market and production uncertainties. These business risks necessarily must be factored into the lender's decision-making process. When the higher foreclosure and insolvency costs are added on to these inherent business risks, the lender may either cease lending all together, reduce the pace of its loans until market conditions become clearer, or restrict its lending activities to projects in only a few selected sectors. In each case, the rate of growth and development in the economy is adversely impacted. Moreover, if international lenders perceive that investment risks may be lower in one country as opposed to another, perhaps because the formal legal systems in one country are more advanced, then the flow of capital to the riskier country will be reduced and then diverted to other countries, perhaps creating an imbalance among countries in the same geographic region.

Inefficient legal system costs critically affect development. The greater the inefficiency and costs associated with doing business, the greater the handicap to development. Where inefficiency is reduced and costs minimized, development is more likely to take place. This does not guarantee that it will, since there are too many other variables which must be taken into account. We can only say that the more handicaps removed, the more likely it is to take place. The following passage from a recent issue of the Wall Street Journal illustrates how important the legal system infrastructure can be to a country's ability to attract foreign investment:

> "Recently, Vietnam's top planning ministry called in an international consulting firm to discuss why U.S. investment had been so far below its expectations. Hanoi had hoped for a gold rush after the U.S. economic embargo ended a year and a half ago. Instead, there was an initial flurry of deals, followed by a cautious appraisal. Part of the consulting firm's answer was that most U.S. multinationals have big legal departments that prefer the comfort of established, stable law and tax systems. Vietnam has neither."[366]

The is substantial support for the view of many economists that an inefficient legal system with uncertain property rights retards development. As North wrote, "the institutional framework plays a major role in the performance of an economy ... [it] is the critical key to the relative success of economies, both cross-sectionally as well as through time.[367] Inefficient legal system costs do much more than increase the costs of doing business. Where inefficiency exists, uncertainty prevails. This means profit maximization no longer controls decision making and success is often determined not by skill but by luck. Obviously, an economy which is no longer built on profit maximization and skill is not resting on the best foundation. The following passage is enlightening:

> "Basically, what property rights, contracts, and extracontractual liability do is reduce uncertainty for people who want to invest their labor or capital in the development of existing resources ... In any country, uncertainty or legal instability reduces the volume of long-term investment and investment in plant and equipment. People save less and invest the little they do save in such socially unproductive goods as jewelry, gold, or luxury property. The flight of capital from countries ... is only one more result of the desire to avoid uncertainty ... Thus, the actual value of an economic opportunity is not the value it would have it if could be realized without cost, but rather its estimated value, taking into account the cost of the red tap, the degree to which it can be protected against third party appropriation, and the ease with which it can be sold. The less costly the transaction and the more secure the right to enjoy the fruits of investment, the greater the real value of an economic activity. A law that is efficient in dealing with these elements will encourage people to identify and seize existing opportunities and will systematically increase the value of economic activity."[368]

While the developed countries have insisted that developing countries adopt intellectual property law regimes that resemble those which exist in the developed world, the work of economists suggests that this may often place form over substance, at least to the extent that one seeks a set of laws and regulations which recognize and balance the needs of both the investors and local firms. A better solution may be to seek legal

[366] Greenberger, "Vietnam," *Wall Street Journal*, February 3, 1996, A-11.

[367] North, supra note 365, at 69.

[368] De Soto, *Path*, 180, 181, 182–3. De Soto also found that the costs of not having property rights for market participants in Peru included inefficient use of resources; reduced investment; a bias toward multiple, small-scale operations rather than larger, more efficient, firms; limited use of property as collateral; stagnation resulting from the lack of a system for transferring and organizing property rights; and, not unimportantly, limited technological progress since inventors are less likely to expend efforts on research unless they are certain of being able to reap benefits under well-established property rights in the invention.

solutions which will reduce the uncertainties and transactional costs for foreign investors in a developing country, while providing incentives for local entrepreneurs to protect the investments of technology and capital made by developed country firms. This will occur only if intellectual property rights are framed within the larger scope of competition and foreign investment policies in the developing countries.

C. Competition Law

Competition law is designed to promote fair and effective competition among autonomous enterprises and ensure that the interests of consumers are protected without the need for direct state management of the enterprises or competition. Among the specific objectives of competition laws in developing countries is the prevention of unfair competition in enterprise operations, particularly in market behavior; ensuring that enterprise mergers do not create barriers to entry into the market; regulation of monopolies, which in developing countries can include not only natural monopolies but also sectors which remain controlled by the state or its affiliates; and protection of consumers against cartel-like behavior of enterprises.

While competition law is certainly concerned with the regulation of competition among enterprises at the macroeconomic level, developing countries must also adopt laws and regulations which address unfair competitive practices which enterprise might engage in the course of day-to-day activities in the marketplace. Unfair competition laws address the general rules of trade in the market, and sanction such things as trademark infringement, fraudulent and improper sales practices (e.g. bribery or false advertising), misappropriation of trade secrets and confidential information, and unethical interference with the activities or reputation of competitors.

Many developing countries lack a comprehensive system of laws and regulations relating to competition, although they may have some general laws referring to the need for fair competition among enterprises and addressing concerns regarding behavior that is perceived as monopolistic. Developing countries interested in broadening the scope and sophistication of its competition laws will need to consider establishing a strong relationship between the agency with responsibility for enforcing such laws and the government's economic policy departments, particularly since a number of competition-related issues can be expected to arise in the context of mergers and combinations of enterprises. Also, reforms may be required in company laws, at least to the extent that the government continues to control new entrants through its authority with respect to registration of enterprises.

Developing country policies regarding recognition and enforcement of intellectual property rights must often be read in conjunction with laws and regulations with respect to inbound technology transfers (e.g. licensing agreements). Many nations have enacted specific legislation regulating the content of technology transfer agreements. These regulations are intended to serve a number of objectives, most of which are generally related to the overall economic development of the local country, although some statutes are also drafted in a manner which serves to protect the relative bargaining position of the transferee. For example, a survey of the various regulations regarding technology transfers which have been adopted in South America indicated the following principal objectives:[369]

[369] See G. Cabanellas, *Antitrust and Direct Regulation of International Transfer of Technology Transactions* (1984), 30–47.

(i) increasing the bargaining power of local transferees, such as by prohibiting the use of package licensing and other similar practices;
(ii) increasing the amount of information available to local parties with respect to possible sources of technology;
(iii) controlling the nature of imported technology and improving the quality and local assimilation of any new technology which is to be transferred;
(iv) protecting local innovation and technology;
(v) regulating remuneration for transferred technology in order to protect and improve the balance of payments position of the local economy, control foreign exchange operations, and prevent tax avoidance;
(vi) limiting the protection of industrial property by restricting the prohibitions which might be imposed upon disclosure and use of such information;
(vii) promoting other local public policy goals, such as the use and training of local employees and protection of the environment.

As a general rule, technology transfer regulations operate by requiring the international technology agreement be submitted to, and approved by, a national administrative authority in order for the agreement to be enforceable. Governmental bodies charged with reviewing technology transfer agreements often refer to statutory lists of objectionable business practices which must be excised from any agreement as a condition of approval. Among the most common areas of concern are royalty rates and other forms of remuneration, the scope and content of controls the technology provider seeks to impose on the local transferee, the nature of any implied representations and warranties regarding the quality and performance of the transferred technology, the term of the agreement, governing law, and dispute resolution procedures.

As with "inadequate" intellectual property protections, companies from developed countries often determined that the effect of the technology transfer laws in developing countries substantially dampened the incentives for those firms to invest in those countries. Even if some investments were made, the quality of the technology transferred was rarely adequate to provide the local economy with the knowledge that the laws were intended to appropriate. Many developing countries began to realize that they would be unable to attract foreign investment unless they liberalized, if not eliminated, their technology transfer restrictions. For example, Mexico, which had long had one of the stringent technology transfer regimes, removed almost all of these restrictions in a very short period as a prelude to the North American Free Trade Agreement.

A number of developing countries have restructured their technology transfer laws to encourage and promote imports of technology for use in specified industrial areas. For example, a country may provide that the basic technology transfer rules will be modified in relation to certain technologies, such as by providing for expedited reviews, liberalizing restrictions on the level of foreign investment participation, creating incentives for local manufacturing, permitting components needed for use in local manufacturing to be imported at concessional rates, and authorizing development subsidies, tariff protections, and tax credits. Industry promotion may be accomplished by announcing areas in which incentives and concessions will be available or by the use of "lists" which identify the type of restrictions, if any, on foreign investment in industries specified thereon.[370]

[370] For example, South Korea originally used a "positive list system" which enumerated those industries in which foreign investment was permitted. See Foreign Capital Inducement Law, April 12, 1973. Subsequently,

D. Lessons for Regional Economic Integration

Introduction As the volume of international trade has increased, their has been heightened pressure to reduce trade barriers of all types around the world. This has led to the creation of various regional trading blocs around the world, including the EC, the North American Free Trade Agreement, and the United States–Israel Free Trade Agreement. In addition, as many of the developing countries make the transition toward market-oriented economies, interest has grown in regional alliances among these countries to take advantage of geographic proximity and the efficiencies from economic integration. For example, several trade agreements among countries in Latin America and the Caribbean are now in various stages of development,[371] and the dynamic growth in the Asia-Pacific region has led to a number of proposals for some form of economic coordination and cooperation.

Lessons from comparing the United States and the EC There are number of lessons which can be learned from comparing the United States and the EC. Certainly, we are struck by the differences which exist with respect to the analytical tools applied to antitrust and competition law issues, including the comfort of courts and regulators in the United States with the expansive "rule of reason" doctrine. However, there are also differences between the United States and the EC which may have direct relevance to the policies implemented by other regional economic blocs around the world.

First, and foremost, the United States has an integrated, federal, system of patents and antitrust laws and regulations, while the EC remains an evolving economic union with separate national patent and competition laws. As we have, perhaps the most difficult issue for EC competition law regulators has been reconciling the existence of national patent rights with the overriding EC Treaty objectives of reducing and eliminating barriers to trade among the member states. While some progress has been made regarding harmonization of national laws, it is unlikely that national laws with respect to intellectual property and competition will be supplanted by a universal EC legal regime in the near term.

Second, while there is no denying the diversity of culture and economic background which exists in the various regions of the United States, it does not begin to approach the continuing nationalism which exists in the EC. In spite of the single market objective, the EC will for some time remain a bundle of different markets demarcated by historical experience, location, culture, and language. This means that companies will likely continue to need the flexibility to develop production, sales, and distribution strategies which are tailored to what will be only a portion of the EC market. This will likely con-

cont
as the economy matured, South Korea changed to a "negative list system" under which all industries other than those on the "negative list" would be open to foreign investment. See Foreign Capital Inducement Law, effective July 1, 1984.

371 Mercosur would create a free trade area among Argentina, Brazil, Uruguay, and Paraguay, while the "G-3 Agreement" would create the second largest free trade zone in the region among Mexico, Colombia, and Venezuela. The Andean Pact countries (i.e. Bolivia, Colombia, Ecuador, and Venezuela) hope to establish a common external tariff system. The Caribbean Common Market ("CARICOM"), comprising the English speaking Caribbean countries, is seeking to lower its common external tariff and liberalize trade amongst its members, while pushing to enter into free trade agreements with Mexico, Colombia, and Venezuela. See, generally, Joseph E. Garten, "American Trade Law in a Changing World Economy," *The International Lawyer* 29 (1995) 15, 26.

tinue to cause difficulties for EC competition law regulators which are not significant in the United States.

Another important factor in the technology area is the mobility of the human resources and capital are necessary for the effective exploitation of new inventions. Technology licensing in the United States is done in an environment where scientists and engineers can move relatively freely, law relating to the raising of capital (i.e. securities laws) are largely federal in nature, and differences in state laws with respect to labor and business entities (e.g. corporation and partnership laws) are rarely significant. In contrast, innovation in the EC will be relatively more complicated until agreement is reached relating to free movement of labor and capital across the old national borders.

Economic integration in East Asia The most important economic and political association in Southeast Asia is the Association of Southeast Asian Nations ("ASEAN"). ASEAN was founded in 1967 by Indonesia, Malaysia, the Philippines, Singapore, and Thailand in order to increase the level of economic cooperation in various business sectors, including food production, industry and commerce, shipping and telecommunications.[372] Since that time, membership has been extended to Brunei and Vietnam, and the group is expected to expand even further over the next several years to include Laos, which is current an observer, Cambodia, and Myanmar. ASEAN is headquartered in Djakarta, Indonesia, and its principal organs include the ASEAN Heads of Government; the ASEAN Ministerial Meeting, which is an annual meeting among the foreign ministers of the member states to set general policies; the Standing Committee; nine Permanent Committees, five of which deal with economic cooperation in the areas of financing and banking, food, agriculture, and forestry, industry, minerals, and energy, trade and tourism, and transportation and communication; Ad Hoc Committees and Working Groups, and the ASEAN Secretariat.

ASEAN has spawned a variety of subsidiary agreements, including the ASEAN Free Trade Area ("AFTA"), a free trade zone formed in 1992 and made up of the members of ASEAN, and the ASEAN Preferential Trading Agreement, which is designed to ensure that the member states of ASEAN afford each other reciprocal trade treatment. AFTA is designed to reduce intraregional tariffs, and to offer foreign investors and manufacturers simplified access to business opportunities in the region. Originally, the agreement was to be phased in over fifteen years, since countries like the Philippines and Indonesia had concerns about the competitiveness of their economies in relation to their fellow member states. However, the target date for final implementation of the AFTA has subsequently been moved up first to the year 2003, and now to the year 2000.

Another important association in the region is the Asia-Pacific Economic Cooperation group ("APEC"), which was founded in 1989 and which includes members on both sides of the Pacific.[373] Among other things, APEC is designed to be a forum in which the member states may discuss a broad range of economic issues. In addition, it is hoped

[372] The purpose of the association is "to accelerate economic growth, social progress and cultural development in the region ... to promote social collaboration and mutual assistance on matters of common interest in the economic, social, cultural, technical, scientific, and administrative fields ... to collaborate more effectively for the greater utilization of [the member states'] agriculture and industries, the expansion of their trade, including the study of problems of international commodity trade, the improvement of their transport and communication facilities and raising the living standards of their people."

[373] Members include Australia, Brunei, Canada, Chile, China, Hong Kong, Indonesia, Japan, Korea, Malaysia, Mexico, New Zealand, Papua New Guinea, Philippines, Singapore, Taiwan, Thailand, and the United States.

APEC can promote increased multilateral cooperation among the market-oriented economies of the region. APEC members have shown an interest in studying methods that might be used to harmonize custom procedures and reduce various impediments to "market access" among members. Of note is the so-called Bogor Declaration of late 1994, where the members adopted a "declaration of common resolve" to pursue "free and open trade and investment" over the period extending to the year 2020. The Declaration contemplates that the region's developed nations would dismantle their trade barriers by 2010, and that the other member states would follow suit by 2020.

Another interesting association is the East Asian Economic Caucus ("EAEC"), which was formed as an alternative to what was to be referred to as the East Asian Economic Grouping ("EAEG"). The EAEG was to be an alternative economic bloc including the ASEAN members, Hong Kong, Myanmar, Korea, and possibly Japan and Vietnam, but not those nations with a predominantly non-Asian population.[374] It was originally contemplated that the EAEC would operate under the "umbrella" of the APEC, but would be directly coordinated by the ASEAN economic ministers. However, progress has been somewhat slow. For example, Japan has appeared to be hesitant to move forward, and has called for including Australia and New Zealand in the association in an effort to reduce some of the apprehensions expressed by the United States. Problems also exist as to how to reconcile the interests of China, Hong Kong, and Taiwan, each of which would be important participants in such an association.

The declining United States military presence in the region has lead to the creation of the ASEAN Regional Forum ("ARF"), which is to be a regional security forum designed, among other things, to facilitate the peaceful resolution of disputes and promote regional stability. Members of the ARF include the ASEAN members, Australia, Canada, China, the European Union, Japan, Korea, Laos, New Zealand, Papua New Guinea, Russia, the United States, and Vietnam. The ARF has considered a variety of issues, including North Korea's nuclear posture, the escalating arms race in the region, and certain political and military actions taken by China; however, there has not yet been any formal agreement regarding conflict-resolution mechanisms that might be used throughout the region.

ASEAN, and the various other associations and forums referred to above, provide a framework for regional economic and political integration that will continue to grow in importance in the years to come. ASEAN is closely observing the efforts to develop regional cooperative efforts in the EC and in North America, and has begun to focus on projects to harmonize its laws and regulations relating to trade and business. For example, ASEAN is considering the implementation of a "regional patent," and similar initiatives can be expected in the near future.

It will be interesting to see the form of legal integration, if any, which ultimately emerges in the Pacific Rim area. What has become clear in recent years, particularly as countries like Japan and Korea have developed their own unique blend of law and policy to fuel economic growth, is that developing countries in the region will no longer blindly accept Anglo/Western legal ideas and systems, particularly since commercial practices in the region are based on different cultural values. Moreover, other differences are likely to flow from the fact that individual countries in the region have followed economic growth models that differ from the classic "export-oriented/open market" models that

[374] Initial reaction to the formation of the EAEG was mixed, and the United States warned that such an association might weaken key established alliances, including those with the United States.

are often preached by the United States, and which are more in tune with the strategies of regional success stories like Japan and Korea.

Striking characteristics of the Japanese economic development model include governmental and political structures which are characterized by the presence of a talented and elite bureaucracy with broad discretion regarding economic policy, an executive branch with only limited political opposition, a relatively weak legislative branch, a central economic planning function, and integrated regulation of fiscal, monetary, credit, and foreign investment policies through the finance and trade ministries. In addition, many of the industrial leaders in the private sector have served in the bureaucracy, and these long-standing relationships mean that consensus regarding the nation's economic goals and objectives is often achieved outside of the political process.

Another important growth and development strategy for emerging countries in the region has been targeting those industries which they believe are essential to national economic security.[375] Having identified those industries, a series of tariff and non-tariff barriers are established to "protect" the domestic industries from foreign firms. Also, domestic firms chosen to be active in target industries become the beneficiaries of a wide variety of additional policy measures to provide, without substantial governmental subsidies, a number of competitors with the capital, technology, and training necessary for rapid development of the capacity to satisfy domestic demand and gain entry into the global marketplace. The foregoing illustrates an attitude toward cooperation and competition which differs substantially from that which underlies antitrust and competition law in the United States and in the EC.

The areas which might be covered in a comprehensive system of regional economic laws and regulations are not difficult to identify. However, although the value of such laws is readily accepted, their effectiveness in a given situation is more problematic, since it depends on a variety of factors such as the following:

(i) the clarity of the laws and regulations and the availability of information regarding such laws and regulations;
(ii) the enforceability of legal contracts and instruments;
(iii) the general level of experience, education, and competence in economic, commercial, and legal matters;
(iv) the predictability and stability of laws, regulations, and enforcement mechanisms;
(v) the moral and ethical behavior norms which exist in the community; and
(vi) the perceptions in the community regarding the content and legitimacy of the prevailing legal system.

One obvious hurdle for the success of economic integration throughout wide parts of Asia is that the successful development models adopted in Japan, Korea, and other

[375] For example, in recent years, China has sought economic development and inbound foreign investment by attempting to enhance its resources in the areas of agriculture, industry, defense, and science and technology. Generally, this has led to the establishment of an administrative and legal framework that would support foreign investment activities, including laws and regulations relating to economic contracts, joint ventures, sales contracts, and intellectual property rights. The more developed countries in the area have been forced to turn their attention to high technology industries. One reason for this has been the fact that the wage rates in those countries have risen substantially, undermining the competitiveness of low-end manufacturers. In some cases, such as in Taiwan, firms have actually begun to move their manufacturing activities into areas where labor costs are much lower, such as mainland China. These countries now look to high technology industries to create new jobs for workers who have been displaced from the manufacturing sector.

countries cannot be fully adapted to a regional coalition. For example, while it may be possible to reach agreement as to those industries which might be best suited for cooperative exploitation, it would be far more difficult to reach consensus about structure of the enterprise coalition that would spearhead the innovation efforts. Problems might also be expected from the fact that the region, just like the EC, is one in which there is tremendous national and ethnic diversity. In addition, there are significant differences among the countries with respect to infrastructure conditions and their stages of legal development.

East Asia is a geographic and economic region of tremendous diversity, with a number of large and important ethnic groups. For example, while in Malaysia the largest ethnic group is the so-called "bumiputera," the indigenous Malays, the country also has a significant number of Chinese and Indian inhabitants. In many cases, the various ethnic groups have become involved in different types of economic activity within the country, as has been the case where the native Malays focused primarily on farming while the immigrant Chinese and Indians concentrated on the development of the industrial and manufacturing sectors. In some cases, this has led to internal discord, and governments have been forced to undertake programs designed to redistribute capital, employment opportunities and managerial resources and opportunities. There is also significant religious diversity in the region. For example, religious groups in Singapore include Buddhist (Chinese), Muslim (Malays), Christian, Hindu, Sikh, Taoist, and Confucianism.

Countries are beginning to focus on developing and improving the infrastructure necessary for efficient economic growth. For example, Taiwan embarked upon a $300 billion program dedicated to improvement of its telecommunications, housing, roads, and system of mass transit, although the scope of the program was reduced in light of budgetary constraints. Singapore prides itself on the scope and quality of its infrastructure, noting the extent of its paved highways and railroads, as well as it excellent airports and ports. In fact, Singapore objective of becoming the major airport hub of the region by the end of the current decade has facilitated the development of more than 50 local firms dedicated to aircraft component manufacturing, maintenance and aerospace activities.

The development of commercial laws throughout the region is impacted by the tremendous historical and social diversity among the various countries. For example, several countries, such as Hong Kong, Singapore and Malaysia, base a number of precedents on their experiences with the English common law. However, other countries, such as Indonesia, Vietnam and Cambodia, have been profoundly effected by long periods of civil law influence, and the commercial laws of the Philippines were, for a long period of time, based on the Spanish Code of Commerce. The situation is even more complicated by the fact that portions of national laws are based on other bodies of law, such as Islamic law in the case of Malaysia. Finally, rules of commerce and trade in various communities often follow native laws, many of which are unwritten and unavailable to persons living outside of these communities.

There has been a substantial amount of activity in the region with respect to reform and revision of commercial laws and regulations. Most countries in the region has moved to make substantial modifications to the their laws and practices in the areas of economic contracts (e.g. sales of goods), foreign investment, business entities, and intellectual property laws. However, mere changes in the form and content of the laws have not always been sufficient. For example, while China has made a substantial number of changes in its economic laws over the last few years, concerns remain regarding the transparency of the new legal framework in China. Many of the new laws, while now widely published, have not yet been subjected to critical analysis, and judicial and administrative interpretations are also lacking. Foreign investors also remain concerned about the inaccessibility of local judicial systems, a problem which has led to development of international conventions and agreements relating to dispute resolution for commercial contracts.

Regional economic blocs and global trade There is an obvious tension between regional economic trading blocs and a truly free multilateral trading system, since regional trading areas, by definition, are based on preferences for its parties over nonmember contracting states. It is perhaps too early to tell what effect, if any, regional economic groups might have on global trade in general. However, it may well be that these regional arrangements may serve as models for various disciplines, liberalizations, and procedures which ultimately could find wider application and improve the process of trade amongest all the countries of the world.

VI. Topics for Future Research

Whether or not one accepts or rejects, in whole or in part, any of the incentive theories advanced as justifications for the creation of the patent monopoly, it is clear that a good deal of work is still required in order to reach agreement on the socially optimal level of invention, disclosure and innovation, as well as the means by which optimality might be properly measured through accepted empirical procedures. If anything, the criticisms which have been made regarding each of the important incentive theories makes it clear that we are still far from agreement regarding the underlying policy objectives of the patent system, whether it is considered to be a form of property right enforced by the state or an intrusion of statutory law into the market mechanism.

Economists have arrived at varying conclusions regarding the rule to be followed in establishing the optimal patent system. For example, Kaplow analyzed restrictive patent licensing practices and argued that the propriety of permitting a given practice should be evaluated in light of the "net social benefits (the excess of total benefits over total costs)" of granting a patent and allowing its use by the patent holder with the subject restrictions.[376] Gilbert and Shapiro advocate infinitely-lived patents with enough "breadth to attain the required reward level" for the inventor.[377] Still another proposed solution ignores concerns about perfectly competitive provision of a new product and concen-

[376] Kaplow, "The Patent Anti-Trust Intersection: A Reappraisal," *Harvard Law Review* 97 (1984), 1813. Kaplow's emphasis on "net social benefits" can be contrasted with the views of Bowman and Baxter that the individual should be given a patent and allowed to use restrictive licensing practices so long as the inventor's reward is less than or equal to the value of the invention to society. Ibid. at 1849–54. Kaplow is critical of this view, arguing that "[S]uch a view incorrectly focuses on total social benefits, rather than the net social benefits (the excess of total benefits over total costs)." Ibid. at 1828.

[377] See Gilbert and Shapiro, *Optimal Patent Length and Breadth*, University of California, Berkeley, Department of Economics, Working Paper No. 89-102 (1989), at 2: "Increasing the breadth of the patent typically is increasingly costly, in terms of deadweight loss, as the patentee's market power grows. When the increasing the length of the patent, by contrast, there is a constant tradeoff between the additional reward to the patentee and the increment to deadweight loss ... So, the socially cost-effective way to achieve a given reward to innovators is to have infinitely-lived patents with enough breadth to attain the required reward level."

trates upon a choice of patent length and breadth which will "minimize the social cost stemming from the resulting monopoly provision."[378]

Even if one could agree on the overall social and economic objectives of the patent system, the next problem is how one might refine or amend the current system in a way which actually meets those objectives. For example, it might be assumed, all other things being equal, that a reduction in the "strength" of patent rights would reduce the incentive to engage in the desired activity and that increasing the "breadth" of patent rights would have the opposite effect of providing further encouragement for inventors, disclosers and innovators. However, the magnitude of any of these effects is generally extremely difficult to measure,[379] and may vary substantially depending on the specific technological field. Moreover, it is also likely that incentives for basic research will differ from those which might be needed in order to induce applied research on commercial applications of new technology,[380] and a case could be made for distinguishing the amount of protection afforded to individual inventors and small firms on the one hand and large firms on the other hand.[381]

Some of the most important and challenging questions lie in examining the actual and potential relationship between providing an incentive for the development of new ideas and knowledge, the process which we generally refer to as "invention," and the subsequent activities which might be required in order to create a finished product that can be made available to consumers at a cost which achieves the optimal level of social gain and utility; however that might be defined. It would appear that the current patent system does not always achieved the appropriate result, but it is unclear that any of the other suggestions will be any more successful. Moreover, it seems likely that there may well be a number of different "solutions," each of which will depend on the characteristics of the specific market, competitive factors and the type of technology involved.[382]

Economic analysis has provided a wide variety of useful insights into the basis for the patent system itself, as well as some of the important issues surrounding decisions regarding the scope and duration of the exclusive rights which a patent represents. Eco-

[378] Klemperer, cited in notes 2 and 124 in Merges and Nelson, supra note 287 ("Since any single prize ... will induce the same R&D activities, we can equivalently think of choosing [patent length and breadth] to minimize the social cost stemming from the resulting monopoly provision rather than the perfectly competitive provision of a new product"). It is important to note that changes in scope and changes in duration may not comparable in terms of the effect which the change might have upon incentives. See also ibid. at 869 n128 ("broad claims influence who will be involved in further work in the technology and on what terms. This is different from giving a patent holder a long time to control a particular invention"); Gilbert and Shapiro, supra note 378 at 2 (explaining that their model, which assumes the grant of infinitely-lived patents, focuses on patent scope because increases in scope have greater preclusive effect than increases in length).

[379] See, e.g. Machlup, *An Economic Review of the Patent System* (Subcommittee on Patents, Trademarks, and Copyrights of the Senate Committee on the Judiciary, Study No. 15, 85th Congress 2nd Session) (1958), at 62–73 (discussing difficulties involved in analyzing impact on incentives of one-year change in patent term); Baxter, "Legal Restrictions on Exploitation of the Patent Monopoly: An Economic Analysis," *Yale Law Journal* 76 (1966) 267, 268–9; McGee, "Patent Exploitation: Some Economic and Legal Problems," *Journal of Law and Economics* 9 (1966), 135, 136.

[380] "Basic research" has been defined as research directed solely toward expanding human knowledge and "applied research" has been defined as research directed toward solving practical problems. See Eisenberg, "Patents and the Progress of Science: Exclusive Rights and Experimental Use," *University of Chicago Law Review* 56 (1989), 1017. For the most part, cases and commentary on the various theories relating to the patent system have focused on applied research and many commentators have distinguished applied research from basic research and explicitly excluded the latter from their analysis. Ibid. at 1030. See, e.g. Nelson, "The Economics of Invention: A Survey of the Literature," *Journal of Business* 32 (1959), 101, 105; Machlup, supra note 387, at 1; Polanyi, "Patent Reform," *Review of Economic Studies* 11–12 (1943), 61, 73.

[381] See Machlup, *Production and Distribution of Knowledge in the United States* (1962), at 168 and 169.

[382] If one focuses on innovation, it is also necessary to deal with the laws and regulations effecting each of the steps in the commercialization process (e.g. production licensing practices and distribution procedures).

nomics also suggests a number of different areas for further research, particularly with regard to the role that patents or some other system of exclusive private property rights can play in promoting "innovation," which is taken to mean the process of capturing the social gains of creative activity through market mechanisms by developing and improving inventions to the point where they can be fully commercialized.

One set of topics that would benefit from further economic analysis is the possibility of modifying the patent system, or implementing new complementary legal schemes, to focus upon groups or activities which appear to be important to the process of innovation. For example, should special patent-like rights be created for specific groups of inventors (e.g. individuals or smaller firms), and should new exclusivity schemes, perhaps in the form of a license or franchise, be developed to protect and encourage post-invention investments required to commercialize new products. Also, additional work is need on encouraging work on so-called "minor inventions" which, while not always bringing the far-reaching changes associated with pioneering ideas, are nonetheless an important element of ongoing technical advance.

A second set of interesting topics is based on the realization that the current patent system may not be ideally suited to deal with the variations in technical advance and competition which exist among different industries. Consideration should be given to whether industry-specific frameworks for intellectual property rights or other forms of exclusivity can, or should, be developed in order to take into account their unique characteristics. If this is not feasible, perhaps due to the substantial difficulty in establishing industry boundaries, then perhaps some effort can be made to take technological change, licensing practices, and non-patent barriers to competition into account in the course of making decisions regarding patent scope. Any changes in this area would also require the development of training programs to educate regulators regarding the complex interplay between technological and economic factors in making patent decisions.

A third set of issues are concerned largely with a variety of potential modifications to the operation of the present patent system to enhance its overall utility and reduce waste from the duplication of research efforts. Clearly, there is a need for further work on improving the current system of disclosures, both as to content and access. Changes might run the gambit from requiring enhanced detail to developing sophisticated information systems which can be easily accessed by inventors and interested parties. Also, additional study might be given to developing a system of pre-invention disclosures of product concepts in order to reduce the wasteful effects of "patent races." In addition, the utility of patent system could well be enhanced by allowing limited experimental usage of patented inventions, provided that sufficient guarantees can be developed to avoid any reduction in the incentives for inventors.

Finally, if it is recognized that patents are to be, at least to some extent, instruments for encouraging innovation, then further decisions regarding the focus and operation of the patent system must be coordinated with law and regulation and government policy in other related areas. The biggest challenge lies in the continuing pursuit of the optimal relationship of antitrust and competition law and policy to intellectual property law. A number of the potential issues in this area already have been highlighted, but none is more interesting that reconciling the respective roles of intellectual property laws and antitrust and competition laws in providing or denying incentives in the marketplace for enhanced innovation. Among other things, it is important to understand the tradeoffs that must be made between the "incentive to innovate" created by intellectual property protection and any disincentive to innovate which might result from stringent antitrust and competition law policies based on the need for achieving allocative efficiencies.

Bibliography

Abramovitz, "Resource and Output Trends in the U.S. Since 1870," *American Economic Review* 46 (1956), 5.

Adamo, "Problems Connected With Acquisition, Licensing and Enforcement of Intellectual Property," *Albanian Law Review* 50 (1986), 475.

Adelman, "Property Rights Theory and Patent-Antitrust: The Role of Compulsory Licensing," *New York University Law Review* 52 (1977), 977.

Andewelt, "Antitrust Perspective on Intellectual Property Protection," *Patent, Trademark and Copyright Journal* 30 (July 25, 1985), 319.

Arrow, *Economic Welfare and the Allocation of Resources for Invention*, in National Bureau of Economic Research, Rate ad Direction of Inventive Activity (1962), 609.

"Asia-Pacific Countries Agree to Cooperate on Patent Issues," *World Intellectual Property Report* 4 (March 1990), 53.

Aubrey, "A Justification of the Patent System," in Phillips (ed.), *Patents in Perspective* (1985) at 1.

Austern, "Fish Traps, Indians, and Patents: The Antitrust Validity of Patent License Restrictions on Sales Price, Field of Use, Quantity, and Territory," *University of Pittsburgh Law Review* 28 (1966), 181.

Bainbridge, *Intellectual Property* (1992).

Baldwin and Childs, "The Fast Second and Rivalry in Research and Development," *Southern Economic Journal* 36 (1969), 36.

Baldwin and Scott, *Market Structure and Technological Change* (1987).

Banner, "The Patent System and the Antitrust Laws: Stabilization of the Tensions?" *Antitrust Law Journal* 49 (1980), 41.

Barrett, *The Theory of Microeconomic Policy* (1974).

Barzel, "Optimal Timing of Innovations," *Review of Economics and Statistics* 40 (1968), 348.

Bator, "The Anatomy of Market Failure," *Quarterly Journal of Economics* 72 (1958), 351.

Baumol, "Contestable Markets: An Uprising in Theory of Industry Structure," *American Economic Review* 72 (March 1982), 1.

Baumol and Ordover, *Antitrust: Source of Dynamic and Static Inefficiencies?*, Working Paper (July 1988).

Baxter, "Legal Restrictions on Exploitation of the Patent Monopoly: An Economic Analysis," *Yale Law Journal* 76 (1966), 267.

Beck, "Patents, Property Rights and Social Welfare: Search for a Restricted Optimum," *Southern Economic Journal* (October 1976).

Beck, "The Prospect Theory of the Patent System and Unproductive Competition," *Research in Law and Economics* 5 (1983), 193.

Bellamy and Child, *Common Market Law of Competition,* 4th ed. (1993).

Berg, Duncan and Friedman, *Joint Venture Strategies and Corporate Innovation* (1982).
Berg and Friedman, "Joint Ventures, Competition and Technological Complementaries: Evidence from Chemicals," *Southern Economic Journal* (1977), 1330.
Bidault, *Technology Pricing: From Principles to Strategy* (B. Page and P. Sherwood translation 1989).
Blair and Kaserman, *Law and Economics of Vertical Integration and Control* (1983).
Bock, "Notes on the Future of Competition," *RECORD* (March 1975), 28.
Bock, "Rediscovering Economic Realism in Defining Competition," *RECORD* (June 1974), 6.
Bosworth, *Intellectual Property Rights* (1986).
Bouckaert, "What is Property?," *Harvard Journal of Law* and *Public Policy* 13 (1990), 775.
Bowman, *Patent and Antitrust Law* (1973).
Bowman, *Patent and Antitrust Law: A Legal and Economic Appraisal* (1974).
Bowman, "Tying Arrangements and the Leverage Problem," *Yale Law Journal* 67 (1957), 19.
Branch, "Research and Development Activity and Profitability: A Distributed Lag Analysis," *Journal of Political Economics* 82 (1974), 999.
Brenner, *Rivalry: In Business, Science, Among Nations* (1987).
Brinson, "Proof of Economic Power in a Sherman Act Tying Arrangement Case: Should Economic Power by Presumed When the Tying Product is Patented or Copyrighted," *Louisiana Law Review* 48 (1987), 29.
British Patent System, *Report of the Committee to Examine the Patent System*, Cmnd 4407, HMSO (1970).
Brodley, "Antitrust Law and Innovation Cooperation," in *Collaboration*, Jorde and Teece (1990), 911.
Brodley, *The Economic Goals of Antitrust: Efficiency, Consumer Welfare and Technological Progress* (1987).
Brozen, "Bain's Concentration and Rates of Return Revisited," *Journal of Law and Economics* 14 (1971), 351.
Brozen, "The Antitrust Task Force Demonstration Recommendation," *Journal of Law and Economics* 13 (1970), 279.
Brozen, "The Persistence of "High Rates of Return," in High-Stable Concentration Industries," *Journal of Law and Economics* 14 (1971), 501.
Brunet, "Eights Session (Second Part) of Committee of Experts on the Harmonization of Certain Provisions in Laws for the Protection of Inventions," *Patents, Trademark Copyright News* 9 (Winter 1990) at 15.
Buckley, "A Creative Alternative to the Private Inventor's Patent," in Phillips (ed.), *Patents in Perspective* (1985), 88.
Cabanellas, *Antitrust and Direct Regulation of International Transfer of Technology Transactions* (1984).
Carter, "Owning What Doesn't Exist," *Harvard Journal of Law and Public Policy* 13 (1990), 99.
Caves *et al*, "The Imperfect Market for Technology Licenses," *Oxford Bulletin for Economics and Stategies* 45 (1983), 249.
Chafee, "Unfair Competition," *Harvard Law Review* 53 (1940), 1289.
Chevigny, "Validity of Grant-Back Agreements Under the Antitrust Laws," *Fordham Law Review* 34 (1966), 569.
Chisum, *Patents* (1984).
Coase, "The Lighthouse in Economics," *Journal of Law and Economics* 17 (1974), 157.

Coase, *The Firm, The Market, and the Law* (1988).

Coase, "The Nature of the Firm," *Economica* 4 (1937), 386.

Coase, "The Problem of Social Cost," *Journal of Law and Economics* 3 (1960), 1.

Cohen, *Awards to Inventors*, address at University of Birmingham (1960).

Coleman, "Protection of Innovation Under the Common Law and In Equity," in Phillips (ed.), *Patents in Perspective* (1985) 18.

Comanor and Wilson, "Advertising Market Structure and Performance," *Review of Economics and Stategies* 49 (1967), 423.

Comment, "Transparent Wrap: The Continuing Case for Grant-Back Agreements," *Catholic University Law Review* 17 (1967), 228.

Commission on Competition Policy, *Joint Venture Notice*.

Conference Board Antitrust Conferences, publications for 1979 through 1983.

Conference Report, *BNA Antitrust and Trade Regulation Report* 47 (1984), 590.

Contractor, *International Technology Licensing: Compensation, Costs and Negotiation* (1981).

Cooper, "Patents and Innovation," in Teeling-Smith (ed.), *Innovation and the Balance of Payments: The Experience in the Pharmaceutical Industry* (1967).

Cooter and Ulen, *Law and Economics* (1988).

Cornish, *Intellectual Property: Patents, Copyright, Trade Marks and Allied Rights* (1989) 79.

Curley, "Patent Acquisitions and Restricted Licenses Under the Antitrust Laws," *Seton Hall Law Review* 8 (1977), 645.

Dasgupta, "The Theory of Technological Competition," in Stiglitz and Mathewson (eds.), *New Development in the Analysis of Market Structure* (1986) 399.

Dasgupta and Maskin, "The Simple Economics of Research Portfolios," *Economic Journal* 97 (1987), 581.

David, *R&D Consortia, High Technology* (October 1985).

Dembo, Dias and Morehouse, "Technology to Aid the Poor: Constraints to Access Resulting from the Privatization – The Case of Biotechnology," in *The International Context of Rural Property in the Third World* (1986) 103.

Demsetz, "Industry Structure, Market Rivalry, and Public Policy," *Journal of Law and Economics* 16 (April 1973), 1.

Demsetz, *The Market Concentration Doctrine* (1973).

Demsetz, "Toward a Theory of Property Rights," *American Economic Review* 57 (1967), 347.

Denison, *Accounting for United States Economic Growth, 1929–1969* (1974).

Denison, *The Sources of Economic Growth in the United States and the Alternatives Before Us* (1962).

Department of Consumer and Corporate Affairs, *Working Paper on Patent Law Revision* (1972).

Department of Justice, *Antitrust Guide for International Operations* (1988).

Department of Justice, *Centrifugal Pump Industry*, DOJ Business Review Letter (June 5, 1985).

Dertouzos, Lester and Solow, *Made in America: Regaining the Productive Edge* (1986).

De Soto, *Path*.

Dias, Dembo and Morehouse, "Product Displacement: Biotechnology's Impacts on Developing Countries," in *The International Context of Rural Poverty in the Third World* (1986) 129.

Dosi, "Sources, Procedures, and Microeconomic Effects of Innovation," *Journal of Economic Literature* (September 1988), 26.
Dunne, "Anti-Competitive Considerations of Patent Accumulation by License Grant-Back Provisions," *Journal of the Patent Office Society* 57 (1975), 124.
Dutton, *The Patent System and Inventive Activity During the Industrial Revolution 1750–1852* (1984).
Easterbrook, "Intellectual Property is Still Property," *Harvard Journal of Law and Public Policy* 13 (1990), 108.
Easterbrook, "The Limits of Antitrust," *Texas Law Review*, 63 (1984), 1.
Eighth Report on Competition Policy (1978).
Eisenberg, "Patents and the Progress of Science: Exclusive Rights and Experimental Use," *University of Chicago Law Review* 56 (1989), 1017.
Eisenberg, "Proprietary Rights and the Norms of Science in Biotechnology Research," *Yale Law Journal* 97 (1987), 177.
Eisenschitz, "The Value of Patent Information," in Phillips (ed.), *Patents in Perspective* (1985) at 47.
Ensley, "Intellectual Property Rights in the GATT," *New Matter* 15 (1990), 1.
Farrell and Shapiro, *Horizontal Mergers: An Equilibrium Analysis*, Working Paper (May 1988).
Faull, "Joint Ventures under the EEC Competition Rules," *EC Law Review* (1984) 358.
Federal Trade Commission, *Hearings on the Changing Nature of Competition* (June 29, 1995).
Federico, *Renewal Fees and Other Patent Fees in Foreign Countries*, Study of Sub-Committee on Patents, Trade Marks and Copyrights (1958).
Fifth Report on Competition Policy (1975).
Fifteenth Report on Competition Policy (1985).
Finnis, *Natural Law and Natural Rights* (1980).
Fisher, "Diagnosing Monopoly," *Quarterly Review of Economics and Business* 19 (1979), 7.
Fisher, "The Misuse Doctrine and Post-Expiration-Discriminatory and Exorbitant Patent Royalties," *Industry Law Journal* 43 (1967–68), 106.
Fisher and Lande, "Efficiency Considerations and Merger Enforcement," *California Law Review* 71 (1983), 1580.
Fondasol Techniques Louis Menard, Ninth Report on Competition Policy (1979).
"Foreign Trade: China, India, and Thailand Named in First 'Priority Foreign Country' List," *Patents, Trademark and Copyright Journal* 42 (1991), 7.
Forst, "Legal Incidents of Non-Use of Patented Inventions Reconsidered," 14 Geo Wash L Rev (1946) 273.
Fourteenth Report on Competition Policy (1984).
Fourth Report on Competition Policy (1974).
Fox, "Maize Seed: A Comparative Comment," in B. Hawk (ed.) (1982) 151.
Frame, "National Commitment to Intellectual Property Protection: An Empirical Investigation," *Journal of Law and Technology* 2 (1987), 209.
French State/Suralmo, *Ninth Report on Competition Policy* (1979).
Frost, "Tying Clauses and Package Licensing," *University of Pittsburgh Law Review* 28 (1966), 207.
Fudenberg and Tirole, "Understanding Rent Dissipation: On the Use of Game Theory in Industrial Organization," *American Economic Review Papers and Proceedings*, 77 (1987), 176.

Fuscone, "The Exploitation of Individual Invention," in Phillips (ed.), *Patents in Perspective* (1985), 81.
Garten, "American Trade Law in a Changing World Economy," *The International Lawyer* 29 (Spring 1955), 15.
Gibbons, "Domestic Territorial Restrictions in Patent Transactions and the Antitrust Laws," *George Washington Law Review* 34 (1966), 893.
Gilbert, "An Interview with Richard Gilbert," *Competition: The Journal of the Antitrust and Trade Regulation Law Section of the State Bar of California* 4 (Fall 1994), 1.
Gilbert and Shapiro, *Optimal Patent Length and Breadth* University of California, Berkeley, Department of Economics, Working Paper No. 89-102 (January 1989), 2.
Gilfillan, *Invention and the Patent System* (1964).
Gilfillan, "The Root of Patents, or Squaring Patents by Their Roots," *Journal of the Patent Office Society* 31 (1949), 611.
Goldschmid *et al.* (eds.), *Industrial Concentration: The New Learning* (1974).
Goyder, *EC Competition Law* 2nd ed. (1993).
Greenberger, "Vietnam," *Wall Street Journal* (1993), A-11.
Griliches, "Productivity, R&D, and Basic Research at the Firm Level in the 1970's," *American Economic Review* 76 (March 1986), 141.
Grossman and Shapiro, "Dynamic R&D Competition," *Economic Journal* (June 1987), 97.
Grossman and Shapiro, "Research Joint Ventures: An Antitrust Analysis," *Journal of Law and Economics* (Fall 1986), 2.
Grossman and Shapiro, *The Sociology of Invention* (1935).
"Guidelines for Standards for the Protection and Enforcement of Patents of the U.S. Chamber of Commerce Intellectual Property Task Force".
Hamilton, *Investigation of Concentration of Economic Power: Patents and Free Enterprise 21* TNEC Monograph No. 31, (1941).
Harberger, "Monopoly and Resource Allocation," *American Economic Review* 44 (1954), 77.
Hardin, "The Tragedy of the Commons," *Science* 162 (1968), 1243.
Hardin and Baden (eds.), *Managing the Commons* (1977).
Hawke, *United States, Common Market and International Antitrust: A Comparative Guide*, Vol. 2 (1986).
Hay and Vickers, *The Economics of Market Dominance* (1987).
Hettinger, "Justifying Intellectual Property," *Philosophy and Public Affairs* 18 (1989), 31.
Higgins in collaboration with Silberston, *Invention and Economic Growth in Britain and the United States 1853–1963* (1969).
Hirshleifer, "The Private and Social Value of Information and the Reward to Inventive Activity," *American Economic Review* 61 (1971), 561.
Holmes, *Intellectual Property and Antitrust Law* (1983).
Hornick, "The Per Se Rule in Tying Contexts: A Critical View," *Delaware Journal of Corporate Law* 10 (1985), 703.
Horstmann *et al.*, "Patents as Information Transfer Mechanisms: To Patent or (Maybe) Not to Patent," *Journal of Political Economy* 93 (1985), 837.
"IIPA Submission to U.S. Trade Representative on 'Priority Foreign Countries'," *World Intellectual Property Report* 4 (April 1990), 84.
"Innovation, Rivalry and Competitive Advantage: Interview with Professor Michael E. Porter," *Antitrust* 2 (Spring 1991), 5.

Jalloq, *Use of Patent Literature by Academics in Chemistry and Physics, Electrical and Mechanical Engineering*, British Library R&D Report (1983).

Jefferson, Letter to Isaac McPherson, Monticello, August 13, 1813, in XIII *The Writings of Thomas Jefferson* (A. Lipscomb ed. 1904).

Jewkes, *The Sources of Invention* (1958).

Jewkes, Sawers and Stillerman, *The Sources of Invention* (1969).

Jorde and Teece, *Antitrust, Innovation and Competitiveness* (1992).

Jorde and Teece, *Innovation and Cooperation: Implications for Competition and Antitrust, Collaboration Among Competitors* (1990) 887.

Jorde and Teece, "Innovation, Cooperation and Antitrust: Balancing Competition and Cooperation," *High Technology Law Journal* (1989), 4.

Jorde and Teece, "Discussion of National Cooperative Research Act," *High Technology Law Journal* 4:1 (1989) 62.

Kahn, "Fundamental Deficiencies of the American Patent Law," *American Economic Review* 30 (1940), 475.

Kamien and Schwartz, *Market Structure and Innovation* (1982).

Kaplow, "The Patent-Antitrust Intersection: A Reappraisal," *Harvard Law Review* 97 (1984), 1813.

Katz, "An Analysis of Cooperative Research and Development," *Rand Journal of Economics* 17 (1986), 527.

Kelman, "Consumption Theory, Production Theory, and Ideology in the Coase Theorem," *Southern California Law Review* 52 (1979), 669.

Kemp, "The Follow-On Development Process and the Market for Diuretics," in Helms (ed.), *Drug Development and Marketing* (1975).

Kenney, "Schumpeterian Innovation and Entrepreneurs in Capitalism: A Case Study of the U.S. Biotechnology Industry," *Res Policy 21* 15 (1958), 21.

Kingston, "Innovation Patents and Warrants," in Phillips (ed.), *Patents in Perspective* (1985) at 80.

Kingston, *The Political Economy of Innovation* (1984).

Kitch, "Patents: Monopolies or Property Rights?," *Research in Law and Economics* 8 (1986), 31.

Kitch, "Patents, Prospects, and Economic Surplus: A Reply," *Journal of Law and Economics* 23 (1980), 205.

Kitch, "The Antitrust Economics of Joint Ventures," in *Collaboration*, Jorde and Teece (1990) 865.

Kitch, "The Nature and Function of the Patent System," *Journal of Law and Economics* 20 (1977), 265.

Klitzke, "Patents and Section 7 of the Clayton Act: The Significance of Patents in Corporate Acquisitions," *Loyola University Law Journal* 12 (1981), 401.

Ko, "An Economic Analysis of Biotechnology Patent Protection," *Yale Law Journal* 102 (1992), 777.

Korah, *Patent Licensing and EEC Competition Rules: Regulation 2349/84* (1985).

Korah, *R&D and the EEC Competition Rules: Regulation 418/85* (1986).

Korah, "The Preliminary Draft of a New EC Group Exemption for Technology Licensing," *European Intellectual Property Report* 7 (1994), 262.

Kronz, "Patent Protection for Innovations: A Model," *EIPR* (1983), 178.

Krueger, "The Political Economy of the Rent-Seeking Society," *American Economic Review* 64 (1974), 291.

Kwoka and White (eds.), *The Antitrust Revolution* (1989).

Landes and Posner, "An Economic Analysis of Copyright Law," *Journal of Legal Studides* 17 (1989), 325.

Landes and Posner, "Market Power in Antitrust Cases," *Harvard Law Review* 94 (1981), 937.

Langrish, Gibbons, Evans and Jevons, *Wealth from Knowledge* (1972).

Leibenstein, "Allocate Efficiency vs. 'X-Efficiency'," *American Economic Review* 56 (1966), 392.

Levin, "A New Look at the Patent System," *AEA Papers and Proceedings* Vol 2, 76 (1986), 201.

Levin, Cohen and Mowery, "R&D Appropriability, Opportunity, and Market Structure: New Evidence on Some Schumpeterian Hypotheses," *American Economic Review* 75 (1985), 20.

Levin, Klevorick, Nelson and Winter, "Appropriating the Returns from Industrial Research and Development," 1987 Brookings Papers on Economic Activity (1987) 783.

Liebeler, "Market Power and Competitive Superiority in Concentrated Industries," *University of California Law Review* 25 (1978), 1231.

Lippman and McCardle, "Dropout Behavior in R&D Races with Learning," *Rand Journal of Economics* 18 (1987), 287.

Lipsky, "Remarks to ABA Antitrust Section", reprinted in *Antitrust Law Journal* 50 (1981), 518.

Livesley, "Patent Law Reform – A Patent Agent's Personal View", in Phillips (ed.), *Patents in Perspective* (1985) at 111.

Locke, *Two Treatises of Government* (P. Laslett ed. 1970).

Machlup, *An Economic Review of the Patent System* (1958).

Machlup, *The Production and Distribution of Knowledge in the United States* (1962), 162.

Machlup and Penrose, "The Patent Controversy in the Nineteenth Century," 10 *Journal of Economic History* 1 (1950) 1.

Mackaay, "Economic Incentives in Markets for Information and Innovation," *Harvard Journal of Law and Public Policy* 13 (1990), 867.

Mandeville, Lamberton and Bishop, *Economic Effects of the Australian Patent System*, Report to the Industrial Property Advisory Committee (1982).

Mansfield, Industrial Research and Technological Progress (1968).

Mansfield, "Patents and Innovation: An Empirical Study," *Management Science* 32 (1984), 173.

Mansfield, *The Economics of Industrial Change* (1968).

Mansfield *et al.*, *Research and Innovation in the Modern Corporation* (1971).

Mansfield *et al.*, "Social and Private Rates of Return from Industrial Innovations," *Quarterly Journal of Economics* 91 (1977), 221.

Mansfield *et al.*, *The Production and Application of New Industrial Technology* (1977).

Mansfield, Schwartz and Wagner, "Imitation Costs and Patents: An Empirical Study," *The Economic Journal* 91 (1981), 907.

Marquis, "Limitations on Patent License Restrictions: Some Observations," 58 *Iowa Law Review* 41 (1972) 54.

Marshall, *Principles of Economics* (1890).

Martyn, *Literature Searching by Research Scientists* (1964).

McConnell, *Joint Venture: An Alternative to Merger* Purdue University (March 1983).

McFetridge and Rafiquzzaman, "The Scope and Duration of the Patent Right and the Nature of Research Rivalry," *Research in Law and Economics* 8 (1986), 91.

McFetridge and Smith, "Patents, Prospects, and Economic Surplus: A Comment," *Journal of Law* 23 (1980), 202.

McGee, *In Defense of Industrial Concentration* (1971).

McGee, "Patent Exploitation: Some Economic and Legal Problems," *Journal of Law and Economics* 9 (1966), 135.

McGrath, "Patent Licensing: A Fresh Look at Antitrust Principles in a Changing Economic Environment", *Trade Regulation Report* 4 (April 5, 1984) 130.

Meiners and Staaf, "Patents, Copyrights, and Trademarks: Property or Monopoly?", *Harvard Journal of Law and Public Policy* 13 (1990), 911.

Meinhardt, *Inventions, Patents and Monopoly* (1950).

Merges, "Battle of Lateralisms: Intellectual Property and Trade," *B. U. International Law Journal* 8 (1990), 239.

Merges, "Commercial Success and Patent Standards: Economic Perspectives on Innovation," *California Law Review* 76 (1988), 803.

Merges and Nelson, "On the Complex Economics of Patent Scope," *Columbia Law Review* 90 (1990), 839.

Merges and Nelson, "Market Structure and Technical Advance: The Role of Patent Scope Decisions", in Jorde and Teece (eds.), *Antitrust, Innovation and Competitiveness* (1992), 185.

Meurer, "The Settlement of Patent Litigation," *Rand Journal of Economics* 20 (1989), 77.

Millstein, "The Role of Antitrust in an Age of Technology", *Cardoza Law Review* 9 (1988), 1175.

Mowery (ed.), *International Collaborative Ventures in U.S. Manufacturing*, Ballinger Publishing Company, Cambridge (1988).

Nalebuff and Stiglitz, "Information, Competition and Markets", *American Economic Review* (May 1983), 72.

Nance, "Forward: Owning Ideas," *Harvard Journal of Law and Public Policy* 13 (1990), 757.

National Science Board, *University-Industry Research Relationships: Myths, Realities and Potentials*, Fourteenth Annual Report (October 1, 1982), 11.

National Society of Professional Engineers, *Engineering Stages of New Product Development* (1990).

Neilson-Hordell/Richmark, *Twelfth Report* (1982).

Nelkin, *Science as Intellectual Property* (1984).

Nelson, "The Economics of Invention: A Survey of the Literature," *Journal of Business* 32 (1959), 101.

Nelson, "The Simple Economics of Basic Scientific Research," *Journal of Political Economics* 67 (1959), 297.

Nicholas, *French Law of Contract* (1982).

Nodet-Gougis/Lamazou, *Tenth Report on Competition Policy* (1980).

Nordhaus, *Invention, Growth and Economic Welfare: A Theoretical Treatment of Technological Change* (1969).

North, *Institutions*.

Note, "Characterizing Horizontal Market-Division Agreements Under the Ancillary Restraints Doctrine," *Loyola University of Chicago Law Journal* 16 (1985), 775.

Note, "Mandatory Package Licensing: A New Patent Misuse," *Virginia Law Review* 44 (1958), 727.

Note, "Post Expiration Royalty Payments and Mandatory Package Licensing as Patent Misuses," *Vill. Law Review* 11 (1966), 382.
Note, "Standard Antitrust Analysis and the Doctrine of Patent Misuse: A Unification Under the Rule of Reason," *University of Pittsburgh Law Review* 46 (1984), 209.
Note, "The Presumption of Economic Power for Patented and Copyrighted Products in Tying Arrangements," *Columbia Law Review* 85 (1985), 1140.
Note, "Total Sales Royalties Under the Patent Misuse Doctrine: A Critique of Zenith," *Michigan Law Review* 76 (1978), 1144.
O'Brien, "Patents: An Economist's View", in Phillips (ed.), *Patents in Perspective* (1985), 32.
O'Driscoll and Rizzo, *The Economics of Time and Ignorance* (1985), 155.
Ogburn, *Social Change* (1923), 86.
Oppenheim, "Information Aspects of Patents," in Phillips (ed.), *Patents in Perspective* (1985), at 54.
Ordover and Baumol, "Antitrust Policy and High-Technology Industries," *Oxford Review of Economic Policy: Technical Progress* (Winter 1988).
Ordover and Saloner, "Predation, Monopolization and Antitrust" in Schmalensee and Willig (eds.), *Handbook of Industrial Organization* (1988).
Ordover and Willig, "Antitrust for High-Technology Industries: Assessing Research Joint Ventures and Mergers," *Journal of Law and Economics* 28 (1985), 311.
Pakes, "On Patents, R&D, and the Stock Market Rate of Return", *Journal of Political Economics* 93 (1985), 390.
Palmer, "Intellectual Property Rights: A Non-Posnerian Law and Economics Approach," *Hamline Law Review* 12 (1989), 261.
Phillips (ed.), *Patents in Perspective* (1985).
Pickford, "A New Test for Manufacturing Industry Efficiency", *International Journal of the Industrial Organization* 3 (1985), 153.
Pitofsky, "A Framework for Antitrust Analysis of Joint Ventures," in *Collaboration*, Jorde and Teece (1990), 839.
Pitofsky, "Joint Ventures Under the Antitrust Laws: Some Reflections on the Significance of Penn-Olin," *Harvard Law Review* 82 (1969), 1007.
Plant, *Selected Economic Essays and Addresses* (1974), 39.
Plant, "The Economic Theory Concerning Patents for Inventions," *Economica* 30 (1934), 1.
Plant, "The Economic Theory Concerning Patents for Inventions," *Economica* 47 (1934).
Plant Royalty Bureau, *Ninth Report on Competition Policy* (1979), 120.
Polanyi, "Patent Reform," *Review of Economic Studies* 11 (1944), 61.
"Portion of U.S. Trade Representative's Fact Sheet on 1988 Trade Act's 'Special 301' Provision on Intellectual Property, Released 25 May 1989," *World Intellectual Property Report* 3 (July 1989), 162.
Posner, "Antitrust Policy and the Supreme Court: An Analysis of the Restricted Distribution, Horizontal Merger and Potential Competition Decisions," *Columbia Law Review* 75 (1975), 282.
Posner, "The Next Step in Antitrust Treatment of Restricted Distribution: Per Se Legality," *University of Chicago Law Review* 48 (1980), 6.
Posner, "The Social Costs of Money and Regulation," *Journal of Political Economics* 8 (1975), 807.
Posner, "Utilitarianism, Economics, and Legal Theory," *Journal of Legal Studies* 8 (1979), 325.

Preflex and Lipski, Tenth Report on Competition Policy (1980).
Preovolos, "Intellectual Property Licensing: The Practitioner's Perspective," *Competition: The Journal of the Antitrust and Trade Regulation Law Section of the State Bar of California* No. 2 (Fall 1994), 4.
President's Commission on Industrial Competitiveness Report, *Global Competition: The New Reality* (1985).
Rachels, "What People Deserve," in J. Arthur and W. Shaw (eds.), *Justice and Economic Distribution* (1978), 150.
Rahl, "The Right to 'Appropriate' Trade Values," *Ohio State Law Journal* 23 (1962), 56.
Rapp and Rozek, "Benefits and Costs of Intellectual Property Protection in Developing Countries," *Journal of World Trade* 24 (October 1990), 75.
Rawls, *A Theory of Justice* (1971).
Reiniganum, "The Timing of Innovation: Research, Development and Diffusion," in Schmalensee and Willig (eds.), *Handbook of Industrial Organization* (1988).
Report of the Banks Committee on the British Patent System 1970 (Cmnd. 4407, 1970, § 1).
Richardson, "The Orphan Drug Tax Credit: An Inadequate Response to an Ill-Defined Problem," *American Journal of Tax Policy* 6 (1987), 135.
Rin-Laures and Janofsky, "Recent Developments Concerning the Orphan Drug Act," *Harvard Journal of Law and Technology* 4 (1991), 291.
Roberts, "Antitrust and Patent Misuse Considerations of Royalty Rate Differences Among Patent Licensees," *Journal of the Patent Office Society* 65 (1983), 344.
Roguin, *La Regle de Droit: Systeme des Rapports de Droit Prive* (1989).
Rule, "Antitrust Implications of International Licensing: After the Nine No-Nos," reprinted in *Trade Regulation Report* 4 (October 21, 1986), 131.
Ruttan, "Usher and Schumpeter on Invention, Innovation, and Technological Change," *Quarterly Journal of Economics* 73 (1959), 596.
Samuelson, "The Pure Theory of Public Expenditure," *Review of Economic Stategies* 36 (1954), 387.
Sanders, "Speedy Entry of Patented Inventions into Commercial Use," *Patent, Trademark and Copyright Journal of Research and Education* 6 (1962), 87.
Sanders, Rossman and Harris, "Attitudes of Assignees Toward Patented Inventions," *Patent, Trademark and Copyright Journal of Research and Education* 2, 463.
Say, *Traité d'Economie Politique* (1st Ed., 1803), 263.
Scherer, "Antitrust, Efficiency, and Progress," in *Collaboration*, Jorde and Teece (1990), 33.
Scherer, *Industrial Market Structure and Economic Performance* (1970).
Scherer, *Innovation and Growth: Schumpeterian Perspectives* (1984).
Scherer, "Inter-Industry Technology Flows and Productivity Growth," *Review of Economics and Stategies* 64 (1982), 627.
Scherer, "Nordhaus's Theory of Optimal Patent Life: A Geometric Reinterpretation," *American Economic Review* 62 (1972), 422.
Scherer, "Research and Development Resource Allocation Under Rivalry," *The Quarterly Journal of Economics* 81:3 (August 1967), 359.
Scherer, *The Economic Effects of Compulsory Licensing* (1977).
Schiff, *Industrialization Without National Patents* (1971).
Schmalbeck, "The Validity of Grant-Back Clauses in Patent Licensing Agreements," *University of Chicago Law Review* 42 (1975), 733.
Schmookler, *Invention and Economic Growth* (1966).

Schumpeter, *Capitalism, Socialism and Democracy* (1950).

Schumpeter, *Business Cycles* 1 (1939).

Schumpeter, *The Theory of Economic Development* (Transaction reprint, Redvers Opie translated 1983).

Schwartzman, *Innovation in the Pharmaceutical Industry* (1976).

Scott, "Diversification versus Cooperation in R&D Investment," *Managerial Decision Economics* (1988), 9.

Selinger, "Patent Licensing in the Afterglow of Sylvania: Practicalities of Life Under the Rule of Reason," *JPOS* 63 (1981), 353.

Senate Report No. 427, 98th Congress, 2d Session (1984).

Shapiro, "Patent Licensing and R&D Rivalry," *American Economic Review* 75 (1985), 25.

Shapiro and Willig, "On the Antitrust Treatment of Production Joint Ventures," in *Collaboration*, Jorde and Teece (1990), 929.

Siegfried and Wheeler, "Cost Efficiency and Monopoly: A Survey," *Quarterly Review of Economics and Business* 21 (1981), 11.

Siekman, "Post Expiration Royalty Payments and Mandatory Package Licensing as Patent Misuses," *Vill. Law Review* 11 (1966), 382.

Singer, *Antitrust Economics* (1968).

Sixth Report on Competition Policy (1976).

Slawson, "A New Concept of Competition: Reanalyzing Tie-In Doctrine After Hyde," *Antitrust Bulletin* 30 (1985), 257.

Smit, *The Social Creation of a Legal Reality: A Study of the Emergence and Acceptance of the British Patent System as a Legal Instrument for the Control of New Technology*, unpublished Ph.D. thesis, University of Edinburgh (1980), quoted in Bainbridge (1992).

Smith, *An Inquiry into the Nature and Causes of the Wealth of Nations* (E. Cannon ed. 1937) (5th Ed. 1789).

Smith and Parr, *Valuation of Intellectual Property and Intangible Assets*, John Wiley, New York (1989).

Solo, "Innovation in the Capitalist Process: A Critique of the Schumpeterian Theory," *Quarterly Journal of Economics* 65 (1951), 417.

Solow, "Technological Change and the Aggregate Production Function," *Review of Economics and Strategies* 39 (1957), 312.

"Something Old, Something New, Special Survey on World Trade," *The Economist* (September 22, 1990), 34.

Spitzer and Van Hool, *Twelfth Report on Competition Policy* (1982).

Staines, "Some Thoughts on the 1983 Green Paper," in Phillips (ed.), *Patents in Perspective* (1985) at 100.

Stedman, "Acquisition of Patents and Know-How by Grant, Fraud, Purchase and Grant-Back," *University of Pittsburgh Law Review* 28 (1966), 161.

Stephenson, "The Use of Patent Information in Industry," *World Patent Information* 4 (1982), 164.

Stephenson and Riley, *Use of Patent Information in Industry, Library and Information* Research Report 4 (1982).

Strasser, "An Antitrust Policy for Tying Arrangements," *Emory Law Journal* 34 (1985), 253.

Stroup and Baden, *National Resources: Bureaucratic Myths and Environmental Management* (1983).

Subcommittee on Patents, Trademarks, and Copyrights of the Senate Committee on the Judiciary, Study No. 15, 85th Congress, 2d Session (1958).

Swann, O'Brien *et al.*, *Competition in British Industry* (1974).

Tandon, "Optimal Patents with Compulsory Licensing," *Journal of Political Economics* 90 (1982), 470.

Taylor, "Counterpoint: Will the Guidelines Work?," 4 *Competition: The Journal of the Antitrust and Trade Regulation Law Sectin of the State Bar of California* 2 (Fall 1994), 1.

Taylor, "Do We Still Need a Patent System?" *CIPA* (1972–73), 292.

Taylor and Silberston, *The Economic Impact of the Patent System: A Study of the British Experience* (1973).

Teece, "Economies of Scope and the Scope of the Enterprise," *Journal of Economic Behavior and Organization* (1980), 223.

Teece, *Inter-organizational Requirements of the Innovation Process, Managerial and Decision Economics*.

Teece, "Profiting from Technological Innovation," *Research Policy* (December 1986), 285.

Teece, *The Multinational Corporation and the Resource Cost of International Technology Transfer* (1976), 44.

Teece, "Towards an Economic Theory of the Multiproduct Firm," *Journal of Economic Behavior and Organization* (1982), 3.

Thirteenth Report on Competition Policy (1983).

Thomas, "Re-Assessing the Orphan Drug Act," *Columbia Journal of Law and Social Problems* 23 (1990), 413.

Timberg, "Antitrust and Industrial Property: Tie-Ins and Grant-Backs," *Antitrust Law Journal* 42 (1973), 651.

Tirole, *The Theory of Industrial Organization*, MIT Press, Cambridge (1988).

Tullock, "The Welfare Costs of Tariffs, Monopolies and Theft," *World Economic Journal* 5 (1967) 224.

Turner, "The Patent System and Competitive Policy," *New York University Law Review* 44 (1969), 450.

Twentieth Report on Competition Policy (1990).

Twenty-First Report on Competition Policy (1991).

Twenty-Second Report on Competition Policy (1992).

UARCO, *Fourteenth Report* (1984).

United Kingdom Green Paper, *Intellectual Property Rights and Innovation*, Cmnd. HMSO (1983).

United Kingdom White Paper on Patent Law Reform, Cmnd. 6000 (1975).

US National Science Board, National Science Foundation, *Science Indicators: The 1985 Report* (1985).

Usher, *A History of Mechanical Inventions* (1929).

Usher, "The Welfare Economics of Invention," *Economica* 31 (1964), 279.

Van Hise, *Concentration and Control* (1973).

Villard, "Competition, Oligopoly and Research," *Journal of Political Economics* 66 (1958), 483.

Von Hippel, *Increasing Innovators' Returns from Innovation*, Working Paper No. 1192-1181 under National Science Foundation Grant No. PRA 80-119244.

Von Hippel, *The Sources of Innovation* (1988).

Waite, "The Draft Technology Transfer Block Exemption," *European Intellectual Property Report* 7 (1994), 259.

Wallace, "Proper Use of the Patent Misuse Doctrine," *Mercer Law Review* 26 (1975), 813.

Weston, "Concentration and Efficiency: The Other Side of the Monopoly Issue," *Special Issues in the Public Interest* 4 (May 1978).

Weston and Ornstein, "Efficiency Considerations in Joint Ventures," in *Collaboration*, Jorde and Teece (1990), 875.

Wetzel, "Did Line Material Really Rule Out Patent Price Fixing?" *Mercer Law Review* 26 (1975), 471.

Wheeler, "A Reexamination of Antitrust Law and Exclusive Territorial Grants by Patentees," *University of Philadelphia Law Review* 199 (1971), 642.

Whish, *Competition Law* (1989), 262.

White, *Research and Development Joint Ventures under EEC Competition Law* (1985).

White, Jeffs, Jacob, Cornish and Vitoria, *Encyclopeda of United Kingdom and European Patent Law* (1977).

Williamson, "Economics as an Antitrust Defense: The Welfare Tradeoffs," *American Economic Review* 58 (1968), 18.

Williamson, *Markets and Hierarchies, Analysis and Antitrust Implications: A Study in the Economics of Internal Organization* (1975).

Williamson, *The Economic Institutions of Capitalism: Firms, Markets, Relational Contracting* (1985).

Winter, "Knowledge and Competence as Strategic Assets," in Teece (ed.), *The Competitive Challenge* (1987).

Zoller and Frohlich/Telemecanique, *Ninth Report on Competition Policy* (1979).

Index